DATE			

NITROGEN CONTROL AND PHOSPHORUS REMOVAL IN SEWAGE TREATMENT

NITROGEN CONTROL AND PHOSPHORUS REMOVAL IN SEWAGE TREATMENT

Edited by D.J. De Renzo

NOYES DATA CORPORATION
Park Ridge, New Jersey, U.S.A.
1978

Library of Congress Catalog Card Number: 78-59820
ISBN: 0-8155-0711-9
Printed in the United States

Published in the United States of America by
Noyes Data Corporation
Noyes Building, Park Ridge, New Jersey 07656

This book is based on two reports entitled "Process Design Manual for Nitrogen Control" by Brown and Caldwell, Walnut Creek, California and "Process Design Manual for Phosphorus Removal" by Shimek, Roming, Jacobs and Finklea, Dallas, Texas. The original reports were prepared for the Office of Technology Transfer of the U.S. Environmental Protection Agency as reports EPA/625/1-75/007 and EPA/625/1-76/001a. These reports were made available by NTIS as PB-259149 and PB-259150.

Foreword

Part I of this book presents theoretical and process design criteria for the implementation of nitrogen control technology in municipal wastewater treatment facilities. Design concepts are emphasized as much as possible through examination of data from full-scale and pilot installations. Design data are included on biological nitrification and denitrification, breakpoint chlorination, ion exchange and air stripping. A special chapter presents the concepts involved in assembling various unit processes into rational treatment trains and presents actual case examples of specific treatment systems that incorporate nitrogen control processes.

Part II discusses phosphorus removal methods that have been found effective and practical for use in treatment plants. All the methods included involve chemical precipitation of the phosphorus and removal of the resultant precipitate. Precipitants include lime and salts of aluminum or iron. The practical points of addition are before the primary settler, in the aerator of an activated sludge plant, before the final settler, or in a tertiary process.

Included in the discussion of each treatment is a description of the method, pilot or full scale performance data, equipment requirements, design parameters, and cost estimates. This information should be of value to designers, municipal officials, regulatory agencies, city planners, and treatment plant operators.

This book is based on the original reports entitled "Process Design Manual for Nitrogen Control" by Brown and Caldwell, Walnut Creek, CA and "Process Design Manual for Phosphorus Removal" by Shimek, Roming, Jacobs and Finklea, Dallas, TX. The reports were prepared for the Office of Technology Transfer of the U.S. Environmental Protection Agency as Report No. EPA/625/1-75/007 and EPA/625/1-76/001a.

The table of contents is organized in such a way as to serve as a subject index.

In order to keep the price of this large and extensive book to a reasonable level, it has been reproduced by photo offset directly from the original typeset report.

Table of Contents

PART I: NITROGEN CONTROL

PART II: PHOSPHORUS REMOVAL

PART I

NITROGEN CONTROL

ACKNOWLEDGEMENTS

This design manual was prepared for the Office of Technology Transfer of the U.S. Environmental Protection Agency. Coordination and preparation of the manual was carried out by the firm of Brown and Caldwell, Walnut Creek, California, under the direction of Denny S. Parker with assistance from Richard W. Stone and Richard J. Stenquist. Chapters 7, 8, and portions of Chapter 9 were prepared by Gordon Culp of Culp/Wesner/Culp. Clair N. Sawyer and Perry L. McCarty served as consultants to the U.S. EPA for the purpose of reviewing portions of the text. U.S. EPA reviewers were Edwin F. Barth and Irwin J. Kugelman of the U.S. EPA National Environmental Research Center, Cincinnati, Ohio, and Robert S. Madancy of the Office of Technology Transfer, Washington, D.C.

NOTICE

CHAPTER 1

INTRODUCTION

1.1 Background and Purpose

Man's influence on the environment is receiving increasing public and scientific attention. The quality of some of the nation's water bodies has been subjected to continuing degradation as a result of man's activities. While there has been considerable success in reversing this trend, one roadblock to greater progress often has been the lack of the necessary technology to reliably and economically remove the pollutants which are the cause of degradation of receiving waters. While conventional technology is well developed for removing organics from wastewater, the processes for the control of nitrogen in wastewater effluents have been developed only recently.

The beginnings of the implementation of nitrification on a significant scale occurred in the U.S. as late as the 1960's. The practice of nitrification was widespread in England much earlier. The *first* implementation of full nitrogen removal was as late as 1969 at South Lake Tahoe in California and even this installation encountered many problems. A flurry of research and development activity on the various nitrogen control methods occurred very recently beginning in the late 1960's and continues to date. Recent legislation and state regulatory activities have spurred many localities into nitrogen control projects.

Nitrogen control techniques are divided into two broad categories. The first group of nitrogen control processes is involved with the conversion of organic and ammonia nitrogen to nitrate nitrogen, a less objectionable form. These processes are termed nitrification processes. The second category involves processes which result in the removal of nitrogen from the wastewater, not just merely the conversion of nitrogen from one form to another form in the wastewater. This latter group includes biological nitrification-denitrification, ion exchange, ammonia stripping and breakpoint chlorination.

The purpose of this manual is the dissemination of the available data on the nitrogen control techniques developed to date. Further, this manual is not simply an assembly of data, rather, data from a variety of sources has been scrutinized and reasonable design criteria drawn on the basis of all available sources. Where design procedures come directly from a single investigator, appropriate reference is made to the work.

This manual could not have been prepared five years ago because of the state of nitrogen control technology at that time. It may well be that continuing research will require an update of this manual in the future. Nonetheless, the body of knowledge on nitrogen control techniques is now well developed and municipalities and local agencies have a firm basis upon which to plan those wastewater treatment facilities which require nitrogen control techniques.

1.2 Scope of the Manual

This manual presents theoretical and process design information on a number of nitrogen control processes. While all of the possible nitrogen removal processes are discussed, details are presented only on those general methods which are most technically and economically feasible, as evidenced by their actual or planned full-scale application. One exception to this is nitrogen control in oxidation ponds; material on nitrogen control in oxidation pond systems was not included because of the paucity of generally applicable design information. Another exception is land treatment; nitrogen removal by land treatment systems is beyond the scope of this manual.

The information in this manual was developed from the following sources: (1) the experience of the individuals involved in the preparation of the manual, (2) the EPA research, development and demonstration program, (3) the literature, (4) from progress reports on on-going projects, (5) from private communication with investigators active in the field, and (6) from operating personnel at existing wastewater treatment plants.

1.3 Guide to the User

A perusal of the table of contents will give the reader a fairly complete picture of the subject matter contained in this manual. The following chapter-by-chapter description is oriented toward providing a general description of the contents of each chapter.

Chapter 2, *Nitrogenous Materials in the Environment and the Need for Control in Wastewater Effluents,* describes the sources of nitrogen compounds entering water bodies, the nitrogen transformations which take place in the environment, and the effects of nitrogen compounds as pollutants. Also given in Chapter 2 is a general introduction into the various types of nitrogen control methods and their applicability to the individual chemical forms of nitrogen. Chapter 2 is useful for establishing the rationale for nitrogen removal.

Chapter 3, *Process Chemistry and Biochemistry of Biological Nitrification and Denitrification,* is a presentation of the basic factors affecting the growth of nitrifying and denitrifying organisms. With an understanding of these factors on a fundamental level, the design concepts evolved in Chapters 4 and 5 can be better appreciated. However, should the reader decide not to involve himself in basic theory, Chapters 4 and 5 are designed to stand by themselves without requiring reference to Chapter 3 except when detailed explanations of individual points are required.

Chapter 4, *Biological Nitrification,* presents design criteria for a wide variety of nitrification processes. Since it has been anticipated that the greatest number of manual users will be concerned with ammonia oxidation, as opposed to nitrogen removal, Chapter 4 presents more material than any other chapter. Both combined carbon oxidation-nitrification and separate stage nitrification systems are described with details, whether given on attached growth or suspended growth processes. The alternative methods for pretreatment for organic carbon removal prior to separate stage nitrification systems are presented. Sections are included on aeration, pH control, and solids-liquid separation.

Chapter 5, *Biological Denitrification,* completes the sequence of the three chapters on the biological approach to nitrogen removal. Design information is provided for both attached growth and suspended growth denitrification systems. For those systems using methanol as the carbon source for denitrification, a section is included describing the methods for chemical handling. The increasingly popular systems using wastewater carbon sources are described in detail. Chapter 5 concludes with a section on solids-liquid separation and a qualitative comparison of the alternative denitrification techniques.

Chapter 6, *Breakpoint Chlorination,* is the first of a set of three chapters on physical-chemical techniques for nitrogen removal. Basic process chemistry is presented along with a host of process design considerations for breakpoint chlorination applications. Because of the importance of process control, details of methods are given. Information is presented on the removal of toxic chlorine residuals.

Chapter 7, *Selective Ion Exchange for Ammonium Removal,* is a presentation of the design concepts involved in the use of clinoptilolite, a natural zeolite exchange material, for ammonium removal from wastewater. Ion exchange fundamentals are discussed along with clinoptilolite properties. Process loading and regeneration relationships are presented. Alternative methods of regenerant recovery are described.

Chapter 8, *Air Stripping for Ammonia Nitrogen Removal,* describes the application of ammonia stripping to wastewater treatment. The air pollution aspects of the method are discussed and general conclusions drawn. The major factors affecting design and process performance are described. The problem of equipment scaling and its control is given detailed consideration. Methods of removing ammonia and controlling the carbon dioxide levels in the stripping air are described.

Chapter 9, *Total System Design,* describes the concepts involved in assembling various unit processes into rational treatment trains that can accomplish not only nitrogen removal, but organics removal and phosphorus removal (where it is required). The main thrust of Chapter 9 is to present actual examples of treatment systems that incorporate the nitrogen control processes described in the previous chapters of this manual. Design concepts that evolved to suit local circumstances are given emphasis.

CHAPTER 2

NITROGENOUS MATERIALS IN THE ENVIRONMENT AND THE NEED FOR CONTROL IN WASTEWATER EFFLUENTS

2.1 Introduction

Various compounds containing the element nitrogen are becoming increasingly important in wastewater management programs because of the many effects that nitrogenous materials in wastewater effluent can have on the environment. Nitrogen, in its various forms, can deplete dissolved oxygen levels in receiving waters, stimulate aquatic growth, exhibit toxicity toward aquatic life, affect chlorine disinfection efficiency, present a public health hazard, and affect the suitability of wastewater for reuse. Biological and chemical processes which occur in wastewater treatment plants and in the natural environment can change the chemical form in which nitrogen exists. Such change may eliminate one deleterious effect of nitrogen while producing, or leaving unchanged, another effect. For example, by converting ammonia in raw wastewater to nitrate, the oxygen-depleting and toxic effects of ammonia are eliminated, but the biostimulatory effects may not be changed significantly.

It is important, therefore, prior to the detailed discussions of nitrogen removal processes which form the principal content of this manual, to review the chemistry of nitrogen and the effects that the various compounds can have. Several specific aspects are discussed in this chapter. First, the nitrogen cycle for both surface water and soil/groundwater environments is described, with emphasis on the important compounds and reactions associated with each. Second, sources of nitrogen, both natural and man-caused, are discussed. Important elements of the latter category include domestic and industrial wastewater, urban and suburban runoff, surface and subsurface agricultural drainage, and emissions to the atmosphere which may eventually enter the aquatic environment through precipitation or dustfall. Then, the effects of nitrogen discharge to surface water, groundwater, and land are summarized. And finally, introductory to the following chapters, a brief discussion is presented on the relationship between the various nitrogen compounds and process removal efficiency.

2.2. The Nitrogen Cycle

Nitrogen exists in many compounds because of the high number of oxidation states it can assume. In ammonia or organic compounds, the form most closely associated with plants and animals, its oxidation state is minus 3. At the other extreme its oxidation state is plus 5 when in the nitrate form. In the environment, changes from one oxidation state to another can be brought about biologically by living organisms. The relationship between the various compounds and the transformations which can occur are often presented schematically in a diagram known as the nitrogen cycle. Figure 2-1 shows a common manner of presentation.[1] The atmosphere serves as a reservoir of N_2 gas from which nitrogen is removed naturally by electrical discharge and nitrogen-fixing organisms and artificially by chemical manufacturing. Nitrogen gas is returned to the atmosphere by the action of denitrifying organisms. In the fixed state, nitrogen can undergo the various reactions shown. A general description of the nitrogen cycle is presented here, and aspects of particular importance to surface water and soil/groundwater environments are discussed in the following sections.

FIGURE 2-1

THE NITROGEN CYCLE (AFTER REFERENCE 1)

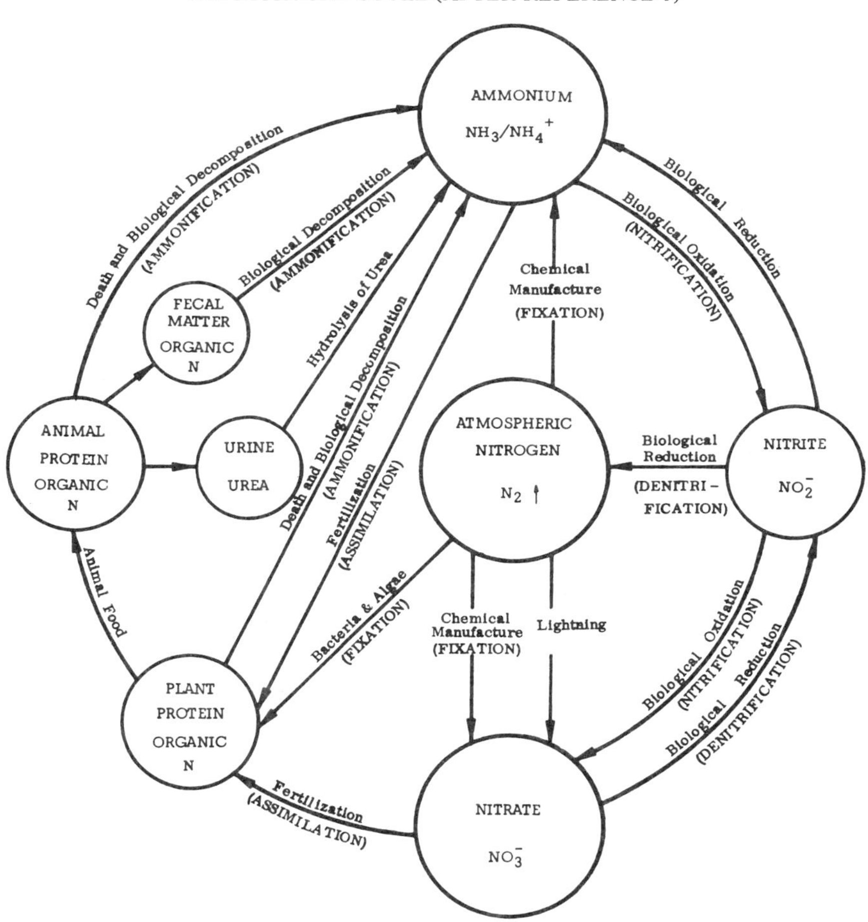

Transformation reactions of importance include fixation, ammonification, assimilation, nitrification and denitrification.[2] These reactions can be carried out by particular microorganisms with either a net gain or loss of energy; energy considerations often play an important role in determining the reaction which occurs. The principal compounds of concern in the nitrogen cycle are nitrogen gas, ammonium, organic nitrogen, and nitrate. These compounds and their oxidation states are shown below:

$$\overset{-3}{NH_3 / NH_4^+} - \overset{0}{N_2} - \overset{+3}{NO_2^-} - \overset{+5}{NO_3^-}$$

$$\updownarrow$$

Organic
Derivatives

It is important to note that at neutral pH values there is very little molecular ammonia (NH_3) in wastewater as most is in the form of the ammonium ion (NH_4^+). The distribution of ammonia and ammonium as a function of pH is discussed in Section 6.1.1.

Fixation of nitrogen from N_2 gas to organic nitrogen is accomplished biologically by specialized microorganisms. This reaction requires an investment of energy. Biological fixation accounts for most of the natural transformation of nitrogen to compounds which can be used by plant and animal life. Lightning fixation has been estimated to account for approximately 15 percent of the total which occurs naturally.[3] Industrial fixation was initially developed in the early 20th Century for manufacture of both fertilizer and explosives. Presently, nitrogen fixed by industry is about half the amount that is removed from the atmosphere by natural means.

Ammonification is the change from organic nitrogen to the ammonium (NH_3/NH_4^+) form. This occurs to dead animal and plant tissue and to animal fecal matter.

$$\text{Protein (organic N)} + \text{microorganisms} \longrightarrow NH_3/NH_4^+$$

Nitrogen in urine exists principally as urea. Urea is hydrolyzed by the enzyme urease to ammonium carbonate.

$$H_2NCONH_2 + 2H_2O \xrightarrow[\text{Urease}]{\text{Enzyme}} (NH_4)_2CO_3$$

Assimilation is the use of ammonium or nitrate compounds to form plant protein and other nitrogen-containing compounds.

$$NO_3^- + CO_2 + \text{green plants} + \text{sunlight} \longrightarrow \text{protein}$$

$$NH_3/NH_4^+ + CO_2 + \text{green plants} + \text{sunlight} \longrightarrow \text{protein}$$

Animals require protein from plants or from other animals. With certain specific exceptions, they are incapable of converting inorganic nitrogen forms into organic forms.

The term "nitrification" is applied to the biological oxidation of ammonium, first to the nitrite, then to the nitrate, form. The bacteria responsible for these reactions are termed chemoautotrophic because they use inorganic chemicals as their source of energy. Generally, the *Nitrosomonas* genera are involved in conversion of ammonium to nitrite under aerobic conditions as follows:

$$2NH_4^+ + 3O_2 \xrightarrow{\text{bacteria}} 2NO_2^- + 4H^+ + 2H_2O$$

The nitrites are in turn oxidized to nitrate generally by *Nitrobacter* according to the following reaction:

$$2NO_2^- + O_2 \xrightarrow{\text{bacteria}} 2NO_3^-$$

The overall nitrification reaction is as follows:

$$NH_4^+ + 2O_2 \longrightarrow NO_3^- + 2N^+ + H_2O$$

To oxidize 1 mg/l of ammonia-nitrogen requires about 4.6 mg/l of oxygen when synthesis of nitrifiers is neglected. The nitrate thus formed may be used in assimilation as described above to promote plant growth, or it may be used in denitrification, wherein through biological reduction, first nitrite and then nitrogen gas are formed. A fairly broad range of bacteria can accomplish denitrification, including *Psuedomonas, Micrococcus, Achromobacter,* and *Bacillus.* In simplified form, the reaction steps are as follows:

$$NO_3^- + 0.33\ \underset{\substack{\text{(organic carbon}\\ \text{source)}}}{CH_3OH} \longrightarrow NO_2^- + 0.33\ CO_2 + 0.67\ H_2O$$

$$NO_2^- + 0.5\ \underset{\substack{\text{(organic carbon}\\ \text{source)}}}{CH_3OH} \longrightarrow 0.5N_2 + 0.5\ H_2O + OH^- + 0.5\ CO_2$$

Here methanol is used as the example organic carbon source, although many natural and synthetic organic compounds can serve as the carbon source for denitrification.

Oxidation of organic matter to carbon dioxide and water furnishes energy for bacteria. Either oxygen or nitrate may be used for the oxidation, but the use of oxygen results in the release of more energy. When both oxygen and nitrate are present, bacteria preferentially use oxygen. Therefore, use of nitrate for denitrification can only occur under anoxic conditions, an important consideration when attempting to remove nitrate from wastewater.

Nitrite, since it is an intermediate in the nitrification and denitrification processes, can link the nitrification and denitrification steps directly without passing through nitrate. First, nitrite is formed from oxidation of ammonium by *Nitrosomonas*, then nitrite can be denitrified to nitrogen gas. By this route less oxygen is required for nitrification and less organic matter (energy) is required for denitrification. This is a special case, however, and not broadly applicable to municipal wastewater treatment.

In discussing the nitrogen cycle, it is useful to differentiate between the surface water and sediment environment and the soil/groundwater environment. This aids in understanding the roles that nitrogenous compounds play in each and the problems which can be encountered.

2.2.1 The Nitrogen Cycle in Surface Waters and Sediments

A modified representation of the nitrogen cycle applicable to the surface water environment is presented in Figure 2-2.[4] Nitrogen can be added by precipitation and dustfall, surface runoff, subsurface groundwater entry, and direct discharge of wastewater effluent. In addition, nitrogen from the atmosphere can be fixed by certain photosynthetic blue-green algae and some bacterial species.

Within the aquatic environment ammonification, nitrification, assimilation, and denitrification can occur as shown in Figure 2-2. Ammonification of organic matter is carried out by microorganisms. The ammonium thus formed, along with nitrate, can be assimilated by algae and aquatic plants; such growths may create water quality problems.

Nitrification of ammonium can occur with a resulting depletion of the dissolved oxygen content of the water. To oxidize 1.0 mg/l of ammonia-nitrogen, 4.6 mg/l of oxygen is required.

Denitrification produces nitrogen gas which may escape to the atmosphere. Because anoxic conditions are required, the oxygen-deficient hypolimnion (or lower layer) of lakes and the sediment zone of streams and lakes are important zones of denitrification action.[4]

2.2.2 The Nitrogen Cycle in Soil Groundwater

Figure 2-3 shows the major aspects of the nitrogen cycle associated with the soil/groundwater environment.[5] Nitrogen can enter the soil from wastewater or wastewater effluent, artificial fertilizers, plant and animal matter, precipitation, and dustfall. In addition,

FIGURE 2-2
THE NITROGEN CYCLE IN SURFACE WATER (AFTER REFERENCE 4)

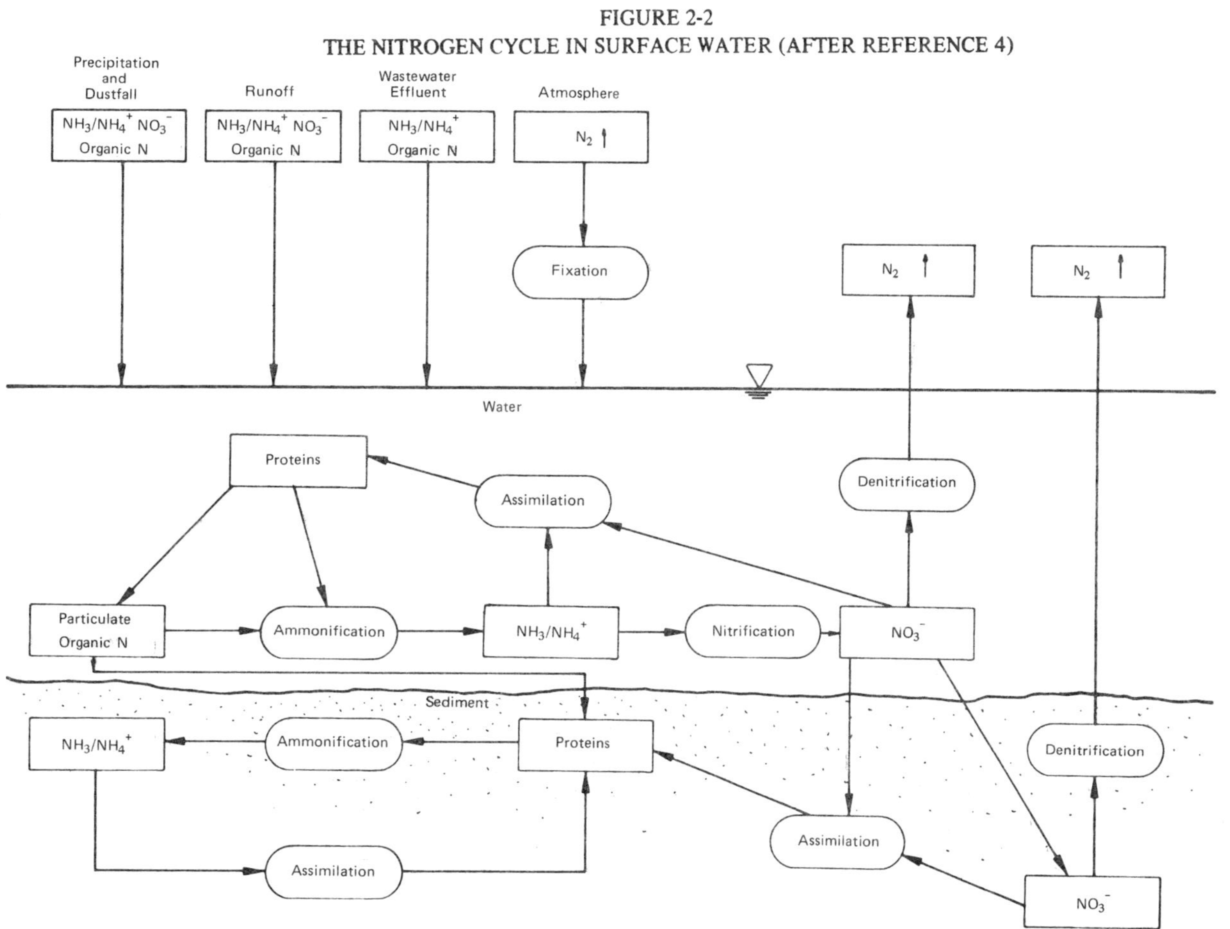

FIGURE 2-3

THE NITROGEN CYCLE IN SOIL AND GROUNDWATER (AFTER REFERENCE 5)

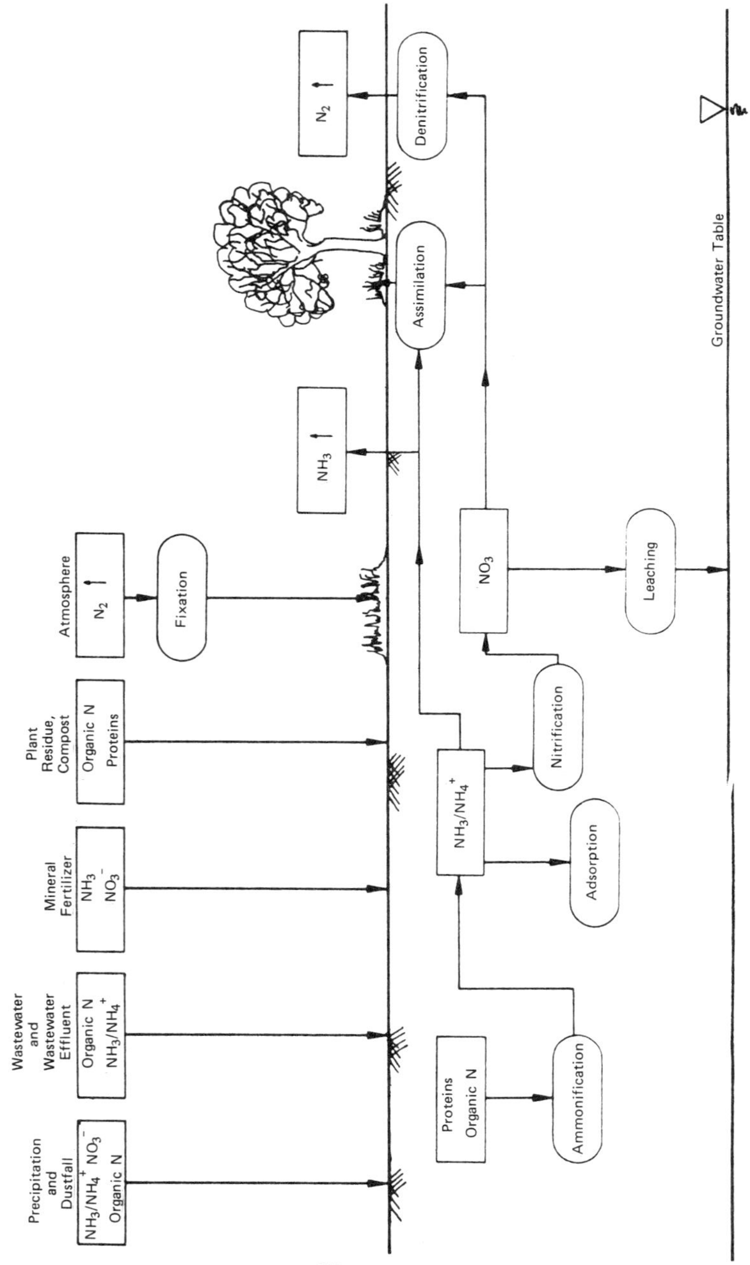

Precipitation and Dustfall
NH_3/NH_4^+ NO_3^-
Organic N

Wastewater and Wastewater Effluent
Organic N
NH_3/NH_4^+

Mineral Fertilizer
NH_3
NO_3^-

Plant Residue, Compost
Organic N
Proteins

Atmosphere
N_2 ↑

Fixation

NH_3 ↑

N_2 ↑

Proteins
Organic N

Ammonification

NH_3/NH_4^+

Adsorption

Nitrification

NO_3

Leaching

Assimilation

Denitrification

Groundwater Table

nitrogen-fixing bacteria convert nitrogen gas into forms available to plant life. Man has increased the amount of nitrogen fixed biologically by cultivation of leguminous crops (e.g., peas and beans). It is estimated that nitrogen fixed by legumes now accounts for approximately 25 percent of the total fixed.[3]

Usually more than 90 percent of the nitrogen present in soil is organic, either in living plants and animals or in humus originating from decomposition of plant and animal residues. Most of the remainder is ammonium (NH_4^+), which is tightly bound to soil particles.

The nitrate content is generally low due to assimilation by plant roots and leaching by water percolating through the soil. Nitrate pollution is the principal groundwater quality problem in many areas. Denitrification, which is the dominating reaction below the aerobic top layer of soil, rarely removes all nitrates added to the soil from fertilizers or wastewater effluents. Thus, most of the nitrogen which is not assimilated by plant growth eventually enters the groundwater table in the nitrate form.

2.3 Sources of Nitrogen

Nitrogenous materials may enter the aquatic environment from either natural or man-caused sources. Further, the quantities from natural sources are often increased by man's activity. For example, while some nitrogen may be expected in rainfall, the combustion of fossil fuels or the application of liquid ammonia agricultural fertilizers with subsequent release to the air through volatilization can increase rainfall concentrations of nitrogen substantially. It is useful to have an understanding of the various sources of nitrogenous materials and to have an appreciation of the quantities of nitrogen which may be expected from each.

Although the source of nitrogen causing a specific pollution problem is often obvious, difficulty may be encountered in determining which of several possible sources is most important. As an example, if a stream with excessive aquatic growths due to nitrogen receives effluent from a sewage treatment plant, drainage from fertilized cropland, and runoff from pastures or feedlots, the contribution of nitrogen from the treatment plant may be a small fraction of that from the other two sources. Thus, in analyzing a nitrogen pollution problem, care must be taken to ensure that all possible sources are investigated and that the amount to be expected from each is accurately estimated. Once an estimate is made, nitrogen control measures can be oriented toward the more significant sources.

2.3.1 Natural Sources

Natural sources of nitrogenous substances include precipitation, dustfall, nonurban runoff, and biological fixation. Amounts from all may be increased in some way by man. It may be quite difficult to determine quantities which might be expected under completely natural conditions.

In order to find levels of nitrogenous substances in precipitation which are as close to "natural" as possible, it is necessary to take samples far from urban or agricultural areas.

Even these values may be suspect, however. In one review of nutrient levels in precipitation, total nitrogen in rainfall in Sweden was cited as 0.2 mg/l.[6] The average concentration of nitrogen in western snow samples, mainly in the Sierra Nevada Mountains, was 0.15 ppm of ammonia-nitrogen, 0.01 ppm of nitrite-nitrogen and 0.02 ppm of nitrate-nitrogen. How representative such values are of "natural" conditions cannot be determined with any certainty.

The quantities of nitrogen in nonurban runoff from non-fertilized land may be expected to vary greatly, depending on the erosive characteristics of the soil. One study found that runoff from forested land in Washington contained 0.13 mg/l of nitrate-nitrogen and 0.20 mg/l of total nitrogen.[7]

Biological fixation may add nitrogen to both soil and surface water environments. Of particular interest is the role of fixation in eutrophication of lakes. Certain photosynthetic blue-green algae, such as the species of *Nostoc, Anabaena, Gleotrichia* and *Calothrix*, are common nitrogen fixers.[4]

As much as 14 percent of the total nitrogen entering eutrophic Lake Mendota, Wisconsin, was added by fixation.[4] The role of nitrogen fixation in oligotrophic lakes has not been established.

2.3.2 Man-caused Sources

The activities of man may increase quantities of nitrogen added to the aquatic environment from three of the sources discussed above: precipitation, dustfall, and nonurban runoff. These sources are increased principally by fertilization of agricultural land and the combustion of fossil fuels.

Other man-related sources include runoff from urban areas and livestock feedlots, municipal wastewater effluents, subsurface drainage from agricultural lands and from septic tank leach fields, and industrial wastewaters.

Nitrogen concentrations in raw municipal wastewaters are well documented.[4,8,9] Values generally range from 15 to 50 mg/l, of which approximately 60 percent is ammonia-nitrogen, 40 percent is organic nitrogen, and a negligible amount (one percent) is nitrite- and nitrate-nitrogen. Unless wastewater treatment facilities are designed to remove nitrogen specifically, most will pass through the treatment works to the receiving waters or land disposal site. An estimate for the total amount of nitrogen discharged into sewerage systems in domestic wastewater is 0.84 million metric tons per year in the United States.[9]

Nitrogen discharged into individual septic tank systems can also create pollution problems. It has been estimated that up to 25 percent of the national population utilizes individual systems,[9] contributing up to 0.23 million metric tons of nitrogen annually. In a well-operating septic tank system, most of the nitrogen leaving the tank will be converted to nitrate in the leaching field. This may then percolate downward to a groundwater table.

Problems from high nitrate concentrations occasionally occur when septic tank waste disposal is located near shallow wells used for water supply, particularly on the fringes of urban areas where the population density may be fairly high.

The nitrogen content of industrial wastes varies dramatically from one industry to the next. Among those industries whose wastewater nitrogen contents may be quite high are meat processing plants, milk processing plants, petroleum refineries, ice plants, fertilizer manufacturers, certain synthetic fiber plants, and industries using ammonia for scouring and cleaning operations.[4]

Feedlot runoff constitutes a source of nitrogen which has become significant as a result of the increased number of concentrated, centralized feedlots. Ammonium is a major constituent of feedlot waste as a result of urea hydrolysis. Ammonia-nitrogen concentrations may reach 300 mg/l,[4,8,10] and organic nitrogen concentrations of up to 600 mg/l have been reported.[8,10] The total annual nitrogen load from livestock in the U.S. is estimated to be 6.0 million metric tons.[4] While the majority of the animals are apparently still raised on small farms, the trend toward feedlot operations is continuing, and unless steps are taken to prevent drainage and runoff, serious localized problems can occur.

Urban runoff can contribute significant quantities of nitrogen to receiving waters during and after periods of precipitation. Average concentrations which have been reported are 2.7 mg/l total nitrogen in Cincinnati,[11] 2.1 mg/l total nitrogen in Washington, D.C.,[12] 2.5 mg/l total nitrogen in Ann Arbor, Michigan,[13] and 0.85 mg/l organic nitrogen in Tulsa, Oklahoma.[14] Sanitary or combined sewer overflows can also add to the nitrogen load.

The use of artificial fertilizers has increased the nitrogen concentrations which can be expected in nonurban runoff. In rural Ohio, runoff from a 1.45 acre field planted in winter wheat contained an average of 9 mg/l total nitrogen.[15] For agricultural land in Washington, the nitrate-nitrogen concentration was 1.25 mg/l.[7] On a 75-acre site in North Carolina which consisted of grassed pasture, wooded pasture, corn field, and orchard, the mean nitrogen concentration in the runoff was 1.2 mg/l.[16]

Subsurface irrigation drainage from fertilized cropland can contain high concentrations of nitrates. In agricultural areas of California's San Joaquin Valley, monitoring of subsurface tile drainage systems between 1966 and 1968 showed average nitrate-nitrogen concentrations of 19.3 mg/l.[17]

In the same way that increased nitrogen concentrations in nonurban runoff and subsurface drainage have been caused by man's activities, increased nitrogen levels in precipitation and dustfall have also resulted. For example, high ammonium concentrations in spring rains in California are due to the use of liquid ammonium fertilizers there.[6] Most atmospheric nitrogen (other than nitrogen gas), however, is associated with soil picked up by the wind and can be returned to earth by gravitational settling (dry fallout) or in precipitation, and several studies have been conducted to determine the quantities to be expected from such sources. The 10-year average of ammonia- plus nitrate-nitrogen concentrations in rainfall at Geneva, New York, was 1.1 mg/l.[6] Snow samples from Ottawa, Canada, over 17 years contained an average of 0.85 ppm inorganic nitrogen.[6] Rainwater from the same area for the same period had concentrations of 1.8 mg/l ammonia-nitrogen and 0.35 mg/l nitrate-nitrogen. In rainfall measurements at Cincinnati, Ohio, total and inorganic nitrogen concentrations were 1.27 and 0.69 mg/l, respectively.[15] For a rural area near Coshocton, Ohio, the respective concentrations were 1.17 and 0.80 mg/l.[15]

A study made near Hamilton, Ontario, was cited[6] which related dustfall to rainfall. It was found that the nitrogen fall totaled 5.8 lb per acre per year. Approximately 61 percent of the nitrogen came down on rainy days, which constituted 25 percent of the days monitored during the test.

In a study on dustfall in Seattle[18] the fall rate for soluble nitrate-nitrogen was 0.63 lb per acre per year. The concentration of nitrate-nitrogen in the total dustfall was 700 ppm.

As a summary to this discussion of sources of nitrogen, Table 2-1 shows estimates of nitrogen quantities discharged from various sources in the San Francisco Bay Basin, California.[19] The bay basin has a population of about 4,500,000 people, a land area of 4,300 square miles, and a water surface area of about 450 square miles. Because of the high population density, the greatest amount of nitrogen discharged is from municipal and industrial sources. This table is presented only as an example. Care must be taken for each case to accurately evaluate the significance of each source.

TABLE 2-1

ESTIMATED NITROGEN LOADINGS FOR THE SAN FRANCISCO BAY BASIN

Identified Nitrogen Source[a]	Nitrogen mass emission, thousand lb per.year	(thousand kg per year)	Percent of total
Municipal wastewater, before treatment	55,000	(26,000)	49
Industrial wastewater, before treatment	35,000	(16,000)	30
Vessel wastes, before treatment	130	(60)	0.1
Dustfall directly on Bay	1,300	(590)	1.1
Rainfall directly on Bay	870	(390)	0.8
Urban runoff	3,100	(1,400)	2.7
Non-urban runoff	4,100	(1,900)	3.6
Nitrogen applied to irrigated agricultural land[b]	2,000	(900)	1.7
Nitrogen from dairies and feedlots[b]	13,000	(6,000)	11
Total	118,000	(53,000)	100

From Reference 19

[a]A major source not included is biological fixation

[b]An estimated 50 percent percolates to groundwater

2.4 Effects of Nitrogen Discharge

It was previously noted that nitrogenous compounds discharged from wastewater treatment facilities can have several deleterious effects. Although biostimulation of receiving waters has generated the most concern in recent years, other less well publicized impacts can be of major importance in particular situations. These impacts include toxicity to fish life, reduction of chlorine disinfection efficiency, an increase in the dissolved oxygen depletion in receiving waters, adverse public health effects – principally in groundwater, and a reduction in the suitability for reuse.

2.4.1 Biostimulation of Surface Waters

A major problem in the field of water pollution is eutrophication, excessive plant growth and/or algae "blooms" resulting from over-fertilization of rivers, lakes, and estuaries. Results of eutrophication include deterioration in the appearance of previously clear waters, odor problems from decomposing algae, and a lower dissolved oxygen level which can adversely affect fish life.

Four basic factors are required for algal growth: nitrogen, phosphorus, carbon dioxide, and light energy. The absence of any one will limit growth. In special cases, trace micronutrients such as cobalt, iron, molybdenum and manganese may be limiting factors under natural conditions.

Good generalizations concerning which factor is growth limiting and at what concentration are difficult to make. Light and carbon dioxide are essentially impossible to control. Both nitrogen and phosphorus are present in waste discharges and hence subject to control. The questions which must usually be answered when faced with a eutrophication problem are: is nitrogen or phosphorus (or neither) the limiting nutrient, and if either one is, can the amount entering the receiving water be significantly reduced by removing that nutrient from the waste stream? In some cases algal assay procedures may allow a conclusion as to which nutrient is limiting. Under some circumstances, however, removal of both nitrogen and phosphorus may be undertaken to limit algal growth.

Eutrophication is of most concern in lakes because nutrients which enter tend to be recycled within the lake and build up over a period of time.[9] A river, by contrast, is a flowing system. Nutrients are always entering or leaving any given section. Accumulations tend to occur only in sediment or in slack water, and the effects of these accumulations are normally moderated by periodic flushing by floods.

In estuaries and oceans, nitrogen compounds are often present in very low concentrations and may limit the total biomass and the types of species it contains.[9] Thus, upwelling, which brings nutrient-rich waters to the surface, may result in periodic blooms of algae or other aquatic life. While in some estuaries discharges from wastewater treatment plants may increase nitrogen concentrations to the level where blooms occur, the high dilution provided by a direct ocean discharge probably eliminates the danger of algae blooms caused by such discharges. In summary, while nitrogen in wastewater treatment plant effluents can in particular cases cause undesirable aquatic growths, determination of the limiting constituent and other sources of that constituent (such as feedlot runoff or fixation) should be made before the decision is made to require nitrogen removal from municipal wastewaters.

2.4.2 Toxicity

The principal toxicity problem is from ammonia in the molecular form (NH_3) which can adversely affect fish life in receiving waters. A slight increase in pH may cause a great increase in toxicity as the ammonium ion (NH_4^+) is transformed to ammonia in accordance with the following equation.

$$NH_4^+ + OH^- \rightleftharpoons NH_3 + H_2O$$

Factors which may increase ammonia toxicity at a given pH are: greater concentrations of dissolved oxygen and carbon dioxide; elevated temperatures; and bicarbonate alkalinity.[9] Reported levels at which acute toxicity is detectable have ranged from 0.01 mg/l[9] to over 2.0 mg/l[20] of molecular ammonia-nitrogen.

2.4.3 Effect on Disinfection Efficiency

When chlorine, in the form of chlorine gas or hypochlorite salt, is added to wastewater containing ammonium, chloramines, which are less effective disinfectants, are formed. The major reactions are as follows:

$$NH_4^+ + HOCl \rightleftharpoons NH_2Cl \text{ (monochloramine)} + H_2O + H^+$$

$$NH_2Cl + HOCl \rightleftharpoons NHCl_2 \text{ (dichloramine)} + H_2O$$

$$NHCl_2 + HOCl \rightleftharpoons NCl_3 \text{ (nitrogen trichloride)} + H_2O$$

Only after the addition of large quantities of chlorine does free available chlorine exist. If the effluent ammonia-nitrogen concentration were 20 mg/l, about 200 mg/l of chlorine would be required to complete the reactions with ammonium and organic compounds. Only rarely in wastewater treatment is this level of chlorine addition ("breakpoint" chlorination) used. Therefore, as a practical matter, the less effective combined chlorine residuals (monochloramine and dichloramine) must be relied upon for disinfection. This results in increased chlorine dose requirements for the same level of disinfection. Further information on the relative effectiveness of free chlorine and combined residuals is presented in Section 6.2.7.

2.4.4 Dissolved Oxygen Depletion in Receiving Waters

Ammonium can be biologically oxidized to nitrite and then to nitrate in receiving waters and thereby add to the oxygen demand imparted by carbonaceous materials. Table 2-2 shows a typical example of the removal of total oxygen demand obtainable with varying degrees of treatment. If either conventional biological treatment or physical-chemical treatment is utilized to provide 90 percent BOD_5 removal, an effluent will be discharged which still contains over 100 mg/l of oxygen demand. This high level of oxygen demand may cause significant oxygen depletion in the receiving water if insufficient dilution is available. Nitrification (or ammonia nitrogen removal) will reduce the total oxygen demand of the effluent to less than 40 mg/l.

The Potomac Estuary in the United States[22] and the Thames Estuary in Great Britain[23] are examples of estuaries which are greatly affected by nitrification. Figure 2-4 shows, as a function of the degree of nitrification provided by wastewater treatment facilities, the estimated discharge into the Thames Estuary which will cause the maximum oxygen depletion to be 10 percent of saturation. The calculation assumes an effluent BOD_5 of 20 mg/l, an effluent organic plus ammonia-nitrogen concentration of 19 mg/l, and discharge at a point 10 miles above London Bridge. From the figure, the allowable discharge for non-nitrified effluent is about 12 mgd, while for completely nitrified effluent, over 40 mgd can be discharged.

2.4.5 Public Health

The public health hazard from nitrogen is associated with the nitrate form and is limited principally to groundwater where high concentrations can occur. Nitrate in drinking water

TABLE 2-2

EFFECT OF AMMONIUM REMOVAL ON TOTAL OXYGEN DEMAND OF WASTEWATER TREATMENT PLANT EFFLUENT

Parameter	Raw wastewater[a]	Final effluent: Organic carbon removal only[a]	Final effluent: With ammonium and organic carbon removal
Organic matter, mg/l	250	25	20
Organic oxygen demand, mg/l	375[b]	37	30
Organic and ammonia nitrogen, mg/l	25	20	1.5
Nitrogenous oxygen demand (NOD) mg/l	115[c]	92[c]	7[c]
Total oxygen demand (TOD) mg/l	490	129	37
Percent of TOD due to nitrogen	23.5	71.3	18.9
Percent organic oxygen demand removed	-	90	92
Percent of TOD removed	-	73.7	92.5

[a]After Reference 21
[b]Taken as 1.5 times organic matter
[c]Taken as 4.6 times the nitrogen level

was first associated in 1945 with methemoglobinemia, a sometimes fatal blood disorder which affects infants less than three months old. When water high in nitrate is used for preparing infant formulas, nitrate is reduced to nitrite in the stomach after ingestion. The nitrites react with hemoglobin in the blood to form methemoglobin, which is incapable of carrying oxygen. The result is suffocation accompanied by a bluish tinge to the skin, which accounts for the use of the term "blue babies" in conjunction with methemoglobinemia. In suspect areas water should be analyzed for both nitrite and nitrate since either form will cause methemoglobinemia.

FIGURE 2-4

ALLOWABLE EFFLUENT DISCHARGE INTO THE THAMES ESTUARY (AFTER REFERENCE 23)

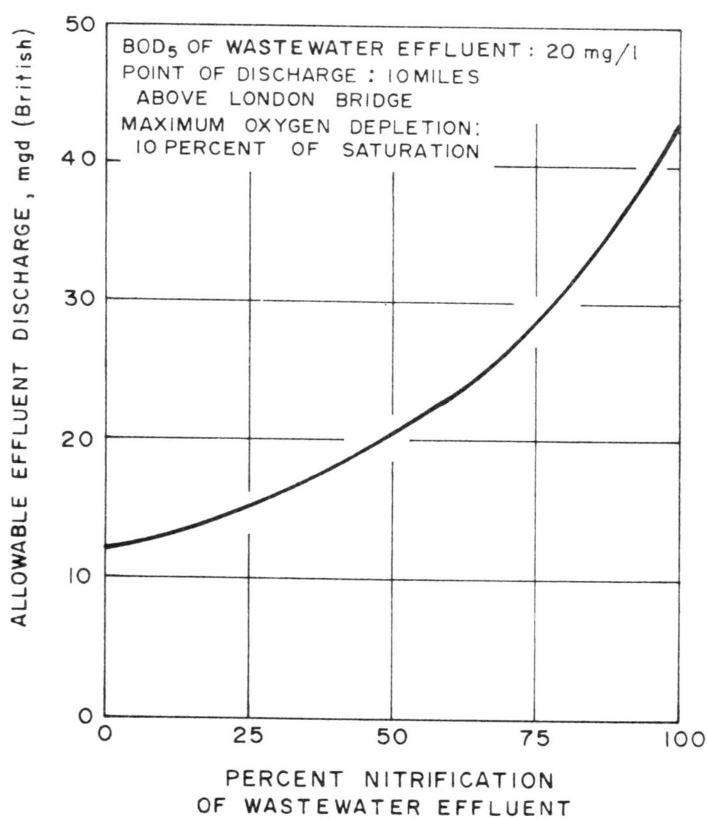

Since 1945 about 2,000 cases of methemoglobinemia have been reported in the U.S. and Europe, with a mortality rate of seven to eight percent. Because of difficulty in diagnosing the disease and because no reporting is required, the actual incidence may be many times higher.[10]

The EPA's interim primary drinking water standard (40 CFR Part 141) for nitrate is 10 mg/l as nitrogen. This standard is exceeded most often in shallow wells in rural areas.

2.4.6 Water Reuse

While direct wastewater reuse for domestic water supply is not yet a reality because of public health considerations, plans for industrial reuse are being carried out in several areas. When reclaiming wastewater for industrial purposes, ammonia may need to be removed in order to prevent corrosion. Further, nitrogen compounds can cause biostimulation in cooling towers and distribution structures.

2.5 Treatment Processes for Nitrogen Removal

In the past several years the number of processes utilized in wastewater treatment has increased rapidly. Many of these processes have been developed with the specific purpose of transforming nitrogen compounds or removing nitrogen from the wastewater stream. Others can remove several compounds, including significant amounts of nitrogen. Still others may remove only a small amount of nitrogen or a particular form of nitrogen which is a small fraction of the total.

In determining which method is most suitable for a particular application, consideration must be given to six principal aspects: (1) form and concentration of nitrogen compounds in the process influent, (2) required effluent quality, (3) other treatment processes to be employed, (4) cost, (5) reliability, and (6) flexibility. Great care must be taken in developing and evaluating alternatives.

Presented below are brief descriptions of the various processes employed in wastewater treatment facilities which, to varying degrees, remove nitrogen from the waste stream. Process characteristics, compound selectivity, and normal range of efficiency are presented. It is stressed that this discussion is descriptive and is intended only to provide an introduction to the following chapters of this manual.

2.5.1 Conventional Treatment Processes

Nitrogen in raw domestic wastewaters is principally in the form of organic nitrogen, both soluble and particulate, and ammonia. The soluble organic nitrogen is mainly in the form of urea and amino acids. Primary sedimentation acts to remove a portion of the particulate organic matter. This generally will amount to less than 20 percent of the total nitrogen entering the plant.

Biological treatment will remove more particulate organic nitrogen and transform some to ammonium and other inorganic forms. A fraction of the ammonium present in the waste will be assimilated into organic materials of cells formed by the biological process. Soluble organic nitrogen is partially transformed to ammonium by microorganisms, but concentrations of 1 to 3 mg/l are usually found in biological treatment effluents.[24] Through these processes, an additional 10 to 20 percent of the total nitrogen is removed when biological treatment and secondary sedimentation follows primary sedimentation. Thus, total nitrogen removal for a conventional primary-secondary facility will generally be less than about 30 percent.

2.5.2 Advanced Wastewater Treatment Processes

Advanced treatment processes designed to remove wastewater constituents other than nitrogen often remove some nitrogen compounds as well. Removal is often restricted to particulate forms, and overall removal efficiency is rarely high.

Tertiary filtration can remove a significant fraction of the organic nitrogen present. Overall removal depends on the amount of nitrogen in the suspended organic form. As noted above, most of the organic nitrogen in secondary effluent is insoluble, but ammonium usually accounts for the majority of the total nitrogen. Carbon adsorption, used to remove residual organics, will also remove organic nitrogen. The amount of organic nitrogen remaining at that point in the treatment scheme will generally be quite small.

Electrodialysis and reverse osmosis are tertiary processes used primarily for reduction of total dissolved solids. Nitrogen entering such systems is mainly in the ammonium or nitrate form. Electrodialysis can be expected to remove about 40 percent of these forms; reverse osmosis, 80 percent. However, these processes are not currently in use for treatment of municipal wastewater.

Chemical coagulation, often utilized for phosphate removal, also aids in removal of particulate matter, including particulate organic nitrogen. While chemical coagulation does not remove ammonium directly, lime addition is used prior to ammonia stripping (discussed in Section 2.5.3.4) in order to raise the pH and allow the process to proceed.

Land disposal may be used to remove nitrogen. Removal occurs when the effluent is used for irrigation purposes with the nitrogen assimilated by growing crops which are subsequently harvested. However, nitrogen removal by land treatment systems is not within the scope of this manual.

2.5.3 Major Nitrogen Removal Processes

The major processes considered in this manual are nitrification-denitrification, breakpoint chlorination (or superchlorination), selective ion exchange for ammonium removal, and air stripping for ammonia removal (ammonia stripping). These are the processes which are technically and economically most viable at the present time.

2.5.3.1 Biological Nitrification-Denitrification

Biological nitrification does not increase the removal of nitrogen from the waste stream over that achieved by conventional biological treatment. The principal effect of the nitrification treatment process is to transform ammonia-nitrogen to nitrate. The nitrified effluent can then be denitrified biologically. Nitrification is also used without subsequent biological denitrification when treatment requirements call for oxidation of ammonia-nitrogen. Oxidation of ammonium can be as high as 98 percent. Overall transformation to nitrate depends on the extent to which organic nitrogen is transformed to ammonia-nitrogen in the secondary stage or is removed by another process. Nitrification can be carried out in conjunction with secondary treatment or in a tertiary stage; in both cases, either suspended growth reactors (activated sludge) or attached growth reactors (such as trickling filters) can be used.

Biological denitrification can also be carried out in either suspended growth or attached growth reactors. As previously noted, an anoxic environment is required for the reactions to proceed. Overall removal efficiency in a nitrification-denitrification plant can range from 70 to 95 percent.

2.5.3.2 Breakpoint Chlorination

Breakpoint chlorination (or superchlorination) is accomplished by the addition of chlorine to the waste stream in an amount sufficient to oxidize ammonia-nitrogen to nitrogen gas. After sufficient chlorine is added to oxidize the organic matter and other readily oxidizable substances present, a stepwise reaction of chlorine with ammonium takes place. The overall theoretical reaction is as follows:

$$3Cl_2 + 2NH_4^+ \longrightarrow N_2 + 6HCl + 2H^+$$

In practice, approximately 10 mg/l of chlorine is required for every 1 mg/l of ammonia-nitrogen. In addition, acidity produced by the reaction must be neutralized by the addition of caustic soda or lime. These chemicals add greatly to the total dissolved solids and result in a substantial operating expense. Often dechlorination is utilized following breakpoint chlorination in order to reduce the toxicity of the chlorine residual in the effluent.

An important advantage of this method is that ammonia-nitrogen concentrations can be reduced to near zero in the effluent. The effect of breakpoint chlorination on organic nitrogen is uncertain, with contradictory results presented in the literature. Nitrite and nitrate are not removed by this method.

2.5.3.3 Selective Ion Exchange for Ammonium Removal

Selective ion exchange for removal of ammonium from wastewater can be accomplished by passing the wastewater through a column of clinoptilolite, a naturally occurring zeolite which has a high selectivity for ammonium ion. The first extensive study was undertaken in 1969 by Battelle Northwest in a federally sponsored demonstration project. Regeneration of the clinoptilolite is undertaken when all the exchange sites are utilized and breakthrough occurs.

Filtration prior to ion exchange is usually required to prevent fouling of the zeolite. Ammonium removals of 90-97 percent can be expected. Nitrite, nitrate, and organic nitrogen are not affected by this process.

2.5.3.4 Air Stripping for Ammonia Removal

Ammonia in the molecular form is a gas which dissolves in water to an extent controlled by the partial pressure of the ammonia in the air adjacent to the water. Reducing the partial pressure causes ammonia to leave the water phase and enter the air. Ammonia removal from wastewater can be effected by bringing small drops of water in contact with a large amount of ammonia-free air. This physical process is termed desorption, but the common name is "ammonia stripping."

In order to strip ammonia from wastewater, it must be in the molecular form (NH_3) rather than the ammonium ion (NH_4^+) form. This is accomplished by raising the pH of the wastewater to 10 or 11, usually by the addition of lime. Because lime addition is often used for phosphate removal, it can serve a dual role. Again, nitrite, nitrate, and organic nitrogen are not affected.

The principal problems associated with ammonia stripping are its inefficiency in cold weather, required shutdown during freezing conditions, and formation of calcium carbonate scale in the air stripping tower.

The effect of cold weather has been well documented at the South Lake Tahoe Public Utility District where ammonia stripping is used for a 3.75 mgd tertiary facility. The stripping tower is designed to remover 90 percent of the incoming ammonium during warm weather. During freezing conditions, the tower is shut down. One mechanism of scale formation is attributed to the carbon dioxide in the air reacting with the alkaline wastewater and precipitating as calcium carbonate.[22] In some instances, removal with a water jet has been possible; in other applications the scale has been extremely difficult to remove. Some factors which may affect the nature of the scale are: orientation of air flow, recirculation of sludge, pH of the wastewater, and chemical makeup of the wastewater.[22]

2.5.4 Other Nitrogen Removal Processes

In addition to the processes listed above, there are other methods for nitrogen removal which might usefully be discussed. Most are in the experimental stage of development or occur coincidentally with another process.

Use of anionic exchange resins for removal of nitrate was developed principally for treatment of irrigation return waters.[22] Two major unsolved problems are the lack of resins which have a high selectivity for nitrate over chloride and disposal of nitrogen-laden regenerants.

Oxidation ponds can remove nitrogen through microbial denitrification in the anaerobic bottom layer or by ammonia emission to the atmosphere. The latter effect is essentially ammonia stripping but is relatively inefficent due to a low surface-volume ratio and low pH. In a study of raw wastewater lagoons in California, removals of 35-85 percent were reported for well-operated lagoons.[5]

Nitrogen in oxidation ponds is assimilated by algal cultures. If the algal cells are removed from the pond effluent stream, nitrogen removal is thereby effected. Methods for removal of algae are summarized in the EPA Technology Transfer Publication, *Upgrading Lagoons.*[25]

It was noted previously that in secondary biological treatment and in nitrification, some nitrogen is incorporated in bacterial cells and is removed from the waste stream with the sludge. If an organic carbon source such as ethanol or glucose is added to the wastewater, the solids production will be increased and a greater nitrogen removal will be effected. Disadvantages are that large quantities of sludge are produced and that difficulties occur in regulating the addition of the carbon source, with high effluent BOD_5 values or high nitrogen levels resulting.[2]

2.5.5 Summary

Table 2-3 summarizes the effect of various treatment processes on nitrogen removal. Shown is the effect that the process has on each of the three major forms: organic nitrogen, ammonium, and nitrate. In the last column is shown normal removal percentages which can be expected from that process. Overall removal for a particular treatment plant will depend on the types of unit processes and their relation to each other. For example, while many processes developed for nitrogen removal are ineffective in removing organic nitrogen, incorporation of chemical coagulation or multimedia filtration into the overall flowsheet can result in a low concentration of organic nitrogen in the plant effluent. Thus, the interrelationship between processes must be carefully analyzed in designing for nitrogen removal. Further discussion of process interrelationships is presented in Chapter 9.

TABLE 2-3

EFFECT OF VARIOUS TREATMENT PROCESSES ON NITROGEN COMPOUNDS

Treatment Process	Effect on Constituent: Organic N	NH_3/NH_4^+	NO_3^-	Removal of Total Nitrogen Entering Process (%)*
Conventional treatment process				
Primary	10-20% removed	no effect	no effect	5-10
Secondary	15-25% removed** urea → NH_3/NH_4^+	<10% removed	nil	10-20
Advanced wastewater treatment processes				
Filtration***	30-95% removed	nil	nil	20-40
Carbon sorption	30-50% removed	nil	nil	10-20
Electrodialysis	100% of suspend organic N removed	40% removed	40% removed	35-45
Reverse osmosis	100% of suspend organic N removed	85% removed	85% removed	80-90
Chemical coagulation***	50-70% removed	nil	nil	20-30
Land application				
Irrigation	→ NH_3/NH_4^+	→ NO_3^- → plant N	→ plant N	40-90
Infiltration/percolation	→ NH_3/NH_4^+	→ NO_3^-	→ N_2	0-50
Major nitrogen removal processes				
Nitrification	limited effect	→ NO_3^-	no effect	5-10
Denitrification	no effect	no effect	80-98% removed	70-95
Breakpoint chlorination	uncertain	90-100% removed	no effect	80-95
Selective ion exchange for ammonium	some removal, uncertain	90-97% removed	no effect	80-95
Ammonia stripping	no effect	60-95% removed	no effect	50-90
Other nitrogen removal processes				
Selective ion exchange for nitrate	nil	nil	75-90% removed	70-90
Oxidation ponds	partial transformation to NH_3/NH_4^+	partial removal by stripping	partial removal by nitrification-denitrification	20-90
Algae stripping	partial transformation to NH_3/NH_4^+	→ cells	→ cells	50-80
Bacterial assimilation	no effect	40-70% removed	limited effect	30-70

*Will depend on the fraction of influent nitrogen for which the process is effective, which may depend on other processes in the treatment plant.

**Soluble organic nitrogen, in the form of urea and amino acids, is substantially reduced by secondary treatment.

***May be used to remove particulate organic carbon in plants where ammonia or nitrate are removed by other processes.

2.6 References

1. Sawyer, C.N., and P.L. McCarty, *Chemistry for Sanitary Engineers*. New York, McGraw-Hill Book Co., 1967.
2. Christensen, M.H., and P. Harremoes, *Biological Denitrification in Wastewater Treatment*. Report 2-72, Department of Sanitary Engineering, Technical University of Denmark, 1972.
3. Delwiche, C.C., *The Nitrogen Cycle*. Scientific American, 223, No. 3, pp 137-146 (1970).

4. Martin, D.M., and D.R. Goff, *The Role of Nitrogen in the Aquatic Environment.* Report No. 2, Department of Limnology, Academy of Natural Sciences of Philadelphia, 1972.

5. Sepp, E., *Nitrogen Cycle in Groundwater.* Bureau of Sanitary Engineering, State of California Department of Public Health, 1970.

6. McCarty, P.L., *et al., Sources of Nitrogen and Phosphorus in Water Supplies.* JAWWA, 59, pp 344 (1967).

7. Sylvester, R.O., *Nutrient Content of Drainage Water for Forested, Urban, and Agricultural Areas.* Algae and Metropolitan Wastes, Robert A. Taft Sanitary Engineering Center, Tech. Rep. W61-3, 1963.

8. Reeves, T.G., *Nitrogen Removal: A Literature Review.* JWPCF, 44, No. 10, pp 1896-1908 (1972).

9. *Nitrogenous Compounds in the Environment.* Hazardous Materials Advisory Committee (to the EPA), EPA-ASB-73-001, December, 1973.

10. Kaufman, W.J., *Chemical Pollution of Ground Waters.* JAWWA, 66, No. 3, pp 152-159 (1974).

11. Weibel, S.R., Anderson, R.J., and R.L. Woodward, *Urban-Land Runoff as a Factor in Stream Pollution.* JWPCF, 43, p 2033 (1971).

12. American Public Works Association, *Water Pollution Aspects of Urban Runoff.* FWPCA Report No. WP-20-15, January, 1969.

13. Burn, R.J., Krawezyk, D.F., and G.T. Harlow, *Chemical and Physical Comparison of Combined and Separated Sewer Discharges.* JWPCF, 40, pp 112 (1968).

14. Avco Economic Systems Corporation, *Storm Water Pollution From Urban Land Activity.* EPA Report No. 11034 FKL 07/70, July, 1970.

15. Weibel, S.R., *et al., Pesticides and Other Contaminants in Rainfall and Runoff.* JAWWA, 58, pp 1075 (1966).

16. Dept. of Biological and Agricultural Engineering, North Carolina State University at Raleigh, *Role of Animal Wastes in Agricultural Land Runoff.* EPA Report No. 13020 DGX 08/71, August 1971.

17. California Department of Water Resources, *Nutrients From Tile Drainage Systems.* EPA Report No. 13030 ELY 5/71-3, May 1971.

18. Johnson, R.E., Rossano, A.T., Jr., and R.O. Sylvester, *Dustfall as a Source of Water Quality Impairment,* ASCE, JSED, 92, No. SA1, pp 145 (1966).

19. California State Water Resources Control Board, *Tentative Water Quality Control Plan, San Francisco Bay Basin.* November, 1974.

20. Brown and Caldwell/Dewante and Stowell, *Feasibility Study for the Northeast-Central Sewerage Service Area.* Prepared for County of Sacramento, Department of Public Works, November 1972.

21. Ehreth, D.J., and E. Barth, *Control of Nitrogen in Wastewater Effluents.* Prepared for the EPA Technology Transfer Program, March, 1974.

22. *Nitrogen Removal from Wastewaters.* EPA Advanced Waste Treatment Research Laboratory, ORD-17010, October, 1970.

23. *Effects of Pollution Discharges on the Thames Estuary.* Department of Scientific and Industrial Research, Water Pollution Research Technical Paper No. 11, Her Majesty's Stationery Office, 1964.

24. Parkin, G.F., and P.L. McCarty, *The Nature, Ecological Significance, and Removal of Soluble Organic Nitrogen in Treated Agricultural Wastewaters.* Stanford University, prepared for the Bureau of Reclamation, Contract USDI- 14-06-200-6090-A, September, 1973.

25. Caldwell, D.H., Parker, D.S., and W.R. Uhte, *Upgrading Lagoons.* Prepared for the EPA Technology Transfer Program, August, 1973.

CHAPTER 3

PROCESS CHEMISTRY AND BIOCHEMISTRY OF NITRIFICATION AND DENITRIFICATION

3.1 Introduction

The purpose of this chapter is to present a treatment process-oriented review of the chemistry and biochemistry of nitrification and denitrification. An understanding of this subject is useful for developing an appreciation of the factors affecting the performance, design, and operation of nitrification and denitrification processes. Subsequent chapters deal with design aspects of nitrification (Chapter 4) and denitrification (Chapter 5). Since these latter chapters are laid out to be used without reference to this chapter, review of the theoretical material in this chapter is not mandatory.

Biological processes for control of nitrogenous residuals in effluents can be classified in two broad areas. First, a process designed to produce an effluent where influent nitrogen (ammonia and organic nitrogen) is substantially converted to nitrate nitrogen can be considered. This process, nitrification, is carried out by bacterial populations that sequentially oxidize ammonia to nitrate with intermediate formation of nitrite. Nitrification will satisfy effluent or receiving water standards where reduction of residual nitrogenous oxygen demand due to ammonia is mandated or where ammonia reduction for other reasons is required for the treatment system. The second type of process, denitrification, reduces nitrate to nitrogen gas and can be used following nitrification when the total nitrogenous content of the effluent must be reduced.

3.2 Nitrification

The two principal genera of importance in biological nitrification processes are *Nitrosomonas* and *Nitrobacter*. Both of these groups are classed as autotrophic organisms. These organisms are distinguished from heterotrophic bacteria in that they derive energy for growth from the oxidation of inorganic nitrogen compounds, rather than from the oxidation of organic matter. Another feature of these organisms is that inorganic carbon (carbon dioxide) is used for synthesis rather than organic carbon. Each group is limited to the oxidation of specific species of nitrogen compounds. *Nitrosomonas* can oxidize ammonia to nitrite, but cannot complete the oxidation to nitrate. On the other hand, *Nitrobacter* is limited to the oxidation of nitrite to nitrate. Since complete nitrification is a sequential reaction, treatment processes must be designed to provide an environment suitable to the growth of both groups of nitrifying bacteria.

3.2.1 Biochemical Pathways

On a biochemical level, the nitrification process is more complex than simply the sequential oxidation of ammonia to nitrite by *Nitrosomonas* and the subsequent oxidation of nitrite to nitrate by *Nitrobacter*. Various reaction intermediates and enzymes are involved.[1] More important than an understanding of these pathways is knowledge of the response of nitrification organisms to environmental conditions.

3.2.2. Energy and Synthesis Relationships

The stoichiometric reaction for oxidation of ammonium to nitrite by *Nitrosomonas* is:

$$NH_4^+ + 1.5\,O_2 \longrightarrow 2H^+ + H_2O + NO_2^- \qquad (3\text{-}1)$$

The loss of free energy by this reaction at physiological concentrations of the reactants has been estimated by various investigators to be between 58 and 84 kcal per mole of ammonia.[1,2]

The reaction for oxidation of nitrite to nitrate by *Nitrobacter* is:

$$NO_2^- + 0.5\,O_2 \longrightarrow NO_3^- \qquad (3\text{-}2)$$

This reaction has been estimated to release between 15.4 to 20.9 kcal per mole of nitrite under *in vivo* conditions.[2] Thus, *Nitrosomonas* obtains more energy per mole of nitrogen oxidized than *Nitrobacter*. If it assumed that the cell synthesis per unit of energy produced is equal, there should be greater mass of *Nitrosomonas* formed than *Nitrobacter* per mole of nitrogen oxidized. As will be seen, this is in fact the case.

The overall oxidation of ammonium by both groups is obtained by adding Equations 3-1 and 3-2:

$$NH_4^+ + 2\,O_2 \longrightarrow NO_3^- + 2H^+ + H_2O \qquad (3\text{-}3)$$

As previously mentioned, these reactions furnish the energy required for growth of the nitrifying organisms. Assuming that the empirical formulation of bacterial cells is $C_5H_7NO_2$, the equations for the growth of *Nitrosomonas* and *Nitrobacter* are shown in Equations 3-4 and 3-5, respectively:

$$15CO_2 + 13NH_4^+ \longrightarrow 10NO_2^- + \underset{\textit{Nitrosomonas}}{3C_5H_7NO_2} + 23H^+ + 4H_2O \qquad (3\text{-}4)$$

$$5CO_2 + NH_4^+ + 10NO_2^- + 2H_2O \longrightarrow 10NO_3^- + \underset{\textit{Nitrobacter}}{C_5H_7NO_2} + H^+ \qquad (3\text{-}5)$$

Equations 3-1, 3-4 and 3-5 have terms showing the production of free acid (H^+) and the consumption of gaseous carbon dioxide (CO_2). In actual fact, these reactions take place in aqueous systems in the context of the carbonic acid system. These reactions usually take place at pH levels less than 8.3. Under this circumstance, the production of acid results in immediate reaction with bicarbonate ion (HCO_3) with the production of carbonic acid (H_2CO_3). The consumption of carbon dioxide by the organisms results in some depletion of the dissolved form of carbon dioxide, carbonic acid (H_2CO_3). Table 3-1 presents the modified forms of Equations 3-1 to 3-5 to reflect the changes in the carbonic acid system. As will be later described in Sections 3.2.3 and 3.2.5.6, the variations occurring in pH resulting from changes in the carbonic acid system can significantly affect nitrification process performance.

The equations for energy yielding reactions (Equations 3-1 and 3-2) can be combined with the equations for organism synthesis (Equations 3-4 and 3-5) to form overall synthesis-oxidation relations by knowledge of the yield coefficients for the nitrifying organisms. Experimental yield values for *Nitrosomonas* range from 0.04 to 0.13 lb VSS grown per lb ammonia nitrogen oxidized.[1] Experimental yields for *Nitrobacter* are in the range from 0.02 to 0.07 lb VSS grown per lb of nitrite nitrogen oxidized.[1] Values based on thermodynamic theory are 0.29 and 0.084 for *Nitrosomonas* and *Nitrobacter*, respectively.[2] The experimentally based yield may be lower than theoretical values due to the diversion of a portion of the free energy released by oxidation to microorganism maintenance functions.[2]

Equations for synthesis-oxidation using representative measurements of yields and oxygen consumption for *Nitrosomonas* and *Nitrobacter* are as follows:[3,4]

$$55NH_4^+ + 76\,O_2 + 109HCO_3^- \longrightarrow \underset{Nitrosomonas}{C_5H_7NO_2} + 54NO_2^- + 57H_2O + 104H_2CO_3 \tag{3-6}$$

$$400NO_2^- + NH_4^+ + 4H_2CO_3 + HCO_3^- + 195\,O_2 \longrightarrow \underset{Nitrobacter}{C_5H_7NO_2} + 3H_2O + 400NO_3^- \tag{3-7}$$

Using Equations 3-6 and 3-7, the overall synthesis and oxidation reaction is:

$$NH_4^+ + 1.83\,O_2 + 1.98HCO_3^- \longrightarrow 0.021C_5H_7NO_2 + 1.041H_2O + 0.98NO_3^- + 1.88H_2CO_3 \tag{3-8}$$

In these equations, yields for *Nitrosomonas* and *Nitrobacter* are 0.15 mg cells/mg NH_4^+-N and 0.02 mg cells/mg NO_2-N, respectively. On this basis, the removal of twenty mg/l of ammonia nitrogen would yield only 1.8 mg/l of nitrifying organisms. This relatively low yield has some far reaching implications, as will be seen in Section 3.2.7. Oxygen consumption ratios in the equations are 3.22 mg O_2/mg NH_4^+-N oxidized and 1.11 mg O_2/mg NO_2^--N oxidized, which is in agreement with measured values.[4]

3.2.3. Alkalinity and pH Relationships

Equation 3-3A (Table 3-1) shows that alkalinity is destroyed by the oxidation of ammonia and carbon dioxide (H_2CO_3 in the aqueous phase) is produced. When synthesis is neglected, it can be calculated that 7.14 mg of alkalinity as $CaCO_3$ is destroyed per mg of ammonia nitrogen oxidized. The effect of synthesis is relatively small; in Equation 3-8, the ratio is 7.07 mg of alkalinity per mg of ammonia nitrogen oxidized. Experimentally determined ratios are presented in Table 3-2; differences between the experimental and theoretical ratios are due either to errors in alkalinity or nitrogen analyses or the inadequacy of theory

TABLE 3-1

RELATIONSHIPS FOR OXIDATION AND GROWTH IN NITRIFICATION REACTIONS IN RELATIONSHIP TO THE CARBONIC ACID SYSTEM

Reaction	Equation	Equation No.
Oxidation - Nitrosomonas	$NH_4^+ + 1.5O_2 + 2HCO_3^- \rightarrow NO_2^- + 2H_2CO_3 + H_2O$	3-1A
Oxidation - Nitrobacter	$NO_2^- + 0.5O_2 \rightarrow NO_3^-$	3-2
Oxidation - overall	$NH_4^+ + 2O_2 + 2HCO_3^- \rightarrow NO_3^- + 2H_2CO_3 + H_2O$	3-3A
Synthesis - Nitrosomonas	$13NH_4^+ + 23HCO_3^- \rightarrow 8H_2CO_3 + 10NO_2^- + 3C_5H_2NO_2 + 19H_2O$	3-4A
Synthesis - Nitrobacter	$NH_4^+ + 10NO_2^- + 4H_2CO_3 + HCO_3^- \rightarrow 10NO_3^- + 3H_2O + C_5H_7NO_2$	3-5A

TABLE 3-2

ALKALINITY DESTRUCTION RATIOS IN EXPERIMENTAL STUDIES

System	$\frac{\text{mg alkalinity destroyed}^a}{\text{mg } NH_4^+\text{-N oxidized}}$	Reference
Suspended growth	6.4	5
Suspended growth	6.0	6
Suspended growth	7.1	7
Attached growth	6.5	8
Attached growth	6.3 to 7.4	9
Attached growth	7.3	2

[a]As $CaCO_3$; the theoretical value is 7.1

to completely explain the phenomenon. A ratio of 7.14 mg alkalinity destroyed per mg of ammonia nitrogen oxidized may be used for engineering calculations.

These changes may have a depressing effect on pH in the nitrification system, as the relationship for pH in the system is:

$$pH = pK_1 - \log \frac{(H_2CO_3)}{(HCO_3^-)} \tag{3-9}$$

Since nitrification reduces the HCO_3^- level and increases the H_2CO_3 level, it is obvious that the pH would tend to be reduced. The effect is mediated by stripping of carbon dioxide from the liquid by the process of aeration and the pH is elevated upwards. If the carbon dioxide is not stripped from the liquid, such as in enclosed high purity oxygen systems, the pH can be depressed as low as 6.0. It has been calculated that to maintain the pH greater than 6.0 in an enclosed system, the alkalinity of the wastewater must be 10 times greater than the amount of ammonia nitrified.[2]

Even in open systems where the carbon dioxide is continually stripped from the liquid, severe pH depression can occur when the alkalinity in the wastewater approaches depletion by the acid produced in the nitrification process. For example, if in a wastewater 20 mg/l of ammonia nitrogen is nitrified, 143 mg/l of alkalinity as $CaCO_3$ will be destroyed. In many wastewaters there is insufficient alkalinity initially present to leave a sufficient residual for buffering the wastewater during the nitrification process. The significance of pH depression in the process is that nitrification rates are rapidly depressed as the pH is reduced below 7.0 (see Section 3.2.5.6). Procedures for calculating the operating pH in aeration systems are presented in Section 4.9.

3.2.4 Oxygen Requirements

The theoretical oxygen requirement for nitrification, neglecting synthesis, is 4.57 mg O_2/mg NH_4^+-N (Equation 3-3). Synthesis has an effect on oxygen requirements; the oxygen requirement is calculated to be, from Equation 3-8, 4.19 mg O_2/mg NH_4^+-N. An oxygen requirement sufficiently accurate to be used in engineering calculations for aeration requirements is 4.6 mg O_2/mg NH_4^+-N.

The oxygen demand for nitrification is significant; for instance if 30 mg/l of ammonia nitrogen is oxidized by the nitrification system, about 138 mg/l of oxygen will be required. Caution: in virtually all practical nitrification systems, oxygen demanding materials other than ammonia are present in the wastewater, raising the total oxygen requirements of nitrification systems even higher (see Section 4.8).

3.2.5 Kinetics of Nitrification

A complete review of the kinetics of biological systems is beyond the scope of the manual; however, several excellent reviews are available.[10,11,12] Rather, the basics of biological kinetics are drawn upon to usefully portray a mathematical description of the oxidation of ammonia and nitrite. In the succeeding portions of this section, the impact of a variety of environmental factors on the rates of growth and nitrification are considered. A combined kinetic expression is then formulated which accounts for the effects of ammonia concentration, temperature, pH and dissolved oxygen concentration.

At several points, reference is made to data developed from various types of nitrification processes. Comprehensive descriptions of the various nitrification processes are presented in Chapter 4 and will not be reproduced herein. One distinction that needs to be clearly understood in discussions in this chapter is the difference between combined carbon oxidation-nitrification processes and separate stage nitrification processes. The combined carbon oxidation-nitrification processes oxidize a high proportion of influent organics (BOD) relative to the ammonia nitrogen content. This causes relatively low populations of nitrifiers to be present in the biomass. Separate stage nitrification systems, on the other hand, have a relatively low BOD_5 load relative to the influent ammonia load. As a result, higher proportions of nitrifiers are obtained. Separate stage nitrification can be provided in municipal treatment applications when a high level of organic carbon removal is provided prior to the nitrification stage. This level of treatment is generally greater than provided by primary treatment. Other differences between these classes of processes can be drawn, but these are left for detailed discussion in Section 4.2.

3.2.5.1 Effect of Ammonia Concentration on Kinetics

A description of ammonia and nitrite oxidation can be derived from an examination of the growth kinetics of *Nitrosomonas* and *Nitrobacter*. *Nitrosomonas'* growth is limited by the concentration of ammonia nitrogen, while *Nitrobacter's* growth is limited by the concentration of nitrite.

The kinetic equation proposed by Monod[13] is used to describe the kinetics of biological growth of either *Nitrosomonas* or *Nitrobacter*:

$$\mu = \hat{\mu} \frac{S}{K_s + S} \tag{3-10}$$

where:
μ = growth rate of microorganisms, day^{-1},
$\hat{\mu}$ = maximum growth rate of microorganisms, day^{-1},
K_s = half velocity constant = substrate concentration, mg/l, at half the maximum growth rate and
S = growth limiting substrate concentration, mg/l.

Since the maximum growth rate of *Nitrobacter* is considerably larger than the maximum growth rate of *Nitrosomonas*, and since K_S values for both organisms are less than 1 mg/l-N at temperatures below 20 C, nitrite does not accumulate in large amounts in biological treatment systems under steady state conditions. For this reason, the rate of nitrifier growth can be modeled with Equation 3-10 using the rate limiting step, ammonia conversion to nitrite. For cases where nitrite accumulation does occur, other approaches are available.[14,15,16]

3.2.5.2 Relationship of Growth Rate to Oxidation Rate

The ammonia oxidation rate can be related to the *Nitrosomonas* growth rate, as follows:

$$q_N = \frac{\mu_N}{Y_N} = \hat{q}_N \frac{N}{K_N + N} \tag{3-11}$$

where:
μ_N = *Nitrosomonas* growth rate, day^{-1},
$\hat{\mu}_N$ = peak *Nitrosomonas* growth rate, day^{-1},
$\hat{q}_N = \frac{\hat{\mu}_N}{Y_N}$ = peak ammonia oxidation rate, lb NH_4^+ - N oxidized/lb VSS/day,
q_N = ammonia oxidation rate, lb NH_4^+ - N oxidized/lb VSS/day
Y_N = organism yield coefficient, lb *Nitrosomonas* grown (VSS) per lb NH_4^+ - N removed,
N = NH_4^+ - N concentration, mg/l, and
K_N = half-saturation constant, mg/l NH_4^+ - N, mg/l.

In Equations 3-10 and 3-11, only the effect of ammonia concentration is considered; in later sections, the effects of temperature, pH, and dissolved oxygen are considered.

3.2.5.3. Relationship of Growth Rate to Solids Retention Time

The growth rate of organisms can be related to the design of activated sludge systems by noting the inverse relationship between solids retention time and growth rate of nitrifiers:

$$\theta_c = \frac{1}{\mu} \qquad (3\text{-}12)$$

where: θ_c = solids retention time, days.

The solids retention time can be calculated from system operating data by dividing the inventory of microbial mass in the treatment system by the quantity of biological mass wasted daily. Equations applicable for this calculation are presented in Section 4.3.3.

3.2.5.4 Kinetic Rate Constants for Temperature and Nitrogen Concentration

The most widely accepted kinetic constants for the nitrifers are those presented by Downing and coworkers.[17,18] Their results are presented in Figures 3-1 and 3-2. As can be seen, both the maximum growth rate, $\hat{\mu}$, and the half saturation constants, K_s for *Nitrosomonas* and *Nitrobacter* are markedly affected by temperature. Further, the maximum growth rate for *Nitrosomonas* in activated sludge was found to be considerably less than for *Nitrosomonas* in pure culture.

Kinetic constants found by other investigators are summarized in Tables 3-3 and 3-4. The observations of maximum growth rates of *Nitrosomonas* of Gujer and Jenkins[4], Wuhrmann[19], Loehr, *et al.*[22], Poduska and Andrews[15], and Lawrence and Brown[24], are closer to the pure culture values of *Nitrosomonas* rather than the activated sludge values in Figure 3-1. This suggests that some additional parameter such as dissolved oxygen (DO) may have been limiting Downing's activated sludge measurements.[4] The influence of DO is discussed in the next section. For illustrative use in this manual the pure culture values of Downing, *et al.* for *Nitrosomonas* are used for considering the effect of temperature and ammonia on growth and nitrification rates. The expressions for the half saturation constant for oxidation of ammonia-nitrogen, K_N, is:

$$K_N = 10^{0.051T - 1.158} \text{ mg/l, as N} \qquad (3\text{-}13)$$

T = temperature, C.

The expression for the effect of temperature on the maximum growth rate of *Nitrosomonas* is:

$$\hat{\mu}_N = 0.47e^{0.098(T-15)} \text{ day}^{-1} \qquad (3\text{-}14)$$

FIGURE 3-1

TEMPERATURE DEPENDENCE OF THE MAXIMUM GROWTH RATES OF NITRIFIERS

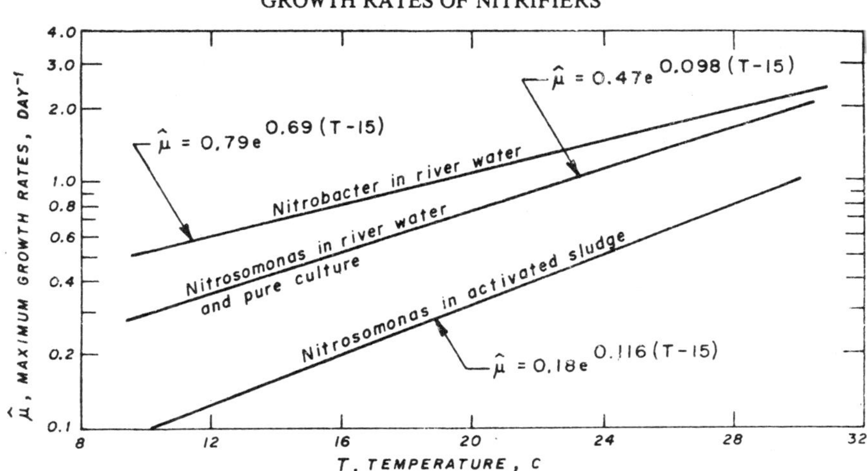

FIGURE 3-2

TEMPERATURE DEPENDENCE OF THE HALF SATURATION CONSTANTS OF NITRIFIERS

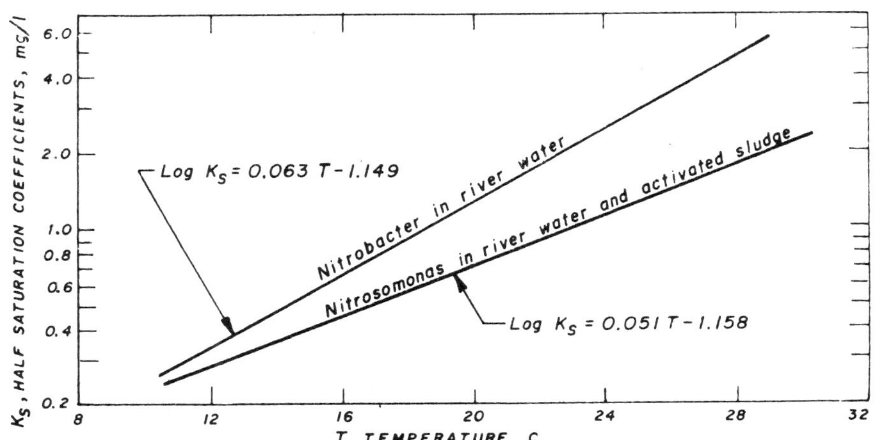

TABLE 3-3

MAXIMUM GROWTH RATES FOR NITRIFIERS IN VARIOUS ENVIRONMENTS

Organism	$\hat{\mu}_N$, day-1 at stated temperature, C								Ref.	Environment
	8	12	15	16	20	21	23	25		
Nitrosomonas		0.40				0.85			4	Activated sludge, wash out
		0.34				0.65			4	Activated sludge, math model
				0.57					19	Activated sludge
								0.17	20	Activated sludge
							0.37		21	Activated sludge
					0.71				22	Activated sludge
			0.21		0.48			0.55	11 23	Synthetic river water
							1.08		15	Activated sludge
	0.25				0.5				24	Activated sludge
Nitrobacter			0.28			0.34		0.53	11,23	Synthetic river water
							1.44		15	Activated sludge

TABLE 3-4

HALF-SATURATION CONSTANTS FOR NITRIFIERS IN VARIOUS ENVIRONMENTS

Organism	K_S, mg/l-N at stated temperature, C						Ref.	Environment
	15	20	25	28	30	32		
Nitrosomonas			0.37				20	Activated sludge
	2.8	3.6	3.4				11,23	Synthetic river water
					10		1,25	Lab culture
	0.5 to 1.0	0.5 to 1.0					2	Warburg analyses
			3.5				1,26	Lab culture
		1.0					1,27	Lab culture and activated sludge
		0.5					28	Lab culture
Nitrobacter			0.25				20	Activated sludge
	0.7	1.1	0.7				11,23	Synthetic river water
					6		1,29	Lab culture
				5			1,30	Lab culture
						8.4	1,31	Lab culture
			5				1,26	Lab culture
		0.07					28	Lab culture

Somewhat differing temperature effects have been found for attached growth systems. Huang and Hopson's summary, with some modifications, is shown in Figure 3-3 for attached growth systems.[34] Downing and coworkers' relationship for *Nitrosomonas* (Equation 3-14) is also shown for comparison purposes. Comparing the suspended growth and attached nitrification data, one can conclude that attached growth systems have an advantage in withstanding low temperatures (<15 C) without as severe losses in nitrification rates. However, measurement of nitrification rates for suspended growth systems are not normally made on the same basis as attached growth systems. In suspended growth systems, rates are expressed on a per unit of biomass basis (MLVSS is used). Precise measurement of biomass is normally not possible in attached growth systems so other parameters are used such as reaction rate per unit surface or volume. Therefore, attached growth systems can compensate for colder temperature conditions by the effective slime growth growing thicker. Thus, if rates could be expressed on a unit biomass basis for both system types, reaction rate variation with temperature might be more similar.

It could be argued that compensation for low temperature in suspended growth systems could be provided by an increase in mixed liquor level, much as an increase in slime growth occurs in attached growth systems. However, suspended growth systems are limited by reactor-sedimentation tank interactions which at cold temperatures might prohibit this due to reduction of thickening rates of the sludge (cf. Section 4.10).

Other differences in reaction rates shown in Figure 3-3 may arise from the fact that some determinations were on separate stage nitrification systems while others were made on combined carbon oxidation-nitrification processes.

FIGURE 3-3

COMPARISON OF EFFECT OF TEMPERATURE ON NITRIFICATION IN SUSPENDED GROWTH AND ATTACHED GROWTH SYSTEMS (REFERENCE 34)

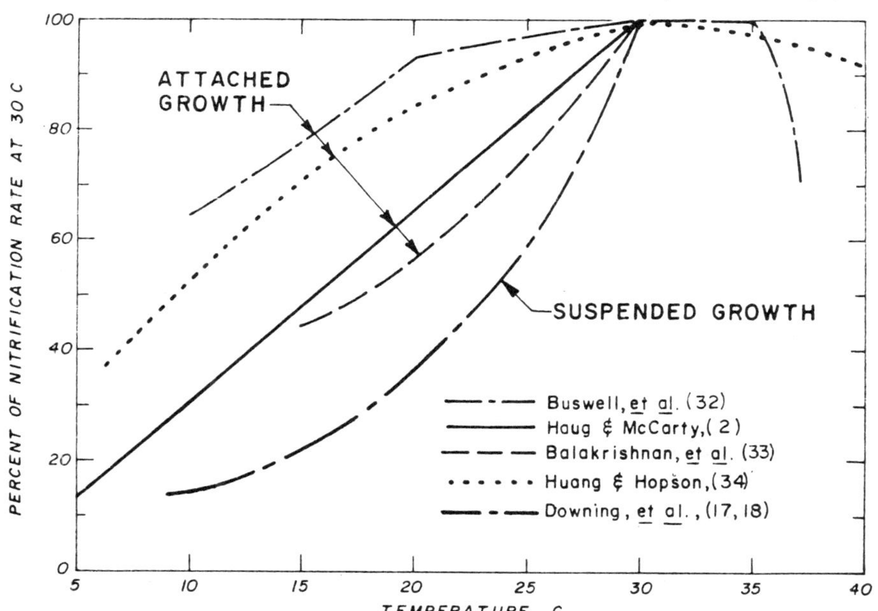

3.2.5.5 Effect of Dissolved Oxygen on Kinetics

The concentration of dissolved oxygen (DO) has a significant effect on the rates of nitrifier growth and nitrification in biological waste treatment systems. The Monod relationship has been used to model the effect of dissolved oxygen, considering oxygen to be a growth limiting substrate, as follows:

$$\mu_N = \hat{\mu}_N \frac{DO}{K_{O_2} + DO} \qquad (3\text{-}15)$$

where: DO = dissolved oxygen, mg/l, and
K_{O_2} = half-saturation constant for oxygen, mg/l.

British investigators found that the K_{O_2} value was about 1.3 mg/l at an unspecified temperature.[35] One U.S. investigator has suggested a relationship that would indicate half-saturation constants of 0.15 mg/l at 15 C and 0.42 mg/l at 25 C, but did not provide supporting data.[14] Studies conducted by the Los Angeles County Sanitation Districts at its Pomona Water Renovation Plant represent one of the most careful attempts to evaluate the effect of DO on nitrification rate.[36] This facility is a combined carbon oxidation-nitrification plant. Sludge samples were withdrawn, dosed with ammonia, and aerated at various DO levels. Nitrification rates determined from the data collected are shown in Figure 3-4.[36] Fitted to this data is a Monod expression for nitrification rate as a function of DO. The K_{O_2} determined from this data is 2.0 mg/l. Temperature was not specified, but indicated to be above 20 C.

Several investigations have provided indirect evidence of the importance of the effect of DO on nitrification rate. A treatment plant operated continuously at a DO near 1 mg/l gave lower degrees of nitrification than plants held at 4 and 7 mg/l.[37] When small scale activated sludge plants were held at 1, 2, 4, and 8 mg/l. British investigators found that the nitrification rates at 2.0 mg/l were about 10 percent lower than at higher levels of DO, although nitrification was complete.[35] Pilot investigations at the Metro Sewer District of Cincinnati, Ohio, showed that when the DO was held at 2 mg/l, only about 40 percent nitrification occurred, but when the DO was increased to 4 mg/l, about 80 percent nitrification took place.[38] Murphy found that in two parallel activated sludge plants, that nitrification was enhanced in the plant maintaining the DO at 8 mg/l compared to the plant where the DO was maintained at 1 mg/l.[39]

The influence of DO on nitrification rates has been somewhat controversial, as examples of plants can be found with completely nitrified effluents with operating DO levels of 0.5 mg/l. However, this type of evidence does not indicate that nitrification *rate* was unaffected, merely that nitrification could be completed in the presence of a low DO level. Low nitrification rates, depressed by low DO levels, can still be sufficient to cause complete nitrification if the aeration tank detention time is large enough.

FIGURE 3-4

EFFECT OF DISSOLVED OXYGEN ON NITRIFICATION RATE

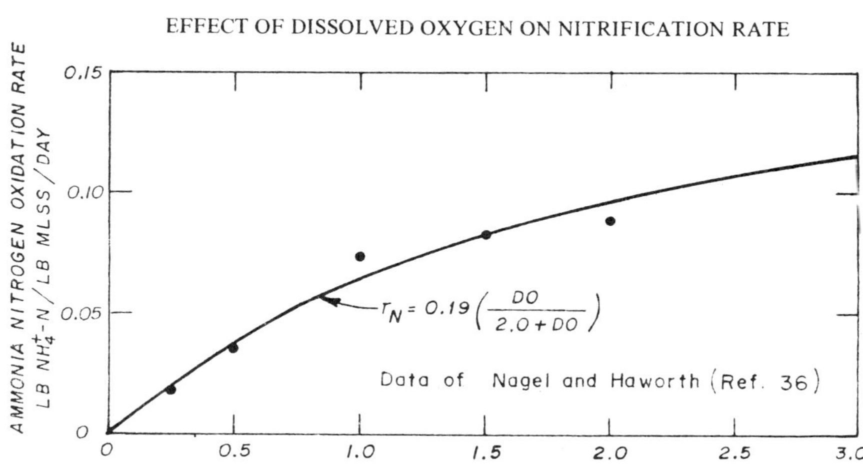

While the general effect of DO on kinetics is firmly established, there needs to be further study to determine the factors affecting the values of K_{O_2}. All of the various estimates are from systems where combined carbon oxidation-nitrification is practiced and no measurements have been made on separate stage nitrification systems. K_{O_2} values for separate stage nitrification systems may very well be different than those for combined carbon oxidation-nitrification systems. Further refinement of K_{O_2} values can be expected. For illustrative use in this manual, a value of K_{O_2} of 1.3 mg/l has been assumed. This value falls in the middle of the range of K_{O_2} observations (0.15 to 2.0 mg/l) and is of a magnitude such that if the operating DO is 2.0 mg/l or less, the nitrification (or nitrifier) growth rate is 60.6 percent (or less) of the peak rate. This order of reduction in rate could account for most of the difference in growth rate observed by Downing, *et al.* (see Section 3.2.5.4) between river water values at high DO levels and activated sludge operating data at DO levels of 2.0 mg/l of less.

3.2.5.6 Effect of pH on Kinetics

The hydrogen ion concentration (pH) has been generally found to have a strong effect on the rate of nitrification. Figure 3-5 presents typical pH relationships from a number of investigations. The results of other investigations have been summarized in the literature.[1,44] There is a wide range in reported pH optima; the almost universal finding is that as the pH moves to the acid range, the rate of ammonia oxidation declines. This has been found to be true for both unacclimated and acclimated cultures, although acclimation tends to moderate pH effects. The findings for an attached growth reactor (Curve E, Figure 3-5) are very similar to the findings for an activated sludge (Curve C). In neither case were the cultures acclimated to each pH value prior to determining nitrification rates. When a three-week acclimation period was provided for the attached growth reactor, it was found that the rate at pH 6.6 rose to 85 percent of the optimum rate at pH 8.4 to 9.0.[34]

Various investigators have reported the effects of pH depression on nitrification. For instance, in an activated sludge with insufficient wastewater alkalinity, pH values of 5 to 5.5 were attained. This high acid concentration resulted in a cessation of nitrification; at the same time sludge bulking occurred. The point at which the rate of nitrification decreased was pH 6.3-6.7.[1] Parallel investigations on air and high purity oxygen-activated sludge systems at the Blue Plains Treatment Plant, Washington, D.C., have shown that depressed pH values in the last oxygen activated sludge stage produced slightly lower nitrification rates than when the system was operated at higher pH.[45] Further information on the EPA investigations at the Blue Plains Treatment Plant is presented later in Section 4.6.5.

In a study of the effect of abrupt changes in pH, it was found that an abrupt change in reactor pH from 7.2 to 6.4 caused no adverse effects. However, when the pH was abruptly changed from 7.2 to 5.8, nitrification performance deteriorated markedly as effluent ammonia levels rose from approximately zero to 11 mg/l NH_4^+-N. A return to pH 7.2 caused rapid improvement indicating that the lowered pH was only inhibitory and not toxic.[15] Haug and McCarty showed that nitrifiers could adapt to nitrify at pH levels as low as 5.5 to 6.0.[2] However, since the concentration of biomass in their column was not defined at each pH level, no conclusions can be drawn from their work as to the effect of pH on the peak ammonia oxidation rate, $\hat{q}_N$.

For illustrative use in this manual, the equation of Downing, *et al.* 43, showing the effect of pH on nitrification is adopted. For pH values < 7.2:

$$\mu_N = \hat{\mu}_N \, (1 - 0.833(7.2 - \mathrm{pH})) \qquad (3\text{-}16)$$

For pH levels between 7.2 and 8.0, the rate is assumed constant. This expression was developed for combined carbon oxidation-nitrification systems, but its application to separate stage nitrification systems is probably conservative.

Because of the effect of pH on nitrification rate, it is especially important that there be sufficient alkalinity in the wastewater to balance the acid produced by nitrification. Equation 3-3A (Table 3-1) indicates that 7.14 mg of alkalinity are destroyed per mg of ammonia nitrogen. Caustic or lime addition may be required to supplement moderately alkaline wastewaters. Design considerations for pH control are presented in Section 4.9.

3.2.5.7 Combined Kinetic Expressions

In previous sections, the effect of ammonia level, temperature, pH and dissolved oxygen on nitrification rate has been presented. In all practical systems, these parameters act to affect

FIGURE 3-5

EFFECT OF pH ON NITRIFICATION RATE
(AFTER SAWYER, ET AL.)

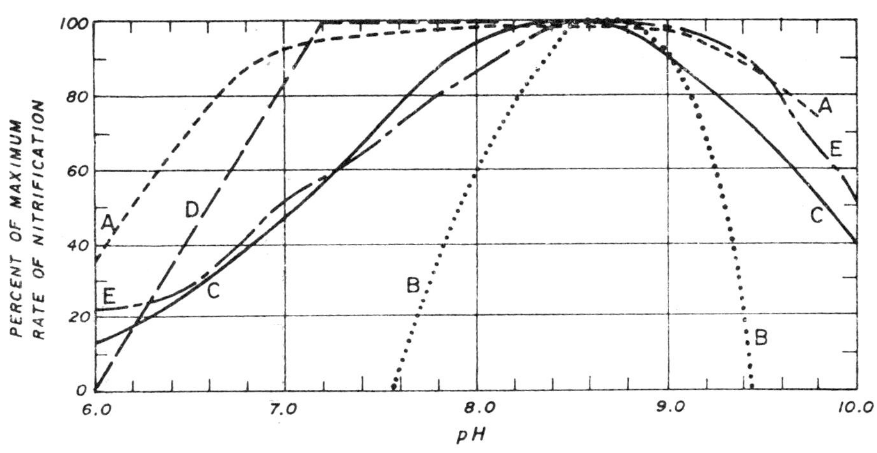

KEY

SYMBOL	ENVIRONMENT	REFERENCE
A	Nitrosomonas - pure culture	Engle and Alexander (40)
B	Nitrosomonas - pure culture	Myerhof (41)
C	Activated sludge at 20 C	Sawyer, et al. (42)
D	Activated sludge	Downing, et al. (43)
E	Attached growth reactor at 22 C	Huang and Hopson (34)

the nitrification rate simultaneously. It has been shown that the combined effect of several limiting factors on biological growth can be introduced as a product of Monod-type factors:[46]

$$\mu_N = \hat{\mu}_N \left(\frac{L}{K_L + L}\right) \left(\frac{N}{K_N + N}\right) \left(\frac{P}{K_P + P}\right) \qquad (3\text{-}17)$$

where: L = concentration of growth limiting substance L,

N = concentration of growth limiting substance N,

P = concentration of growth limiting substance P, and

K_L, K_N, and K_P = half-saturation constants for substances L, N, and P, respectively.

This concept has also been applied to the analysis of algal growth kinetic data[47] and to denitrification kinetics.[48]

Taking this approach for nitrification, the combined kinetic expression for nitrifier growth would take the form:

$$\mu_N = \hat{\mu}_N \left(\frac{N}{K_N + N}\right)\left(\frac{DO}{K_{O_2} + DO}\right)(1 - 0.833(7.2 - pH)) \qquad (3\text{-}18)$$

where: $\hat{\mu}_N$ = maximum nitrifier growth rate at temperature, T, and pH.

Downing, *et al.*[43] used this procedure to describe nitrifier growth rates, excepting that no term for DO was included. Using specific values for temperature, pH, ammonia and oxygen, adopted in this manual in Sections 3.2.5.4, 3.2.5.5., and 3.2.5.6, the following expression results for pH < 7.2 for *Nitrosomonas* valid for temperatures between 8 and 30 C:

$$\mu_N = 0.47\left[e^{0.098(T-15)}\right]\left[1 - 0.833(7.2 - pH)\right]$$

$$\times\left[\frac{N}{N + 10^{0.051T - 1.158}}\right]\left[\frac{DO}{DO + 1.3}\right] \text{day}^{-1} \qquad (3\text{-}19)$$

The first term in brackets accommodates the effect of temperature. The second term in brackets considers the effect of pH. For pH > 7.2, the second quantity in brackets is taken to be unity. The third term in brackets is the Monod expression for the effect of ammonia nitrogen concentration. Similarly, the fourth term in brackets accounts for the effect of DO on nitrification rate. Equation 3-19 has been adopted for illustrative use in this manual. As more reliable data becomes available, Equation 3-19 can be modified to suit particular circumstances.

An example evaluation of Equation 3-19 at T = 20 C, pH = 7.0, N = 2.5 mg/l, and DO equal to 2.0 mg/l is as follows:

$$\mu_N = 0.47\ [1.63]\ [.833]\ [0.775]\ [0.606] = 0.300 \text{ day}^{-1}$$

The ammonia removal rate is defined as done previously (Equation 3-11):

$$q_N = \frac{\mu}{Y_N} = \hat{q}_N \left(\frac{N}{K_N + N}\right)\left(\frac{DO}{K_{O_2} + DO}\right)(1 - 0.833(7.2 - pH) \qquad (3\text{-}20)$$

where: $$\hat{q}_N = \frac{\hat{\mu}_N}{Y_N} \tag{3-21}$$

In the numerical example above, with a yield coefficient of 0.15 lb VSS per lb NH_4^+-N removed, the nitrification rate is 2 lb NH_4^+-N oxidized per lb VSS per day. This rate is expressed per unit of nitrifiers, assuming that there are no other types of bacteria in the population. Nitrification rates of a comparable magnitude have been found experimentally by a number of investigators for laboratory enrichment cultures of nitrifiers.[2,23,28] As will be seen in Section 3.2.7, nitrification rates in mixed cultures are much lower.

3.2.6 Population Dynamics

In previous sections, the kinetics of the growth of nitrifiers have been presented. In all practical applications in wastewater treatment, nitrifier growth takes place in waste treatment processes where other types of biological growth occurs. In no case are there opportunities for pure cultures to develop. This fact has significant implications in process design for nitrification.

In both combined carbon oxidation-nitrification systems and in separate stage nitrification systems, there is sufficient organic matter in the wastewater to enable the growth of heterotrophic bacteria. In this situation, the yield of heterotrophic bacteria growth is greater than the yield of the autotrophic nitrifying bacteria. Because of this dominance of the culture, there is the danger that the growth rate of the heterotrophic organisms will be established at a value exceeding the maximum possible growth rate of the nitrifying organisms. When this occurs, the slower growing nitrifiers will gradually diminish in proportion to the total population and be washed out of the system.[43]

Thus, for consistent nitrification to occur, the following design condition must be satisfied, assuming pH and DO do not limit nitrifier growth:

$$\hat{\mu}_N > \mu_b \tag{3-22}$$

where: $\hat{\mu}_N$ = maximum growth rate of the nitrifying population,

μ_b = growth rate of the heterotrophic population.

Reduced DO or pH can act to depress the peak nitrifier growth rate and cause a washout condition. An expression for this possible reduced rate of growth is:

$$\dot{\mu}_N = \hat{\mu}_N \left(\frac{DO}{K_{O_2} + DO}\right) (1 - 0.833(7.2 - pH)) \tag{3-23}$$

where: $\dot{\mu}_N$ = maximum possible nitrifier growth rate under environmental conditions of T, pH, DO, and $N \gg K_N$.

As before, the last expression in brackets is taken to be unity above a pH of 7.2. The relationship between actual nitrifier growth rate and maximum possible nitrifier growth rate can easily be seen from Equations 3-8 and 3-23:

$$\mu_N = \dot{\mu}_N \left(\frac{N}{K_N + N} \right) \qquad (3\text{-}24)$$

A more rigorous condition for prevention of nitrifier washout than Equation 3-22 is:

$$\dot{\mu}_N \geq \mu_b \qquad (3\text{-}25)$$

More typically, Equation 3-25 is expressed in reciprocal form in terms of solids retention time as:[11]

$$\theta_c^d \geq \theta_c^m \qquad (3\text{-}26)$$

where: θ_c^d = solids retention time of design, days, and

θ_c^m = minimum solids retention time, days, for nitrification at given pH, temperature and DO.

Since $\dot{\mu}$ or θ_c^m is fixed by the environmental conditions (T, pH and DO), Equations 3-25 or 3-26 is satisfied by modifying μ_b or θ_c^d. The various ways of satisfying these relationships can be established by examining the following growth relationship for the heterotrophic population:[10,11,12]

$$\mu_b = \frac{1}{\theta_c^d} = Y_b q_b - K_d \qquad (3\text{-}27)$$

where: Y_b = heterotrophic yield coefficient, lb VSS grown per lb of substrate (BOD or COD) removed, and

q_b = rate of substrate removal, lb BOD (or COD) removed/lb VSS/day,

K_d = "decay" coefficient, day^{-1}, and

μ_b = net growth rate for heterotrophic population.

The rate of substrate removal is defined as:

$$q_b = \frac{S_0 - S_1}{X_1 \, HT} \tag{3-28}$$

where:
S_0 = influent total BOD (or COD), mg/l,
S_1 = effluent soluble BOD (or COD), mg/l,
HT = hydraulic detention time, days, and
X_1 = MLVSS, mg/l.

Since both Y_b and K_d are assumed to be constant, the only way μ_b or θ_c^d can be manipulated is by altering q_b. One way the substrate removal rate can be reduced is to place an organic carbon removal step ahead of the nitrification stage, creating a "separate stage" nitrification process. The result of this procedure is to reduce the food available to the heterotrophic bacteria and to lessen their dominance in controlling the solids retention time. Separate nitrification stages can have very long solids retention times (θ_c^d = 15 to 25 days). Another procedure for reducing the substrate removal rate, without separating the carbon oxidation and nitrification processes, is to increase the biological solids in the system. This can be done by increasing the concentration of biological solids under aeration (the MLVSS in the activated sludge system) or by increasing the volume of the oxidation tank while maintaining the concentration of biological solids at the same concentration.

The level set for biological solids retention time, θ_c^d , establishes the biological solids retention time (or growth rate) of the nitrifiers, since selective wasting of the heterotrophic population is not practical. Therefore, the design solids retention time can be related to the effluent ammonia level through the Monod relationship and the inverse relationship between nitrifier growth rate and solids retention time (Equations 3-12 and 3-24). With specific values of θ_c^d , T, pH and DO, Equations 3-12 and 3-24 can be solved for the effluent ammonia level. Figure 3-6 was developed by such a procedure, with the assumption of T = 15 C, DO = 2 mg/l, and pH $>7.2 <8$. Also plotted is the nitrification efficiency, assuming an influent Total Kjeldahl Nitrogen (TKN) concentration of 25 mg/l and that all of the influent nitrogen is available for nitrification. As can be seen, significant breakthrough of ammonia from the system does not occur unless the solids retention time is reduced below 5 days. At that point, ammonia nitrogen breakthrough is very abrupt, rising from 1 mg/l at θ_c = 4.9 days to 15 mg/l at θ_c = 3.6 days. The principal cause of the sharpness of the ammonia breakthrough is due to the low value of K_N (0.40 mg/l NH_4^+ – N in this case). A number of investigators have experimentally determined very similar relationships to that shown in Figure 3-6.[19,20,21]

Lawrence and McCarty[11] have introduced the concept of a safety factor (SF) in the application of biological treatment process kinetics to design. The safety factor was defined as the ratio of the design solids retention time to the minimum solids retention time; the safety factor can also be related to the nitrifier growth rates through Equation 3-12. The expression for the SF is:

$$SF = \frac{\theta_c^d}{\theta_c^m} = \frac{\hat{\mu}_N}{\mu_N} \qquad (3\text{-}29)$$

A conservative safety factor was recommended to minimize process variations caused by pH extremes, low DO concentrations and toxicants.[11] Also, the SF can be used to ensure that ammonia breakthrough does not occur during diurnal peaks in load (see Section 4.3.3).

Interestingly, Equation 3-24 can be manipulated to show that the specification of a SF of 2 will establish an ammonia level equal to K_N in the effluent of a complete mix activated sludge plant, if there is a high DO level (DO not limiting). This is because the process will be operating at one-half its maximum growth rate. Recall that the half-saturation constant, K_N, in this Monod expression is defined as the level of substrate which will cause the organism to

FIGURE 3-6

EFFECT OF SOLIDS RETENTION TIME ON EFFLUENT AMMONIA CONCENTRATION AND NITRIFICATION EFFICIENCY

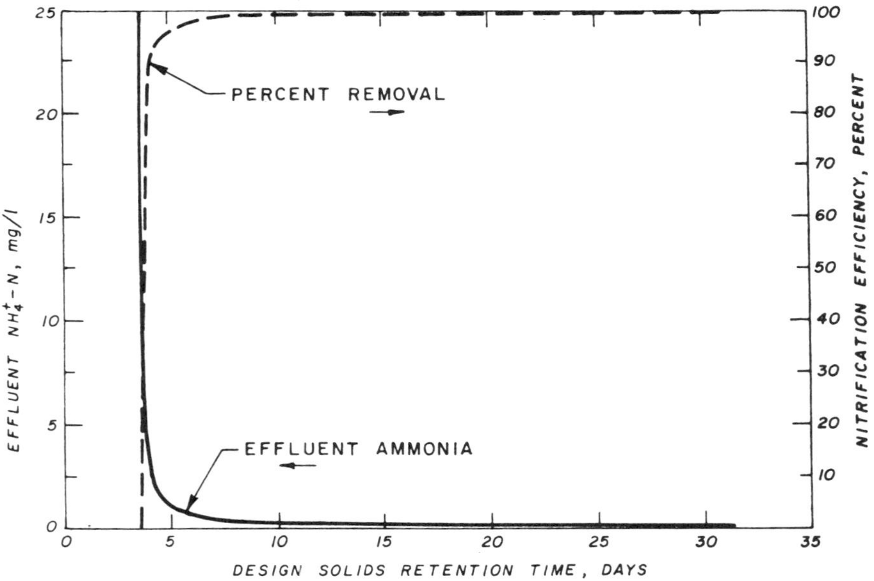

grow at half its maximum rate. Many nitrifying activated sludge plants have been observed to have 1 to 2 mg/l of ammonia in their effluents, values close to the theoretical value of K_N.

Criteria for establishing the safety factor are presented in Chapter 4. Furthermore, specific examples of the use of the kinetic expressions developed in this section are presented in Sections 4.3.3. and 4.3.5.

3.2.7 Nitrification Rates in Activated Sludge

The basic design approach for separate stage nitrification systems (activated sludge type) has been on a different basis than for combined carbon oxidation-nitrification systems. Rather than use the sludge growth rate or solids retention time approach described in Section 3.2.6, the practice frequently has been to base reactor sizing for separate stage systems on the basis of nitrification rates in terms of lb NH_4^+-N oxidized/lb MLVSS/day.[5,42,49] However, it will be shown that this parameter is fundamentally related to the nitrification kinetics previously presented in this chapter.

The nitrification rate can be calculated from the ammonia oxidation rate, q_N, by recognizing that the nitrifiers are only a fraction of the total mass of biological solids in a nitrification system. The other biological solids in the system result from the growth of the heterotrophic population. On this basis, the nitrification rate, r_N, is as follows:

$$r_N = q_N \cdot f \tag{3-30}$$

where: f = nitrifier fraction of the mixed liquor solids, and

r_N = nitrification rate, lb NH_4^+ - N oxidized/lb MLVSS/day.

A similar expression for the peak nitrification rates (in activated sludge) is:

$$\hat{r}_N = \hat{q}_N \cdot f \tag{3-31}$$

where: $\hat{r}_N$ = peak nitrification rate, lb NH_4^+ - N oxidized/lb MLVSS/day.

This latter rate is normally determined experimentally in activated sludge systems, as will be described in Section 4.6.3.

Specific analytical techniques for determination of the nitrifier fraction have not as yet been developed. However, f can be estimated from knowledge of the biological yields of the autotrophic and heterotrophic populations, as follows:

$$f = \frac{M_N}{M_N + M_c} \tag{3-32}$$

where: M_N = nitrifiers grown through oxidation of ammonia, and
M_c = heterotrophs grown through oxidation of organic carbon;

M_N and M_c can be estimated as follows:

$$M_N = Y_N(N_0 - N_1) \tag{3-33}$$

where: N_0 = TKN in the influent, mg/l, and
N_1 = NH_4^+ - N in the effluent, mg/l.

$$M_c = \bar{Y}_b (S_0 - S_1) \tag{3-34}$$

where: S_0 = carbon (BOD_5 or COD) in the influent, mg/l
S_1 = carbon (BOD_5 or COD) in the effluent, mg/l, and
$\bar{Y}_b$ = net yield of VSS of heterotrophs per unit of carbon (BOD_5 or COD) removed.

This procedure neglects the ammonia assimilated by heterotrophic growth and therefore is approximate. A further approximation is that the net yield of the heterotrophs has been assumed constant, whereas it is known that it varies with solids retention time.

Several examples can be drawn for separate stage nitrification systems as compared to combined carbon oxidation-nitrification systems. In a separate stage system, illustrative values of BOD_5 removed and TKN oxidized would be 50 mg/l and 25 mg/l, respectively. A reasonable estimate for Y_N is 0.15 lb/lb NH_4^+-N rem. (Section 3.2.2) and for $\bar{Y}_b$ (BOD), 0.55 lb/lb BOD_5 removed. Thus, for this separate stage example, f can be calculated to be:

$$f = \frac{0.15(25)}{0.55(50 + 0.15(25)} = 0.12$$

A similar example can be drawn for combined carbon oxidation-nitrification systems. Assuming 200 mg/l of BOD_5 removed and 25 mg/l of TKN oxidized, f can be calculated to be:

$$f = \frac{0.15(25)}{0.55(200) + 0.15(25)} = 0.033$$

Thus, it can be seen that the fraction of nitrifiers is lower in combined carbon oxidation-nitrification systems that in separate stage nitrification systems.

Equation 3-32 can be reexpressed in terms of the BOD_5/TKN ratio in the influent, by assuming effluent BOD and ammonia are negligibly small:

$$f = \frac{1}{\dfrac{S_0}{N_0}\dfrac{\bar{Y}_b}{Y_N} + 1} \qquad (3\text{-}35)$$

Table 3-5 presents numerical values for the fraction of nitrifiers, using Equation 3-35, and a condition where $Y_N = 0.15$ and $\bar{Y}_b$ (BOD_5) = 0.55. For most separate stage nitrification systems, the BOD_5/TKN ratio is greater than 1.0 and less than 3.0 (Section 4-2). Thus even in separate stage systems, the fraction of nitrifiers is relatively low. For the assumed yield values, the fraction is less than 20 percent and greater than 8 percent. It must be emphasized that the values of nitrifier fraction given in Table 3-5 are estimates only, and not supported by actual measurements of nitrifier fractions. Table 3-5 does have value in that it shows, at least qualitatively, the impact of influent BOD_5/TKN ratio on the fraction of nitrifiers in a nitrification system. The influence of this ratio on nitrifier fraction and nitrification rates was recognized as early as 1940, when Sawyer showed that the BOD_5/NH_3 ratio correlated with the nitrifying ability of various activated sludges.[50]

The effect of ammonia concentration on the nitrification rate is portrayed in Figure 3-7. The assumptions made were: f = 0.1, pH > 7.2 < 8, and the DO = 2.0 mg/l. Equations 3-20, 3-21, and 3-27 were employed to construct the figure. As can be seen, the nitrification rates

TABLE 3-5

RELATIONSHIP BETWEEN NITRIFIER FRACTION AND THE BOD_5/TKN RATIO

BOD_5/TKN ratio	Nitrifier fraction[a]	BOD_5/TKN ratio	Nitrifier fraction[a]
0.5	0.35	5	0.054
1	0.21	6	0.043
2	0.12	7	0.037
3	0.083	8	0.033
4	0.064	9	0.029

[a] Using Equation 3-35 and Y_N=0.15, $\bar{Y}_b$ = 0.55.

are relatively unaffected by ammonia concentration above 2.5 mg/l ammonia N. As the nitrification rates approach their plateau values, nitrification approaches a zero order rate, uninfluenced by ammonia level. It has been shown that for suspended growth processes, the rate of removal approximates a zero order reaction.[42,45,51] However, in none of these cases were nitrification rates determined at ammonia concentrations at or below the value of K_N where non-zero order rates effects would be evident.

FIGURE 3-7

EFFECT OF AMMONIA CONCENTRATION ON NITRIFICATION RATE

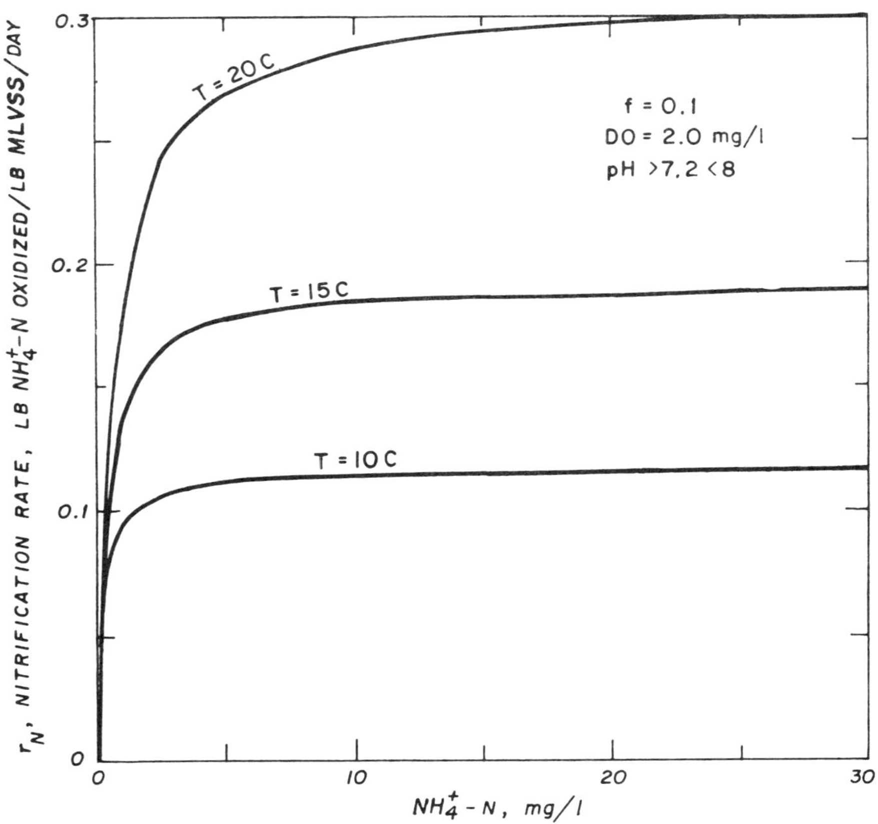

The fraction of nitrifiers has a marked effect on the nitrification rates. Figure 3-8, demonstrating this effect, was developed similarly to Figure 3-7, excepting that the effluent ammonia nitrogen concentration was assumed to be 2.5 mg/l. A principal means of increasing the nitrification rate is to increase the fraction of nitrifiers. From Table 3-5, it can be seen that this can be accomplished by lowering the BOD_5/TKN ratio. In terms of plant

FIGURE 3-8

EFFECT OF TEMPERATURE AND FRACTION OF NITRIFIERS ON NITRIFICATION RATE

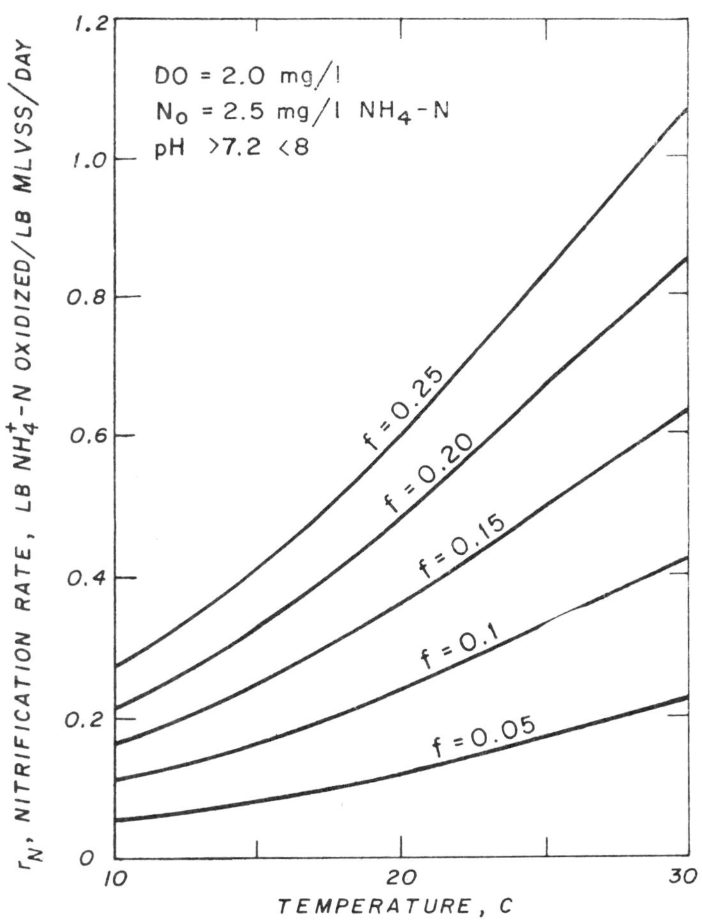

design, the BOD_5/TKN ratio can be altered by increasing the organic carbon removal ahead of the nitrification unit.

It must be emphasized that the nitrification rates developed in this section are only estimated relationships based on theoretical considerations. Actual measured values are presented in Section 4.6.3.

As a practical example of the effect of the BOD_5/TKN ratio on nitrification rates, rate data for an attached growth system[44] are plotted against the BOD_5/TKN ratio in Figure 3-9. The effect of BOD_5 in the synthetic waste was to displace nitrifiers with heterotrophic bacteria in the bacterial film, thereby reducing the nitrifier fraction and the nitrification rate. Interestingly, a small amount of BOD_5 (~10 mg/l) was found to enhance the nitrification rate.

FIGURE 3-9

EFFECT OF BOD_5/TKN RATIO ON NITRIFICATION RATE - EXPERIMENTAL ATTACHED GROWTH SYSTEM

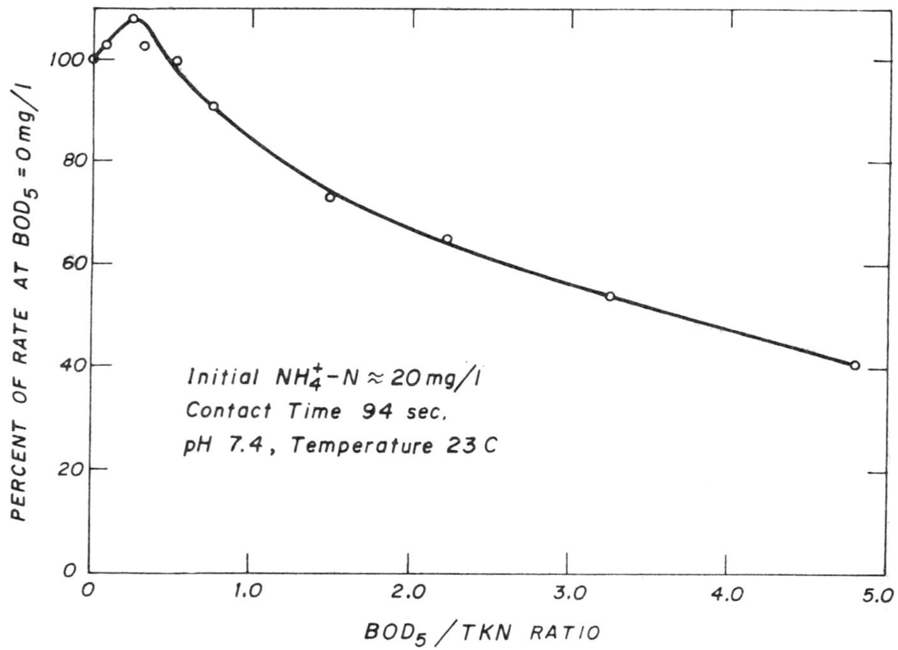

3.2.8 Nitrification Rates in Trickling Filters and Other Attached Growth Systems

Discussion of kinetic rates in the previous sections has been primarily oriented to nitrification in activated sludge type (suspended growth) nitrification systems, although some comparisons have been drawn with attached growth system measurements for comparative purposes. The growth and oxidation rate relationships presented in Sections 3.2.5.7 and 3.2.6 are directly applicable to design of suspended growth systems, as will be shown in Chapter 4. While these relationships are operative in attached growth systems, their application is complicated by the fact that oxygen mass transfer limitations through the bacterial slimes may limit reaction rates in some situations. A biofilm model has been developed which yields insight into which factors are controlling[52,53] but it has not yet been extended to the point where it can be applied directly to design applications. As a consequence, the design relationships presented for attached growth nitrification in Chapter 4 are empirically based and therefore, less theoretically precise than those developed for suspended growth sytems. Though the design relationships presented are empirical, where possible the loading relationships are presented on a basis that is at least consistent with the biofilm model. For instance, ammonia nitrogen oxidation rates are expressed on a surface area basis when describing separate stage nitrification in trickling filters and the rotating biological disc process (Sections 4.7.1 and 4.7.2).

Some of the conclusions that can be drawn from the biofilm model are of interest in considering surface ammonia removal rates in attached growth systems. The biofilm model shows that the ammonia oxidation rate in attached growth systems should not be decreased as drastically under adverse environmental conditions as in suspended growth systems.[52] This finding is consistent with the observation made in Section 3.2.5.4, namely that attached growth systems have an advantage over suspended growth systems in withstanding lower temperatures without as severe losses in nitrification rates.

The biofilm model also shows that the dissolved oxygen concentration must be 2.7 times the ammonia nitrogen concentration to prevent oxygen transfer from limiting nitrification rates in attached growth systems.[52] Two operational procedures have been suggested to overcome this limitation: (1) dilution of the ammonia nitrogen through recirculation and (2) increasing the oxygen transfer through the use of high purity oxygen.[52] The first recommendation has been made in this manual in Sections 4.4.1.4 and 4.7.1.3.

3.2.9 Effect of Inhibitors on Nitrification

Certain heavy metals and organic compounds are toxic to nitrifiers. To date, these effects have not been quantitatively incorporated into the kinetic description of nitrifier growth, although such approaches have been used to describe toxicity in other biological systems. A listing of substances toxic to unacclimated nitrifying organisms is presented in Table 3-6, which is drawn primarily from the review by Painter.[1]

Sawyer, on reviewing the English literature, suggests that 10 to 20 mg/l of heavy metals can be tolerated due to the low ionic concentrations at high pH values of 7.5 to 8.0.[56] It has also been pointed out that precipitated metals (such as hydroxides) that concentrate in the sludge can be disastrous to the sludge if the pH falls and the precipitate dissolves. Such conditions may occur in the sludge collection zone of the secondary clarifier where continuing organism activity may cause low pH values. Alternatively, low pH values may occur when pH control systems fail.

Davis[57] found that silver (Ag) was extremely toxic to nitrification of secondary effluent on a plastic media trickling filter. Levels of 2 ppb in the influent to the filter were concentrated to 5 ppm in the biomass on the media. This inhibitory effect was found to severely reduce allowable loading rates and result in only partial nitrification.

TABLE 3-6

COMPOUNDS TOXIC TO NITRIFIERS
(AFTER PAINTER (1))

Organics	Inorganics[a]
Thiourea	Zn
Allyl-thiourea	OCN^{-1}
8-hydroxyquinoline	ClO_4^{-1}
Salicyladoxine	Cu
Histidine	Hg
Amino acids	Cr
Mercaptobenzthiazole	Ni
Perchloroethylene[b]	Ag
Trichloroethylene[c]	
Abietec acid[b]	

[a] Also reference 54
[b] Reference 5
[c] Reference 55

The rate and change of magnitude of environmental conditions are nearly as critical to the biomass as the conditions themselves. It has been shown that nitrifiers can adapt to toxic substances when they are consistently present at concentrations higher than cause toxic effects in slug discharges.[1,58] Unfortunately, slug discharges are often present in municipal systems and can result from industrial dumps or from urban stormwater inflow.

Under the unusual conditions of discharge of highly concentrated industrial wastes into municipal systems that contain either nitrite or ammonia, the resulting high concentrations of ammonia or nitrite in the municipal waste can be temporarily toxic to the nitrifying population.[59] When these conditions are suspected, the reader is referred to reference 59 which contains charts which allow the identification of the regions of toxicity. Under normal municipal conditions of pH and concentrations complete nitrification will occur. When concentrated industrial wastes are present, slug discharges should be avoided; rather, storage facilities should be provided so that wastes can be metered into the collection system at a rate sufficient to ensure dilution to safe loads.

In sum, the possibility of toxic inhibition must be recognized in the design of nitrification systems. Either implementation of source control programs or inclusion of upstream toxicity removal processes may be required, particularly in those cases where significant industrial dischargers are tributary to the collection system.

3.3 Denitrification

The biological process of denitrification involves the conversion of nitrate nitrogen to a gaseous nitrogen species. The gaseous product is primarily nitrogen gas but also may be nitrous oxide or nitric oxide. Gaseous nitrogen is relatively unavailable for biological growth, thus denitrification converts nitrogen which may be in an objectionable form to one which has no significant effect on environmental quality.

As opposed to nitrification, a relatively broad range of bacteria can accomplish denitrification, including *Psuedomonas, Micrococcus, Achromobacter* and *Bacillus*. These groups accomplish nitrate reduction by what is known as a process of nitrate dissimilation whereby nitrate or nitrite replaces oxygen in the respiratory processes of the organism under anoxic conditions. Because of the ability of these organisms to use either nitrate or oxygen as the terminal electron acceptors while oxidizing organic matter, these organisms are termed facultative heterotrophic bacteria.

Confusion has arisen in the literature in terminology; the process has been termed anaerobic denitrification. However, the principal biochemical pathways are not anaerobic, but merely minor modifications of aerobic biochemical pathways. The term anoxic denitrification is preferred, since it describes the environmental condition of the absence of oxygen, without implying the nature of the biochemical pathways.

3.3.1 Biochemical Pathways

Specific information on the specific biochemical reaction intermediates involved in denitrification are available in the literature[60] and only certain concepts are of interest in process design applications. Denitrification is a two-step process in which the first step is a conversion of nitrate to nitrite. The second step carries nitrite through two intermediates to nitrogen gas. This two-step process is normally termed "dissimilation."

Denitrifiers are also capable of an assimilation process whereby nitrate (through nitrite) is converted to ammonia. Ammonia is then used for the bacterial cell's nitrogen requirements.

If ammonia is already present, assimilation of nitrate need not occur to satisfy cell requirements.

As will be shown in Section 3.3.2, electrons pass from the carbon source (the electron donor) to nitrate or nitrite (the electron acceptor) to promote the conversion to nitrogen gas. This involves the nitrifiers "electron transport system" and is involved with the release of energy from the carbon source for use in organism growth. It happens that this electron transport system is *identical* to that used for respiration by organisms oxidizing organic matter aerobically, except for one enzyme. Because of this close relationship, many facultative bacteria can shift between using oxygen or nitrate (or nitrite) rapidly and without difficulty.

3.3.2 Energy and Synthesis Relationships

The use of oxygen as the final electron acceptor is more energetically favored than the use of nitrate. Table 3-7 compares the energy yields per mole of glucose when oxygen and nitrate are used as electron acceptors.[61] The greater free energy released for oxygen favors its use whenever it is available. Therefore, denitrification must be conducted in an anoxic environment to ensure that nitrate, rather than oxygen, serves as the final electron acceptor.

TABLE 3-7

COMPARISON OF ENERGY YIELDS OF NITRATE DISSIMILATION VS OXYGEN RESPIRATION FOR GLUCOSE

Reaction	Energy yield[a] per mole glucose, kilocalories
Nitrate dissimilation	
$5\,C_6H_{12}O_6 + 24\,KNO_3 \rightarrow 30\,CO_2 + 18\,H_2O + 24\,KOH + 12\,N_2$	570
Oxygen respiration	
$C_6H_{12}O_6 + 6\,O_2 \rightarrow 6\,CO_2 + 6\,H_2O$	686

[a] Reference 61

Methanol, rather than glucose or any other organic, has seen widest use as the electron donor in the U.S. (see Section 3.3.4). Using methanol as an electron donor and neglecting synthesis for the moment, denitrification can be represented as a two-step process as shown in Equations 3-36 and 3-37:

First Step

$$NO_3^- + 0.33\ CH_3OH = NO_2^- + 0.67\ H_2O \qquad (3\text{-}36)$$

Second Step

$$NO_2^- + 0.5\ CH_3OH = 0.5\ N_2 + 0.5\ CO_2 + 0.5\ H_2O + OH^- \qquad (3\text{-}37)$$

The overall transformation is obtained by addition of Equations 3-36 and 3-37 to yield Equation 3-38:

$$NO_3^- + 0.833\ CH_3OH = 0.5\ N_2 + 0.833\ CO_2 + 1.167\ H_2O + OH^- \qquad (3\text{-}38)$$

In this equation, methanol serves as the electron donor and nitrate as the electron acceptor. This can be shown by splitting up Equation 3-38 into the following oxidation-reduction half reactions:

$$NO_3^- + 6H^+ + 5e^- \longrightarrow 0.5\ N_2 + 3\ H_2O \qquad (3\text{-}39)$$

(electron acceptor)

$$0.833\ CH_3OH + 0.833\ H_2O \longrightarrow 0.833\ CO_2 + 5H^+ + 5e^- \qquad (3\text{-}40)$$

(electron donor)

Equation 3-38 can be obtained by adding Equations 3-39 and 3-40 and the following equation for water:

$$H_2O = H^+ + OH^-$$

From Equations 3-39 and 3-40 the meaning of the terms of electron donor and acceptor are clear. Nitrate gains electrons and is reduced to nitrogen gas, hence it is the electron acceptor. The carbon source, methanol, loses electrons and is oxidized to carbon dioxide, hence it is the electron donor.

As mentioned in Section 3.2.2, these reactions take place in the context of the carbonic acid system. Equations 3-36 and 3-38 have been modified in Table 3-8 to reflect the fact that hydroxide (OH^-) produced reacts with carbonic acid (carbon dioxide) to produce bicarbonate alkalinity. Also shown in Table 3-8 is the equation of synthesis for those organisms deriving energy through nitrate respiration.[62]

The equations for energy yielding reactions (Equations 3-36 and 3-37) have been combined with the equation for oganism synthesis (Equation 3-41, Table 3-8) through knowledge of organism yields and are summarized in Table 3-9. Also, shown for completeness is the combined expression for oxygen respiration (Equation 3-44) since, if any oxygen is present, it will be used preferentially. Similar expressions can be developed for other organic sources serving as electron donors if organism yields are known.[63]

TABLE 3-8

RELATIONSHIPS FOR NITRATE DISSIMILATION AND GROWTH IN DENITRIFICATION REACTIONS

Reaction	Equation	Equation No.
Nitrogen dissimilation		
Nitrate to nitrite	$NO_3^- + 0.33CH_3OH =$ $NO_2^- + 0.33H_2O + 0.33H_2CO_3$	3-36A
Nitrite to nitrogen gas	$NO_2^- + 0.5\ CH_3OH + 0.5H_2CO_3 =$ $0.5N_2 + HCO_3^- + H_2O$	3-37A
Nitrate to nitrogen gas	$NO_3^- + 0.833\ CH_3OH + 0.167H_2CO_3$ $= 0.5\ N_2 + 1.33\ H_2O + HCO_3^-$	3-38A
Synthesis - denitrifiers	$14\,CH_3OH + 3\,NO_3^- + 4\,H_2CO_3 =$ $3\,C_5H_7O_2N + 20\ H_2O + 3\,HCO_3^-$	3-41

The theoretical methanol requirement for nitrate reduction, neglecting synthesis, is 1.9 mg methanol per mg nitrate N (Equation 3-36, Table 3-9). Including synthesis (Equation 3-42), the requirement is increased to 2.47. Similarly, Equations 3-43 and 3-44 in Table 3-9 allow calculation of methanol requirements for nitrite reduction and deoxygenation to allow a combined expression to be formulated for the methanol requirement:[62]

$$C_m = 2.47\ NO_3^- - N + 1.53\ NO_2^- - N + 0.87\ DO \qquad (3\text{-}45)$$

where:

C_m = required methanol concentration, mg/l,

$NO_3^- - N$ = nitrate concentration removed, mg/l,

$NO_2^- - N$ = nitrite concentration removed, mg/l, and

DO = dissolved oxygen removed, mg/l.

Biomass production can be calculated similarly:

$$C_b = 0.53\,NO_3^- - N + 0.32\,NO_2^- - N + 0.19\,DO \qquad (3\text{-}46)$$

where: C_b = biomass production, mg/l.

TABLE 3-9

COMBINED DISSIMILATION-SYNTHESIS EQUATIONS FOR DENITRIFICATION (AFTER MC CARTY, ET AL. (62))

Transformation	Equation	Equation No.
Overall nitrate removal	$NO_3^- + 1.08\,CH_3OH + 0.24\,H_2CO_3 =$ $0.056\,C_5H_7NO_2 + 0.47\,N_2 + 1.68\,H_2O + HCO_3^-$	3-42
Overall nitrite removal	$NO_2^- + 0.53\,H_2CO_3 + 0.67\,CH_3OH =$ $0.04\,C_5H_7NO_2 + 1.23\,H_2O + 0.48\,N_2 + HCO_3^-$	3-43
Overall deoxygenation	$O_2 + 0.93\,CH_3OH + 0.056\,NO_3^- =$ $0.056\,C_5H_7NO_2 + 1.04\,H_2O + 0.59\,H_2CO_3 + 0.056\,HCO_3^-$	3-44

Most experimental data is expressed in terms of the "M/N ratio," which is the mg of methanol per mg of initial nitrate nitrogen concentration. The ratio includes the requirements for nitrite and oxygen, which are usually small relative to the nitrate requirement. For instance, for a NO_3^- value of 25 mg/l of nitrate -N, 0.5 mg/l nitrite -N and 3.0 mg/l dissolved oxygen, the methanol requirement can be calculated to be 64.1 mg/l from Equation 3-45. The M/N ratio is therefore 2.57 (64.4/25), which is only 4 percent greater than the requirement for nitrate alone (2.47).

Values of the "M/N" ratio required for complete denitrification range from the levels estimated from Equation 3-45 at 2.5 lb methanol per lb of nitrate nitrogen removed up to 3.0 lb methanol per lb of nitrate -N removed.[5,6,64,65,66] Departures of methanol requirements from Equation 3-45 are most likely due to variations in sludge yields among experimental systems. It has been suggested that column denitrification systems require a lower M/N ratio than suspended growth systems due to the higher concentration of biomass maintained in the column systems.[67] Higher biomass levels produce longer solids retention times and reduce organism yields due to increased endogenous metabolism. In turn, this lower yield would result in less methanol required for synthesis and reduce the "M/N" ratio.[67]

In general, an M/N ratio of 3.0 will enable "complete" denitrification (95 percent removal of nitrate) and this value may be used for design purposes when methanol is employed as the carbon source for denitrification.

3.3.3 Alkalinity and pH Relationships

Equations 3-42 and 3-43 (Table 3-9) show that bicarbonate is produced and carbonic acid concentration is reduced whenever nitrate or nitrite is denitrified to nitrogen gas. The stoichiometric quantity of alkalinity produced is 3.57 mg alkalinity as $CaCO_3$ produced per mg of nitrate or nitrite -N reduced to nitrogen gas.

Since both the alkalinity concentration is increased and the carbonic acid concentration is reduced, the tendency of denitrification is to at least partially reverse the effects of nitrification and raise the pH of the biological reaction (Equation 3-9). Denitrification only partially offsets the alkalinity loss caused by nitrification, since the alkalinity gain per mg of nitrogen is only one-half the loss caused by nitrification (see Section 3.2.3).

Measured alkalinity production has been reported to be somewhat lower than indicated theoretically. Experiments with an attached growth process showed that the alkalinity produced averaged 2.95 mg as $CaCO_3$ per mg of nitrogen reduced.[65] Similarly, the ratio for a suspended growth system was 2.89.[6] Departures from theory may be due to the fact that Equations 3-42 and 3-43 (Table 3-9) represent over-simplications of the biological transformations taking place and do not include all factors affecting alkalinity production.

A value for alkalinity production suitable for engineering calculations would be 3.0 mg alkalinity as $CaCO_3$ produced per mg nitrogen reduced.

3.3.4 Alternative Electron Donors

Although methanol has found a predominance in U.S. practice as the electron donor of choice, the significance of the cost of the organic chosen for the process has led to the consideration of alternative electron donors available, particularly those from waste sources. Considering alternate commercial sources, methanol seems to continue to be the most economic choice, because price increases in alternate sources have paralleled those for methanol.

A variety of compounds that can substitute for methanol have been experimentally evaluated,[60] but design data are available only for municipal wastewater organics, volatile acids, brewery wastes, and molasses. The use of wastewater organics for denitrification is discussed extensively in Section 5.5.2. Denitrification rates with wastewater oganics are approximately one-third of those when methanol is employed. Therefore, denitrification reactors must be proportionately larger. Since using wastewater organics adds ammonia and organic nitrogen to the wastewater, the sequence of nitrification-denitrification steps must be modified to ensure that these compounds do not escape from the system. Thus, wastewater organics are not completely interchangeable with methanol; their attraction, however, is the possible reduction in operating costs with the elimination of the need for methanol in the treatment plant.

In studies conducted for the development of the City of Tampa, Florida's treatment plant, it was shown that brewery wastes could substitute for methanol when used in both suspended growth and column denitrification systems.[68] Bench scale studies exhibited denitrification rates of 0.25 to 0.22 lb NO_3^- -N rem./lb MLVSS/day with brewery wastes compared to 0.18 lb NO_3^- -N rem./lb MLVSS/day with methanol at a temperature in the range of 19 to 24 C. Solids production was found to be greater with brewery wastes than methanol, but values were not given. Removal efficiencies were similar in a parallel test of brewery wastes and methanol using columnar denitrification.[68]

Volatile acids have also been used as a carbon source for denitrification. In studies of nitrate reduction in wastewaters generated in the manufacture of nylon intermediates, it was found that a mixture of C_1 to C_5 volatile acids was very effective as a carbon source for denitrification.[69] Denitrification rates with this mixture were 0.36 lb. NO_3^- -N rem./lb MLVSS/day at 20 C and 0.10 lb NO_3^- -N rem./lb MLVSS/day at 10 C. These rates compare favorably with those measured for use with methanol (see Section 5.2.1). Volatile acids can be produced from wastewater organics by anaerobic fermentation or by low temperature wet oxidation. In either case, the product will contain varying amounts of ammonia nitrogen which may have to be removed in the process (as described in Section 5.5.2) or removed prior to use by ammonia stripping.

Molasses was tested at the Central Contra Costa Sanitary District's Advanced Treatment Test facility as a substitute for methanol.[70] Peak denitrification rates at 16 C in a suspended growth reaction were found to be only 0.036 lb NO_3^- -N rem./lb MLVSS/day. In addition to having a slower reaction rate with molasses, the sludge tended to bulk to a greater degree than with methanol, rising from a sludge volume index (SVI) averaging 164 ml/gram to one having a SVI averaging 257 ml/gram. This caused a decrease in the settling rates of the sludge when molasses was employed.

Some of the alternatives cause greater sludge production than others. For instance, about twice as much sludge is produced per mg of nitrogen reduced when saccharose is used than when methanol is employed. On the other hand, acetone, acetate and ethanol produced similar quantities of sludge to that produced when methanol is employed.[62]

Methanol has certain advantages over wastewater carbon sources. It is free of contaminants, such as nitrogen, and therefore can be used directly in the process without taking the special precautions that must be made for use of with a waste carbon source. Second, the product is of consistent quality while wastewater sources may vary in strength and composition either daily or seasonally, complicating process control and optimization. Use of wastewater sources will require regular assaying of the source to check its purity, strength and biological availability. Methanol also has the advantage of being nationally distributed while suitable waste carbon sources may not be geographically close to the point of use. Nonetheless, the significant disadvantage of methanol is its cost and this alone mandates the necessity of economic comparisons of alternate carbon sources.

3.3.5 Kinetics of Denitrification

Just as in the case of nitrification (Section 3.2.5), environmental factors have a significant effect on the kinetic rates of denitrifier growth and nitrate removal. Factors considered in subsequent sections are temperature, pH, carbon concentration and nitrate concentration. A combined kinetic expression incorporating all these factors is presented.

3.3.5.1 Effect of Nitrate on Kinetics

The absence of significant quantities of nitrite in denitrification systems[60,71] has led to the description of the kinetics of denitrification as a one-step process from nitrate to nitrogen gas. The Monod expression is employed to describe the influence of nitrate on growth rate:

$$\mu_D = \hat{\mu}_D \frac{D}{K_D + D} \tag{3-47}$$

where:
- μ_D = growth rate, day^{-1}
- $\hat{\mu}_D$ = maximum denitrifier growth rate, day^{-1}
- D = concentration of nitrate nitrogen, mg/l, and
- K_D = half saturation constant, mg/l NO_3^- - N.

3.3.5.2 Relationship of Growth Rate to Removal Rate

Denitrification rates can be related to the oganism growth rates by the following relationship:

$$q_D = \mu_D/Y_D \tag{3-48}$$

where:
- q_D = nitrate removal rate, lb NO_3^- - N rem./lb VSS/day, and
- Y_D = denitrifier gross yield, lb VSS grown/lb NO_3^- - N removed.

Similarly, peak denitrification rates are related to maximum denitrifier growth rates as follows:

$$\hat{q}_D = \hat{\mu}_D / Y_D \tag{3-49}$$

3.3.5.3 Solids Retention Time

Consideration of solids production and solids retention time is an important design consideration. A mass balance of the biomass in a completely mixed reactor yields the relationship.[10,11]

$$\frac{1}{\theta_c} = Y_D q_D - K_d \tag{3-50}$$

where: θ_c = solids retention time, days, and

K_d = decay coefficient, day^{-1}.

3.3.5.4 Kinetic Constants for Denitrification

The value of the half saturation constant, K_D, is very low. Investigators at the University of California at Davis, found K_D for suspended growth systems to be 0.08 mg/l NO_3^--N without solids recycle and 0.16 mg/l NO_3^-- N with solids recycle at 20 C.[71,48] For attached growth systems the value of K_D was found to be 0.06 mg/l NO_3^--N at 25 C.[72,73] It can be seen from examination of Equation 3-47 using these values of K_D that nitrate has almost no effect on denitrification above 1-2 mg/l nitrate -N and approaches a zero order rate. The observations of several investigators suport these low values of K_D, as they have reported zero order rates above 1-2 mg/l nitrate -N.[60,74,75,76]

Yield and decay coefficients from data of a number of investigations are shown in Table 3-10. In most cases only net yields are reported or can be calculated from the data reported. The relationship between the gross yield in Equation 3-50 and the net yield is:

$$\bar{Y}_D = \frac{Y_D q_D - K_d}{q_D} = \frac{Y_D}{(1 + \theta_c K_d)} \tag{3-51}$$

where: $\bar{Y}_D$ = denitrifier net yield, lb VSS/lb NO_3^- – N rem.

The data of Stensel, *et al.*[74] for K_d has been used in some cases to derive calculated Y values for those cases where none was reported (Table 3-10). Data from another study[48] allow calculation of both Y_D and K_d at 20 C. The value of K_d of 0.04 day^{-1} is consistent with Stensel's findings at 20 C. Values suitable for use in engineering calculations are Y_D = 0.6 to 1.20 and K_d = 0.04 day^{-1}.

It is notable that when an aerobic stabilization step was incorporated into the process after anoxic denitrification, net yields reduced by almost an order of magnitude.[78,80] This effect was attributed to enhanced endogenous metabolism where oxygen is provided as an electron-acceptor.[80] Table 3-10 shows a calculated value for K_d under these conditions of 0.19 $days^{-1}$, almost five times the rate when nitrate serves as the electron acceptor for endogenous metabolism. This concept is supported by the results of other investigators whose data show that the endogenous respiration rates when expressed on an equivalent basis are significantly greater when oxygen serves as the electron acceptor than when nitrate does.[81,82]

Both organism growth rate and nitrate removal rate are significantly affected by temperature. Only one investigator has reported growth rates,[74] all others have reported removal rates. To show the effect of temperature on growth and denitrification rates, the available data have been summarized in Figure 3-10 on a basis that is normalized with respect to the rate at 20 C. It can be seen that denitrification proceeds at a reduced rate as

TABLE 3-10

VALUES OF DENITRIFICATION YIELD AND DECAY COEFFICIENTS FOR VARIOUS INVESTIGATIONS USING METHANOL

Process description	Ref.	q_D, $days^{-1}$	$\overline{Y}_D$, lb VSS / lb NO_3^--N rem.	Y_D, lb VSS / lb NO_3^--N rem	K_d, $days^{-1}$	Temp, C
Suspended growth, no solids recycle, continuous	74	Variable	Variable	0.57	0.05	10
	74	Variable	Variable	0.63	0.04	20
	74	Variable	Variable	0.67	0.02	30
	71	0.12 to 0.32	0.55 to 1.4	-	-	20
	77	0.16 to 0.9	0.57 to 0.73	-	-	20
Suspended growth, batch	75[a]	0.24 to 3.8	0.45 to 1.43	-	-	5 to 27
	62	Variable	0.53	-	-	20
Suspended growth, solids recycle, continuous	5	-[b]	0.58	0.77[c]	0.04[d]	10 to 20
	48	0.131 to 0.347	0.542 to 0.703	0.84	0.04	20
	78	0.25	0.49	0.65[c]	0.04[d]	16 to 18
	79	e	0.7 to 1.4	0.83 to 1.67	0.04[d]	
Suspended growth, solids recycle, continuous, aerated stabilization	78	0.30	0.061	0.65[d]	0.19[c]	19 to 20

[a] Substrate was sodium citrate

[b] q not given, but Θ_c = 8.0

[c] Calculated

[d] Assumed

[e] Θ_c = 3 to 6 days

FIGURE 3-10
EFFECT OF TEMPERATURE ON DENITRIFICATION RATE

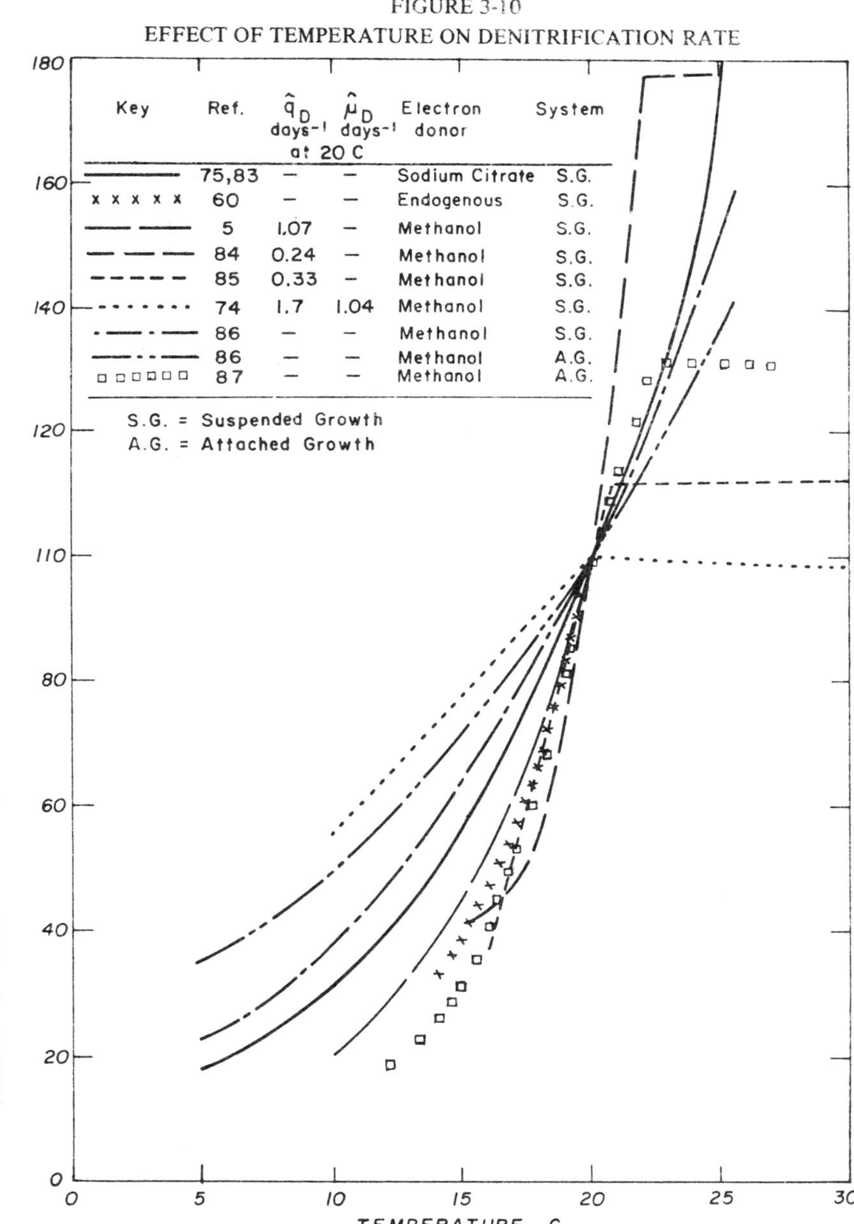

low as 5 C. Above 20 C, four out of seven sets of data indicate that the denitrification rates find plateau values at some temperature and do not keep climbing. The parallel systems study of Murphy *et al.*[86] is interesting in that it shows attached growth systems to be less affected by cold temperatures than suspended growth systems. Differences between attached growth systems and suspended growth systems may reflect differences in the method of measurement rather than differences in organism reaction rate. Attached growth system removal rates were expressed on a unit surface basis while suspended growth systems were expressed per unit of biomass (MLVSS) in Murphy's study.[86] It is probable that surface slimes in attached growth systems expand in cold weather and compensate for reduced reaction rates. If the biomass level could be measured the rate per unit of biomass may very well be similar. For instance, in one study parallel tests of suspended and attached growth systems were at 30 C. Biomass measurements were made in both systems and peak denitrification rates were found to be comparable, 0.38 lb NO_3^--N rem./lb MLVSS/day for the suspended growth system and 0.45 lb NO_3^--N rem./lb MLVSS/day for the attached growth system.[88]

Further information on the effect of temperature on denitrification rates is presented in Chapter 5.

3.3.5.5 Effect of Carbon Concentration on Kinetics

The effect of carbon concentration on the rate of denitrification has been modeled in terms of a Monod type of expression. When methanol serves as the carbon source, the expression is:[74,89]

$$\mu_D = \hat{\mu}_D \frac{M}{K_M + M} \tag{3-52}$$

where:

M = methanol concentration, mg/l

K_M = half saturation constant for methanol, mg/l of methanol.

The earliest investigators used a nonspecific test for methanol, COD.[74] As a result the initial evaluation of K_M was somewhat obscured. A later more definitive investigation evaluating K_M used a specific test for methanol.[89] A chemostat system operating at a solids retention time of five days and a temperature of 20 C was operated in a manner whereby reaction rates were limited by methanol and not nitrate. The value of K_M was found to be very low, 0.1 mg/l as methanol. The practical implication of this finding is that to achieve 90 percent of the maximum denitrification rate in a suspended growth reactor, only about 1 mg/l of methanol need be in the effluent. In other words, great excesses of methanol above stoichiometric requirements need not be in the effluent from a suspended growth denitrification process to achieve nearly the maximum denitrification rates.

3.3.5.6 Effect of pH on Kinetics

Representative observations of the effect of pH on denitrification rates are shown on Figure 3-11. While there are some anomalies, it is apparent that denitrification rates are depressed below pH 6.0 and above pH 8.0. There is some disagreement about the pH of the optima, but the data show the highest rates of denitrification are at least within the range of pH 7.0 to 7.5.

FIGURE 3-11

EFFECT OF pH ON DENITRIFICATION RATE

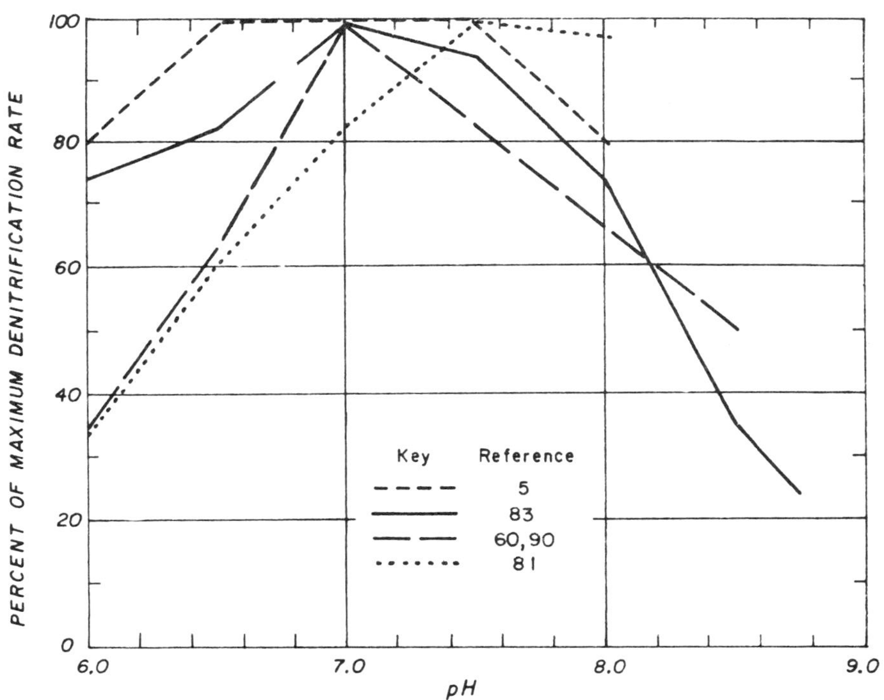

3.3.5.7 Combined Kinetic Expression

The same approach as employed for nitrification can be used for denitrification to establish the effects of environmental conditions on the rates of denitrifier growth (and nitrate removal):

$$\mu = \hat{\mu}_D \left(\frac{D}{K_D + D} \right) \left(\frac{M}{K_M + M} \right) \qquad (3\text{-}53)$$

where: $\hat{\mu}_D$ = peak rate of denitrifier growth at given temperature, T, and pH, and

μ_D = actual rate of denitrifier growth affected by nitrate, methanol, T, and pH.

Relationships for temperature, pH, nitrate and methanol established in Sections 3.3.5.3. 3.3.5.4, 3.3.5.5, and 3.3.5.6 can be employed when using this equation to predict growth rates or removal rates. Ordinarily, the term for methanol can be neglected (Section 3.3.5.5). Removal rates can be related to growth rates through Equation 3-48.

The safety factor concept presented in Section 3.2.6 can be applied to denitrification as well as to nitrification, as the concept has general validity for biological systems. Restating the concept for denitrification:

$$SF = \frac{\theta_c^d}{\theta_c^m} \qquad (3\text{-}29)$$

In the case of denitrification, the safety factor can be related to nitrate removal rates through Equation 3-50 and the following similar equation for the minimum solids retention time:

$$\frac{1}{\theta_c^m} = Y\hat{q}_D - k_d \qquad (3\text{-}54)$$

The use of these equations for design of suspended growth systems is given in Section 5.2.2 in terms of illustrative examples.

The above equations cannot be directly applied to attached growth denitrification because the reactions take place in a more complex environment than is present in suspended growth systems. Rates of nitrate removal in the bacterial films developed in denitrification systems may be affected by the mass transfer of nitrate or methanol through the bacterial film. A biofilm model has been developed[52,53] that may be used to describe denitrification in bacterial slimes, but its use has not yet been extended to the point where it can be used in system design. However, the model indicates that removal rates are most usefully expressed on a unit surface area basis and this is the procedure adopted in Section 5.3.1 to describe denitrification in the various attached growth systems.

The biofilm model usefully predicts certain properties of attached growth denitrification that are significant in design. The model shows that the nitrate removal rate in attached growth sytems should not be drastically affected by adverse environmental conditions compared to effects in suspended growth systems.[52] In Section 3.3.5.4, for instance, it was shown that attached growth systems are less affected by cold temperatures than suspended growth systems. Another interesting prediction of the biofilm model is that methanol will normally be film transfer limiting rather than nitrate, unless the methanol is supplied in concentrations five times as large as the nitrate concentration[52] (an impractical situation).

3.3.6 Effect of DO on Denitrification Inhibition

The role of dissolved oxygen in denitrification is generally to suppress denitrification. This has been explained on the basis that the rate of dissimilatory nitrate reduction is considerably slower than the rate of aerobic respiration.[71] While it has been observed that denitrification can occur in the presence of low levels of DO,[6,66] the mechanism of denitrification is attributed to an oxygen gradient in the system whereby some cells are at zero dissolved oxygen and thus able to reduce nitrate.[1,6,91]

3.4 References

1. Painter, H.A., *A Review of Literature on Inorganic Nitrogen Metabolism in Microorganisms.* Water Research, 4, No. 6, pp 393-450 (1970).

2. Haug, R. T., and P. L. McCarty, *Nitrification with the Submerged Filter.* Report prepared by the Department of Civil Engineering, Stanford University for the Environmental Protection Agency, Research Grant No. 17010 EPM, August, 1971.

3. *Notes on Water Pollution No. 52,* Water Pollution Research Laboratory, Stevenage, England, 1971.

4. Gujer, W. and D. Jenkins, *The Contact Stabilization Process-Oxygen and Nitrogen Mass Balances.* University of California, Sanitary Engineering Research Lab, SERL Report 74-2, February, 1974.

5. Mulbarger, M. C., *The Three Sludge System for Nitrogen and Phosphorus Removal.* Presented at the 44th Annual Conference of the Water Pollution Control Federation, San Francisco, California, October, 1971.

6. Horstkotte, G. A., Niles, D.G., Parker, D.S., and D.H. Caldwell, *Full-Scale Testing of A Water Reclamation System.* JWPCF, 46, No. 1, pp 181-197 (1974).

7. Newton, D., and T.E. Wilson, *Oxygen Nitrification Process at Tampa.* In *Applications of Commercial Oxygen to Water and Wastewater Systems,* Ed. by R.E. Speece and J.F. Malena, Jr., Austin, Texas: The Center for Research in Water Resources, 1973.

8. Gasser, J.A., Chen, C.L., and R.P. Miele, *Fixed-Film Nitrification of Secondary Effluent.* Presented at the EED-ASCE Specialty Conference, Penn. State University, Pa., July, 1974.

9. Osborn, D.E., *Operating Experiences with Double Filtration in Johannesburg.* J. Inst. of Sew. Purif., Part 3, pp 272-281 (1965).

10. Pearson, E.A., *Kinetics of Biological Treatment.* In *Advances in Water Quality Improvement.* Ed. by E.F. Gloyna and W.W. Eckenfielder, Austin, Texas: University of Texas Press, 1968.

11. Lawrence, A.W., and P.L. McCarty, *Unified Basis for Biological Treatment Design and Operation.* JSED, Proc. ASCE, 96, No. SA3, pp 757-778 (1970).

12. Jenkins, D., and W.E. Garrison, *Control of Activated Sludge by Mean Cell Residence Time.* JWPCF, 40, No. 11, pp 1904-1919 (1968).

13. Monod, J., *Researches sur la croissance des cultures.* Hermann and Cie, Paris, 1942.

14. Stankewich, M. J., *Biological Nitrification with the High Purity Oxygenation Process.* Presented at 27th Annual Purdue Industrial Waste Conference, Lafayette, Indiana, May, 1972 .

15. Poduska, R.A. and J.F. Andrews, *Dynamics of Nitrification in the Activated Sludge Process.* Presented at the 29th Industrial Waste Conference, Purdue University, Lafayette, Indiana, May 7-9, 1974.

16. Christensen, D.R. and P.L. McCarty, *Bio-treat: A Multi-Process Biological Treatment Model.* Presented at the 47th Annual Conference of the Water Pollution Control Federation, Denver, Colorado, October, 1974.

17. Knowles, G., Downing, A.L., and M.J. Barrett, *Determination of Kinetic Constants for Nitrifying Bacteria in Mixed Culture, with the Aid of An Electronic Computer.* J. Gen. Microbiology, 38, p 263 (1965).

18. Downing, A.L., and A.P. Hopwood, *Some Observations on the Kinetics of Nitrifying Activated Sludge Plants.* Schweiz. Zeitsch f Hydrol., 26, No. 2, p 271 (1964).

19. Wuhrmann, K., *Research Developments in Regard to Concept and Base Values of the Activated Sludge System.* Ed. by E.F. Gloyna and W.W. Eckenfielder, Austin, Texas: University of Texas Press, 1968.

20. Melamed, A., Saliternik, C., and A.M. Wachs, *BOD Removal and Nitrification of Anaerobic Effluent by Activated Sludge.* Presented at the 5th IAWPR Conference, San Francisco, California, July-August, 1970.

21. Balakrishnan, S., and W.W. Eckenfielder, *Nitrogen Relationships in Biological Waste Treatment Processes – I, Nitrification in the Activated Sludge Process.* Water Research, 3, pp 73-81 (1969).

22. Loehr, R.C., Prakasam, T.B.S., Srinath, E.G., and Y.D. Joo, *Development and Demonstration of Nutrient Removal from Animal Wastes.* Report prepared for the Environmental Protection Agency, EPA-R2-73-095, January, 1973.

23. Stratton, F.E., and P.L. McCarty, *Prediction of Nitrification Effects on the Dissolved Oxygen Balance of Streams.* Environmental Science and Technology, 1, No. 5, pp 405-410 (1967).

24. Lawrence, A.M., and C.G. Brown, *Biokinetic Approach to Optimal Design and Control of Nitrifying Activated Sludge Systems.* Presented at the Annual Meeting of the New York Water Pollution Control Association, New York City, January 23, 1973.

25. Hoffman, T., and H. Lees, *The Biochemistry of Nitrifying Organisms 4, The Respiration and Intermediate Metabolism of Nitrosomonas.* Biochemistry Journal, 54, 575-583 (1953).

26. Ulken, A., *Die Herkunft des Nitrits in der Elbe.* Arch. Hydrobiol., 59, pp 486-501, 1963.

27. Loveless, J.E., and J.E. Painter, *The Influent of Metal Ion Concentration and pH Value on the Growth of a Nitrosomonas Strain Isolated From Activated Sludge.* J. Gen. Microbiology, 52, pp 1-14 (1968).

28. Williamson, K.J. and P.L. McCarty, *Rapid Measurement of Monod Half-Velocity Coefficients for Bacterial Kinetics.* Unpublished paper, Stanford University, May, 1974.

29. Lees, H., and J.R. Simpson, *The Biochemistry of the Nitrifying Organisms. 5, Nitrate Oxidation by Nitrobacter.* Biochemistry Journal, 65, pp 297-305 (1957).

30. Gould, G.W., and H. Lees, *The Isolation and Culture of the Nitrifying Organisms. Part I. Nitrobacter.* Can. J. Microbiology, 6, pp 299-307 (1960).

31. Landelot, H., and L. Van Tichilen, *Kinetics of Nitrite Oxidation by Nitrobacter Winogradskyi*, J. Bact., 79, pp 39-42 (1960).

32. Buswell, A.H., Shiota, T., Lawrence, N., and I. Van Meter, *Laboratory Studies on the Kinetics of the Growth of Nitrosomonas with Relation to the Nitrification Phase of the BOD Test.* Applied Microbiology, 2, pp 21-25 (1954).

33. Balakrishnan, S., and W.W. Eckenfielder, *Nitrogen Relationships in Biological Waste Treatment Processes – II, Nitrification in Trickling Filters.* Water Research, 3, pp 167-174 (1969).

34. Huang, C.S., and N.E. Hopson, *Temperature and pH Effect on the Biological Nitrification Process.* Presented at the Annual Winter Meeting, New York Water Pollution Control Association, New York City, January, 1974.

35. *Water Pollution Research*, 1964. London: HMSO, 1965.

36. Nagel, C.A. and Haworth, J.G., *Operational Factors Affecting Nitrification in the Activated Sludge Process.* Presented at the 42nd Annual Conference of the Water Polluation Control Federation, Dallas, Texas, October (1969), (available as a reprint from the County Sanitation Districts of Los Angeles County).

37. Wuhrman, K., *Effects of Oxygen Tension on Biochemical Reactions in Sewage Purification.* In *Advances in Biological Treatment, Proc. 3rd Conf. Biological Waste Treatment*, Ed. by Eckenfielder, W.W., and J. McCabe. (New York, 1963), Pergamon Press.

38. Schwer, A.D., Letter communication to D.S. Parker – Metropolitan Sewer District of Greater Cincinnati, March 9, 1971.

39. Murphy, K.L., Personal communication to D.S. Parker, McMasters University, Ontario, July, 1974.

40. Engel, M.S., and M. Alexander, *Growth and Metabolism of Nitrosomonas europaea.* Journal Bacteriology, 76, pp 217-222 (1958).

41. Meyerhof, O., *Untersuchungen uber den Atmungsvorgany Nitrifizierenden Bakterien.* Pflugers Archges Physiol., 166, pp 240-280, 1917.

42. Sawyer, C.N., Wild, H.E., Jr., and T.C. McMahon, *Nitrification and Denitrification Facilities, Wastewater Treatment.* Prepared for the EPA Technology Transfer Program, August, 1973.

43. Downing, A.L., and G. Knowles, *Population Dynamics in Biological Treatment Plants.* Presented at the 3rd Conference of the IAWPR, Munich, 1966.

44. Huang, C.S., *Kinetics and Process Factors of Nitrification on a Biological Film Reactor.* Thesis submitted in partial satisfaction of the requirements for the degree of Doctor of Philosophy, University of New York at Buffalo, 1973.

45. Heidman, J.A., *An Experimental Evaluation of Oxygen and Air Activated Sludge Nitrification Systems With and Without pH Control.* EPA report for Contract No. 68-03-0349, 1975.

46. Chen, C.W., *Concepts and Utilities of Ecological Model.* JSED, Proc. ASCE, 96, No. SA5, pp 1085-1097 (1970).

47. Middlebrooks, E.J., and D.B. Porcella, *Rational Multivariate Algal Growth Kinetics.* JSED, Proc. ASCE, 97, SA1, pp 135-140 (1971).

48. Engberg, D.J., and E.D. Schroeder, *Kinetics and Stoichiometry of Bacterial Denitrification as a Function of Cell Residence Time.* University of California at Davis, unpublished paper, 1974.

49. Stamberg, J.B., Hais, A.B., Bishop, D.F., and J. Heidman, *Nitrification in Oxygen Activated Sludge.* Unpublished paper, Environmental Protection Agency, 1974.

50. Sawyer, C.N., *Activated Sludge Oxidations, V. The Influence of Nutrition in Determining Activated Sludge Characteristics.* Sewage Works Journal, 12, No. 1, pp 3-17 (1940).

51. Kiff, R.J., *The Ecology of Nitrification/Denitrification in Activated Sludge.* Water Pollution Control, 71, p 475 (1972).

52. Williamson, K.L. and P.L. McCarty, *A Model of Substrate Utilization by Bacterial Films.* Presented at the 46th Annual Conference of the Water Pollution Control Federation, Cincinnati, Ohio, October, 1973.

53. Williamson, K.L. and P.L. McCarty, *Verification Studies of the Biofilm Model for Bacterial Substrate Utilization.* Presented at the 46th Annual Conference of the Water Pollution Control Federation, Cincinnati, Ohio, October, 1973.

54. *Interaction of Heavy Metals and Biological Sewage Treatment Processes.* Division of Water Supply and Pollution Control, USPHS, May, 1965.

55. Environmental Quality Analysts, Inc., Letter Report to Valley Community Services District, March, 1974.

56. Sawyer, C.N., Letter communication to D.S. Parker, January 24, 1975.

57. Davis, G., Personal communication with D.S. Parker, B.F. Goodrich, Co., August, 1974.

58. Tomlinson, T.G., Boon, A.G., and C.N.A. Trotman, *Inhibition of Nitrification in the Activated Sludge Process of Sewage Disposal.* J. Appl. Bact., 29, pp 266-291 (1966).

59. Anthonisen, A.C., Loehr, R.C., Prakasam, T.B.S., and E.G. Srinath, *Inhibition of Nitrification by Un-ionized Ammonia and Un-ionized Nitrous Acid.* Presented at the 47th Annual Conference, Water Pollution Control Federation, Denver, Colorado, October, 1974.

60. Christensen, M.H., and P. Harremoes, *Biological Denitrification in Wastewater Treatment.* Report 2-72, Department of Sanitary Engineering, Technical University of Denmark, 1972.

61. Delwiche, C.C., *The Nitrogen Cycle.* Scientific American, 223, No. 3, pp 137-146 (1970).

62. McCarty, P.L., Beck, L., and P. St. Amant, *Biological Denitrification of Wastewaters by Addition of Organic Materials.* In *Proc. of the 24th Industrial Waste Conference,* May 6, 7, and 8, 1969. Lafayette, Indiana: Purdue University, 1969.

63. McCarty, P.L., *Stoichiometry of Biological Reactions.* Presented at the Summer Institute in Water Pollution Control, Biological Waste Treatment, Manhattan College, New York, N.Y., May, 1974.

64. Dholakia, S.G., Stone, J.H., and H.P. Burchfield, *Methanol Requirement and Temperature Effects in Wastewater Denitrification.* U.S. Environmental Protection Agency, Washington, D.C., WPCRS 17010 DHT 09/70, August, 1970.

65. Jeris, John S., and R.W. Owens, *Pilot Scale High Rate Biological Denitrification at Nassau County, N.Y.* Presented at the Winter Meeting of the New York Water Pollution Control Association, January, 1974.

66. Smith, J.M., Masse, A.N., Feige, W.A., and L.J. Kamphake, *Nitrogen Removal from Municipal Wastewater by Columnar Denitrification.* Environmental Science and Technology, 6, p 260 (1972).

67. Michael, R.P., and W.J. Jewell, *Optimization of the Denitrification Process.* Journal of the Environmental Engineering Division, Proc. ASCE, in press.

68. Wilson, T.E., and D. Newton, *Brewery Wastes as a Carbon Source for Denitrification at Tampa, Florida.* Presented at the 28th Annual Purdue Industrial Waste Conference, May, 1973.

69. Climenhage, D.C., *Biological Denitrification of Nylon Intermediates Waste Water.* Presented at the 2nd Canadian Chemical Engineering Conference, September, 1972.

70. Central Contra Costa Sanitary District, *Operating Report, Advanced Treatment Test Facility.* January, 1974.

71. Moore, S.F., and E.D. Schroeder, *The Effect of Nitrate Feed Rate on Denitrification.* Water Research, 5, pp 445-452 (1971).

72. Requa, D.A., *Kinetics of Packed Bed Denitrification.* Thesis submitted in partial satisfaction of the requirements for the degree of Master of Science in Engineering, University of California at Davis, 1970.

73. Requa, D.A., and E.D. Schroeder, *Kinetics of Packed Bed Denitrification.* JWPCF, 45, No. 8, pp 1696-1707 (1973).

74. Stensel, H.D., Loehr, R.C., and A.W. Lawrence, *Biological Kinetics of the Suspended Growth Denitrification.* JWPCF, 45, No. 2, pp 249-261 (1973).

75. Murphy, R.L., and R.N. Dawson, *The Temperature Dependency of Biological Denitrification.* Water Research, 6, pp 71-83 (1972).

76. Ericsson, et al., *Nutrient Reduction at Sewage Treatment Plants.* KTH Publication 67:5, Stockholm, 1967.

77. Moore, S., and E.D. Schroeder, *An Investigation of the Effects of Residence Time on Anaerobic Bacterial Denitrification.* Water Research, 4, pp 685-694 (1970).

78. Parker, D.S., Zadick, F.J., and K.E. Train, *Sludge Processing for Combined Physical-Chemical-Biological Sludges.* Prepared for the EPA, Report No. R2-73-250, July, 1973.

79. Sutton, P.M., Murphy, K.L., and R.N. Dawson, *Low Temperature Biological Denitrification of Wastewater.* JWPCF, 47, No. 1, pp 122-134 (1975).

80. Parker, D.S., Aberley, R.C., and D.H. Caldwell, *Development and Implementation of Biological Denitrification for Two Large Plants.* Presented at the Conference on Nitrogen as a Water Pollutant, sponsored by the IAWPR, Copenhagen, Denmark, August, 1975.

81. Clayfield, G.W., *Respiration and Denitrification Studies on Laboratory and Works Activated Sludges.* Water Pollution Control, London, 73, No. 1, pp 51-76 (1974).

82. Balakrishnan, S., and W.W. Eckenfielder, *Nitrogen Relationships in Biological Waste Treatment Processes – III, Denitrification in the Modified Activated Sludge Process.* Water Research, 3, pp 177-188 (1969).

83. Dawson, R.N., and K.L. Murphy, *Factors Affecting Biological Denitrification in Wastewater.* In *Advances in Water Pollution Research,* S.H. Jenkins, Ed., Oxford, England: Pergamon Press, 1973.

84. Parker, D.S., *Case Histories of Nitrification and Denitrification Facilities.* Prepared for the EPA Technology Transfer Program, May, 1974.

85. Bishop, D.F., Personal communication to D.S. Parker. Environmental Protection Agency, Washington, D.C., April, 1974.

86. Murphy, K.L., and P.M. Sutton, *Pilot Scale Studies on Biological Denitrification.* Presented at the 7th International Conference on Water Pollution Research, Paris, September, 1974.

87. Ecotrol, Inc., *Biological Denitrification Using Fluidized Bed Technology.* August, 1974.

88. Jewell, W.J., and R.J. Cummings, *Denitrification of Concentrated Wastewaters.* Presented at the Water Pollution Control Federation, Cleveland, October, 1973.

89. Kaufman, G., and E. Schroeder, Personal communication to D.S. Parker. University of California at Davis, July, 1974.

90. Renner, *Production of Nitric Oxide and Nitrous Oxide During Denitrification by Cornybacterium nephridii.* J. Bacterial., 101, pp 821-826 (1970).

91. Skerman, V.B.D., and I.C. MacRae, *The Influence of Oxygen Availability on the Degree of Nitrate Reduction by Pseudomonas Denitrificans.* Can. J. Microbiology, 3, pp 505-530 (1957).

CHAPTER 4

BIOLOGICAL NITRIFICATION

4.1 Introduction

The application of biological nitrification in municipal wastewater treatment is particularly applicable to those cases where an ammonia removal requirement exists, without need for complete nitrogen removal. Biological nitrification is also the first step of the biological nitrification-denitrification approach to nitrogen removal.

4.2 Classification of Nitrification Processes

The first means of categorizing nitrification systems concerns the degree of separation of the carbon removal and nitrification processes. The first nitrification processes developed combined the functions of carbon oxidation and nitrification in one process. The extended aeration modification of the activated sludge process is an example of a combined carbon oxidation-nitrification process. Combined carbon oxidation-nitrification processes generally have low populations of nitrifiers due to a high ratio of BOD_5 to Total Kjeldahl Nitrogen (TKN) in the influent (see Section 3.2.7 for a discussion of this effect). The bulk of the oxygen requirement for this process comes from the oxidation of organics.

Separate stage nitrification is the other category of nitrification processes. In this process, there is a lower BOD_5 load relative to the influent ammonia load. As a result, a higher proportion of nitrifiers is obtained, resulting in higher rates of nitrification. The bulk of the oxygen requirements in the nitrification stage derive from ammonia oxidation. To obtain separate stage nitrification, pretreatment is required to lower the organic load or BOD_5/TKN ratio in the influent to the nitrification stage.

Both the combined carbon oxidation-nitrification and separate stage nitrification processes can be further subdivided into *suspended growth* and *attached growth* processes. Suspended growth processes are those which suspend the biological solids in a mixed liquor by some mixing mechanism. A subsequent clarification stage is required for returning these solids to the nitrification stage. Attached growth processes, on the other hand, retain the bulk of the biomass on the media and therefore do not require a solids separation step for returning the solids to the nitrification reactor. In separate stage processes operated in the attached growth mode, a clarification step may not be required since solids synthesis is low and the sloughed solids are often low in concentration.

There are many different configurations of suspended and attached growth reactors; these are described in subsequent sections of this manual. For suspended growth reactors refer to Sections 4.3 and 4.6; for attached growth reactors refer to Sections 4.4 and 4.7.

Using the classification described above, representative nitrification processes have been classified in Table 4-1 according to the degree of separation of the carbon removal and nitrification processes. Many of these facilities are described in further detail in either this chapter or Chapter 9. Each facility listed has been categorized according to the BOD_5/TKN ratio of the wastewater influent to the nitrification process. Interestingly, the processes listed can all be categorized according to whether the BOD_5/TKN ratio is less than 3.0 or greater than 5.0. If the BOD_5/TKN ratio is less than 3.0, the system can be classified as a separate stage nitrification process. If the BOD_5/TKN is greater than 5.0, the process can be classed as a combined carbon oxidation-nitrification process. Also shown in Table 4-1 is the distribution of total oxygen demand in the process between carbonaceous sources (BOD_5) and nitrogenous sources (NOD). It can be seen that in separate stage processes, the proportion of nitrogenous oxygen demand is at least 60 percent of the total. In combined carbon oxidation-nitrification processes, the proportion of nitrogenous oxygen demand is lower than 50 percent.

There exists a range of BOD_5/TKN ratios between 3.0 and 5.0 where no practical examples currently exist. Facilities in this range could be considered to provide an intermediate degree of separation of carbon removal and nitrification.

4.3 Combined Carbon Oxidation-Nitrification in Suspended Growth Reactors

The conventional activated sludge process has seen relatively wide application in Great Britain for use in obtaining effluents low in ammonia nitrogen. Much of the U.S. practice derives from that experience. Recent U.S. design practice has provided amplifying information.

General design concepts for the activated sludge process are covered in the Technology Transfer publication, *Process Design Manual for Upgrading Existing Wastewater Treatment Plants.*[25] The following sections provide an extension of these concepts to combined carbon oxidation-nitrification applications.

4.3.1 Activated Sludge Modifications

Not all of the various modifications of the activated sludge process are appropriate for nitrification applications, although some see use where only partial ammonia removal is required. Figure 4-1 gives pictorial representations of four common modifications.

4.3.1.1 Complete Mix Plants

Many plants are designed to operate on the complete mix principle. Shown on Figure 4-1 is an example of the feed and withdrawal arrangement for a complete mix plant. The complete mix design provides uniformity of load to all points within the aeration tank, easing the problems of oxygen transfer presented in the head end of the conventional plants. Complete mix plants can be designed for complete nitrification at loading rates comparable to

TABLE 4-1

CLASSIFICATION OF NITRIFICATION FACILITIES

Type and Location	Scale, mgd[a]	BOD_5/TKN Ratio	Oxygen Demand Distribution in percentage		Ref.	Classification - Degree of separation		Pretreatment
			BOD_5	NOD		Combined oxidation - nitrification	Separate stage	
Suspended Growth								
Manassas, Va.	0.2	1.2	20	80	1		x	Activated sludge
Hyperion, Los Angeles, Ca.	46	7.3[b]	61	39[b]	2	x		Primary treatment
Central Contra Costa Sanitary	1.0, design 30	2.4	34	66	3		x	Lime primary treatment
District, Ca.	pilot	1.0	18	82	4		x	Activated sludge
Livermore, Ca.	3.3	2.8	38	62	5		x	Roughing filter
Flint, Michigan	34	5.5	65	35[b]	6	x		Primary treatment
Valley Community Services District, Ca.	3.8	10.8[b]	70	30[b]	3	x		Primary treatment
Blue Plains, D.C.	pilot, design 309	1.3 to 3.0	22 to 39	61 to 78	7, 8		x	Activated sludge
Whittier Narrows, LACSD, Ca.	12	6.6	61	39	9	x		Primary treatment
Jackson, Michigan	13.5	9	66	34	10	x		Primary treatment
Tampa, Florida	pilot, design 60	3.0	40	60	11		x	Activated sludge
South Bend, Indiana	pilot	1.8	28	72	12		x	Activated sludge
New Market, Ontario, Canada	2.4	2.6	36	64	13		x	Lime primary treatment
Cincinnati, Ohio	pilot	7.2	61	39	14	x		Primary treatment
	pilot	1.0[c]	18[c]	82[c]	15		x	Activated sludge
Fitchburg, Mass.	pilot	1.0	18	82	16		x	Activated sludge
Marlboro, Mass.	pilot	3.6[d]	40	60	17		x	Trickling filter
Amherst, N. Y.	pilot	0.8 to 2.0	22	78	18		x	Activated sludge
Denver, Col.	pilot	2.7	37	63	19, 20		x	Activated sludge
Attached Growth								
Stockton, Ca.	pilot, design 58	5.3	54	46	21	x		Primary treatment
Midland, Mich.	pilot	1.1	19	81	22		x	Trickling filter
Union City, Ca.	pilot	1.7	27	73	23, 24		x	Activated sludge
Allentown, Pa.	40	1.9	30	70	25		x	Trickling filter
Lima, Ohio	pilot	0.79	15	85	26, 27		x	Activated sludge

[a] 1 mgd = 0.044 m^3/sec

[b] Calculated from effluent

[c] Approximate, calculated from COD data

[d] BOD/NH_4^+ - N ratio; BOD/TKN would be about 3.0

conventional plants. As will be shown in Section 4.3.3.2, complete mix plants may have slightly higher effluent ammonia contents than conventional plants due to increased short circuiting of the influent to the effluent. Design procedures for nitrification with complete mix plants are presented in Section 4.3.3.

4.3.1.2 Extended Aeration Plants

Extended aeration plants are similar to complete mix plants excepting that hydraulic retention times range from 24 to 48 hr instead of the 2 to 8 hr used in complete mix plants. Extended aeration plants are operated to maximize endogenous respiration, consequently solid retention times of 25 to 35 days are not uncommon. Because of their long aeration periods, they suffer from unusual heat losses and low temperatures. Extended aeration plants, because of their low net growth rate, can be expected to nitrify except at the coldest of temperatures (< 10 C). Unless the sludge inventory is kept under control via intentional sludge wasting, solids are periodically lost in the effluent and nitrification efficiency wanes. Section 4.3.4 includes a discussion of design procedures for extended aeration plants.

4.3.1.3 Conventional or Plug Flow Plants

Conventional plants consist of a series of rectangular tanks or passes with the total tank length to width ratio of 5 to 50.[25] The hydraulics of the system have been loosely termed a plug flow configuration, so called because the influent wastewater and return activated sludge are returned to the head end of the process and the combined flow must pass along a long narrow aeration tank prior to exiting from the system. The degree to which the process actually approaches plug flow is dependent on the amount of longitudinal mixing in the process. Conventional plants can be designed to dependably nitrify using the design approach presented in Section 4.3.5.

4.3.1.4 Contact Stabilization Plants

The contact stabilization modification of the activated sludge process derives from the alteration of the feed pattern to the process. Instead of mixing the influent wastewater with the return sludge, the return activated sludge is separately aerated in a sludge reaeration tank prior to mixing with the influent wastewater. Backmixing between the contact tank and the sludge reaeration tank is prevented by providing overflow weirs or pumps between the tanks. BOD_5 removal can take place in the contact tank which has a relatively short detention time, 0.5 to 1 hr based on average dry weather flow (ADWF). BOD_5 removals can be fairly high because the bulk of the organics in domestic wastewater are particulate or colloidal and can be adsorbed to the biological solids for later oxidation in the sludge reaeration (or stabilization) tank.

The process is not well suited for complete nitrification, even though relatively high solids retention times can be maintained in the process because of the inventory of solids in the sludge reaeration tank. Nonetheless, insufficient biological mass is present in the contact

FIGURE 4-1

MODIFICATIONS OF THE ACTIVATED SLUDGE PROCESS

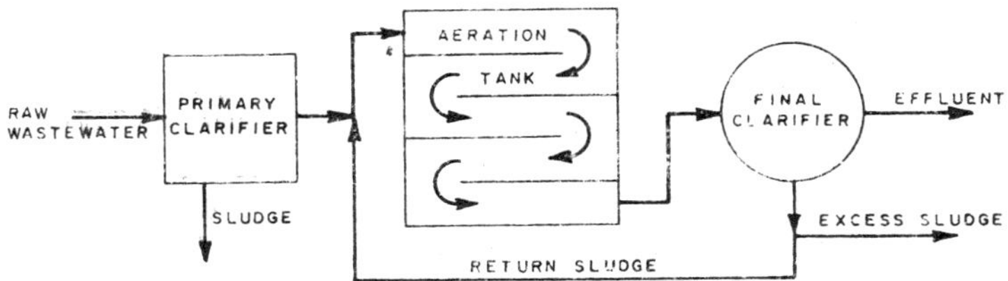

CONVENTIONAL ACTIVATED SLUDGE PLANT

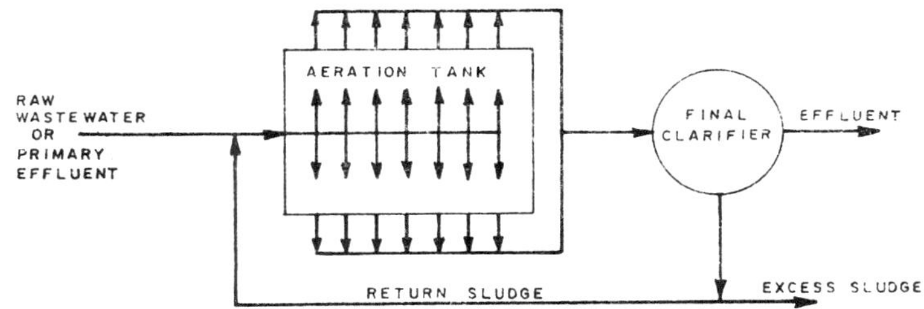

COMPLETE MIX PLANT

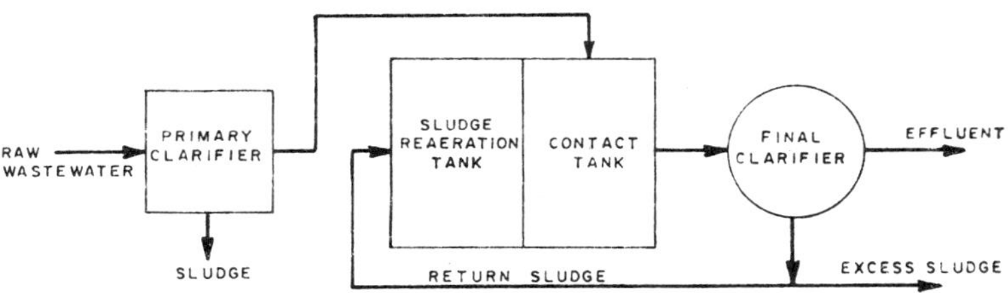

CONTACT STABILIZATION PLANT

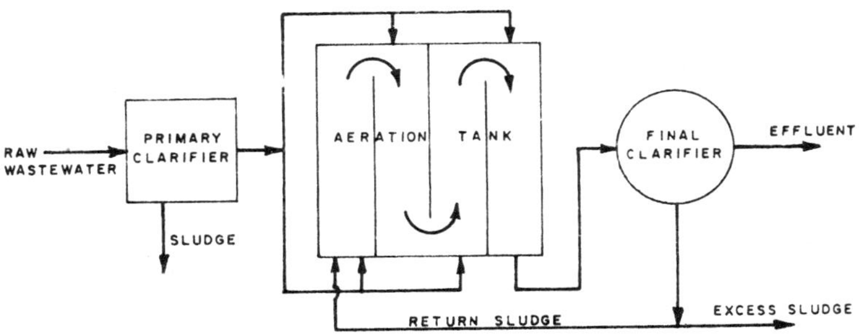

STEP AERATION PLANT

tank to completely nitrify the ammonia and since ammonia is not adsorbed on the biological floc, ammonia will bleed through to the effluent. Partial nitrification can be obtained at levels which can be predicted by methods presented in Section 4.3.6.

4.3.1.5 Step Aeration and Sludge Reaeration Plants

A typical step aeration plant is illustrated on Figure 4-1. Like the conventional plant, the return sludge is introduced at the head end of the aeration tank. However, the step aeration plant differs from the conventional plant in that influent wastewater is introduced at several points along the aeration tank. This distribution of influent flow reduces the initial oxygen demand usually experienced in the conventional plant.[25]

A variation on the step aeration plant that has been popular on the West Coast is to introduce no feed into the first pass while directing the flow into the remaining downstream passes. A sludge reaeration zone is established in the first pass and this variation has become known as a "sludge reaeration plant." Normally, no provision is made to prevent back mixing between the sludge reaeration pass and the downstream passes.

The ammonia bleedthrough characterizing contact stabilization plants is avoided in a step aeration plant because of the greater contact times employed and backmixing of influent occurs. Nonetheless, some bleedthrough of ammonia as well as organic nitrogen can occur. This breakthrough results from short circuiting of influent to the effluent and insufficient contact time for complete organic nitrogen hydrolysis (ammonification) and oxidation of ammonia.

4.3.1.6 High Rate and Modified Activated Sludge

High rate activated sludge processes (high MLSS) and modified activated sludge (low MLSS) processes are characterized by low solids retention times (0.5 days). Under these conditions, a nitrifying activated sludge cannot be developed. The high rate and modified activated sludge processes are acceptable pretreatment techniques for separate stage nitrification processes (Section 4.5).

4.3.1.7 High Purity Oxygen Activated Sludge Plants

Both covered and uncovered reactors have been used with pure oxygen activated sludge, but only the former technique has seen actual implementation in full-scale plants.[28] The covered reactor approach involves the recirculation of reactor off-gases to achieve efficient oxygen utilization. As a consequence, the carbon dioxide which is present in the off-gas is returned to the liquid. The end result is that high carbon dioxide concentrations build up in the mixed liquor and recycle gases, depressing the mixed liquor pH. pH levels as low as 6.0 are not uncommon. This effect can have a depressing effect on nitrification rates (cf. Sections 3.2.5.6 and 4.6.3), resulting in the requirement for somewhat longer solids retention times for nitrification than would otherwise be the case.

Virtually all applications of the high purity oxygen activated sludge process to nitrification have been for separate stage nitrification applications (Section 4.6), rather than for combined carbon oxidation-nitrification applications.

4.3.2 Utility of Nitrification Kinetic Theory in Design

The nitrification kinetic theory presented in Chapter 3 may be directly applied to the design of those activated sludge modifications compatible with nitrification. The equations must be adapted to the hydraulic configuration under consideration, but in all cases this adaptation is relatively straightforward.

Nitrification kinetic theory can be very usefully applied to define the following parameters:

1. The safety factor required to handle diurnal transients in loading to prevent significant ammonia bleedthrough under peak load conditions.

2. The design solids retention time under the most adverse conditions of pH, DO and temperature.

3. The allowable organic loading on the combined carbon oxidation-nitrification stage.

4. The required hydraulic detention time in the aeration tank at ADWF.

5. The excess sludge wasting schedule.

The following sections present the design procedures in terms of a number of specific examples. The procedure developed for each case has often been termed the "solids retention time" design approach.

4.3.3 Complete Mix Activated Sludge Kinetics

As a design example, consider a 1 mgd treatment plant that must achieve complete nitrification at 15 C. The plant incorporates primary treatment. Primary effluent BOD_5 is 150 mg/l, including solids handling return streams to the primary. Total Kjeldahl Nitrogen (TKN) is 25 mg/l as N. As a simplifying assumption, neglect that portion of the TKN that is assimilated into biomass or associated with refractory organics. The wastewater has an alkalinity of 280 mg/l as $CaCO_3$. The procedure is as follows:

1. Establish the safety factor, SF. The SF is affected by the desired effluent quality. Assume a minimum SF of 2.5 is required due to transient loading conditions at this particular plant (see Section 4.3.3.2).

2. Establish the minimum mixed liquor dissolved oxygen (DO) concentration.

Consideration of aeration efficiency at the peak hourly load is required (see Section 4.8). Assume a minimum DO of 2.0 mg/l is selected as a compromise between power requirements and a consideration of the depressing effects of low DO levels on the rate of nitrification as discussed in Section 3.2.5.5.

3. Estimate the process operating pH (see Section 4.9.2). Approximately 7.14 mg/l of alkalinity as $CaCO_3$ is destroyed per mg/l of NH_4^+-N oxidized. Neglecting the incorporation of nitrogen into biomass, the alkalinity remaining after nitrification will be at least:

$$280 - [7.14(25)] = 102 \text{ mg/l}$$

If a coarse bubble aeration system is chosen, the pH should remain above pH 7.2 and chemical addition is not required for pH control (see Section 4.9.2).

4. Calculate the maximum growth rate of nitrifiers at 15 C, DO = 2 mg/l, and pH > 7.2. The appropriate equation to be used was presented in Section 3.2.6 and is as follows:

$$\dot{\mu}_N = \hat{\mu}_N \left(\frac{DO}{K_{O_2} + DO}\right)\left(1 - 8.33(7.2 - pH)\right) \qquad (3\text{-}23)$$

where: $\dot{\mu}_N$ = maximum possible nitrifier growth rate, day^{-1}, environmental conditions of pH, temperature, and DO,

$\hat{\mu}_N$ = maximum nitrifier growth rate, day^{-1}, and

K_{O_2} = half-saturation constant for oxygen, mg/l.

The last bracketed term is taken as unity at a pH above 7.2. Using the specific values adopted in Section 3.2.5 for $\hat{\mu}_N$ and K_{O_2} leads to the following expression:

$$\dot{\mu}_N = 0.47 \left[e^{0.098(T-15)}\right]\left[\frac{DO}{DO + 1.3}\right]\left[1 - 0.833(7.2 - pH)\right] \qquad (4\text{-}1)$$

Using the numbers given above:

$$\dot{\mu}_N = (0.47)(0.61) = 0.285 \text{ day}^{-1}$$

5. Calculate the minimum solids retention time for nitrification. From Equation 3-15, the correct expression is:

$$\theta_c^m = \frac{1}{\hat{\mu}_N} \tag{4-2}$$

where: θ_c^m = minimum solids retention time, days, for nitrification at pH, temperature and DO.

For this example:

$$\theta_c^m = \frac{1}{0.285} = 3.51 \text{ days}$$

6. Calculate the design solids retention time. From Equation 3-29, the correct expression is:

$$\theta_c^d = SF \cdot \theta_c^m \tag{4-3}$$

where: θ_c^d = solids retention time of design, days.

For this example:

$$\theta_c^d = 2.5(3.51) = 8.78 \text{ days.}$$

7. Calculate the design nitrifier growth rate. From Equation 3-12, the correct expression is:

$$\mu_N = \frac{1}{\theta_c^d} \tag{4-4}$$

where: μ_N = nitrifier growth rate *Nitrosomonas* , day^{-1}.

For this example:

$$\mu_N = \frac{1}{8.78} = 0.114 \text{ day}^{-1}$$

8. Calculate the half-saturation constant for ammonia oxidation at 15 C. The proper expression is:

$$K_N = 10^{0.051T - 1.158} \tag{3-13}$$

where: K_N = half-saturation constant for NH_4^+ – N, mg/l, and

T = Temperature, C

For this example:

$$K_N = 10^{-0.393} = 0.405 \text{ mg/l}$$

9. Calculate the steady state ammonia content of the effluent. Equation 3-24 is directly applicable to complete mix activated sludge systems, where N_1 is the effluent ammonia-nitrogen content:

$$\mu_N = \hat{\mu}_N \frac{N_1}{K_N + N_1} \qquad (3\text{-}24)$$

where: N_1 = effluent NH_4^+ - N, mg/l

For this case:

$$\mu_N = 0.114 = 0.285 \frac{N_1}{N_1 + 0.405}$$

$$\therefore \quad N_1 = 0.27 \text{ mg/l}$$

Transient loading effects on effluent quality are presented in Section 4.3.3.2.

10. Calculate the organic removal rate. The design solids retention time θ_c^d applies to both the nitrifier population and the heterotrophic population. Equation 3-27 can be applied to determine substrate removal rates:

$$\mu_b = \frac{1}{\theta_c^d} = Y_b q_b - K_d \qquad (3\text{-}27)$$

where: Y_b = heterotrophic yield coefficient, lb VSS grown per lb BOD_5 removed,

q_b = rate of substrate removal, lb BOD_5 removed/lb VSS/day, and

K_d = "decay" coefficient, day^{-1}.

Assume representative values for Y_b and K_d:[29]

Y_b = 0.65 lb VSS/lb BOD rem.

K_d = 0.05 day^{-1}

Therefore:

$0.114 = 0.65 q_b - 0.05$

q_b = 0.252 lb BOD rem./lb MLVSS/day

In the above calculation of q_b, it is assumed that the fraction of nitrifiers is low and can be neglected (see Section 4.6.1 for a discussion of this point).

11. Determine the hydraulic detention time at ADWF. In this analysis, the MLVSS content and effluent soluble BOD must be known. The effluent soluble BOD_5 can be assumed to be very low (say 2 mg/l). The MLVSS content is dependent on the mixed liquor total suspended solids, which is in turn dependent on the operation of the nitrification sedimentation tank (Section 4.10). Assume for the purposes of this example that the design mixed liquor content at 15 C is 2500 mg/l. At a volatile content of 75 percent, the MLVSS is 0.75 (2500) = 1875 mg/l. From Equation 3-28, the expression for hydraulic detention time is:

$$HT = \frac{S_o - S_1}{X_1 q_b} \qquad (4\text{-}5)$$

where:
- HT = hydraulic detention time, days,
- X_1 = mixed liquor volatile suspended solids, MLVSS, mg/l,
- S_o = influent total BOD_5, mg/l, and
- S_1 = effluent soluble BOD_5, mg/l.

For this example, the hydraulic detention time at ADWF is:

$$HT = \frac{148}{(1875)(0.252)} = 0.313 \text{ days}$$

$$= 7.5 \text{ hours}$$

12. Determine the organic loading per unit volume. The volume required in the aeration basin for 1 mgd flow is:

$$\text{Volume} = Q \cdot HT = 1(0.313) = 0.313 \text{ mil gal} = 41{,}844 \text{ cu ft}$$

where: Q = influent flow rate, mgd

The BOD_5 loading is:

$$(1)(8.33)(150) = 1249 \text{ lb/day}$$

The BOD_5 load per 1000 cu ft is:

$$\frac{1249}{41.84} = 29.9 \text{ lb } BOD_5/1000 \text{ cu ft/day}$$

13. Determine the sludge wasting schedule. Sludge is wasted from the system from two sources: (1) solids contained in the effluent from the secondary sedimentation tank, and (2) intentional sludge wasting from the return sludge or mixed liquor. The sludge to be wasted under steady state conditions can be calculated from the solids retention time. The total sludge wasted per day is:

$$S = 8.33(Q \cdot X_2 + W \cdot X_w) \qquad (4\text{-}6)$$

where: S = total sludge wasted in lb/day,

W = waste sludge flow rate, mgd

X_2 = effluent volatile suspended solids, mg/l, and

X_w = waste sludge volatile suspended solids, mg/l

The inventory of sludge in the system is:

$$I = 8.33(X_1 \cdot V) \qquad (4\text{-}7)$$

where: I = inventory of VSS under aeration, lb, and

V = volume of aeration tank, mil gal

The solids retention time is defined as:

$$\theta_c^d = \frac{I}{S} \qquad (4\text{-}8)$$

In this case, application of Equation 4-7 yields:

$$I = 8.33(1875)(0.313) = 4889 \text{ lb VSS}$$

Using Equation 4-8 and a design θ_c^d of 8.78 days, the sludge wasted from the system is:

$$S = 4889/8.78 = 557 \text{ lb/VSS day}$$

The sludge contained in the effluent at 1 mgd can be calculated assuming that the effluent volatile suspended solids is equal to 12 mg/l:

$$8.33\,(1)\,(12) = 100 \text{ lb VSS/day}$$

By difference, the lb of MLVSS to be wasted from the mixed liquor or return sludge is:

$$557 - 100 = 457 \text{ lb VSS/day}$$

4.3.3.1 Effect of Temperature and Safety Factor on Design

The design example presented in the previous section provided one solution to a set of stated conditions. Alteration of the lowest temperature at which nitrification will be supported, or the design safety factor, or the wastewater strength, or the assumption of different kinetic constants can materially alter the design.

To give one illustration, Table 4-2 has been prepared using differing safety factors (2.0 to 3.0) and differing minimum wastewater temperatures with design calculations to derive the computed quantities shown. Assumptions have been made for illustrative purposes as to the allowable MLSS. Allowable mixed liquor levels are a function of sedimentation tank operation. The mixed liquor level that can be maintained will be affected by reduced sedimentation efficiency at lower temperatures. Consideration of aeration tank-secondary sedimentation tank interactions is presented in Section 4.10.

As can be seen from Table 4-2, low temperature applications (10 C) of combined carbon oxidation-nitrification in complete mix activated sludge systems require very long hydraulic residence times to achieve favorable conditions for nitrification. This factor was one of the reasons for the development of separate stage nitrification systems. As temperatures rise, required residence times are materially reduced. At 20 C, less than five hours is required for virtually complete nitrification in the specific case examined. While it is possible to design for nitrification using the relatively low detention times given in Table 4-2 for 20 C, special attention must be given to oxygen transfer as a very high oxygen demand is expressed per unit volume. Considerations for oxygen transfer are given in Section 4.8.

4.3.3.2 Consideration in the Selection of SF

In introducing the safety factor concept to the design of biological treatment systems, Lawrence and McCarty[29] noted that the SF was necessary to achieve high efficiency of treatment, to insure process stability and to provide resistance to toxic upsets. Excessively high safety factors resulted in higher operating and capital costs. It was noted that the safety factor concept had been implicitly incorporated into treatment plant design practice by the selection of solids retention times in excess of θ_c^m

TABLE 4-2

CALCULATED DESIGN PARAMETERS FOR A 1 MGD COMPLETE MIX ACTIVATED SLUDGE PLANT

Minimum temp. for nitrification, C	Maximum possible nitrifier growth rate, $\hat{\mu}_N$, day-1	Assumed allowable MLSS/MLVSS mg/l	Safety Factor, SF	Design solids retention time, days θ_c^d	Steady state effluent NH_4^+-N, mg/l	Organic removal rate, lb BODrem/ lb MLVSS-day	Hydraulic retention time,[a] hours	BOD_5 loading (volumetric) lb/1000/cf/day[b]
10	0.175	2,000	2.0	11.5	0.23	0.21	11.0	20.5
			2.5	14.3	0.15	0.19	12.8	17.5
		1,500	3.0	17.2	0.11	0.17	14.0	15.8
15	0.285	2,500	2.0	7.0	0.40	0.29	6.4	34.9
			2.5	8.8	0.27	0.25	7.5	29.9
		1,875	3.0	10.5	0.20	0.22	8.5	26.5
20	0.465	3,000	2.0	4.3	0.73	0.44	4.4	51.5
			2.5	5.4	0.49	0.36	5.2	43.0
		2,250	3.0	6.4	0.36	0.32	6.0	37.3

[a] At ADWF

[b] 62.4 lb/1000 cf/day = kg/m^3/day

Because the SF concept is relatively new, there is no plant scale experience with its application accumulated as yet on which to base broad recommendations. Rather, kinetic theory itself is used in this section to establish *minimum* factors of safety considering the desired degree of nitrification under steady state and transient load conditions. It must be emphasized that these are minimum values and individual designs may exceed these values for a variety of reasons. For instance, the presence of industrial wastes may adversely affect nitrification rates, requiring conservatism in the selection of the SF.

Figure 4-2 provides a wider array of safety factors for the design example presented in Table 4-2. As may be seen, the selection of the SF has a marked effect on the calculated steady state values of ammonia in the effluent. If relatively complete nitrification is to be obtained (at steady-state) resulting in 0.5-2 mg/l of ammonia nitrogen in the effluent, a minimum SF of 1.5 is appropriate for application to complete mix activated sludge systems. Further, effluent values for a comparable plug flow system are also shown in Figure 4-2 (see Section 4.3.5 for plug flow data). As may be seen, complete mix systems have higher effluent ammonia levels than plug flow systems at the same SF.

In all practical applications, waste treatment plants do not operate at "steady state." Significant diurnal variation in the nitrogen loading on such systems occurs. Figure 4-3 shows the diurnal variations in influent flow and TKN loading experienced at the Chapel Hill, N.C. treatment plant. The ratio of the maximum TKN loading to the average was 2.17, while the ratio of the maximum to minimum was 6.72. The Chapel Hill system is a relatively small system (1.8 mgd) with high peak to average ratios for all constituents.[30] The variation in load for each community will be a function of the unique characteristics of that community (see Section 4.8), and data must be individually developed for each situation.

TKN load variations have a significant impact on nitrification kinetics, and ammonia bleedthrough can occur under peak load situations.[31,32] Kinetic theory can be applied to these situations, however, and the safety factor established at levels which will prevent ammonia bleedthrough from causing significant deterioration of effluent quality.

A mass balance on nitrogen in the organic and ammonia form can be made at any time during a diurnal cycle which states that the influent TKN load is equal to the effluent ammonia load plus that nitrified in the complete-mix reactor during any time, Δt:

$$N_o Q \Delta t = q_N f X_1 V \Delta t + N_1 Q \Delta t \qquad (4\text{-}9)$$

where N_o = influent TKN concentration, mg/l,

N_1 = effluent ammonia nitrogen concentration, mg/l,

Q = influent or effluent flow rate, mgd,

Δt = time increment,

V = volume of aeration basin, mil gal,

f = nitrifier fraction of the mixed liquor solids

FIGURE 4-2

EFFECT OF THE SAFETY FACTOR ON STEADY STATE EFFLUENT AMMONIA LEVELS IN SUSPENDED GROWTH SYSTEMS

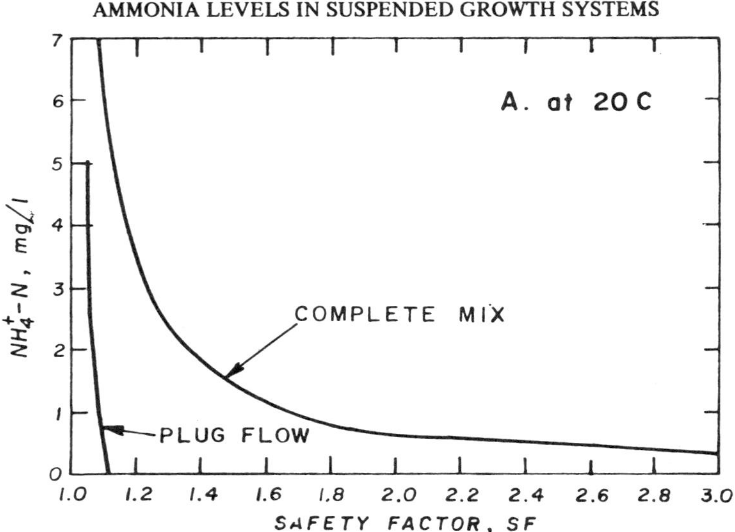

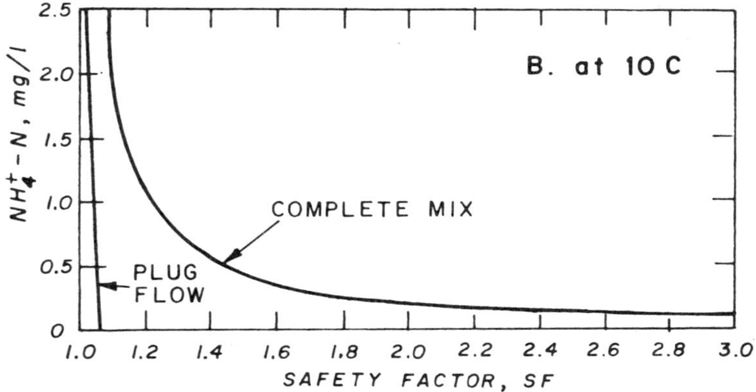

FIGURE 4-3
DIURNAL VARIATIONS AT THE CHAPEL HILL, N.C.
TREATMENT PLANT (AFTER HANSON, ET AL. (30))

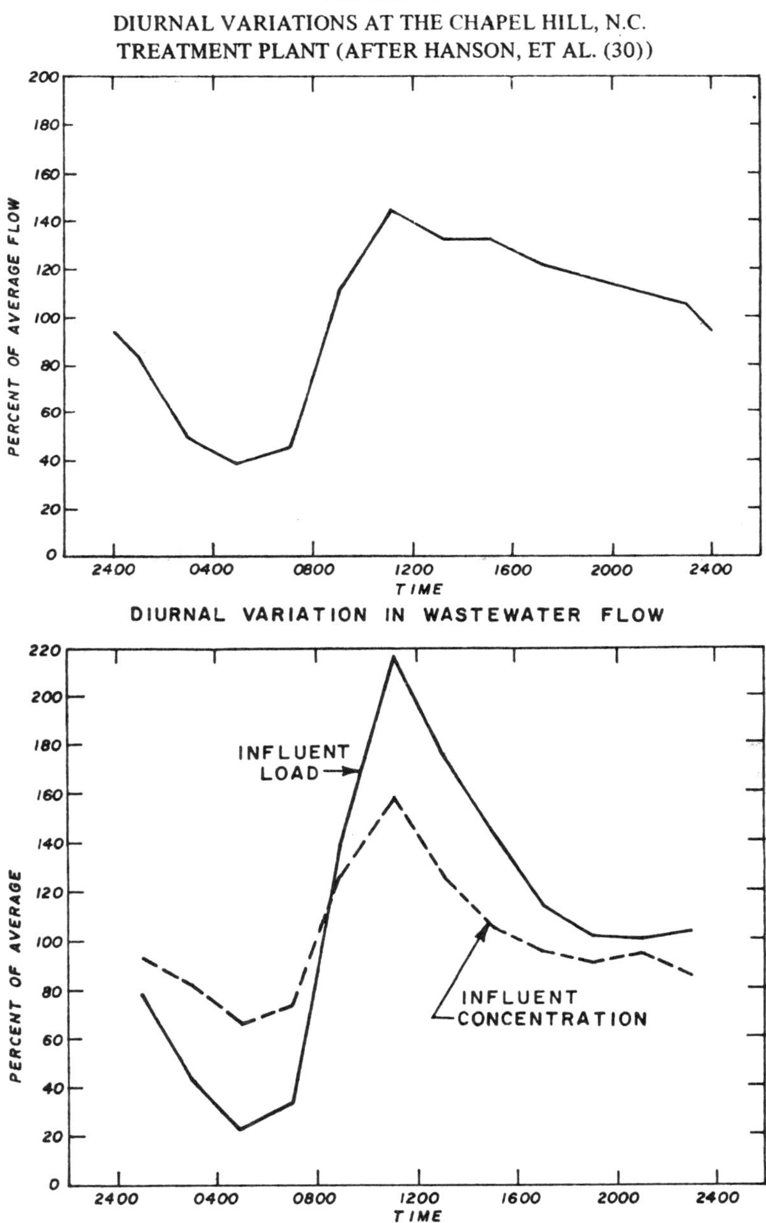

This equation neglects synthesis terms, assumes all influent organic N is hydrolyzed, and neglects terms relating to the rate of change of ammonia concentration in the reactor. Numerical solution techniques are available to handle transient load effects more exactly.[32] Equation 4-9, however, is useful for approximating the effects of transient loads.

Equation 4-9 may be solved for N_1, by substitution for the terms for nitrification rate, q_N and the term fX_1V, representing the inventory of nitrifying organisms. The inventory of nitrifying organisms can be related to the solids retention time through the following equation:

$$\theta_c^d = \frac{fX_1 V}{\bar{Q}Y_N(\bar{N}_o - \bar{N}_1)} \tag{4-10}$$

where:

$\bar{N}_o$	=	24 hr-average influent TKN, mg/l
$\bar{N}_1$	=	24 hr-average effluent NH_4^+ - N, mg/l
$\bar{Q}$	=	mean flow rate (ADWF), mgd, and
Y_N	=	nitrifier yield coefficient, lb VSS/lb NH_4^+ - N removed.

The term $\bar{Q}Y_N(\bar{N}_o-\bar{N}_1)$ represents the quantity of nitrifiers grown per day, which must be wasted each day to establish a steady-stage solids retentioh time, θ_c^d. The average terms, $\bar{N}_o$ and $\bar{N}_1$, are flow weighted averages of nitrogen concentration of an entire day (the equivalent of composite samples). $\bar{Q}$ represents the average dry weather flow (ADWF).

The nitrification rate from Equations 3-20 and 3-24 is:

$$q_N = \frac{\hat{\mu}_N}{Y_N}\left(\frac{N_1}{K_N + N_1}\right) \tag{4-11}$$

Substitution of Equations 4-10, 4-11 and Equation 3-29 into Equation 4-9 yields:

$$N_o = (\bar{N}_o - \bar{N}_1)\ \frac{\bar{Q}}{Q}\ SF\left(\frac{N_1}{K_N + N_1}\right) + N_1 \tag{4-12}$$

Equation 4-12 can be used to solve for N_1 over a 24-hr cycle since all other quantities in Equation 4-12 are known or can be estimated. Initially, $\bar{N}_1$ can be estimated to be the calculated steady-state value. Once Equation 4-12 has been applied to generate a 24-hr cycle of N_1 values, a new value of $\bar{N}_1$ may be calculated. If $\bar{N}_1$ differs significantly from the initial assumption, the calculation process can be repeated.

Equation 4-12 has been applied to the variations in load observed at Chapel Hill, and using the design information used to generate Table 4-2 at a temperature of 15 C. The results of this analysis are plotted in Figure 4-4 for three different assumed safety factors, 1.5, 2.0, and 2.5. As may be seen, the assumption of the safety factor has a marked effect on the average effluent ammonia content, $\bar{N}_1$. For this particular case, the ratio of peak to average TKN loading was 2.2; the SF had to exceed this ratio (2.5) to produce an effluent that had, on the average, less than 1 mg/l of ammonia-N.

The application of Equation 4-12 to several other such cases showed the same effect; namely, the minimum safety factor should equal or exceed the ratio of peak ammonia load to average load to prevent high ammonia bleedthrough at peak loads. This statement may be used as "a rule of thumb" for designing suspended growth nitrification systems operated in the complete mix mode.

A flow equalization procedure applicable to reducing diurnal peaking on nitrification systems is presented in Chapter 3 of the *Process Design Manual for Upgrading Existing Wastewater Treatment Plants.*[25] By incorporating flow equalization into treatment plants,

FIGURE 4-4

EFFECT OF SF ON DIURNAL VARIATION IN EFFLUENT AMMONIA

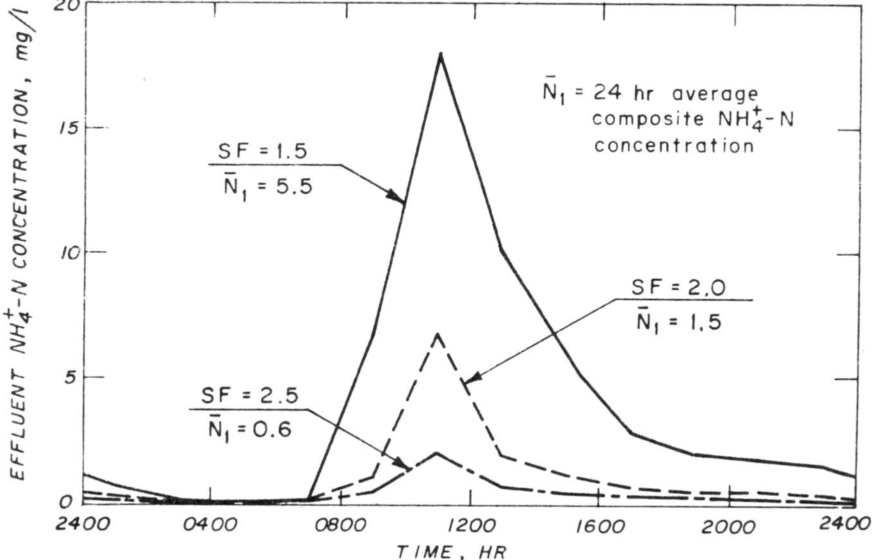

the safety factor used in kinetic design of the nitrification tanks may be reduced. Case examples for treatment plants incorporating flow equalization are presented in Sections 9.5.1.1, 9.5.1.2 and 9.5.2.1.

4.3.4 Extended Aeration Activated Sludge Kinetics

The procedure presented in Section 4.3.3 for complete mix activated sludge kinetics is directly applicable to extended aeration activated sludge. Extended aeration systems are usually operated at such long solids retention times that except during cold temperatures (5-10 C) nitrification is usually obtained in properly operated systems.

4.3.5 Conventional Activated Sludge (Plug Flow) Kinetics

The approach for conventional activated sludge plants is similar to that for complete mix plants with the exception of the equations used to predict effluent quality. The plug flow model may be applied to approximate the hydraulic regime in these plants. The Monod expression for substrate removal rate (Equation 3-24) must be integrated over the period of time an element of liquid remains in the nitrification tank. The following is a solution for plug flow kinetics that can be adapted to this problem as shown:[29]

$$\frac{1}{\theta_c^d} = \frac{\hat{\mu}_N(N_o - N_1)}{(N_o - N_1) + K_N \ln \frac{N_o}{N_1}} \quad \text{for } r < 1 \qquad (4\text{-}13)$$

where: r = recycle ratio (or return sludge ratio).

or

$$\frac{1}{SF} = \frac{(N_o - N_1)}{(N_o - N_1) + K_N \ln \frac{N_o}{N_1}} \quad \text{for } r < 1 \qquad (4\text{-}14)$$

Equation 4-14 is evaluated in Figure 4-2 for the design example presented in Section 4.3.3. Comparing the safety factor to the safety factor producing the same effluent ammonia in the complete mix case, it can be seen that lower values of the SF are required for plug flow nitrification processes than for complete mix nitrification processes. This means that plug flow processes theoretically can be more efficient at the same SF, or alternatively, require less aeration tank volume for the same level of nitrification efficiency. However, plug flow type reactors have the disadvantage that the carbonaceous oxygen demand is concentrated at the head end of the tank, making it difficult to supply enough air in that area for both carbonaceous oxidation and nitrification. Air diffusion systems must be specifically designed to handle this concentrated load. Otherwise, the first portion of the tank will not be available for nitrification and thus effective volume for nitrification will be reduced.

A typical DO and nitrification pattern for plug flow tanks where aeration capability is limited in the front end of the tank is presented in Figure 4-5, where an aeration tank profile for DO and ammonia nitrogen is plotted. As may be seen, nitrification is inhibited in the first portion of the tank, because of the DO suppression due to carbon oxidation. Once the DO rises, the ammonia level falls at a reaction rate that approximates zero order, a reactor order predicted by kinetic theory (Section 3.2.7). It is notable that if sufficient aeration capability had been available in the head end of the tank, virtually complete nitrification probably would have been obtained.

Thus, the first portion of plug flow tanks may be ineffective for nitrification, reducing the effective contact time for nitrification. If oxygen supply limitations are present in the head end of the tank, the plug flow type reactor's advantage over the complete mix reactor is reduced.

The degree to which full-scale nitrification tanks approach plug-flow operation can be examined through reactor diffusion theory.[34,35] Reactors can be characterized by an axial disperson number, D/uL, where D is the axial disperson coefficient in square ft per hr, u is the mean displacement velocity along the tank length, in feet per hr, and L is the tank length, ft. In the calculation of the axial disperson number, u and L are known for any

FIGURE 4-5

DO AND AMMONIA NITROGEN PROFILE IN A PLUG-FLOW SYSTEM
(AFTER NAGEL AND HAWORTH (33))

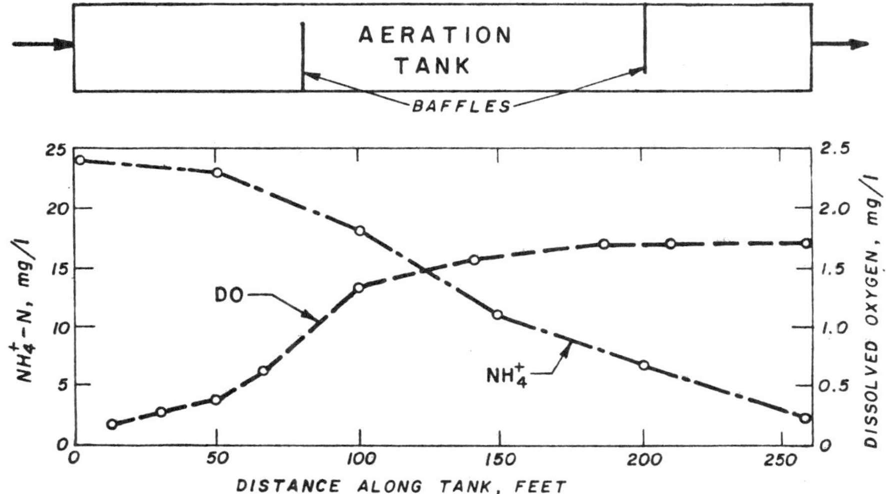

particular plant design and D must be measured. A valid empirical relationship for D for both fine and coarse bubble diffused air plants is as follows:[35]

$$D = 3.118\ W^2 (A)^{0.346} \tag{4-15}$$

where: W = tank width, ft and

A = air flow per unit tank volume, in standard cubic feet per minute per 1,000 cu ft.

The axial disperson coefficient, D/uL is zero for true plug flow plants and infinite (∞) for true complete mix plants. Plants with D/uL between 0 to 0.2 are usually classed as plug flow reactors, while for complete mix systems, D/uL is usually in the range from 4.0 to ∞.[36]

As an example calculation, the Central Contra Costa Sanitary District (CCCSD) plant's nitrification tanks (Section 9.5.2.1) have the following characteristics:

Air flow (av.) = 51.1 SCFM/1000 CF
Width = 35 ft
Area of tank = 525 sf
Length (all 4 passes) = 1080 ft
Flow each tank (4 passes) @ 50% recycle = 22.5 mgd

From the above data, the mean displacement velocity can be calculated to be 239 ft/hr. From Equation 4-15, the diffusion coefficient is:

$$D = 3.118\ (35)^2\ (51.1)^{0.346} = 14{,}989\ ft^2/hr$$

Therefore:

$$D/uL = 14{,}898/239\ (1080) = 0.058$$

Therefore, the CCCSD nitrification tanks closely approach a plug flow reactor.

Equation 4-15 can be utilized to evaluate mixing in actual plant designs to determine whether they approach plug flow closely enough to allow use of Equation 4-15 to describe nitrification. It is probable that most plants operated in the conventional mode do approach plug flow. For those plants with intermediate values of D/uL, complete mix kinetics can be employed which yield conservative answers.

The hydraulic configuration of nitrification tanks can also be designed to discourage back mixing. A series of complete mix tanks can approximate a plug flow reactor. In the case example for Canberra, Australia (Section 9.5.2.2) complete mix reactors are used in series for nitrification. Absolute prevention of back mixing is provided by virtue of the mixed liquor overflowing weirs between reactors. Available head at the site was utilized, eliminating the need for mixed liquor pumping.

4.3.5.1 Considerations in the Selection of the Safety Factor

The factors affecting the choice of the SF for plug flow activated sludge are similar to those for complete mix applications. Diurnal peaking in load has an important influence on the choice of the SF, although kinetic models have not been extended to handle diurnal loads in plug flow systems at the present time. It can be expected that the effects of diurnal loads on plug flow systems will be similar to those for complete mix systems as when the effluent ammonia nitrogen level rises to 2 to 3 mg/l or above, the rate of removal becomes a zero order reaction (unaffected by ammonia nitrogen concentration). In zero order reaction situations, differences between plug flow and complete mix kinetics are negligible. Therefore, the adoption of the criteria advanced for complete mix systems (that the minimum SF equal or exceed the ratio of peak ammonia load to average daily load), should prevent high ammonia bleedthrough during diurnal peak loads. The problem of low dissolved oxygen due to carbonaceous load in the head end of plug flow systems should be considered in aeration design for combined carbon oxidation-nitrification applications; indeed, this factor alone would justify a conservative safety factor.

4.3.5.2 Kinetic Design Approach

The kinetic design approach for plug flow (conventional) plants is identical to that presented in Section 4.3.3, excepting in Step 9, where Equation 4-13 is used instead of Equation 3-24. If a portion of the nitrification tank is rendered ineffective by DO suppression at its head end, then only the sludge inventory maintained under adequate DO conditions should be used in the calculation of θ_c^d or the SF.

4.3.6 Contact Stabilization Activated Sludge Kinetics

Gujer and Jenkins[37,38] have developed the kinetic design procedure for nitrification in contact-stabilization activated sludge plants. The procedure described herein is a summary of their approach, and the reader is referred to their publications for theoretical bases.

The overall nitrifier growth rate in the contact stabilization process is the weighted mean of their growth rate in the contact tank and in the stablization (sludge reaeration) tank:

$$\mu_N = C\mu_c + B\mu_s \tag{4-16}$$

where: μ_N, μ_c, μ_s = growth rate of the nitrifiers in the overall process, in the contact and stabilization tanks respectively (day^{-1}).

C, B = the fractions of total sludge in the contact and stabilization basins respectively.

Gùjer and Jenkins used Equation 4-16, and Monod type expressions for complete mix tanks to develop a graphical solution for nitrification in contact stabilization (Figure 4-6). In Figure 4-6, the efficiency of nitrification, η_{nit}, is defined as a fraction by:

$$\eta_{nit} = \frac{(NO_3^-)_c}{(NO_3^-)_s} \tag{4-17}$$

where: $(NO_3^-)_c$ = NO_3^- - N level in the contact tank, mg/l, and

$(NO_3^-)_s$ = NO_3^- - N level in the stabilization tank, mg/l.

4.3.6.1 Design Example

As a design example, consider a 1 mgd contact stabilization plant operated at a minimum temperature of 15 C. Influent BOD_5 is 150 mg/l, including solids handling returns to the primary. Total Kjeldahl Nitrogen is 30 mg/l, of which 21 mg/l is ammonia -N and 9 mg/l is organic -N. The wastewater has an alkalinity of 210 mg/l. The effluent requirement is not more than 10 mg/l reduced soluble nitrogen (organic and ammonia). The procedure is as follows:

1. Establish a reasonable safety factor for nitrification, say 2.5 as in Section 4.3.3.

2. Establish the minimum mixed liquor DO; assume 2.0 mg/l as in Section 4.3.3.

3. Establish the maximum growth rate of nitrifiers, assuming for the moment that there is sufficient alkalinity in the wastewater to buffer the nitrification pH to greater than 7.2 (see step 14). Therefore,

   $\hat{\mu}_N = 0.285\ day^{-1}$ as in Section 4.3.3, step 4.

4. Calculate the minimum solids retention time, the design solids retention time and the actual nitrifier growth rate (as in Section 4.3.3, Steps 5, 6, 7):

$$\theta_c^m = 3.51\ days$$

$$\theta_c^d = 8.78\ days$$

$$\mu_N = 0.114\ day^{-1}$$

FIGURE 4-6

NITRIFICATION EFFICIENCY AS A FUNCTION OF PROCESS PARAMETERS (AFTER GUJER AND JENKINS (37))

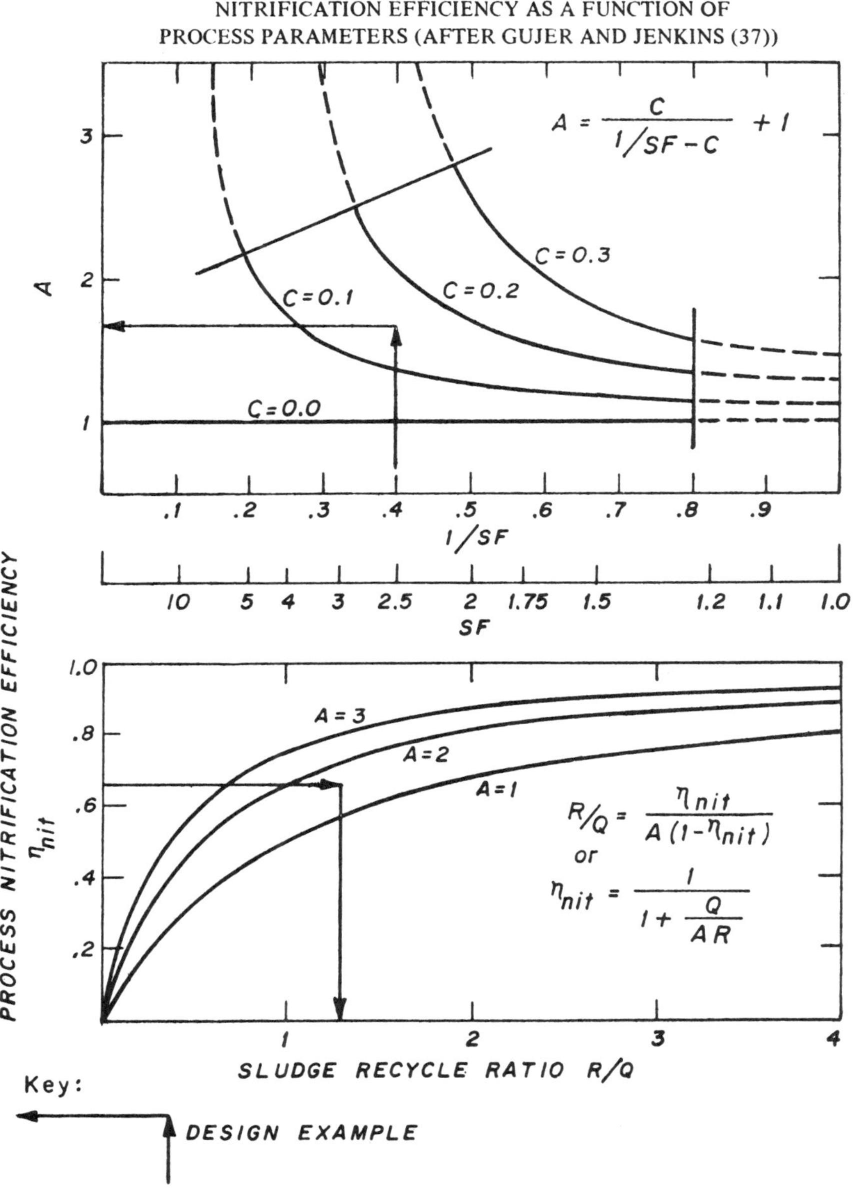

5. Calculate the organic removal rate, as in Step 10, Section 4.3.3:

$$q_b = 0.252 \text{ lb BOD rem/lb MLVSS/day}$$

6. Calculate the VSS produced per unit of wastewater treated:

$$(S_o - S_1)\,\mu_N/q_b = (150 - 2)(0.114)/0.252 = 67.0 \text{ mg/l}$$

7. Compute the nitrogen incorporated into VSS, assuming 12 percent nitrogen incorporated into VSS:

$$0.12\,(67.0) = 8.0 \text{ mg/l N}$$

8. Compute the soluble N content of the effluent, assuming no denitrification. The effluent soluble N = the total influent N minus N incorporated into VSS:

$$30 - 8.0 = 22.0 \text{ mg/l soluble N}$$

9. Calculate the soluble organic N in the effluent. Gujer and Jenkins found that 40 percent of the influent organic N appeared in soluble form in the effluent:

$$0.4\,(9) = 3.6 \text{ mg/l organic N}$$

$$\approx 4.0 \text{ mg/l organic N}$$

10. Calculate ammonia nitrogen in the effluent under steady state conditions:

$$\text{Total reduced N} - \text{organic N} = 10 - 4 = 6 \text{ mg/l } NH_4^+\text{-N}$$

11. Calculate nitrate nitrogen in the effluent and in the contact tank under steady state conditions:

$$\text{Total soluble N} - \text{total reduced N} = (NO_3^-)_c$$

$$(NO_3^-)_c = 22 - 10 = 12 \text{ mg/l}$$

12. Calculate the required nitrification efficiency from Equation 4-17:

$$\eta_{nit} = 12/(12 + 6) = 0.667$$

In this calculation, it is assumed that the concentration of nitrate nitrogen in the stabilization tank $(NO_3^-)_s$ totals 18 mg/l, since the contact tank concentration is 12 mg/l and with the assumption that the 6 mg/l of ammonia nitrogen is completely nitrified in the stablization tank.

13. Calculate the required sludge recycle ratio. Assume the fraction of biomass in the contact tank is 15 percent (C= 0.15). The dimensionless number A, is used in the calculation; A is defined as follows:

$$A = \frac{C}{\frac{1}{SF} - C} + 1 \qquad (4\text{-}18)$$

For this example:

$$A = \frac{0.15}{\frac{1}{2.5} - 0.15} + 1 = 1.60$$

The required sludge recycle ratio, R/Q, depends on the value of A and the required nitrification efficiency as follows:

$$R/Q = \frac{\eta_{nit}}{A(1 - \eta_{nit})} \qquad (4\text{-}19)$$

where:
R = recycle flow rate, mgd
Q = influent flow rate, mgd

For this example:

$$R/Q = \frac{0.667}{1.60(1 - 0.667)} = 1.25$$

Since Q has been assumed to be 1 mgd, the return activated sludge rate is 1.25 mgd.

This example is also worked graphically in Figure 4-6. The top part of the figure is used first by entering the abscissa with the value of the SF and rising vertically to the chosen value of C and then reading the value of A on the ordinate. The bottom part of Figure 4-6 is then used; the nitrification efficiency, η_{nit}, is entered on the ordinate and traveling horizontally to the value of A just determined and then finding the required recycle ratio on the abscissa. Figure 4-6 also demonstrates a general result; in order to obtain high nitrification efficiency, a higher than normal sludge recycle ratio must be employed.

14. Check the buffering of the wastewater. The quantity of ammonia nitrified is reflected by the level of nitrate in the process effluent. Approximately 7.14 mg/l of alkalinity as $CaCO_3$ is destroyed per mg/l of NH_4^+-N oxidized. The alkalinity remaining after nitrification would be at least:

$$210 - 7.14\,(12) = 124 \text{ mg/l as } CaCO_3$$

This should be sufficient residual alkalinity to maintain the pH above 7.2 for coarse bubble aeration systems. If a fine bubble aeration system were chosen, chemical addition would be required and the dose estimated from the procedures discussed in Section 4.9.2. Alternatively, a lower operating pH could be used with a longer aeration period.

15. Calculate required reactor volumes. As in Section 4.3.3, assume the mixed liquor content in the contact tank is 2500 mg/l at a volatile content of 75 percent. The mixed liquor volatile suspended solids in stabilization can be obtained from the balance:

$$(Q + R)\, X_c = RX_s \tag{4-20}$$

where: X_c = contact MLVSS, mg/l, and

X_s = stabilization MLVSS, mg/l.

Therefore: $X_s = 1875 \dfrac{1 + 1.25}{1.25} = 3375$ mg/l

The total sludge inventory can be calculated from the following equation for substrate removal rate:

$$q_b = \frac{Q(S_o - S_1)}{\Sigma XV} \tag{4-21}$$

where: ΣXV = total inventory of MLVSS in the contact and stabilization tanks, lb

therefore: $\Sigma XV = \dfrac{1(148)(8.33)}{(.252)} = 4889$ lb

Of this inventory, 15 percent is in the contact tank:

$$0.15(4889) = 733 \text{ lb MLVSS}$$

The remainder is in the stabilization tank:

$$4889 - 733 = 4156 \text{ lb MLVSS}$$

The volume in contact is:

$$V_c = 733/(8.33)(1875) = 0.047 \text{ mil gal}$$

The volume in stabilization is:

$$V_s = 4156/(8.33)(3375) = 0.148 \text{ mil gal}$$

The total volume is 0.148 + 0.047 = 0.195 mil gal

16. Calculate the residence time in contact.

$$\theta_c = V_c/Q = 0.047 \text{ days} = 1.1 \text{ hr}$$

17. Calculate the sludge wasting schedule. See Section 4.3.3, Step 13.

As can be seen from Figure 4-6, the design of contact stabilization for nitrification is highly sensitive to the safety factor chosen and the sludge recirculation ratio. A wide range of alternate designs can be derived from variation in these parameters. Required reactor volumes are also sensitive to the assumed growth rate of nitrifiers, creating a need for kinetic data of high accuracy when designing for contact stabilization.

The design procedure described is based on the assumption of steady state operation. Diurnal variations in load will cause average effluent ammonia levels to exceed those calculated above. To compensate for this, it would be necessary to use an even higher SF than assumed in the above example. Regardless of the safety factor chosen, contact stabilization plants cannot be expected to completely nitrify except under the normally impractical condition of very high recycle rates (R/Q = 4.0 and above). However, at high recycle ratios the major advantage of contact stabilization is lost because the sludge in the stabilization basin becomes more dilute and the overall basin volume requirements approach those of the conventional process. This limits the application of contact stabilization to situations where only partial nitrification is required.

A further limitation on nitrification in contact stabilization plants is the incomplete hydrolysis of organic nitrogen occurring in the short detention time contact tank. As noted under step 9 above, about 40 percent of the influent organic nitrogen appears in the process effluent. Conventional or complete mix activated sludge plants, on the other hand, have sufficient contact time to hydrolyze the bulk of the organic nitrogen to ammonia thus making the nitrogen available to nitrifiers and leaving very little organic nitrogen in the effluent.

The short contact time of the contact tank can create problems in the sedimentation tank. The mixed liquor solids are not well stabilized in the contact tank prior to sedimentation.

Denitrification activity in the sedimentation tank is therefore greater than in conventional or complete mix plants and floating sludge may be the result. Procedures for circumventing the floating sludge problem are discussed in Section 4.10.

4.3.7 Step Aeration Activated Sludge Kinetics

Because of backmixing, the step feed pattern of step aeration plants causes the kinetics of such plants to more closely approach complete mix than plug flow. As a result, the design approach developed for complete mix (Section 4.3.3) can usually be employed for step aeration plants as a reasonable approximation. In those step aeration plants where influent is fed to the last pass (as in Figure 4-1), there is the danger that there will be insufficient time for the organic nitrogen to be hydrolyzed prior to discharge, resulting in elevated quantities of organic nitrogen in the effluent. Further discussion of this effect is presented in Section 4.3.8.2 which is a description of an operating step aeration plant.

4.3.8 Operating Experience with Combined Carbon Oxidation-Nitrification in Suspended Growth Reactors

While activated sludge-type systems are commonly used in England to obtain dependable nitrification, their use in the U.S. has not been widespread. Early U.S. activated sludge plants of the conventional design nitrified in the warmer months of the year or if they were underloaded. But nitrification became unpopular because of the additional aeration power cost and the propensity of some sludges to float in the sedimentation tank when nitrifying, and it was questioned whether the added expense was worth it in many cases.[39,40] As a consequence, ways and means were sought to prevent nitrification rather than to encourage it through increasing organic loading or through tapered aeration or by picking modifications of the process which were less favorable for nitrification. This early experience with the process may have led to uncertainty about its reliability.

Nonetheless, there have been several plant-scale operations in the U.S. which have demonstrated the viability of the process. The purpose of this section is to review some of these cases. Other case examples are presented in Sections 9.5.1 and 9.5.2.

4.3.8.1 Step Aeration Activated Sludge In a Moderate Climate

The Whittier Narrows Water Reclamation Plant is a 12 mgd activated sludge plant designed and operated by the Los Angeles County Sanitation Districts. The basic purpose of the plant is to reclaim water for groundwater recharge; the entire effluent of the plant is discharged to spreading basins for recharging groundwater aquifers.

Design data for the plant are summarized in Table 4-3.[41] The plant operates at either a constant flow rate or a constant oxygen demand load by pumping wastewater from a trunk sewer and returning grit, skimmings, primary and waste activated sludge back to the trunk. No solids handling facilities are provided as the solids returned to the trunk are processed at a downstream primary treatment plant. The plant was constructed in 1961 at a cost of $1,700,000; this cost includes influent pumping, foam fractionation and effluent pipelines in addition to those items shown in Table 4-3.

Recently, the plant has been operated in a manner promoting nitrification. The three aeration tanks are operated in a 3 pass series configuration; two-thirds of the primary effluent is added along the first pass, with the head end of the first pass operating as sludge reaeration. One-third of the primary effluent is added to the second pass. The plant

TABLE 4-3

DESIGN DATA
WHITTIER NARROWS WATER RECLAMATION PLANT

Plant Flow	12 mgd (0.53 m^3/sec)
Raw Wastewater Loadings	
Biochemical Oxygen Demand (BOD)	270 mg/l
Suspended Solids (SS)	280 mg/l
Primary Sedimentation Tanks	
Number	2 (1 stand-by)
Overflow Rate	2000 gpd/ft^2 (82 m^3/m^2/day)
Detention Time	1.1 hr
BOD Removal	35%
SS Removal	60%
Air Blowers	
Number	3
Discharge Pressure	6.5 psig (0.46 kgf/cm^2)
Capacity - Total	29,500 cfm (840 m^3/min)
Aeration Tanks	
Number	3
Detention Time (@ 12 mgd)	6.0 hrs
BOD_5 loading	45 lb/1000 cf/day) (0.18 kg/m^3/day)
Final Sedimentation Tanks	
Number	5
Overflow Rate (@ 12 mgd)	790 gpd/ft^2 (32.2 m^3/m^2/day)
Detention Time (@ 12 mgd + 33% return)	1.7 hrs
Weir Rate (@ 12 mgd)	12,000 gpd/ft (150 m^3/m/day)
Chlorine Contact Chambers	
Number	1
Detention Time (@ 12 mgd) including time in Foam Fractionation Tank & Effluent Pipe	43 min
Chlorine	600 lb/day (272 kg/day)

performance reflects its very careful control and operation; operating data for a one-year period are summarized in Table 4-4.[9] While organic nitrogen data are not available, the data indicate that year-round complete nitrification has been obtained. Climatic conditions for this California treatment plant are very favorable for nitrification as average monthly wastewater temperatures did not fall below 21 C for the year examined.

4.3.8.2 Step Aeration Activated Sludge in a Rigorous Climate

The Flint, Michigan sewage treatment plant is being upgraded to comply with requirements of the Michigan Water Resources Commission which mandate nitrification for the purpose of preventing DO depletion in the Flint River. In connection with this upgrading, a large scale test of combined carbon oxidation-nitrification was conducted with the existing activated sludge plant over a ten-month period to determine design conditions for the plant upgrading. The minimum wastewater temperature tested was 7 C.[6] During the test, ferric chloride and polymer were added to the primary treatment stage for phosphorus removal. This also had the effect of reducing the organic loading to the aeration tank.

The existing plant had three aeration tanks, each with four passes providing a 750,000 cu ft (21,240 cu m) capacity. With an average design BOD_5 loading of 24,500 lb/day (11,110 kg/day) to the aeration tanks at a 20 mgd (75,700 cu m/day) flow, the aeration tank load was 32.7 lb/1000 cu ft/day (523 kg/1000 cu m/day). Flows were varied to the facility, however, to provide variation in loads. Three secondary sedimentation tanks were provided having a design overflow rate of 678 gal/sq ft/day ($27.6 m^3/m^2$/day) at ADWF. The plant was usually operated in a step aeration mode, with one-half the influent directed to the head ends of the second and third passes.

Average effluent qualities for eight months of the test are shown in Table 4-5. While nitrate and nitrite are not shown, it is reported that a relatively good balance between ammonia removal and nitrate production was obtained.[6] Nitrite nitrogen was always less than 0.1 mg/l. The appearance of high concentrations of organic nitrogen was attributed to the low rate of hydrolysis of organic nitrogen compounds.[6] It is probable that the provision of feeding wastewater to the last pass exacerbated the problem by causing insufficient contact time for that portion of the wastewater to complete the hydrolysis of organic nitrogen to ammonia.

The effect of temperature and solids retention are considered in Table 4-6. Effluent qualities deteriorated somewhat with colder temperatures, with only 75 percent ammonia removal being obtained at 10 C. This ammonia bleedthrough may have been due to diurnal peaking in ammonia at the relatively low solids retention time employed (c.f. Section 4.3.3.2).

4.3.8.3 Conventional Activated Sludge In a Rigorous Climate

The Jackson, Michigan wastewater treatment plant is a 17 mgd conventional activated sludge plant designed for year-round complete nitrification.[10] The existing plant was upgraded in

TABLE 4-4

NITRIFICATION PERFORMANCE AT THE WHITTIER NARROWS WATER RECLAMATION PLANT (REFERENCE 9)

Calendar Month	Year	Flow, mgd (m^3/s)	Recycle ratio	Temp., C	MLVSS, mg/l		SVI, ml/g	θ_c, days	HT[a], hours	Air Use, MCF/day (l/s)	COD, mg/l		Ammonia-N, mg/l		
					1st pass	3rd pass					Primary effluent	Secondary effluent	Primary effluent	Secondary effluent	Percent Removal
4	1973	10.4 (0.46)	0.39	22	2177	1474	78	13.1	7.0	29.3 (9600)	241	62	21.6	2.0	91
5	1973	11.7 (0.51)	0.40	24	2337	1778	64	15.4	6.2	27.9 (9145)	243	47	23.5	3.2	86
6	1973	11.9 (0.52)	0.45	26	2390	2319	66	17.2	6.1	29.3 (9600)	239	39	20.8	0.8	96
7	1973	11.9 (0.52)	0.46	26	2603	2092	78	37.5	6.1	28.1 (9400)	227	32	20.1	0.6	97
8	1973	12.8 (0.56)	0.45	27	2889	2005	77	33.4	5.7	27.5 (9010)	223	30	18.8	0.6	97
9	1973	13.4 (0.59)	0.43	25	2850	1938	64	40.2	5.4	27.2 (8920)	216	34	19.2	1.4	93
10	1973	13.5 (0.59)	0.41	25	2958	2094	77	20.5	5.4	27.6 (9050)	228	34	21.5	1.5	93
11	1973	13.2 (0.58)	0.45	24	2791	2000	62	9.4	5.5	27.3 (8950)	235	33	21.9	2.8[b]	87
12	1973	11.1 (0.49)	0.52	22	2724	1986	55	10.7	6.6	25.5 (8360)	233	35	21.8	1.9	91
1	1974	9.9 (0.43)	0.60	21	2675	1971	114	16.1	7.4	25.6 (8400)	227	27	21.4	0.4	98
2	1974	12.1 (0.53)	0.49	21	2857	2097	88	9.8	6.0	28.9 (9470)	242	31	22.1	0.9	96
3	1974	12.2 (0.54)	0.48	22	2888	1982	75	9.4	6.0	28.8 (9440)	229	41	21.2	1.0	95

[a] Based on influent flow and entire aeration tank

[b] Blower trouble this month

1973 in response to an order to improve treatment to a point where a minimum dissolved oxygen of 4.0 mg/l could be maintained in the Grand River. An analysis of the assimilative capacity of the reach indicated that this could only be done if the effluent were completely nitrified to prevent discharge of NOD to the river.

TABLE 4-5

AVERAGE NITRIFICATION PERFORMANCE AT FLINT, MICHIGAN FOR 8 MONTHS (REFERENCE 6)

Parameter (all values in mg/l except Temp.)	Raw wastewater	Settled wastewater	Secondary effluent
BOD_5	250	131	13.6
SS	300	140	24.1
Total Kjeldahl nitrogen	27.6	23.3	7.8
Organic nitrogen	13.3	9.9	6.1
Ammonia nitrogen	14.3	13.4	1.7
Phosphorus	15.4	2.7	2.3
Temp., C	7.2 to 18.3		

TABLE 4-6

EFFECT OF TEMPERATURE AND SOLIDS RETENTION TIME ON NITRIFICATION EFFICIENCY AT FLINT, MICHIGAN (REFERENCE 6)

Temperature, C	Solids retention time, days	NH_4^+ removal, percent
18 and greater	4	95
13	4 - 5	87
10	6	75
7	10 - 12	50 (Lab) [a]

[a]Based on bench scale test results

Pilot studies[10] indicated that a combined carbon oxidation-nitrification system was more economical than a two-stage activated sludge system. Design data for this plant are contained in Section 9.5.1.1. The plant is operated in the conventional (or plug flow) mode. Table 4-7 summarizes the plant operating data; since start-up the plant has obtained complete nitrification at daily temperatures as low as 8 C.[42] A characteristic of this wastewater is that the primary effluent is weak, both in terms of BOD_5 and ammonia -N. The mixed liquor concentrates very well due to the high inert concentration of the raw wastewater, allowing high mixed liquor levels under aeration. Both the weak wastewater and the ability to maintain the mixed liquor at a high concentration allow nitrification to be obtained in hydraulic retention times of less than eight hours even at temperatures of 8-10 C. Nitrate and organic nitrogen values are not available.

This case history clearly demonstrates that combined carbon oxidation-nitrification can be dependably accomplished at temperatures as low as 10 C.

4.4 Combined Carbon Oxidation-Nitrification In Attached Growth Reactors

The two attached growth reactor systems seeing application for combined carbon oxidation-nitrification in the U.S. are the trickling filter process and the rotating biological disc process. Procedures for designing nitrification with these two systems are described in this section.

4.4.1 Nitrification with Trickling Filters in Combined Carbon Oxidation-Nitrification Applications

Trickling filter design concepts are discussed extensively in the Technology Transfer publication, *Process Design Manual for Upgrading Existing Wastewater Treatment Plants*.[25] Therefore, the following discussion is limited to the loading ranges that are applicable for nitrification in trickling filters used in combined carbon oxidation-nitrification applications.

As is the case for the activated sludge system, the development and maintenace of nitrifying organisms in a trickling filter is dependent on a variety of factors including organic loading, temperature, pH, dissolved oxygen and the presence of toxicants. However, in the case of the trickling filter, there has been no comparable development of kinetic theory for combined carbon oxidation-nitrification that can be directly applied with any degree of confidence. The approach applied to date has largely been empirical and relied mostly on specification of an organic loading rate suitable for application to each media type.[21]

4.4.1.1 Media Selection

The types of media currently available are summarized in Table 4-8. Rock applications are generally limited to four to ten feet in depth; the plastic and redwood media may be built in towers of 15 to 25 ft in height due to their lighter weight and greater void space for ventilation, affording considerable space saving economies. Loading capabilities of trickling

TABLE 4-7

NITRIFICATION PERFORMANCE AT THE JACKSON, MICHIGAN WASTEWATER TREATMENT PLANT (REFERENCE 42)

Month	Year	Flow, mgd (m^3/sec)	Recycle ratio	Temp., C	MLSS, mg/l	SVI, ml/g	θ_c, days	HT,[a] hours	Air use, MCF/day[b]	BOD_5, mg/l		Ammonia-N, mg/l		
										Primary effluent	Secondary effluent	Primary effluent	Secondary effluent	Percent Removal
8	1973	14.5 (0.63)	.38	21.7	4320	42	15.6	7.5	14	75	2.5	8.4	0.6	93
9	1973	12.4 (0.54)	.42	20.0	4110	45	16.4	8.8	14	82	2.6	9.9	0.7	93
10	1973	13.2 (0.58)	.38	17.2	4390	47	16.7	8.3	14	84	3	11.6	1.2	90
11	1973	12.2 (0.53)	.40	15.6	4480	47	18.6	9.0	14	94	4	11.0	0.8	93
12	1973	11.4 (0.50)	.42	12.2	4560	45	16.4	9.9	14	85	3	11.5	0.6	95
1	1974	14.0 (0.61)	.43	11.1	4630	43	10.3	7.9	14	97	4	9.2	0.7	92
2	1974	14.2 (0.62)	.49	10.6	4800	42	11.0	8.2	14	104	5	9.3	0.6	94
3	1974	18.4 (0.81)	.43	11.1	4930	38	11.1	6.3	14	90	4	7.1	0.5	93

[a] Based on influent flow

[b] 14 MCF = 4590 l/sec

filter media are known to be related to the available surface area for biological slime growth. Specific surface, or the amount of media surface contained in a unit volume, is a gross measure of the available surface for growth of organisms. Plastic media is available in higher specific surfaces than that shown in Table 4-8. Design practice has been to avoid the use of media with higher specific surface and lower voids due to the danger of clogging in combined carbon oxidation-nitrification applications. However, there has been recent experience which indicates that medias with specific surfaces exceeding 35 sq ft/cu ft (115 m^2/m^3) have been used to treat domestic wastewaters without media clogging.[43]

TABLE 4-8

COMPARATIVE PHYSICAL PROPERTIES OF TRICKLING FILTER MEDIA

Media[a]	Nominal Size in. (cm)	Unit Weight lb/cu ft (kg/m^3)	Specific Surface Area sq ft/cu ft m^2/m^3	Void Space percent
Plastic[b]	24 x 24 x 48 (61 x 61 x 122)	2-6 (32-96)	25-35[c] (82-115)	94-97
Redwood[d]	47-1/2 x 47-1/2 x 35-3/4 (121 x 121 x 51)	10.3 (165)	14 (46)	76
Granite	1-3 (2.5 x 6.5)	90 (1440)	19 (62)	46
Granite	4 (10.2)	-	13 (47)	60
Blast Furnace Slag	2-3 (5.1 x 7.6)	68 (1090)	20 (67)	49

[a]Reference 25
[b]Currently manufactured in the U.S. by: the Envirotech Corp., Brisbane, Ca.; the Munters Corp., Fort Meyers, Fla., and the B.F. Goodrich Co., Marietta, Ohio
[c]Denser media may be used for separate stage applications, see Section 4.7.1.1
[d]Currently manufactured in the U.S. by Neptune-Microfloc, Corvalis, Or.

4.4.1.2 Organic Loading Criteria

Observations of the effect of organic loading on nitrification efficiency in rock media and trickling filters are summarized in Figure 4-7. The data are from the following full-scale and pilot-scale plants: Lakefield, Minn.,[25] Allentown, Pa.,[25] Gainesville, Fla.,[44,45] Corvallis, Or.,[46] Fitchburg, Mass.,[25] Ft. Benjamin Harrison, Ind.,[25] Johannesburg, South Africa,[47] and Salford, England.[48] Several interesting factors affecting design are evident. First, organic loading significantly affects nitrification efficiency. This is principally caused by the fact that the bacterial film in the rock becomes dominated by heterotrophic bacteria. The relative high bacterial yield when BOD is removed causes displacement of the nitrifiers from the film by heterotrophic organisms at high organic loadings.

As opposed to nitrification with activated sludge, the breakthrough of ammonia in a trickling filter is not abruptly affected by loading rate. For rock media, attainment of 75 percent nitrification or better requires the organic loading to be limited to 10-12 lb BOD_5/1000 cu ft/day (0.16 to 0.19 kg/m^3/day). At higher organic loading rates the degree of nitrification diminishes, such that above 30 to 40 lb BOD_5/1000 cu ft/day (0.48 to 0.64 kg/m^3/day) very little nitrification occurs. These findings are consistent with those of the National Research Council whose evaluation of World War II military installations indicates that the organic loading should not exceed 12 lb BOD_5/1000 cu ft/day (0.19 kg/m^3/day) for rock media.[49]

The partial nitrification occurring at intermediate loading rates can cause confusion when attempting to analyze organic carbon removals across trickling filters with the BOD_5 test. Samples from the effluent of these partially nitrifying trickling filters will contain a

FIGURE 4-7

EFFECT OF ORGANIC LOAD ON NITRIFICATION EFFICIENCY OF ROCK TRICKLING FILTERS

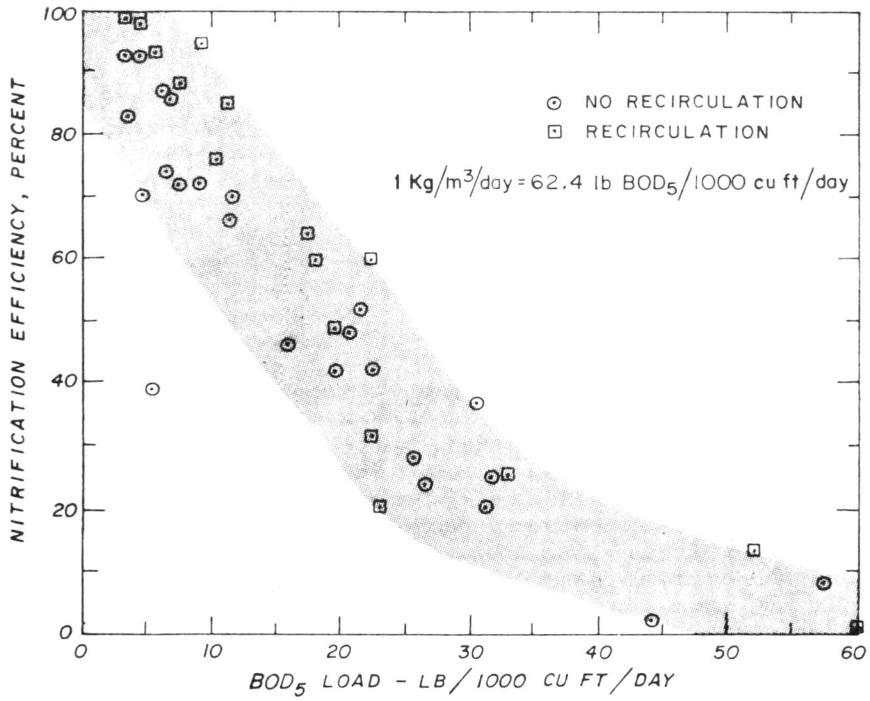

significant quantity of nitrifiers that could act as seed for promoting nitrification within the 5-day incubation period of the BOD_5 test.[21,50] About 1.5 mg/l of ammonia nitrogen is added to the dilution water in the BOD_5 test and will also be nitrified. This will result in unexpectedly high oxygen demands. If the BOD_5 test is to measure organics in effluents from trickling filters, then nitrification must be suppressed. The same is true for activated sludge, but to a lesser degree as partial nitrification is less prevalent.

As opposed to the relatively efficient removal of ammonia in trickling filters, it appears that reductions in organic nitrogen are variable and range between 20 and 80 percent. (Table 4-9). Organic nitrogen reduction can be obtained through employing effluent filtration to remove particulate organic nitrogen. However, all treatment systems are limited to about 1 to 2 mg/l of soluble organic nitrogen, contained in refractory organics, and therefore there are limits to the improvements that can be obtained with effluent filtration.

4.4.1.3 Effect of Media Type on Allowable Organic Loading

The specific surface of media selected has a substantial effect on the allowable organic loading rate for trickling filters. Greater specific surface in the media allows greater biological film development and therefore a greater concentration of organisms within a unit volume. Therefore, the organic loading may be higher in cases where the specific surface of

TABLE 4-9

ORGANIC NITROGEN REDUCTIONS IN NITRIFYING TRICKLING FILTERS

Facility Location	Organic load, lb BOD_5/1000 cu ft/day (kg BOD_5/ m^3/day)	Depth, ft (m)	Media	Influent organic-N, mg/l	Effluent organic-N, mg/l	Organic-N removal, percent
Gainesville, Fla.[a] (pilot)	31.5 (0.50)	6 (1.8)	1-1/2 - 2-1/2 in. (3.8 - 6.4 cm) rock	16.6	7.3	56
Johannesburg, S.A.[b] (full-scale)	31.2 (0.50	12 (3.7)	2 in. (5 cm) rock	9.8	4.7	52
	19.6 (0.31)	6 (1.8)	1-1/2 in. (3.8 cm) rock	6.3	2.5	60
	13.1 (0.21)	12 (3.7)	2-3 in. (5.1 - 7.6 cm) rock	8.2	3.6	56
	10.5 (0.17)			9.9	2.2	78
	6.8 (0.11)			13.9	2.1	85
Stockton, Ca.[c] (pilot)	14 (0.22)	21.5 (6.6)	plastic 27 sf/cu ft ($86\ m^2/m^3$)	11.3	8.9	21
	22 (0.35)			11.4	9.0	21

[a]References 44, 45
[b]Reference 47
[c]References 21, 51

the media is increased over that of rock media. An example is the work of Stenquist, *et al.*[21] who showed that plastic media (27 sq ft/cu ft) could be loaded at about 25 lb/1000 cu ft/day and still achieve good nitrification (Table 4-10). The higher allowable loading attributable to plastic trickling filters was attributed to be at least partly due to the greater specific surface of plastic media when compared to rock media. Another factor favoring greater capacity of the plastic media filters may be oxygen supply. Rock filters often have poor ventilation, particularly when water and air temperatures are close or the same.

4.4.1.4 Effect of Recirculation on Nitrification

The beneficial effects of recirculation on enhancing nitrification in trickling filters is evident in the data for Salford, England in Table 4-11. The imposition of a 1:1 recycle ratio consistently improved ammonia removals compared to when no recirculation was the rule. Trickling filter plants designed for nitrification should incorporate provision for recirculation.

The minimum hydraulic application rate for plastic media trickling filters is in the range of 0.5 to 1.0 gpm/sf (0.020 to 0.041 $m^3/m^2/min.$). This minimum rate must be supplied to ensure uniform wetting of the media. Without recirculation, nitrification design loadings may result in applied hydraulic loads lower than the minimum hydraulic application rate. Recirculation provides the means for preventing drying out of portions of the media by ensuring that at least the minimum hydraulic application rate is applied at all times.

4.4.1.5 Effect of Temperature on Nitrification

Available data for nitrification are largely for warm liquid temperatures, and the practical effects of reduced temperatures (< 20C) on allowable organic loads for combined carbon oxidation-nitrification applications are not known at this time. However, the kinetic rate data given in Section 3.2.5.4 would indicate that organic loads would have to be reduced below those shown in Figure 4-7 for cold weather operation. This reduction in organic load

TABLE 4-10

LOADING CRITERIA FOR NITRIFICATION WITH PLASTIC MEDIA AT STOCKTON

BOD_5 load, lb/1000 cu ft/day ($kg/m^2/day$)	Temp, C	Influent BOD_5, mg/l	Depth, ft (m)	Media	Recycle ratio[b]	Influent NH_4^+-N, mg/l	Effluent NH_4^+-N, mg/l	Percent nitrification (or ammonia removal)	Reference
14 (0.22)	26	155	21.5 (6.6)	plastic (Surfpac)[a]	5.5	16.5	1.0	94	21,51
22 (0.35)	24	131	21.5 (6.6)	Same	2.25	17.5	2.0	89	21,51

[a]27 sq ft/cu ft (86 m^2/m^3)

[b]Recycle ratio is the ratio of recycled effluent to influent. Effluent was recycled prior to sedimentation.

would reduce the loadings to such low values as to cause capital costs to be higher than other available ammonia removal techniques, such as separate stage nitrification or a physical chemical technique.

4.4.1.6 Effect of Diurnal Loading on Performance

While it is known that diurnal variations in nitrogen loading will cause variations in effluent quality, no information is available which would allow quantitative guidelines to be formulated. In cases where large peak to average flow ratios are experienced, flow equalization before the nitrification step may be appropriate.

4.4.2 Nitrification with the Rotating Biological Disc Process in Combined Carbon Oxidation-Nitrification Applications

The rotating biological disc (RBD) process is beginning to see use in the U.S. in combined carbon oxidation-nitrification applications. The following discussion is abstracted from Antonie.[52]

The RBD process consists of a series of large-diameter plastic discs, which are mounted on a horizontal shaft and placed in a concrete tank. The discs are slowly rotated while approximately 40 percent of the surface area is immersed in the wastewater to be treated.

TABLE 4-11

EFFECT OF RECIRCULATION ON NITRIFICATION IN ROCK TRICKLING FILTERS AT SALFORD, ENGLAND (REFERENCE 48, 25)[a]

BOD_5 Load lb/1,000 ft^3/day (kg/m^2/day)	Influent BOD_5 (mg/l)	Influent NH_4^+-N (mg/l)	. . Effluent NH_4^+-N (mg/l). . . Without Recirculation	With Recirculation	. . . Percent Nitrification. . . . Without Recirculation	With Recirculation
22.6 (0.36)	266	33.9	19.7	13.6	42	60
16.3 (0.26)	235	31.3	16.9	11.8	46	62
11.8 (0.19)	191	32.0	9.7	4.8	70	85
9.2 (0.15)	239	43.9	125	2.2	72	95
7.7 (0.12)	165	40.5	11.4	4.9	72	88
5.9 (0.095)	192	40.7	5.7	2.8	86	93
4.6 (0.074)	199	38.3	2.8	0.9	93	98
3.2 (0.051)	206	36.6	0.7	0.4	93	99

[a]Media was blast slag, 8 ft (2.4 m) deep. With recirculation a 1:1 ratio was employed.

Shortly after start-up, organisms present in the wastewater begin to adhere to the rotating surfaces and grow until in about one week, the entire surface area is covered with a layer of aerobic biomass. In rotation, the discs pick up a thin film of wastewater, which flows down the surface of the discs and absorbs oxygen from the air. Microorganisms remove both dissolved oxygen and organic materials from this thin film of wastewater. Shearing forces exerted on the biomass as it passes through the wastewater strip excess growth from the discs into the mixed liquor. The mixing action of the rotating discs keeps the sloughed solids in suspension, and the wastewater flow carries them out of the disc sections into a secondary clarifier for separation and disposal. The discs also serve to mix the contents of each treatment stage. Treated wastewater and sloughed solids flow to a secondary clarifier where the solids settle out and the effluent passes on for further treatment or disinfection. The settled solids, which can thicken up to 4 percent solids content in the secondary clarifier, are removed for treatment and disposal. A flow diagram for a typical application of the RBD process is shown in Figure 4-8.

FIGURE 4-8

A TYPICAL ROTATING BIOLOGICAL DISC PROCESS
(COURTESY OF THE AUTOTROL CORP.)

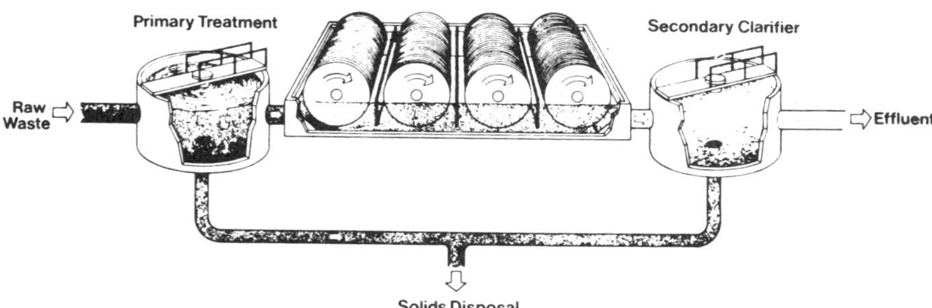

One current disc design consists of vacuum formed polyethylene sheets formed into concentric corrugations which provide a high density of surface area. The corrugated sheets are then welded together to form a stack of discs with approximately 1¼ in. (3.2 cm) center-to-center spacing. This type of construction has a surface area density of approximately 37 ft^2/ft^3 (121 m^2/m^3). A key feature of this disc design is the provision of radial passages extending from the shaft to the outer perimeter of the discs. This assures that wastewater, air, and stripped biomass can pass freely into and out of the disc assembly. In the twelve-ft (3.7 m) diameter size, radial passages are provided every thirty degrees.

The disc units are normally housed to avoid temperature drops across the process, to prevent algae growth on the disc surface, and to protect the surface from hail or rain which can wash the slimes off. Information on disc types and on the general design of these facilities can be obtained from the various disc manufacturers.

4.4.2.1 Loading Criteria for Nitrification

As the rotating discs operate in series, organic matter is removed in the first disc stages and subsequent disc stages are used for nitrification. This separation of function occurs without the need for intermediate clarification. Nitrification does not commence until the bulk of the BOD_5 is oxidized. When low levels of BOD_5 are reached, the disc stage is no longer dominated by heterotrophs, and nitrification can proceed

Antonie[53] has summarized the effects of process operating conditions on nitrification. Test data from a number of locations are shown in Figure 4-9B. The degree of ammonia oxidation is related to the hydraulic loading on the rotating media as gallons per day per unit surface of available surface area. Also important, as it affects population dynamics, is the influent BOD strength. The data in Figure 4-9B was used to arrive at the design criteria shown in Figure 4-9A. Two changes have been made to allow use of the relationships in design. One is that there is a maximum ammonia nitrogen concentration for which the data is considered valid. This concentration is generally 1/5 to 1/10 the influent BOD_5 concentration. When the ammonia concentration exceeds these maximum ammonia nitrogen concentrations on the appropriate BOD_5 curve, it has been recommended that the design curve be used which is rated for that ammonia concentration.[53] The second change made in Figure 4-9A is the identification of a region of unstable nitrification; that is, a region of hydraulic loading where either a slight change in the hydraulic loading or influent BOD strength could result in a displacement of the nitrifying population. It is considered advisable to stay out of that region even during daily peak flow conditions.[53]

4.4.2.2 Effect of Temperature

Investigators at Rutgers University[54] found that the nitrifying capability of the discs was relatively constant in a temperature range from 15 to 26 C. Simliar results were obtained by Antonie[53] who has found no effect of temperature above 13 C. Temperature correction factors derived from pilot data are shown in Figure 4-10.[53] These correction factors should be used to reduce the design hydraulic loading determined in Figure 4-9A for any wastewater temperature lower than 13 C.

4.4.2.3 Effect of Diurnal Load Variations

Precise design criteria for handling load variations in RBD units have not been formulated. The concern has been that high BOD_5 concentrations would break through to the last disc stage during peak loads and would cause displacement of nitrifiers from that stage.[53] While firm design recommendations have not been made, there are two general approaches available. One is to arbitrarily derate the surface hydraulic loading to the disc to ensure low BOD loadings at all times. This will result in a larger amount of rotating surface area. The other is to install flow equalization to reduce the peak to average flow ratio.[53]

FIGURE 4-9

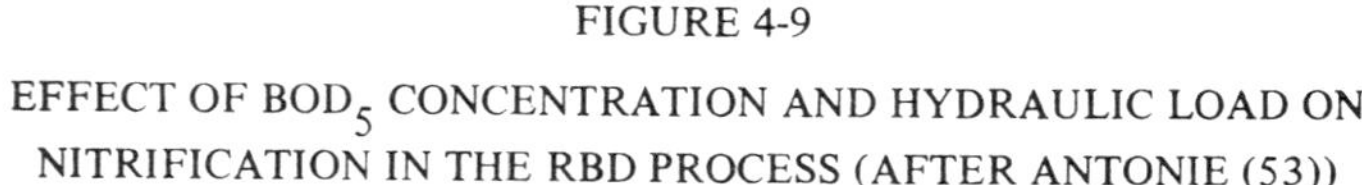

EFFECT OF BOD_5 CONCENTRATION AND HYDRAULIC LOAD ON NITRIFICATION IN THE RBD PROCESS (AFTER ANTONIE (53))

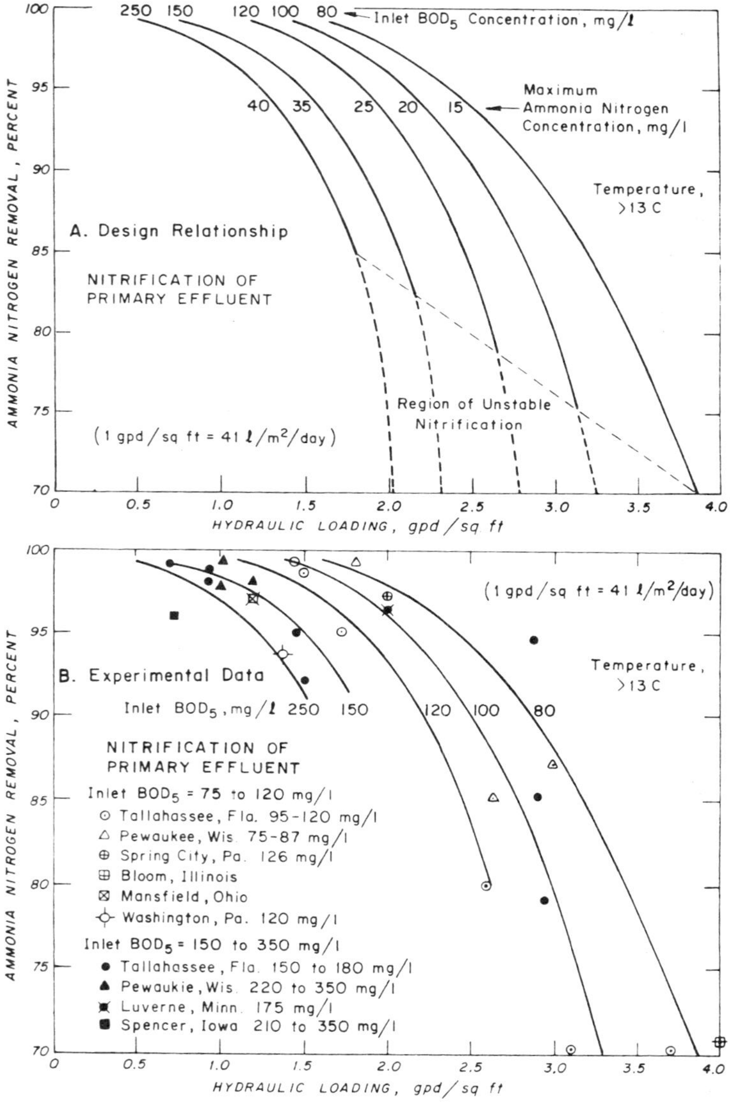

4.5 Pretreatment for Separate Stage Nitrification

Nitrification facilities have been previously classified in Table 4-1. As can be seen from that table, in order to obtain a separate stage nitrification process, the influent to that process must be pretreated to remove organic carbon. Further, the pretreatment must achieve a degree of carbon removal greater than is obtained by primary treatment alone, in order to reduce the BOD_5/TKN ratio to a sufficiently low level to ensure a significant fraction of nitrifiers in the biomass. Alternatives listed in Table 4-1 include chemical treatment in the primary, activated sludge, roughing filters, and trickling filters. Not listed, but possible, is activated carbon treatment in conjunction with primary chemical addition. These alternatives are summarized in Figure 4-11. This set of pretreatment alternatives is not meant to be exhaustive, but merely illustrative; other flowsheets are possible. Detailed design of pretreatment steps is beyond the scope of this manual; however, design procedures may be found in the publications referenced in Figure 4-11 and the sources cited on Table 4-1.

The pretreatment alternative adopted may have significant effects on the downstream nitrification stage. In this section, some of these possible effects are considered.

FIGURE 4-10

TEMPERATURE CORRECTION FACTOR FOR NITRIFICATION IN THE RBD PROCESS (AFTER ANTONIE (53))

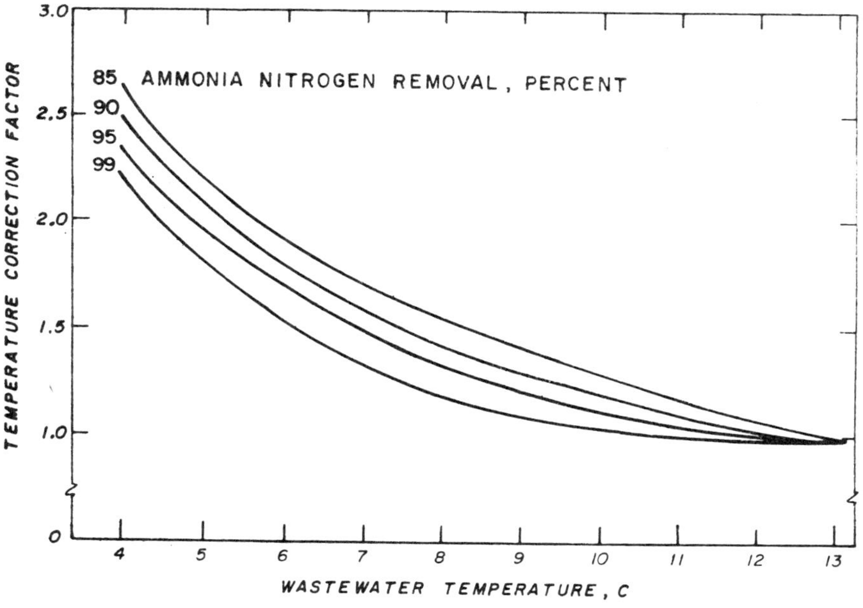

4.5.1 Effects of Pretreatment by Chemical Addition

The chemical treatment step may cause significant changes in alkalinity and pH in the downstream nitrification stage. Several alternate chemicals are available and their results differ.

FIGURE 4-11

PRETREATMENT ALTERNATIVES FOR SEPARATE STAGE NITRIFICATION

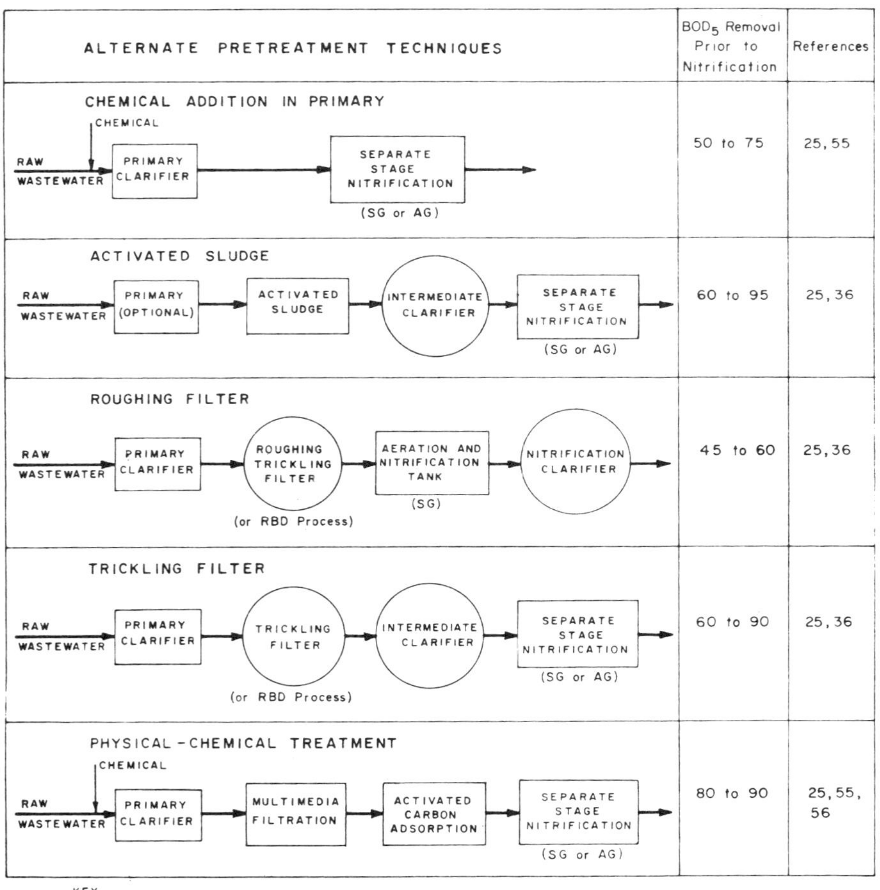

KEY
SG = Suspended growth
AG = Attached growth

When alum or ferric chloride is used in the primary treatment stage, both the carbonate and phosphate components of the alkalinity in wastewater are changed. Table 4-12 summarizes the changes occurring when alum is added to wastewater. Effects of iron addition are similar. Since orthophosphate is present in wastewater as $HPO_4^=$ and $H_2PO_4^-$ between pH 4.5 and 9.3,[57] Table 4-12 shows reactions with aluminum for both forms. The compound $HPO_4^=$ is measured as part of the alkalinity in wastewaters because of the following equilibria:

$$H^+ + HPO_4^= \rightarrow H_2PO_4^-$$

This reaction is shifted completely to the right at pH 4.5 which is approximately the endpoint titration pH for the conversion of bicarbonate (HCO_3) to carbonic acid (H_2CO_3) in the standard alkalinity determination. Table 4-12 shows that the same amount of total alkalinity is lost regardless of the form of the inorganic phosphorus ($HPO_4^=$ or $H_2PO_4^-$).

TABLE 4-12

EFFECT OF ALUM ADDITION TO WASTEWATER ON ALKALINITY

1. Hydrolysis

$$Al_2(SO_4)_3 + 6\ HCO_3^- + 6\ H_2O \rightarrow 2\ Al(OH)_3 + 3\ SO_4^= + 6\ H_2CO_3$$

(54 as Al^{+3}) (300 alkalinity as $CaCo_3$)

or 5.6 mg of alkalinity as $CaCO_3$ lost per mg Al^{+3} added

2. Precipitation of inorganic phosphorus

$HPO_4^=$ form

$$Al_2(SO_4)_3 + 2\ HCO_3^- + 2\ HPO_4^= \rightarrow 2\ AlPO_4 + 3\ SO_4^= + 2\ H_2CO_3$$

(54 as Al^{+3}) (100 alkalinity as $CaCO_3$) (100 alkalinity as $CaCO_3$) (62 as P)

or 3.7 mg of alkalinity as $CaCO_3$ lost per mg Al^{+3} added

and .87 mg Al^{+++} required per mg P

$H_2PO_4^-$ form

$$Al_2(SO_4)_3 + 4\ HCO_3^- + 2\ H_2PO_4^- \rightarrow 2\ AlPO_4 + 3\ SO_4^= + 4\ H_2CO_3$$

(54 as Al^{+3}) (200 alkalinity as $CaCO_3$) (62 as P) (0 alkalinity as $CaCO_3$) (62 as P)

or 3.7 mg of alkalinity as $CaCO_3$ per mg Al^{+3} added

and 0.87 mg of Al^{+++} required per mg P

As an illustrative example of the results of alum addition, consider a case where 10 mg/l of inorganic phosphorus is precipitated in the primary stage of treatment. Using a value of 0.64 mg P removed per mg of aluminum added results in the following aluminum use:[1]

$$\frac{10 \text{ mg/l}}{0.64 \dfrac{\text{mg P}}{\text{mg Al}^{+++}}} = 15.5 \text{ mg/l Al}^{+++} \text{ required (total)}$$

The aluminum used in phosphorus precipitation is (Table 4-12):

$$10 \text{ mg/l P}(0.87 \frac{\text{mg Al}^{+++}}{\text{mg P}}) = 8.7 \text{ mg/l Al}^{+++} \text{ for precipitation}$$

By difference the aluminum used in hydrolysis is:

$$15.5 - 8.7 = 6.8 \text{ mg/l Al}^{+++} \text{ for hydrolysis}$$

The alkalinity loss due to hydrolysis is (Table 4-12):

$$6.8 \frac{(5.6 \text{ mg CaCO}_3)}{\text{mg Al}^{+++}} = 37.4 \text{ as CaCO}_3$$

The alkalinity loss due to precipitation is (Table 4-12):

$$8.7 \frac{(3.7 \text{ mg CaCO}_3)}{\text{mg Al}^{+++}} = 32.2 \text{ as CaCO}_3$$

Thus, a total alkalinity loss of 70 mg/l will occur when 10 mg/l of P is removed. This loss, depending on the initial wastewater alkalinity, may have to be made up with downstream chemical addition to prevent adverse effects on the operating pH of nitrification (see Section 4.9.2.).

On the other hand, lime addition to the primary may have quite the opposite effect of alum. The changes in alkalinity occurring with lime addition are dependent on lime dose (or pH) and the quality of the raw wastewater. Cases of both increases and decreases in alkalinity with lime treatment may be found.[55]

Lime addition may also cause elevation of the operating pH of the nitrification stage.[3] Primary effluent after lime treatment will typically have a pH between 9.5 and 11.0, depending on lime dose and treatment requirements.[55] This pH is higher than can normally be discharged or introduced into downstream treatment units. To reduce the pH, normal practice is to "recarbonate" the high pH primary effluent. Conventionally, this involves the introduction of gaseous carbon dioxide (CO_2) into the high pH primary effluent in a reaction basin of at least 20 minutes detention time. Typically the carbon dioxide is either drawn from refrigerated storage or furnace stack gases containing carbon dioxide are used. The recarbonation step can be thought of as the conversion of alkalinity in the hydroxide form (OH^-) to that in the bicarbonate form (HCO_3^-) as follows:[55]

$$Ca(OH)_2 + 2CO_2 \longrightarrow Ca(HCO_3)_2$$

In many cases there is sufficient carbon dioxide produced from the oxidation of organic carbon and from nitrification to completely satisfy recarbonation requirements. In a lime precipitation-nitrification sequence in California,[3] it was found that external carbon dioxide requirements were minimal and only occasionally required. In this case, it was calculated that approximately two-thirds of the carbon dioxide produced was derived from nitrification, while the remaining one-third was derived from the oxidation of the organic carbon remaining in the primary effluent. When the same process was tried at the EPA-DC pilot plant at Blue Plains, it was found that supplemental carbon dioxide was continuously required to maintain a neutral pH. This is because wastewater in the Washington area is weaker than in the California case. There is less production of carbon dioxide because there are lower concentrations of oxidizable substances entering the nitrification stage at Blue Plains.

The tendency is for the high pH effluent of the lime primary process to elevate the pH in the nitrification reactor. Often this effect enhances nitrification rates (see Section 3.2.5.6.).

When lime primary treatment is employed, care must be taken not to mix the primary effluent with the return activated sludge prior to entry into the nitrification tank. The high pH of the primary effluent return activated sludge mixture would be toxic to both the nitrifiers and heterotrophic bacteria in the return sludge.

A concern often expressed is that the phosphorus removal obtained in a chemical primary treatment step will be so great as to starve the downstream nitrifying biomass for phosphorus as a nutrient for growth. Actually, the requirements for phosphorus in a separate stage nitrification system are very low. Typically, organism biomass contains about 2.6 percent phosphorus. This number can be used to calculate phosphorus requirements, as in the following example. Assume a case where 36 mg/l of TKN are nitrified and 60 mg/l of BOD_5 are removed in a separate nitrification stage. The quantity of biomass grown can be conservatively estimated as follows: (see Section 3.2.7)

Nitrifiers:	0.15 (36)	=	5.4 mg/l VSS
Heterotrophs:	0.55 (60)	=	33.0 mg/l VSS
Total biomass:			38.4 mg/l VSS

The phosphorus required is 2.6 percent of the VSS grown:

$$0.026(38.4) = 1.0 \text{ mg/l as P}$$

Thus, the phosphorus requirement in this example is only 1.0 mg/l P. In most cases, the BOD_5 and TKN levels will be less than given in the example, so that the 1 mg/l requirement can be considered a maximum requirement for separate stage nitrification.

It is a relatively simple matter to manipulate chemical doses in the primary so that phosphorus residuals are sufficient to support biological growth.[3] It may be that low levels of phosphorus or some trace micronutrient removed by the primary may cause "bulking" in a suspended growth type nitrification stage.[58] Very high SVI measurements were observed in the nitrification stage following lime treatment at the Central Contra Costa Sanitary District's Advanced Treatment Test Facility.[3] However, it has been found that this bulking could be very effectively controlled by a continuous low dose of chlorine to the return sludge, without impairment of nitrification efficiency. A dose of 2 to 3 lb. per 1,000 lb. MLVSS per day (2 to 3 g/kg/day) in the inventory reduced the SVI from 160-200 to 45-80 ml/gram. Overdosing at 5 to 7 lb./1,000 lb./day (5 to 7 g/kg/day) caused impairment of nitrification and higher effluent solids.[59]

4.5.2 Effects of Degree of Organic Carbon Removal

The pretreatment alternatives arrayed in Figure 4-11 provide varying degrees of organic carbon removal ahead of the nitrification step. Not only is there variation among the alternatives, but there is possible range of intermediate effluent qualities within each alternative. It was shown in Chapter 3 that high degrees of organic carbon removals led to the highest nitrification rates. This implies that reactor requirements diminish with increasing degrees of carbon removal in the pretreatment stage.

Very low levels of organic carbon in the nitrification process influent has differing implications for suspended growth and attached growth nitrification reactors. The effluent solids from the sedimentation step in a suspended growth system can exceed the solids synthesized in the process when the level of organics is low. This leads to a requirement of continuously wasting solids from other suspended growth stages (carbon removal or denitrification) to the nitrification stage or to increase the BOD of the influent in order to maintain the inventory of biological solids in the system. This point is discussed further in Section 4.6.4.

Low levels of organics in the influent to attached growth reactors can be advantageous. The synthesis of solids occurring with low levels of influent organics results in very low levels of solids in the effluent from an attached growth reactor. In some cases this can eliminate the need for a clarification step, especially if multimedia filtration follows or if there is some other downstream treatment unit such as denitrification.

4.5.3 Protection Against Toxicants

All of the pretreatment alternatives portrayed in Figure 4-11 provide a degree of removal of the toxicants present in raw wastewater. However, the types of toxicants removed by each pretreatment stage vary among the alternatives. Chemical primary treatment can be used where toxicity from heavy metals is the major problem. Lime primary treatment is one of the most effective processes for removal of a wide range of metals.[55] Chemical treatment is usually not effective for removal of organic toxicants, unless it is coupled with a carbon adsorption step such as in the physical-chemical treatment sequence. The biological pretreatment alternatives (activated sludge, trickling filters, and roughing filters) provide a degree of protection against both organic and heavy metal toxicants. An exception would be organics resistant to biological oxidation, such as the solvents perchloroethylene and trichloroethylene which have been identified as toxicants which can upset nitrification.[1,60]

When materials toxic to nitrifiers are present in the influent raw wastewater on a regular basis, the pretreatment technique most suitable for their removal can be used in the plant design to safeguard the nitrifying population. The determination of the most suitable system configuration need not involve an extensive sampling or elaborate pilot program. A recently developed bench top analysis can be used to screen alternates.[61] The test procedure involves batch oxygen uptake tests using a respirometer to measure oxygen utilization. Composite wastewater samples are subjected to various pretreatments, e.g., alum or powdered activated carbon via a jar test procedure or to biological oxidation by batch aeration. Each treated sample is then split and placed into two respirometers. One respirometer is used as a non-nitrifying control by treatment with a nitrification inhibitor such as Allythiourea (ATU).[61] The other respirometer is inoculated with a small amount of mixed liquor from a nitrifying activated sludge plant. Differences between the oxygen used in the control and in the seeded sample can be used to establish batch nitrification rates. At the end of the test, the respirometer contents are sampled and analyzed for the nitrogen species to confirm whether nitrification took place in the inoculated samples as well as to check the control. The adequacy of the seed used can also be checked by running an inoculated, but uninhibited sample, known to contain ammonia and organics, but no toxicants.

The batch nitrification rates, determined by the batch procedure, can be examined to determine the pretreatment technique apparently most suitable among the options examined. Often, some of the pretreatment techniques will result in little or no nitrification in the inoculated sample, indicating inadequate removal of the toxicant(s). In other cases, the pretreatment techniques will allow vigorous nitrification in the sample indicating good removal of the toxicant(s).[61] The particular pretreatment technique that is effective may also indicate the type of toxicant that is interfering with nitrification and may permit identification and elimination of the source tributary to the system. For instance, if lime treatment is effective, the problem may be a heavy metal that can be precipitated by the lime. Alternatively, if biological oxidation is ineffective but activated carbon treatment allows nitrification to proceed, then a nonbiodegradable organic is suspect. Subsequent specific analyses can then be run in the identified category of compounds. If the toxicants cannot be eliminated by a source control program, often a pilot study of the process identified by the bench scale procedure can be justified to confirm the process selection. Pilot studies also have value in determining the ability of the nitrifiers to adapt to the toxicants, something the batch test is not capable of doing.

4.6 Separate Stage Nitrification with Suspended Growth Processes

There are many examples of separate stage nitrification processes in the U.S. (Table 4-1). The initial development of the suspended growth process in separate stage application was oriented to the isolation of the operation of the carbonaceous removal and nitrification processes so that each could be separately controlled and optimized.[15] By placing a carbon removal system (originally conceived as a high rate activated sludge system) ahead of the separate nitrification stage, the sludge would be enriched with nitrifiers as opposed to the marginal population present in combined carbon oxidation-nitrification systems. By this enrichment process, nitrification would be expected to be less temperature sensitive than in a combined carbon oxidation-nitrification system.

First applications of the process were in the northern portion of the U.S. where low (< 10 C) liquid temperatures were obtained in the wintertime. Subsequently, the system has been applied in moderate climates such as Florida and California where some of the other advantages of the process have made its application desirable.

4.6.1 Application of Nitrification Kinetic Theory to Design

The kinetic approach for design of separate stage nitrification in suspended growth systems is fundamentally related to the kinetic design approach used for combined carbon oxidation-nitrification, as was shown in Section 3.2.7. To date, however, the practice has been to adopt the "solids retention time" design approach for combined carbon oxidation-nitrification applications and to use the nitrification rate approach for separate stage design. The nitrification rate approach has been based on experimentally measured rates,[1,62,63] rather than attempting to relate the rates to fundamental kinetic theory (see Section 3.2.7). The theoretically determined nitrification rates are limited in their applicability chiefly because the nitrifier fraction of the mixed liquor cannot be accurately assessed. Nonetheless, the concepts developed from kinetic theory are applicable even in the absence of information about the nitrifier fraction.

The solids retention time design approach is also directly applicable to many separate stage nitrification design problems. The limit of its applicability is related to the difficulties with the nitrification rate approach, that is,the yield of nitrifiers grown through nitrification is not accurately known. However, for designs where the BOD_5 level in the nitrification influent is 30 to 60 mg/l (or BOD_5/TKN ratio is 2 to 3) the growth of nitrifiers will generally be a small fraction relative to the heterotrophic population. In this case, the errors in assumption of the yield of nitrifiers will be masked by the growth of heterotrophs. For these cases, the contribution of the nitrifiers to the overall process growth rate may be neglected with the assurance that their contribution is less than 12 percent (Section 3.2.7). A model which explicitly considers the nitrifier contribution to the system growth rate is available which can be used when accurate yields for nitrification become known.[38]

In situations where the organic carbon is low in the influent (BOD_5/TKN ratio 0.5 to 2.0) to the nitrification stage, both the assumptions for heterotrophic and nitrifier yields become uncertain. These low BOD_5/TKN ratio situations are usually cases where a well stabilized secondary effluent is being nitrified. Residual BOD_5 in these effluents is often biological solids rather than residual raw wastewater organic matter. The biomass yield in the nitrification stage for these cases is less well defined. This leads to uncertainty in estimating design sludge inventories and wasting schedules when using the solids retention time design approach.

In this section, both the solids retention time approach and the nitrification rate approach are presented. The direct theoretical interrelationships of the approaches, as developed in Section 3.2.7, should not be overlooked.

4.6.2 Solids Retention Time Approach

The design procedures developed in Section 4.3 are directly applicable to separate stage nitrification design when the BOD_5/TKN ratio equals or exceeds 2.0. A summary of these design concepts follows.

4.6.2.1 Choice of Process Configuration

In general, the favored system for separate stage nitrification is the plug flow system. It was already shown in Section 4.3.5 that the plug flow process results in lower effluent ammonia than a complete mix process at the same SF, or alternately, the same ammonia level at a lower SF. The only disadvantage of the process in combined carbon oxidation-nitrification applications is the difficulty in supplying adequate DO in the head end of the system, rendering that zone ineffective for nitrification in cases of low DO. With a carbon removal step ahead of the separate nitrification stage, the high oxygen demand at the head end of the system is minimized, and less difficulty is found in designing aeration systems that ensure adequate DO levels throughout the nitrification tanks. With adequate DO levels throughout the tanks, the full advantage of plug flow kinetics over the kinetics of any other configuration is obtained.

In cases where a lime precipitation step precedes the nitrification step, it is often desirable to use the carbon dioxide (CO_2) produced by the nitrification process for recarbonation (c.f. Section 4.5.1). Since the process of nitrification will be spread throughout the plug flow reactor, and recarbonation is desired at its head end to avoid pH toxicity, only a portion of the carbon dioxide produced in the process is available for recarbonation at the head end of the process. A solution to this problem is to increase the use of external carbon dioxide. On the other hand, external carbon dioxide usage may be minimized by adopting one of the other process configurations, such as complete mix, that more evenly spread out the load throughout the aeration tank, thereby taking full advantage of the in-process carbon dioxide generation for recarbonation.

4.6.2.2 Choice of the Safety Factor

The same considerations discussed in Section 4.3.3.2 are applicable to separate sludge applications. Diurnal variations in nitrogen load will effect effluent quality in the same manner as for combined carbon oxidation-nitrification applications. However, upstream treatment steps may moderate the fluctuations in nitrogen load experienced by separate stage nitrification processes. While primary treatment and roughing filters have so little liquid holdup that little or no nitrogen load equalization is provided, this is not true of either the activated sludge or trickling filter pretreatment alternatives. Both of these alternatives provide a degree of nitrogen load equalization.

A typical ammonia load curve for the secondary effluent from the Rancho Cordova, California treatment plant is shown in Figure 4-12. The plant provides activated sludge treatment. On the date monitored, the average flow was 1.9 mgd with a peak hourly flow of

FIGURE 4-12

RANCHO CORDOVA WASTEWATER TREATMENT FACILITY
EFFLUENT AMMONIA CHARACTERISTICS, MARCH 19-20, 1974

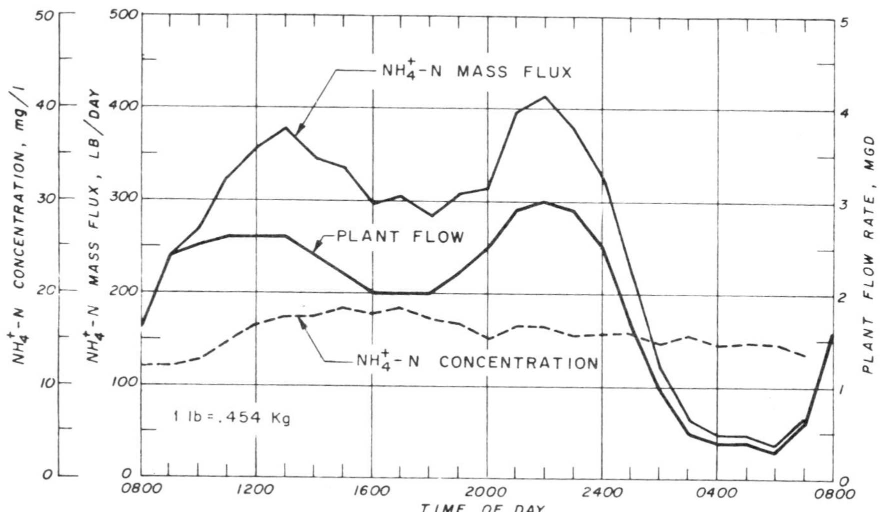

3.0 mgd, for a flow peaking factor of 1.6. The ammonia loading behaved similarly with an average load of 252 lb/day and a peak hourly load of 379 lb/day for a nitrogen load peaking factor of 1.6. Using the concept developed in Section 4.3.3.2, the minimum safety factor adopted for a separate stage nitrification process serving this plant should be 1.6, to prevent significant ammonia leakage during the peak hour.

4.6.3 Nitrification Rate Approach

The kinetic design approach using nitrification rates places reliance on experimentally determined rates obtained from pilot studies. Available data are summarized in Figure 4-13, plotted against temperature. Also shown are two other principal variables that can effect nitrification rates, the BOD_5/TKN ratio and the mixed liquor pH.

In general, the nitrification rates follow the predictions of the theory (Section 3.2.7). As the temperature rises, nitrification rates increase. The BOD_5/TKN ratio strongly influences the nitrification rates. Comparing the Manassas, Blue Plains and Marlborough data, it can be seen that the lower the BOD_5/TKN ratio (and the higher the nitrifier fraction) the higher the nitrification rates. Also the effects of pH depression on nitrification rates is apparent. Particularly interesting is the data from Blue Plains for the air and oxygen system run at approximately the same pH, but at differing BOD_5/TKN ratios. The air nitrification system, running at a lower influent BOD_5/TKN ratio exhibited higher nitrification rates than the oxygen system running at a relatively higher BOD_5/TKN ratio. It is notable that when oxygen and air nitrification systems were run in parallel at the same pH and influent BOD_5/TKN ratios, the same nitrification rates were obtained.[8] A further comparison of oxygen and air nitrification is presented in Section 4.6.5.

In the absence of pilot data specific to a particular situation, Figure 4-13 can be used to approximate design nitrification rates. What must be known or estimated is the BOD_5/TKN ratio in the influent, the minimum temperature for nitrification and the mixed liquor pH. In essence, Fig. 4-13 is a plot of experimentally determined values of the peak nitrification rate, as defined in Section 3.2.7 as follows:

$$\hat{r}_N = \hat{q}_N \cdot f \qquad (3\text{-}31)$$

where:

$\hat{r}_N$ = peak nitrification rate, lb NH_4^+ - N oxidized/lb MLVSS/day,

f = nitrifier fraction, and

$\hat{q}_N$ = peak ammonia oxidation rate, lb NH_4^+ - N rem/lb VSS/day.

In other words, these rates are determined at values where the DO and the ammonia level are not limiting the rate of nitrification. However, to be useful for design purposes, the effects of ammonia content and DO should be considered. The effect of operating DO can

FIGURE 4-13

OBSERVED NITRIFICATION RATES AT VARIOUS LOCATIONS

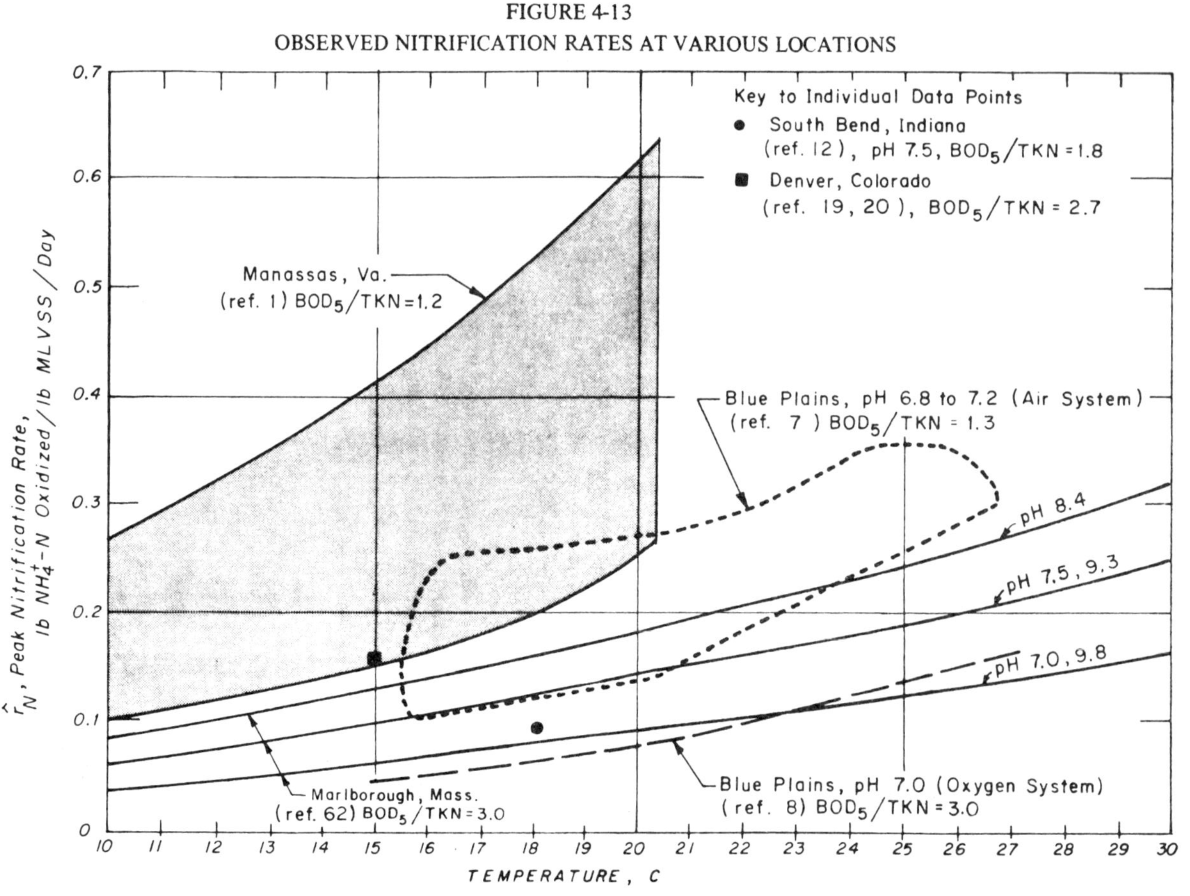

be incorporated through the Monod expression for DO. The effect of desired effluent ammonia content can be considered through the safety factor concept. The use of Equations 3-31, 3-20, and 3-29 yields the following expression for either a complete mix, or plug flow system:

$$r_N = \frac{\hat{r}_N}{SF}\left(\frac{DO}{K_{O_2} + DO}\right) \tag{4-22}$$

Effluent ammonia nitrogen content can be estimated for a complete mix system at steady state from the equation:

$$\frac{1}{SF} = \frac{N_1}{N_1 + K_N} \tag{4-23}$$

For plug flow reactors, the effluent ammonia content can be estimated from Equation 4-14. Criteria for establishing the safety factor are discussed in Sections 4.3.3.1, 4.3.3.2, and 4.6.2.2.

Once the design value of the nitrification rate is established, the design can proceed in a manner similar to the F/M design approach adopted for activated sludge design. The total nitrogen load per day and the nitrification rate are used to establish the mass of solids that must be maintained in the nitrification reactor. The volume of reactor is determined from the allowable mixed liquor solids level and the inventory of solids required. The allowable mixed liquor level is influenced primarily by the efficiency of solids-liquid separation (Section 4.10).

4.6.4 Effect of the BOD_5/TKN Ratio on Sludge Inventory Control

At Manassas, Virginia, Jackson, Michigan, and Contra Costa, California, difficulty was experienced in maintaining a nitrifying sludge inventory when the influent contained low amounts of organics (low BOD_5/TKN ratios).[64,10,4] Typically, effluent solids fluctuated between 10 and 50 mg/l and the effluents contained a good deal of dispersed solids that were not captured in the secondary clarifier. It has been suggested that a high fraction of the mixed liquor must be heterotrophic to maintain good bioflocculation in a separate stage nitrification system.[12] Since the synthesis of solids in these systems is often less than the solids appearing in the effluent from the system, an unstable condition can result. Several remedies are available. At the three locations mentioned, solids from the upstream activated sludge process were periodically transferred to the separate stage nitrification process to maintain the solids inventory.

In other cases, pretreatment steps have been purposely chosen which do not provide as high a degree of carbon removal as activated sludge and therefore cause greater synthesis of heterotrophic biomass in the nitrification stage. Examples are lime precipitation in the primary[3] and modified aeration activated sludge with alum addition.[65] In still another case, 10 percent of the primary effluent was bypassed around the activated sludge carbon removal step to a separate stage nitrification step.[11] The amount of primary effluent bypassed can be varied to precisely control the solids retention time and details of a recommended procedure for accomplishing this control can be found in reference 66.

4.6.5 Comparison of the Use of Conventional Aeration to the Use of High Purity Oxygen

A comprehensive study of the effect of pH on covered high purity oxygen nitrification systems with comparisons to conventionally aerated systems has recently been completed.[8] High purity oxygen systems typically operate at somewhat lower pH levels than conventionally aerated systems, and it has been intimated that this will result in lower nitrification rates and efficiency than conventionally aerated systems. The reason for the lower pH is because of the method of oxygenation in covered high purity oxygen systems. As shown in Figure 4-14, the system uses a covered and staged oxygenation basin for contact of gases and mixed liquor. High purity oxygen (90+ percent purity) enters the first stage and flows concurrently with the wastewater being treated. The gas is reused in successive stages, resulting in the buildup of carbon dioxide released by biological activity in the gas and in the liquid. This results in a depression of pH. While the pH is also depressed by carbon dioxide release in conventionally aerated systems (c.f. Sec. 4.9), the pH depression is less than occurs in the high purity oxygen system because evolved carbon dioxide is continually stripped from the system by the aeration air.

The work at the EPA-DC Blue Plains treatment plant consisted of two carefully controlled pilot investigations as follows: (1) separate stage nitrification with high purity oxygen with and without pH control and (2) separate stage nitrification by conventional aeration and

FIGURE 4-14

COVERED HIGH PURITY OXYGEN REACTOR WITH THREE STAGES AND MECHANICAL AERATORS

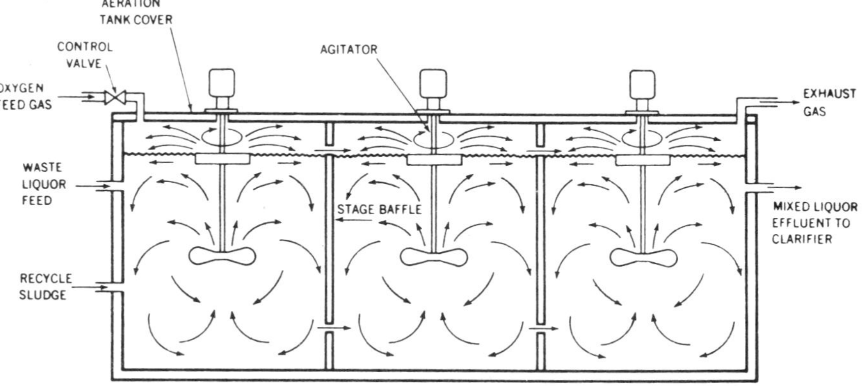

high purity oxygenation controlled at the same pH level. The purpose of this section is to review the salient features of this work as it affects nitrification design with high purity oxygen. No attempt will be made to present general high purity oxygen design concepts as these are covered in the EPA Office of Technology Transfer's publication, *Oxygen Activated Sludge Wastewater Treatment Systems Design Criteria and Operating Experience.*[67]

4.6.5.1 High Purity Oxygen Nitrification With and Without pH Control

Operational results are shown in Table 4-13 for a high purity oxygen pilot plant without pH control and a plant with the last of four stages held at pH 7.0 by lime addition to the first stage. As may be seen from the table, the system with pH control provided an effluent with somewhat better quality. Lime addition to the pH controlled reactor caused buildup of inerts which resulted in better thickening sludge and better clarification efficiency at the expense of greater sludge production.[8] The effluent ammonia level was somewhat lower with pH control, but the difference is not significant when it is considered that a chlorine dose as little as 10 mg/l would be sufficient to remove all traces of ammonia in both wastewaters (c.f. Chapter 6). It should be noted that both systems were operated at high solids retention times and therefore had high SF values. Therefore, differences in effluent ammonia levels would be expected to be small.

Comparisons of nitrification rates did not present a clear picture of differences between the two systems. However, it was shown that when the pH drops below 6.0 nitrification rates did decline. However, the pH generally did not drop below 6.0 in the system without pH control until the last reactor stage, but there was so little ammonia remaining to be oxidized there that essentially no effect of pH on nitrification performance could be discerned.

In sum, the work at Blue Plains demonstrates that the pH of the nitrification reactor can drop as low as 6.0 and allow acclimation of the nitrification organisms with attainment of complete nitrification. The pH should not be allowed to drop below 6.0, except perhaps in the last reactor stage, and in those cases where the carbon dioxide evolution is sufficient to cause the pH to drop below 6.0, pH control should be implemented. From an organics standpoint, effluent qualities are superior in the pH controlled system because of lime addition, and this should be kept in mind when designing for stringent effluent requirements.

4.6.5.2 Comparison of Conventional Aeration and High Purity Oxygen at the Same pH

Table 4-14 shows the results of a parallel study of a conventional aerated nitrification system with a high purity oxygen system; both systems were held at pH 7.0 in the last stage of a four-stage system.[8] As may be seen from the table, the concentration of organics and nitrogen species were virtually identical in the two systems. Greater lime was required in the high purity oxygen system than in the conventional system to maintain the same pH level.

This greater lime dose resulted in greater sludge production in the oxygen system but also improved the sludge thickening properties, allowing the same MLVSS level to be maintained in both systems at a higher MLSS level in the oxygen system. Nitrification kinetic rates were found to be the same in both systems. Choice between the two pH controlled systems should be based on economic considerations as the systems are equivalent in other respects.

TABLE 4-13

COMPARISON OF PROCESS CHARACTERISTICS FOR OXYGEN NITRIFICATION SYSTEMS WITH AND WITHOUT pH CONTROL AT BLUE PLAINS, WASHINGTON, D.C.

	With pH control, days 51-150[a]	Without pH control, days 26-85[a]
Operating Parameters[b]		
Oxygenation time, hrs[c]	4.0	3.9
F/M ratio[d]	0.15	0.16
Solids retention time, days	13	17
MLSS, mg/l	5,660	3,520
MLVSS, mg/l	2,620	2,780
Return sludge SS, mg/l	38,650	13,640
Sedimentation tank overflow rate, gpd/sq ft	620	645
$m^3/m^2/day$	25.3	26.3
Sludge production, mg/l	69	35.4
Lime dose (CaO), mg/l	126	0

Effluent Qualities[b]

Concentration, mg/l	With pH control: Influent mean	Effluent mean	Effluent standard deviation	Without pH control: Influent mean	Effluent mean	Effluent standard deviation
BOD_5	67	6.3[e]	3.1[e]	70	8.6[e]	3.0[e]
COD	152	26.4	6.2	152	42.5	8.2
SS	77	10	4.8	75	24.0	10.2
VSS	58	6.0	4.2	58	17.1	8.6
TKN	21.7	1.3	0.6	22.7	2.4	0.7
NH_4^+-N	14.9	0.15	0.11	150	0.64	0.64
NO_2^--NO_3^--N	-	13.4	1.1	-	12.8	12

[a]Day 1 = January 1, 1974

[b]Data from reference 8 ; while the data are not from the same time period, data from a common time period (days 56-85) showed the same trends.

[c]Based on influent flow

[d]F/M ratio is the ratio of the lb BOD_5 in the influent to the activated sludge process and the lb of MLVSS inventory under aeration or oxygenation.

[e]Nitrification inhibited

4.7 Separate Stage Nitrification with Attached Growth Processes

Three types of attached growth processes have been employed for separate stage nitrification. The differences lie in the type of medium provided for biological growth. The three types of processes are the trickling filter, the rotating biological disc and the packed bed reactor.

TABLE 4-14

COMPARISON OF PROCESS CHARACTERISTICS OF CONVENTIONALLY AERATED AND HIGH PURITY OXYGEN SYSTEMS WITH pH CONTROL AT BLUE PLAINS, WASHINGTON, D.C.

	High purity oxygen, days 186-265[a]	Conventional diffused aeration, days 186-265[a]
Operating Parameters[b]		
Oxygenation time, hr[c]	3.8	3.7
F/M ratio[d]	0.16	0.16
Solids retention time, days	10.0	9.9
MLSS, mg/l	6,355	3,890
MLVSS, mg/l	2,260	2,240
Return sludge SS, mg/l	40,850	16,460
Sedimentation tank overflow rate,		
gpd/sq ft	656	678
$m^3/m^2/day$	26.7	27.6
Sludge production, mg/l	97.6	59.2
Lime dose (CaO), mg/l	128	47

Effluent Qualities[b]

Concentration, mg/l	High purity oxygen: Influent mean	Effluent mean	Effluent standard deviation	Conventional: Influent mean	Effluent mean	Effluent standard deviation
BOD_5	56	3.5[c]	1.1[e]	56	3.4	1.3[e]
COD	130	20.6	2.6	130	20.6	3.4
SS	65	8.1	2.6	65	6.2	3.4
VSS	49	4.8	2.1	49	4.2	2.5
TKN	19.5	0.94	0.23	19.5	1.0	0.3
NH_4^+-N	15.2	0.19	0.20	15.2	0.19	0.16
NO_2^-+ NO_3^--N	-	13.6	1.2	-	12.5	1.2

[a]Day 1 = January 1, 1974
[b]Data from reference 8
[c]Based on influent flow
[d]F/M ratio is the ratio of the lb BOD_5 in the influent to the activated sludge process and the lb of MLVSS inventory under aeration or oxygenation.
[e]Nitrification inhibited

4.7.1 Nitrification with Trickling Filters

The development of two-stage trickling filtration or double filtration preceded the development of two-stage suspended growth systems for nitrification. In fact, two-stage filtration was in operation at several military installations during World War II.[49] Initially, the two-stage trickling filtration process was developed to increase the removal of organics in the effluents from the high rate trickling filters. Later, it was observed that under some operating conditions, the second stage produced a well nitrified effluent.[68]

In separate stage nitrification application, trickling filters can follow a high rate trickling filter plant with intermediate clarification, or an activated sludge process or any of the other alternatives listed in Fig. 4-11.

4.7.1.1 Media Type and Specific Surface

It has been shown that in separate stage nitrification applications, the rate of nitrification is proportional to the surface area exposed to the liquid being nitrified.[69,70] In other words, when all other factors are held constant, the allowable loading rates can be expected to be related to the media surface area, rather than to the media volume.

Very little biological film development has been observed in separate stage applications.[71,72] As a consequence, pluggage of voids in the media and ponding becomes of less concern than in combined carbon oxidation-nitrification applications. Media of higher specific surface than normally employed may be used. Plastic media is characterized by having very high specific surface available while maintaining a high void ratio (> 90 percent). The high specific surface area of plastic media allows the trickling filter volume to be reduced, significantly reducing the cost of the distributor arms and the structure.

Available types of plastic media are summarized in Table 4-15. Most experience in the U.S. has been with the corrugated sheet module type, rather than with the dumped media which has just become available. Media applicable to nitrification applications is commercially available in specific surfaces ranging from 27 to 68 sq ft/cu ft (89 to 223 m^2/m^3).

4.7.1.2 Loading Criteria

As previously stated, nitrification rates in trickling filters are related to the wetted surface area of the media. Therefore, the most rational criterion would be in terms of surface area. Unfortunately, information on specific surface is not always available, and volumetric loading criteria must occasionally be resorted to.

The pilot study at the Midland, Michigan wastewater treatment plant provides the most comprehensive set of data currently available on nitrification with trickling filters.[22,71] The influent to the pilot plant was well treated trickling filter effluent with BOD_5, SS and ammonia-N values ranging from 15-20, 15-20 and 8-18 respectively. The BOD_5/TKN ratio

TABLE 4-15

COMMERCIAL TYPES OF PLASTIC MEDIA FOR SEPARATE STAGE NITRIFICATION APPLICATIONS

Manufacturer	Trade Name	Type	Specific surface available sf/cu ft	(m^2 / m^3)
Envirotech Corp., Brisbane, Ca.[a]	Surfpac	Corrugated sheet modules	27	(89)
B.F.Goodrich, Marietta, Ohio	Vinyl Core	Corrugated sheet modules	30.5 45	(100) (148)
Enviro Development Co., Inc. Palo Alto, Ca.[b]	Flocor	Corrugated sheet modules	27 40	(89) (131)
Mass Transfer, Ltd., Houston, Texas	Filterpack	Dumped rings	36 57	(118) (189)
Norton Co., Akron, Ohio	Actifil	Dumped rings	27 42	(89) (138)
Munters Corp., Ft. Meyers, Fla.	PLASdek	Corrugated sheet modules	42 68	(138) (223)

[a]Formerly available from the Dow Chemical Co., Midland, Mich.

[b]Under license from ICI, Great Britain; formerly available from the Ethyl Corp., Baton Rouge, La.

was 1.1, indicating a high degree of BOD removal in the pretreatment stage. The pilot unit was a 21.5 ft (6.55 m) unit filled with Surfpac media. During the 18 month project period, a variety of climatic conditions were experienced with wastewater temperatures in the pilot unit as low as 7 C and as high as 19 C encountered.

The data from the various operating periods for the project have been reexpressed in Figure 4-15 in terms of the surface area required for nitrification and the desired effluent ammonia nitrogen content. As may be seen, greater surface area is required at low temperature (7 to 11 C) than high temperatures (13 to 19 C). Further, to obtain ammonia-N contents below 2.5 to 3.0 mg/l, greater surface area is required than for effluent ammonia contents above 2.5 to 3.0 mg/l ammonia-N.

When effluent ammonia-N levels less than 2.5 mg/l are desired, consideration should be given to using breakpoint chlorination (Chapter 6) for removing ammonia residuals rather than increasing the surface area of the filter, as the cost of removing the last 1-3 mg/l of ammonia becomes very high because of the very much larger trickling filters required.

Figure 4-15 was developed solely from the Midland, Michigan data. Data are available from two other locations which allow calculation of surface requirements for nitrification.[27,73] Figure 4-16 shows surface reaction rates for Lima, Ohio data[27] compared with the trend lines developed from the Midland, Michigan data. Lesser surface area is required at the Lima location for the same degree of nitrification. This lower surface requirement is chiefly due to the higher wastewater temperature, but the fact that the influent BOD_5 levels were lower may also have caused a higher proportion of nitrifiers to be present in the trickling filter's surface film. In the case of Lima, Ohio, the influent to the nitrification stage is produced by a step aeration activated sludge plant.

Surface requirements for nitrification of oxidation pond effluent at Sunnyvale, California are shown in Figure 4-17.[73] In this case, large quantities of algae were present in the trickling filter influent. While the bulk of the algae passed through the unit unaffected, at

FIGURE 4-15

SURFACE AREA REQUIREMENTS FOR NITRIFICATION - MIDLAND MICHIGAN

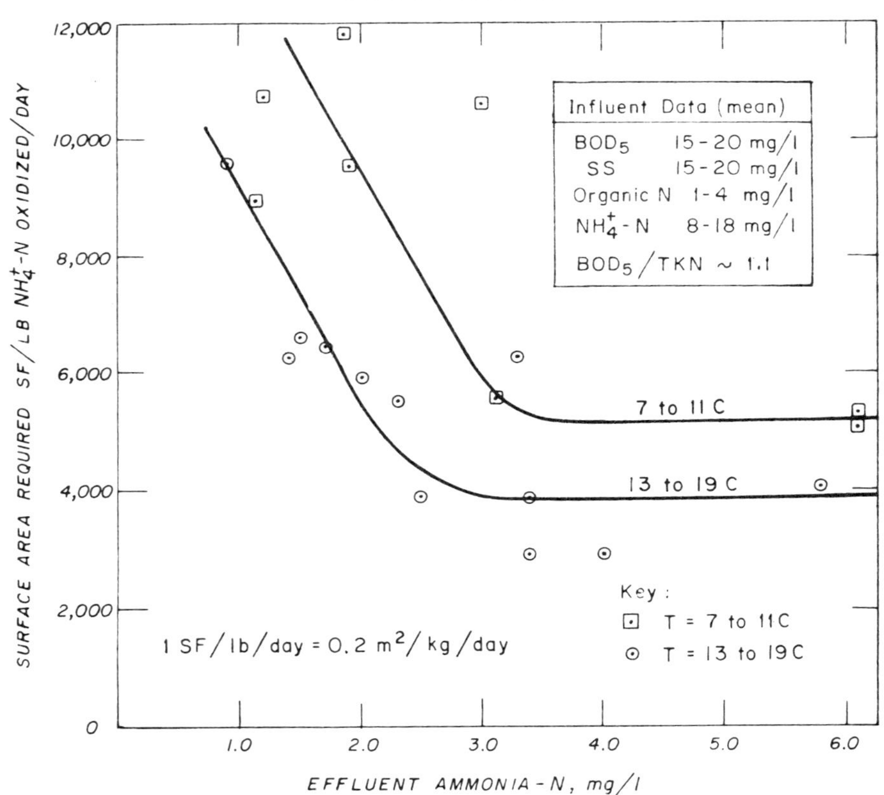

least 20 to 40 percent were trapped and eventually oxidized. This would have affected the proportion of heterotrophic bacteria in the bacterial film, causing higher surface requirements for nitrification at Sunnyvale, California than at Midland, Michigan.

Available data with rock media is more sparse and is summarized in Table 4-16 in terms of ammonia oxidized per unit volume. Rock media is capable of ammonia oxidation at only 15 to 50 percent of the plastic media rate, on a volumetric load basis. The principal reason for this is undoubtedly the rock media's lower specific surface, although the lower depth of the typical rock filter may also have a role to play.

FIGURE 4-16

SURFACE AREA REQUIREMENTS FOR NITRIFICATION - LIMA, OHIO

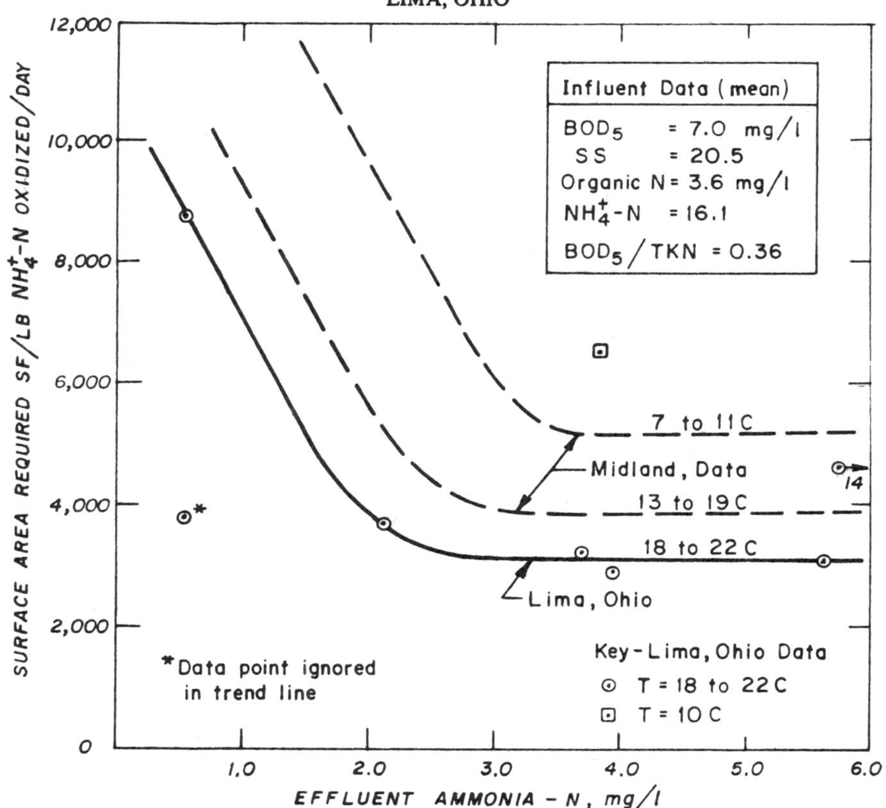

TABLE 4-16

NITRIFICATION IN SEPARATE STAGE ROCK TRICKLING FILTERS

Facility location	Ref.	Depth, ft (m)	Media	Influent		Effluent			Ammonia - N oxidized lb/1000 cu ft/day (kg/m^3/day)
				BOD_5 mg/l	NH_4^+-N, mg/l	BOD_5, mg/l	NH_4^+-N mg/l	NH_4^+-N Percent removed	
Johannesburg, S.A. (full-scale)	47	12 (3.7)	2-3 in. (5.1 to 7.6 cm) rock	28	23.9	14	8.3	65	3.5 (0.055)
		12 (3.7)	1.5 in. (3.8 cm) rock	32	25.2	13	4.4	83	2.2 (0.035)
		9 (2.7)	1 in. (2.5 cm) rock	23	22	10	9.1	59	2.4 (0.038)
Northhampton, England (pilot-scale)	72	6 (1.8)	1.5 in. (3.8 cm) rock	80	33	10	11.2	66	1.0 (0.016)

4.7.1.3 Effect of Recirculation

An analysis of the Midland, Michigan data and Lima, Ohio, data has led to the conclusion that while recirculation improved nitrification efficiency only marginally on an average basis, the periods with recirculation demonstrated greater consistency (less fluctuations) than when no recirculation was employed.[26,27] This conclusion, together with the improvements seen with recirculation in combined carbon oxidation-nitrification applications (Section 4.4.1.4), leads to a general recommendation for the provision of recirculation. A 1:1 recirculation ratio is considered adequate at average dry weather flow for most applications.

4.7.1.4 Effluent Clarification

Since the organisms are attached to the media in attached growth systems, effluent clarification steps are not required in all cases. In the case of Midland, Michigan it was found that the effluent solids were approximately equal to the influent solids at 9 to 28 mg/l.[22] This is because influent BOD_5 levels were low (15 to 20 mg/l). When influent BOD_5 loads were increased above previous low levels, trickling filter effluent solids rose to 58 mg/l. The insertion of clarifier allowed this to be reduced to 19 mg/l. Subsequent multimedia filtration allowed further reduction to about 4 mg/l.

4.7.1.5 Effect of Diurnal Load Variations

Trickling filters used for nitrification, like any other nitrification process, are affected by diurnal variations in nitrogen load. The rule of thumb developed in Section 4.3.3.2 can likely be applied to trickling filters to prevent high ammonia bleed through during diurnal peaks in load. Thus, the amount of surface area determined from Figures 4-15, 4-16 or 4-17 under average daily loading conditions should be multiplied by the ratio of peak ammonia load to average load to establish design surface area. An alternative would be to provide flow equalization.

FIGURE 4-17

SURFACE AREA REQUIREMENTS FOR NITRIFICATION - SUNNYVALE, CALIFORNIA

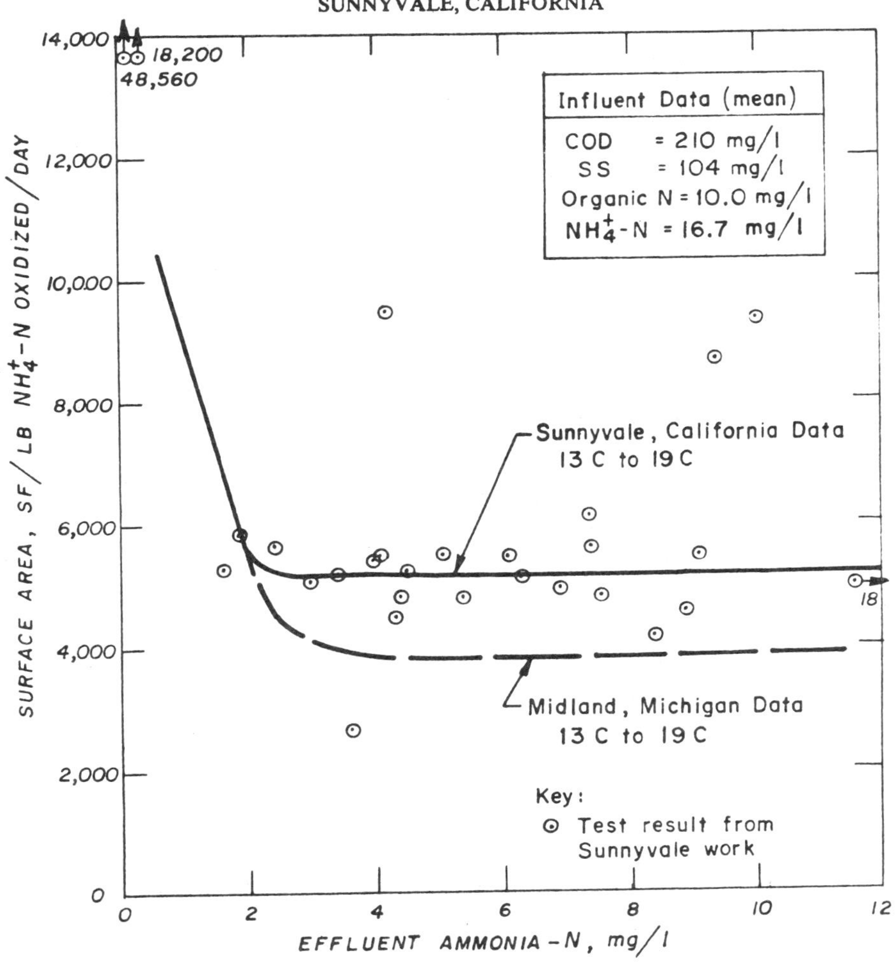

4.7.1.6 Design Example

As an example consider a 10 mgd conventional activated sludge plant that must be upgraded to meet effluent requirements of 4 mg/l ammonia nitrogen and 10 mg/l suspended solids on an average basis. The plant is located in a temperate zone, and the minimum wastewater temperature is 15 C. Present average effluent qualities are 1 mg/l organic nitrogen, 20 mg/l ammonia nitrogen, 15 mg/l of suspended solids, and a BOD_5 of 25 mg/l. The peak to average nitrogen load ratio is 1.9. Consider as one alternative a plastic media trickling filter.

1. Calculate the BOD_5/TKN ratio:

$$BOD_5/TKN = 25/21 = 1.19$$

2. The closest set of data based on BOD_5/TKN ratio and temperature is that for Midland, Michigan (Figure 4-15). For an effluent ammonia nitrogen concentration of 4 mg/l at 15 C, the unit surface area requirement is about 3,800 sf/lb NH_4^+-N oxidized/day.

3. Calculate the ammonia nitrogen oxidized daily. The following equation is appropriate:

$$NT = 8.33 \cdot Q(N_o - N_1) \qquad (4\text{-}24)$$

where:

NT	=	ammonia nitrogen oxidized, lb per day
Q	=	average daily flow, mgd
N_o	=	influent NH_4^+ - N, mg/l
N_1	=	effluent NH_4^+ - N, mg/l

For this example, assuming no change in the organic nitrogen level, the result is:

$$NT = 8.33\ (10)\ (20-4) = 1{,}332 \text{ lb/day}$$

4. Find the total surface area requirement under average load conditions. Multiplying the nitrogen oxidized per day (step 3) by the unit surface area requirement (step 2) results in:

$$(1{,}332)\ (3{,}800) = 5{,}061{,}000 \text{ sf of media}$$

5. Consider diurnal peak loading. One approach would be to provide flow equalization. Assume that in this case site restrictions prevent this. Therefore, increase the surface requirement by the peak to average nitrogen load ratio as follows (Section 4.7.1.5):

1.5 (5,061,000) = 7,592,000 sf

6. Choose a media type and establish media volume requirements. Effluent BOD_5 and SS are low enough so that fairly high density media can be employed. In this instance, a corrugated sheet module media having a specific surface of 42 sf/cu ft is chosen. Media volume requirements are determined by dividing the total surface requirement by the specific surface as follows:

   7,592,000/42 = 180,750 cu ft

   This volume could be provided by a variety of configurations; for instance two 75 ft diameter trickling filters with a media height of 21 ft would have the necessary volume of media. Whatever configuration is chosen, the filter shouldn't be less than about 12 to 15 ft in height because of the danger of short circuiting. Usual practice is to consult with the media manufacturer(s) prior to final selection of media configuration.

7. Establish recirculation rate. At 10 mgd ADWF, a 1:1 recycle is adequate (Section 4.7.1.3); therefore 10 mgd of recirculation capacity is recommended.

8. Establish clarification requirements. Effluent solids in the nitrification process effluent will be approximately at the influent SS level, 15 mg/l. Therefore, to meet a 10 mg/l requirement, some form of effluent clarification is required such as dual or multimedia filtration.

4.7.2 Nitrification with the Rotating Biological Disc Process

The rotating biological disc (RBD) process, discussed in Section 4.4.2 for combined carbon oxidation-nitrification applications may also be applied to nitrifying secondary effluents. The process is constructed as shown in Fig. 4-8, excepting that it may be possible to eliminate the secondary clarifier when the secondary effluent being treated has a BOD_5 and suspended solids less than about 20 mg/l.[74] Under this circumstance the very low net growth occurring in the nitrification stage causes the RBD process effluent suspended solids to approximately equal the influent solids level. If lower levels of suspended solids are required, the RBD process could be followed directly by tertiary filtration without the need for intermediate clarification.[74]

One manufacturer has announced the availability of media especially adapted to nitrification. The minimal biomass film development in separate stage nitrification applications has allowed a 50 percent increase in surface area of the corrugated polyethylene media. Standard shafts were 100,000 sq ft (9300 m^2) of available surface area; the new media is available at 150,000 sq ft (13,900 m^2) of surface per shaft. This results in a reduction of one-third in the number of shaft assemblies required for nitrification with the RBD process.[74]

4.7.2.1 Kinetics

The reaction rates occurring in each stage of the RBD process treating secondary effluents have been analyzed; the correlation between surface reaction rates and stage (effluent) concentration is shown in Figure 4-18.[74] The trend line does not reach a plateau value but keeps gradually rising because the biomass developed per unit surface is not constant. Antonie found that the amount of culture developed on the rotating surface increased with increasing ammonia nitrogen concentration.[74]

A stage-by-stage application of Fig. 4-18 allowed the construction of Fig. 4-19 to be used for design of 4 stage nitrification systems, the most commonly employed configuration.[74] It can also be employed for other numbers of stages using the relative capacities shown in the figure. The relative capacity factor should be applied to the hydraulic loading to obtain design values for situations where other than 4 stages are employed.

Very little test data is available for temperatures below 13 C. For applications below 13 C, the provisional recommendation has been made that the temperature correction factors

FIGURE 4-18

NITRIFICATION RATES AS A FUNCTION OF STAGE EFFLUENT CONCENTRATION (AFTER ANTONIE (74))

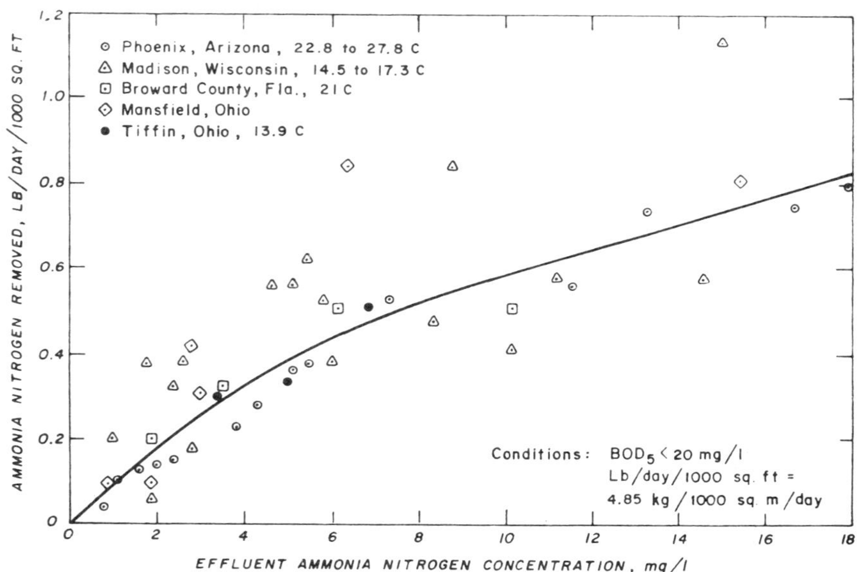

developed for combined carbon oxidation-nitrification (Section 4.4.2.2) be applied for separate stage nitrification.[74]

Figure 4-19 may also be used for hour-by-hour analysis of the effects of diurnal variations in flow on effluent quality.[74] This may tend to overestimate effluent quality during peaking periods, however. To ensure that severe ammonia bleedthrough does not occur during peak load periods, it would appear prudent to adopt the rule formulated in Section 4.7.1.5 for trickling filters. Namely, the surface area determined from Figure 4-19 should be multiplied by the ammonia nitrogen peaking ratio to establish the design surface area.

FIGURE 4-19

DESIGN RELATIONSHIPS FOR A 4-STAGE RBD PROCESS TREATING SECONDARY EFFLUENT (AFTER ANTONIE (74))

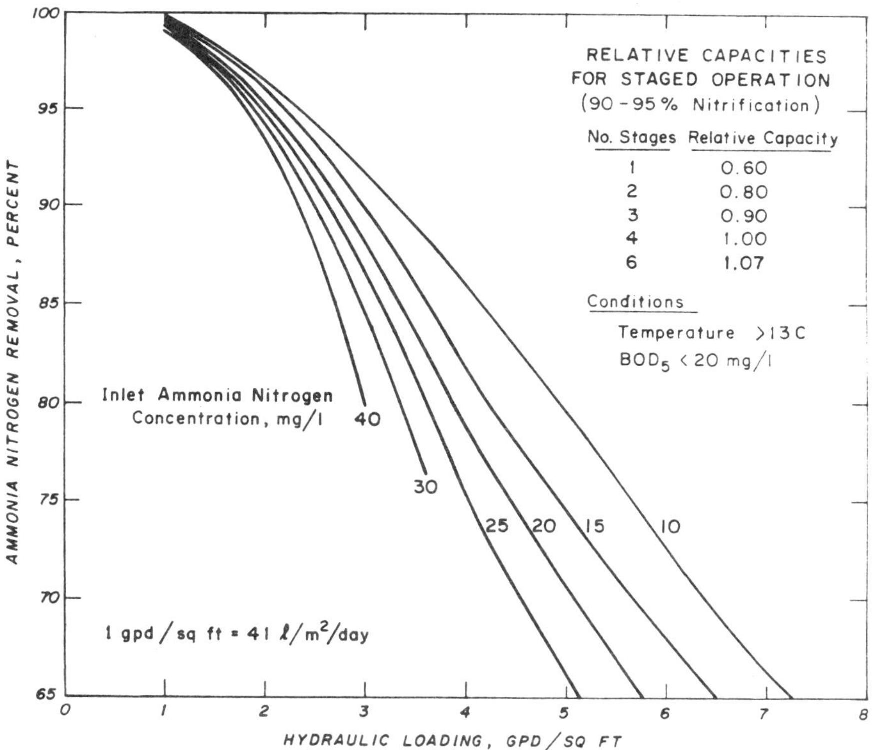

4.7.3 Nitrification with Packed-Bed Reactors

Packed bed reactors (PBR) for nitrification are a comparatively recent development, having progressed from the laboratory stage to pilot-scale and commercial availability in a period of only 5 years.[23,24,75,76,77,78,79]

Figure 4-20 shows one design.[77,80] A PBR consists of a bed of media upon which biological growth occurs overlaying an inlet chamber, much as in an upflow carbon column or filter. Wastewater is distributed evenly across the floor of the PBR by baffles, nozzles or strainers, similar to the way backwash water is distributed in down flow rapid sand filters. The wastewater flow is upward, and a nitrifying biological mass is developed on the large surface area of the media.

FIGURE 4-20

SCHEMATIC DIAGRAM OF A PACKED-BED REACTOR (PBR). (AFTER YOUNG, ET AL., REF 77)

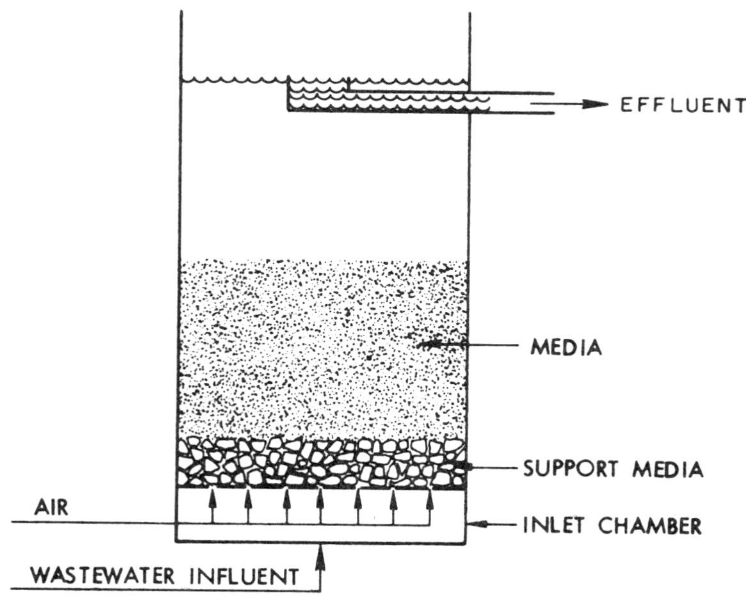

4.7.3.1 Oxygenation Techniques

Several means have been employed for supplying the necessary oxygen for nitrification. The earliest work used injection of air into the feed line entering the chamber.[79] A subsequent pilot-scale investigation used a similar procedure, excepting that the air was distributed across the PBR floor, as shown in Figure 4-20.[77] High purity oxygen has been used in two alternative procedures.[23,75,76] In one the oxygen was bubbled directly into the PBR. In the second procedure, the liquid was preoxygenated in a reaction chamber prior to entry into the PBR. Since preoxygenation is limited to satisfying the oxidation of about 10 mg/l NH_4^+-N due to the solubility of oxygen in water, effluent was recycled at a 2 to 3:1 ratio to provide sufficient oxygen for nitrification.

4.7.3.2 Media Type, Backwashing and Loading Criteria

Several types of media have successfully performed in the PBR including 1-1.5 in. (2.5 to 3.8 cm) stones, 0.5 cm gravel, 1.8 mm (effective size) anthracite and 9 cm "Maspac," a plastic dumped media manufactured by the Dow Chemical Company, Midland, Michigan.[76,75,23,81,77]

In the studies using the relatively light density anthracite and Maspac where air was injected directly into the PBR, no backwashing was found to be necessary due to the turbulence developed in the bed.[77] Despite this, the General Filter Company recommends that when anthracite is used, provision be made for increasing the hydraulic loading in surges for 1 to 2 hours to about four times the average rate, with air at a rate of 0.5 to 1.0 scfm/sq ft (2.54 to 5.08 $l/s/m^2$). The frequency of the surging will vary, depending on influent quality and flow rate. With the plastic media, the frequency of the surging can be reduced considerably because of the high void volume, and in most cases excess solids can be withdrawn simply by draining or backflushing the unit on a monthly or less frequent schedule.[82]

With the gravel media, standard practice was to backwash the reactor at 25 gpm/ft^2 (127 $l/s/m^2$) at least three times per week and in some cases daily.[81] In the studies with the stone media, backwashing was required with both direct and pre-oxygenation. Gravity draining at 6 to 20 gpm/sq ft (30 to 102 $l/s/m^2$) once or twice per week was sufficient to prevent clogging.[23,24]

Data available for formulation of design criteria for PBR units are summarized in Table 4-17. Oxidation rates fall in the range of 4 to 27 lb NH_4^+-N oxidized per 1000 cu ft/day (0.06 to 0.43 $kg/m^3/day$). Factors affecting the oxidation rate are the influent quality (BOD_5, TKN and NH_4^+-N), temperature, and the type of media selected as a biological growth surface. Oxidation rates at Pomona, Ca. were much greater than those at Ames, Iowa at the same temperature, which is very probably due to the higher BOD_5 and lower ammonia content of the Ames secondary effluent. Very likely, there was a higher fraction of nitrifiers in the Pomona biofilm. Interestingly, chemically clarified raw sewage (BOD = 93 mg/l) was compared to secondary effluent (COD = 46) at Pomona and only 60 percent nitrification was achieved with the chemically clarified feed at the detention time sufficient to produce virtually complete nitrification of the secondary effluent.[81] It is very probable that this reduced efficiency was caused by a higher fraction of heterotrophs being present when chemically clarified wastewater was being treated.

Temperature has a strong effect on the PBR process. Figure 4-21 shows the detention time required for relatively complete nitrification (< 2 mg/l NH_4^+-N) at steady state as a function of temperature. If Figure 4-21 is used for sizing a PBR, attention must also be given to diurnal variation in nitrogen loads. It would be prudent to multiply the detention time determined from Figure 4-21 by the peak to average nitrogen load ratio to establish the design detention time. This should present extremes in ammonia bleedthrough during diurnal peak conditions.

The media type chosen affects the amount of surface available for nitrifier growth. For instance, Table 4-17 shows that anthracite was superior to Maspac media in terms of oxidation rate at Ames, Iowa; this may be due to the higher surface area of anthracite media compared to Maspac.

FIGURE 4-21

TEMPERATURE DEPENDENCE OF DETENTION TIME FOR COMPLETE NITRIFICATION, (<2 mg/l NH_4^+-N) AT STEADY STATE IN THE PBR

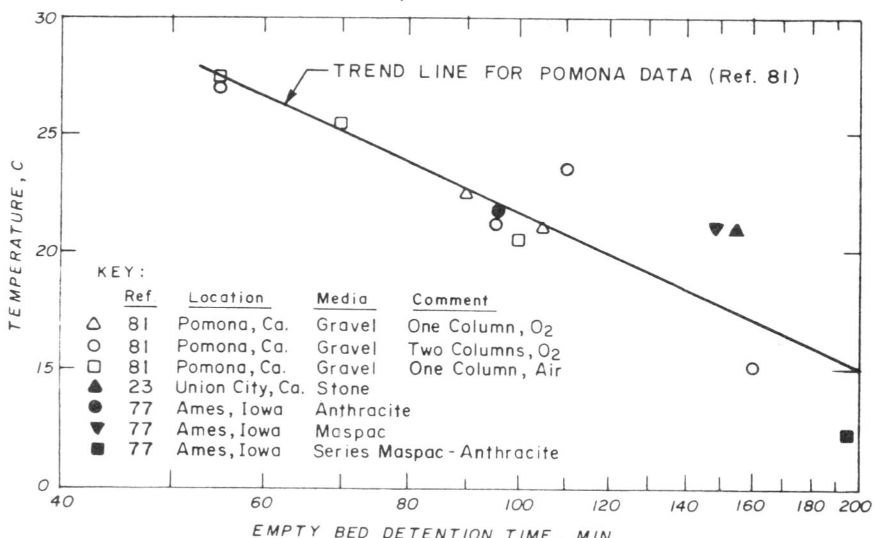

TABLE 4-17

PACKED BED REACTOR PERFORMANCE WHEN TREATING SECONDARY EFFLUENTS

Location and type of aeration	Ref.	Media depth, ft (m)	Media type	Surface loading gpm/sf[a] (m³/m²/day)	Empty bed hydraulic detention time, min[a]	Ammonia-N oxidation rate lb/1000 cu ft/day (kg/m³/day)	Temp., C	Influent Quality, mg/l				Effluent Quality, mg/l						Removals, percent			
								BOD_5	SS	Organic-N	Ammonia-N	BOD_5	SS	Organic-N	Ammonia-N	Nitrite-N	Nitrate-N	BOD_5	SS	Organic-N	NH_4^+-N
Union City, Ca. Preoxygenation (oxygen)	23, 24	3.0	1 to 1.5 in	.15 (8.8)	154	7.7 (.12)	21 to 27	35	27	3.6	14.3	5	4	1.5	1.0	0.6	15.9	86	87	58	93
		(.91)	(2.5 to 3.8 cm) stone	.21 (12)	103	12.1 (.19)	21 to 27	38	38	-	19.6	3	7	-	5.6	4.1	6.9	91	83	-	71
				.29 (17)	77	9.8 (.15)	21 to 27	37	25	5.7	15.2	10	7	2.5	6.9	1.7	6.0	74	74	56	55
Bubble (oxygen)	23, 24	3.0 (.91)	1 to 1.5 in (2.5 to 3.8 cm) stone	.15 (8.8)	154	9.7 (.15)	16 to 30	37	30	5	18.3	10	16	2.2 - 4.7	1.8	0.4	18.3	74	48	-	91
				.29 (17)	77	-	16 to 30	43	48	-	-	25	51	-	-	-	-	41	-6	-	-
Ames, Iowa Bubble (air)	77, 83	5.0 (1.5)	1.8 mm (D_{10}) U.C. = 1.7 anthracite	1.0 (59)	37	9.4 (.15)	21 to 23	39	43	-	8.4	19	-	-	5	-	-	51	20	-	46
				0.4 (23)	94	5.9 (0.09)		20	26	-	6.8	5	-	-	1	-	-	77	32	-	90
		8.0 (2.4)	Maspac[b]	1.0 (59)	60	5.1 (0.08)	21 to 23	39	43	-	8.4	19	-	-	5	-	-	51	42	-	40
				0.4 (23)	150	3.5 (0.06)		20	26	-	6.8	8	-	-	1	-	-	59	40	-	87
		13 (4.0)	Series anthracite Maspac[b]	0.5 (29)	195	6.1 (0.10)	9 to 14	26	47	-	14.4	8	-	-	1	-	-	68	49	-	92
				0.75[b] (44)	130	4.6 (0.07)	12	37	63	-	11.2	16	-	-	5	-	-	56	39	-	59
Pomona, Ca. Bubble (oxygen or air)	81	5.5 (1.7)	5 mm gravel	0.75 (44)	55	26.5 (0.42)	27 to 28	7[c]	9[c]	-	18.1	-	-	-	1.9	0.6	16.6	-	-	-	90
				0.59 (35)	70	20.1 (0.32)	25 to 26			-	17.6	-	-	-	1.4	0.6	16.3	-	-	-	92
				0.41 (24)	100	13.3 (0.21)	19 to 22			-	16.8	-	-	-	2.0	0.9	16.9	-	-	-	88
				0.46 (27)	90	14.7 (0.24)	20 to 25			-	16.4	-	-	-	1.7	0.4	15.6	-	-	-	90
				0.39 (23)	105	16.4 (0.26)	20 to 22			-	20.7	-	-	-	1.5	0.5	20.7	-	-	-	93
		11.0 (3.4)	5 mm, gravel, two columns in series	1.49 (87)	55	26.7 (0.43)	26 to 28			-	17.6	-	-	-	1.3	0.4	17.1	-	-	-	93
				0.75 (44)	110	13.8 (0.22)	22 to 25			-	18.9	-	-	-	1.9	0.2	18.2	-	-	-	90

[a] Basis: influent flow

[b] A product of the Dow Chemical Co., Midland, Mich.

[c] Average for test series treating activated sludge effluent

Effluent BOD and SS levels are affected by the type of aeration (see Table 4-17). Preoxygenation allows the PBR to produce effluents of similar quality to tertiary multimedia filtration. Bubble aeration, however, causes continuous shearing of the biological film from the media, resulting in lower reductions of BOD and suspended solids. When very low levels of effluent solids are required, effluent filtration may be required when bubble aeration is used in the PBR.

4.8 Aeration Requirements

Care must be exercised in designing aeration systems for nitrification. Unlike BOD, ammonia is not adsorbed to the biological floc for later oxidation, and therefore the ammonia must be oxidized during the relatively short period it is in the nitrification reactor. Therefore, sufficient oxygen must be provided to handle the load impressed on the nitrification process at all times. This problem is particularly critical when either activated sludge or a packed bed reactor system is used for nitrification; in the other attached growth systems the aeration is provided as the liquid spills over the media and the design considerations relate to proper ventilation rather than oxygen transfer.

Very significant diurnal changes in nitrogen load have been observed. Load variations at the Chapel Hill treatment plant are shown in Fig. 4-3 and the load pattern would be representative of systems provided with no significant in-process flow equalization. In this case, peak to average nitrogen load rose to nearly 2.2, considerably above the peak to average flow ratio of 1.44. As an example of a plant with some in-process flow equalization, Fig. 4-12 shows the load variations observed in the activated sludge effluent of the Rancho Cordova Wastewater Treatment Plant. In this case the nitrogen peaking is moderated by the equalization in quality provided by the activated sludge aeration tanks and secondary clarifiers. The ammonia peak to average ratio at 1.63 approximates the flow peak to average ratio of 1.57.

In addition to being affected by in-process flow equalization, the diurnal variation in nitrogen loading is also very significantly affected by equalization in the wastewater collection system. Large collection systems serving spreadout urban areas have high built-in storage providing unintentional flow and quality equalization. This relationship is indicated in Fig. 4-22 where the nitrogen load peaking (expressed as the ratio of the maximum hourly load to average load) is plotted for eight treatment plants having no significant in-process equalization. There is an interesting relationship between flow peaking and ammonia load peaking shown in Figure 4-22. In large plants such as Blue Plains plant at Washington, D.C. and Sacramento, California a spread out collection system causes moderation of both flow and nitrogen load peaking. In the smaller systems, however, without such "flow equalization," ammonia load peaking can be substantial; for example at the Central Contra Costa Sanitary District's (CCCSD) plant an hourly peaking factor of 2.4 has been measured. The aeration system must accommodate these changes in loads to avoid ammonia bleeding through during the peak load period. The diurnal variations in load can be quite extreme; in Figure 4-23 the peak to minimum hourly loads are plotted against the flow peaking factor. Ratios as high as 10:1 have been observed.

FIGURE 4-22

RELATION BETWEEN AMMONIA PEAKING AND HYDRAULIC PEAKING LOADS FOR TREATMENT PLANTS WITH NO IN-PROCESS EQUALIZATION

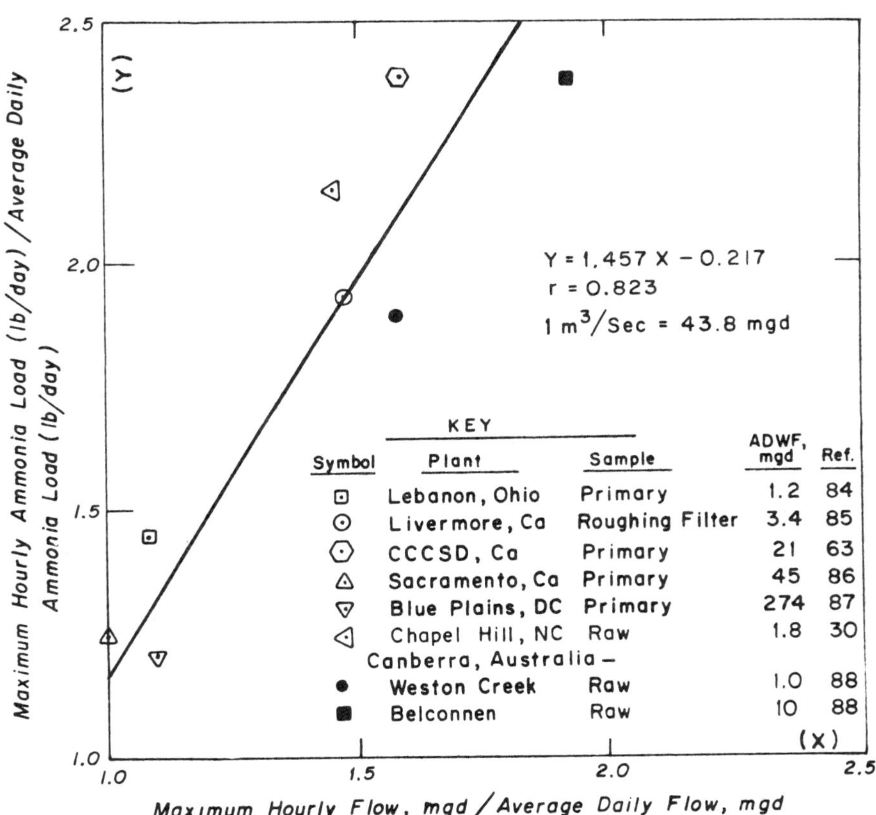

An early decision must be made during the design process as to what level of peaking of oxygen demanding substances will be designed for. In addition to peaking of ammonia or organic nitrogen, a concurrent peak may also occur in the loading of organic substances. If very low levels of ammonia nitrogen are required at all times care must be used to develop a statistical base whereby the frequency of peak oxygen loads can be identified. Not only should daily peaks be identified, but possibly those occurring on weekly or monthly bases. Table 4-18 presents an example of such an analysis for the primary effluent from two plants in St. Louis, Missouri using COD as a measure of oxygen demanding substances since in that particular instance nitrification was not required. As may be seen, significant departures from average conditions occur on a fairly frequent basis. Similar analyses may be justified when designing for nitrification.

The extra aeration capacity required for handling diurnal variations in nitrogen load, coupled with the extra tankage and equipment required, may dictate in-plant flow equalization in many instances. The reductions in capital and operating cost of aeration tankage and aeration facilities must be compared with the cost of flow equalization to determine applicability to specific cases. Design procedures for flow equalization are contained in Chapter 3 of the *Process Design Manual for Upgrading Existing Wastewater Treatment Plants.*[25]

FIGURE 4-23
RELATIONSHIP OF MAXIMUM/MINIMUM NITROGEN LOAD RATIO TO MAXIMUM/AVERAGE FLOWS

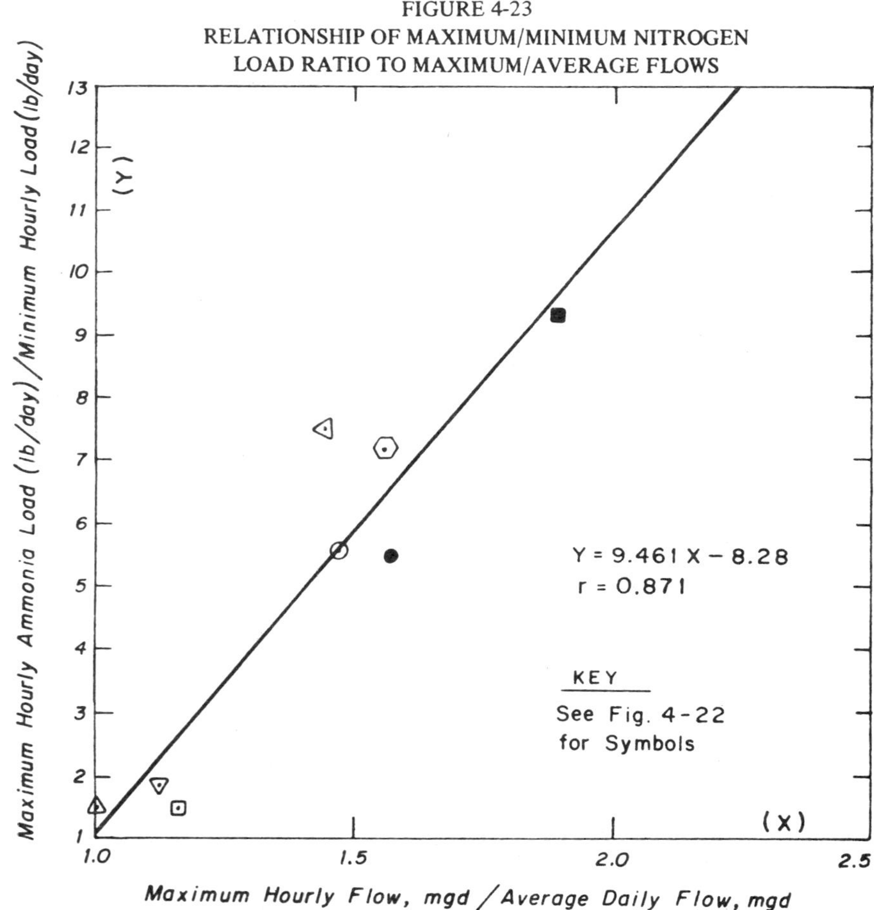

TABLE 4-18

PEAKING FACTORS VERSUS FREQUENCY OF OCCURRENCE FOR PRIMARY TREATMENT PLANT EFFLUENT

Frequency of Occurrence	Bissel Point Treatment Plant[a]		Lemay Treatment Plant[a]	
	COD load peaking factor[b]	Flow peaking factor[b]	COD load peaking factor[b]	Flow peaking factor[b]
4 hours/day	1.30	1.18	1.40	1.20
4 hours/week	1.72	1.40	1.85	1.44
4 hours/month	2.02	1.60	2.35	1.70
4 hours/3 months	2.25	1.72	2.62	1.88
4 hours/6 months	2.40	1.80	2.80	1.96

[a]Data is from reference 89; both plants serve the St. Louis area in Missouri and process about 100 mgd each.

[b]Peaking factor is defined in each case as the ratio of the 4 hour load listed to the average daily load.

4.8.1 Adaptability of Alternative Aeration Systems to Diurnal Variations in Load

Careful consideration should be given to maximizing oxygen utilization per unit power input. In the face of significant load variation, the aeration system should be designed to match the load variation while economizing on power input. Obviously, designing the aeration system to provide for the maximum hourly demand 24 hours a day would provide over aeration the majority of the time with wasteful losses of power.

The available means for aeration are summarized in Section 5.3.4 of the *Process Design Manual for Upgrading Existing Wastewater Treatment Plants*, a Technology Transfer publication.[25] Of the available aeration devices, the mechanical surface aerator is least well suited to nitrification applications. This is because they are normally designed to operate at fixed speed and therefore must overaerate the majority of the day to satisfy the peak oxygen demands. Even when equipped with variable submergence of the blade, the units are limited to matching less than a 2:1 variation in load, at best. Therefore, unless flow equalization is provided somewhere in the system, mechanical surface aerators are not capable of matching variations in nitrogen loads without overaerating the mixed liquor during a significant portion of the day.

Using diffused air aeration, air rates can be easily modulated to closely match the load, merely by turning down or shutting off individual blowers. Thus, the diurnal load variations can be matched without the necessity of over aerating the mixed liquor and wasting power. Fine bubble diffusers can be arranged across the tank floor,[90] allowing fairly even distribution of energy input. Gentler mixing is provided than with mechanical aeration plants, providing less tendency for floc breakup.

Submerged turbine aeration systems are intermediate in terms of their responsiveness to the problem of aeration in nitrification systems. Because of their capability to vary the air rate to the sparger, they may be designed to match the load variation in oxygen demand. A drawback, however, is that the impeller normally operates at fixed speed, imparting no turn down capability for a significant part of the power draw. In fact, in some impeller designs, the power draw of the ungassed impeller is actually greater than when gas is fed to the unit.

4.8.2 Oxygen Transfer Requirements

Oxygen requirements for nitrification alone were discussed in Section 3.2.4. Oxygen requirements in all practical cases are compounded by the oxygen required for stabilization of organics.

Reasonably exact expressions for oxygen requirements for heterotrophic organisms and nitrifiers have been developed.[38] The approach, however, requires pilot plant data to provide COD balances and sludge yields. In general, this information is not available and a simpler approach may be adopted.

In normal activated sludge treatment when nitrification is not required, the amount of oxygen needed to oxidize the BOD_5 can be calculated by the following equation:

$$B = X(BOD_5) \tag{4-25}$$

where: B = oxygen required for carbonaceous oxidation, mg/l
X = a coefficient

The coefficient X relates to the amount of endogenous respiration taking place and to the type of waste being treated. For normal municipal wastewater, the X value would range from .5-.7 for high rate activated sludge systems to 1.5 for extended aeration. For conventional activated sludge systems X can be taken as 1.0.

In the case of nitrification, the oxygen requirement for oxidizing ammonia must be added to the requirement for BOD removal. The coefficient for nitrogen to be oxidized can be conservatively taken as 4.6 times the TKN content of the influent (Section 3.2.4) to obtain the nitrogen oxygen demand (NOD) and the value of X in Equation 4-25 can be assumed to be approximately 1.0. In actual fact, some of the influent nitrogen will be assimilated into the biomass or is associated with refractory organics and will not be oxidized. These assumptions lead to the following oxygen requirement:

$$W = BOD_5 + NOD \qquad (4\text{-}26)$$

where: W = the total oxygen demand mg/l, and
NOD = oxygen required to oxidize a unit of TKN taken as 4.6 times the TKN

Since aeration devices are rated using tap water at standard conditions, the rated performance of the aerator must be converted to actual process conditions by the application of temperature corrections and by factors designated α and β which relate waste characteristics to tap water characteristics.

Temperature corrections are made by the relationship:

$$1.024^{(T-20)} \text{ where } T = \text{process temperature in degrees C.}$$

The α factor is the ratio of oxygen transfer in wastewater to that in tap water and is represented by the following:

$$\alpha = \frac{K_L a \text{ (process conditions)}}{K_L a \text{ (standard conditions)}} \qquad (4\text{-}27)$$

Values of α can vary widely in industrial waste treatment applications, but for most municipal plants, it will range from 0.40 to 0.90.

The β factor is the ratio of oxygen saturation in waste to that in tap water at the same temperature. A value of 0.95 is commonly used. Thus, the actual amount of oxygen required to be transferred (W) can be determined from the amount transferred under test conditions (W_O) by the equation:

$$W = W_O \alpha (1.024)^{(T-20)} \frac{\beta C_s - C_1}{9.2} \qquad (4\text{-}28)$$

where: W = oxygen transferred at process conditions, lb/day
W_O = oxygen transferred at standard conditions (T = 20 C, DO = .01 mg/l, tap water), lb/day
T = process temperature, C.
C_s = oxygen saturation in water at temperature T, mg/l
C_1 = process dissolved oxygen level, mg/l

The process dissolved oxygen level, C_1, must be set high enough to prevent inhibition of nitrification rates (Section 3.2.5.5). For this purpose, a minimum value of 2.0 mg/l is recommended. This value is also applicable under peak diurnal load conditions, and the practice of allowing the DO to drop below 2.0 mg/l under peak load is not recommended. If the DO were to be allowed to drop below 2.0 mg/l during peak load conditions, excessive bleedthrough of ammonia could be expected.

Using example values for domestic sewage ($\alpha = 0.9$, $\beta = 0.95$, $C_1 = 2.0$, T = 20 C and $C_s = 9.2$) in Equation 4-28, the relationship between oxygen required under test conditions and that required under actual process conditions is:

$$W_O = 1.5\ W$$

W_O can then easily be converted to horsepower (hp) requirements for mechanical aerators or to volumetric air flow rates for diffused air plants. The latter is accomplished by the equation:

$$Q_A = W_O \times \frac{100}{23} \times \frac{1}{0.75} \times \frac{1}{1440} \times \frac{100}{e_o} \qquad (4\text{-}29)$$

where: Q_A = air flow, cfm,

e_o = aerator rated oxygen transfer efficiency at standard conditions, percent,

air composition = 23 percent oxygen (weight basis), and
air density = 0.75 lb/cf.

Applying the above equation and the numbers in the previously stated example to diffusers of various efficiencies produces air rates of from 1500 to 600 cu ft per lb BOD_5 + NOD (99 to 43 m^3/kg) corresponding to diffuser efficiencies of 6 to 15 percent as shown in Figure 4-24.

Diffuser placement within the aeration tank has been shown in full-scale tests to have a large effect on oxygen transfer efficiency.[91] For oxygen transfer efficiencies, diffusion system design and layout, the reader should refer to the literature as well as to air diffusion equipment manufacturers.

Mechanical aerators are commonly rated in terms of pounds of oxygen transferred at standard conditions per brake hp (bhp) hour. Diffusers are rated by their oxygen transfer efficiency expressed as a percentage of oxygen supplied. The two can be approximately compared by the values given in Table 4-19.[92] Mechanical surface aerators typically have aerator power efficiencies of about 3.2 O_2lb/bhp/hr.

Actually measured values of diffused air requirements are summarized in Table 4-20 for eight treatment plants. As may be seen, air usage is within the range expected from

FIGURE 4-24

RELATIONSHIP OF AERATION AIR REQUIREMENTS FOR OXIDATION OF CARBONACEOUS BOD AND NITROGEN

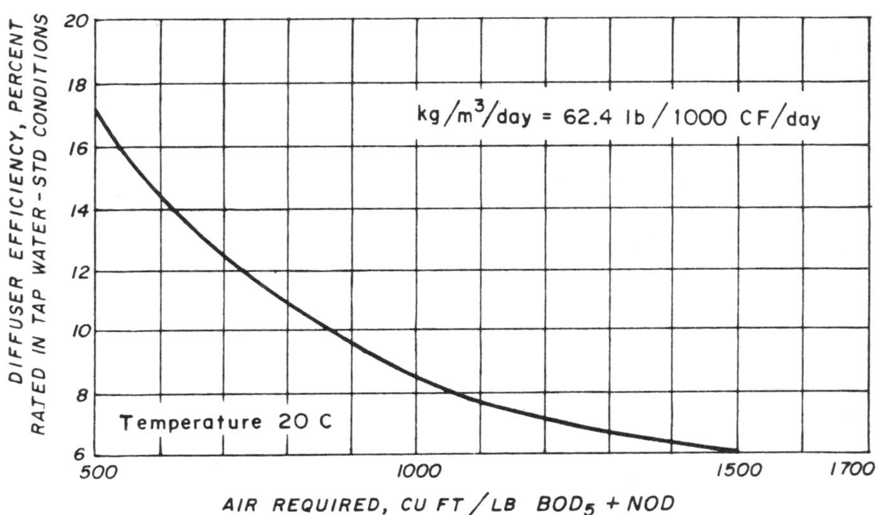

theoretical considerations for coarse bubble aeration systems. It should be noted that none of the plants listed in Table 4-20 have automated blower control systems linked to DO probes. When the Livermore, California plant staff manually adjusted the blower output according to the DO probe reading, they were able to reduce the air requirement to an average for the day of 680 CF/lb BOD_5 + lb NOD (42 m^3/kg) based on hourly variations in

TABLE 4-19

RELATION OF OXYGEN TRANSFER EFFICIENCY TO AERATOR POWER EFFICIENCY (AFTER NOGAJ (92))

Diffuser oxygen transfer efficiency, percent (at standard conditions)	Aerator power efficiency lb O_2/bhp/hr
4	1.23
6	1.85
8	2.46
10	3.08
12	3.70

load.[85] Thus, it is possible that automated blower operation may reduce aeration requirements.

In addition to determining the total air requirement, attention should also be given to air distribution within the aeration tanks. If the conventional (or plug flow) mode of operation is established as the normal operating procedure, the air requirements will be greatest at the head end of the aeration tanks (see Section 4.3.5) .

TABLE 4-20

AIR REQUIREMENTS FOR NITRIFICATION ACTIVATED SLUDGE PLANTS

Treatment plant	Configuration	Diffuser type	Oxygen demand distribution, %		cu ft/lb of BOD_5 and NOD[a]	Data reference
			BOD_5	NOD		
Medford, Oregon	Plug flow or step aeration	Coarse bubble	73	27	1390	93
Flint, Michigan	Plug flow or step aeration	--	65	35	1280	6
Livermore, Calif.	Separate stage: roughing bio-filter followed by nitrification	Coarse bubble	50	50	1250	5
Central Contra Costa Sanitary District, Calif.	Separate stage: lime followed by nitrification	Coarse bubble	21	79	1700	3
Jackson, Mich.	Plug flow	Coarse bubble	66	34	910	42
Hyperion Treatment Plant West Battery, Los Angeles, Ca.	Plug flow	Coarse bubble	61	39	1160	2
Whittier Narrows, County Sanitation Districts of Los Angeles County, Ca.	Step aeration	Coarse bubble	59[b]	41	1180[b]	9
San Pablo Sanitary District Treatment Plant, Ca.	Plug flow; roughing filter followed by nitrification	Coarse bubble	59	41	1410[c]	94

[a] 1 cu ft/lb = .062 m^3/kg

[b] Assuming primary effluent BOD_5/COD ratio of 0.6

[c] During June 1974 when nitrification run at design loads

4.8.3 Example Sizing of Aeration Capacity

As a design example, consider a 10 mgd plant with lime clarification in the primary sedimentation stage:

Given the following primary effluent properties:

Design flow = 10 mgd
Average BOD_5 load = 4,170 lb/day
Average TKN load = 2,100 lb/day
Peak/average TKN load ratio = 1.8
Peak/average BOD_5 load ratio = 1.5 (coincident with peak TKN load)

Average load condition:

BOD_5 = 4,170 lb/day
NOD = 4.6 x 2,100 = 9,660 lb/day
BOD_5 + NOD = 13,830 lb/day

Peak hourly load condition:

BOD_5 = 4,170 x 1.5 = 6,260 lb/day
NOD = 9,660 x 1.8 = 17,390 lb/day
BOD_5 + NOD = 23,550 lb/day

Ratio peak hour to average load: 1.7

For the purposes of this example assume a fine bubble diffused air aeration with a design figure of 725 cf/lb BOD_5 + NOD.

Average aeration requirement:

725 x 13,830/1,440 = 6,963 CFM

Peak aeration requirement:

725 x 23,550/1,440 = 11,860 CFM

In sizing plant air requirements, separate provision must be made for preaeration, air in mixed liquor and return sludge channels, and air requirements for aerated stabilization in downstream denitrification processes.

4.9 pH Control

The implications of adverse operating pH and its causes have been discussed previously in Sections 3.2.3 and 3.2.5.6. In cases where the alkalinity of the wastewater will be depleted by the acid produced by nitrification, the natural alkalinity of the wastewater will have to be supplemented by chemical addition. As discussed in Section 4.5.1, the effects of operation of upstream processes on alkalinity and pH must be considered. Alum or iron addition tend to deplete wastewater alkalinity, whereas in some instances lime addition increases the alkalinity.

4.9.1 Chemical Addition and Dose Control

Two alternative chemicals, caustic (NaOH) and lime (CaO or $Ca(OH)_2$) are in predominate use for pH control. As lime is less expensive than caustic for the same change in alkalinity, lime will generally be favored, except in smaller plants. Procedures for the feeding and storage of these chemicals are described in two EPA Technology Transfer publications, *Process Design Manual for Phosphorus Removal*[95] and *Process Design Manual for Suspended Solids Removal.*[96]

The need for pH adjustment may vary diurnally. In one case it was found that an alkalinity deficit occurred daily only for several hours at the peak nitrogen load condition. At other times, sufficient alkalinity was available. This condition caused a cyclic variation in pH. In situations where diurnal variation in the pH depression may be encountered, continuous on-line monitoring of pH for the purpose of controlling chemical addition seems justified. In the case of suspended growth systems operated in the plug flow mode, probes may be positioned at several points in the aeration basin, with provision for addition of chemical at several points.[62] In the case of attached growth sytems and suspended growth systems operated in the complete mix mode, effluent monitoring of pH would be the usual choice for controlling chemical addition to the influent.

4.9.2 Effect of Aeration Method on Chemical Requirements

The type of oxygen transfer device chosen can have a marked effect on the chemical dose required for pH control. As an example, the differences between coarse bubble and fine bubble air diffusion systems will be examined. The following equation from bicarbonate-carbonic acid equilibrium is useful for estimating the pH level in aeration tanks:

$$pH = pK_1 - \log \frac{(H_2CO_3)}{(HCO_3^-)} \tag{3-9}$$

At 20 C, the value of pK_1 is approximately 6.38. In using the equation, the alkalinity in the aeration tank can be used to estimate the bicarbonate level (HCO_3^-) and the value used should reflect any alkalinity depletion resulting from nitrification. The level of H_2CO_3 (carbon dioxide in solution) can be estimated from Henry's Law as follows:

$$C_{eq} = H\,P_{gas} \tag{4-30}$$

where:

C_{eq} = concentration of gas dissolved in liquid at equilibrium, mg/l,

H = Henry's Law constant, mg/l/atmosphere

P_{gas} = partial pressure of gas in equilibrium with the liquid, atmosphere

The Henry's law constant for carbon dioxide at 20 C is 1688 mg/l/atmosphere. To establish C_{eq} for use in Equation 3-9, the level of carbon dioxide in the aeration air must be estimated. In unpolluted atmospheric air, the partial pressure of carbon dioxide is only about 0.0003 atm. However in aeration air, carbon dioxide is generated from the oxidation of organics and from nitrification. For the mix of organics in municipal wastewaters, about one mole of carbon dioxide is produced per mole of oxygen consumed. Similarly for nitrification, about one mole of carbon dioxide is produced per mole of oxygen consumed (c.f. Section 3.2.2). On a weight basis, for every lb of oxygen consumed about 1.38 lb of carbon dioxide is produced. Equations 3-9 and 4-30 can be solved simultaneously to determine the amount of dissolved carbon dioxide in the liquid and the amount present in the gas as well as the process operating pH.

The following example is illustrative of the procedure. Assume a residual alkalinity of 150 mg/l as $CaCO_3$ (3×10^{-3} moles/liter of HCO_3^-) and an oxygen transfer efficiency of 12 percent, under standard conditions. Equation 4-29 would indicate that 725 cu ft/lb BOD_5 + NOD is required. Assume also that the example conditions given in Section 4.8.3 are used, namely that 13,830 lb of oxygen demand are contained in 10 million gallons of wastewater; this allows the calculation that 1 lb of oxygen demand is contained in 725 gallons. Therefore the aeration rate is 1 cu ft/gal of wastewater nitrified (725 cu ft/725 gal). Assuming a pH of 7.1 (by trial and error), Equation 3-9 allows calculation of the dissolved carbon dioxide concentration, and for this case it is found to be 25 mg/l. Therefore at this concentration, 725 gallons contain 0.15 lb of carbon dioxide (CO_2) of the total evolved (1.38 lb of CO_2 evolved/lb of oxygen). Equation 4-30 allows calculation of the partial pressure of CO_2 in the off gases; the calculated result is 0.0149 atm. This concentration, plus the volume of gas required allows calculation that the gas contains 1.23 lb of carbon dioxide. The total is then 1.38 lb which checks with the expected result. If the total had not been 1.38 lb, a new pH value level would be assumed and the procedure tried again until a balance is obtained.

Table 4-21 presents the results for a number of oxygen transfer efficiency values, using the procedure outlined above. The trends shown are valid for municipal wastewaters, although the absolute value of the pH may differ slightly from those indicated due to variation from the assumed conditions.

Examining Table 4-21, it can be seen that fine bubble diffuser systems, at high oxygen transfer efficiency, will operate at lower operating pH levels than coarse bubble diffusers operated at lower oxygen transfer efficiencies. If the same operating pH is to be maintained, this will translate into higher chemical doses for fine bubble aeration systems. For instance, Table 4-21 shows that if the residual alkalinity is 100 mg/l, a pH of 7.0 can be maintained with a coarse bubble aeration system rated at 9 percent efficiency under standard conditions. For a comparable fine bubble system rated at 18 percent under standard conditions, the residual alkalinity would have to be raised to 150 mg/l as $CaCO_3$ to maintain the pH at 7.0. This would require an extra lime dose of at least 28 mg/l as CaO.

TABLE 4-21

EFFECT OF OXYGEN TRANSFER EFFICIENCY AND RESIDUAL ALKALINITY ON OPERATING pH

Residual alkalinity as $CaCO_3$, mg/l	pH at stated Operating transfer efficiency, percent[a]			
	Coarse bubble range		Fine bubble range	
	6	9	12	18
50	6.9	6.7	6.6	6.5
75	7.1	6.9	6.8	6.7
100	7.2	7.0	6.9	6.8
125	7.3	7.1	7.0	6.9
150	7.4	7.2	7.1	7.0
175	7.4	7.3	7.2	7.0
200	7.5	7.3	7.2	7.1

[a]at standard conditions

As an example of the use of Table 4-21, presume that it is desired to maintain the operating pH level at 7.2. Assume that an aeration system has been chosen which has an oxygen transfer efficiency, e_o, of 12 percent and residual alkalinity of 75 mg/l as $CaCO_3$. If it is desired to maintain the process pH at 7.2 to prevent inhibition of nitrification rates, Table 4-21 indicates that a residual alkalinity of 175 mg/l is required. The difference between the available alkalinity and the required alkalinity is 100 mg/l as $CaCO_3$ and this could be supplied by a lime dose of 56 mg/l (as CaO).

In plug flow systems, the pH will steadily decline from the influent end to the effluent end, following the similar decline in alkalinity occurring due to nitrification. The most severe pH depression will be at the effluent end, after the bulk of the nitrogen has been oxidized. Complete mix systems, on the other hand, will be uniformly depressed in pH throughout the tank because of the uniformity of aeration tank contents. This is a disadvantage for nitrification with complete mix activated sludge compared to plug flow as the pH of the entire complete mix tank will be the same as the pH at the effluent end of the plug flow tank at the same oxygen transfer efficiency.

The preceding discussions about the effect of carbon dioxide evolution in operating pH in the aeration tank are not applicable to cases where nitrification follows lime treatment. In these cases, the carbon dioxide that is evolved is used in recarbonation reactions and very little enters the aeration air (c.f. Section 4.5.1).

4.10 Solids-Liquid Separation

In all suspended growth systems and in most attached growth systems, the nitrification stage must be followed with a solids removal stage. Because of the complexity of the solids-liquid separation problem, full consideration cannot be given to it within the scope of this manual. Rather, a brief review of the problem is given with reference to the pertinent literature.

In some attached growth nitrification systems, multimedia filtration is provided following the nitrification stage as the level of effluent solids is low. Design criteria for multimedia filtration are presented in the EPA Technology Transfer Publication, *Wastewater Filtration, Design Considerations.*[97] General design criteria for secondary sedimentation as well as multimedia filtration applicable to nitrification processes are presented in the *Process Design Manual for Upgrading Existing Wastewater Treatment Plants.*[25]

In suspended growth systems, there is a strong interrelationship between operation of the secondary clarifier and operation of the aeration basin. The degree to which the return sludge can be thickened will affect the allowable mixed liquor in the aeration tank and therefore the size of the aeration tank. Dick and coworkers have presented design relationships which are useful for analyzing clarifier-aeration tank interactions.[98,99,100]

Consideration of several factors is required when designing secondary sedimentation tanks. Mechanical design of the tank is very important. Inlet turbulence must not upset the tank; an energy dissipating inlet well serves the purposes of distributing the flow equally across the tank and of providing for some mild turbulence to help aggregate the finely divided solids into floc and increase the separation efficiency. Effluent launders must be sized and placed so that currents are not created that will upset the tank and cause short-circuiting of the solids to the effluent. The tank must also have sufficient depth to allow a sludge blanket to form under all conditions, especially under those conditions which occur when the system operation is limited by the ability to obtain a specific return activated sludge concentration. Overflow rates must not be so great that sludge is suspended in the "upflow" and carried over the weirs. Lastly, the sludge must be quickly removed from the tank to minimize the ocurrence and duration of anoxic (zero DO) conditions.

The nature of the biological solids developed in suspended growth systems plays an important role in the operation and design of the secondary sedimentation tank. First, the ability of the mixed liquor to be clarified will affect the size of the required sedimentation tank. This property is dependent on the extent of disperson of the biological solids. If a large proportion of the biological solids are finely divided, the separation efficiency of these solids will be poor. If, however, only small portions of the biological solids are dispersed and the bulk is incorporated into floc, the separation efficiency of the mixed liquor should be high. Consolidation of the small dispersed particles into the large floc to improve effluent clarity can be encouraged by a number of techniques. Chemical coagulation of the dispersed solids has been successfully employed to enhance the flocculation of biological solids.[101] Physical conditioning of the activated sludge floc prior to secondary sedimentation is another technique which can be used with or without chemicals. It has been found that by incorporating mild turbulence beweeen the oxidation tank and the secondary sedimentation tank, effluent clarification can be enhanced.[102] Mild turbulence can be conveniently accomplished by using mild aeration in transfer channels and by incorporating energy-dissipating inlet wells in circular secondary sedimentation tanks.

The thickening qualities of the sludge will determine the required area for sludge thickening in the tank, depending on the desired or optimal MLSS or Return Activated Sludge (RAS) levels. The limiting design consideration for thickening is usually at peak wet weather flow (PWWF), for it is under these conditions that the solids loading on the secondary sedimentation tanks are the greatest. Secondary sedimentation tanks must be sized such that mixed liquor solids are not lost in the effluent during PWWF conditions. Polymer feed facilities are appropriate to allow short-term polymer dosing during wet weak peak load conditions to enhance the thickening qualities of the sludge.

Thickening qualities will decrease and sedimentation tank area requirements will increase when wastewater temperatures decline. Figures 4-25 and 4-26 show the effect of temperature on three different oxygen activated sludges and the temperature trends would be expected to be similar for air activated sludges. In sedimentation tank design, solids loadings should reflect the minimum wastewater temperatures expected. When temperatures decline the mixed liquor levels that can be maintained under aeration also decline due to the decreased level of thickening that can be obtained in the sedimentation tank.

Rising sludge caused by denitrification in secondary clarifiers has occasionally plagued nitrification operations. Denitrification occurs because the organisms in the biological sludge in the secondary clarifier can utilize nitrate and nitrite as electron acceptors to oxidize organic compounds in the sludge layer. The formation of bubbles from nitrogen gas evolved in denitrification and other gasses, such as carbon dioxide, can then cause flotation. Early experimental work by Sawyer and Bradney[104] and subsequent experimental and theoretical work by Clayfield [105] have identified the important parameters affecting denitrification. A factor of foremost importance is the presence of adequate quantities of nitrites or nitrates in the wastewater to cause bubble formation. In essence, the dissolved gases in the wastewater (nitrogen, carbon dioxide, and oxygen) must be in sufficient concentration so that the sum of their partial pressures equals or exceeds the ambient liquid pressure. In the case of Sawyer and Bradney's work, as little as 4-6 mg/l of nitrate-N or nitrite-N would cause flotation in a graduated cylinder, while Clayfield found that 16-18 mg/l nitrate-N were needed to cause flotation in full scale sedimentation tanks. Differences between investigations are in part due to the greater liquid pressures in deep sedimentation tanks compared to graduated cylinders and may also reflect differing levels of concentration of other gases, such as carbon dioxide.

The degree of stabilization of the sludge also has a profound effect on denitrification. It has been shown that sludges incorporating unoxidized feed organics float more readily than well oxidized sludges.[104] Temperature is also important as it affects the rate of denitrification and therefore affects the rate of gas and bubble formation.[104,105]

FIGURE 4-25

EFFECT OF TEMPERATURE ON THICKENING PROPERTIES OF OXYGEN ACTIVATED SLUDGE AT MLSS = 4000 mg/l (REFERENCES 86 AND 103)

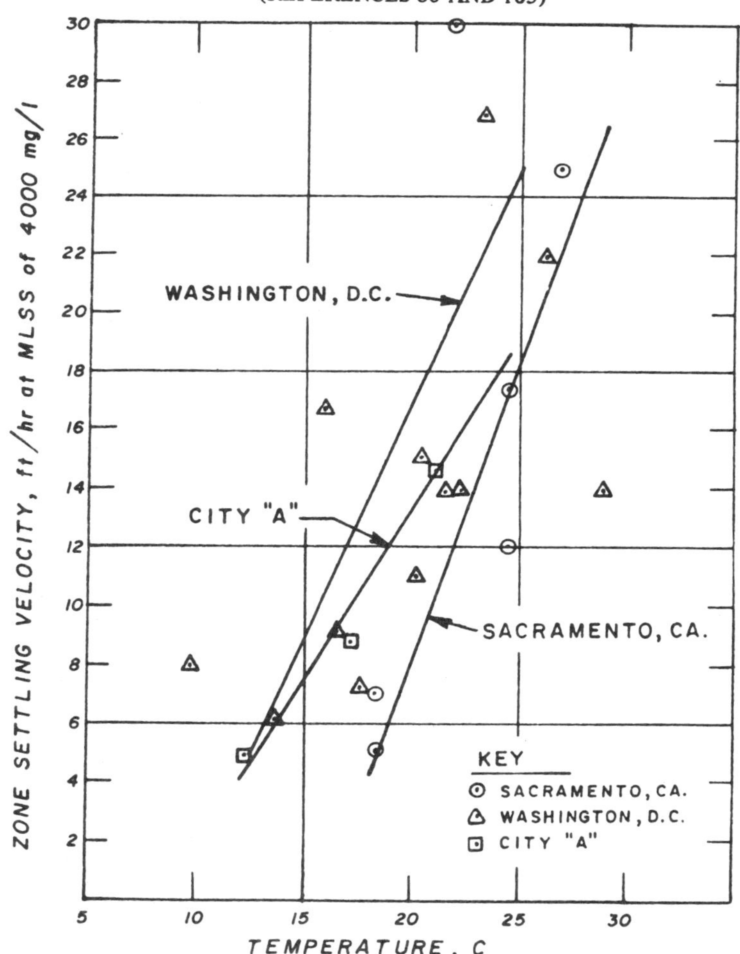

FIGURE 4-26

EFFECT OF TEMPERATURE ON THICKENING PROPERTIES OF OXYGEN ACTIVATED SLUDGE AT MLSS = 7000 mg/l (REFERENCES 86 AND 103)

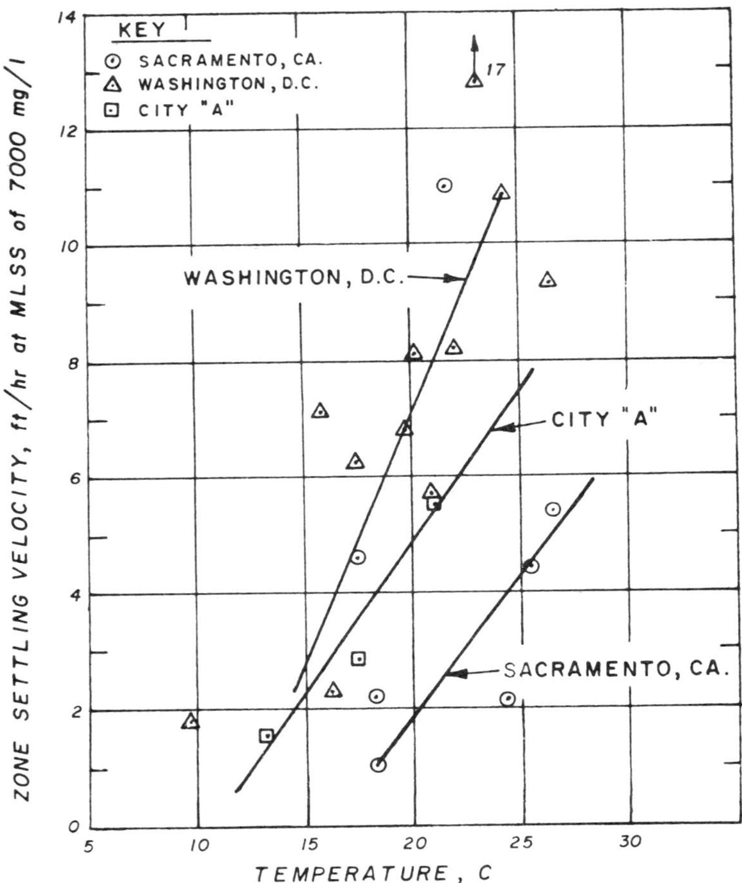

The following conclusions can be drawn about denitrification in secondary clarifiers, based on work done to date:[33,104,105] (1) Rapid sludge removal can prevent sufficient time being available for nitrogen bubble formation; (2) sludges with low SVI values are preferable as they can be withdrawn faster; (3) since the saturation level of nitrogen is greater in deep tanks than laboratory cylinders, bubbles will form and sludges will float faster in the laboratory than in the field; (4) there is a minimum concentration of nitrate nitrogen below which there is insufficient nitrogen to cause flotation. In weak wastewaters or for those plants in which nitrification is suppressed, sludge flotation will not occur; (5) a drop in temperature will reduce denitrification rates and may render rising sludge a problem only under warm conditions; (6) an equivalent amount of nitrite will produce flotation faster than nitrate, because denitrification is more rapid when nitrite serves as the electron acceptor rather than nitrate; (7) a sludge with low activity or low rate of denitrification should be less susceptible to flotation. Sludge activity will vary among the types of suspended growth processes. For instance, separate stage nitrification sludge is less susceptible to flotation than combined carbon oxidation-nitrification sludges due to the low organic loading on the former process and resultant lower activity with respect to carbon oxidation and denitrification in sedimentation tanks. Among combined carbon oxidation-nitrification alternatives, the contact stabilization and step aeration alternatives are most susceptible to sludge flotation due to the enhanced possibility in those modifications of influent organics being incorporated into the mixed liquor without being oxidized prior to the clarification step. To a lesser degree, complete mix systems are also prone to sludge flotation for the same reason.

Control measures for preventing floating sludge should be incorporated into the initial plant design. Provision of rapid sludge removal (vacuum pickup type) in sedimentation tank design can prevent there being sufficient contact time for bubble formation to occur and cause flotation. Flexibility in influent feed points (e.g., allowing switching from step aeration to plug flow in warm weather periods) can provide the operator with options in process operation that allow him to get out of difficult operating situations. Provision for chlorination of the return activated sludge is recommended for all suspended growth applications. Recent work done in California has shown that continuous low dose chlorination can be used for controlling sludge bulking and reducing the sludge volume index, apparently without impairing nitrification (see Section 4.5.1).[59] This allows more rapid sludge withdrawal from the sedimentation tanks. When nitrification is not required, higher doses of chlorine can be used to suppress nitrification and thus avoid flotation. Overdosing chlorine on a slug dose or continuous dose basis should be avoided, however, as it can cause increases in the level of organics in the process effluent.[104]

Control of solids retention time to values below that which will support nitrification has been a procedure that has been occasionally recommended for cases where nitrification is not required and sludge flotation is to be prevented. However, when the DO level or pH is not limiting the nitrification rate, nitrification will proceed at solids detention times as low as 1.30 days at temperatures equal to or greater than 20 C (Section 3.2.5.4). This renders the control of nitrification impractical with solids retention time control in warm weather as stable operation at such a low value for the solids retention time would not be possible. It may be possible to suppress nitrification through control of DO to levels <0.5 mg/l and thereby limit the nitrifier growth rate to levels which result in washout of the nitrifying organisms, but this will require accurate around-the-clock control of DO. Further, maintenance of low DO can cause another operational problem, sludge bulking.

4.11 Considerations for Process Selection

In selecting nitrification as the process for ammonia removal, two kinds of comparisons can be made. First, the process can be compared to the physical-chemical alternatives. Second, alternative nitrification processes can be compared. It is emphasized that no single alternative will be the best choice for all situations.

4.11.1 Comparison to Physical-Chemical Alternatives

Several factors dictate the choice between biological and physical-chemical techniques for ammonia removal. Cost is often the single most influential factor in process choice. Ammonia removal via the nitrification process is widely recognized to be the least costly ammonia removal alternative. Unless phosphorus removal is also required, the combined cost of lime precipitation-air stripping is normally greater than the cost for nitrification. Likewise, breakpoint chlorination and ion exchange are normally more costly than nitrification.[106]

In the majority of situations existing facilities are utilized when treatment is upgraded, rather than construction of wholly new facilities. The layout of the existing facility may be more adaptable to one specific alternative or another. In many instances, it has been found that biological nitrification has been the process most easily incorporated into the upgraded treatment system.

Very low temperature operation (< 10 C) may favor a physical-chemical process rather than a biological process as reaction rates become very low, requiring very large reactors. Physical chemical processes are also affected by low temperatures, but to a lesser degree.

The presence of compounds toxic to nitrifiers may also dictate against the choice of nitrification. Some toxicants are resistant (e.g., nonbiodegradable solvents) to most forms of pretreatment. Unless a very effective source control program can be implemented for these compounds, dependable operation of nitrification may become impractical.

4.11.2 Choice Among Alternative Nitrification Systems

All of the factors described in Section 4.11.1 are also factors to be considered in selection among nitrification systems. Other factors affecting choice among nitrification alternatives are summarized in Table 4-22 as a guide for process selection. Each of these factors is considered earlier in this chapter.

Higher effluent ammonia (1-3 mg/l) in the attached growth effluents than suspended growth effluent is cited as a disadvantage of attached growth systems in Table 4-22. However, breakpoint chlorination is easily appended to attached growth systems, as the chlorine dose for breakpoint is low (< 30 mg/l). The addition of breakpoint chlorination puts attached growth systems on an equal footing with suspended growth systems with respect to ammonia control.

TABLE 4-22

COMPARISON OF NITRIFICATION ALTERNATIVES

System Type	Advantages	Disadvantages
Combined carbon oxidation - nitrification		
Suspended growth	Combined treatment of carbon and ammonia in a single stage Very low effluent ammonia possible Inventory control of mixed liquor stable due to high BOD_5/TKN ratio	No protection against toxicants Only moderate stability of operation Stability linked to operation of secondary clarifier for biomass return Large reactors required in cold weather
Attached growth	Combined treatment of carbon and ammonia in a single stage Stability not linked to secondary clarifier as organisms on media	No protection against toxicants Only moderate stability of operation Effluent ammonia normally 1-3 mg/l (except RBD) Cold weather operation impractical in most cases
Separate stage nitrification		
Suspended growth	Good protection against most toxicants Stable operation Very low effluent ammonia possible	Sludge inventory requires careful control when low BOD_5/TKN ratio Stability of operation linked to operation of secondary clarifier for biomass return Greater number of unit processes required than for combined carbon oxidation - nitrification
Attached growth	Good protection against most toxicants Stable operation Less sensitive to low temperatures Stability not linked to secondary clarifier as organisms on media	Effluent ammonia normally 1-3 mg/l Greater number of unit processes required than for combined carbon oxidation - nitrification

Refinement of process choice may require pilot studies. This is particularly true where wastewater toxicity may affect the efficacy of nitrification (see Section 4.5.3).

A common issue faced by the engineer when dealing with suspended growth system design is whether to separate the carbon oxidation stage from the nitrification stage or whether to provide a combined carbon oxidation-nitrification system. In a recently conducted pilot study using these systems in parallel at two wastewater temperatures, 8 C and 20C, for the town of Cheektowaga, New York, it was shown that when the separate nitrification stage system was operated at the same solids retention time as the combined carbon oxidation-nitrification system and the same temperature, nitrification effluents of essentially identical quality were produced.[107]

The investigators concluded that in most cases the combined carbon-oxidation nitrification system should be chosen for the following reasons:[107]

1. Use of the biological solids retention time concept and controlled sludge wasting make the combined carbon oxidation-nitrification system as controllable as a two stage suspended growth system. The investigators did not agree with the often stated concept that separating the stages leads to more positive control of the carbon oxidation and nitrification functions, as their experimental study demonstrated quite the opposite.

2. The use of combined carbon oxidation-nitrification results in lower sludge quantities to be wasted than in a two stage suspended growth system. This is because the first stage (carbon oxidation) operates at a low solids retention time (say θ_c = 2 days) which results in less solids destruction than when 10 or 20 days are used in the combined carbon oxidation-nitrification system. This phenomenon has also been observed by others.[108,109]

3. The longer sludge retention times employed in separate stage nitrification systems (θ_c = 10 to 20 days) results in improved sludge settling characteristics as compared to high rate activated sludge systems (at $\theta_c \sim 2$ days).

4. A two stage suspended growth system appears to be more prone to control problems relating to sludge inventory control. The separate stage reactor's sludge inventory must be maintained by shifting inventory from the first stage, or by some other means (Section 4.5.2). Further, two sets of sedimentation tanks are required. Sedimentation tanks are often the most vulnerable components of the activated sludge system. "Thus it is difficult to envision that the path to increased controllability of nitrifying activated sludge should lead to a doubling of the least stable element in the process configuration, i.e.; the clarifier. Rather, if a two stage nitrification system is required, it appears more reasonable to explore the capabilities of a fixed film nitrification reactor . . ."[107]

5. Toxicants affecting nitrifiers present in the raw sewage or primary effluent are often cited as a reason to provide a high rate activated sludge effluent to act as a toxicant removal step ahead of a separate stage nitrification unit. First, a detailed evaluation may show that in fact toxicity is not a problem. "Secondly, toxic materials might better be excluded from wastewater systems by regulation rather than relying on 'sacrificial' biosystems to protect the nitrifying capability of the system. Thirdly, in most cases it may become attractive to remove phosphorus in primary treatment by the addition of coagulants."[107]

These conclusions are presented in this manual as an excellent basis for consideration of the reasons for process selection. In many cases effective counter arguments can be presented. In the last analysis, the process choice must be made by the local agency and its engineering consultant or staff. As an example, counter-arguments are listed in the same order as the arguments previously presented:

1. The parallel study of carbon oxidation-nitrification and two sludge systems at Cheektowaga was based primarily on municipal sewage, with only 10 percent industrial load.[107] Situations do exist where significant industrial contributors of organic load are tributary to the municipal system. For instance, seasonal canning industries tributary to California municipalities treatment plants in some instances cause 3 to 4 times the non-canning season organic load at the peak of the canning season. Further, industrial waste production may vary from year to year depending on factors beyond control or accurate prediction. In the face of this unpredictable variation it may be difficult or uneconomic to design a system for combined carbon oxidation-nitrification due to the sensitivity of nitrification to solids retention time. Less difficulty is experienced in designing for high rate activated sludge in a two stage system, as production of effluent of a quality suitable for separate stage nitrification is not as sensitive to load or solids retention time as is nitrification.

2. The phenomenon of lower sludge production from a two stage suspended growth system compared to a combined carbon oxidation-nitrification system is well established and cannot be contested. However, from an energy standpoint, the greater sludge production from the two stage system may be an advantage for two reasons. First, the lower sludge production of the combined carbon oxidation-nitrification system is obtained at the expense of greater power requirements because greater amounts of oxygen must be supplied for the oxidation of solids. Second, if anaerobic digestion is employed and the gas recovered for useful energy purposes, less energy is available from the digestion process when less solids are produced. However, these two factors may be outweighed by the increased cost of ultimate disposal in some cases.

3. It has occasionally been found that the longer solids retention times (10 to 20 days) also result in sludge settling difficulties.

4. With careful monitoring, the separate nitrification stage's inventory requirements can be managed. In some situations where industries are tributary, the inventory control problem may be easier with a two stage system (see counter-argument 1 above).

   Two sets of sedimentation tanks do present more control requirements than one set. However, if the conditions for tank upset are present when two sets are provided, they also can be present for one set. Careful control of sedimentation tank operation is mandatory in either case.

5. In the case of the parallel study at Cheektowaga, there was no evidence of significant toxicants in the primarily domestic sewage processed.[107] So perhaps it is natural to discount the difficulty in dealing with this problem.

The unfortunate aspect of the presence of toxicity in wastewater is that it is often ephemeral in nature; i.e., it's there and then it's gone. Toxicants discharged on a continuous basis are handled with relative ease compared to the occasional dump. Unless the problem is recognized in its earliest stage, the causative agent may not even be sampled by plant personnel, rendering it impossible to trace. Even if a sample is caught, tracing it back through the system is not always possible until the next occurrence. Another aspect of the problem is that the dumps offer no opportunity for the biomass to adapt to the toxicant, whereas if it were continuously present, adaptation of the nitrifiers might be possible (see Section 3.2.9).

In cases where toxicants are occasionally present, the issue boils down to the need for plant reliability. In cases of discharge to an estuary or groundwater where mixing in the environment causes dilution of the effluent, occasional process failures may be accepted. However, where stringent regulatory requirements exist or where the water is reused and the water user demands the consistent performance expected of a water utility, some compensation must be made to handle the problem of toxic upsets. This may be done by any number of means, one of which is to provide pretreatment via chemical addition or by a biological treatment stage (Section 4.5.3). Another is to provide supplemental breakpoint chlorination at the end of the system. The cost of the latter facility is very much affected by the degree of upset in nitrification expected.

In the last analysis, the parallel study at Cheektowaga showed that combined carbon oxidation-nitrification could be just as reliable as separate stage nitrification at low temperatures (8 C) with a primarily domestic wastewater.[107] The study provides further proof dispelling the poor reputation that the combined carbon oxidation-nitrification system has acquired in the U.S. Its wide application in England where it is coupled with a toxicity source control program offers additional testimony to the efficacy of the process.

Similar lengthy discussions could be prepared which present the pros and cons of other systems: e.g., rotating biological discs versus plastic media trickling filters. These comparisons not only are beyond the scope of this manual but would not have general validity. It is a hazardous task to make such an attempt as the specifics of individual circumstances affect the decisions to a large degree. There is no universally best nitrification approach. Rather, the broad variety of alternatives should be viewed as a positive situation. The fact that there are many alternatives makes the task of adapting nitrification into waste treatment easier, not harder. A myriad of flowsheets incorporating nitrification are not only possible, but economically feasible and with proper design and operation, quite reliable.

4.12 References

1. Mulbarger, M.C., *The Three Sludge System for Nitrogen and Phosphorus Removal.* Presented at the 44th Annual Conference of the Water Pollution Control Federation, San Francisco, California, October, 1971.

2. City of Los Angeles, California, *Hyperion Treatment Plant West Battery Operating Reports.* January through March, 1969.

3. Horstkotte, G.A., Niles, D.G., Parker, D.S., and D.H. Caldwell, *Full-Scale Testing of a Water Reclamation System.* JWPCF, 46, No. 1, pp 181-197 (1974).

4. Weddle, C.L., Niles, D.G., Goldman, E., and J.W. Porter, *Studies of Municipal Wastewater Renovation for Industrial Water.* Presented at the 44th Annual Conference of the Water Pollution Control Federation, San Francisco, California, October, 1971.

5. Loftin, W.E., *Annual Report, Livermore Water Reclamation Plant, 1970.* City of Livermore, California, March, 1971.

6. Beckman, W.J., Avendt, R.J., Mulligan, T.J., and G.J. Kehrberger, *Combined Carbon Oxidation-Nitrification.* JWPCF, 44, No. 10, pp 1916-1931 (1972).

7. Stamberg, J.B., Hais, A.B., Bishop, D.F., and J.A. Heidman, *Nitrification in Oxygen Activated Sludge.* Unpublished paper, Environmental Protection Agency, 1974.

8. Heidman, J.A., *An Experimental Evaluation of Oxygen and Air Activated Sludge Nitrification Systems With and Without pH Control.* EPA report for Contract No. 68-03-0349, 1975.

9. County Sanitation Districts of Los Angeles County, *Monthly Operating Reports, Whittier Narrows Water Reclamation Plant.* April, 1973 to March, 1974.

10. Greene, R.A., *Complete Nitrification by Single Stage Activated Sludge.* Presented at the 46th Annual Conference of the Water Pollution Control Federation, Cleveland, Ohio, October, 1973.

11. Newton, D., and T.E. Wilson, *Oxygen Nitrification Process at Tampa.* In *Applications of Commercial Oxygen to Water and Wastewater Systems,* Ed. by R.E. Speece and J.F. Malena, Jr., Austin, Texas: The Center for Research in Water Resources, 1973.

12. Tenney, M.W., and W.F. Echelberger, *Removal of Organic and Eutrophying Pollutants by Chemical-Biological Treatment.* Prepared for the EPA, Report No. R2-72-076 (NTIS PB-214628), April, 1972.

13. Black, S.A., *Lime Treatment for Phosphorus Removal at the New Market/East Gwillimbury WPCF.* Paper No. W3032, Ontario Ministry of the Environment, Research Branch, May, 1972.

14. Schwer, A.D., Letter communication to D.S. Parker – Metropolitan Sewer District of Greater Cincinnati, March 9, 1971.

15. Barth, E.F., Brenner, R.C., and R.F. Lewis, *Chemical-Biological Control of Nitrogen in Wastewater Effluent.* JWPCF, 40, No. 12, pp 2040 – 2054 (1968).

16. Rimer, A.E., and R.L. Woodward, *Two Stage Activated Sludge Pilot Plant Operations,* Fitchburg, Massachusetts. JWPCF, 44, No. 1, pp 101-116 (1972).

17. Wild, H.E., Jr., Letter communication to D.S. Parker. Briley, Wild and Associates, Ormond Beach, Florida, September 9, 1974.

18. Union Carbide Corporation, *"UNOX" System Study at Town of Amherst, New York.* 1972.

19. Linstedt, K.D., and E.R. Bennett, *Evaluation of Treatment for Urban Wastewater Reuse.* Report prepared for the Environmental Protection Agency, EPA-R2-73-122, July, 1973.

20. Linstedt, K.D., Letter communication to D.S. Parker. University of Colorado, Boulder, Colorado, August 5, 1974.

21. Stenquist, R.J., Parker, D.S., and T.J. Dosh, *Carbon Oxidation-Nitrification In Synthetic Media Trickling Filters.* JWPCF, 46, No. 10, pp 2327-2339 (1974).

22. Duddles, G.A., Richardson, S.E., and E.F. Barth, *Plastic Medium Trickling Filters for Biological Nitrogen Control.* JWPCF, 46, No. 5, pp 937-946 (1974).

23. McHarness, D.D., Haug, R.T., and P.L. McCarty, *Field Studies of Nitrification with Submerged Filters.* JWPCF, 47, No. 2, pp 291-309 (1975).

24. McHarness, D.D., and P.L. McCarty, *Field Study of Nitrification with Submerged Filter.* Report prepared for the EPA, EPA-R2-73-158, February, 1973.

25. *Process Design Manual for Upgrading Existing Wastewater Treatment Plants.* U.S. EPA, Office of Technology Transfer, Washington, D.C. (1974).

26. Richardson, S.E., *Pilot Plants Define Parameters for Plastic Media Trickling Filter Nitrification.* Presented at the 46th Annual Conference of the Water Pollution Control Federation, Cleveland, Ohio, October, 1973.

27. Sampayo, Felix F., *The Use of Nitrification Towers at Lima, Ohio.* Presented at the Second Annual Conference Water Management Association of Ohio, Columbus, Ohio, October, 1973.

28. Brenner, R.C., *EPA Experiences in Oxygen Activated Sludge.* Prepared for the EPA Technology Transfer Program, October, 1973.

29. Lawrence, A.W., and P.L. McCarty, *Unified Basis for Biological Treatment Design and Operation.* JSED, Proc. ASCE, 96, No. SA3, pp 757-778 (1970).

30. Hanson, R.L., Walker, W.C., and J.C. Brown, *Variations in Characteristics of Wastewater Influent at the Mason Farm Wastewater Treatment Plant, Chapel Hill, No. Carolina.* Report No. 13, UNC Wastewater Research Center, Chapel Hill, N.C., February, 1971.

31. Lijklema, L., *A Model for Nitrification in the Activated Sludge Process.* ESE Publication No. 303, Department of Environmental Sciences and Engineering, University of North Carolina, June, 1972.

32. Poduska, R.A. and J.F. Andrews, *Dynamics of Nitrification in the Activated Sludge Process.* Presented at the 29th Industrial Waste Conference, Purdue University, Lafayette, Indiana, May 7-9, 1974.

33. Nagel, C.A. and J.G. Haworth, *Operational Factors Affecting Nitrification in the Activated Sludge Process.* Presented at the 42nd Annual Conference of the Water Pollution Control Federation, Dallas, Texas, October (1969), (available as a reprint from the County Sanitation Districts of Los Angeles County).

34. Murphy, K.L., and P.L. Timpany, *Design and Analysis of Mixing for an Aeration Tank.* JSED, Proc. ASCE, 93, No. SA5, pp 1-15 (1967).

35. Murphy, K.D., and B.I. Boyko, *Longitudinal Mixing in Spiral Flow Aeration Tanks;* JSED, Proc. ASCE, 96, No. SA2, pp 211-221 (1970).

36. Metcalf and Eddy, Inc., *Wastewater Engineering.* New York, McGraw Hill Book Co., 1972.

37. Gujer, W. and D. Jenkins, *A Nitrification Model for the Contact Stabilization Activated Sludge Process.* Water Research, in press.

38. Gujer, W. and D. Jenkins, *The Contact Stabilization Process-Oxygen and Nitrogen Mass Balances.* University of California, Sanitary Engineering Research Lab, SERL Report 74-2, February, 1974.

39. Sawyer, C.N., Letter communication to D.S. Parker; January 24, 1975.

40. Sawyer, C.N., *Activated Sludge Oxidations, III. Factors Involved in Prolonging the High Initial Rate of Oxygen Utilization by Activated Sludge-Mixtures.* Sewage Works Journal, 11, No. 4, pp 595-606 (1939).

41. County Sanitation Districts of Los Angeles County, *A Plan for Water Reuse.* Report prepared for the members of the Boards of Directors, July, 1963.

42. City of Jackson, Michigan, *Sewage Treatment Plant Operating Reports.* August, 1973 to March, 1974.

43. Bruce, A.M., and J.C. Merkins, *Further Studies of Partial Treatment of Sewage by High Rate Biological Filtration.* Water Pollution Control (London), pp 499-527, 1973.

44. Grantham, G.R., Phelps, E.B., Calaway, W.T., and D.L. Emerson, *Progress Report on Trickling Filter Studies.* Sewage and Industrial Wastes, 22, No. 7, pp 867-874 (1950).

45. Grantham, G.R., *Trickling Filter Performance at Intermediate Loading Rates.* Sewage and Industrial Wastes, 23, No. 10, pp 127-1234 (1951).

46. Burgess F.J., Gilmour, C.M., Merryfield, F., and J.K. Carswell, *Evaluation Criteria for Deep Trickling Filters.* JWPCF, 33, No. 8, pp 787-799 (1961).

47. Osborn, D.E., *Operating Experiences with Double Filtration in Johannesburg.* J. Inst. of Sew. Purif., Part 3, pp 272-281 (1965).

48. Stones, T., *Investigation on Biological Filtration at Salford.* Journal of the Institute of Sewage Purification, No. 5, pp 406-417 (1961).

49. Mohlman, F.W., Norgaard, J.T., Fair, G.M., Fuhrman, R.E., Gilbert, J.J., Heacox, R.E., and C.C. Ruchoft, *Sewage Treatment at Military Installations.* Sewage Works Journal, 18, No. 4, pp 789-1028 (1946).

50. Heukelekian, H., *The Relationship Between Accumulation, Biochemical and Biological Characteristics of Film and Purification Capacity of a Biofilter and a Standard Filter.* Sewage Works Journal, 17, No. 3, pp 516-524 (1945).

51. Brown and Caldwell, *Report on Pilot Trickling Filter Studies at the Main Water Quality Control Plant.* Prepared for the City of Stockon, California, March, 1973.

52. Antonie, R.L., *Three Step Biological Treatment with the Bio-Disc Process.* Presented at the New York Water Pollution Control Association, Spring Meeting, Mantauk, New York, June, 1972.

53. Antonie, R.L., *Nitrification and Denitrification with the Bio-Surf Process.* Presented at the Annual Meeting of the New England W.P.C. Association in Kennebunkport, Maine, June 10-12, 1974.

54. *Rotating Biological Disk Wastewater Treatment Process – Pilot Plant Evaluation.* Report by the Department of Environmental Sciences, Rutgers University, prepared for the Environmental Protection Agency, Project No. 17010 EBM, August, 1972.

55. Brown and Caldwell, Consulting Engineers, *Lime Use in Wastewater Treatment: Design and Cost Data.* Report submitted to the U.S. Environmental Protection Agency, 1975.

56. *Process Design Manual for Carbon Adsorption.* U.S. EPA, Office of Technology Transfer, Washington, D.C. (1974).

57. Rainwater, F.H. and L.L. Thatcher, *Methods for Collection and Analysis of Water Samples.* Geological Survey Water-Supply Paper 1454, USGPO, 1960.

58. Wood, D.K. and G. Tchbanoglous, *Trace Elements in Biological Waste Treatment with Specific Reference to the Activated Sludge Process.* Presented at the 29th Industrial Waste Conference, Purdue University, May, 1974.

59. Eisenhauer, D.L., Sieger, R.B. and D.S. Parker, *Design of an Integrated Approach to Nutrient Removal.* Presented at the EED- ASCE Specialty Conference, Penn. State University, Pa., July, 1974.

60. Environmental Quality Analysts, Inc., Letter Report to Valley Community Services District, March, 1974.

61. Stensil, H.D., *Oases Wastewater Characterization Study, Chattanooga Moccasin Bend Wastewater Treatment Plant.* Report prepared by Air Products and Chemicals, Inc., 1975.

62. Sawyer, C.N., Wild, H.E., Jr., and T.C. McMahon, *Nitrification and Denitrification Facilities, Wastewater Treatment.* Prepared for the EPA Technology Transfer Program, August, 1973.

63. Parker, D.S., *Case Histories of Nitrification and Denitrification Facilities.* Prepared for the EPA Technology Transfer Program, May, 1974.

64. Mulbarger, M.C., Private communication to D.H. Caldwell, 1971.

65. Schwinn, D.E., *Treatment Plant Designed for Anticipated Standards.* Public Works, 104, No. 1, pp 54-57 (1973).

66. Wilson, T.E., and M.D.R. Riddel, *Nitrogen Removal: Where Do We Stand?* Water and Wastes Engineering, 11, No. 10, pp 56-61 (1974).

67. Wilcox, E.A., and A.A. Thomas, *Oxygen Activated Sludge Wastewater Treatment Systems, Design Criteria and Operating Experience.* Prepared for the EPA Technology Transfer Program, August, 1973.

68. Sorrels, J.H., and P.J.A. Zeller, *Two-Stage Trickling Filter Performance.* Sewage and Industrial Wastes, 18, No. 8, pp 943-954 (1956).

69. Huang, C.S., *Kinetics and Process Factors of Nitrification On a Biological Film Reactor.* Thesis submitted in partial satisfaction of the requirements for the degree of Doctor of Philosophy, University of New York at Buffalo, 1973.

70. Williamson, K.L. and P.L. McCarty, *A Model of Substrate Utilization by Bacterial Films.* Presented at the 46th Annual Conference of the Water Pollution Control Federation, Cincinnati, Ohio, October, 1973.

71. Duddles, G.A. and S.E. Richardson, *Application of Plastic Media Trickling Filters for Biological Nitrification.* Report prepared for the Environmental Protection Agency, EPA- R2-73-199, June, 1973.

72. Bruce, A.M., and J.C. Merkens, *Recent Studies of High-Rate Biological Filtration.* Water Pollution Control, pp 113-148 (1970).

73. Brown and Caldwell, *Report on Tertiary Treatment Pilot Plant Studies.* Prepared for the City of Sunnyvale, California, February, 1975.

74. Antonie, R.L., *Nitrification of Activated Sludge Effluent with the Bio-Surf Process.* Presented at the Annual Conference of the Ohio Water Pollution Control Association, Toledo, Ohio, June 7-13, 1974.

75. Haug, R.T., and P.L. McCarty, *Nitrification with the Submerged Filter.* JWPCF, 44, p 2086 (1972).

76. Haug, R.T., and P.L. McCarty, *Nitrification with the Submerged Filter.* Report prepared by the Department of Civil Engineering, Stanford University, for the Environmental Protection Agency, Research Grant No. 17010 EPM, August, 1971.

77. Young, J.C., Baumann, E.R., and D.J. Wall, *Packed-Bed Reactors for Secondary Effluent BOD and Ammonia Removal.* JWPCF, 47, No. 1, pp 46-56 (1975).

78. General Filter Co., *Paktor, Packed Bed Reactor,* Bulletin No. 7305, 1974.

79. Mechalas, B.J., Allen, P.M. and W.W. Matyskiela, *A Study of Nitrification and Denitrification.* A report prepared for the Federal Water Quality Administration, WPCRS 17010 DRD 07/70, July, 1970.

80. Young, J.C. and M.C. Stewart, *Advanced Wastewater Treatment with Packed Bed Reactors.* Report of the Engineering Research Institute, Iowa State University, No. ERI-73108, May, 1973.

81. Gasser, J.A., Chen, C.L., and R.P. Miele, *Fixed-Film Nitrification of Secondary Effluent.* Presented at the EED-ASCE Specialty Conference, Penn. State University, Pa., July, 1974.

82. Kenney, F.R., Letter communication to D.S. Parker. General Filter Co., Ames, Iowa, September, 1974.

83. Young, J.C., Unpublished data, September 27, 1974.

84. Smith, J., Personal communication to D.S. Parker. Environmental Protection Agency, Cincinnati, Ohio, March, 1974.

85. Loftin, W.E., Personal communication to D.S. Parker. City of Livermore, California, April, 1972.

86. Sacramento Area Consultants, *Oxygen Activated Sludge Pilot Studies for the Sacramento Regional Treatment Plant.* Report prepared for the Sacramento Regional County Sanitation District of Sacramento County, California, July, 1974.

87. Bishop, D.F., Personal communication to D.S. Parker. Environmental Protection Agency, Washington, D.C., April, 1974.

88. Commonwealth Department of Works, Australian Capital Territory, Letter communication to R.C. Aberley, dated June 1 and July 20, 1972.

89. Metropolitan St. Louis Sewer District and Havens and Emerson, Ltd., *Cost-Effective Design of Wastewater Treatment Facilities Based on Field Derived Parameters.* Prepared for the EPA, Report No. EPA- 670/2-74-062 (PB-234 356), July, 1974.

90. Aberley, R.C., Rattray, G.B. and P.P. Dougas, *Air Diffusion Unit.* JWPCF, 46, No. 5, pp 895-910 (1974).

91. Leary, R.D., Ernest, L.A., and W.J. Katz, *Full Scale Oxygen Transfer Studies of Seven Diffuser Systems.* JWPCF, 41, No. 3, pp 459-473 (1969).

92. Nogaj, R.J., *Selecting Wastewater Aeration Equipment.* Chemical Engineering, April, 1972.

93. City of Medford, Oregon, *Sewage Treatment Plant Operating Reports.* July, 1974.

94. San Pablo Sanitary District, California, *Wastewater Treatment Plant Operating Reports.* June, 1973, to July, 1974.

95. *Process Design Manual for Phosphorus Removal.* U.S. EPA, Office of Technology Transfer, Washington, D.C. (1971).

96. *Process Design Manual for Suspended Solids Removal.* U.S. EPA, Office of Technology Transfer, Washington, D.C., January, 1975.

97. Cleasby, J.L., and E.R. Baumann, *Wastewater Filtration, Design Considerations.* Prepared for the EPA Technology Transfer Program, July, 1974.

98. Dick, R.I., *Role of Activated Sludge Final Settling Tanks.* JSED, Proc. ASCE, 96, No. SA2, pp 423-436 (1970).

99. Dick, R.I. and A.R. Javaheri, *Discussion of Unified Basis for Biological Treatment Design and Operation by A.W. Lawrence and P.L. McCarty.* JSED, Proc. ASCE, 97, SA2, pp 234-238 (1971).

100. Dick, R.I. and K.W. Young, *Analysis of Thickening Performance of Final Settling Tanks.* Presented at the 27th Industrial Waste Conference, Purdue University, Lafayette, Indiana, May 7-9, 1974.

101. Tenney, M.W. and W. Stumm, *Chemical Flocculation of Microorganisms in Biological Waste Treatment.* JWPCF, 37, p. 1370 (1965).

102. Parker, D.S., Kaufman, W.J., and D. Jenkins, *Physical Conditioning of Activated Sludge Floc.* JWPCF, 43, No. 9, pp 1817-1833 (1971).

103. Stamberg, J.B., Bishop, D.F., Hais, A.B., and S.M. Bennet, *System Alternatives in Oxygen Activated Sludge.* Presented at the 45th Annual Conference of the WPCF, Atlanta, Ga., 1972.

104. Sawyer, C.N., and L. Bradney, *Rising of Activated Sludge in Final Settling Tanks.* Sewage Works Journal, 17, No. 6, pp. 1191-1209 (1945).

105. Clayfield, G.W., *Respiration and Denitrification Studies on Laboratory and Works Activated Sludges.* Water Pollution Control, London, 73, No. 1, pp 51-76 (1974).

106. Stone, R.W., Parker, D.S., and J.A. Cotteral, *Upgrading Lagoon Effluent to Meet Best Practicable Treatment.* Presented at the 47th Annual Conference of the Water Pollution Control Federation, Denver, Colorado, October, 1974.

107. Lawrence, A.M., and C.G. Brown, *Biokinetic Approach to Optimal Design and Control of Nitrifying Activated Sludge Systems.* Presented at the Annual Meeting of the New York Water Pollution Control Association, New York City, January 23, 1973.

108. Stall, T.R., and R.H. Sherwood, *One Sludge or Two Sludge?* Water and Wastes Engineering, p 41-44, April, 1974.

109. Sutton, P.M., Murphy, K.L., and B.E. Jank, *Biological Nitrogen Removal – The Efficacy of the Nitrification Step.* Presented at the WPCF Conference, Denver, October, 1974.

CHAPTER 5

BIOLOGICAL DENITRIFICATION

5.1 Introduction

The process of biological denitrification is applicable to the removal of nitrogen from wastewater when the nitrogen is predominately in the nitrate or nitrite form. In municipal applications, the nitrogen in the raw wastewater is primarily present as organic and ammonia-nitrogen and first must be converted to an oxidized form (nitrite or nitrate) prior to biological denitrification. The biological oxidation process used for this conversion, nitrification, was described in Chapters 3 and 4.

This chapter presents design criteria for several alternative denitrification systems including suspended growth and attached growth systems using methanol as the carbon source and combined carbon oxidation-nitrification-denitrification systems using wastewater or endogenous carbon sources. The basic chemistry of denitrification was described in Section 3.3.

5.2 Denitrification in Suspended Growth Reactors Using Methanol as the Carbon Source

The suspended growth denitrification process is a form of the activated sludge process. There are several differences between its typical application for organic carbon removal and in its use for denitrification. In common is the provision of a reactor, in which the biomass is kept in suspension in the liquid by mixing. Also provided in both applications is a sedimentation tank for separation of the mixed liquor solids from the effluent, allowing the biomass to be recycled in the system and also allowing the production of a clear effluent for discharge or subsequent treatment. Two typical suspended growth denitrification systems are illustrated in Figure 5-1.

There are other analogies between suspended growth systems used for denitrification and organic carbon removal. In organic carbon removal applications, dissolved oxygen is introduced into the reactor by aeration so that biological oxidation of the organic matter can take place. In the process of carbon oxidation, oxygen is consumed as the electron acceptor in the oxidation process. In the process of denitrification, carbon (usually methanol) is oxidized with nitrate or nitrite serving as the electron acceptor (see Section 3.3.2). In denitrification *as opposed to organics removal,* it is the nitrate that is the pollutant that is to be removed and the carbon source that is added. In organics removal, it is the carbon that is the pollutant that is to be removed and the oxygen that is added. Needless to say, only sufficient carbon (such as methanol) is added in denitrification to accomplish the nitrate removal, as excess dosing causes organics to appear in the effluent unless control measures are undertaken. These residual organics, if left in the effluent, would exert BOD_5 and might cause violation of effluent requirements.

FIGURE 5-1

SUSPENDED GROWTH DENITRIFICATION SYSTEMS USING METHANOL

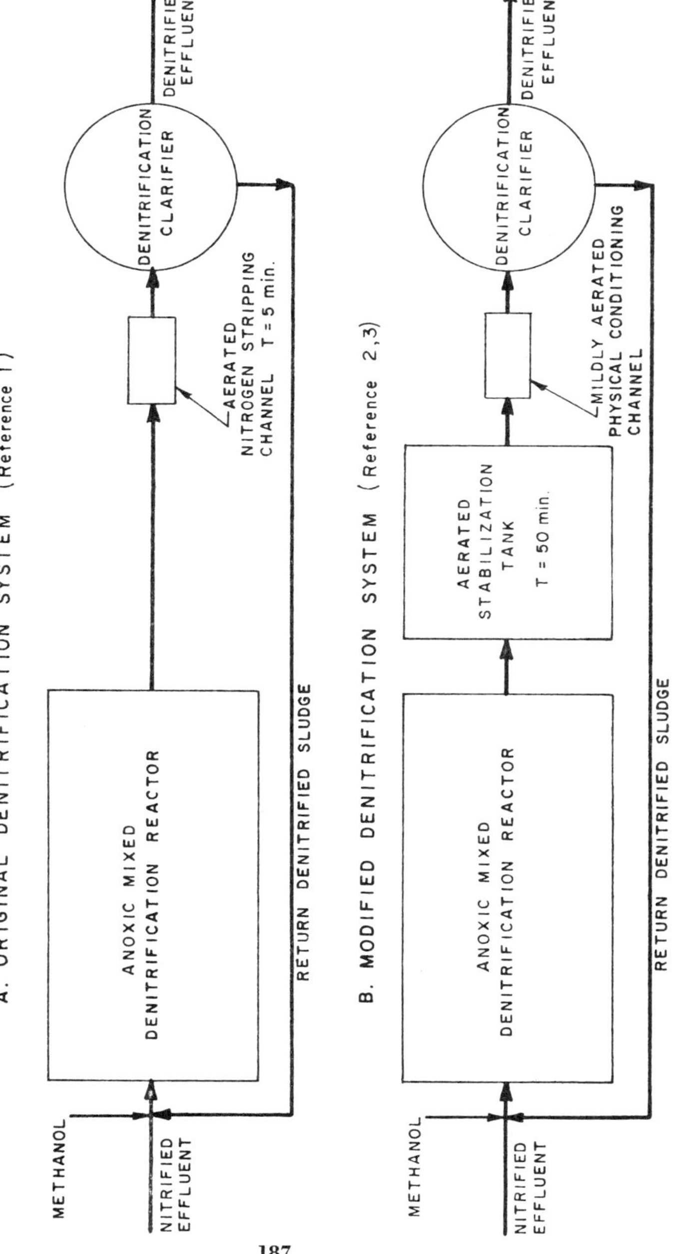

5.2.1 Denitrification Rates

Currently used denitrification rate data for design of denitrification systems are based upon work described in references 4, 5, 6, and 7. Data from these investigations are summarized in Figure 5-2. Rather than show individual data points, or trend lines, Figure 5-2 shows boundaries for the data so that the range in variation of denitrification rates can be inspected. Earliest available measurements were those from Manassas, Va.[4] which have been found to be considerably higher than subsequent observations at three other locations. The data from CCCSD, Ca., Blue Plains, D.C., and Burlington, Canada, all are in reasonable agreement with each other, and are all well below the Manassas rates. A possible reason for

FIGURE 5-2

OBSERVED DENITRIFICATION RATES FOR SUSPENDED GROWTH SYSTEMS USING METHANOL

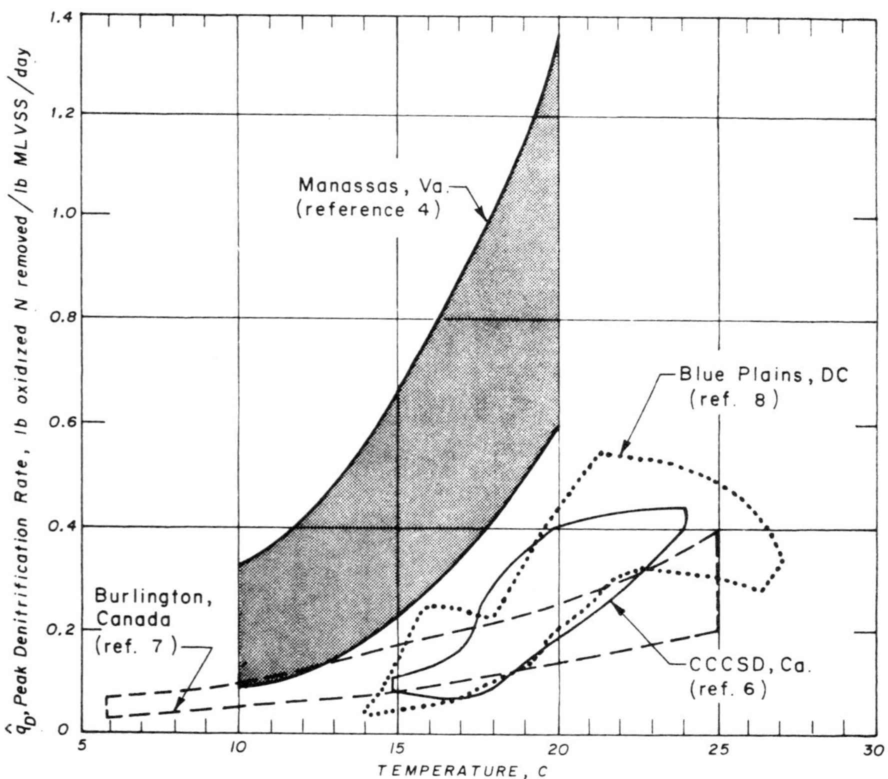

the higher measurements at Manassas is that an acid solubilization procedure was employed prior to the volatile solids determination, which may have acid hydrolized organic matter resulting in low measurements of volatile solids (and higher apparent denitrification rates). The earlier work at Manassas[9] prior to implementation of the acid step showed denitrification rates closer to the observations of other locations.

Laboratory studies on synthetic nitrate containing wastes have shown much higher denitrification rates than are found in Figure 5-2.[10,11] However, the biological solids developed in such laboratory systems do not contain the levels of refractory solids that build up in practical systems operated under field conditions. Therefore, the data developed from such laboratory studies are not directly useful in establishing accurate design parameters.

Conditions maintained during the field studies may influence field measured denitrification rates considerably. When the denitrification reactor is continuously operated close to the maximum growth rate of the denitrifying organisms, it is probable that the denitrifying activity of the biomass is higher than when the system is operated with a high safety factor. For instance, in studies in Ontario it was found that measured peak denitrification rates were approximately 30 percent greater at a solids retention time of 3 days than rates measured at a solids retention time of 6 days.[12] Thus, differences in measured rates may reflect variations in operating conditions among the various locations.

Observed rates in Figure 5-2 are essentially the experimentally determined values of the term $\hat{q}_D$ using the notation presented in Section 3.3.5.2. The peak nitrate removal rate, $\hat{q}_D$, is the reaction rate when neither methanol nor nitrate is limiting the reaction rate. Subsequent sections show how these peak nitrate removal rates are used in design calculations.

5.2.2 Complete Mix Denitrification Kinetics

The equations presented in Section 3.3.5 are directly applicable to the design of complete mix denitrification systems. The design procedure for denitrification uses the safety factor concept to relate peak nitrate removal rates, $\hat{q}_D$, to design nitrate removal rates, q_D. Expressed in terms of solids retention time, the safety factor concept is:

$$SF = \frac{\theta_c^d}{\theta_c^m} \tag{3-29}$$

where:

SF = safety factor,

θ_c^d = solids retention time of design, days, and

θ_c^m = minimum solids retention time, days, for denitrification.

The design nitrate removal rate, q_D, can be related to the safety factor and the peak nitrate removal rate by using the following equations in conjunction with Equation 3-53:

$$\frac{1}{\theta_c^d} = Y_D q_D - K_d \tag{3-50}$$

$$\frac{1}{\theta_c^m} = Y_D \hat{q}_D - K_d \tag{3-54}$$

where: Y_D = denitrifier gross yield, lb VSS grown/lb NO_3^- - N removed,

K_d = decay coefficient, day^{-1},

q_D = nitrate removal rate, lb NO_3^--N rem./lb VSS-day, and

$\hat{q}_D$ = peak nitrate removal rate, lb NO_3^--N rem./lb VSS-day.

In evaluating these equations in design calculations, the specific values of Y and K_d given in Section 3.3.5.4 may be used. Figure 5-2 may be used to arrive at estimated values of $\hat{q}_D$. Considering the range in the data in Figure 5-2, conservative practice in the absence of pilot data would be to pick the lowest denitrification rates observed for $\hat{q}_D$, e.g. at 10 C $\hat{q}_D$ = 0.05, at 15 C $\hat{q}_D$ = 0.08, at 20 C $\hat{q}_D$ = 0.15, and at 25 C $\hat{q}_D$ = 0.20. Use of these minimum values of $\hat{q}_D$ will result in very conservative reactor designs. Pilot plant studies may be useful to define applicable values of $\hat{q}_D$, as the potential for establishing higher denitrification rates for a particular location is good; evidence of this is the range of denitrification rates among the various locations shown in Figure 5-2.

As a design example consider a case where the temperature is 25 C, $\hat{q}_D$ = 0.2 lb NO_3 rem./lb MLVSS/day, Y = 0.9 lb VSS/lb NO_3^- rem., K_d = 0.04 day^{-1}, and K_D = 0.15 mg/l. Assume that due to diurnal variations in load (Section 5.2.2.2), a minimum safety factor of 2.0 is adopted. Consider a 30 mgd treatment plant, where 25 mg/l of nitrate-N must be removed.

1. Using Equation 3-54, calculate the minimum solids retention time for denitrification:

$$\frac{1}{\theta_c^m} = 0.9(0.2) - 0.04 = 0.14$$

$$\therefore \quad \theta_c^m = 7.14 \text{ days}$$

2. Calculate the design solids retention time (Equation 3-29):

$$\theta_c^d = 2.0(7.14) = 14.3 \text{ days}$$

3. Calculate the design nitrate removal rate (Equation 3-50):

$$\frac{1}{14.3} = 0.9q_D - 0.04$$

$$\therefore \quad q_D = 0.12 \text{ lb } NO_3^- \text{-N rem./lb MLVSS/day}$$

4. Calculate the steady state nitrate content of the effluent. The expression relating removal rates to nitrate level, from Equations 3-47, 3-48 and 3-49, is as follows:

$$q_D = \hat{q}_D \frac{D_1}{K_D + D_1} \tag{5-1}$$

where: K_D = half saturation constant, mg/l NO_3^--N, and

D_1 = effluent concentration of nitrate nitrogen mg/l.

Evaluation of this equation for this example yields:

$$0.12 = 0.20 \frac{D_1}{0.15 + D_1}$$

$$D_1 = 0.23 \text{ mg/l } NO_3^- \text{-N}$$

5. Determine the hydraulic detention time at average dry weather flow. The equation for nitrate removal rate is useful in this calculation.

$$q_D = \frac{D_o - D_1}{X_1 \cdot HT} \tag{5-2}$$

where: D_o = influent NO_3^--N, mg/l

D_1 = effluent NO_3^--N, mg/l

X_1 = MLVSS, mg/l, and

HT = hydraulic detention time, days.

The mixed liquor volatile suspended solids (MLVSS) level is dependent on the mixed liquor total suspended solids, which is in turn dependent on the operation of the denitrification sedimentation tank (see Sections 4.10 and 5.6). Assume for the purposes of this example that the design mixed liquor content at 25 C is 3000 mg/l. At a volatile content of 80 percent, the MLVSS is 2400 mg/l. From Equation 5-2, the hydraulic detention time is:

$$HT = \frac{(25 - 0.23)}{(0.12)(2500)} = 0.083 \text{ days}$$

$$= 1.99 \text{ hr}$$

6. Determine the sludge wasting schedule. The equations developed for wasting in the nitrification system are directly applicable here. The necessary sludge inventory is:

$$I = 8.33(X_1 \cdot V) \tag{4-7}$$

where: I = inventory of VSS in the anoxic denitrification reactor, lb,

X_1 = MLVSS in the reactor, mg/l, and

V = volume of the reactor, mil gal.

In the example at hand:

$$I = 8.33(2400)(0.083)(30) = 49{,}780 \text{ lb VSS}$$

From Equation 4-8, the sludge wasting from all sources is defined by the following equation:

$$S = \frac{I}{\theta_c^d} \tag{5-3}$$

where: S = total sludge wasted in lb/day

For this example:

$$S = 49{,}780/14.3 = 3{,}481 \text{ lb VSS/day}$$

The total sludge to be wasted each day is made up of two components, as shown:

$$S = 8.33(Q \cdot X_2 + W \cdot X_w) \tag{4-6}$$

where: Q = influent (or effluent) flow rate, mgd,

W = waste sludge flow rate, mgd,

X_2 = effluent volatile suspended solids, mg/l, and

X_w = waste sludge volatile suspended solids, mg/l.

The sludge contained in the effluent (the term $Q \cdot X_2$ above) can be calculated assuming that the effluent VSS is 10 mg/l:

$$8.33(10)(30) = 2{,}499 \text{ lb VSS/day}$$

By difference, the lb of MLVSS to be wasted from the mixed liquor or return sludge is:

$$3{,}481 - 2{,}499 = 982 \text{ lb VSS/day}$$

7. Methanol requirement. From Section 3.3.2, an estimate of 3.0 lb per lb of nitrate N removed is reasonable. The methanol requirement is:

$$3.0(25 - 0.23)(8.33)(30) = 18{,}570 \text{ lb/day}$$

The sludge yield and decay values used above are for a case where only a short aeration period is used prior to clarification. When an aerated stabilization step is employed, very much lower sludge wasting is required than presented in the above example. In cases where an aerated stabilization tank is employed, only the sludge inventory under anoxic conditions should be considered in the sizing of the anoxic reactor for denitrification (Step 5).

5.2.2.1 Effect of Safety Factor on Steady-State Effluent Quality

In the design example previously presented, the safety factor was assumed to be 2.0, based on considerations presented in Section 5.2.2.2. The effect of alternative assumptions on the effluent nitrate level are presented in Figure 5-3. As may be seen, the assumption of the safety factor has a marked effect on the effluent quality of complete mix denitrification systems.

5.2.2.2 Effect of Diurnal Load Variations on Effluent Quality

As is the case for nitrification, load variations have a significant effect on effluent quality. Since upstream treatment units to some extent equalize load variations, the peak to average load ratio is generally lower than for the nitrification stage. The effect of load variations can be analyzed in a similar manner to that used for nitrification (Section 4.3.3.2). By analogy to the nitrification case, the mass balance yields the following expression for nitrate at any time:

FIGURE 5-3
EFFECT OF SAFETY FACTOR ON EFFLUENT NITRATE LEVEL IN SUSPENDED GROWTH SYSTEM

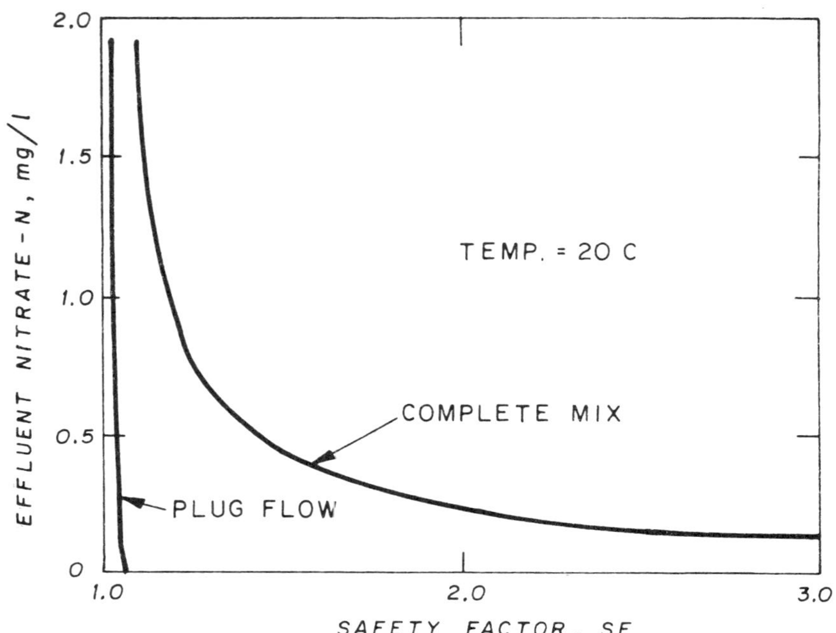

$$D_o = (\bar{D}_o - \bar{D}_1)\,\frac{\bar{Q}}{Q}\,\frac{\hat{\mu}_D}{\bar{\mu}_D}\,\frac{D_1}{K_D + D_1} + D_1 \qquad (5\text{-}4)$$

where:

D_o = influent nitrate -N level at any time, mg/l,

$\bar{D}_o$ = mass average influent nitrate -N level over 24 hours, mg/l,

D_1 = mass average effluent nitrate -N level at any time, mg/l,

$\bar{D}_1$ = mass average effluent nitrate -N level over 24 hours, mg/l,

Q = influent flow rate at any time,

$\bar{Q}$ = average daily influent flow rate,

$\bar{\mu}_D$ = design denitrifier growth rate, day^{-1}, and

$\hat{\mu}_D$ = maximum denitrifier growth rate, day^{-1}.

Equation 5-4 has been used to evaluate the effect of the diurnal load variations shown on Figure 5-4 using the design example conditions given in Section 5.2.2.1. The influent nitrate-nitrogen concentration was assumed constant at 25 mg/l, and the load variation was assumed to be due to variation in flow only. Several trial calculations using Equation 5-4 over a 24 hour cycle were necessary to derive values of D_1. Results of these calculations for several values of the SF are also shown on Figure 5-4. As may be seen, the safety factor, SF, has a marked effect on the nitrate bleedthrough occurring during peak load conditions. In the case examined, a safety factor of 2.0 was sufficient to prevent excessive nitrate leakage. The peak to average load for this case was 1.5. In summary, it would appear that as a minimum, the safety factor should exceed the peak to average load ratio to prevent excessive nitrate leakage during peak load conditions.

5.2.3 Plug Flow Denitrification Kinetics

The design approach for plug flow denitrification reactors is similar to the approach developed for complete mix reactors, with the exception of the equations used to predict effluent quality. Lawrence and McCarty's[13] solution for plug flow kinetics is applicable:

$$\frac{1}{\theta_c^d} = \frac{Y_D \hat{q}_D (D_o - D_1)}{(D_o - D_1) + K_D \ln \frac{D_o}{D_1}} - K_d \qquad (5\text{-}5)$$

All these terms are as defined previously in Section 5.2.2.

The kinetic design approach for plug flow follows that used for complete mix systems in Section 5.2.2, excepting that at step 4, Equation 5-5 is used instead of Equation 5-1 to find the effluent nitrate level.

Equation 5-5 can also be used to find the safety factor required to obtain any desired nitrate level. This has been done for the example presented in Section 5.2.2.1 and plotted in Figure 5-3. A comparison of the safety factor required to obtain the same nitrate level in a complete mix system yields the conclusion that plug flow systems can be designed with considerably lower safety factors while obtaining the same effluent quality.

While kinetic models have not been extended to the point where they can be expected to describe the effect of diurnal variations on plug flow systems, it can be expected that the effects of these loads will be similar to those experienced in complete mix systems. This is a result of the fact that once the effluent level rises above 1 mg/l nitrate –N, the denitrification rate becomes essentially zero order. For zero order reactions, there is little difference between plug flow and complete mix reaction kinetics. Therefore, the nitrate bleedthrough in a plug flow reactor can be expected to closely approach that in a complete mix reactor under diurnal peak load conditions. To prevent excessive nitrate leakage during peak load conditions, the recommendation in Section 5.2.2.2 should be adopted; as a minimum, the safety factor should exceed the peak to average load ratio.

The plug flow hydraulic regime can be approximated by a series of complete mix denitrification tanks, in which backmixing is prevented. An example is provided by the case history of the CCCSD Water Reclamation Plant, presented in Section 9.5.2.1.

FIGURE 5-4

EFFECT OF DIURNAL VARIATION IN LOAD ON EFFLUENT NITRATE LEVEL IN COMPLETE MIX SUSPENDED GROWTH SYSTEM

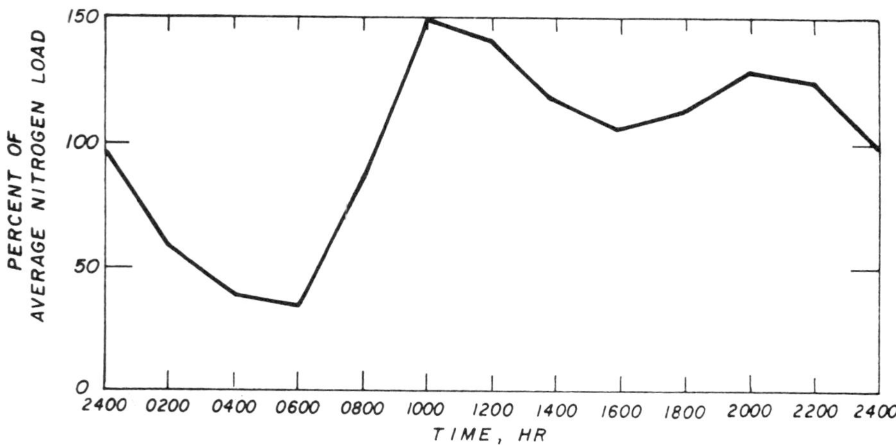

A. ASSUMED VARIATION IN INFLUENT NITROGEN LOAD

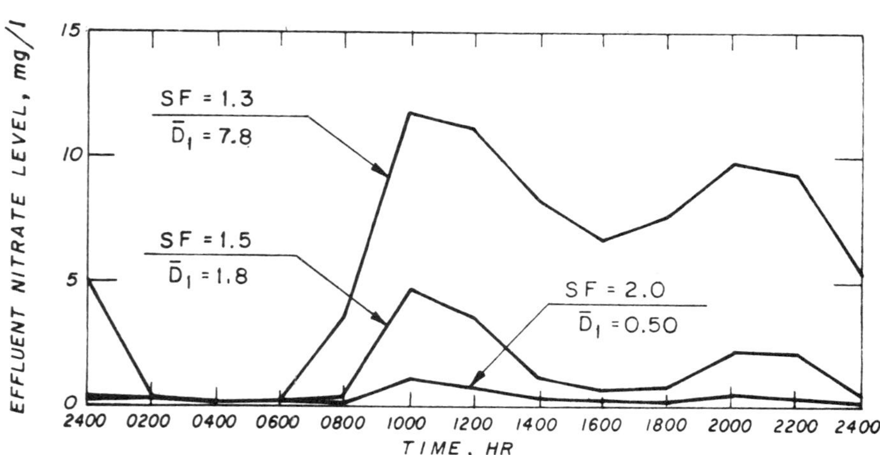

B. CALCULATED EFFLUENT NITRATE LEVEL

5.2.4 Effluent Quality from Suspended Growth Denitrification Processes

Since denitrification technology is new, there is a concern on the part of some design engineers that biological nitrification-denitrification systems are unstable and produce results of high variability. However, large scale tests of biological nitrogen removal have demonstrated over relatively long periods that a consistently low nitrogen level can be obtained.

5.2.4.1 Experience at Manassas, Va.

The EPA conducted a 0.2 mgd (760 cu m/day) test of the "three sludge" system for eight months at Manassas, Va. The system consisted of primary treatment and three separate suspended growth stages for organic carbon oxidation, nitrification and denitrification followed by filtration, as shown in Figure 5-5. Alum was added to the first and third suspended growth stages for phosphorus removal. A dose of polymer was added to the third reactor effluent. Performance data presented in the form of frequency distribution diagrams show that the performance of the closely monitored system was very stable.[4] A tabular summary of denitrification effluent quality is shown in Table 5-1 for the last four months of operation. As may be seen, an effluent very low in total phosphorus and total nitrogen was obtained from the denitrification system and filtration provided further reductions. Further details are available in the papers produced from the project.[4,9]

FIGURE 5-5

THE THREE SLUDGE SYSTEM AS TESTED AT MANASSAS, VA (REFERENCE 4)

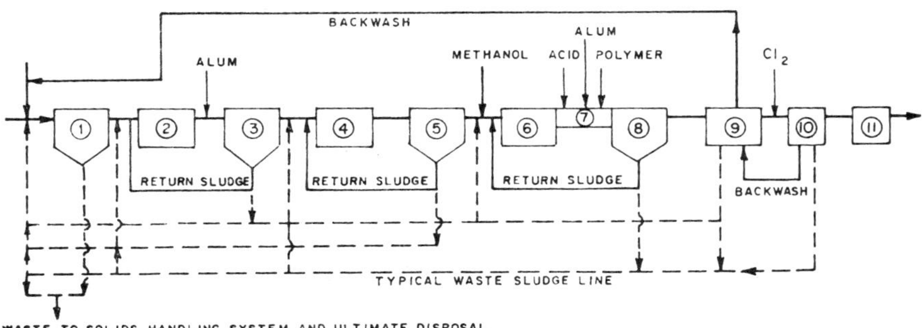

PRIMARY TREATMENT	HIGH RATE ACTIVATED SLUDGE	NITRIFYING ACTIVATED SLUDGE	DENITRIFYING ACTIVATED SLUDGE	POST TREATMENT
(1) SEDIMENTATION TANK	(2) AERATION TANK (3) SEDIMENTATION TANK	(4) AERATION TANK (5) SEDIMENTATION TANK	(6) ANOXIC REACTORS (7) AERATED CHANNEL (8) SEDIMENTATION TANK	(9) MIXED MEDIA FILTERS (10) CHLORINE CONTACT (11) POST AERATION

TABLE 5-1

DENITRIFICATION PERFORMANCE: FINAL FOUR MONTHS OF OPERATION AT MANASASS, VIRGINIA (REFERENCE 4)

Parameter		After final clarification, mg/l		After mixed media filtration, mg/l	
SS		2		0	
COD		21		16	
BOD_5		4.0		0.8	
Total P		0.6		0.3	
Organic N	TOTAL N	1.0	1.8	0.8	1.5
NH_4^+-N		0.0		0.0	
NO_2^--N		0.0		0.0	
NO_3^--N		0.8		0.7	

5.2.4.2 Experience at the CCCSD's Advanced Treatment Test Facility

In November, 1971, the Central Contra Costa Sanitary District (CCCSD) began the operation of a full-scale Advanced Treatment Test Facility (ATTF) at its existing wastewater treatment plant in California.[3] Operation of the facility ultimately extended over 23 months. A purpose of the test facility was to obtain data on the ATTF System sequence of processes (Figure 5-6) that had been proposed for the CCCSD Water Reclamation Plant.[14] Another purpose was to dispel the notion that the nitrification and denitrification processes were unstable due to the erratic nitrification-denitrification results previously obtained in a small-scale pilot study.[3,15,16]

The ATTF process units had capacities ranging from 2.5 mgd (9,464 cu m/day) in the primary step to 0.5 mgd (1,990 cu m/day) in the denitrification facilities. Primary clarification follows lime addition and preaeration and is coupled with a separate stage nitrification step. The use of lime in the primary stage removes much of the organic carbon load from the nitrification stage, thus allowing stable oxidation of ammonia to nitrate. Addition of lime also enhances the removal of phosphorus, heavy metals and viruses. Addition of lime in the initial stage of treatment, in contrast to the use of lime after conventional secondary treatment, enables the achievement of better stability in succeeding

FIGURE 5-6

ATTF SYSTEM FOR NITROGEN AND PHOSPHORUS REMOVAL

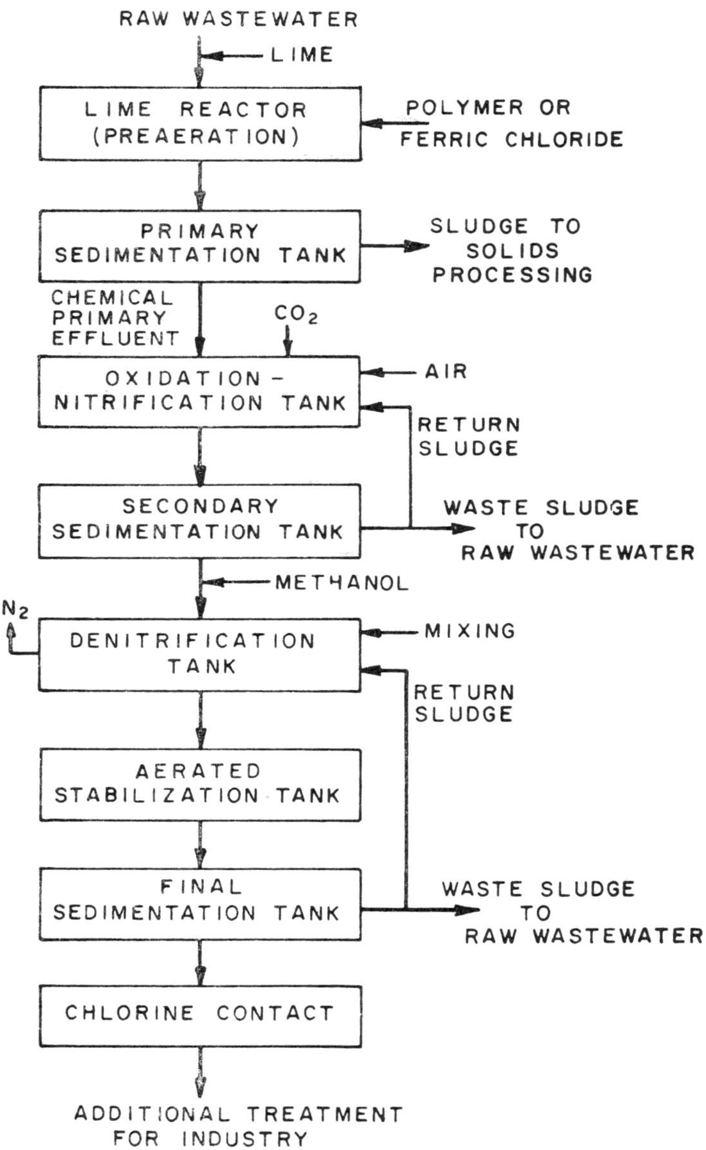

treatment processes and also allows the elimination of the need for a biological treatment step for organic carbon removal ahead of the nitrification stage. Biological denitrification follows nitrification, converting nitrate to nitrogen gas.

Performance data for a representative three months of operation of the ATTF are shown in Table 5-2.[3] Of particular interest is the fact that the 90 percentile performance level did not vary widely from the median performance level for the various constituents. This provides statistical confirmation that the nitrification and denitrification performance of the ATTF system was quite stable. The concentration of organics in the nitrified and denitrified effluents was low, as measured by BOD and organic carbon. Operation for complete nitrification also resulted in high organic removals. Similarly, suspended matter in the nitrified and denitrified effluents were also exceptionally low (Table 5-2). Nutrients are effectively removed in the ATTF system. Total nitrogen in the denitrified effluent averages less than 2 mg/l. Total phosphorus averaged 0.5 mg/l.

5.3 Denitrification in Attached Growth Reactors Using Methanol as the Carbon Source

Denitrification in attached growth reactors has been accomplished in a wide variety of denitrification column configurations using various media to support the growth of denitrifiers. In part because of this variability among systems, it is difficult to set forth generally useful design criteria at the present time. Several useful approaches are suggested for characterizing denitrification in attached growth systems and presently available data (1975) are analyzed by these procedures.

5.3.1 Kinetic Design of Attached Growth Denitrification Systems

In order to size an attached denitrification reactor, knowledge is required of the reaction rates taking place in the reactor volume. In estimating reaction rates, the level of biomass effective in denitrification must also be known. One approach is to estimate the level of biomass on the media surface and then use measured reaction rates per unit of biomass to obtain the nitrogen removal capability of a column containing the estimated amount of biomass.[17,18,19] This approach is limited in its usefulness in design applications because there is insufficient data available at the present time to predict in advance the level of biomass that will develop on the media. Biomass development will be dependent on hydraulic regime, type of media, loading, means for promoting sloughing and possibly the temperature of operation.

An indirect procedure for consideration of denitrification rates in design is to adopt the approach in which denitrification rates are expressed in terms of surface nitrate removal rates, e.g. lb nitrate –N removed per sq ft per day.[7,18,20] On this basis, high surface removal rates would reflect extensive biological film development, whereas low surface removal rates would reflect minimal surface film development. The surface denitrification rate varies considerably among the various denitrification column configurations and is affected by the loadings under which the process is operated.

TABLE 5-2

ATTF PERFORMANCE SUMMARY, APRIL 16 TO JULY 15, 1972
(CENTRAL CONTRA COSTA SANITARY DISTRICT, CA., REF. 3)

Constituent	Raw wastewater, mg/l			Chemical primary,[b] mg/l			Nitrified effluent, mg/l			Denitrified effluent, mg/l		
	mean	median	90%[a]	mean	median	90%[a]	mean	median	90%[a]	mean	median	90%[a]
BOD_5	203	199	235	57	54	79	3.6	3.5	6.8	3.2	3.0	4.8
TOC[c]	122	120	152	42	41	55	8.9	8.5	11.5	9.5	9.0	11
SOC[c]	22	23	25	27	27	31	5.6	5.5	6.7	5.9	6.0	6.6
SS	214	212	295	26	23	45	4.5	4.0	7.8	4.5	4.0	7.7
Turbidity (JTU)	-	-	-	12.8	12.0	23	1.4	1.3	2.2	1.4	1.3	1.8
Settleable solids (ml/l)	13.5	13.0	16.5	.084	0	.37	-	-	-	-	-	-
Organic N	-	-	-	-	-	-	.26	0	.67	1.1	1.1	2.5
NH_4^+ - N	-	-	-	24.0	23.8	28.5	.48	.30	.58	.31	.30	.40
NO_3^- - N	-	-	-	-	-	-	27	27	32	.48	0	.79
NO_2^- - N	-	-	-	-	-	-	.015	.013	.027	.009	.008	.018
Total P	9.86	9.57	11.19	.86	.59	1.85	1.04	.72	2.11	.50	.36	.98
Ortho P	9.74	9.90	11.24	.61	.46	1.75	1.00	.71	1.80	.52	.36	.90
TDS	583	561	750	-	-	-	634	636	724	537	551	616
Conductivity (10^5 mho/cm)	1230	1210	1366	-	-	-	1226	1223	1394	1160	1147	1318
Alkalinity[d]	215	218	237	254	249	293	105	106	127	183	187	217
Ca[d]	67.9	66.0	78.5	155	150	184	159	161	187	174	172	190
Mg[d]	84.0	84.0	95.6	38.6	34.0	74.0	57.3	61.0	73.0	30.1	30.0	48.4

[a]90% of observations are equal to or less than stated value.

[b]pH 10.2 operation to June 1, pH 11.0 thereafter.

[c]TOC = total organic carbon, SOC = soluble organic carbon

[d]as $CaCO_3$

5.3.2 Classification of Column Configurations

The various types of denitrification columns currently available are summarized in Table 5-3 along with calculated peak surface denitrification rates. The first type of categorization is with respect to the condition of the void space in the column. Until very recently, all denitrification work has been conducted on submerged columns wherein the voids were filled with the fluid being denitrified. Very recently, a new type of column has been developed in which the voids are filled with nitrogen gas, a product of denitrification.[6,21] The submerged columns can be further subdivided into packed bed and fluidized bed operations.

The varieties of media being employed for commercial application are also shown in Table 5-3; the listing is not meant to exclude commercial products which may be equivalent to those listed. For instance, other vendors of plastic media are listed in Table 4-15.

Details of design construction and operation of each column type are presented in the following sections. Also included is a comparison of column systems.

5.3.2.1 Nitrogen Gas Filled Denitrification Columns – Packed Bed

The nitrogen gas filled column was recently developed for use at the Lower Molonglo Water Quality Control Centre (LMWQCC), currently under construction.[6,21] Details of the column design used for this Canberra, Australia installation are shown in Figure 5-7. The column media consists of corrugated plastic sheet modules of the same type used in plastic media trickling filters (Table 4-15). As opposed to previously developed attached growth processes, the present column system is not submerged with liquid; rather, the column's void spaces are filled with nitrogen gas, a reaction product of the denitrification process.

In the denitrification column the influent wastewater is spread out over the top of the media and then the liquid flows in a thin film over the media on which the organisms grow. These organisms maintain a balance so that an active biological film develops. The balance is maintained by sloughing of biomass from the media, either by death or by hydraulic erosion or both. Sufficient voids are present in the media to prevent clogging and ponding. The denitrification column must be followed by a clarification step to remove sloughed solids. Pilot studies for the LMWQCC facility indicated that effluent solids should be sufficiently low so that the effluent can go directly to a tertiary multi-media filter.

Oxygen must be excluded from the denitrification column since its presence would prevent nitrate or nitrite in the applied liquor from serving as the electron acceptor in the biological oxidation of the applied carbon (methanol). Therefore, the denitrification column is sealed to prevent intrusion of air. The units are vented to allow outflow of nitrogen gas while preventing inflow of atmospheric air. Soon after start-up, the nitrogen gas displaces the air or other gases initially present, leaving a nitrogen gas atmosphere in the voids. This ensures the anoxic environment that is required for denitrification.

TABLE 5-3

TYPES OF DENITRIFICATION COLUMNS AND MEASURED DENITRIFICATION RATES

Void Space	Type	Nature of Surface	Ref. No.	Media trade name	Specific surface sf/cu ft (m^2/m^3)	Voids, percent	Surface or volume denitrification rate at stated temperature, C lb N rem/sf/day x 10^{4} [a] (lb N rem/1000 cf/day [b])																				
							5	10	11	12	13	14	15	16	17	18	19	20	21	22	23	24	25	26	27	Other
Nitrogen gas	Packed bed	High porosity corrugated sheet modules	6, 21	Munters	68 (223)	~95									6.2 (42)		18.6 (126)									
			6, 21	Plasdek	42 (138)	~95															12.1 (82)					
Liquid	Packed bed	High porosity corrugated sheet modules or dumped media	22	Koch Flexirings	65 (213)	96					.43 (2.8)		.50 (3.3)		.53 (3.4			1.3 (8.5)	1.1 (7.2)		.59 (3.8)					
			23	Envirotech Surfpac	27 (89)	94																				.93[c] (2.5)
			24	Koch Flexirings	105 (344)	92																			2.02 (21)	
			7, 20	Intalox saddles	142 to 274 (466 to 899)	70 to 78	.32 (4.5 to 8.8)	.37 (5.3 to 10.1)					.26 (3.7 to 7.1)					.95 (13.5 to 26)					1.1 (15.6 to 30.1)			
			17, 25	Raschig rings	79 (259)	80																	2.4[d] (19) 1.45[e] (12)			
		Low porosity fine media	18	Gravel d_{50}=3.4 to 14.5 mm	245 to 85 (804 to 279)	28 to 37																			2.3 -	
			26	Sand d_{50}=0.9mm[f]	450[f] (1,476)	-																				3.9 (176)
			24	Sand d_{50}=3 to 4 mm	250 (825)	40																			2.9 (73)	
			27, 28	Sand d_{10}=2.9mm U.C.= 1.13	270[f] (886)						0.6 (16)														4.6 (124)	
	Fluidized bed	High porosity fine media, sand	29	d_{50}=0.85mm[f]	130[f] (426)	-										55 (720)				98 (1275)						
		Activated carbon	30	d_{10}=0.65mm	390[g] (1,279)	-								- (257)				- (308)		- (292) - (388)		- (424)				

[a] 1 kg/m^2/day = 0.205 lb/sf/day
[b] 1 kg/m^3/day = 62.4 lb/1000 CF/day
[c] 10 - 23C
[d] 1 hour detention
[e] 2 hour detention
[f] Estimated from data in publication
[g] Prior to bed expansion

Note: d_{10} = effective size, d_{50} = median particle size, d_{60} = particle size at which 60 percent of the material is smaller and U.C. = uniformity coefficient d_{60}/d_{10}

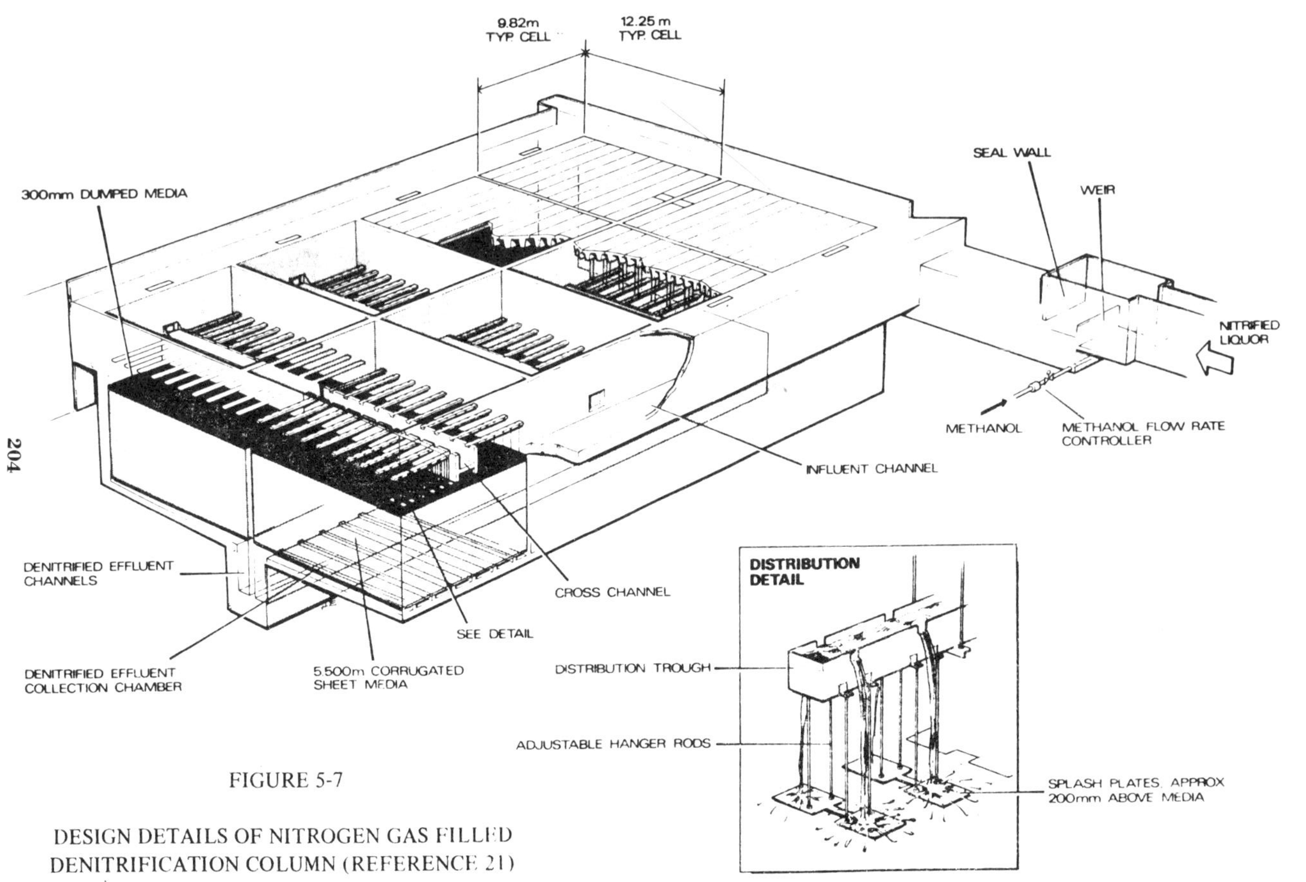

FIGURE 5-7

DESIGN DETAILS OF NITROGEN GAS FILLED DENITRIFICATION COLUMN (REFERENCE 21)

Media specific surface and configuration must be carefully selected to ensure that clogging does not occur. Clogging can occur under high loadings if the restrictions in the media are too small. Pilot studies may be warranted for media selection when design conditions depart from those previously tested. Once the proper media is selected, the process has the advantage that backwashing is not required, considerably simplifying design, construction, and operation. Further, the column construction is simplified since it need not be a pressure vessel as the column is at approximately atmospheric pressure.

The denitrification column was first tested at the Central Contra Costa Sanitary District's Advanced Treatment Test Facility (ATTF).[6,21] The test program had three primary objectives: the first was to develop confidence in the ability of the attached growth reactor to function consistently and predictably; second, to determine the optimum specific area of the plastic media; third, to develop criteria upon which design of the prototype columns for the LMWQCC could be based.

The test denitrification column consisted of a sealed vertical 24-inch (610 mm) diameter reinforced concrete pipe 12 ft (3.66 m) in height and filled with 10 ft (3.05 m) of media. Plastic media consisted of PVC corrugated sheet modules, supplied by Munters Corporation. The top of the column was sealed by a gasketed cover and the nitrified liquor was applied to the top of the tower by a round pattern nozzle. After passing through the column, the denitrified effluent was collected in a sump from where it was discharged. Provision was made to allow pumping of column effluent back to the influent of the column to test the merits of recirculation. A brief description of the test program and a summary of the results is presented on Table 5-4. The initial media tested had a specific surface of 68 sq ft per cu ft (223 m^2/m^3). This media clogged at the high application rates applied in May and June 1972 and was subsequently replaced with media having a specific surface of 42 sq ft per cu ft (138 m^2/m^3). Recirculation of effluent was not found to be required and was dropped from the test program because of the high energy costs that would be incurred if recirculation was used in the full-scale plant.

Experience with the pilot column indicated that an application rate of 5 gpm appeared to be a reasonable flow that could be applied continuously without an objectionable buildup of growth on the media. That value is equivalent to a rate of 245 gallons per cubic foot per day (33,300 l/m^3/day). Because the Contra Costa wastewater has a lower nitrogen concentration than the wastewater at the LMWQCC, the hydraulic loading rate was adjusted to ensure an equivalent nitrogen load was applied. The value was then further reduced to permit lower loadings at low wastewater temperature. Taking these factors into account, an ADWF loading rate of 144 gallons per cubic foot per day (12,562 l/m^3/day) was selected for design purposes. In terms of nitrate removal rate on the basis of media surface, this corresponds to 9.9×10^{-4} lb NO_3^-–N rem/sq ft/day (4.83×10^{-3} kg/m^2/day).

While media with a specific surface of 42 sq ft per cu ft (12.8 m^2/m^3) is believed to be not susceptible to clogging in municipal applications based on the limited experience to date, it would appear prudent to provide chlorine addition capability to aid sloughing should clogging occur.

TABLE 5-4
SUMMARY OF OPERATION - NITROGEN GAS FILLED DENITRIFICATION COLUMN

Date, 1972	Effluent flow rate, gpm[a]	Recirculation rate gpm[a]	NO_3^--N, mg/l		TOC, mg/l		Comment
			In	Out	In	Out	
April 24	4.7	9.4	27	4.5	26	15.5	Media installed had 68 sq ft/cu ft. (20.7 m^2/m^3)
April 27	5.3	9.4	22	1.6	29	19	
May 2	2.3	9.4	28	<0.2	37	25	
May 8	2.1	9.4	30	<0.4	36	11	
May 11	2.0	8.0	24.5	Nil	40	16	
May 16	2.5	0	28	1.6	10	13	Eliminate recirculation and increase flow rate.
May 18	2.5/5.0	0	27	<0.1	9	57	
May 23	5.0	0	27	0.6	8	14	
May 26	10.0	0	-	-	-	-	
May 31	10.0	0	22	<1	9	16	
June 1	15.0	0	25.5	3.4	-	-	Column being overloaded.
June 5	15.0	0	-	-	-	-	Severe foaming. Feed rate decreased.
June 9	10.0	0	27	1	7	15	
June 12	10.0	0	27	5	8	17	
June 15	7.0	0	27	6.8	7.5	12	
June 16	-	-	-	-	-	-	Mechanical breakdowns and operation difficulties.
July 27	-	-	-	-	-	-	Test results unreliable.
July 28	5.0	0	15.3	5.4	46	38	
August 1	5.0	0	13.5	2.5	10	10	
August 4	5.0	0	11.8	2.7	73	45	
August 9	5.0	0	11.8	1.0	11	14	
August 15	5.0	0	9.8	Nil	-	-	
August 21	5.0	0	13.2	4.0	8	18	
August 25	5.0	0	12.5	5.0	8	13	Media removed.
August 28	5.0	0	23	17	8	12.5	New Media used was 42 sq. ft/cu ft.
August 30	5.0	0	30	22	8	13	(12.8 m^2/m^3)
September 1	5.0	0	18	3.2	8	13	
September 7	5.0	0	27	0.5	-	-	Denitrification reestablished.
September 12	5.0	0	17	Nil	8	18.5	
September 19	5.0	0	24.5	Nil	8.5	12	
September 27	5.0	0	18	0.8	8	12	
October 3	5.0	0	19	Nil	8	15.5	
October 25	5.0	0	20	1.4	6.5	11	

[a]1 gpm 0.065 l/s

The design of the denitrification column for the LMWQCC is portrayed in Figure 5-7 and design data are given in Section 9.5.2.2. Operation will be by gravity flow, as the LMWQCC is located on a very steep site. Fixed distribution troughs with splash plates were used to handle the wide range in flows expected and to minimize the restrictions in the distribution system which might be clogged by the growth of denitrifiers. Further, a top layer of dumped media is placed over the corrugated media to ensure good distribution and growth at the top of the column.

5.3.2.2. Submerged High Porosity Media Columns – Packed Bed

Submerged denitrification columns packed with high porosity media have been piloted at several locations[7,17,20,22,23] and tested full-scale at El Lago, Texas.[24]

A typical schematic illustrating essential process elements is shown on Figure 5-8. The media is normally contained in pressure vessels. To obtain sufficient contact time, a series configuration of 2 or 3 vessels is employed.[7,20,24] Either an upflow or downflow column operation may be used. While a variety of media types have been used (Table 5-3), a

FIGURE 5-8

TYPICAL PROCESS SCHEMATIC FOR SUBMERGED HIGH POROSITY MEDIA COLUMNS

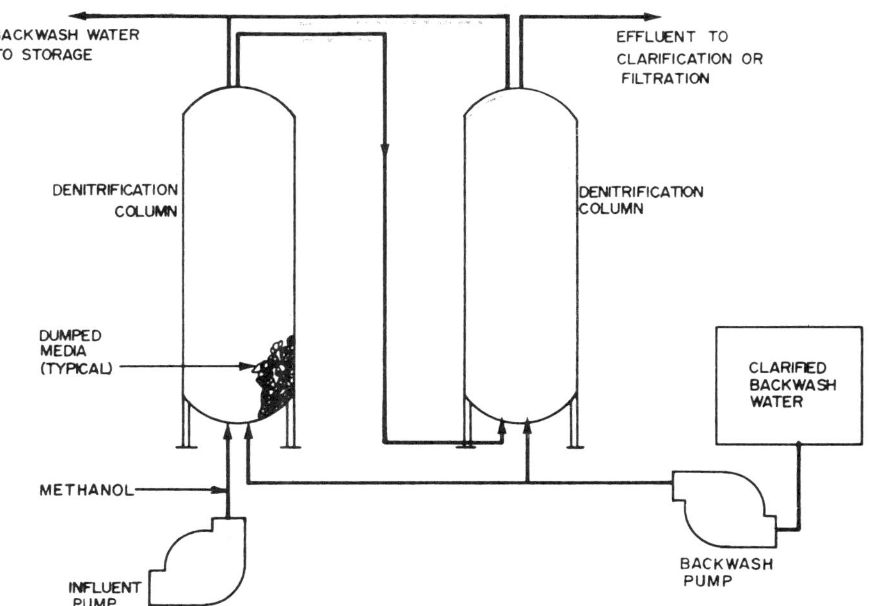

common characteristic is that a high void volume is maintained in the unit. As a consequence, biomass is allowed to continuously slough from the media, minimizing the requirements for backwashing. A corollary is that the media does not build up the layers of biomass that would develop if the void fraction were smaller such as with sand or pebble media.[23] The lower surface denitrification rates for this type of media compared to sand or rock (Table 5-3) reflect this difference in attached biomass development. Most often dumped media have been used, though there is one instance when corrugated sheet media has been tried.[23]

Backwashing, though infrequently used, is still required. At El Lago, Texas, where the media was Koch Flexirings, the water backwash rate was 10 gpm/sf (13.5 $l/s/m^2$) coupled with an air backwash rate of 10 cfm/sf (3.6 $m^3/m^2/min$). Backwashing was routinely done every four weeks.[24] Backwashing in this type of column is not required due to excessive head losses in the column; rather, it is required to prevent the accumulated solids in the column from continuously sloughing into the effluent and causing high effluent suspended solids. Others have used backwash rates up to 44 gpm/sf (29.7 $l/s/m^2$) but did not use an air backwash procedure.[7,20] The El Lago air-water backwash procedure is the recommended approach for design purposes. As opposed to the situation with respect to other column designs, a fairly broad data base exists for this column type. Surface removal rates observed at various locations are summarized in Figure 5-9. As may be seen, most data points fit the data correlation of Sutton, *et al.*[20] for Intalox saddles.

Figure 5-9 may be used to size the denitrification column. First, peak diurnal nitrate loading and minimum wastewater temperature must be known. From Figure 5-9, surface removal rate can be determined. Then from the loading, the media surface area can be calculated. Finally, a specific media is selected and column volume requirements are calculated.

5.3.2.3 Submerged Low Porosity Fine Media Columns – Packed Bed Configuration

The submerged low porosity column using fine media (Table 5-3) is the column system seeing widest commercial application at the present time. One manufacturer's concept (Dravo) of how to incorporate this type of column into a treatment plant is shown in Figure 5-10.[27] In this flowsheet combined carbon oxidation-nitrification is accomplished in an activated sludge step. Of course, other nitrification processes could be employed for producing a nitrified effluent. Clarified nitrified effluent then flows to the denitrification column. The concept employed in this flowsheet is that the column combines two functions in one. First, it serves the purpose of denitrifying the wastewater; second, the column serves the purpose of effluent filtration that normally would be required in many plants anyway.[27] A discussion of the cost-effectiveness of combining the denitrification and filtration functions is presented at the end of this section.

The units manufactured by the Dravo Corp. typically consist of 6 ft (1.83 m) of uniformly graded sand 2 to 4 mm in size. Filtration rates normally recommended by the Dravo Corp. when removing 20 mg/l NO_3^- –N from municipal wastewaters are 2.5 and 1.0 gpm/sf for

FIGURE 5-9

SURFACE DENITRIFICATION RATE FOR SUBMERGED HIGH POROSITY MEDIA COLUMNS

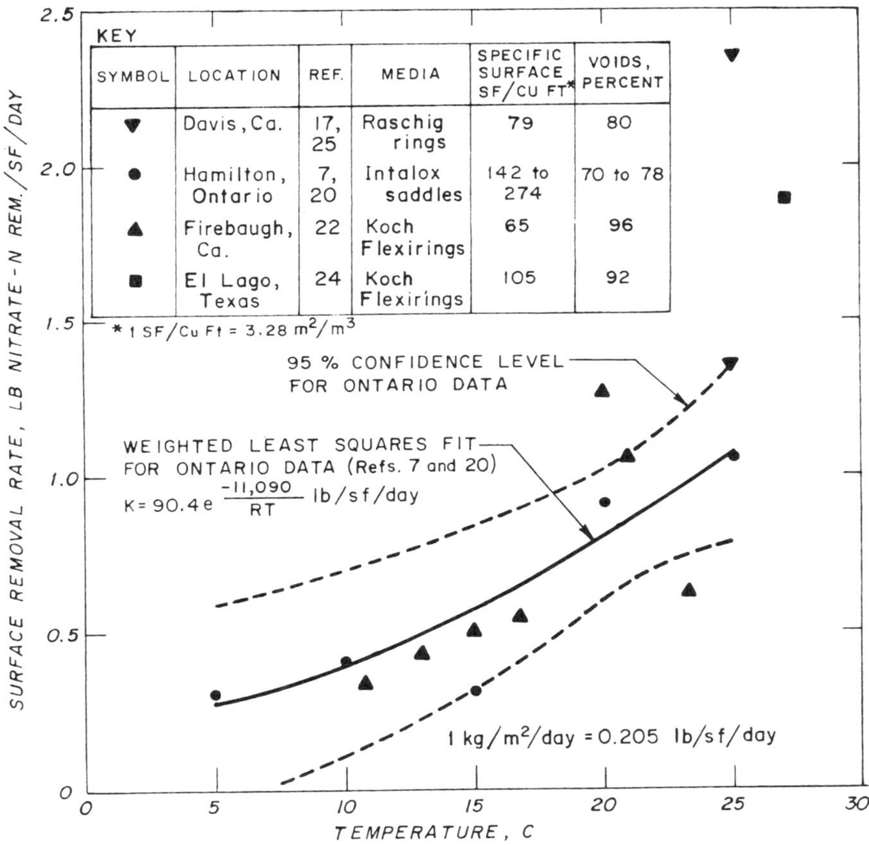

minimum wastewater temperatures of 21 C and 10 C respectively.[31]

The procedure for backwashing the Dravo filter begins with one or two minutes of air agitation followed by 10 to 15 minutes of air and water scouring and finally five minutes of water rinse. Air and water backwash rates recommended by Dravo are 6 cfm/sf (1.83 m^3/m^2/min) and 8 gpm/sf (5.41 l/s/m^2), respectively.[31] In addition, it has been found that nitrogen gas accumulates in the filter during a filter run. This imposes a loss of head on the

FIGURE 5-10

NITRIFICATION-DENITRIFICATION FLOW SHEET UTILIZING LOW POROSITY FINE MEDIA IN COLUMNS (REFERENCE 27)

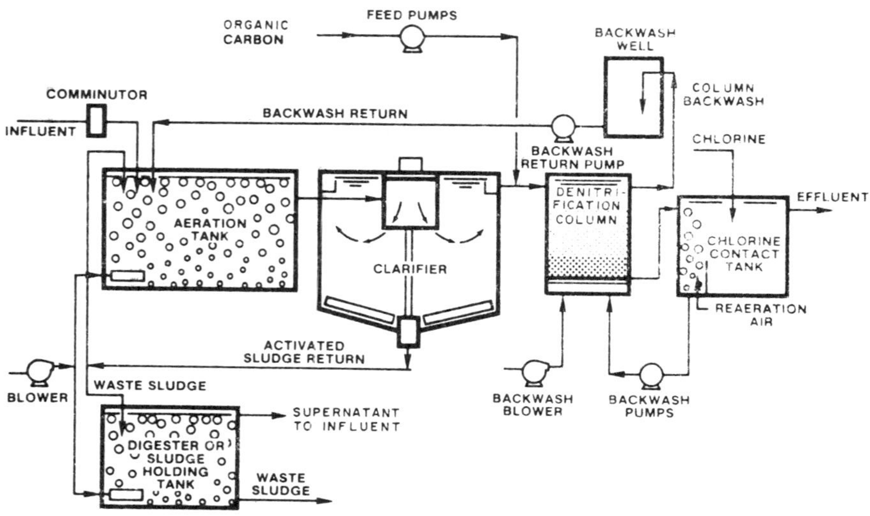

filter and requires periodic removal of the trapped gas bubbles in the media. A "bumping" procedure was evolved whereby a filter was taken out of service for what amounts to a short backwash cycle. In the Dravo design, "bumping" backwash rates of 8 to 16 gpm/sf (5.4 to 10.8 $l/s/m^2$) for one or two minutes is required every four to twelve hours.[27]

The period between regular backwashes in the Dravo filter is dependent on the rate of headloss buildup. At El Lago, Texas it was found that daily backwashing was required whereas in a pilot study in Tampa, Florida the time between regular backwashes ranged from 4 to 40 days.[24,32] During the Tampa study, the air backwashing was found to cause a temporary partial inhibition of denitrification that was not present when only a water backwash was used. For instance, with an influent nitrate-nitrogen level of 15 mg/l, the effluent nitrate nitrogen level was 10 mg/l one-half hour after the air-water backwashing and reached 0 mg/l seven and one-half hours after backwashing.[32] In multiple filter installations, the effluents from the recently backwashed filter would be blended with other normally operating filters, so the impact of this nitrate leakage would be expected to be moderated. Generally speaking, even the smallest of plants will require multiple filters so that an effluent can be continously produced, otherwise a filter influent storage basin will be required.

Neptune-Microfloc, Inc. has made available suggested design guidelines for their media designs.[33] Four media designs were tested on a nitrified effluent containing 20-30 mg/l of NO_3^--N from an extended aeration plant. Best overall performance was obtained from the two media designs shown in Table 5-5. Basic conclusions of the study were as follows:

> "Utilizing a 36-inch (0.92m) mixed-media filter (F-III), essentially complete denitrification of a highly nitfified wastewater can be achieved at filtration rates of 1.5 gpm/sq ft (1.01/s/m^2) for temperatures of 10 C, and at 3 gpm/sf (2.01/s/m^2) at temperatures of 20 C. The methanol to nitrate nitrogen ratio was found to be between 2.0 and 2.5. At applied nitrate nitrogen concentrations of 10 mg/l, filter run times between 16 and 24 hours to 8 feet of headloss were realized at a filtration rate of 3 gpm/sq ft (2.01/s/m^2). At higher applied nitrogen levels, filter runs were reduced in direct relation to nitrogen concentration."[33]

In another study of a Neptune-Microfloc filter, a fully nitrified effluent from a trickling filter was denitrified. No attempt was made to determine limiting filter loadings, however.

TABLE 5-5

NEPTUNE-MICROFLOC MEDIA DESIGNS
FOR DENITRIFICATION (REFERENCE 33)

Filter Material	Layer depths, inches (cm)	
	F-II	F-III
Garnet Sand d_{10} = 0.27 mm[a]	3 (7.6)	3 (7.6)
Silica Sand d_{10} = 0.5 mm[a]	9 (22.9)	9 (22.9)
Anthracite Coal d_{10} = 1.05 mm[a]	18 (45.1)	8 (20.3)
Anthracite Coal d_{10} = 1.75 mm[a]	-	16 (40.6)

[a] d_{10} = effective size

The filter had three layers of media: anthracite, silica sand and garnet sand with sizes ranging from 1.2 mm at the top to 0.2 mm on the bottom. At a surface application rate of 2.3 gpm/sf (1.61/s/m^2) and influent nitrate levels averaging 8 to 9 mg/l, better than 95 percent removals were obtained. Operating temperature ranged from 16 to 18 C during this study.[34,35]

Media size is important in establishing denitrification column requirements. The relationship between specific surface and column size was established in a pilot study at Lebanon, Ohio using the following media sizes for three columns: 3.4 mm, 5.9 mm, and 14.5 mm.[18] Biological film development per unit surface area was shown to be approximately the same for each size media. Therefore, the smaller the media, the higher the media surface per unit volume and the smaller the column as shown in Figure 5-11.

FIGURE 5-11

COLUMN DEPTH VS SPECIFIC SURFACE AREA (REFERENCE 18)

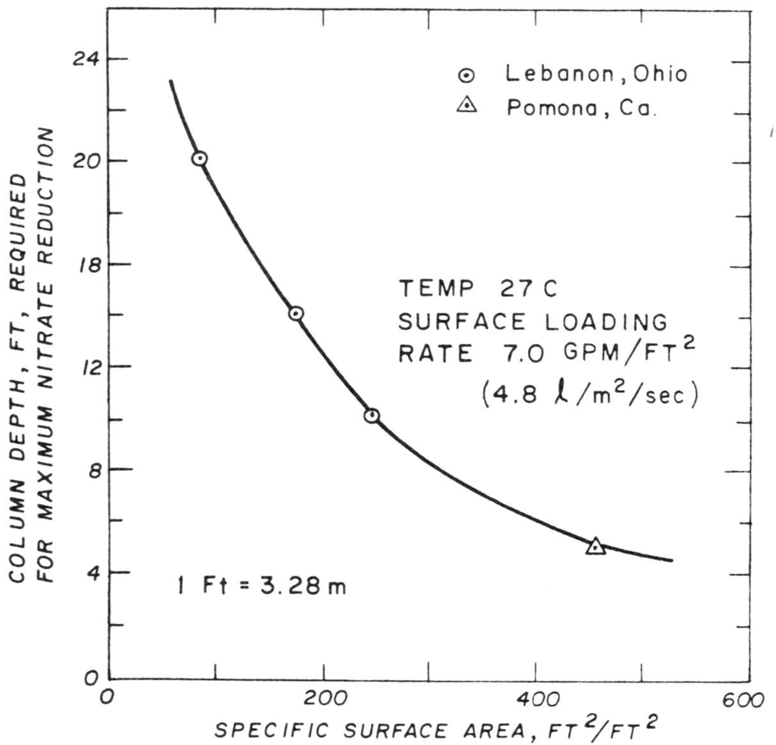

Care should be exercised in the design of the column underdrain system. Neptune-Microfloc recommends that an extremely open underdrain system be employed (pipe lateral, Leopold tile, etc.) to avoid the very real possibility that an overfeed of methanol will cause denitrifier growth to clog the underdrain as was experienced in one test with a nozzle-type underdrain.[33]

Since it has been proposed to use this type of column for both filtration of suspended solids and nitrate removal, it would be well to examine the performance of the columnar system for suspended solids removal. Various observations of suspended solids removal are shown in Table 5-6. With the exception of the El Lago, Texas data, the performance of these columns as tertiary filters falls within the range normally expected for tertiary filtration (for comparative data see Section 9.3.2.3 of the *Process Design Manual for Suspended Solids Removal,* an EPA Technology Transfer publication).[36] Suspended solids removals will be affected by filter design and the fact that the filter is operating as a biological treatment system as opposed to a purely physical separation process.

In considering the use of this process as both an effluent filter and a denitrification system, an important design factor should be borne in mind that has considerable implications on

TABLE 5-6

COMPARISON OF SUSPENDED SOLIDS REMOVAL EFFICIENCY FOR SUBMERGED FINE MEDIA DENITRIFICATION COLUMNS

Location	Media type	Reference	Surface loading gal/min/sf (l/s/m²)	Depth, ft (m)	Influent SS, mg/l	Effluent SS, mg/l	SS removal efficiency, percent
El Lago, Texas	Dravo d_{50} = 3 to 4mm[a]	24	6.27 (4.23)	13 (4)	37	17	54
North Huntington Township, Pa	Dravo d_{10} = 2.9mm	27,28,31	0.72 (0.49)	6.0 (1.8)	16	7	56
Tampa, Fla.	Dravo d_{10} = 2.9 mm	27,31	2.5 (1.70)	a	20	5	75
Lebanon, Ohio	d_{50} = 3.4mm[a]	18	7.0 (4.75)	10 (3.1)	13	4	69
	d_{50} = 5.9mm[a]	18	7.0 (4.75)	20 (6.1)	13	2	85
	d_{50} = 14.5mm[a]	18	7.0 (4.75)	20 (6.1)	13	1	92
Corvallis, Or.	Neptune Media F-II [b]	33	3.0 (2.04)	2.5 (0.76)	25-65	8	68-88
	Media F-III [b]	34,35	3.0 (2.04)	3.0 (0.91)	25-65	4	84-93
Midland, Mi.	Neptune		2.5 (1.70)	5.0 (1.50)	13-30	2-10	67-93

[a] uniformly graded

[b] Table 5-5

cost. It has been claimed that combining the functions of filtration and denitrification reduces tankage and equipment requirements and therefore yields cost savings in plants requiring filtration.[27] However, it should be recognized that the column loading criteria are different for the functions of filtration and nitrogen removal. For effluent filtration, fairly high hydraulic loadings can be applied (4 to 6 gpm/sf or 2.7 to 4.1 l/s/m^2). However, for filters 3 to 6 ft (0.9 to 1.8 m) deep acting as denitrification columns, available data indicates that hydraulic loading should be between 0.5 to 1.5 gpm/sf (0.34 to 1.02 l/s/m^2) at a wastewater temperature of 10 C. Thus, to accomplish denitrification at 10 C, it would be necessary to have column surface areas five times as large as required for filtration alone. Thus, an economic analysis must be done in each case to determine the most economic process configuration.

5.3.2.4 Submerged High Porosity Fine Media Columns – Fluidized Bed

The introduction of fluidized bed technology into the field of columnar denitrification is a comparatively recent development.[29,30,37,38] Figure 5-12 depicts a typical fluidized bed reactor with its ancillary facilities. In the fluidized bed unit wastewater passes upwards vertically through a bed of small media such as activated carbon or sand at a sufficient velocity to cause motion or fluidization of the media. The small media provides a large surface for growth of denitrifiers.

High surface application rates were recently employed in a pilot study of the process at Nassau County, New York (15 gpm/sf or 10.2 l/s/m^2).[38] The column had a fluidized bed depth of 12 ft (3.7 m). The bed settled to about 6 ft (1.8 m) when the influent was shut off

FIGURE 5-12

FLUIDIZED BED DENITRIFICATION SYSTEM

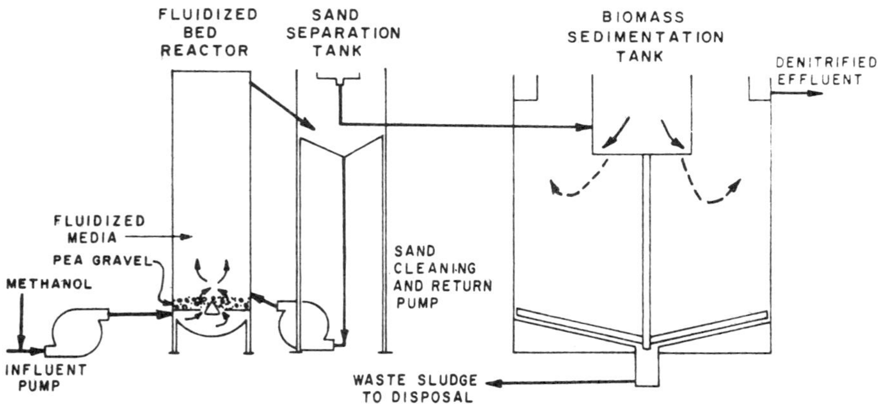

so bed expansion during operation was 100 percent. Initially the clean bed in the column contained 1.5 in. (38 mm) of pea gravel and 3 ft (0.91 m) of silica sand with an effective size of 0.6 mm and a uniformity coefficient of 1.5. During operation, the media became completely covered with denitrifier growth and the individual particles grew in size. During the initial lab-scale test for this process, the media grew from 0.65 mm particles, to particles 3 to 4 mm in size.[30] The attached growth accounts for the greater depth of media in the non-fluidized bed after the column had been in operation.

In a packed bed, this growth of particle size would result in high headloss, channeling, and a loss in efficiency. In an expanded bed, however, there are sufficient voids between the denitrifier-sand particles to provide good liquid contact at modest headlosses. This greater biological film development allows higher surface reaction rates (expressed per unit of media surface) than for any other type of column configuration as shown in Table 5-3. Since the surface contained in a unit volume is high, higher volumetric loadings are also possible as compared to any other column configuration (Table 5-3). Empty bed detention time during the recent pilot test at Nassau County was only 6.5 min. Maximum nitrate removal rates as a function of temperature are shown in Figure 5-13 and are based on the Nassau County data.[39] If diurnal variations in nitrogen load are to be accommodated by the column without nitrate bleedthrough, then column volume requirements will be greater than that determined in Figure 5-13. A provisional recommendation would be to increase the reactor requirements by the ratio of the peak to average nitrogen loads.

The process has been shown to be responsive to both diurnal load variations and to cold temperature operation.[29] Methanol feed was not automatically controlled, so periods of nitrate bleedthrough occurred.[29] When methanol feed was under control, 99 percent removal of influent nitrate and nitrite was demonstrated. Total nitrogen reductions were not given.[29,38]

While backwashing facilities are not required in this type of column, facilities must be provided for managing the column media inventory. During operation, the denitrification column increases in depth due to biological growth causing a continuous small loss of media from the system. Further, diurnal flow variations cause height variations which may contribute to media loss. This loss can be minimized by provision of flow equalization facilities (see Chapter 3, Flow Equalization, *Process Design Manual for Upgrading Existing Wastewater Treatment Plants,* an EPA Technology Transfer Publication).[40]

The manufacturer of the system suggests that for most plants subject to diurnal flow variations, media losses and effluent solids levels can be controlled by placing two tanks in series with the column as shown in Figure 5-12. The first tank would be a sand separation tank followed by a biomass sedimentation tank for biological solids removal. The sand separation tank might be very small, as a tank with an overflow rate of 13,600 gpd/sf (554 m^3/m^2/day) served satisfactorily during the pilot study. It has been suggested by Ecolotrol, the manufacturer of the fluidized bed system, that the swirl concentrator which was developed for grit removal from combined stormwater and wastewaters could serve as the

FIGURE 5-13

VOLUME DENITRIFICATION RATE FOR SUBMERGED HIGH POROSITY FINE MEDIA COLUMNS (REFERENCE 39)

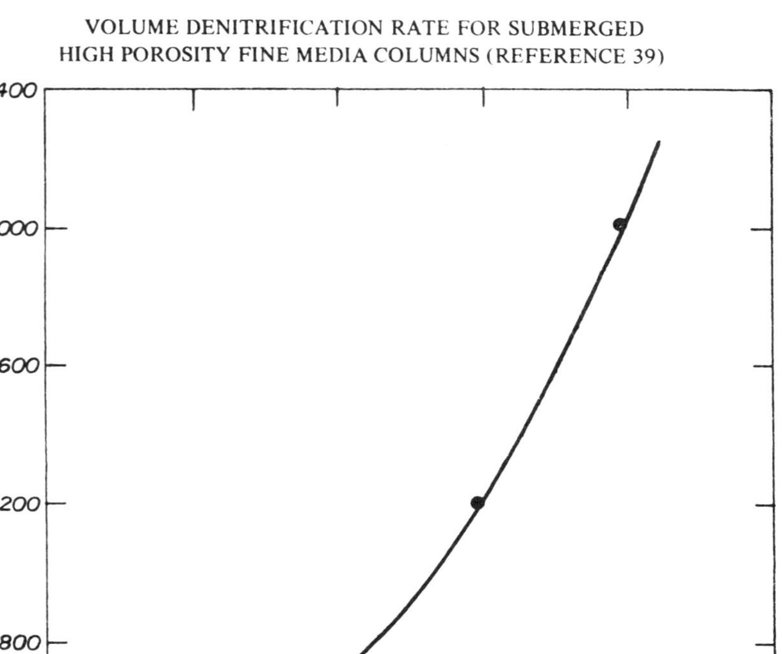

sand separation device.[41,42] The media settling in the sand separation tank would be pumped back to the column with the pumping action shearing the denitrifiers from the media. This sheared biomass would pass through the column and sand separating tank and then settle in the biomass sedimentation tank.[29] If very low levels of suspended solids are required, the system would have to be followed with tertiary filtration.

The manufacturer's cost estimate indicates that the fluidized bed system is competitive with suspended growth or packed bed columnar systems, but it was also stated that these must be confirmed by larger scale tests on the fluidized bed system.[29]

5.3.2.5 Comparison of Attached Growth Denitrification Systems

The three column systems previously described in Sections 5.3.2.1, 5.3.2.2, and 5.3.2.3 are seeing commercial applications at this time and the fluidized bed system described in Section 5.3.2.4 will likely see application in the near future. Therefore, the design engineer has four alternative column systems to consider.

Where low treatment plant effluent solids are required, tertiary filtration will have to follow all column systems except the low porosity fine media system described in Section 5.3.2.3. In small plants, the elimination of a unit process may favor combining the denitrification and filtration functions. In larger plants the cost trade-offs between alternatives need to be considered.

Where space restrictions exist at a plant site, there is an incentive to pick those systems requiring the least land area possible. While from Table 5-3 it might appear that the fluidized bed had the distinct advantage because of highest volumetric loading rates, the ancillary sand separation and biomass sedimentation tanks diminish its advantage over the submerged fine media column and the nitrogen gas filled column.

The submerged high porosity column configuration appears to offer the least attractive alternative. Both surface and volumetric removal rates are low, requiring comparatively large reactors (Table 5-3). Further, the unit must incorporate the design features of a filter, without having the advantage of low effluent suspended solids of the submerged low porosity columns. The system has an advantage for small treatment plants in that backwashing is only infrequently required and can be scheduled to coincide with plant staff availability.

The advantages of the nitrogen gas filled column are: (1) similar space requirements to low porosity submerged columns (2) column walls need not be designed to handle hydrostatic loads and (3) with proper media selection, sloughing occurs naturally and backwashing is not required.

5.4 Methanol Handling, Storage, Feed Control, and Excess Methanol Removal

Methanol is a chemical not normally dealt with in wastewater treatment plant operation and care must be exerted in the design and operation of methanol handling, storage and feeding facilities to ensure its safe and proper use.

5.4.1 Properties of Methanol

Methanol, CH_3OH, has a variety of names such as methyl alcohol, carbinol and wood alcohol and is normally supplied pure (99.90 percent). It is a colorless liquid noncorrosive (except to aluminum and lead) at normal atmospheric temperature. Some important properties of methanol are shown in Table 5-7. Additional data is available in references 43 and 44 and manufacturer's information.

TABLE 5-7

PROPERTIES OF METHANOL

Property		Value
Density		0.7913 g/ml @ 20C (6.59 lb/gal)
Vapor Density (air = 1.00)		1.11
Vapor Pressure	0 C	29 mm Hg
	10 C	52 mm Hg
	20 C	96 mm Hg
	30 C	159 mm Hg
	40 C	258 mm Hg
	50 C	410 mm Hg
Solubility		Miscible in all proportions with water
Viscosity @ 20 C		0.614 cps
Combustible Limits, percent by volume in air at STP		7.3 to 36
Flash Point Tag Open Cup		16 C (61 F)
Tag Closed Cup		12 C (54 F)

Taken internally, methanol is highly toxic. It is harmful if the vapors are inhaled or if skin contact by liquid or vapors is prolonged or repeated. Fire and explosion are primary dangers of methanol. Persons involved in handling methanol should be aware of these hazards. Federal, state and local regulations for safety should be posted along with information from references 43, 45 and manufacturer's data.

5.4.2 Standards for Shipping, Unloading, Storage and Handling

The shipping, unloading, storage and handling of any flammable chemical including methanol is governed by a multitude of stringent regulations which include: Federal, such as the Department of Transportation (DOT) and the Occupational Safety and Health Act (OSHA); State, which has various safety orders and codes; municipal ordinances; independent associations such as the National Fire Protection Association (NFPA) and the Manufacturing Chemists Association (MCA); and insurance requirements. It is necessary that all of these regulations be reviewed and studied before the design of any methanol facilities, and all such regulations must be followed.

5.4.3 Methanol Delivery and Unloading

Methanol may be received in 55 gallon (208 l) metal drums, tank wagon, tank truck or tank cars. Other methods of shipping, not discussed herein, are by barge, metal drums (less than 55 gal) and glass and metal cans. Tank wagons are normally 1,000 to 4,000 gal (3,785 to 15,142 l) in size, tank trucks range from 4,000 to 9,000 gal (15,142 to 34,069 l) while tank cars are shipped in 6,000, 8,000 and 10,000 gal (22,713, 30,283, 37,854 l) capacities. Tank cars and tank trucks are the most economical shipping mode for most plants. However, for pilot work and small plants, 55 gallon metal drums may be feasible. Since methanol is classified as a flammable liquid by the DOT, all shipping containers must be approved and labeled in accordance with applicable DOT regulations.

The recommended method of unloading methanol from any container is by pumping. Some barges and tank wagons have their own pumps for unloading. Tank cars and trucks can be unloaded from the top or bottom and be pumped or conveyed by gravity or syphoning. The preferred method of unloading is pumping from the top via an eductor tube. Syphoning and gravity unloading are only permitted when the top of the storage vessel is below the bottom of the shipping container. Due to the increased spillage probability using bottom unloading, it is only permitted on cars or trucks approved for bottom unloading which include valving approved by the Association of American Railroads (AAR) and in agreement with the DOT requirements. This valving helps contain the product by safely controlling flow. Additional precautions such as fusible link valves and excess flow valves may be used.

Air pressurization of the tank ("air padding") must never be used for methanol unloading. However, top unloading using the water displacement method or inert gas padding, i.e. carbon dioxide, nitrogen, etc., may be used if the exact unloading procedures are as provided by the chemical manufacturer. Unloading procedures can be found in references 43 and 45 and the supplier's data.

General requirements for the design of unloading facilities for methanol are applicable to both tank car and truck. The unloading area should be arranged to avoid traffic areas. Also, all facilities should be outside due to the fire hazard and all equipment in the vapor area must be explosion-proof, Class I, Group D, Division 1 or 2 per the National Electrical Code. Tools should be "non-sparking." Unloading should occur during daylight hours since the safety and lighting requirements for night operation are very extensive. Ample fire extinguishers, safety blankets, deluge showers, eye washes, no smoking signs and unloading signs are also required.

If top unloading will be practiced, approach platforms are required for access to the top of the tank.[46] The approach platforms used at the CCCSD Water Reclamation Plant are steel, swinging type with pneumatic operated drawbridges. These drawbridges provide for three feet (0.9 m) of horizontal adjustment, 30 in. (76 cm) of vertical adjustment and a 45 degree pivot to either side for ease in unloading both tank cars and trucks. The drawbridges fold to the platforms when not in use providing the required railroad clearances.

In all unloading setups, all equipment must be grounded. This includes the shipping vessel, interconnecting piping, pumps, approach platforms, etc. Also, bonding jumpers must be used to provide a good continuous system. Periodic checks of the grounding system must be made.

Static electricity buildup must be minimized since it is not dangerous until at the spark discharge level. A spark discharge can easily start a fire or cause an explosion. Refer to reference 47 for static electricity accumulation prevention.

Facilities for truck unloading must provide for ease in truck maneuvering, both entering and leaving the area. Consideration must be given to the number of individual unloading spots regarding frequency of use, space and simultaneous unloading. For ease in truck traffic, it is best to have parallel unloading areas for straight-through driving.

Rail unloading creates additional considerations. Unloading areas must have derails or a closed switch a minimum of one car length from the car. A primary concern is who spots the cars, the plant or the railroad. It is preferable to have private sidings so the railroad can drop off or pick up cars at any time without disrupting plant operations. Also, the cost of having the railroad spot cars at night, over weekends or holidays is high and because the railroad cannot guarantee time of shipment, safety provisions and lighting must be provided for night operations. Two sidings, one for empty cars and one for full cars should be provided. The cars can be spotted by the plant personnel with rail car movers which operate both on rail or streets or in the case of short distances, car spotters (winches) can be used.

By paving the railroad yard area, both truck and rail unloading can be practiced. This is advantageous because major strikes affecting either kind of transport cannot cripple the plant. At the CCCSD Water Reclamation Plant (described in Section 9.5.2.1), two sidings are provided with space for ten cars on each track for storage. Unloading sidings with two platforms and two bottom stations will be used for simultaneous spotting and unloading. A rail car mover with a capacity of two cars is also provided.

Unloading equipment is normally steel but many other materials, except aluminum, are acceptable providing they can withstand the pressure and are completely grounded. Pumps must be "non-sparking" such as bronze fitted steel pumps with bronze impellers. Many materials are compatible with methanol so seals and gaskets can be common materials. Pumps may be either centrifugal or positive displacement gear type. However, positive displacement pumps must have relief valves. Because of the widely varying heads encountered during unloading, care must be taken in pump selection. Piping should have as few joints as possible and should be Schedule 40 minimum. Splash guards at joints may be desired in traffic areas. Valves may be gate, plug or diaphragm, iron or steel with bronze trim, and neoprene plugs in the plug valve or a neoprene disc in the diaphragm valves. Refineries on the West Coast have adopted a standard of cast steel valves on all flammable materials to prevent damage during a fire. Couplings must be leak tight and it is preferable to have a valve next to the coupling to limit material leakage and waste during disconnection. If flexible hose connections are used, a coupling with an integral valve can be used. A strainer should be used ahead of any pumping or storage equipment.

Care must be exercised to not overflow the storage vessel. A high level alarm and pump shutoff should be utilized. Due to the increasing cost of methanol, it may be desirable to have a flowmeter in the unloading piping. All vessels must be vented during unloading or loading.

5.4.4 Methanol Storage

In order to provide for possible delays in methanol delivery, a storage capability of two to four weeks supply is recommended. The volume of storage will be determined by various site and cost requirements; however, storage of less than two weeks is too short for expected delivery delays and strikes. Tank truck deliveries require in-plant storage. However, with rail deliveries, the rail cars can be used for storage, but charges (demurrage) are levied by the carriers for time on site in excess of a fixed time. For small plants, demurrage may become cost effective. However, carriers may have a time limit on cars or have excessive demurrage charges.

Methanol may be stored in vertical or horizontal tanks above ground inside or outside, or buried. It is strongly recommended that all methanol equipment and tanks be located outside. If interior storage is required refer to reference 43 and 44 for detailed requirements. An exception to this rule is drum storage which, if not stored indoors, must be shaded from direct sunlight or constantly sprinkled with water.

Layout of methanol tanks should be in accordance with reference 48. There should also be a dike around each aboveground tank or group of tanks to contain 125 percent of the largest tank volume in case of rupture or fire. If the tanks are not of steel, care must be taken so that a fire will not cause a rupture in the group of tanks thereby overflowing the dike. Fire protection is very critical, especially when the tanks are near other structures. For large volumes of methanol storage, low expansion alcohol-type foam is used for fire extinguishing. For very small fires, dry chemical or carbon dioxide extinguishers can be used. Rate-of-rise or ultraviolet detectors may be used for sensing of fire and initiating automatic foam release. Water should not be used, but may be used for plant area fire control.

Storage tanks are normally of steel, but most materials are satisfactory except for aluminum. Tank size is only dependent upon the capacities required and any size limitations imposed by the tank material. Piping, valves, etc. should be as described in Section 5.4.3. Tank fittings should include the following: (1) an inlet with dip tube to prevent splash and static electricity, (2) an anti-siphon valve or hole on the inlet to prevent back siphonage, (3) vent pipe with pressure-vacuum relief valve[48] with flame arrester, (4) an outlet connection, (5) a drain connection, and (6) various openings for depth gauges, sample points, level switches, etc. Manholes for access should also be provided. Also, extreme corrosion will take place if the tank is drained dry. The tank must also be grounded. Due to increasing air pollution requirements, venting must be controlled by conservation type vents or by maintaining a slight negative pressure in the tank using a small ejector.

To maintain a correct inventory of tank contents, a diaphragm level sensor or float should be used. Low and high level alarms are needed for protection against overfill and settled material at the bottom of the tank. The high level alarm should be separate from the tank sensors for a fail-safe design.

5.4.5 Transfer and Feed

Methanol must be transferred and controlled from the storage vessel to the point of feed. Methanol is removed from the tank and is fed by gravity or pumps. Normally, pumping will be utilized for ease of control. The transfer pumps should always have positive suction pressure and should be protected by a strainer. As with all methanol situations, it is desirable to mount all equipment outside. There are three basic pumping arrangements which can be used: (1) diaphragm chemical feed pumps using an adjustable stroke for volume control; (2) positive displacement pumps with variable speed drives controlled by either counting revolutions to obtain flow or using a flow meter; (3) centrifugal or regenerative turbine pumps with variable speed drives controlled by a flow meter. Each arrangement has its own particular problems and must be studied for each installation. To cover the widest range of feed rates, arrangements 1 and 2 are used due to the limited accuracy range of flow meters.

All pumps, piping, etc., should be the same as noted in Section 5.4.3. All piping should be tested for 1.5 times the maximum system pressure for 30 minutes with zero leakage. Methanol addition to the denitrification process is relatively simple. In the CCCSD's Advanced Treatment Test Facility, methanol was pumped into the influent line ahead of the denitrification reactor and the stirring of the reactor was sufficient mixing. In the CCCSD Water Reclamation Plant, a multi-orifice diffuser is used to evenly distribute methanol in the channel ahead of denitrification.

5.4.6 Methanol Feed Control

Because methanol is expensive and a methanol overdose can result in a high effluent BOD_5, it is essential to accurately pace methanol with oxidized nitrogen load. Simply pacing methanol dose against plant flow is inaccurate as it does not account for daily and diurnal variations in the nitrate concentration. Feed forward control utilizing plant flow and nitrification effluent nitrate nitrogen is shown in Figure 5-14. Feed ratio is approximately three parts methanol per part of nitrate nitrogen by weight (see Section 3.3.2). This method requires continuous on-line measurement of nitrate utilizing an automated wet chemistry analyzer. The wet chemistry analyzer (AIT) output is proportional to nitrate concentration in the nitrification effluent. The manual control station (HIK) provides means to select either the analyzer output or to enter a manual concentration value in case of analyzer failure. The output of HIK is multiplied by a signal proportional to flow from the ratio stations (FFIK) to obtain a signal proportional to required methanol flow ratio. This signal may then be fed to a chemical proportioning pump, as shown, or may be the setpoint of a flow control loop. FFIK provides means to adjust the methanol feed ratio. The dependability of this control procedure is predicated on the reliability of the automated wet chemical analyzer. These analyzers require very careful routine maintenance and calibration. In a research laboratory environment methanol was paced with the use of a Technicon Auto Analyzer for one week periods between maintenance checks.[18]

FIGURE 5-14

FEEDFORWARD CONTROL OF METHANOL BASED ON FLOW AND NITRATE NITROGEN

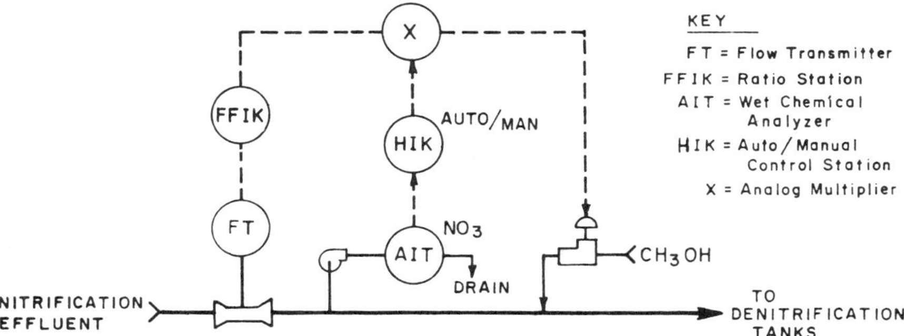

5.4.7 Excess Methanol Removal

Unless specific measures are taken to provide for methanol removal, methanol addition above stoichiometric requirements (Section 3.3.2) will cause methanol to appear in the denitrification process effluent.[3,4,7,15,16,27,29,32,49,50] In one instance, a methanol overdose caused an effluent BOD_5 of 106 mg/l.[3] Placing total reliance on the methanol feed control system to prevent methanol overdoses may be unrealistic in small plants where a trained technician's attention can be expected to be infrequent. The provision of a good methanol control system and a methanol removal system as backup should allow nearly fail-safe operation in terms of preventing effluents containing high levels of organics.

A recent modification of the suspended growth denitrification process,that has as one of its objectives preventing methanol bleedthrough,is shown in Part B of Figure 5-1.[2,51] After denitrification, mixed liquor passes to an aerated stabilization tank. In this tank, facultative organisms "switch over" from using nitrate to dissolved oxygen and oxidize any remaining methanol. While refinements undoubtedly could be made in determining the length of time required for the facultative bacteria to switch over and complete methanol oxidation, it is known that 30 minutes of aeration in an aerated stabilization tank at a Burlington, Ontario pilot plant was insufficient, as high effluent methanol values were periodically observed in the system.[52] A period of about 48 minutes has been found to be sufficient for methanol oxidation.[3] Therefore, a period of one hour aeration is recommended based on experience to date. Further details of this modification concerning solids-liquid separation are presented in Section 5.6.

In attached growth denitrification systems, the provision of an aerated basin after the denitrification column would not ensure oxidation of excess methanol. This is because there is an insufficient mass of facultative organisms in the column effluent to accomplish the biological oxidation of the carbon, since the denitrifying organisms are retained on the media and only a few pass into the column effluent. To date, excess methanol removal systems applicable to attached growth systems have not been developed.

5.5 Combined Carbon Oxidation-Nitrification-Denitrification Systems with Wastewater and Endogenous Carbon Sources

The methanol price increases experienced during late 1973 have caused renewed interest in alternative carbon sources. Alternatives having the least chemical cost for nitrate reduction are the organics present in domestic wastewater or endogenous respiration of the biological sludge. The problems experienced in the past with these sources are lower denitrification rates and contamination of the effluent with the ammonia released when wastewater organics or biological sludge serve as the carbon source for denitrification. The former problem has been mitigated by increasing reactor detention time. The latter problem has been addressed by adapting the suspended growth process configuration in specific ways to expose the sludge to alternating aerobic and anoxic environments so that released ammonia is subjected to nitrification and subsequent denitrification. The alternatives which achieve higher than 80 percent nitrogen removal, while avoiding the use of methanol, combine the carbon oxidation, nitrification and denitrification processes in single sludge systems with no intervening clarification steps.

5.5.1 Systems Using Endogenous Respiration in a Sequential Carbon Oxidation-Nitrification-Denitrification System

When the process uses the endogenous decay of the organisms for denitrification in the system, the rate limiting step becomes the organism's endogenous decay rate in an oxygen-free environment. The process flowsheet for this system is shown on Figure 5-15; it was developed and first tested in Switzerland[53] and subsequently evaluated in other locations.[54,55,56,57] In the initial tests, fairly high mixed liquor solids (5200-5300 mg/l) were employed and detention times of 2.2 to 2.8 hours were used in the denitrification section. Results of the various studies are shown in Table 5-8. Deviations at the other locations from the Swiss results are explained on the basis of insufficient reaction times in either the nitrification and denitrification stages.[59]

FIGURE 5-15

SEQUENTIAL CARBON OXIDATION-NITRIFICATION-DENITRIFICATION

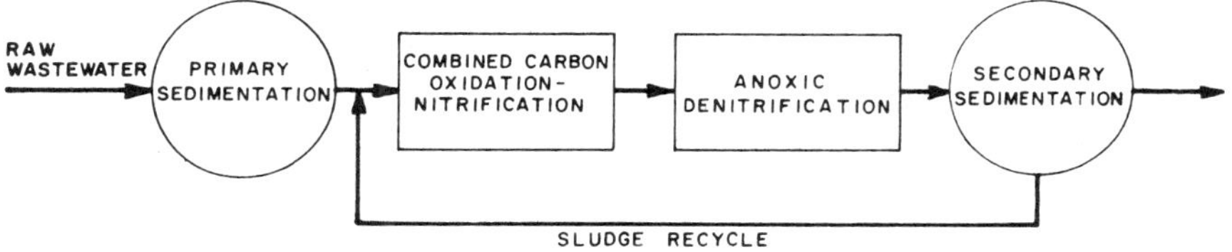

TABLE 5-8

PILOT TESTS OF WUHRMAN'S SEQUENTIAL CARBON OXIDATION-NITRIFICATION DENITRIFICATION SYSTEM (AFTER CHRISTENSEN & HARREMOËS, REF. 59)

Location	Reference	Temperature, C	$\hat{q}_D$, peak denitrification rate lb NO_3^- rem/lb MLVSS/day or kg/kg/day	Effluent NO_3^--N, mg/l	Range of total nitrogen removal, percent
Switzerland	53[a]	13.6	0.0168	2	82-90
		17.1	0.041		
Germany	54[a]	16	0.022	-	40-60
		16	0.026		
Germany	55[a]	12-16	0.038	3	36-88
Germany	56[b]	20	0.048	7.8	46 (average)
Seattle, Wa.	57[a]	20	0.026	-	31-65
New York	58[c]	-	-	1.5 to 3.1	84-89

[a] pilot scale
[b] lab scale
[c] full scale, 95,000 gpd (360 m^3/day)

A modification of the concept tested in New York state involved placing a short anoxic cell prior to the combined carbon oxidation-nitrification tank.[58] An aerobic cell of 0.5 hr detention time was also placed downstream of the anoxic denitrification, presumably to improve settling. Total nitrogen removals of 84 to 89 percent were obtained, which are comparable to the Swiss results.

Denitrification rates determined at the various locations are shown in Figure 5-16. Data for Blue Plains and Pretoria, S.A. are shown also. In these two pilot plants wastewater was used as the principal carbon source, but treatment stages using endogenous respiration were employed also. The data shown for these two pilot plants are for those treatment stages where endogenous respiration was predominant. When data were reported in terms of MLSS only, volatile content was assumed to be 70 percent to allow reporting on a basis consistent with denitrification rates given in Section 5.2.

An envelope has been drawn around all data points in Figure 5-16 to emphasize the variation in measured denitrification rates. In the absence of pilot data, it would be prudent to establish reactor sizes on the basis of the lower envelope line shown in Figure 5-16.

FIGURE 5-16

DENITRIFICATION RATES USING ENDOGENOUS CARBON SOURCES

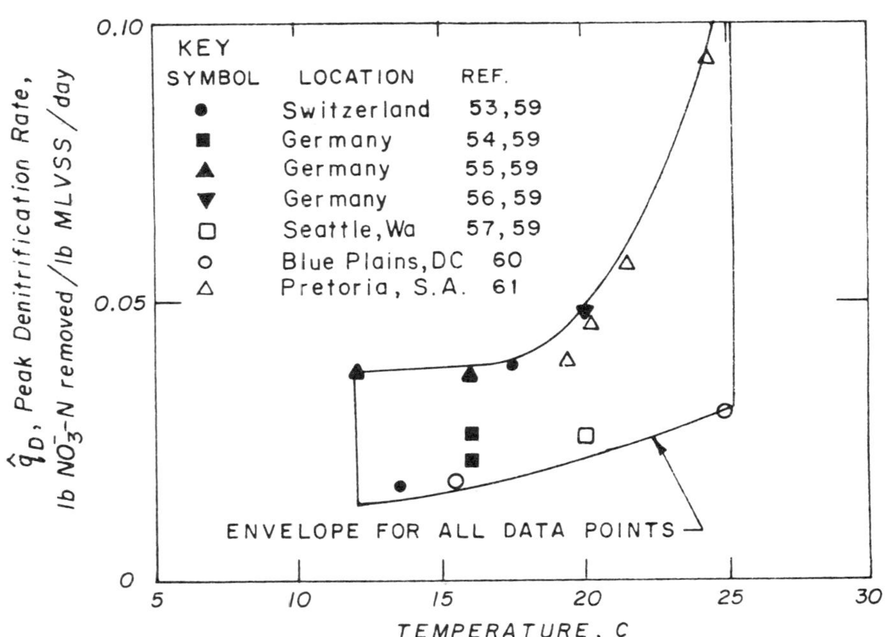

Further, rates shown in Figure 5-16 are peak nitrate removal rates, $\hat{q}_D$, and a safety factor must be employed in design to ensure low nitrate levels in the effluent, as is done in Section 5.2. It is notable that these rates fall considerably below those shown for methanol as the carbon source in Figure 5-2. For instance, median values at 20 C for methanol and endogenous carbon are about 0.25 lb NO_3^- –N rem./lb MLVSS/day (0.25 g/g/day) and 0.04 lb NO_3^- –N rem./lb MLVSS/day (0.04 g/g/day). Therefore, a denitrification reactor using an endogenous carbon source would have to be about six times larger than a reactor with a methanol carbon source at 20 C.

The combined carbon oxidation-nitrification step can be designed with the criteria set forth in Sections 4.3.3 and 4.3.5, because it has been found that the anoxic denitrification step has no impact on sludge activity, if the length of the anoxic period is below 5 hours.[59,62] In the calculations for the carbon oxidation-nitrification function, only the inventory under aeration should be included in the solids retention time, growth rate, and removal rate calculations.

5.5.2 Systems Using Wastewater Carbon in Alternating Aerobic/Anoxic Modes

All of the systems using wastewater as the chief organic carbon source for denitrification use an alternating aerobic-anoxic sequence of stages, without intermediate clarification, to effect total nitrogen removal while attempting to avoid ammonia nitrogen bleedthrough. Some of the demonstrations of these systems have shown that removals of 90 percent were possible with the alternating mode concept.

5.5.2.1 Aerobic/Anoxic Sequences in Oxidation Ditches

Pasveer was the first to investigate denitrification in oxidation ditches and reported total nitrogen removals of 90 percent, but few details have been reported.[59,63] Subsequent investigators have confirmed Pasveer's results and defined the conditions required for denitrification in oxidation ditches.

Ditch design has varied among investigators principally in the means used for aeration. Even though aeration devices have ranged from Kessner brushes and cage aerators to vertical turbine mechanical aerators, all the variations can be termed oxidation ditches because all use the concept of a channel with aeration devices placed at localized points as shown in Figure 5-17. By controlling the level of aeration, the mixed liquor is exposed to alternating aerobic and anoxic zones. The channel operates as an "endless channel" since only a portion of the mixed liquor passing the channel outlet is withdrawn, with the bulk of the flow remaining in the channel. In this way the mixed liquor is recirculated many times through aerobic and anoxic zones prior to discharge from the channel.

The largest scale test of the oxidation ditch employed for nitrogen removal has taken place at the Vienna-Blumenthal plant in Vienna, Austria.[64,65] The plant flowsheet is shown in Figure 5-18 and design data are given in Table 5-9. The plant consists of a pumping station,

FIGURE 5-17

PASVEER DITCH OR ENDLESS CHANNEL SYSTEM FOR NITROGEN REMOVAL

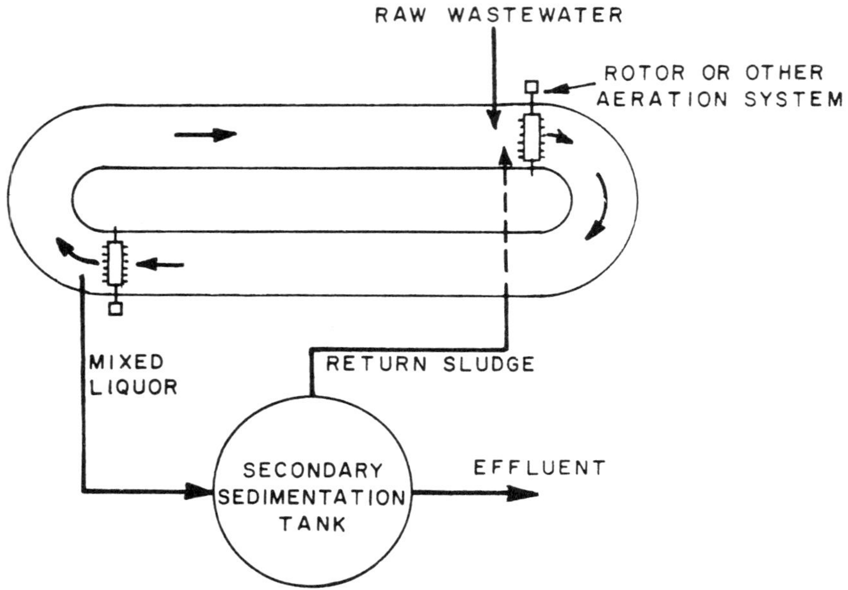

screens, aerated grit chamber, two aeration tanks, two secondary sedimentation tanks and a return sludge pumping station. The plant does not incorporate either primary sedimentation or sludge handling facilities. It was constructed in 1968 at a cost of $1,623,000 (1968 U.S. dollars).

Since the Vienna-Blumenthal plant is currently operating below design capacity, it has been possible to operate it in a manner encouraging nitrogen removal. The two aeration tanks have been connected in series and the number of operational cage aerators varied to encourage nitrogen removal. It has been found that dissolved oxygen could be measured in the mixed liquor immediately after the rotor, however the oxygen demand of the microorganisms caused the oxygen to be depleted prior to contact with the next rotor. This resulted in an alternating contact of the mixed liquor with aerobic and anoxic zones. Nitrification took place in the aerobic zones and denitrification occurred in the anoxic zones.[65]

FIGURE 5-18

VIENNA-BLUMENTHAL WASTEWATER TREATMENT PLANT

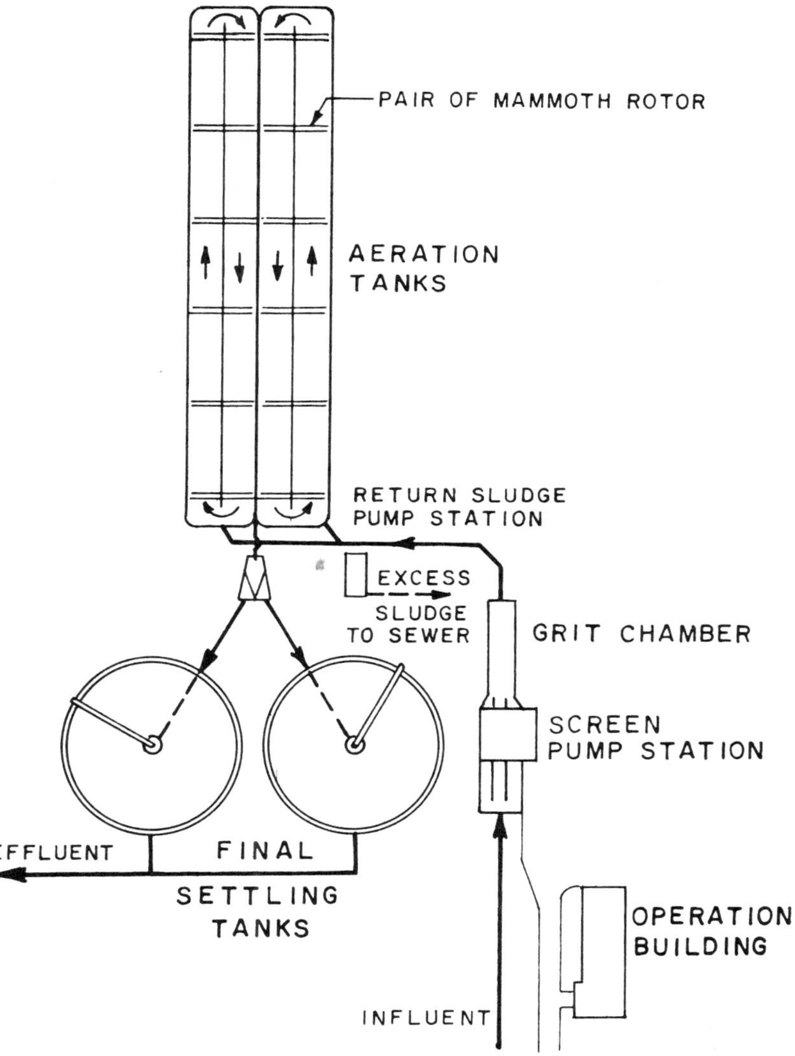

TABLE 5-9

DESIGN DATA FOR THE VIENNA-BLUMENTHAL TREATMENT PLANT (REFERENCE 65)

Population equivalents	150,000[a]
Average dry weather flow (ADWF)	22.8 mgd[a] (1.0 m^3/sec)
Aerated grit removal tanks	
Number	1
Volume	9,500 cu ft (270 m^3)
Detention time, ADWF	4.5 min
Aeration tanks	
Number	2
Passes/tank	2
Length/pass	492 ft (150m)
Width/pass	29 ft (8.5m)
Depth	8.2 ft (2.5m)
Volume - total	420,000 cu ft (12,000m^3)
Detention time, ADWF	3.3 hr
Cage aerator pairs/tank	6
Horsepower each, rotor, pair	56
Secondary sedimentation tanks	
Number	2
Diameter	148 ft (45m)
Average depth	10 ft (3.0m)
Overflow rate, ADWF	670 gpd/sf (27.2m^3/m^2/day)

[a]Estimate current connections, ultimate capacity estimated to be 300,000 population equivalents

The results of special 24-hr tests at the Vienna-Blumenthal plant are shown in Table 5-10. While in some cases the available data on total nitrogen are incomplete, the ammonia and nitrate nitrogen data indicate that nitrogen removals of 80 to 90 percent are obtainable. Special attention should be given to the operating conditions for these tests. While aeration detention time was only 5.5 to 8.4 hr, the mixed liquor solids were maintained at relatively high values, 5,300 to 6,700 mg/l. These high MLSS levels were made possible due to the low overflow rates maintained in the sedimentation tanks (280 to 410 gpd/sf or 11.4 to 16.6 m^3/m^2/day). U.S. practice would be to use higher overflow rates in sedimentation, reducing the ability to maintain high MLSS levels under aeration and consequently requiring higher aeration detention times (see Sections 4.10 and 5.6).

Alternative aeration patterns were evaluated in a recent South African Study.[66] Disc-type aerators were arranged in a series of four concentrically arranged channels. Effluent was monitored over a full year of weather conditions in the South African study.

One problem that was eventually overcome in the South African work was floating sludge in the sedimentation tank. This condition was especially prevalent when an anoxic channel section preceded the sedimentation tank. The condition was corrected by ensuring that each channel had sufficient aeration to maintain DO in the channel after the disc. Before the disc

TABLE 5-10

OPERATION AND PERFORMANCE OF THE VIENNA-BLUMENTHAL PLANT
24-HOUR INVESTIGATIONS (REFERENCE 65)

	Sept. 2, 1971	Feb. 17, 1972	June 6, 1974	July 16, 1974	July 31, 1974	Aug. 7, 1974
Influent flow rate, mgd (m^3/sec)	9.6(0.42)	14.0(0.61)	12.4(0.54)	10.9(0.48)	9.5(0.42)	9.0(0.40)
Aeration detention time, hr	8.0	5.5	6.1	7.0	8.0	8.4
MLSS, mg/l	6,700	6,200	5,300	5,900	5,400	5,400
Return sludge SS, mg/l	10,000	10,300	8,000	9,000	7,900	8,400
Average sedimentation tank						
Overflow rate, gpd/sf	280	410	360	320	280	270
m^3/m^2/day	11.4	16.6	14.7	13.0	11.4	10.9
Aeration tank temperature, C	18	12	18	20	19	20
Operating rotors, tank 1, tank 2	4, 2	3, 3	3, 3	4, 2	4, 3	4, 4
Sludge loading,						
lb BOD_5/lb MLSS/day	0.11	0.13	0.20	0.19	0.16	0.24
Influent concentrations, mg/l						
BOD_5	268	200	268	a	a	a
COD	475	384	463	574	515	778
TOC	155	126	128	143	134	208
Total - N	36	24	a	a	a	a
Ammonia - N	21.8	9.0	21.8	16.6	17.2	21.2
Effluent concentrations, mg/l						
BOD_5	13	13	10	a	a	a
COD	49	50	39	39	35	35
TOC	14	13	12	14	13	15
Total - N	4	4	a	a	a	a
Ammonia - N	3.8	2.7	6.3	3.9	2.4	2.7
Removal efficiencies, percent						
BOD_5	95	93	96	a	a	a
COD	90	87	92	93	93	92
TOC	91	90	91	90	90	93
Total - N	88	82	a	b	c	d
Ammonia - N	83	70	71	76	86	87

[a]Not available

[b]Not available, but no nitrate - N in effluent

[c]Not available, but only 0.3 mg/l Nitrate - N in effluent

[d]Not available, but only 2.4 mg/l Nitrate - N in effluent

aerator, DO was allowed to drop to zero. Since mixed liquor was withdrawn after the disc in a fresh condition, a sparkling clear effluent was produced.

Table 5-11 summarizes the performance data for the successful configurations of aeration in the South African study. Nitrogen removal averages 79 percent. A detailed 4-day analysis during a period when 86 percent nitrogen removal was observed showed that 40 percent of the influent nitrogen was incorporated in the sludge, 45 percent was nitrogen gas lost to the atmosphere and 14 percent appeared in the effluent.

TABLE 5-11

OPERATION AND PERFORMANCE OF OXIDATION DITCH
OPERATED FOR NITROGEN REMOVAL IN SOUTH AFRICA (REFERENCE 66)

	Run 1	Run 2	Run 3	Run 4
Feed flow rate, gpm (l/hr)	21.4 (4,800)	35.2 (8,000)	22 (5,000)	26.5-28.6 (6,000-6,500)
Retention time, hr	23.2	13.9	22.2	17.1-18.5
Return sludge ratio	2:1	1.5:1	2:1	>2:1
MLSS, mg/l	4,330	4,280	3,720	3,660
Water temperature, °C	21-22.5	20-22	14-17	14-15.5
Number of discs used, total	5 (on single shaft)	11 (on two shafts)	5 (on single shaft)	7 (on single shaft, 3-2-1-1)
Dissolved oxygen: mg/l				
Before aeration discs	Virtually always 0	Positive DO with only infrequent zero readings	Usually zero or slightly positive, 0-0.3 in channel 1	0-0.4 in first 3 chs. 0.6-2.0 in ch. 4
After aeration discs	Ranging 0.3-0.8	Higher by only 0.2-0.3 mg/l than before the discs	0.3-0.6 in chs. 2 and 3. 0.6-1.0 in chs. 1 and 4	0.3-1.3 in chs. 1, 2 and 3. 1.2-2.3 in ch. 4
SVI, ml/g	213	212	263	268
COD, lb COD/lb MLVSS/day (g/g/day)	0.155	0.28	0.20	0.26
Influent				
COD, mg/l	749	791	678	723
Kjeldahl-N, mg/l	39.5	39.4	43.1	45.0
Ammonia-N, mg/l	21.3	20.7	28.2	29.2
Effluent				
COD, mg/l	28.2	30.9	34.4	37.4
Kjeldahl-N, mg/l	4.5	4.05	6.4	5.0
Ammonia-N, mg/l	3.4	3.3	3.4	2.1
Nitrite-N, mg/l	trace	0.3	trace	0.6
Nitrate-N, mg/l	1.4	2.65	3.4	7.1
COD removed, %	96.2	96.0	94.6	94.5
N removed, %	85.2	81.6	77.3	72.0

The sludge in the South African study was always in a bulking condition, with SVI levels ranging from 212 to 268 ml/g. This bulking tendency may be a general property of alternating aerobic-anoxic processes, as bulking sludges have developed in other alternating aerobic/anoxic systems (see Sections 5.5.2.3 and 5.5.2.4).

It has been found that the cage aerators which are typically employed in the oxidation ditch are not well suited to nitrogen removal applications.[67] The cage aerator is not capable of simultaneously mixing and maintaining DO control; too much oxygen is imparted to allow development of alternating aerobic and anoxic zones while maintaining sufficient ditch velocities (1 fps or 0.30 m/s) for prevention of settling of solids in the ditch. In one case, the problem was solved by providing separate submerged propellers for mixing which allowed the cage aerators to be managed for DO control alone.[67] An aeration system has been developed that can both control the mixing level and the level of aeration simultaneously. Vertical shaft aerators are placed at the channel turning points in restricted areas so that the fluid rotation generated by the aerator is marshalled to move the flow around a semicircular turn. Power input and the number of on-line aerators can be varied to control DO level while maintaining sufficient mixing in the system. Nitrogen removals of 80 to 87 percent are reported.[68]

5.5.2.2 Denitrification in an Alternating Contact Process

The novel alternating contact process shown in Figure 5-19 has been tested in Denmark in both lab-scale and full-scale tests at wastewater flows up to 1.5 mgd (0.067 m^3/sec).[69] The process is similar to the oxidation ditch approach in that alternating aerobic and anoxic residence periods are provided in contact with raw wastewater. The means of accomplishing this alternating contact are very different from an oxidation ditch, however.

The operational sequence consists of the four phases shown in Figure 5-19. Aeration air is controlled to provide alternating aerobic and anoxic conditions. When anoxic conditions were required, the aeration air was merely turned down to a level sufficient to keep the sludge anoxic but in suspension. Raw wastewater is alternately directed to one or the other of the two tanks, according to a predetermined cycle. The cycle that an individual tank passes through is depicted on Figure 5-20. The operational phasing of wastewater addition and aeration are shown in addition to the phasing of nitrification and denitrification. Multiple liquid exposure to anoxic and aerobic zones is done by limiting the sequence time so that an average of 6 to 15 cycles are completed prior to liquid discharge. The effluent is discharged from the tank to sedimentation only when the tank is under aeration.

The stoichiometric carbon to nitrogen ratio for denitrification was defined in terms of a 7 day BOD and found to be $BOD_7/NO_3^- - N = 5.2$. Denitrification rates are summarized in Section 5.2.6. Effluent nitrate levels of 2.0 to 5.0 mg/l appeared obtainable with proper selection of design and operating mode. A full description of the mathematical model used for process design can be found in reference 69.

FIGURE 5-19

ALTERNATING CONTACT PROCESS

Phase 1 - Denitrification/Nitrification

RAW WASTEWATER

AN

A

S

SLUDGE RETURN

Phase 2 - Intermediate Aeration

RAW WASTEWATER

A

A

S

SLUDGE RETURN

Phase 3 - Nitrification/Denitrification

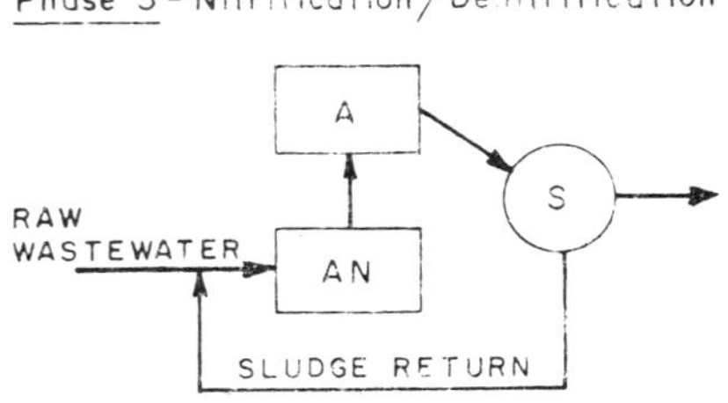

Phase 4 - Intermediate Aeration

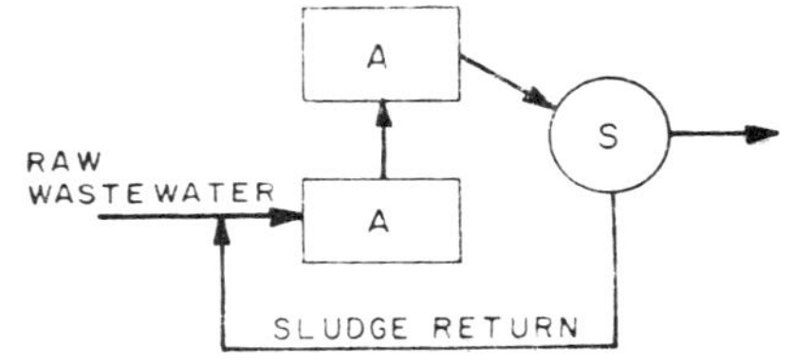

KEY

AN = Anoxic conditions, A = Aerobic conditions, S = Sedimentation

5.5.2.3 The Bardenpho Process

A recent South African development for nitrogen removal using both wastewater and endogenous carbon for denitrification is shown in Figure 5-21. Termed the "Bardenpho" process by its developer, the system is a combination of two previously developed processes.[70] Mixed liquor containing nitrate is recycled from the second (aerobic) tank to the initial (anoxic) tank for denitrification.[71] Appended to the first two tanks is a third (anoxic) tank for removing the nitrate remaining in the effluent from the second (aerobic) tank. In this third tank endogenous respiration is used for denitrification in the manner described in Section 5.5.1. Finally, a period of aeration is provided to improve sedimentation.

Initial lab scale tests of the Bardenpho process showed that nitrogen removals of 93 percent were possible at recycle rates of 4:1 and 1:1 of mixed liquor and return sludge, respectively.

FIGURE 5-20

OPERATIONAL SEQUENCING OF ONE OF TWO AERATION TANKS IN ALTERNATING CONTACT PROCESS

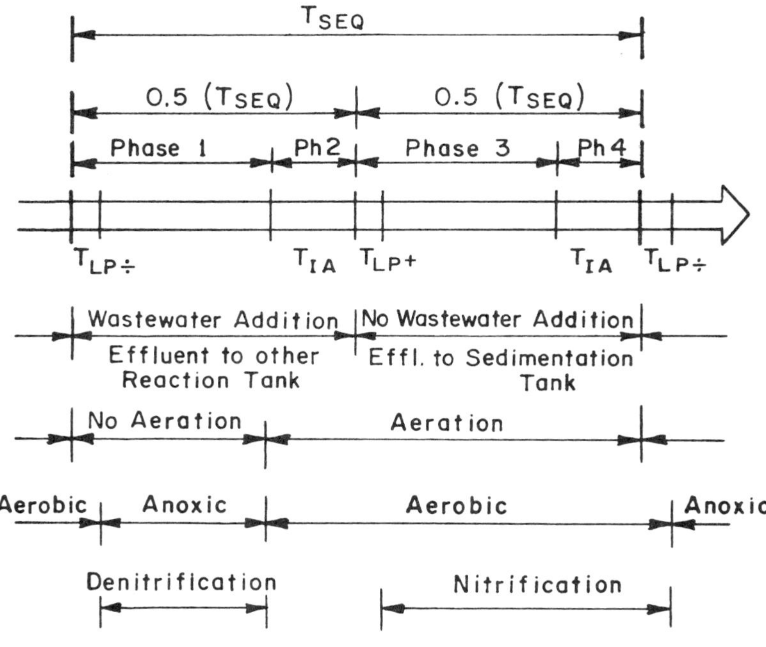

KEY TO SEQUENCING TIMES

T_{SEQ} = Overall Anoxic-Aerobic Sequence Time

$T_{LP\div}$ = Lag phase of Denitrification

T_{LP+} = Lag phase of Nitrification

T_{IA} = Intermediate Aeration Time

Under this condition, 5 to 7 mg/l of total nitrogen appeared in the effluent.[70] A pilot test of 9 months duration was then conducted to determine the long-term performance of the system on a scale of 26,000 gpd (100 m^3/day). During the last three months of the study, nitrogen removals were in the range of 80 to 90 percent.[61] Observations of denitrification rates for the first (anoxic) rank are summarized in Section 5.5.2.5 and denitrification rates for the third (anoxic) tank where endogenous carbon is employed are presented in Section 5.5.1.

FIGURE 5-21

THE BARDENPHO SYSTEM - SEQUENTIAL UTILIZATION OF WASTEWATER CARBON AND ENDOGENOUS CARBON

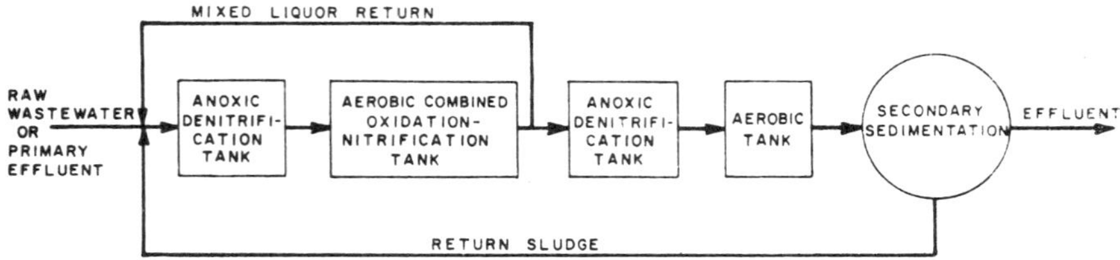

The sludge tended to be in bulking condition, settling very little in the standard one-liter cylinder SVI test. Even with stirring a relatively high SVI was obtained (150 ml/g). This bulking sludge condition appears to be typical for alternating aerobic/anoxic systems, and care must be employed in design and operation to deal with this problem.

The Bardenpho process has been tested in a second 26,000 gpd (100 m^3/day) pilot plant at Pretoria, South Africa.[72] While full test results were not available, test data from the initial few weeks of operation are shown in Table 5-12. These data demonstrate that relatively high nitrogen removals are obtainable with the Bardenpho process.

TABLE 5-12

PERFORMANCE OF THE "BARDENPHO" PROCESS AT PRETORIA, SOUTH AFRICA (REFERENCE 72)

Parameter	Period: Jan. 7 to Jan. 31, 1975	Period: Feb. 2 to Feb. 14, 1975
Influent COD, mg/l	226	176
Effluent COD, mg/l	46	48
Percent COD removal	79	73
Influent TKN[a], mg/l	21.1	15.9
Effluent TKN, mg/l	1.9	1.4
Percent TKN removal	91	91
Effluent nitrate - N, mg/l	2.6	1.7
Influent total nitrogen[b], mg/l	21.1	15.9
Effluent total nitrogen[c], mg/l	4.5	3.1
Percent total nitrogen removal	79	81

[a]TKN = total Kjeldahl nitrogen

[b]Assuming influent oxidized nitrogen is zero

[c]Assuming effluent nitrite nitrogen is zero

5.5.2.4 Alternating Aerobic/Anoxic System Without Internal Recycle

Investigators at the EPA Blue Plains pilot plant conceived of still another way to achieve alternating aerobic and anoxic environments with the system shown in Figure 5-22.[60] A two pass aeration tank was provided with separate aeration and mixing facilities. In each basin, 2 mechanical mixers were employed to keep the mixed liquor in suspension, independent of the aeration system. Air was supplied alternately to each basin; first to one basin and then to the other in a 30 minute cycle.[60] Dissolved oxygen level in the pass under aeration was controlled between 2 and 3 mg/l, while the anoxic pass decreased to zero rapidly after cessation of aeration. The pilot process was typically operated at a flow rate of 50,000 gpd (189 m^3/day).

FIGURE 5-22

BLUE PLAINS ALTERNATING ANOXIC AEROBIC SYSTEM (REF. 60)

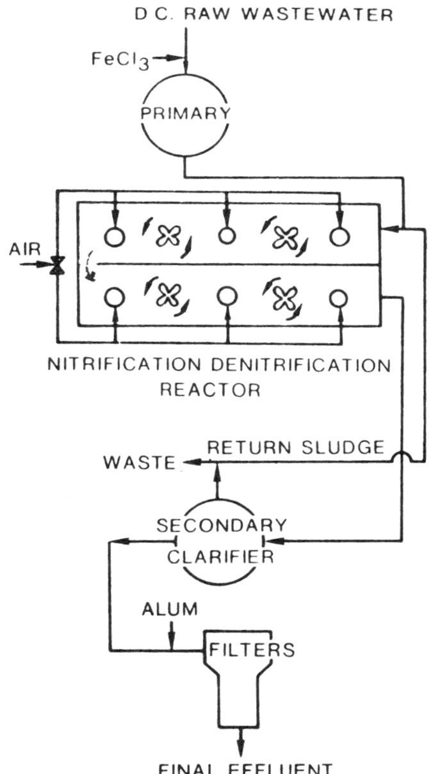

A summary of operating and performance data for the 9 months of test work is shown in Table 5-13. During 9 months of operation, the plant was operated at a F/M of approximately 0.1 lb BOD_5/lb MLVSS/day (0.1 g/g/day), expressed on the basis of the total inventory in the reactor. This F/M was sufficiently low to permit the development of a mixed culture of organisms for carbon oxidation, nitrification and denitrification.

Nitrogen removals during the study varied from 54 to 84 percent, but operational problems contributed to the lower reported removals. These problems may be avoidable in full-scale operation. For instance, in July and August, ferric chloride addition in the primary treatment stage reduced the COD/TKN ratio from 10 to about 7.5 to 8.0, and resulted in a situation in which the lack of a sufficient organic carbon source limited the degree of denitrification obtainable. The lower removals experienced in April and May were due to the fact that during a portion of each month, the alternate aerobic/anoxic regime was altered to a full aerobic mode to rid the system of filamentous growth.

During September, performance deteriorated for an unexpected reason. The flow sheet in Figure 5-22 was modified to include two more reactor stages prior to the clarifier. In the first added stage, an anoxic one, methanol was added to cause denitrification of residual nitrite and nitrate. An aerobic stabilization step was added as the last stage. It was found that methanol addition caused an immediate ammonia increase in the process. Subsequent studies showed that methanol is toxic to nitrifiers.

Since all operating problems can be explained, it can be concluded that the system is capable of 84 percent nitrogen removal in the summer (23 C) and 75 percent nitrogen removal in the winter (14 C).

Like in other alternating aerobic/anoxic studies, it was found that a severe filamentous bulking condition developed in the sludge, limiting wintertime clarifier overflow rates to about 300 gpd/sf.[60] Bulking sludge has been observed at low temperatures at Blue Plains and at low F/M operation at Blue Plains and at other plants. Whatever the cause of the bulking problems, it appears that clarifier operation will limit operation of this aerobic/anoxic system, as it will limit the other aerobic/anoxic systems.

Kinetic data was obtained during the study. On and off aeration of samples of the pilot plant mixed liquor was employed over several cycles to simulate operation of the pilot unit. Nitrification and denitrification rates determined by this procedure are shown in Table 5-14. Nitrification rates were similar through all cycles, while denitrification rates decreased as the batch reaction continued. Denitrification rates measured in the first cycle represent peak rates possible when a readily available carbon source is available during denitrification. By the third or fourth cycle, the rates represent denitrification when the readily available carbon is depleted and an endogenous carbon source is used. Denitrification rates for cycle 1 are also presented in Section 5.5.2.5 with measurements of other observors, while rates for cycle 4 are presented in Section 5.5.1.

TABLE 5-13

SUMMARY OF OPERATION AND PERFORMANCE FOR THE BLUE PLAINS ALTERNATING AEROBIC/ANOXIC SYSTEM (REFERENCE 60)

Month 1973	Det. time hr	F/M lb BOD applied/ lb MLVSS/day or g/g/day	MLSS mg/l (% volatile)	SVI, ml/g	COD/ TKN ratio	Temp., C	Influent quality mg/l			Effluent quality, mg/l [a]				Removals, percent		
							BOD_5	SS	Total Kjeldahl Nitrogen	BOD_5	SS	Total Kjeldahl Nitrogen	NO_3^- and NO_2^- - N	BOD_5	SS	Total Nitrogen
Jan	12.3	0.072	3510 (74)	245	9.6	14.0	96.5	110	25.7	20.4	15.4	2.28	3.99	79	85	76
Feb	12.3	0.066	3980 (73)	250	9.9	14.2	99	108	23.2	14.0	14.3	1.52	4.41	86	87	75
March	12.3	0.10	2950 (73)	330	10.5	15.5	110	128	24.8	6.5[b]	15.0	4.20	2.30	94[b]	88	74
April	12.4	0.081	3540 (67)	277	10.6	-	98.8	120	21.7	5.3[b]	13.0	5.20	6.03	95[b]	89	49
May	10.5	0.089	4170 (69)	227	10.0	-	115	109	23.3	3.3[b]	11.8	1.36	8.25	97[b]	89	59
June	8.8	0.105	4010 (69)	188	10.3	23.0	107	112	24.0	3.2[b]	7.8	1.51	2.30	97[b]	93	84
July	6.8	0.093	3040 (64)	133	7.9	25.0	51	153	15.0	3.8[b]	9.0	2.14	2.72	93[b]	94	68
August	6.6	0.089	3200 (57)	134	7.5	25.5	44.2	197	14.9	2.6[b]	10.0	1.23	3.74	94[b]	95	67
Sept	8.7	0.11	3700 (65)	-	10.0	26.0	99	110	22.6	7.2[b]	16.0	10.2[c]	0.22	93[b]	85	54

[a] Prior to filtration

[b] Nitrification inhibited

[c] Ammonia level was 9.4 mg/l as N (see text)

TABLE 5-14

OBSERVED NITRIFICATION AND DENITRIFICATION RATES FOR BLUE PLAINS ALTERNATING ANOXIC/AEROBIC SYSTEM

Mode	Temp., C	Peak nitrification or denitrification removal rate, lb N/lb MLVSS/day				
		Cycle 1	Cycle 2	Cycle 3	Cycle 4	Cycle 5
Aerobic-Nitrification	15.5	0.032	0.042	0.016	0.026	0.035
	25.0	0.083	0.095	-	-	-
	27.0	0.11	0.11	-	-	-
	26.5	0.12	-	-	-	-
Anoxic-Denitrification	15.5	0.032	0.029	0.021	0.019	-
	25.0	0.055	0.030	0.033	0.030	-
	27.0	0.042	-	-	-	-
	26.5	0.026	0.0075	-	-	-

5.5.2.5 Kinetic Design of Alternating Aerobic/Anoxic Systems

The four factors which can limit denitrification process efficiency in alternating aerobic/anoxic systems using wastewater as the carbon source are as follows:[69]

1. Nitrification

2. Denitrification

3. Carbon-nitrogen ratio

4. Operational mode (process hydraulics)

The third factor has been evaluated by several investigators (see Sections 5.5.2.2 and 5.5.2.4) and further discussion here is unnecessary.

To evaluate nitrification limitations on the system, nitrogen loads and nitrification rates must be taken into account. Most investigators agree that the design of the combined carbon oxidation-nitrification functions of the aerobic phase can be separated from the anoxic phase.[61,69,73] It has been found that anoxic periods up to 5 hours have no impact on aerobic sludge activity.[62,59] Therefore, the carbon oxidation and nitrification calculations for the aerobic periods can be virtually identical to those advanced for combined carbon oxidation-nitrification in Sections 4.3.3 and 4.3.5. In the calculations of nitrifier solids retention time, nitrifier growth rate,and removal rates in the aerobic residence periods, only the solids inventory under aeration is employed. This is because the environment must be aerobic for nitrifier growth to occur. As always, a safety factor must be employed.

Sizing of denitrification steps must consider nitrite load and nitrate removal rates and consideration of the safety factor concept in design. Of the two models formulated for these systems, the safety factor concept is used only for the nitrification step in the Bardenpho design, but for some unapparent reason not for denitrification.[61] The safety factor concept was not used for nitrification or denitrification in the alternating contact process design (Section 5.5.2.2).[69] It must be emphasized that unless a safety factor is incorporated in the design, nitrogen removal will deteriorate under peak load conditions.

Most of the alternating processes employ both wastewater carbon and endogenous carbon for denitrification at some point in the system. Observed denitrification rates for endogenous carbon have been summarized in Section 5.5.1. Experimentally determined denitrification rates in alternating aerobic/anoxic systems with wastewater as the carbon source are shown in Figure 5-23. When data were reported in terms of MLSS only, volatile content was assumed to be 70 percent to allow reporting on a basis consistent with the denitrification rates shown in Section 5.2. These rates are peak nitrate removal rates, and are expressed as $\hat{q}_D$, using the terminology developed in Section 3.3.5.2. As can be seen from Figure 5-23, there is a wide variation in measured denitrification rates in systems using wastewater as the organic carbon source. As a result, it may not be a conservative practice to use the denitrification rates given in Figure 5-23; rather, it would appear prudent to conduct pilot investigations to verify design parameters for denitrification when wastewater is the carbon source for denitrification.

The rates for denitrification with wastewater as the carbon source fall below those found for methanol as the carbon source shown in Figure 5-2. Median rates at 20 C for methanol and wastewater carbon are about 0.25 lb NO_3^- –N removed/lb MLVSS/day (0.25 g/g/day) and 0.07 lb NO_3^- –N removed/lb MLVSS/day (0.07 g/g/day), respectively. A denitrification reactor using a wastewater carbon source would have to be about three and one-half times larger than a denitrification reactor using methanol as the carbon source.

One cause of the difference in reaction rates between methanol and wastewater carbon relates to biological availability. Methanol is a simple, easily degraded compound, whereas wastewater contains a mix of easily degraded and hard to degrade compounds. Wastewaters may vary in the relative distribution of easily degraded and hard to degrade compounds, thus causing variations in denitrification rates between locations.

FIGURE 5-23
EFFECT OF TEMPERATURE ON PEAK DENITRIFICATION RATES WITH WASTEWATER AS CARBON SOURCE

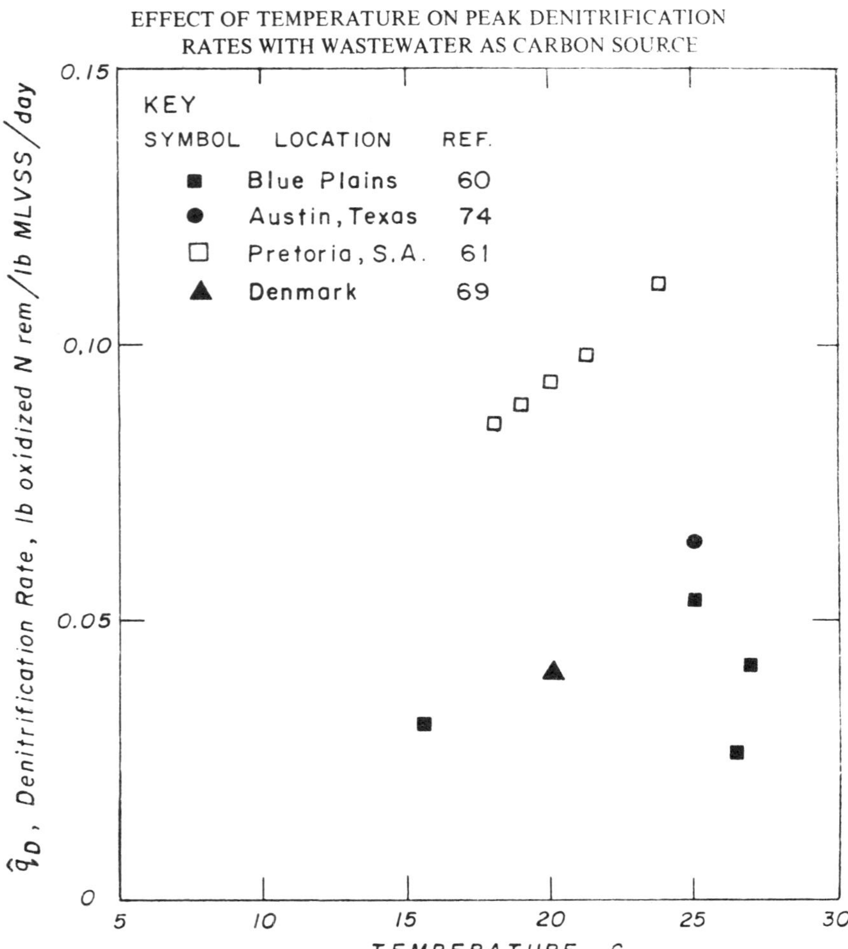

The hydraulic mode of operation significantly affects the kinetic design procedure. In all of these systems, relatively complex and lengthy mass balances are necessary to describe the system. None-the-less, such descriptions are possible and have been developed for two of the alternating aerobic/anoxic systems.[61,69] These models are presented in sufficient detail in the literature to allow their modification for use in design. Iterative solution of equations is required, and the digital computer has proven a useful design tool.[69] The limitation of these models is that generally applicable kinetic rate data are not yet available.

5.6 Solids-Liquid Separation

The considerations for design of sedimentation tanks for denitrification systems are the same as those discussed for nitrification systems in Section 4.10 and those points common to both will not be repeated herein.

Rising sludge has occasionally plagued denitrification systems, depending on system design.[3,4,7,15,49,50] To remedy this and other problems, the original suspended growth denitrification system (using methanol) was modified by placing an aerated stabilization step between the anoxic denitrification reactor and the denitrification clarifier (Figure 5-1B).[2,3,21,51] This step was taken because it was found that the "conventional" design was basically an unstable process. In the conventional system, the methanol: nitrogen ratio (M:N ratio) had to be kept at precisely the optimum level (2.5 to 3.0). When methanol was overfed, the effluent BOD_5 would rise. When methanol was underfed, nitrate would bleed through to the clarifier, and denitrification would proceed in the clarifier using the sludge as the carbon source. Floating sludge, buoyed up by nitrogen gas bubbles, caused a severe deterioration in effluent quality.

The coupling of an anoxic residence period and an aerobic residence period in the modified system (Figure 5-1) is based on the recognition that dissimilatory denitrification is accomplished by facultative bacteria using biochemical pathways that are almost identical to aerobic biochemical pathways. The main difference in the biochemical pathways lies in the electron transport system where the terminal enzyme is changed and nitrate replaces oxygen as the final electron acceptor.[75] These facultative bacteria can shift rapidly from using nitrate to using oxygen and vice versa. In the aerobic tank, the excess methanol is oxidized and the mixed liquor solids are aerobically stabilized. The aerobic tank also serves the purpose of stripping supersaturated nitrogen gas from solution so that nitrogen gas bubbles will not form during sedimentation.

A mildly aerated physical conditioning channel transfers the denitrification mixed liquor to the final clarifier. Recognizing that the very turbulent conditions in the aerated stabilization tank causes floc breakup and dispersed fines in suspension, the purpose of the channel is to allow these dispersed particles to be incorporated into floc under mild turbulence conditions that favor aggregation over breakup.[76]

Tests of the original and modified systems were conducted at the Central Contra Costa Sanitary District's Advanced Treatment Test Facility.[2,3,21,51] Comparing performance with stabilization to performance without it (Tables 5-15 and 5-16), indicates the substantial merits of the aerated stabilization tank. The test period without stabilization was one in which daily adjustments were made in the methanol feed rate, hence there was little if any methanol bled into the effluent. This is reflected in the low soluble BOD_5 of 5 mg/l. During the test period with aerated stabilization, less careful control was exerted in methanol feed which resulted in a fairly high M/N ratio of 3.3. Despite this overfeeding of methanol, the effluent soluble BOD_5 remained 5 mg/l. In other words, the aerated stabilization tank formed a favorable environment for the oxidation of the excess methanol.

One significant factor contributing to less suspended solids and turbidity in the effluent was the development of a ciliate and rotifer population in the culture. Previously these organisms were not abundant. With a significant aerobic residence period, these organisms

TABLE 5-15

EFFECT OF STABILIZATION TANK ON DENITRIFIED EFFLUENT AT THE CENTRAL CONTRA COSTA SANITARY DISTRICT'S ADVANCED TREATMENT TEST FACILITY (REFERENCE 3)

Constituent	Mean effluent quality without stabilization, mg/l (Feb. 13 to Mar. 13, 1972)	Mean effluent quality with stabilization, mg/l (Mar. 28 to April 20, 1972)
Nitrate as N	0.5	0.7
Total BOD_5	37	6
Filtered BOD_5	5	5
Suspended Solids	14	4
Turbidity	5[a]	1.3[a]
Total organic carbon	18	8.6
Soluble organic carbon	7	5
Temperature	16 to 17[b]	16 to 19[b]

[a] JTU [b] degrees C

TABLE 5-16

DENITRIFICATION PROCESS PARAMETERS AT THE CENTRAL CONTRA COSTA SANITARY DISTRICT'S ADVANCED TREATMENT TEST FACILITY (REF. 3)

Parameter	Without stabilization Feb. 13 to Mar. 13, 1972	With stabilization Mar. 28 to April 20, 1972
Flow, mgd	.46	.47
Residence time, hr		
reactor	.85	.82
stabilization tank	0	.79
MLSS, mg/l	3000	2500
SVI, ml/g	143	242
Nitrogen rem lb/lb MLVSS/day	.23	.18
Methanol/nitrate - N ratio	2.8	3.3

could flourish and clarify the liquid. Sulfide odors in the sludge were eliminated by the modification.

In addition to stabilizing the liquid, the sludge is stabilized as well. Without the stabilization tank, solids appearing in the effluent contained about 2.3 lb BOD_5 per lb of SS. With stabilization, effluent solids contain less BOD_5 with a BOD_5/SS ratio of 0.25. The net effect of this is to reduce the effluent BOD_5 from 37 to 6 mg/l. This reduction in BOD_5 value of the solids is very likely due to enhanced endogenous respiration in the stabilization tank.

Another indication of sludge stabilization in the aerated stabilization tank is the reduction in denitrification rates observed in the modified system compared to the conventional system. Prior to the modification, denitrification rates were as high as 0.3 to 0.56 lb Nitrate -N rem./lb MLVSS/day at 16 to 18 C and these rates were below the peak limiting rate, $\hat{q}_D$, as effluent nitrate was always low. This range in rates is roughly twice the range in rates shown for the system employing aerated stabilization at CCCSD as shown in Figure 5-2. While one impact of aerated stabilization is to decrease denitrification rates and therefore increase anoxic reactor requirements, the other effect of this impact is to render the sludge less active and more resistant to the rising sludge problem in the denitrification clarifier due to denitrification. This is in agreement with the conclusions drawn in Section 4.10, where it is shown that the tendency for rising sludge was related to denitrification rates and sludge residence time in the clarifier. Rapid sludge removal equipment, such as the vacuum pickup type, should be provided in all sedimentation tanks to minimize sludge residence time and reduce the likelihood of rising sludge.

Similar to the findings with methanol based denitrification, an aerobic step has been usefully employed in combined carbon oxidation-nitrification-denitrification systems. Generally, a 1 to 2 hr residence period is provided prior to mixed liquor separation in the sedimentation step.

Reactor-clarifier interactions should not be ignored. For instance design examples are presented in the literature for alternating aerobic/anoxic systems where the mixed liquor solids are assumed to be 5000 mg/l (at 14 C) and 7000 mg/l (at 10 C).[61,69] Current U.S. practice is to limit mixed liquor levels below 3000 mg/l unless clarifier overflow rates are reduced to account for the need to thicken and return the sludge.[40] Operation at 5000 to 7000 mg/l would require very large clarifiers to ensure that solids are not lost at peak wet weather flow conditions. Bulking sludge tends to occur in the combined carbon-oxidation-nitrification-denitrification systems which mandates even greater conservatism in allowable mixed liquor levels and clarifier design than would normally be the case.

5.7 Considerations for Process Selection

The choice of denitrification for nitrogen removal mandates the process sequence of nitrification-denitrification for nitrogen removal. Two kinds of comparisons in denitrification process selection can be made. First, the nitrification-denitrification sequence can be compared to the physical-chemical alternatives. Second, if nitrification-denitrification is chosen, comparisons must be made among the denitrification alternatives to make the process selection.

5.7.1 Comparison to Physical-Chemical Alternatives

The comparison made between the nitrification portion of the sequence and the physical-chemical alternatives in Section 4.11.1 need not be repeated here.

Total dissolved solids (TDS) increment in the nitrogen removal system will have a bearing on process selection in many situations. Nitrification-denitrification leads to a net reduction in wastewater alkalinity and no change in the total mineral content of the water. Both the breakpoint chlorination and the selective ion exchange process lead to TDS increases.

5.7.2 Choice Among Alternative Denitrification Systems

Many of the considerations presented in Section 4.11 are applicable to denitrification system selection and are not repeated here. Other factors affecting process choice are summarized in Table 5-17. Most of these factors were considered earlier in this chapter.

In some treatment plants, both nitrogen and phosphorus removal has been mandated. The combined carbon oxidation-nitrification-denitrification systems are somewhat restricted in the sequencing of the phosphorus removal step. Chemical addition in the primary treatment stage cannot be employed, as this would not leave sufficient organic carbon influent to the process to complete denitrification.

Another factor listed in Table 5-17 is stability of operation and degree of nitrogen removal. Several long-term tests of denitrification systems using methanol have successfully demonstrated consistently high levels of nitrogen removal. Equivalent operational experience with the combined carbon oxidation-nitrification-denitrification systems will soon be obtained, as large-scale experimental work is currently underway in the U.S., Denmark, and South Africa. When the results of this work are available, fair comparisons can be made between the two types of systems. Based upon presently available data, it appears that the combined carbon oxidation-nitrification-denitrification systems are capable of 75 to 90 percent nitrogen removal; in comparison, methanol based systems can achieve 90 to 95 percent nitrogen removal.

One other kind of comparison can be made between the systems using methanol and the combined systems. Since the rate of denitrification with wastewater as the carbon source in the combined system is lower than with methanol as the carbon source, greater denitrification reactor sizes are required for the combined system. This issue can best be analyzed with an example comparing the alternative systems. A design example for the Bardenpho process has been presented in the literature that can be usefully employed for the comparison.[61] Specific loading criteria are not important in this example; for these the

TABLE 5-17

COMPARISON OF DENITRIFICATION ALTERNATIVES

System Type	Advantages	Disadvantages
Suspended growth using methanol following a nitrification stage	Denitrification rapid, small structures required Demonstrated stability of operation Few limitations in treatment sequence options Excess methanol oxidation step can be easily incorporated Each process in the system can be separately optimized High degree of nitrogen removal possible	Methanol required Stability of operation linked to clarifier for biomass return Greater number of unit processes required for nitrification-denitrification than in combined systems
Attached growth (column) using methanol following a nitrification stage	Denitrification rapid, small structures required Demonstrated stability of operation Stability not linked to clarifier as organisms on media Few limitations in treatment sequence options High degree of nitrogen removal possible Each process in the system can be separately optimized	Methanol required Excess methanol oxidation process not easily incorporated Greater number of unit processes required for nitrification-denitrification than in combined system
Combined carbon oxidation-nitrification-denitrification in suspended growth reactor using endogenous carbon source	No methanol required Lesser number of unit processes required	Denitrification rates very low; very large structures required Lower nitrogen removal than in methanol based system Stability of operation linked to clarifier for biomass return Treatment sequence options limited when both N and P removal required No protection provided for nitrifiers against toxicants Difficult to optimize nitrification and denitrification separately
Combined carbon oxidation-nitrification-denitrification in suspended growth reactor using wastewater carbon source	No methanol required Lesser number of unit processes required	Denitrification rates low; large structures required Lower nitrogen removal than in methanol based system Stability of operation linked to clarifier for biomass return Tendency for development of sludge bulking Treatment sequence options limited when both N and P removal required No protection provided for nitrifiers against toxicants Difficult to optimize nitrification and denitrification separately

reader is referred to reference 61. Reactor residence times are as provided in reference 61 at a temperature of 14 C excepting that the MLSS value has been downwardly adjusted from 5000 to 3000 mg/l, according to U.S. practice. The effect of this is to increase by the ratio of 5/3 the detention times in the reactors in the design example. The adjusted residence times are shown in Figure 5-24. A methanol-based system useful for comparison purposes is also shown in Figure 5-24. In this sytem, a combined carbon oxidation-nitrification step is chosen, since if a combined operation provides acceptable treatment for the Bardenpho

FIGURE 5-24
COMPARISON OF DENITRIFICATION SYSTEMS

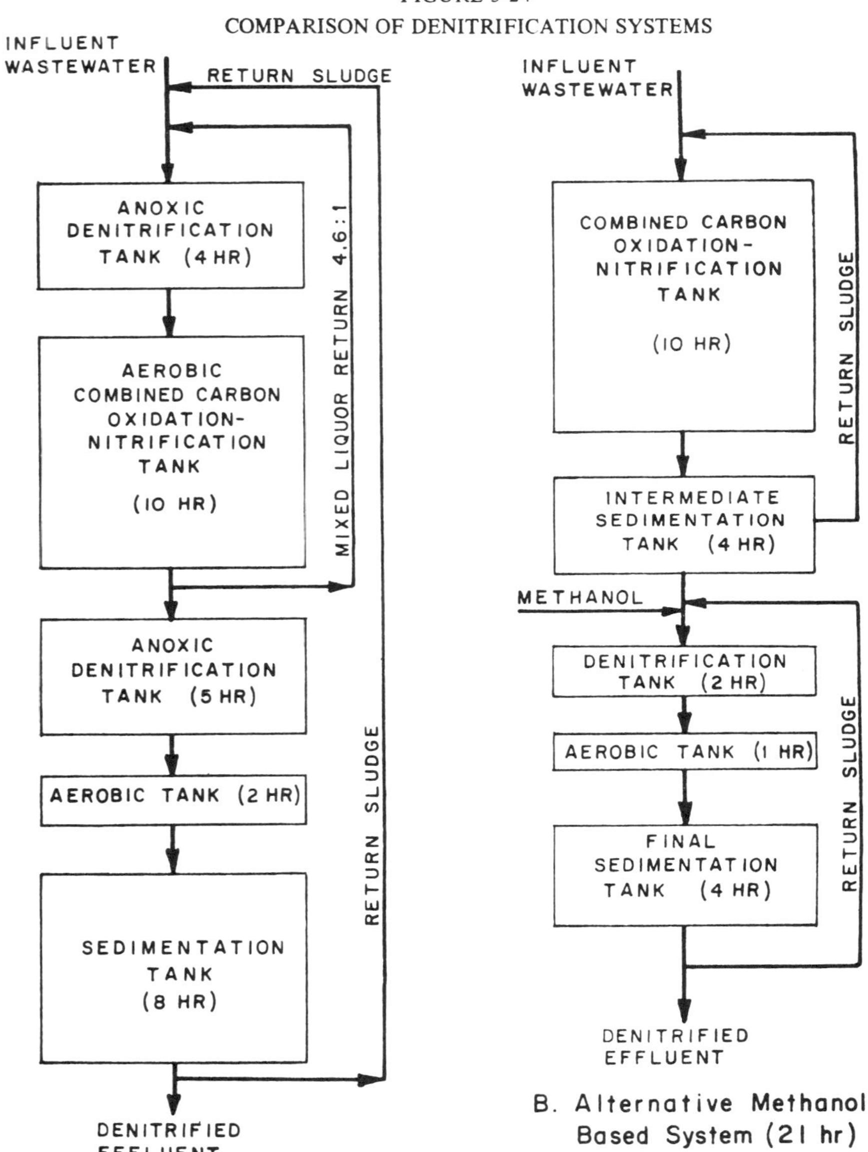

process it would also work effectively for the comparative case. Solids retention time (and hydraulic detention time) of the nitrification step would be the same as for the nitrification tank in the Bardenpho process.

Since denitrification rates are more rapid, the denitrification tank in the methanol based system can be proportionately smaller than in the Bardenpho process. Interpolating denitrification rates from Figures 5-2 and 5-23, about 2 hours would be required. A residence time of 4 hours is assumed for sedimentation tanks in the methanol-based system, whereas 8 hours is assumed for the Bardenpho process due to the bulking tendency of the sludge. Comparing the two alternative systems, it can be seen that the alternating aerobic/anoxic system requires greater tankage than the methanol-based system (29 hours compared to 21 hours). In terms of economics, differences in the systems can be seen as a trade-off of capital cost (tankage) with operating cost (methanol).

5.8 References

1. Barth, E.F., Brenner, R.C., and R.F. Lewis, *Chemical-Biological Control of Nitrogen in Wastewater Effluent.* JWPCF, 40, No. 12, pp 2040-2054 (1968).

2. Parker D.S., Zadick, F.J., and K.E. Train, *Sludge Processing for Combined Physical-Chemical-Biological Sludges.* Prepared for the EPA, Report No. R2-73-250, July, 1973.

3. Horstkotte, G.A., Niles, D.G., Parker, D.S., and D.H. Caldwell, *Full-Scale Testing of a Water Reclamation System.* JWPCF, 46 No. 1, pp 181-197 (1974).

4. Mulbarger, M.C., *The Three Sludge Systems for Nitrogen and Phosphorus Removal.* Presented at the 44th Annual Conference of the Water Pollution Control Federation, San Francisco, California, October, 1971.

5. Sawyer, C.N., Wild, H.E., Jr., and T.C. McMahon, *Nitrification and Denitrification Facilities, Wastewater Treatment.* Prepared for the EPA Technology Transfer Program, August, 1973.

6. Parker D.S., *Case Histories of Nitrification and Denitrification Facilities.* Prepared for the EPA Technology Transfer Program, May, 1974.

7. Murphy, K.L., and P.M. Sutton, *Pilot Scale Studies on Biological Denitrification.* Presented at the 7th International Conference on Water Pollution Research, Paris, September, 1974.

8. Bishop, D.F., Personal communication to D.S. Parker. Environmental Protection Agency, Washington, D.C., April, 1974.

9. Mulbarger, M.C., *Nitrification and Denitrification in Activated Sludge Systems.* JWPCF, 43, No. 10, pp 2059-2070 (1971).

10. Dawson, R.N., and K.L. Murphy, *Factors Affecting Biological Denitrification in Wastewater.* In Advances in Water Pollution Research, S.H. Jenkins, Ed., Oxford, England: Pergamon Press, 1973.

11. Stensel, H.D., Loehr, R.C., and A.W. Lawrence, *Biological Kinetics of Suspended Growth Denitrification.* JWPCF, 43, No. 2, pp 249-261 (1973).

12. Sutton, P.M., Murphy, K.L., and R.N. Dawson, *Continuous Biological Denitrification of Wastewater.* Environmental Protection Service (Canada), Report EPS 4-WP-74-6, August, 1974.

13. Lawrence, A.W., and P.L. McCarty, *Unified Basis for Biological Treatment Design and Operation.* JSED, Proc. ASCE, 96, No. SA3, pp 757-778 (1970).

14. Brown and Caldwell, *Project Report for the Water Reclamation Plant.* Report to the Central Contra Costa Sanitary District, November, 1971.

15. Weddle, C.L., Niles, D.G., Goldman, E., and J.W. Porter, *Studies of Municipal Wastewater Renovation for Industrial Water.* Presented at the 44th Annual Conference of the Water Pollution Control Federation, San Francisco, California, October, 1971.

16. Horstkotte, G.A., Jr., *Pilot Demonstration Project for Industrial Reuse of Renovated Municipal Wastewater.* Prepared for the EPA, EPA-670/2-72-064, August, 1973.

17. Requa, D.A., and E.D. Schroeder, *Kinetics of Packed Bed Denitrification.* JWPCF, 45, No. 8, pp 1696-1707 (1973).

18. Smith, J.M., Masse, A.N., Feige, W.A., and L.J. Kamphake, *Nitrogen Removal from Municipal Wastewater by Columnar Denitrification.* Environmental Science and Technology, 6, p 260 (1972).

19. Jewell, W.J., and R.J. Cummings, *Denitrification of Concentrated Wastewaters.* Presented at the Water Pollution Control Federation, Cleveland, October, 1973.

20. Sutton, P.M., Murphy, K.L., and R.N. Dawson, *Low Temperature Biological Denitrification of Wastewater.* JWPCF, 47, No. 1, pp 122-134 (1975).

21. Parker, D.S., Aberley, R.C., and D.H. Caldwell, *Development and Implementation of Biological Denitrification for Two Large Plants.* Presented at the Conference on Nitrogen as a Water Pollutant, sponsored by the IAWPR, Copenhagen, Denmark, August, 1975.

22. *Denitrification for Anaerobic Filters and Ponds, Phase II.* Robert S. Kerr Water Research Center, EPA WPCRS 13030 ELY 06/71-14, June, 1971.

23. *Denitrification for Anaerobic Filters and Ponds.* Robert S. Kerr Water Research Center, EPA WPCRS 13030 ELY 04171-8, April, 1971.

24. *Description of the El Lago, Texas. Advanced Wastewater Treatment Plant.* Seabrook, Texas: Harris County Water Control and Improvement District Number 50, March, 1974.

25. Requa, D.A., *Kinetics of Packed Bed Denitrification.* Thesis submitted in partial satisfaction of the requirements for the degree of Master of Science in Engineering, University of California at Davis, 1970.

26. English, J.N., Carry, C.W., Masse, A.M., Pitkin, J.B., and F.D. Dryden, *Denitrification in Granular Carbon and Sand Columns.* JWPCF, 46, No. 1, pp 28-42 (1974).

27. Savage, E.S., and J.J. Chen, *Operating Experiences with Columnar Denitrification.* Pittsburgh, Pennsylvania: Dravo Corporation, 1973.

28. Lamb, G.L., *Nitrogen Removal Utilizing Single Stage Activated Sludge and Deep Bed Filtration.* Presented at the Joint Meeting of the Sanitary Engineering Section, ASCE, and the Metropolitan Section, NYWPCA, New York, N.Y., March, 1972.

29. Ecotrol, Inc., *Biological Denitrification Using Fluidized Bed Technology.* August, 1974.

30. Jeris, J., Beer, C., and J.A. Mueller, *High Rate Biological Denitrification Using a Granular Fluidized Bed.* JWPCF, 46, No. 9, pp 2118-2128 (1974).

31. Chen, J.J., Letter communication to D.S. Parker, Dravo Corporation, November, 1974.

32. Kapoor, S.A., and T.E. Wilson, *Biological Denitrification on Deep Bed Filters at Tampa, Florida.* Unpublished paper, Greeley and Hansen, Engineers, Chicago, Ill., (no date).

33. Hansen, S.E., Letter communication to D.S. Parker. Neptune MicroFloc Inc., Corvallis, Oregon, September 11, 1974.

34. Duddles, G.A., Richardson, S.E., and E.F. Barth, *Plastic Medium Trickling Filters for Biological Nitrogen Control.* JWPCF, 46, No. 5, pp 937-946 (1974).

35. Duddles, G.A. and S.E. Richardson, *Application of Plastic Media Trickling Filters for Biological Nitrification.* Report prepared for the Environmental Protection Agency, EPA-R2-73-199, June, 1973.

36. *Process Design Manual for Suspended Solids Removal.* U.S. EPA, Office of Technology Transfer, Washington, D.C., January, 1975.

37. Jeris, J.S., and F.J. Flood, *Plant Gets New Process.* Water and Wastewater Engineering, pp 45-48, March, 1974.

38. Jeris, John S., and R.W. Owens, *Pilot Scale High Rate Biological Denitrification at Nassau County, N.Y.* Presented at the Winter Meeting of the New York Water Pollution Control Association, January, 1974.

39. Owens, R., Letter communication to D.S. Parker, Ecolotrol Corp., October, 1974.

40. *Process Design Manual for Upgrading Existing Wastewater Treatment Plants.* U.S. EPA, Office of Technology Transfer, Washington, D.C. (1974).

41. Ecolotrol, Inc., *Hy-Flo Fluidized Bed Denitrification,* Ecolotrol Technical Bulletin, No. 123-A, November, 1974.

42. Sullivan, R.H., Cohen, M.M., Ure, J.E., and F. Parkinson, *The Swirl Concentrator as a Grit Separator Device.* Prepared for the EPA by the American Public Works Association, EPA-670/2-74-026, June, 1974.

43. Manufacturing Chemists Association, *Properties and Essential Information for Safe Handling and Use of Methanol,* Chemical Safety Data Sheet SD-221, 1970.

44. Austin, George T., *Industrially Significant Organic Chemicals – Part 7.* Chemical Engineering, 81, No. 13, pp 152-153 (1974).

45. Manufacturing Chemists Association, *Recommended Practice-Unloading Flammable Liquids from Tank Cars,* Manual Sheet TC-4. 1969.

46. Manufacturing Chemists Association, *Tank Car Approach Platforms, Manual Sheet TC- 7.* 1965.

47. National Fire Protection Association, *Static Electricity 1966,* NFPA No. 77. 1966.

48. National Fire Protection Association, *Flammable and Combustible Liquids Code 1972, NFPA No. 30,* 1972.

49. Smith, A.G., *Denitrification Reactor Studies in a Lime Treated Sewage Plant.* Paper No. 2029, Ontario Ministry of the Environment, Research Branch, July, 1972.

50. Tenney, M.W., and W.F. Echelberger, *Removal of Organic and Eutrophying Pollutants by Chemical-Biological Treatment.* Prepared for the EPA, Report No. R2-72-076 (NTIS PB-214628), April, 1972.

51. Parker, D.S., and D.G. Niles, *Full-Scale Test Plant at Contra Costa Turns Out Valuable Data On Advanced Treatment.* The Bulletin (of the California Water Pollution Control Association), 9, No. 1, pp 25-27 (1972).

52. Sutton, P.M., Murphy, K.L., and B.E. Jank, *Nitrogen Control: A Basis for Design With Activated Sludge Systems.* Presented at the Conference on Nitrogen as a Water Pollutant, sponsored by the IAWPR, Copenhagen, Denmark, August, 1975.

53. Wuhrmann, K., *Nitrogen Removal In Sewage Treatment Processes.* Verh. Int. Ver. Limnol., 15, pp 580-596 (1964).

54. Hunerberg and Sarfert, *Experiments on the Elimination of Nitrogen from Berlin Sewage.* Gas u. Wasserfach, Wasser-Abwasser, 108, pp 966-969 and 1197-1205 (1967).

55. Hamm, *The Influence on Denitrification of Phosphate Precipitants.* Z. Wasser and Abwasserforschung, 2, pp 180-182 (1969).

56. Schuster, *Laboratory Experiments on Removal of Nitrogen Compounds from Domestic Sewage.* Fortschr. Wasserchem. Greng., 12, pp 139-148 (1970).

57. Carlson, D., *Nitrate Removal from Activated Sludge Systems.* Report for OWRR, Project AO26, July, 1970.

58. Beer, C., Hetling, L.J., and R.E. McKinney, *Nitrogen Removal and Phosphorus Precipitation in a Compartmentalized Aeration Tank.* Technical Paper 32, New York State Department of Environmental Conservation, February, 1974.

59. Christensen, M.H., and P. Harremoës, *Biological Denitrification in Wastewater Treatment.* Report 2-72, Department of Sanitary Engineering, Technical University of Denmark, 1972.

60. Bishop, D.F., Heidman, J.A., and J.B. Stamberg, *Single Stage Nitrification-Denitrification.* Presented at the 47th Annual Conference of the Water Pollution Control Federation, Denver, Colorado, October 6-11, 1974.

61. Barnard, J.L., *Cut P and N Without Chemicals.* Water and Wastes Engineering, 11, No. 7, pp 33-36 (1974) and No. 8, pp 41-94 (1974).

62. Wuhrmann, K., *Effect of Oxygen Tension on Biological Treatment Processes.* Proc. Third Conference Biological Waste Treatment, Manhattan College, N.Y., 1960, pp 27-38.

63. Pasveer, A., *Contribution on Nitrogen Removal from Sewage.* Müncher Bertrage zur Abwasser-Fisherei-and Flussbiologie, Bd 12, pp 197-200 (1965).

64. Matsche, N., *The Elimination of Nitrogen in the Treatment Plant of Vienna-Blumenthal.* Water Research, 6, No. 4/5, pp 485-486 (1972).

65. Matsche, N.F., and G. Spatzierer, *Austrian Plant Knocks Out Nitrogen.* Water and Wastewater Engineering, 12, No. 1, pp 18-24 (1975).

66. Drews, R.J.L.C., and A.M. Greeff, *Nitrogen Elimination by Rapid Alternation of Aerobic/"Anoxic" Conditions in "Orbal" Activated Sludge Plants.* Water Research, 7, pp 1183-1194 (1973).

67. Halvorson, H.O., Irgens, R., and H. Bauer, *Channel Aeration Activated Sludge Treatment at Glenwood, Minnesota.* JWPCF, 44, No. 12, pp 2266-2276 (1972).

68. Zemaitis, W.L., Letter communication to D.S. Parker. Envirobic Systems, Inc., New York, New York, September 27, 1974.

69. Christensen, M.H., *Denitrification of Sewage by Alternating Process Operation.* Presented at the IAWPR, Paris, September, 1974.

70. Barnard, J.L., *Biological Denitrification.* Presented at the monthly meeting of the South African Branch of the Institute of Water Pollution Control, August, 1972.

71. Ludzack, F.J., and M.B. Ettinger, *Controlling Operation to Minimize Activated Sludge Effluent Nitrogen.* JWPCF, 34, No. 9, pp 920-931 (1962).

72. van Vuuren, L.R.J., Letter communication to D.S. Parker, February, 1975.

73. Gujer, W. and D. Jenkins, *The Contact Stabilization Process-Oxygen and Nitrogen Mass Balances.* University of California, Sanitary Engineering Research Lab, SERL Report 74-2, February, 1974.

74. Balakrishnan, S., and W.W. Eckenfielder, *Nitrogen Relationships in Biological Waste Treatment Processes – III, Denitrification in the Modified Activated Sludge Process.* Water Research, 3, pp 177-188 (1969).

75. Schroeder, E.D., and A.W. Busch, *The Role of Nitrate Nitrogen In Bioxidation.* JWPCF, 40, No. 11, pp R445-R457 (1968).

76. Parker, D.S., Kaufman, W.J., and D. Jenkins, *Physical Conditioning of Activated Sludge Floc.* JWPCF, 43, No. 9, pp 1817–1833 (1971).

CHAPTER 6

BREAKPOINT CHLORINATION

6.1 Process Chemistry

When chlorine is added to dilute aqueous solutions containing ammonia nitrogen, reactions occur which may lead ultimately to oxidation of the ammonium ion to end products composed predominantly of nitrogen gas. When such chemical processes are performed in water and wastewater treatment for the purpose of ammonia nitrogen removal, the procedure is termed breakpoint chlorination. This chapter discusses the theoretical stoichiometry of breakpoint chlorination, presents the practical process considerations which influence actual chemical consumption, reaction end products and rate of the reaction, and presents process design criteria.

Recent work at the Blue Plains wastewater treatment pilot plant in Washington, D.C.[1,2,3] has confirmed breakpoint chlorination reaction products. Gas chromatography was used at Blue Plains to identify breakpoint reaction products from buffered aqueous wastewater samples in laboratory tests. Further confirmation of breakpoint reaction end products was obtained in pilot scale investigations with wastewater effluent of different qualities.

Breakpoint chlorination tests on domestic wastewaters at the Blue Plains pilot plant showed that 95 to 99 percent of the ammonia nitrogen in solution is converted to nitrogen gas.[2,3] No breakpoint reaction intermediate compounds of N_2O, NO or NO_2 were detected. The oxidized ammonia nitrogen fraction which did not appear as nitrogen gas was found to be made up of nitrate and nitrogen trichloride.

6.1.1 Chemical Stoichiometry

When chlorine gas is dissolved in water, hydrolysis of the chlorine molecule occurs according to the following relationship:

$$Cl_2 + H_2O \rightleftharpoons \underset{\text{(hypochlorous acid)}}{HOCl} + H^+ + Cl^- \qquad (6\text{-}1)$$

The active (oxidizing) forms of chlorine in solution are hypochlorous acid and its dissociation product, hypochlorite ion.

$$HOCl \rightleftharpoons OCl^- + H^+ \qquad K = 3.3 \times 10^{-8} \text{ at 20 C} \qquad (6\text{-}2)$$

where: K = dissociation constant

The fraction of the total chlorine residual in a sample which is made up of hypochlorous acid and hypochlorite ion is termed the "fiee available" chlorine residual. The rate of dissociation of HOCl is very rapid and equilibrium proportions are maintained even when HOCl is continuously being reacted. The equilibrium relationship between HOCl and OCl^- in relation to solution pH is shown in Figure 6-1.

FIGURE 6-1

RELATIVE AMOUNTS OF HOCl AND OCl^- AT VARIOUS pH LEVELS (REFERENCE 4)

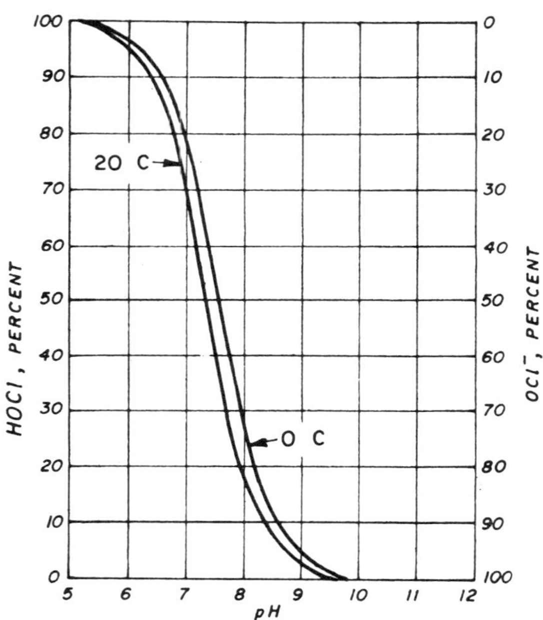

The oxidizing capability of the free available chlorine residual is manifest in the chemical transformation of hypochlorous acid to chloride ion (Cl^-). This transformation involves a gain of two electrons and a valence change of the chlorine atom from "+1" to "-1".

Ammonia nitrogen concentrations of 10 mg/l to 40 mg/l may be found in typical municipal wastewater treatment plant effluents. The source of ammonia nitrogen typically includes direct discharge from industrial processes and release following hydrolysis of urea and biological degradation of amino acids and other organic derivatives of ammonia nitrogen. The actual chemical form of ammonia nitrogen in solution is pH and temperature dependent. The relative distribution of ammonia nitrogen and ammonium ion may be defined according to the equation below:

$$NH_3 + H^+ \rightleftharpoons NH_4^+ \qquad K = 5.0 \times 10^{-10} \text{ at } 20\text{ C} \tag{6-3}$$

This relationship is indicated graphically in Figure 6-2 in relation to pH of the solution. Reactions between chlorine and ammonium in dilute aqueous solution can proceed according to a number of competing pathways. Formation of chloramines, termed "combined residual" and nitrate can occur in the following manner:

$$NH_4^+ + HOCl \longrightarrow NH_2Cl \text{ (monochloramine)} + H_2O + H^+ \tag{6-4}$$

$$NH_2Cl + HOCl \longrightarrow NHCl_2 \text{ (dichloramine)} + H_2O \tag{6-5}$$

FIGURE 6-2

EFFECTS OF pH AND TEMPERATURE ON DISTRIBUTION OF AMMONIA AND AMMONIUM ION IN WATER

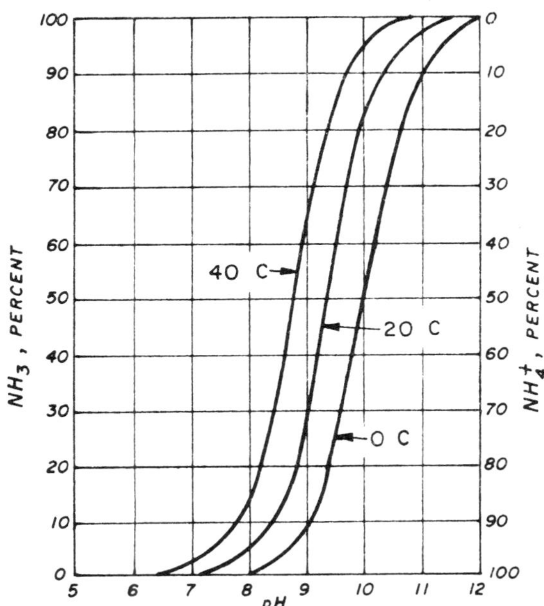

$$NHCl_2 + HOCl \longrightarrow NCl_3 \text{ (nitrogen trichloride)} + H_2O \qquad (6\text{-}6)$$

and

$$NH_4^+ + 4HOCl \longrightarrow HNO_3 + 5H^+ + 4Cl^- + H_2O \qquad (6\text{-}7)$$

The reactions are dependent upon certain process variables, including pH, temperature, contact time, and the initial chlorine to ammonia nitrogen ratio (Cl_2:NH_4^+-N).[5,6,7]

Breakpoint chlorination occurs when sufficient chlorine has been added to a water or wastewater sample to cause the chemical oxidation of the ammonium in solution to nitrogen gas and other end products.

Significant aspects of breakpoint chlorination process chemistry which were studied during the Blue Plains pilot work included identification of the predominant end products of the breakpoint reaction. Tests on municipal wastewater and wastewater treatment plant effluent at Blue Plains indicated that 95 to 99 percent of the ammonia nitrogen in solution was converted to nitrogen gas. Nitrate and nitrogen trichloride account for the remaining fraction. The overall reaction between the ammonium ion and chlorine leading to formation of nitrogen gas may be expressed in terms of the simplified equations below:

$$NH_4^+ + HOCl \longrightarrow NH_2Cl + H_2O + H^+ \qquad (6\text{-}4)$$

$$NH_2Cl + 0.5\ HOCl \longrightarrow 0.5\ N_2 + 0.5\ H_2O + 1.5\ H^+ + 1.5\ Cl^- \qquad (6\text{-}8)$$

$$NH_4^+ + 1.5\ HOCl \longrightarrow 0.5\ N_2 + 1.5\ H_2O + 2.5\ H^+ + 1.5\ Cl^- \qquad (6\text{-}9)$$

Stoichiometrically, the breakpoint reaction of Equation 6-9 requires a weight ratio of chlorine to ammonia nitrogen at the breakpoint of 7.6:1, as shown below:

Molecular Weight HOCl = 70.9 (expressed as Cl_2)

Molecular Weight NH_4^+ = 14.0 (expressed as N)

Therefore, weight ratio of Cl_2:NH_4^+-N at breakpoint:

Cl_2:NH_4^+-N = (1.5)(70.9) : (1)(14.0) = 7.6:1

Therefore 7.6 parts of chlorine are theoretically required to chemically oxidize one part of ammonia nitrogen in aqueous solution. In practice, the actual weight ratio of chlorine to ammonia nitrogen at breakpoint has ranged from about 8:1 to 10:1. Many of the process variables which are known to affect the total chemical requirement for this process have been identified and those factors are discussed in subsequent sections.

6.1.2 The Breakpoint Curve

The breakpoint chlorination curve is a graphical representation of chemical relationships which exist as varying amounts of chlorine are added to dilute solutions of ammonia nitrogen. An investigation in 1939 led to the discovery that increasing the chlorine dose in certain waters resulted in an overall reduction in the chlorine residual measured in the water sample following contact.[8] The point of maximum reduction of chlorine residual was termed the "breakpoint".

The theoretical breakpoint curve shown in Figure 6-3 has several characteristic features. The characteristics of the breakpoint curve shown in Zone 1 include principally the reaction between chlorine and ammonium indicated in Equation 6-4. The hump of the breakpoint curve occurs, theoretically, at a chlorine to ammonia nitrogen weight ratio of 5:1 (molar ratio of 1:1). That ratio corresponds to the point at which the reacting molecules are present in solution in equal numbers.

The chemical equilibria of Zone 2 favor the formation of dichloramine (Equation 6-5) and the oxidation of ammonium according to Equation 6-9. These reactions proceed in competition to, theoretically, a Cl_2:NH_4^+-N weight ratio of 7.6:1. At the breakpoint, the ammonium concentration is minimized.

To the right of breakpoint, Zone 3 chemical equilibria include the build-up of free chlorine residual as well as the presence of small quantities of dichloramine (Equation 6-5), nitrogen trichloride (Equation 6-6), and nitrate (Equation 6-7). The free chlorine residuals which result from dosages beyond breakpoint are known to be considerably more bactericidally potent than the combined residuals found at lower chlorine dosages (See Section 6.2.7).

6.2 Process Application Considerations

The basic theoretical background chemistry must be combined with application fundamentals if proper process designs and operations are to be achieved in full scale wastewater treatment practice. Much background work on breakpoint chlorination has been done on a laboratory "pure system" basis and in potable water. This information is useful, but not always applicable to wastewater treatment considerations. Generally, the presence of high concentrations of ammonia nitrogen and other amino substances as well as the presence of other chemical constituents in wastewater effluents contribute to the discrepancy between some data available from laboratory testing and that collected in actual wastewater treatment applications.

FIGURE 6-3

THEORETICAL BREAKPOINT CHLORINATION CURVE

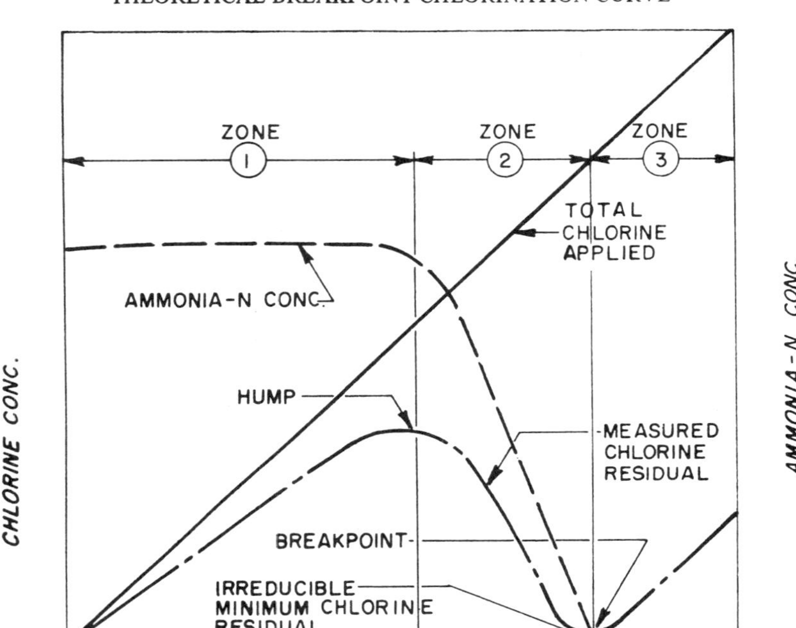

6.2.1 Chlorine Dosage Requirement

The total amount of chlorine which must be added to wastewater to achieve breakpoint is affected by the chemical nature of the wastewater and by the conditions which exist in the zone where the reacting species come into contact. Several important factors which should be considered in breakpoint chlorination process application and design are discussed below.

6.2.1.1 Effect of Pretreatment

The degree of treatment which a wastewater stream receives prior to breakpoint chlorination effects both the chlorine dosage and end-product distribution. The relationship between pretreatment and nitrate and nitrogen trichloride formation is discussed in Section 6.2.2.

The chlorine demand of a wastewater treatment plant effluent sample is the total chlorine oxidative capacity consumed during a given period of time by substances in solution which do not result in measurable chlorine residual or chemical oxidation of ammonia nitrogen. This is the chlorine oxidative capacity which is essentially "lost" from participation in the desired breakpoint reaction. Chlorine demand may be exerted by a number of substances commonly present in wastewater, including S^{2-},$HS-SO_3^{2-}$,NO_2^-,Fe^{2+}, phenols, amino acids, proteins and carbohydrates.[9]

Generally, as the degree of pretreatment of wastewater is increased, the chlorine demand exerted by the substances noted above is reduced. One particularly important factor is that if a treatment process employing anoxic conditions precedes a chlorination facility, substances in solution may be converted from an oxidized to a reduced form and the chlorine demand may be substantially increased.

Laboratory and pilot plant studies of breakpoint chlorination at the Blue Plains pilot plant[1,2] and at Sunnyvale[10,11] have shown that increasing levels of pretreatment decrease the amount of chlorine required to achieve breakpoint. Table 6-1 shows that laboratory tests on buffered distilled water containing only ammonia nitrogen reached breakpoint at a Cl_2:NH_4^+-N ratio of 8:1, a level near that predicted by chemical stoichiometry (7.6:1). In comparison, raw wastewater required a Cl_2:NH_4^+-N of 9:1 - 10:1 to reach breakpoint. Ammonia nitrogen concentrations in the samples following breakpoint were found to be consistently in the range of 0.2 mg/l or less. Pilot plant scale testing of breakpoint chlorination processes has confirmed chlorine dosages predicted through laboratory work.

6.2.1.2 Effect of pH and Temperature

Laboratory studies at Blue Plains[1,2] in which buffered distilled ammonia nitrogen solutions of 20 mg/l concentration were subjected to breakpoint chlorination dosages showed a definite optimum pH for breakpoint in the range of pH 6 to 7. The chlorine dosage at optimum pH levels was found to be Cl_2:NH_4^+-N of 8:1. Breakpoint tests conducted outside the apparent optimum range of pH 6 to 7 showed an appreciably higher chlorine requirement for breakpoint and slower reaction rates.

Comparable tests carried out with filtered secondary effluent did not show a clearly defined relationship between pH and Cl_2:NH_4^+-N to reach breakpoint. Formation of other nitrogenous residuals (NO_3^- and NCl_3) was considered to be the controlling criteria in selection of the optimum pH operating range of pH 6 to 7 (See Section 6.2.2).

There is no evidence that ordinary variations in the temperature of wastewater effluents affect the Cl_2:NH_4^+-N to reach breakpoint.

6.2.1.3 Initial Mixing of Chlorine

The significance of initial mixing in certain unit processes of sanitary engineering has been

TABLE 6-1

EFFECT OF PRETREATMENT ON Cl_2:NH_4^+-N BREAKPOINT RATIO

Sample	Breakpoint pH	Initial NH_4^+-N (mg/l)	Final NH_4^+-N (mg/l)	Irreducible minimum residual (mg/l as Cl_2)	Breakpoint ratio Cl_2:NH_4^+-N (weight basis)	Ref
		Laboratory Tests				
Buffered water	6 - 7	20	0.1	0.6	8:1	2
Raw wastewater	6.5 - 7.5	15	0.2	7	9:1 - 10:1	2
Lime clarified raw wastewater	6.5 - 7.5	11.2	0.1	7	8:1 - 9:1	2
Secondary effluent	6.5 - 7.5	8.1	0.2	3	8:1 - 9:1	2
Lime clarified secondary effluent	6.5 - 7.5	9.2	0.1	4	8:1	2
Ferric chloride clarified raw wastewater-carbon adsorption	3.2	10.2	0.1	20	8.2:1	12
		Pilot Plant Tests				
Filtered secondary effluent	6 - 8	12.9 - 21.0	0.1	2 - 8.5	8.4:1 - 9.2:1	2
Lime clarified raw wastewater-filtered	7.0 - 7.3	9.7 - 12.5	0.4 - 1.2	-	9:1	2
Alum clarified oxidation pond effluent-filtered	6.6	20.6	0.1	7.6	9.6:1	11

amply demonstrated. Tests[13] have shown significantly improved alum coagulation efficiency as a direct result of increasing the level of turbulence of mixing in the zone of alum application to a water sample. Improved disinfection efficiency in laboratory tests was also noted when the application of chlorine to a wastewater effluent was accomplished at increased levels of turbulent mixing.

Recent data from Blue Plains[14] have shown the total chlorine dose required to reach breakpoint is not affected by initial mixing conditions to the degree which was first reported. A number of pilot plant tests with secondary effluent in which the chlorine was dosed into a mechanical mixer showed no difference in the efficiency of chlorine utilization whether or not the mixer was operating. Other laboratory studies have shown the Cl_2:NH_4^+-N ratio at breakpoint to be unaffected by the degree of mixing in the reaction zone.

In plant-scale design of breakpoint chlorination facilities, a quantity of hydraulic or mechanical energy sufficient to facilitate rapid and thorough blending of the chlorine solution, pH adjustment chemical and process influent should be provided. The blending of chemicals with the process influent initiates the breakpoint reactions and allows completion of the breakpoint reaction within the contact zone, provides the basis for containment of process odors and assures process consistency, a necessary prerequisite for the feedback element of process control (Section 6.3).

6.2.2 Residual Nitrogenous Materials

Nitrate (NO_3^-) and nitrogen trichloride (NCl_3) are occasionally found in the effluent from the breakpoint chlorination process. Both compounds can be found in varying concentrations, depending upon the degree of pretreatment and pH in the reaction zone. The total concentration of these residuals seldom exceeds 10 percent of the influent ammonia nitrogen concentration.[2]

Nitrogen trichloride is a particularly volatile compound which exhibits a very strong chlorinous odor. It is an extremely strong oxidizing agent, having been used for many years in the bleaching of flour. Formation of NCl_3 in breakpoint chlorination, even in fairly small concentrations, is undesirable because of the obnoxious and dangerous characteristics of the compound in the gaseous form. Nitrate formation in breakpoint chlorination should be avoided since it represents a reaction product which has consumed considerable amounts of chlorine (Equation 6-7) and because it reduces the overall nitrogen removal capability of the breakpoint process.

Table 6-2 presents data on the effect of wastewater pretreatment on the formation of residual nitrogenous materials at breakpoint.[1,2] Although the investigators concluded that NCl_3 formation decreased with decreasing pretreatment, it appears that very highly treated effluent may be breakpointed with the production of negligible levels of NCl_3. Nitrate production was similar at each treatment level.

TABLE 6-2

EFFECT OF PRETREATMENT ON FORMATION OF NITROGENOUS RESIDUALS AT BREAKPOINT (REFERENCE 1, 2)

Sample	Initial Ammonia-N Conc. (mg/l)	NCl_3 Conc. at Breakpoint (mg/l as N)[a]	NO_3^- Conc. at Breakpoint (mg/l as N)[a]
Raw Wastewater	15.0	0.0	0.3
Lime Clarified and Filtered Raw Wastewater	11.2	0.25	0.45
Secondary Effluent	8.1	0.13	0.24
Lime Clarified and Filtered Secondary Effluent	9.2	0.0	0.2

[a]pH Range = 6.5 to 7.5

Laboratory and pilot-scale tests have confirmed the pH sensitivity of NCl_3 formation in breakpoint chlorination. Pilot-scale tests of filtered secondary effluent showed 0.33 mg/l NCl_3 (as N) after breakpoint chlorination at pH 6; at pH 7 and above, NCl_3 concentration was reduced to about 0.05 mg/l. Nitrate formation was found to be slightly affected by pH at breakpoint. Nitrate concentration in pilot tests of filtered secondary effluent ranged from about 0.7 mg/l (as N) at pH 6 to 1.0 mg/l (as N) at pH 8.

Figure 6-4 shows the consequences of chlorine dosages beyond breakpoint on the formation of nitrogenous residuals. Nitrate production following breakpoint chlorination of lime clarified filtered secondary effluent (breakpoint at Cl_2:NH_4^+-N of 8:1) showed a linear increase in concentration at increased chlorine dosages. Nitrogen trichloride formation was noted beginning at Cl_2:NH_4^+-N ratios exceeding 9:1. Close control of chlorine dosage levels in breakpoint is shown to be an important factor in minimizing production of these residuals.

Pilot testing of breakpoint chlorination at Blue Plains showed that variations in temperature from 5 to 40 C did not affect the final distribution of nitrogenous residuals in the effluent.

Awareness of the pH sensitivity of formation of NO_3^- and NCl_3 has led to formulation of two specific recommendations for design of plant-scale breakpoint chlorination facilities. First, the pH adjustment chemical (Section 6.2.3) should be added to the chlorine solution prior to application of the chlorine solution to the breakpoint chlorination process influent. Without premixing of chemicals, disproportionate mixing of chlorine solution and pH

FIGURE 6-4

EFFECT OF Cl_2:NH_4^+-N ON NITROGEN RESIDUALS IN
LIME CLARIFIED FILTERED SECONDARY EFFLUENT (REF. 1)

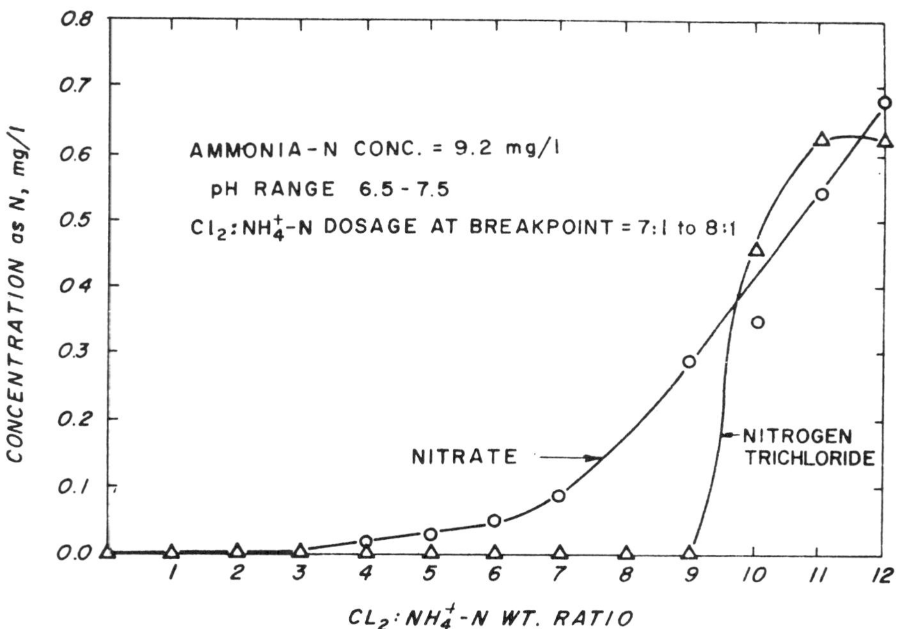

adjustment chemical in the breakpoint reaction zone can result in "pockets" of liquid in the breakpoint reaction zone not at the desired pH level. Occurrence of the breakpoint reaction in such "pockets" could lead to formation of excessive concentrations of NO_3^- and NCl_3.

A second design recommendation is that the operating pH for breakpoint chlorination should be maintained at pH 7. This pH allows optimal Cl_2:NH_4^+-N ratios and reaction rates and also reduces the NCl_3 production and attendant odor production to about 0.2 mg/l or less, as long as careful control over chlorine dosage is maintained.

In summary, the design engineer should be cautioned that a poorly designed or maintained system for breakpoint chlorination pH and chlorine dosage control can result in the generation of chlorinous odors due to the formation of nitrogen trichloride. However, it appears that if appropriate care is exercised in the design and maintenance of pH and chlorine dosage systems, the breakpoint chlorination process can function satisfactorily without the need for odor control facilities.

6.2.3 Alkalinity Supplementation

In the addition of chlorine to a wastewater sample for the purposes of breakpoint chlorination, acidity is generated through the hydrolysis of chlorine gas (Equation 6-1) and oxidation of ammonium (Equation 6-9). Theoretically, four moles of hydrogen ions are generated for every mole of ammonia nitrogen which is oxidized with chlorine gas.

$$1.5\ Cl_2 + 1.5\ H_2O \longrightarrow 1.5\ HOCl + 1.5\ H^+ + 1.5\ Cl^- \quad (6\text{-}1)$$

$$NH_4^+ + 1.5\ HOCl \longrightarrow 0.5\ N_2 + 1.5\ H_2O + 2.5\ H^+ + 1.5\ Cl^- \quad (6\text{-}9)$$

$$1.5\ Cl_2 + NH_4^+ \longrightarrow 0.5\ N_2 + 4H^+ + 3Cl^- \quad (6\text{-}10)$$

The buffering capacity (alkalinity) of the breakpoint process influent is consumed by the acidity of Equation 6-10. Stoichiometrically, 14.3 mg/l of alkalinity (as $CaCO_3$) is consumed for each 1.0 mg/l of ammonia nitrogen which is oxidized in breakpoint chlorination. In actual practice, around 15 mg/l alkalinity is consumed due to the hydrolysis of chlorine needed beyond that predicted by stoichiometry.

It is apparent that if the ammonia nitrogen concentration in the breakpoint process influent is high, or if the wastewater alkalinity is relatively low, insufficient buffering capacity will be available to maintain the process pH at a reasonable level. Any alkaline substance may be used to supplement the natural alkalinity for pH control but sodium hydroxide (NaOH) and lime (CaO) are the compounds most commonly used. If all of the acidity generated in breakpoint must be neutralized through chemical addition, 1.50 parts of sodium hydroxide (NaOH) would be needed for each part of chlorine, or 1.05 parts of lime (CaO) would be needed for each part of chlorine. In practice, the alkalinity of the process influent may supply a portion of the total buffering capacity needed under breakpoint conditions. If a highly alkaline stream is to be treated (such as might result from a lime precipitation process), alkalinity supplementation would not be needed and, perhaps, acid addition would be needed to achieve the recommended operating pH of 7. In situations where breakpoint chlorination is used for removing low ammonia nitrogen residuals (< 3 mg/l) from nitrificaton process effluents, there may be sufficient remaining alkalinity to buffer the water without the need for addition of neutralizing chemicals.

6.2.4 Reaction Rates

The reaction rate for breakpoint chlorination has not been measured quantitatively, but several investigations[2,11] have noted that the reaction is very rapid. According to reaction rates established by Morris,[6,7] the reaction between ammonium and hypochlorous acid (Equation 6-4) to form monochloramine is 99 percent complete in 0.2 seconds at pH 7 and $Cl_2:NH_4^+$-N of 0.2:1. The optimum reaction pH was shown to be pH 8.3, allowing 99 percent conversion to monochloramine in 0.069 seconds. Obviously, the rate of this reaction is important in breakpoint since it is the first of several sequential reactions.

Laboratory studies have shown the breakpoint reaction rate to vary, depending upon pH. At pH levels between 6 and 7, the breakpoint reaction was found to proceed to completion in less than 15 seconds when secondary effluents were tested. When reaction rates at pH levels outside the optimum range of pH 6-7 were tested, the rate was found to be slowed considerably. At pH 3.5, for example, the breakpoint chlorination reactions were not complete following two hours of contact.

The rapid rates noted for breakpoint reactions conducted in the pH range of 6-7 lead to the design recommendation that a breakpoint chlorination contact period of one minute is sufficient for plant-scale applications. The design of the contact basin should provide, as closely as possible, a plug flow contact regime. The blending of breakpoint chemicals with the process influent should be carried out as indicated in Section 6.2.1.3.

6.2.5 Effect on Total Dissolved Solids

In many instances where high levels of wastewater treatment are required, total dissolved solids (TDS) limitations may be a controlling criterion in the selection of alternative treatment processes. Breakpoint chlorination can involve the addition of large quantities of chemicals to solution. The TDS increment attributable to each of several chemicals which may be utilized in breakpoint chlorination is summarized in Table 6-3.

TABLE 6-3
EFFECTS OF CHEMICAL ADDITION ON TOTAL DISSOLVED SOLIDS IN BREAKPOINT CHLORINATION

Chemical Addition	TDS Increase : NH_4^+-N Consumed
Breakpoint with chlorine gas	6.2 : 1
Breakpoint with sodium hypochlorite	7.1 : 1
Breakpoint with chlorine gas - Neutralization of all acidity with lime (CaO)	12.2 : 1
Breakpoint with chlorine gas - Neutralization of all acidity with sodium hydroxide (NaOH)	14.8 : 1

If breakpoint chlorination is contemplated on a wastewater effluent stream which contains 20 mg/l ammonia nitrogen, the increase of TDS in solution following addition of chlorine in the gaseous form would amount to 124 mg/l. If all of the acidity generated in the hydrolysis of chlorine and the oxidation of ammonium is neutralized with lime (CaO), the total increase in TDS would amount to 244 mg/l.

6.2.6 Reactions with Organic Nitrogen

Studies at Blue Plains[3] found only "slight reduction in organic nitrogen within the two hour contact time." The data of Lawrence, *et al.*[15] showed a decrease in organic nitrogen from 3.2 mg/l - 3.5 mg/l to levels of 0.2 mg/l - 0.4 mg/l. More recent data[11] collected in Sunnyvale, California also show an apparent decrease in soluble organic nitrogen following breakpoint (Table 6-4).

TABLE 6-4

EFFECT OF BREAKPOINT CHLORINATION
ON SOLUBLE ORGANIC NITROGEN (REFERENCE 11)[a]

Date, 1973	Soluble Organic Nitrogen before Breakpoint (mg/l as N)	Soluble Organic Nitrogen after Breakpoint (mg/l as N)
8/30	2.7	1.0
9/4	2.8	1.7
9/11	4.6	1.3
9/12	4.7	2.0

[a]Soluble organic nitrogen determination conducted on filtrate from 0.45μ membrane filter. Wastewater treated was tertiary treated oxidation pond effluent

Taras[16] reported in 1953 that the concentration of unsubstituted amino nitrogen of many common amino acids was reduced slowly by reaction with chlorine. Organic nitrogen in the more complex form of proteins was practically unaffected by chlorine over a period of several days. The degree of organic nitrogen reduction through breakpoint chlorination is likely a function of the relative proportion of proteins to the simpler hydrolytic products (including amino acids) of the protein molecules. In summary, the true reductions in organic nitrogen with breakpoint chlorination are difficult to predict.

The presence of organic nitrogen in solution at breakpoint has also been shown to effect the shape of the breakpoint curve.[4] It has been noted that waters containing a mixture of ammonia nitrogen and organic nitrogen did not display the classic "dip" of the breakpoint curve.[17,18,19] The irreducible minimum residual was found to be appreciably greater when organic nitrogen was present than when the sample contained only the inorganic ammonia nitrogen form.

6.2.7 Disinfection

Under normal conditions, disinfection of a non-nitrified wastewater effluent stream with chlorine is mainly accomplished by the form of combined chlorine residual known as monochloramine (NH_2Cl at pH 7.0). When contact for disinfection is provided downstream from the breakpoint chlorination process, the free chlorine residual in solution following breakpoint will provide much higher bactericidal potential than with monochloramine alone.

Figure 6-5 is a comparison of the germicidal efficiency of hypochlorous acid, hypochlorite ion and monochloramine. Figure 6-1 shows the ionization characteristics of the hypochlorous acid at various pH levels. The data of Figure 6-5 show that hypochlorous acid is a far more effective germicidal agent than either hypochlorite ion or monochloramine. Formation of hypochlorous acid following breakpoint chlorination will, therefore, considerably enhance the capability of a wastewater disinfection system if an efficient contacting system is provided.

6.3 Process Control Instrumentation

Both South African[21] and American researchers[2,22] have reported that if a continuously functioning breakpoint chlorination process is to be a consistent and reliable environmental engineering unit process, the system must be capable of responding rapidly to changes in ammonia nitrogen concentration, chlorine demand, pH, alkalinity and flow.

6.3.1 Process Control System

Failure of the chlorine dosage control system to respond adequately to changes in process conditions can result in a substantial loss in nitrogen removal capability as well as potentially adding significant overdoses of chlorine to the process. Overdoses of chlorine to the system are a direct waste of this chemical, they result in increased difficulty of pacing the dechlorination equipment,and can cause the direct discharge of high concentrations of chlorine residuals to the receiving water.

6.3.1.1 Chlorine Dosage Control

A function diagram of the breakpoint chlorination process control system is shown in Figure 6-6. This system is the same as that used in the pilot plant testing of breakpoint chlorination at the Blue Plains pilot plant in Washington, D.C. The control of chlorine dosage is accomplished by a combination feed foreward and feedback control loop. The feed foreward component utilizes ammonia concentration and flow signals together with a manually selected multiplier to establish an approximate chlorine dosage to achieve

FIGURE 6-5

COMPARISON OF GERMICIDAL EFFICIENCY OF HYPOCHLOROUS ACID, HYPOCHLORITE ION, AND MONOCHLORAMINE FOR 99 PERCENT DESTRUCTION OF E. COLI AT 2-6 C (REF. 20)

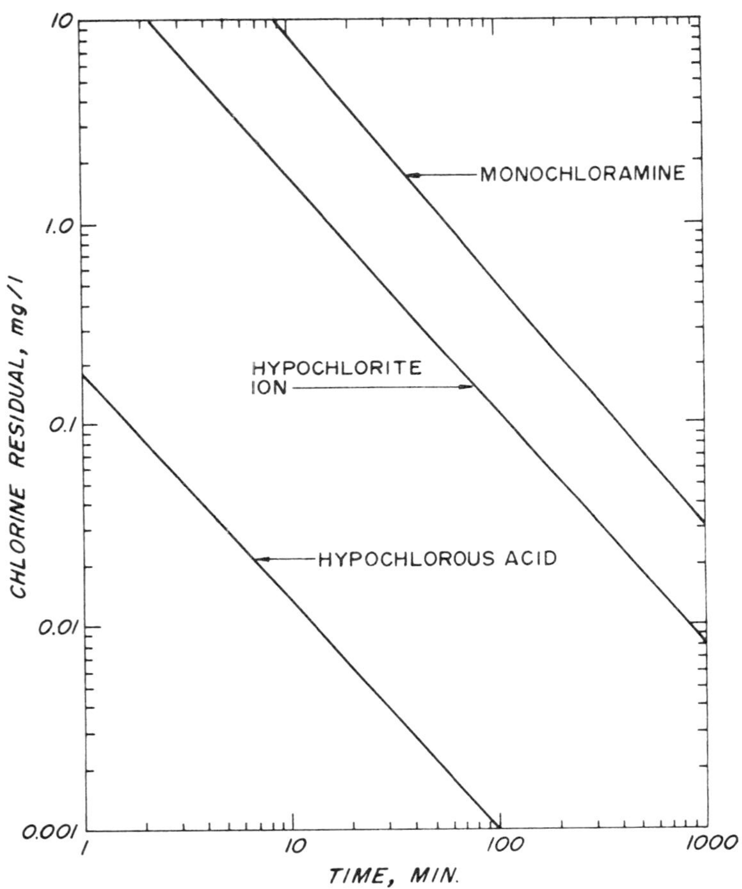

FIGURE 6-6

BREAKPOINT CHLORINATION CONTROL - FUNCTIONAL SCHEMATIC

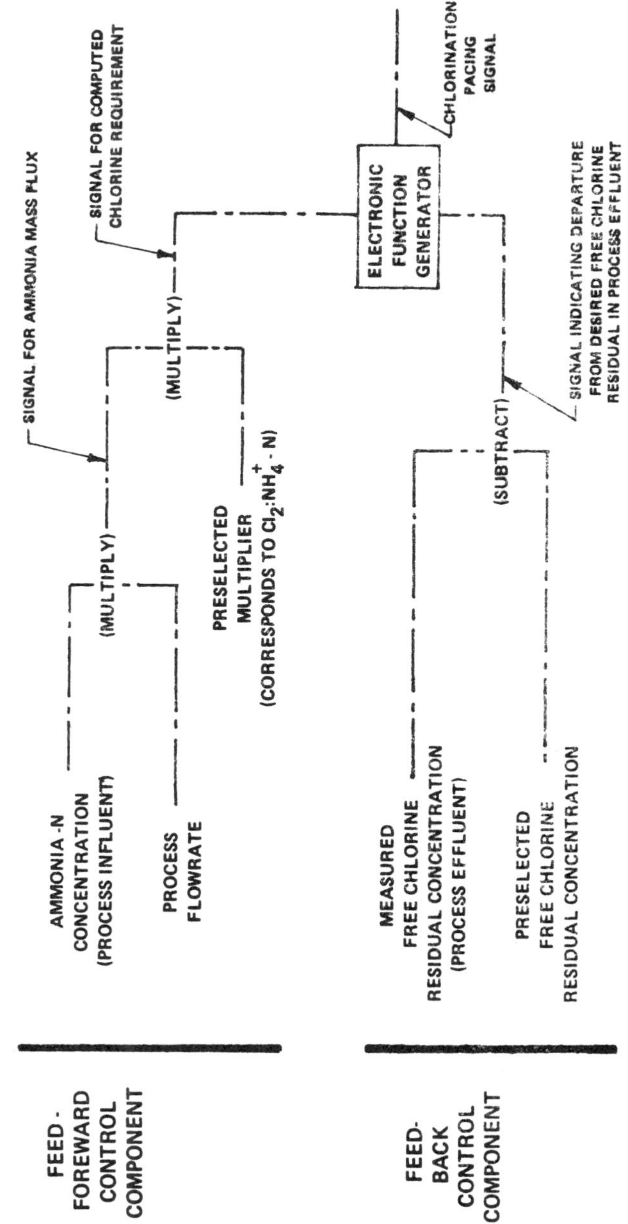

breakpoint. The manually preselected multiplier corresponds to the Cl_2:NH_4^+-N ratio required for breakpoint and varies from values of about 8 to 10.

The feedback control loop involves measurement of the free chlorine residual in the process effluent. The level of measured free chlorine residual is compared to a setpoint value (usually 2-4 mg/l) using a standard process controller and a signal is generated which provides a "trimming" of the chlorine dosage to the system.

6.3.1.2 pH Control

Under most circumstances, a base is added to the breakpoint process to neutralize a portion of the acidity generated in the chlorine addition (Equations 6-1 and 6-9). The base requirements at a particular installation are related to wastewater alkalinity, individual treatment processes employed prior to breakpoint chlorination as well as pH or alkalinity restrictions which might be imposed upon the breakpoint effluent.

pH control in the breakpoint chlorination process may be effectively achieved using a combination feed foreward and feedback system. The feed foreward component of the system involves pacing the base addition directly on the chlorine application rate. This accomplishes neutralization of a preselected portion of the chlorine acidity. A feedback loop is employed to "trim" the system to a designated pH level. Accurate pH control of the process promotes efficient utilization of chlorine and can reduce or eliminate undesirable end products of the reaction.

6.3.2 Process Control Components

Effective control over breakpoint chlorination requires utilization of accurate and reliable sensory equipment. In this regard two components deserve special attention; namely the ammonia nitrogen and free chlorine monitoring elements.

The ammonia nitrogen and free chlorine monitoring device which was used during the Blue Plains pilot testing of breakpoint chlorination was an automated wet chemical analyzer using colorimetric techniques to measure the ammonia nitrogen and free chlorine concentrations in small individual samples of the process influent. A continuous-duty free chlorine residual analyzer was also successfully used at Blue Plains to provide a feedback signal for breakpoint chlorination control.

6.4 Dechlorination Techniques

Dechlorination is a unit process which eliminates the active chlorine residual in solution prior to effluent discharge or additional treatment steps. Current emphasis on elimination of chlorine residual from wastewater treatment effluents has resulted in dechlorination requirements in many cases. Frequently, dechlorination will be required as a companion process to breakpoint chlorination. Dechlorination techniques using sulphur dioxide and activated carbon may be used.

6.4.1 Sulphur Dioxide Dechlorination

Sulphur dioxide is a colorless, odorless gas which hydrolyzes in aqueous solution to form sulfurous acid, a strong reducing agent. When added to a sample containing active chlorine residual (oxidizing agent), the chlorine residual is converted to a non-toxic form, normally chloride ion.

6.4.1.1 Stoichiometry

Sulphur dioxide reacts with both free and combined forms of chlorine residuals according to the expressions noted below:

a. Reaction with free chlorine residual (i.e. HOCl)

$$SO_2 + H_2O \longrightarrow HSO_3^- + H^+ \qquad (6\text{-}11)$$

$$HOCl + HSO_3^- \longrightarrow Cl^- + SO_4^= + 2H^+ \qquad (6\text{-}12)$$

$$SO_2 + HOCl + H_2O \longrightarrow Cl^- + SO_4^= + 3H^+ \qquad (6\text{-}13)$$

b. Reaction with combined chlorine residual (i.e. NH_2Cl)

$$SO_2 + H_2O \longrightarrow HSO_3^- + H^+ \qquad (6\text{-}11)$$

$$NH_2Cl + HSO_3^- + H_2O \longrightarrow Cl^- + SO_4^= + NH_4^+ + H^+ \qquad (6\text{-}14)$$

$$SO_2 + NH_2Cl + 2H_2O \longrightarrow Cl^- + SO_4^= + NH_4^+ + 2H^+ \qquad (6\text{-}15)$$

Several important observations can be made from the chemical reactions presented in Equations 6-11 to 6-15. First, both combined and free chlorine residuals are reduced to chloride ion following the reaction. The sulfite ion is correspondingly oxidized to the sulfate ion. Both chloride and sulfate ions are usually present in abundance in wastewater, so that the additional increment added in dechlorination is usually insignificant.

The predictions of sulphur dioxide dosages required for dechlorination (Equations 6-13 and 6-15) suggest an SO_2:Cl_2 molar ratio of 1:1. This corresponds to a weight ratio of 0.9:1. In practice, 0.9 to 1.0 parts of SO_2 are required for dechlorination of 1.0 part of chlorine residual (expressed as Cl_2).

The acidity generated in dechlorination, while appearing to be significant on a molar basis, is seldom an important factor in practice. Roughly 2 mg/l of alkalinity are consumed for each 1.0 mg/l of sulphur dioxide applied. The chlorine residual remaining in solution following breakpoint chlorination and disinfection contact is usually low enough (approximate range of 1 mg/l to 8 mg/l) so that the alkalinity consumption and pH change resulting from sulphur dioxide addition may usually be neglected.

6.4.1.2 Reaction Rates

The rate of reaction between SO_2 in solution and chlorine residual is practically instantaneous. The high specificity of SO_2 in solution for the chlorine residual and the rapid reaction rate tend to reduce to very low levels competing side reactions which could result in wastage of the sulphur dioxide.

6.4.1.3 Significance of Sulphur Dioxide Overdose

Sulphur dioxide dechlorination is most often accomplished to eliminate the chlorine induced toxicity in a wastewater effluent prior to discharge. Studies on primary and secondary effluents[23,24] have shown that effluent toxicity following chlorination-dechlorination, as measured by fish bioassay, is reduced to a level slightly below that measured on comparable unchlorinated effluent. These studies indicate that sulphur dioxide dechlorination can eliminate acute toxicity related to chlorine residual. Tests in an aerated continuous-flow fish bioassay apparatus at sulphur dioxide overdoses of about 70 mg/l (as SO_2) beyond that necessary for dechlorination showed no adverse effects on the test fish.

Large overdoses of sulphur dioxide should be avoided in plant scale dechlorination systems because of chemical wastage and the oxygen demand exerted by the excess sulphur dioxide in solution. Excess sulphur dioxide reacts slowly with dissolved oxygen in solution according to the following equation:

$$HSO_3^- + 0.5\ O_2 \longrightarrow SO_4^= + H^+ \qquad (6\text{-}16)$$

The net result of this reaction can be a reduction in wastewater effluent dissolved oxygen and an increase in the measured effluent COD and BOD levels. Proper control of the dechlorination system is essential to minimize these adverse effects.

6.4.1.4 Process Application and Control

Chemical dechlorination using sulphur dioxide can be a rather inexpensive process for reducing effluent toxicity related to chlorine residual. Much of the process equipment, chemical handling and chemical addition components of a sulphur dioxide dechlorination system are identical to those used in a chlorination system. For example, gaseous chlorinators are used to feed sulphur dioxide, evaporaters may be used practically interchangeably between chlorine and sulphur dioxide, and piping, valves and gaseous injectors are identical in the two systems. The speed of the reaction precludes the necessity for providing a separate contact basin as in chlorine contact for disinfection. The dosage of sulphur dioxide required is only that amount needed to dechlorinate the effluent following the chlorine contact period, a level substantially less than the total chlorine dose.

There are no analytical instruments currently on the market which have demonstrated a capability to reliably and accurately measure the excess sulphur dioxide in solution following dechlorination. This is why the process control schemes which have been devised do not have the feedback self-corrective aspects which provide a high degree of consistency to effluent quality. Control is usually achieved by pacing the dechlorination equipment on signals of flow and chlorine residual measured directly upstream from the point of sulphur dioxide application. Operator surveillance and frequent manual adjustments to the system are needed to maintain a slight overdose of sulphur dioxide, say 0.5 mg/l.

A more complicated control scheme which does incorporate positive control features involves application of a preselected chlorine dosage to a small side stream of dechlorinated effluent. If, for example, a preselected chlorine dose of 1 mg/l is applied and dechlorination is paced to maintain a chlorine residual of 0.8 mg/l in the "biased" side stream, the net effect of the control system would be to provide a controlled overdose of 0.2 mg/l (expressed as Cl_2) of sulphur dioxide.

With either control scheme, the sulphur dioxide overdose in the final effluent may be controlled to 0.5 mg/l (as Cl_2) or less. Given that the reaction rate of Equation 6-16 is very slow, the effect of the overdose on plant effluent dissolved oxygen concentration would be minimal, probably less than 0.2 mg/l DO depletion. In such a case, aeration following sulphur dioxide dechlorination would seldom be warranted.

6.4.2 Activated Carbon Dechlorination

Dechlorination of wastewater effluent streams using activated carbon can serve several functions other than removal of chlorine residuals. In the case of breakpoint chlorination, the activated carbon can effectively catalyze the chemical reactions, serving as a "reaction bed." Some removal of soluble organics is also accomplished through adsorption.

6.4.2.1 Stoichiometry

Activated carbon (C*) reacts with both free and combined chlorine residuals in the following manner:

a. Reaction with free chlorine residual

$$C^* + 2HOCl \longrightarrow CO_2 + 2H^+ + 2Cl^- \qquad (6\text{-}17)$$

b. Reactions with combined chlorine residual

$$C^* + 2NH_2Cl + 2H_2O \longrightarrow CO_2 + 2NH_4^+ + 2Cl^- \qquad (6\text{-}18)$$

$$C^* + 4NHCl_2 + 2H_2O \longrightarrow CO_2 + 2N_2 + 8H^+ + 8Cl^- \qquad (6\text{-}19)$$

Carbon dioxide is formed in each case following reaction with chlorine residual. Ammonia is returned to solution following reaction of carbon with monochloramine, but dichloramine has been observed to decompose to nitrogen gas following contact with activated carbon.

These chemical pathways have been confirmed by several recent studies.[25,26]

6.4.2.2 Process Application

Studies by Stasiuk *et al.*[25] showed complete dechlorination of both free and combined chlorine residuals following carbon contact times of 10 minutes. Studies on the dechlorination characteristics of granular activated carbon have been conducted[27,28,29] which show a variation in the dechlorination capacity of activated carbon as a function of the hydraulic application rate and particle size. Those data suggest the formation of a dechlorination intermediary compound, nascent oxygen, on the surface of the carbon which builds up and causes a gradual loss of dechlorination efficiency. Regeneration of the carbon was accomplished by heating in the absence of air at 400 C or higher. Smaller activated carbon particles were found to give enhanced dechlorination capacity. One cubic foot of activated carbon was found to dechlorinate 0.55 million gallons at 2 gpm/sq ft hydraulic application rate and 5 mg/l free chlorine residual. Approximately 3.6 million gallons of the same water could be dechlorinated by the activated carbon when applied at a rate of 1 gpm/sq ft.

6.5 Design Example

As an example, consider a 10 mgd conventional activated sludge plant that must be upgraded to meet an effluent nitrogen limitation of 2 mg/l. Present effluent quality is 1 mg/l organic nitrogen, 20 mg/l ammonia nitrogen, 15 mg/l of suspended solids, BOD_5 of 25 mg/l and pH of 7.0. The peak to average nitrogen load ratio is 1.9 The breakpoint chlorination design should provide capacity for removal of all ammonia nitrogen in the plant effluent.

1. Calculate the average ammonia nitrogen oxidized daily.

$$NT = 8.33 \cdot Q(N_o - N_1) \qquad (4\text{-}24)$$

where: NT = ammonia nitrogen oxidized, lb per day

Q = average daily flow, mgd

N_o = influent ammonia nitrogen, mg/l

N_1 = effluent ammonia nitrogen, mg/l

For this example, the result is:

$$NT = 8.33(10)(20\text{–}0) = 1{,}666 \text{ lb/day}$$

2. To calculate the average daily chlorine consumption, a Cl_2:NH_4^+-N of approximately 9:1 would be appropriate. The average daily chlorine consumption would be:

$$\text{Average daily } Cl_2 = (9)(1{,}666) = 14{,}994 \text{ lb/day}$$

3. The peak rate of chlorine utilization, that rate used to size the chlorine feed system, may be computed by multiplying the peak to average nitrogen load ratio by the average daily chlorine consumption as follows:

$$\text{Peak } Cl_2 = (1.9)(14{,}994) = 28{,}489 \text{ lb/day}$$

4. To calculate the average daily quantity of alkalinity supplementation, assume that sodium hydroxide is to be used. Since the effluent pH is at pH 7 prior to breakpoint chlorination,and the breakpoint process itself should be conducted at pH 7, all acidity generated in breakpoint chlorination must be neutralized by the sodium hydroxide added. Since 1.50 lb. of NaOH are required per lb of Cl_2 added in breakpoint, the average daily NaOH added per day would be:

$$\text{Average daily NaOH} = (1.50)(14{,}994) = 22{,}491 \text{ lb/day}$$

5. The TDS increment added to the plant effluent as a result of breakpoint chlorination can be computed from data of Table 6-3. For breakpoint chlorination with chlorine gas and neutralization of all acidity with sodium

hydroxide, the TDS increase per mg/l of ammonia nitrogen consumed is 14.8 mg/l. The total TDS increase in this example would be:

$$\text{TDS increase} = (14.8)(20) = 296 \text{ mg/l.}$$

6. To compute the average daily consumption of sulphur dioxide needed to completely dechlorinate the plant effluent prior to discharge, assume that the average chlorine residual following breakpoint chlorination and disinfection is 5 mg/l. The average daily sulphur dioxide consumption would be:

$$\text{Average daily } SO_2 = (8.33)(10)(5)(1.0) = 416 \text{ lb/day.}$$

6.6 Considerations for Process Selection

The breakpoint chlorination process offers a number of advantages which should be considered when evaluating alternative nitrogen removal processes:

1. Ammonia nitrogen removal may be accomplished in a one-step process to concentrations less than 0.1 mg/l (as N). The major end product of the breakpoint reaction is nitrogen gas which is evolved to the atmosphere.

2. Breakpoint chlorination is free from the toxic upsets and acclimation periods which can affect biological nitrification and denitrification processes. Ammonia nitrogen may be successfully removed from solution regardless of upstream treatment processes. Breakpoint chlorination is also rather insensitive to changes in process temperature.

3. The breakpoint process may be operated intermittently or in split-stream arrangements as needed to meet individual receiving water nitrogen limitations. The process can be used as a polishing step for other ammonia removal processes such as nitrification to provide low ammonia nitrogen concentrations.

4. Breakpoint chlorination is reliable and consistent in terms of process performance.

5. Disinfection of a wastewater effluent is enhanced following breakpoint chlorination due to the presence of free available chlorine residuals ($HOCl$ and OCl^-).

6. The low spacial requirement of the breakpoint process makes it particularly suitable for certain applications, including addition to an existing facility, where nitrogen removal is required but where space constraints exist.

7. The cost of physical facilities for breakpoint chlorination is much less than for biological nitrification-denitrification facilities.

A list of potential disadvantages of the breakpoint chlorination process includes:

1. The breakpoint chlorination process is usually quite high in operating costs.

2. The addition of chorine, pH adjustment chemicals and sulphur dioxide all contribute to the level of total dissolved solids in the wastewater effluent. Some treatment plants have TDS limitations which would limit the applicability of breakpoint chlorination.

3. Dechlorination will be required in many cases to remove the potentially toxic chlorine residual. It should be noted, however, that this is a disadvantage of any chlorination process.

6.7 References

1. Pressley, T.A., Bishop, D.F., and S.G. Roan, *Nitrogen Removal By Breakpoint Chlorination.* Report prepared for the Environmental Protection Agency, September, 1970.

2. Pressley, T.A., Bishop, D.F., Pinto, A.P., and A.F. Cassel, *Ammonia-Nitrogen Removal by Breakpoint Chlorination.* Report prepared for the Environmental Protection Agency, Contract Number 14-12-818, February, 1973.

3. Pressley, T.A., *et al., Ammonia Removal by Breakpoint Chlorination.* Environmental Science and Technology, 6, No. 7, 662, July, 1972.

4. White, G.C., *Handbook of Chlorination.* New York, Van Nostrand Reinhold Company, 1972.

5. Morris, J.C., and I. Weil, *Chlorine-Ammonia Breakpoint Reactions: Model Mechanism and Computer Simulation.* Paper, annual meeting Am. Chem. Soc., Minneapolis, Minn., April 15, 1969.

6. Morris, J.C., Weil, I., and R.H. Culver, *Kinetic Studies on the Break-Point with Ammonia and Glycine.* Unpublished copy from senior author, Harvard Univ. (1952).

7. Morris, J.C., *Kinetic Reactions between Aqueous Chlorine and Nitrogen Compounds.* Fourth Rudolphs Res. Conf., Rutgers Univ. (June 15-18, 1965).

8. Griffin, A.E., and N.S. Chamberlin, *Some Chemical Aspects of Breakpoint Chlorination.* J. NEWWA, 55, pp 371, 1941.

9. Weber, W.J., Jr., *Physiochemical Processes for Water Quality Control.* New York, Wiley-Interscience, 1972.

10. Stone, R.W., Parker, D.S., and J.A. Cotteral, *Upgrading Lagoon Effluent to Meet Best Practicable Treatment.* Presented at the 47th Annual Conference of the Water Pollution Control Federation, Denver, Colorado, October, 1974.

11. Brown and Caldwell, *Report on Tertiary Treatment Pilot Plant Studies.* Prepared for the City of Sunnyvale, California, February, 1975.

12. Stearns and Wheeler, *Wastewater Facilities Report.* Cortland, New York, May, 1973.

13. Stenquist, R.J., and W.J. Kaufman, *Initial Mixing in Coagulation Processes.* Prepared for EPA, Project EPA-R2-72-053, November, 1972.

14. Schuk, W., personal communication with D.S. Parker. EPA-DC Blue Plains Pilot Plant, September 9, 1974.

15. Lawrence, A.W., *et al., Ammonia Nitrogen Removal from Wastewater Effluents by Chlorination.* Presented at the 4th Mid-Atlantic Industrial Waste Conference, University of Delaware, November, 1970.

16. Taras, M.J., *Effect of Free Residual Chlorine on Nitrogen Compounds in Water.* JAWWA, 45, 47 (1953).

17. Griffin, A.E., and N.S. Chamberlin, *Some Chemical Aspects of Break-Point Chlorination.* J. NEWWA, 55, pp 371 (1941).

18. Griffin, A.E., *Chlorine for Ammonia Removal.* Fifth Annual Water Conf. Proc. Engrs. Soc. Western Penn., pp 27 (1944).

19. William, D.B., private communication with G.C. White, Brantford, Ontario, Canada (1967).

20. Clarke, N.A., *et al., Human Enteric Viruses in Water: Source, Survival and Removability.* International Conf. Water Poll. Research, Pergamon Press, London (September, 1962).

21. van Vuuren, L.R.J., *et al., Stander Water Reclamation Plant: Chlorination Unit Process.* Project Report 21, Pretoria, So. Africa, November, 1972.

22. *Pollution: Engineering and Scientific Solutions.* E.S. Barrekette Ed., New York, Plenum Publishing Corp., pp 522-547.

23. Esvelt, L.A., Kaufman, W.J., and R.E. Selleck, *Toxicity Removal from Municipal Wastewater*. University of California, Sanitary Engineering Research Laboratory Report 71-7, 1971.

24. Stone, R.W., Kaufman, W.J., and A.J. Horne, *Long-Term Effects of Toxicants and Biostimulants on the Waters of Central San Francisco Bay*. University of California, Sanitary Engineering Research Laboratory Report 73-1, May, 1973.

25. Stasuik, W.N., Hetling, L.J., and W.W. Shuster, *Removal of Ammonia Nitrogen by Breakpoint Chlorination Using an Activated Carbon Catalyst*. New York State Department of Environmental Conservation Technical Paper No. 26, April, 1973.

27. Hagar, D.G., and M.E. Flentje, *Removal of Organic Contaminants by Granular Carbon Filtration*. JAWWA, 57, 1440 (November, 1965).

28. Kovach, J.L., *Activated Carbon Dechlorination*. Industrial Water Engineering, pp 30-32, October/November, 1973.

29. Calgon Corporation, Bulletin 20-44.

CHAPTER 7

SELECTIVE ION EXCHANGE FOR AMMONIUM REMOVAL

7.1 Chemistry and Engineering Principles

The basic concepts of the ion exchange process apply to its use for ammonium removal and these concepts are discussed in the following paragraphs.

7.1.1 Basic Concept

The use of conventional ion exchange resin for removal of nitrogenous material from wastewaters has not proven attractive because of the preference of these exchangers for ions other than ammonium or nitrate ions. In addition, the regeneration of conventional ion exchange resins results in regenerant wastes which are difficult to handle. The non-selective nature of conventional resins is unfortunate because as a unit process, ion exchange is easily controlled to achieve almost any desired product quality. The efficiency of the process is not significantly impaired at temperatures usually encountered and ion exchange equipment can be automatically controlled, requiring only occasional monitoring, inspection, and maintenance.

The above limitations of conventional resins may be largely overcome by using an exchanger which is selective for ammonium. The exchanger currently favored for this use is clinoptilolite, a zeolite, which occurs naturally in several extensive deposits in the western United States. It is selective for ammonium relative to calcium, magnesium and sodium. The removal of the ammonium from the spent regenerant permits regenerant reuse. The ammonium may be removed from the regenerant and released to the atmosphere as ammonia (in certain situations) or nitrogen gas or it may be recovered as an ammonium solution for use as a fertilizer. Figure 7-1 is a simplified schematic of the process.

The wastewater is passed downward through a bed of clinoptilolite (typically 4-5 ft or 1.2 to 1.5 m of 20 x 50 mesh particles) during the normal service cycle. When the effluent ammonium concentration increases to an objectionable level, the clinoptilolite is regenerated by passing a concentrated salt solution through the exchange bed. By removing the ammonia from the spent regenerant, the regenerant may be reused, eliminating the difficult problem of brine disposal associated with conventional exchange resins. Some of the regenerant recovery techniques, as discussed later in detail, remove the ammonia as nitrogen gas which is discharged to the atmosphere while others remove and recover the ammonia in solution form for potential use as a fertilizer.

FIGURE 7-1

SELECTIVE ION EXCHANGE PROCESS

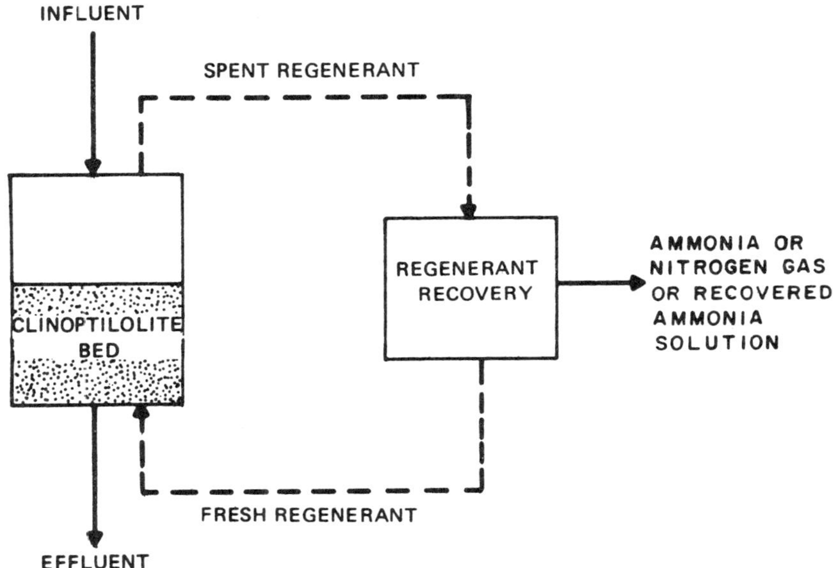

7.1.2 Ion Exchange Principles

Ion exchange is a process in which ions held by electrostatic forces to charged functional groups on the surface of a solid are exchanged for ions of similar charge in a solution in which the solid is immersed. It is a stoichiometric, reversible exchange of ions between a liquid and solid which produces no significant changes in the structure of the solid. The mass action equilibria expression provides a useful model for ion exchange behavior. In a binary system, the reaction,

$$bA^{+a} + aBZ_b \rightleftharpoons bAZ_a + aB^{+b} \tag{7-1}$$

expresses the reversible equilibrium where a and b are the valences of ions A and B, respectively, and Z is the exchange site in the solid.[1] This reaction may be expressed as the equilibrium constant,

$$K = \frac{(a)^b_{AZ_a} \cdot (a)^a_B}{(a)^b_A \cdot (a)^a_{BZ_b}} \tag{7-2}$$

in which $(a)^b_A$, $(a)^b_{AZ_a}$, etc. are the activities of the various species. Because of the difficulty in measuring activities, especially in the solid phase, it is convenient to use concentrations uncorrected for activity. In doing so, the equilibrium constant in Equation 7-2 varies with concentration and has been termed the "selectivity coefficient,"

$$K^A_B = \left(\frac{q}{c}\right)^b_A \left(\frac{c}{q}\right)^a_B \tag{7-3}$$

where q is the solid phase ionic concentration in milliequivalents per gram (meq/g) and c is the solution phase concentration in meq/l. Alternatively, the selectivity coefficient can be expressed in terms of dimensionless concentration,

$$K^A_B \left(\frac{Q}{C_o}\right)^{a-b} = \left(\frac{y}{x}\right)^b_A \left(\frac{x}{y}\right)^a_B \tag{7-4}$$

These variables are expressed in terms of the total solution concentration, C_o, in meq/l and the total exchange capacity, Q, in meq/g. Thus, $x = c/C_o$ and $y = q/Q$.

The preference of an ion exchange for one ion relative to another in binary systems is often expressed as the "separation factor,"

$$\alpha^A_B = \left(\frac{q}{c}\right)_A \left(\frac{c}{q}\right)_B = \left(\frac{y}{x}\right)_A \left(\frac{x}{y}\right)_B \tag{7-5}$$

Because the numerical value of the separation factor is not affected by the choice of concentration units, equilibrium data are often expressed in this way. If the equivalent fraction of ion A in the solid phase, y_A, is plotted against the equivalent fraction of A in the solution, x_A, three cases can be identified corresponding to $\alpha < 1$, $\alpha = 1$, and $\alpha > 1$ as shown in Figure 7-2. Isotherms which are concave upward, $\alpha < 1$, are designated as being "unfavorable" to the uptake of ion A; those which fall along the rising diagonal, $\alpha = 1$, are termed "linear" and exhibit no preference for ion A or B; and curves which are concave downward, $\alpha > 1$, are referred to as "favorable" isotherms since the solid prefers ion A to ion B. Ion exchange operations almost always are concerned with systems in which the ion of concern has a separation factor greater than unity during the service cycle.

The basic principles of ion exchange can be used to determine the capacity of clinoptilolite for ammonium and an excellent example is available in Reference 2. However, such calculations are lengthy and rather complex and a later section (7.2.9) presents a simplified technique for quickly approximating the ammonium capacity of clinoptilolite for varying concentrations of competing cations.

7.1.3 Properties of Clinoptilolite

7.1.3.1 Selectivity

Isotherms demonstrating the selectivity of clinoptilolite for ammonium over other cations have been reported in the literature.[3] An example is the comparison of Hector

FIGURE 7-2

GENERALIZED ION EXCHANGE ISOTHERMS

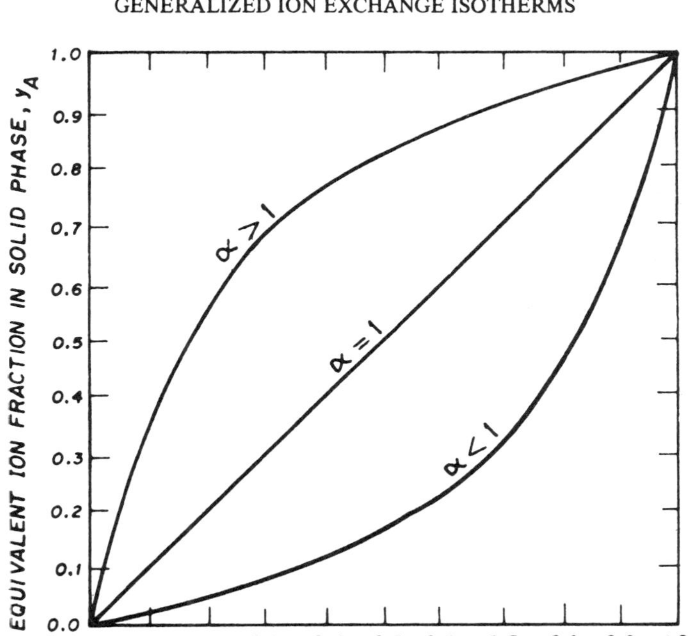

Clinoptilolite and a strong acid polystyrene resin, IR 120, shown in Figure 7-3. The total equilibrium solution normality was constant at 0.1*N*. The terms $(Ca)_Z$, $(NH_4^+)_Z$ = equivalent fraction of calcium or ammonium on the zeolite. The terms $(Ca)_N$, $(NH_4^+)_N$ = normality of calcium or ammonium in the equilibrium solution. The ammonium capacity of IR 120 was 4.29 meq/g of air-dried resin. A comparison of the generalized isotherm presented in Figure 7-2 with Figure 7-3 clearly shows that the IR 120 prefers calcium to ammonium ions. Clinoptilolite, on the other hand, prefers ammonium to calcium ion and is of greater utility for ammonium ion removal from wastewaters containing calcium.

The ion exchange equilibria for the systems NH_4^+-Na^+, NH_4^+-K^+, NH_4^+-Ca^{+2} and NH_4^+-Mg^{+2}, with clinoptilolite and other zeolites, are also available in the literature.[2] Plots of the NH_4^+ selectivity coefficients vs. the solution concentration ratios of the cations are shown in Figures 7-4 and 7-5. Using these data in conjunction with published calculation techniques,[2] it is possible to accurately predict the ammonium capacity of clinoptilolite in the presence of various concentrations of other cations.

FIGURE 7-3

THE 23 C ISOTHERMS FOR THE REACTION, $(Ca)_z + 2(NH_4^+)_N = 2(NH_4^+)_z + (Ca)_N$
WITH HECTOR CLINOPTILOLITE AND IR 120

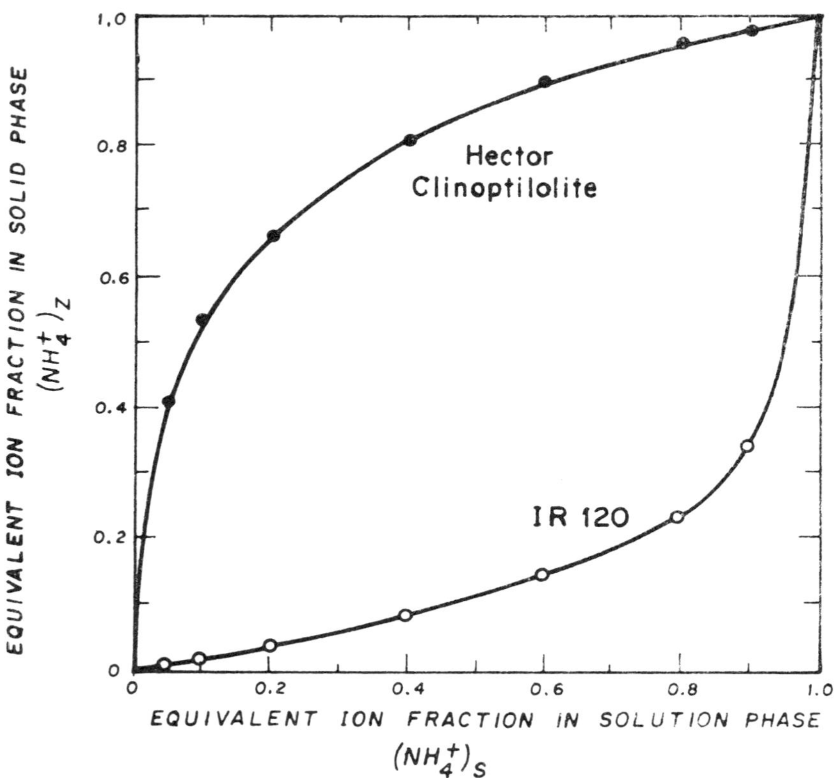

Section 7.2.9 presents a curve useful in approximating the ammonium exchange capacity which can be employed in sizing the ion exchange beds. The equilibrium isotherms for ammonia and other cations which are present as macrocomponents in wastewaters are shown in Figure 7-6.[1] These isotherms illustrate that clinoptilolite is selective for ammonium relative to all of the listed ions except potassium.

7.1.3.2 Mineralogical Classification

The zeolites are classified as a family in the silicate group. They are hydrated alumino-silicates of univalent and bivalent bases which can be reversibly dehydrated to varying

FIGURE 7-4

SELECTIVITY COEFFICIENTS VS. CONCENTRATION RATIOS OF SODIUM OR POTASSIUM AND AMMONIUM IN THE EQUILIBRIUM SOLUTION WITH HECTOR CLINOPTILOLITE AT 23 C FOR THE REACTION

$$(Y)_Z + (NH_4^+)_N = (NH_4^+)_Z + (Y)_N$$

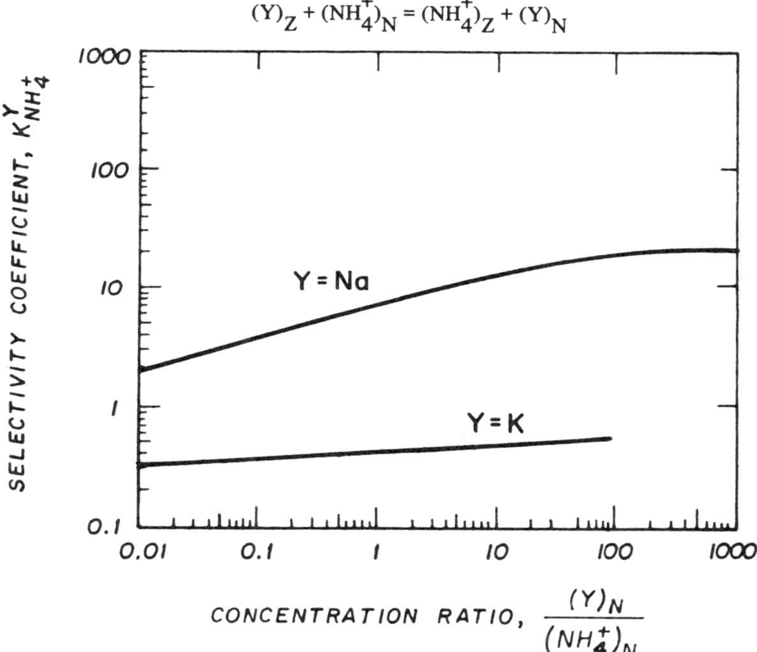

degrees without undergoing a change in crystal structure and are capable of undergoing cation exchange. The general composition of zeolites is given by the formula $(M,N_2)O{\cdot}Al_2O_3{\cdot}nSiO_2{\cdot}mH_2O$ where M and N are, respectively, the alkali metal and alkaline earth counter ions present in the zeolite cavities.

Clinoptilolite is a common material found in bentonite deposits in the western United States. The largest known deposit of clinoptilolite in the United States is found in southern California within a deposit of bentonite called Hectorite because of its proximity to Hector, California. The U.S. deposits are predominantly in the sodium form. Although a widely occurring material, not all deposits produce a clinoptilolite of adequate structural strength to withstand the handling which occurs in a columnar operation.

FIGURE 7-5

SELECTIVITY COEFFICIENTS VS. CONCENTRATION RATIOS OF CALCIUM OR MAGNESIUM AND AMMONIUM IN THE EQUILIBRIUM SOLUTION WITH HECTOR CLINOPTILOLITE AT 23 C FOR THE REACTION

$$(X)_Z + 2(NH_4^+)_N = 2(NH_4^+)_Z + (X)_N$$

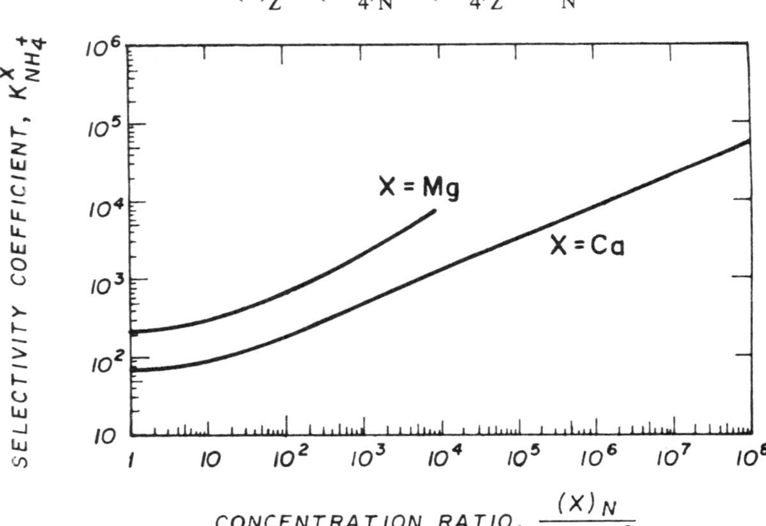

7.1.3.3 Total Exchange Capacity

Although the total ion exchange capacity of a material is by no means a complete description of its ion exchange properties, it is an indication of the applicability of the substance for process use. For example, New Jersey greensand, which was widely used in water softening before the development of organic exchangers, has a total exchange capacity of 0.17 meq/g.[1] In comparison, the exchange capacities of strong acid cation exchangers are usually 4 to 5 meq/g. The total exchange capacity of clinoptilolite as measured by several different investigators ranges from 1.6 to 2.0 meq/g and is slightly lower than the average for zeolites. With typical cation concentrations encountered in municipal wastewaters, the capacity for NH_4^+ is typically 0.4 meq/g (see Section 7.2.9).

7.1.3.4 Chemical Stability

The instability of natural clays and zeolites toward acids and alkalis is known as these materials are widely used in water softening. However, clinoptilolite is considerably more

FIGURE 7-6

ISOTHERMS FOR EXCHANGE OF NH_4^+ FOR K^+, Na^+, Ca^{++}, AND Mg^{++} ON CLINOPTILOLITE

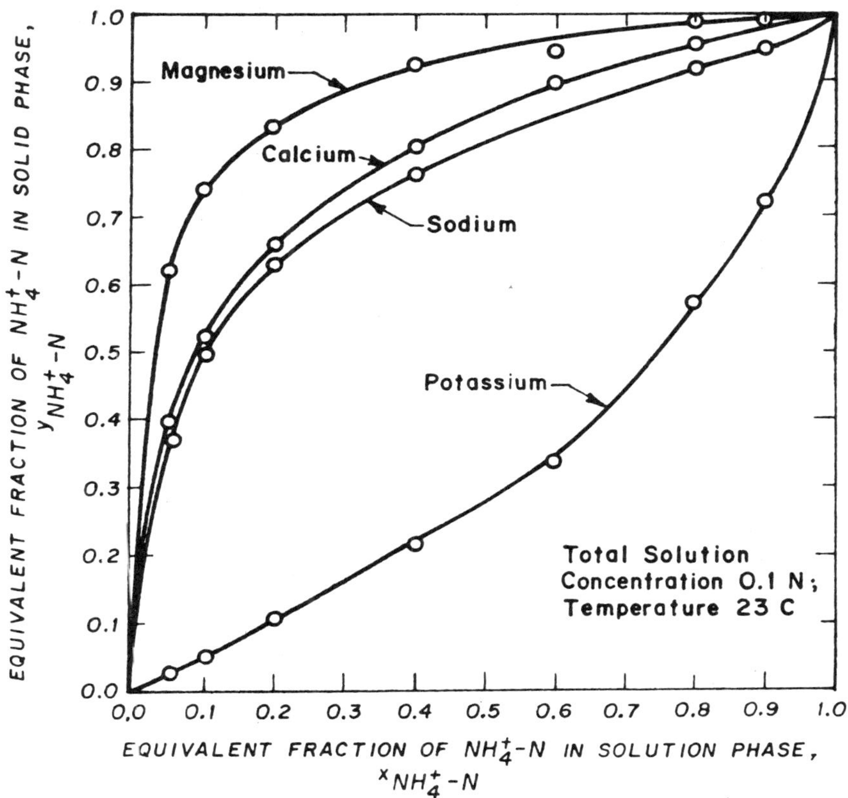

acid resistant than other zeolites.[1] Very high strength (20% NaOH) caustic solutions produce significant chemical attack on the clinoptilolite. However, at the lower solution strengths encountered in systems which use a caustic regenerant, physical attrition is more significant than chemical attack. This will be discussed later in this section.

7.1.3.5 Physical Stability

When crushed, sieved, and thoroughly washed with agitation to remove fines, clay, and other impurities, 20 x 50 mesh Hector clinoptilolite gives a wet attrition test of 3 percent.[4]

The wet attrition test determines the amount of fines (less than 100 mesh) generated by 25 grams of the granular zeolite during rapid mixing with 75 milliliters of water on a paint shaker for 5 minutes. Commercial zeolites, such as erionite and chabazite, which are powdered, mixed with clay binder, extruded, and fined, will generally give a wet attrition test of about 6 percent or twice that of the Hector clinoptilolite. Low wet attrition is important to minimize losses of clinoptilolite in an ion exchange column operation.

7.1.3.6 Density

Clinoptilolite (20 x 50 mesh) has been reported to have a wet particle specific gravity of 1.59 and a bulk density of 0.74-0.79 g/cc[1].

7.2 Major Service Cycle Variables

The factors which have a major effect on process efficiency include: pH, hydraulic loading rate, clinoptilolite size, pretreatment, wastewater composition, and bed depth.

7.2.1 pH

Within an influent pH range of 4 to 8, optimum ammonium exchange occurs.[1] As the pH drops below this range, hydrogen ions begin to compete with ammonium for the available exchange capacity. As the pH values increase above 8, a shift in the NH_3-NH_4^+ equilibrium towards NH_3 begins. Operation outside of the pH range of 4 to 8 results in a rapid decrease of exchange capacity and increased ammonium leakage.

7.2.2 Hydraulic Loading Rate

Variations in column loading rates within the range of 7.5-20 Bed Volumes (BV)/hour (7.5 BV/hr is equivalent to 0.95 gpm/cu ft or 2.15 1/m3/sec) have been shown to produce no significant effects on the ammonium removal efficiency of 20 x 50 mesh clinoptilolite.[3] Ammonium concentrations in the clinoptilolite effluent of 0.22-0.26 mg/l were produced throughout the above range in one set of tests.[1] When rates exceed 20 BV/hour, the exchange kinetics suffer as demonstrated by a significant leakage of ammonium early in the loading cycle. The effects of loading rate as a function of clinoptilolite size are discussed in the next section.

7.2.3 Clinoptilolite Size

Mine-run clinoptilolite is typically 1-2 inch (25 to 51 mm) chunks which must be ground and screened to the size desired for column operation. As would be expected, the smaller the clinoptilolite size, the better the kinetics of the exchange reaction. This effect is illustrated by data that show that 20 x 50 mesh clinoptilolite kinetics begin to suffer (see Section 7.2.2) at rates of 20-30 BV/hour while 50 x 80 mesh kinetics do not suffer until rates of 40 BV/hour are reached. However, the improved rate of exchange is accompanied by the disadvantage of higher head loss. It appears that 20 x 50 mesh clinoptilolite (about the size of typical filter sand) offers an adequate compromise between acceptable headloss and exchange kinetics. At a loading of 15 BV/hour in a 3 ft (0.9m) deep bed (5.6 gpm/sq ft or 3.8 $l/m^2/sec$), the headloss is 2.1 ft (0.64m) with 20 x 50 mesh clinoptilolite. Lower headlosses could be obtained by lower rates. Use of deeper beds would result in greater headloss, i.e., 6 ft (1.8 m) depth would have a headloss of 4.2 ft (1.28 m) at 5.6 gpm/sq ft (3.8 $l/m^2/sec$). These headloss values do not include losses in inlet and outlet piping or in the underdrain system.

7.2.4 Pretreatment

To avoid excessive headloss, the clinoptilolite influent must be relatively free of suspended solids – preferably less than 35 mg/l. Tests with clarified and filtered raw wastewater indicate no problems with organic fouling. Biological growths which occurred were adequately removed in the regeneration cycle.[2] Additional data on pretreatment effects are presented in the next section.

7.2.5 Wastewater Composition

As noted earlier, although clinoptilolite prefers ammonium ions to other cations, it is not absolutely selective and other cations do compete for the available exchange capacity. Pilot tests conducted at several locations illustrate the effects of wastewater composition on the useful capacity of the clinoptilolite.[3] Tests at three locales that span a wide range of wastewater compositions are shown in Table 7-1.

The equilibrium NH_4^+-N bed loading computed for each of the wastewaters listed in Table 7-1 was 4.1 g/l, 3.9 g/l, and 4.3 g/l, respectively, for Tahoe, Pomona, and Blue Plains. Figure 7-7 presents equilibrium bed loading in an alternate way. The minimum bed volumes required to attain equilibrium NH_4^+-N loading are expressed as a function of the NH_4^+-N concentration in the influent wastewater. The equilibrium bed volume values given in Figure 7-7 normally represent the 50 percent breakthrough point (the effluent concentration is 50 percent of the feed concentration). The water lowest in competing cations (Blue Plains) had the greatest ammonium removal capacity. In the 10-20 mg/l influent NH_4^+-N range, the lower competing ion concentrations at Blue Plains resulted in the useful ammonium exchange capacity being about 33 percent greater than that for Pomona with its higher TDS water. The lower degree of pretreatment at Blue Plains (i.e., no biological pretreatment) did not impair the effectiveness of the clinoptilolite for ammonium removal.

7.2.6 Length of Service Cycle

To illustrate the determination of a permissable length service cycle for a given wastewater, ammonia breakthrough curves for a single 6 ft deep (1.8m) bed of clinoptilolite are illustrated in Figure 7-8 for Tahoe tertiary effluent with flow rates varying from 6.5 to 9.7

TABLE 7-1

INFLUENT COMPOSITION FOR SELECTIVE ION EXCHANGE PILOT TESTS AT DIFFERENT LOCALES

Parameter, mg/l	Activated sludge plant effluent		Clarified raw wastewater
	Tahoe Carbon treated	Pomona[a]	Blue Plains (Washington, D.C.)
NH_4^+-N	15	16	12
Na	44	120	35
K	10	18	9
Mg	1	20	0.2
Ca	51	43	30
pH			
Range	7-8[b]	6.5-8.2[b]	7-9[b]
COD	11	10	50
TDS	325	700	250

[a]Approximately half of the runs at Pomona were made with carbon treated secondary effluent and the others with alum coagulated secondary effluent.

[b]pH units

BV/hour (bed volumes per hour) with 15 to 17 mg/l NH_4^+-N in the feed stream. These curves indicate a throughput value of 150 bed volumes should be used for this wastewater for design for effluents requiring a high degree of ammonia removal. Although the effluent concentration had reached 2-3 mg/l at 150 BV, the average concentration produced to this point in the cycle was less than 1 mg/l. Breakpoint chlorination would be more economical for removing the 1 mg/l residual, if required, than would provision of greater ion exchange column capacity. The average ammonium concentration for a breakthrough curve is obtained by integrating the area under the breakthrough curve and dividing by the total flow. For example, integrating the area under the curve for 8.1 BV/hr in Figure 7-8 indicates an average NH_4^+-N concentration of 0.67 mg/l for 150 BV.

7.2.7 Bed Depth

The effect of bed depth on ammonia breakthrough at 9.7 BV/hr is illustrated in Figure 7-9. In general, the 3 ft (0.9 m) bed of clinoptilolite was not as effective for ammonium removal as the 6 ft (1.8 m) bed at the same bed volume rate. The shallow bed has a lower flow velocity because a 9.7 BV/hr flow in a 3 ft (0.9 m) deep bed corresponds to 3.6 gpm/sq ft (2.5 l/m^2/sec) while in a 6 ft (1.8 m) deep bed it corresponds to 7.2 gpm/sq ft (4.9

FIGURE 7-7

MINIMUM BED VOLUMES AS A FUNCTION OF INFLUENT NH_4^+-N CONCENTRATION TO REACH 50 PERCENT BREAKTHROUGH OF AMMONIUM (REFERENCE 2)

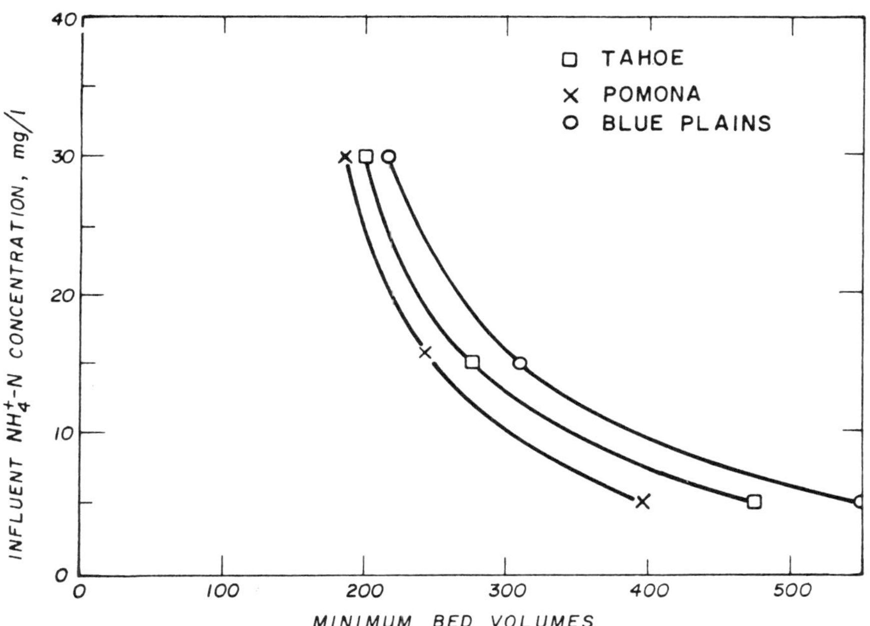

$l/m^2/sec$). The lower velocity might increase the likelihood of plugging of portions of the bed. Plugging would cause poor flow distribution and lower bed efficiency. As discussed in the design examples in Chapter 9, full-scale designs are using bed depths of 4-ft (1.2 m) with a high degree of pretreatment (coagulation and filtration) which will minimize plugging of the clinoptilolite bed.

7.2.8 One Column vs. Series Column Operation

Operation to the 150 bed volume throughput value (Figure 7-9) to maintain an average NH_4^+–N concentration at or below 1 mg/l uses only 55 to 58 percent of the zeolite's equilibrium capacity. The number of bed volumes throughput per bed can be increased while maintaining low NH_4^+–N effluent concentrations with semi-countercurrent operation, using two beds in series. Semi-countercurrent operation is achieved by first operating the

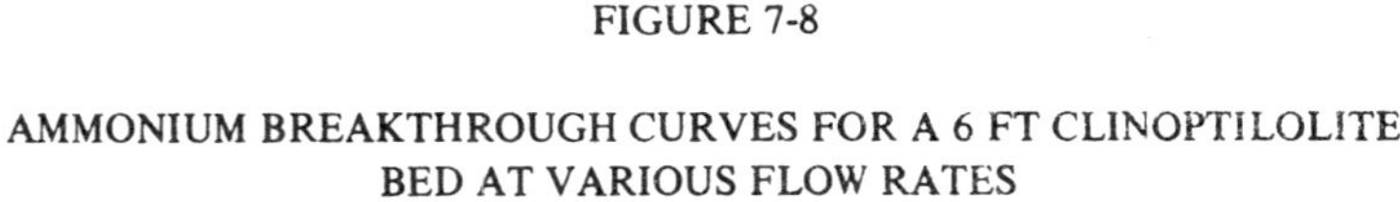

FIGURE 7-8

AMMONIUM BREAKTHROUGH CURVES FOR A 6 FT CLINOPTILOLITE BED AT VARIOUS FLOW RATES

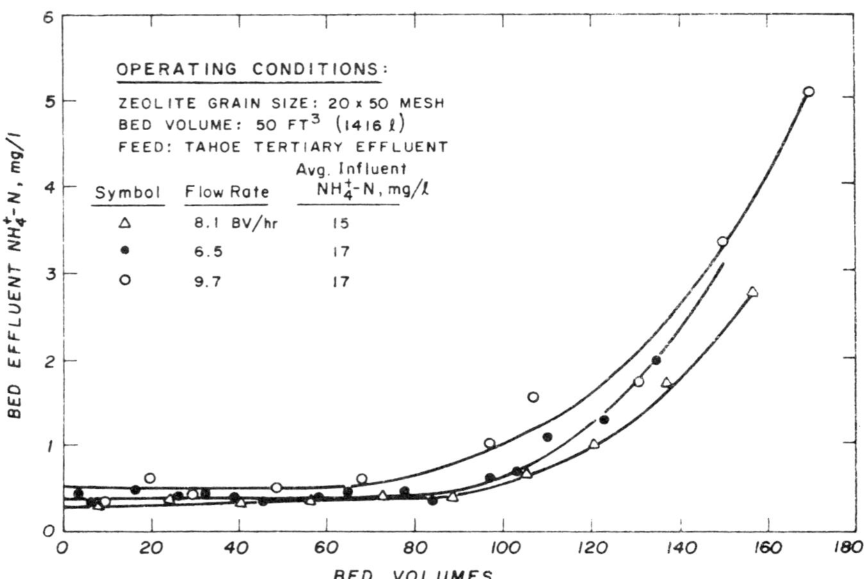

columns in a 1-2 sequence and then placing column 2 into the lead position, after the first regeneration, with column 1 becoming the polishing column. A column is removed from the influent end when it becomes loaded while simultaneously adding a regenerated column at the effluent end. This procedure in effect moves the beds countercurrent to liquid flow by continually shifting the more saturated beds closer to the higher influent concentrations. Beds can be loaded nearer to capacity with this procedure than with single column or parallel feed multi-column operation. The most highly loaded column is always at the influent end backed up by one (if two in series) or more columns having decreasing loadings and NH_4^+-N concentrations at locations progressively nearer the end of the series. Removal of a column is not decided by applying a breakthrough criterion to the column's own effluent but by breakthrough at the end of the series. Tests have indicated that the ammonium loadings could be increased from 55-58 percent of the equilibrium capacity to 85 percent by using two columns in series.[2] Average throughputs for the Tahoe example discussed earlier increased from 150 to 250 BV/cycle. However, such a two column operation requires three columns (two on stream while the third is being regenerated) and more complicated valving and piping than a parallel column operation. Because of the added capital costs involved in a

FIGURE 7-9

EFFECT OF BED DEPTH ON AMMONIUM BREAKTHROUGH AT 9.7 BV/HR

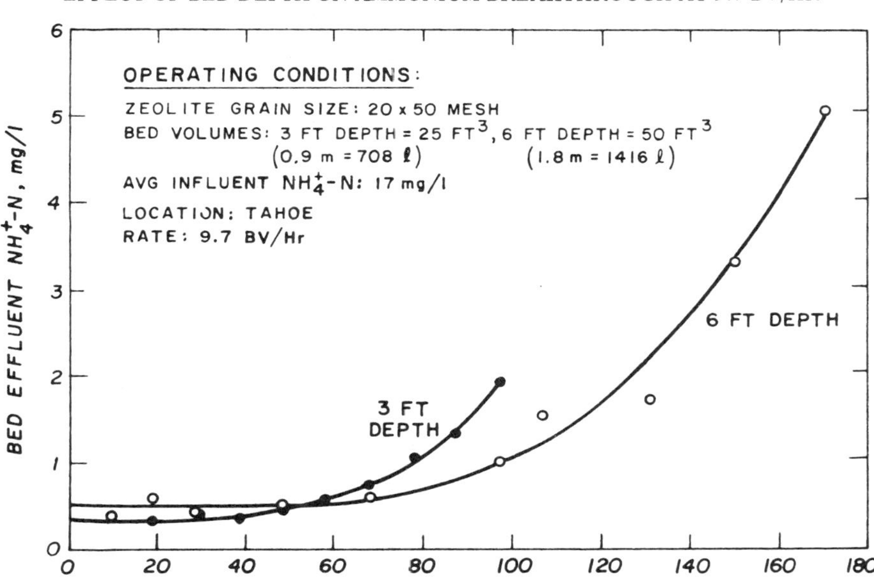

series system, all of the full-scale systems currently under design or in operation utilize parallel single beds. By blending the effluents from several parallel columns, each of which is in a different stage of exhaustion, improved utilization of the available exchange capacity is also achieved. That is, if equal volumes of effluent containing 2 mg/l NH_4^+–N from one column are blended with effluent containing 0.6 mg/l from another, some added throughput through the more heavily loaded column could be achieved while still meeting an overall standard of no more than 2 mg/l NH_4^+–N.

7.2.9 Determination of Ion Exchanger Size

In order to calculate the size of the ion exchange unit needed, the ammonium capacity of the clinoptilolite must be determined from the characteristics of the influent water. The ammonium capacity of clinoptilolite can be estimated from Figure 7-10 if the cationic strength of the wastewater is known. The data used to plot Figure 7-10 were determined in several experimental runs where the influent ammonium nitrogen concentration was 16.4-19.0 mg/l[1]. Although the curve is empirical and is a simplification of the complex effect of competing cation concentrations on ammonium capacity, it illustrates this effect and serves as a useful tool in sizing the exchange bed.

FIGURE 7-10
VARIATION OF AMMONIUM EXCHANGE CAPACITY WITH COMPETING CATION CONCENTRATION FOR A 3 FT DEEP CLINOPTILOLITE BED (REFERENCE 1)

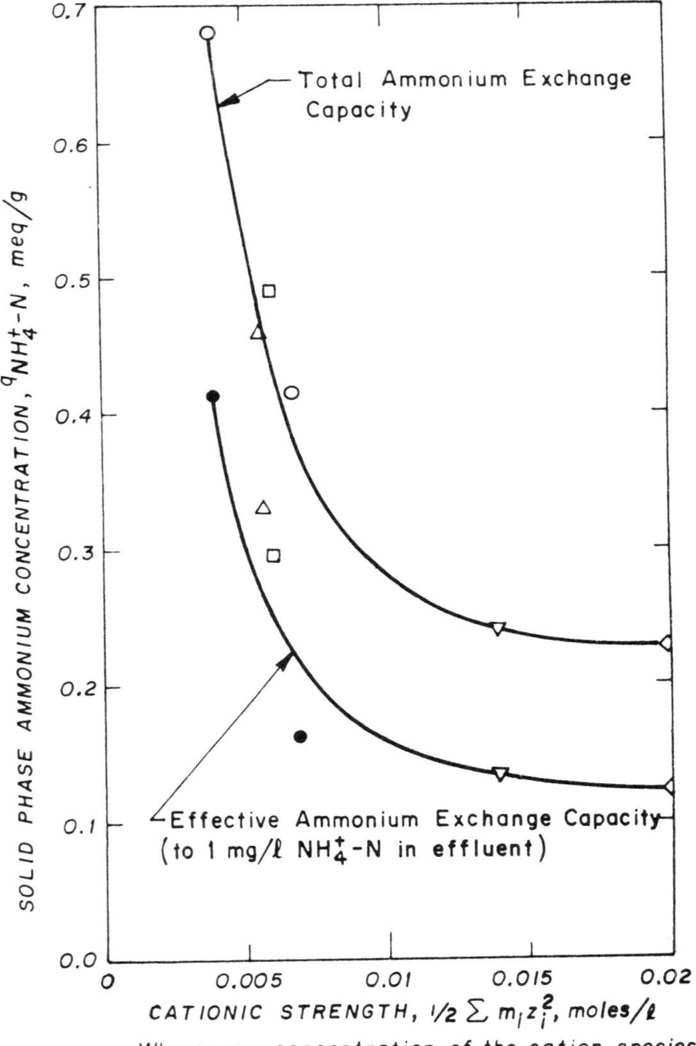

Assuming that the influent water has a cationic strength of 0.006 moles/1, the breakthrough ammonium capacity of the clinoptilolite will be approximately 0.25 meq/g for a 3-ft (0.9 m) bed; the capacity to saturation will be approximately 0.44 meq/g. A greater effective ammonium capacity can be realized by increasing the depth of the zeolite bed. The use of a 6-ft (1.8 m) bed would result in greater ammonia capacity per unit of exchanger and while requiring a deeper structure, the additional cost would be nominal. Assuming that 3 ft (0.9 m) of the zeolite bed will have an ammonium exchange capacity equal to 0.25 meq/g and that the remaining 3 ft (0.9 m) will have a capacity equal to 90 percent of the saturation capacity or 0.40 meq/g, the 6 ft (1.8 m) bed will have an effective capacity of 0.32 meq/g [equivalent to 6.6 eq/cu ft (236 eq/m^3) and 5.1 kgr/cu ft (182 kgr/m^3)].

The zeolite volume required to treat a 10 mgd (0.44 m^3/sec) waste flow at 15 BV/hr (1.9 gpm/cu ft or 4.3 1/sec/m^3) is 3650 cu ft (102 m^3). Assuming complete removal of ammonium, the throughput to ammonium breakthrough would then be 165 BV with a run length of 11 hr. Allowing 2 hr down time per cycle for regeneration and rinsing, the zeolite volume would be increased proportionately to 4300 cu ft (120.4 m^3) to accommodate the total design flow. Using four units, each having the dimensions 12 ft x 15 ft x 6 ft deep (3.66 m x 4.6 m x 1.8 m), the total zeolite volume would be 4320 cu ft (121 m^3).[1]

7.3 Regeneration Alternatives

The key to the applicability of this process is the method of handling the spent regenerant. The following paragraphs discuss available alternates.

7.3.1 Basic Concepts

As noted earlier, after about 150-200 bed volumes of normal strength municipal waste have passed through the bed, the capacity of the clinoptilolite has been used to the point that ammonium begins to leak through the bed. At this point, the clinoptilolite must be regenerated so that its capacity to remove ammonium is restored. The resin is regenerated by passing concentrated salt solutions through the exchange bed when the ammonium concentration in the solid phase has reached the maximum desirable level. The ammonium-laden spent regenerant volume is about 2.5 to 5 percent of the throughput treated prior to regeneration. By removing the ammonium from the spent regenerant, the regenerant may be reused. The alternative approaches available for regenerant recovery are:

- air stripping
- steam stripping
- electrolytic treatment

These alternatives for regenerant recovery will be discussed following a discussion of the regeneration process.

7.3.2 Regeneration Process

The ammonium retained on the clinoptilolite exchange sites may be eluted by either sodium or calcium ions contained in a regenerant solution. While the normal service cycle is downflow, regeneration is carried out by passing the regenerant up through the clinoptilolite bed.

7.3.2.1 High pH Regeneration

The approach originally studied for wastewater applications was to use a lime slurry (5 gm/l) as the regenerant so that the ammonium stripped from the bed during regeneration would be converted to gaseous ammonia which could then be removed from the regenerant by air stripping.[3] It was found that elution with lime could be speeded up by the addition of sufficient NaCl to render the regenerant 0.1*N* with respect to NaCl. [3]

In addition to converting the ammonium ion to ammonia so it can readily be removed from the regenerant, the volume of regenerant required for complete regeneration has been found to decrease with increasing regenerant pH.[1] However, high pH regeneration was found to be accompanied by an operational problem of major significance.[2,3] Precipitation of magnesium hydroxide and calcium carbonate occurs within the exchanger during the regeneration cycle. This leads to plugging of the exchanger inlets and outlets, as well as coating of the clinoptilolite particles. Violent backwashing of the clinoptilolite was found to be necessary to remove these precipitants from the clinoptilolite particles, which resulted in increased mechanical attrition of the clinoptilolite. Chemical attrition also increases at elevated pH values.[1]

Substantial data have been collected on high pH regeneration and are available in references 1, 2 and 3 if this approach is considered. However, the practical problems of scale control are major limitations which can be overcome by using neutral regenerants. The use of closed loop regenerant recovery processes negates the disadvantage of higher regenerant volumes required at lower regenerant pH values (See Section 7.3.2.2).

7.3.2.2 Neutral pH Regeneration

Two of the largest municipal wastewater installations under design which will use clinoptilolite are the Upper Occoquan (Virginia) Regional Plant (15 mgd or 0.66 m^3/sec) as described in Section 9.5.4.1 and the Tahoe-Truckee (California) Sanitation Agency plant (6 mgd or 0.26 m^3/sec), both of which will utilize a regenerant with a pH near neutral. The active portion of the regenerant will be a 2 percent sodium chloride solution. Calcium and potassium will be eluted as well as ammonia and will build up in the regenerant until they reach equilibrium. The typical elution curve for ammonium with a neutral pH regenerant is shown in Figure 7-11. Approximately 25-30 BV were required before the ammonium concentration reached equilibrium.[5] Although greater regenerant volumes are required than with a high pH regenerant (10-30 BV), this is not a major disadvantage if the regenerant is recovered and reused in a closed loop system.

Variations in regenerant flow rates of 4-20 BV/hr do not affect regenerant performance. Higher rates result in less ammonia removed per volume of regenerant. Typical design values are 10 BV/hr which insures efficient use of the regenerant while keeping headloss values at low levels. Provisions should be made for backwash at rates of 8 gpm/ sq ft (3.9 $l/m^2/sec$) and surface wash of the contactor prior to initiation of the regenerant flow. Additional details on neutral pH regeneration are contained in Section 9.5.4.

FIGURE 7-11

AMMONIUM ELUTION WITH 2 PERCENT SODIUM CHLORIDE REGENERANT (REFERENCE 5)

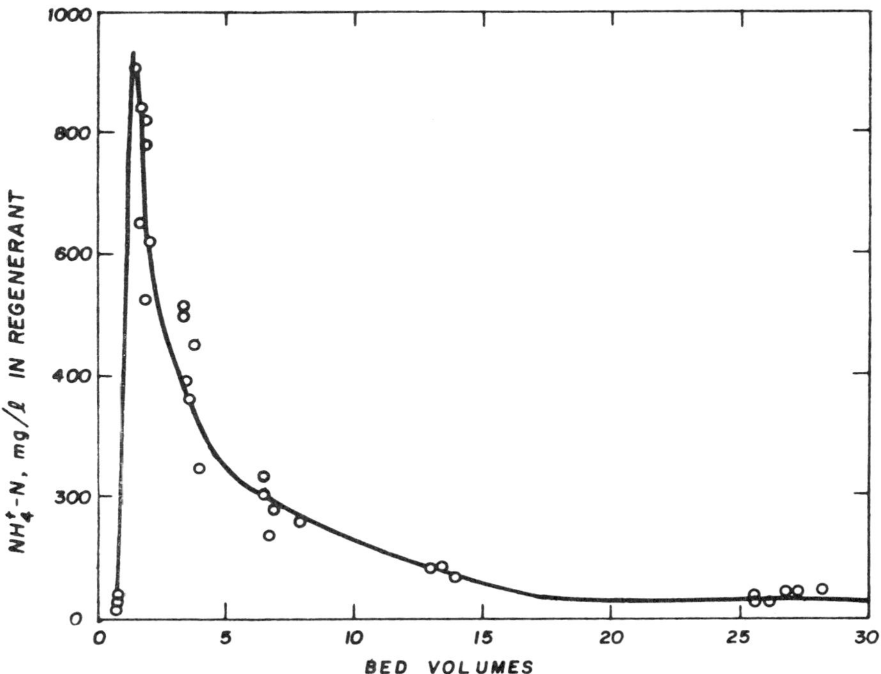

7.3.2.3 Effects on Effluent TDS

Effluent Total Dissolved Solids (TDS) is an important consideration in many plants. When a 2-3 percent solution of salt is used for regeneration, elution of this salt remaining in the bed after regeneration at the start of the service cycle may result in an increase in TDS of about 50 mg/l. The increment would be greater with stronger regenerants. The TDS effect is much less than for the breakpoint process, however.

7.3.3 Regenerant Recovery Systems

Ammonia may be removed from the regenerant so that the regenerant may be reused. Air stripping of a high pH regenerant, air stripping of a neutral regenerant, steam stripping, or electrolytic treatment may be used.

7.3.3.1 Air Stripping of High pH Regenerant

In the original pilot work on this approach to regenerant recovery, a stripping tower packed with 1 inch (2.54 cm) polypropylene saddles was used.[2,3] Because the regenerant volume is only a small portion of the total wastewater flow, it becomes feasible to heat the air used in the stripping process. The regenerant was normally recycled upflow through the zeolite bed at a flow rate of 4.8-7.1 gpm/sq ft (3.3-4.8 $l/m^2/sec$) until the NH_3–N approached a maximum concentration. The regenerant was then recycled through both the zeolite bed and the air stripper until the NH_3–N was reduced to about 10 mg/l. The liquid flow rate to the stripper was normally 2 gpm/sq ft (1.36 $l/m^2/sec$) with an air/liquid ratio of 150 cu ft/gal (1.1 m^3/l). Ammonia removal in the air stripper generally averaged about 40 percent per cycle at 25 C. Calcium carbonate scaling occurred on the polypropylene saddles, but the scale could generally be removed by water sprays. The headloss through the 1 inch (2.54 cm) pilot plant saddles caused the power requirements for the air stripping to be excessive. It was suggested for a full-scale design that the ammonia stripping tower be sized to treat the contents of an elutriant tank in 8 hours, using two passes through the tower at 85 percent removal per pass at an air-to-water ratio of 300 cfm/gpm (2.2 m^3/l), and a loading of 3.5 gpm/ft^2(0.63 $l/m^2/sec$).[2] The tower would be a modified cooling tower with low differential pressure across the tower as discussed in Chapter 8.

An example design of a 7.5 mgd (0.33 m^3/sec) system illustrates how the air stripping system can be integrated into an overall system.[2] A schematic diagram of the ion exchange beds, lime elutriant system, and ammonia air stripping system is shown in Figure 7-12. For design flows, nine beds (12 ft or 3.65 m diameter and 8 ft or 2.4 m deep) would be in service and three beds in regeneration. The direction of flow for the beds in service would be downflow. All beds would operate in parallel. When a given volume of wastewater has passed through a set of three beds, for example, beds 1, 2, and 3, the set of beds would be taken off line for regeneration. At this time elutriant tank A would contain elutriant water from a previous regeneration with a very high ammonia nitrogen content (say 600 mg/l); tank B would contain elutriant water with a low ammonia nitrogen content (say 100 mg/l);

FIGURE 7-12

EXAMPLE ION EXCHANGE - AIR STRIPPING SYSTEM FOR HIGH pH REGENERANT

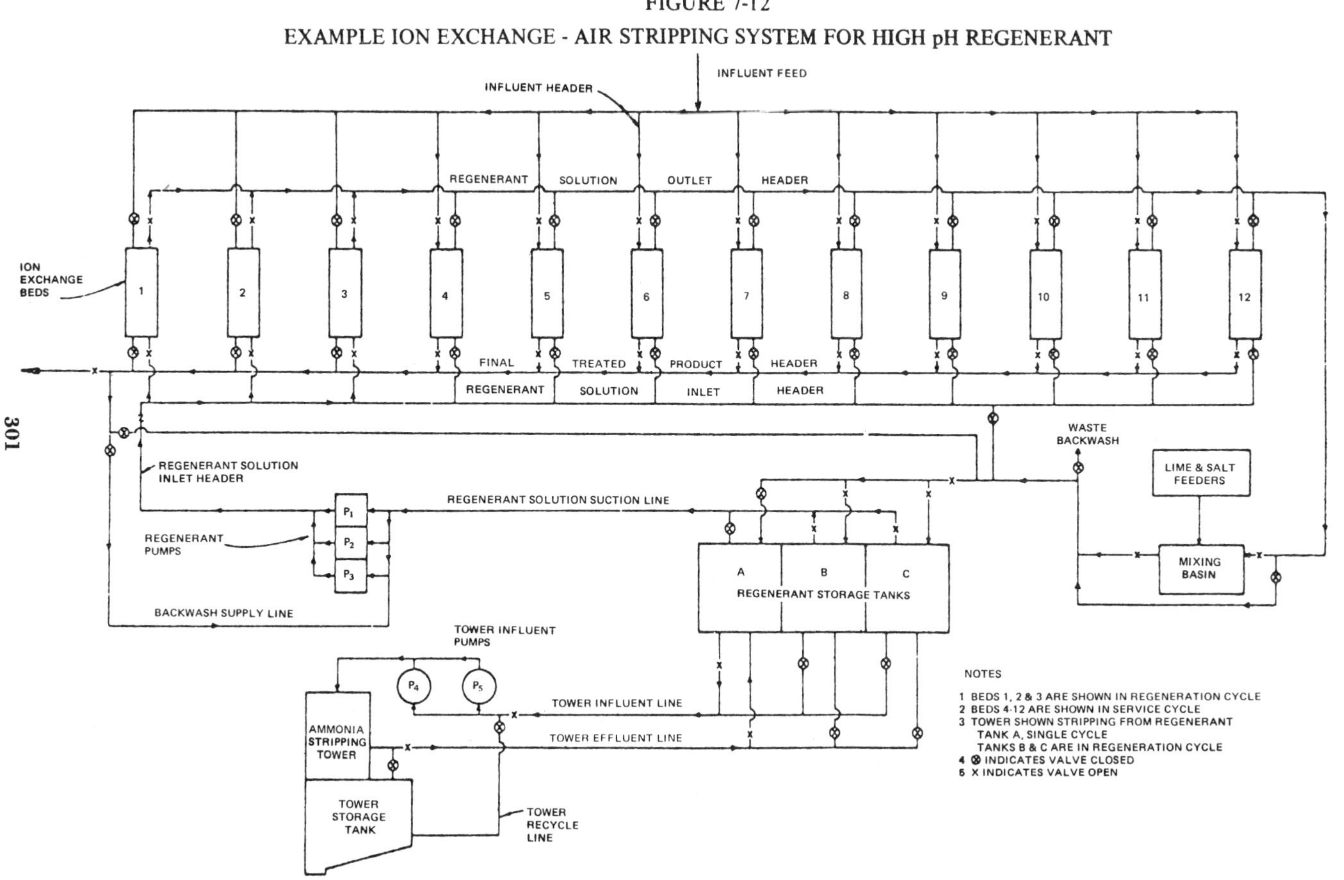

and tank C would contain nearly ammonia-free elutriant water (say 10 mg/l). The contents of tank A would be air stripped during the regeneration of exchange beds 1, 2, and 3. The regeneration would proceed as follows:

1. Exchange beds 1, 2 and 3 would be drained to the final effluent.

2. Low ammonia content elutriant water from tank B (100 mg/l) would be recirculated upflow through the three exchange beds and back through tank B to the exchange beds until the concentration of ammonia in the elutriant began to approach a maximum value (say 600 mg/l). Throughout the recirculation, make-up lime and salt would be added. A pH of about 11.5 would be maintained. About 4 BV will elute 75 percent of the NH_4^+–N.

3. After an allotted time (long enough for elutriant from tank B to approach a maximum ammonia concentration), the elutriant would be changed to recirculation to and from tank C through beds 1, 2, and 3. Tank C with its ammonia-free elutriant water would be recirculated for an allotted time (long enough for elutriant from tank C to reach about 100 mg/l) which will bring the elution up to more than 90 percent. About 4 BV are required. At this stage of the elution, the small amount of ammonia left on the zeolite would be distributed uniformly throughout the bed. Tank A with nearly ammonia-free water (10 mg/l NH_4^+-N, water stripped during the regeneration of beds 1, 2, 3) would be pumped once upflow through the bed to further polish the lower portion of the bed and prevent leakage of ammonia during the downflow service cycle.

   The elutriant tank B (600 mg/l –NH_4^+–N) would be held for air stripping during the regeneration of the next set (say beds 4, 5, and 6) of ion exchange beds. Tank C with 100 mg/l elution water would become the lead tank for this next set of ion exchange beds. Tank A with ammonia-free elution water (10 mg/l-NH_4^+-N, water stripped during the regeneration of beds 1, 2, 3) would be used at the polishing tank for beds 4, 5, and 6.

4. Once the elution of beds 1, 2, and 3 was completed, the three beds would be drained back to tank A.

5. Beds 1, 2, and 3 would then be filled slowly from the bottom to remove trapped air with product water from the other nine beds in service.

6. After the beds were filled with product water, more product water would be pumped at a high rate through beds 1, 2, and 3 in sequence. The backwash water would be returned to the wastewater treatment plant.

7. After backwashing was completed, ion exchange beds 1, 2, and 3 would be placed in service and beds 4, 5, and 6 would be taken off line for regeneration.

Ammonia in the elutriant solution would be removed by air stripping at a pH of about 11.5. In the preceding example, during the regeneration of beds 1, 2, and 3, the very high ammonia nitrogen content (600 mg/l) in the elutriant solution of tank A was to be air stripped. The following procedure would be used:

1. The contents of tank A would pass through the tower down into the recycle basin below the tower.

2. The contents of the recycle basin would then be pumped back up through the tower once again. This time, however, the effluent from the tower would flow back to tank A.

3. The contents of tank A would now contain about 10 mg/l of ammonia, and would be ready to serve as the polishing volume during the regeneration of ion exchange beds 4, 5, and 6.

By using the above batch countercurrent recycle technique, it is possible to achieve complete regeneration with only about 4 BV requiring stripping per cycle. This is a key to making steam stripping (as discussed later) practical.

7.3.3.2 Air Stripping of Neutral pH Regenerant

As previously discussed, the use of high pH regenerant is accompanied by scaling problems within the ion exchange beds. Thus, as discussed in Section 7.3.2.2, a regenerant with 2 percent sodium chloride as the active agent and pH nearer neutral has been used to overcome the scaling problem. This regenerant may also be recovered for reuse by air stripping. Figure 7-13 is a schematic diagram of such a system.[6] In this system, the stripping tower off-gases are not discharged to the atmosphere but are instead passed through an absorption tower where the ammonia in the off-gases is absorbed in sulfuric acid. The stripping gases are recycled to the tower. This approach eliminates discharge of ammonia to the atmosphere and recovers the ammonia in a form suitable for fertilizer usage. The stripping-absorption approach is applicable to high pH regeneration systems as well. It also reduces scaling problems in the stripping tower by limiting the CO_2 content of the stripping air. This system (the Ammonia Removal and Recovery Process – "ARRP") is also discussed in Chapter 8 (see Section 8.4.1 and Figure 8-7).

In the system shown in Figure 7-13, batch-countercurrent regenerant flow similar to that described above for the high pH regenerant is practiced to reduce the amount of regenerant which must be stripped per cycle. The first 11 BV of spent regenerant are discharged to the spent regenerant tank for stripping. The second and third 11 BV batches are stored and used as the first 22 BV of regenerant flow in the next regenerant cycle. The last 6-11 BV batch of regenerant is mixed with the 11 BV of stripped regenerant for use as the final regenerant flow in the next cycle. Thus, although 40-44 BV of regenerant are passed through the exchanger per cycle, only 11 BV are actually renovated by air stripping per cycle.

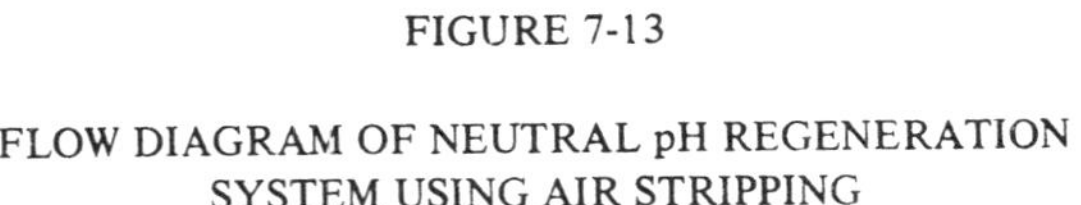

FIGURE 7-13

FLOW DIAGRAM OF NEUTRAL pH REGENERATION SYSTEM USING AIR STRIPPING

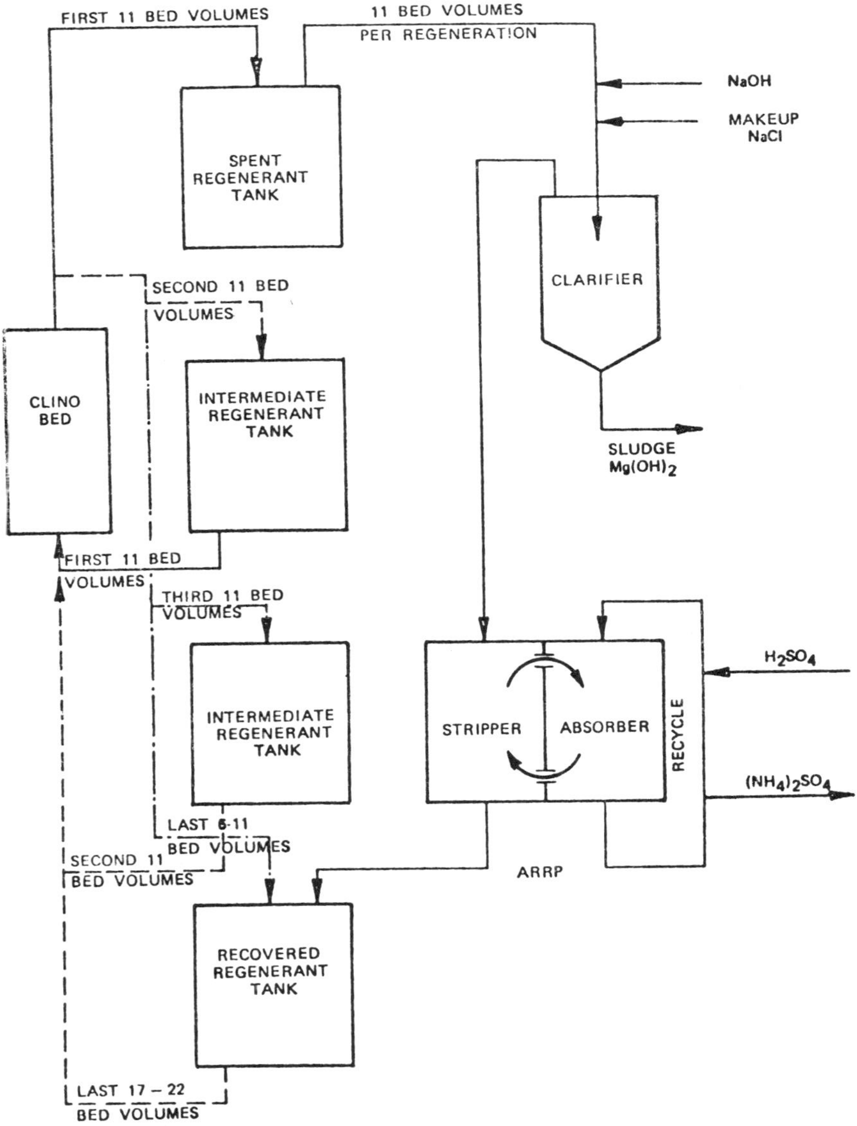

Regenerant stored in the tanks shown in Figure 7-13 varies in ammonium-nitrogen concentration from about 250 mg/l in the spent tank to 50 mg/l in the recovered regenerant tank. The intermediate tanks have intermediate concentrations. The regenerant pH varies from about 9.5 in the recovered regenerant tank to about 7.0 in the spent tank. As discussed earlier, higher pH values produce more efficient regeneration but near-neutral pH levels avoid problems with magnesium hydroxide precipitation in the bed during regeneration and attrition of the clinoptilolite caused by high pH. Media attrition has been insignificant in pilot studies under these pH conditions.[6]

A typical ammonium elution curve is shown on Figure 7-14 with the background concentrations in each regenerant storage tank also shown. At the end of the cycle, the last portion of spent regenerant is discharged to the recovered regenerant tank. This has the effect of neutralizing the alkaline pH from the ARRP process. ARRP effluent is normally at a pH of 10.7 to 11.0. This is reduced to 9.5± by recycling the last portion of spent regenerant. In this manner, pH is controlled without use of acid.

When spent regenerant is accumulated to a predetermined amount, the recovery portion of the process is activated. This system operates at a flow rate of approximately 1/13 of average plant flow since the regenerant concentration is about 13 times as concentrated as plant waste. Initially, sodium hydroxide is added to the spent regenerant to achieve a pH of about 11. Sodium chloride is also added because of some salt loss from the regenerant solution in sludge removal and bed rinsing. Following pH adjustment, the regenerant is clarified and any magnesium hydroxide formed is removed. the clarified regenerant at pH = 11 is then pumped to the ARRP process for ammonia removal and recovery. The ARRP effluent flows to the recovered regenerant tank where it is mixed with the last 6-11 BV of spent regenerant for pH adjustment prior to reuse.

This system is being used in the design of plants for the Tahoe-Truckee Sanitation Agency (California) and for the Upper Occoquan (Virginia) Regional Plant discussed in Section 9.5.4.1.

7.3.3.3 Steam Stripping

Steam Stripping of regenerant is being practiced at the 0.6 mgd (0.026 m^3/sec) Rosemount, Minnesota physical-chemical plant.[7,8] This process is economically feasible only with high pH regenerant.

The higher regenerant volumes resulting from the neutral regenerant approach are not economically treated by this approach. This process is feasible only if the regenerant volume requiring stripping is held to 4 BV per cycle which is achievable with the high pH regenerant batch recycle system discussed in Section 7.3.3.1.[9] In this case, the necessary portion of the spent regenerant is stripped in a distillation tower in which steam is injected countercurrent with the regenerant. An air cooled plate-and-tube condensor condenses the vapor for collection in a covered tank as a one percent aqueous ammonia solution which could be

FIGURE 7-14

TYPICAL ELUTION CURVE

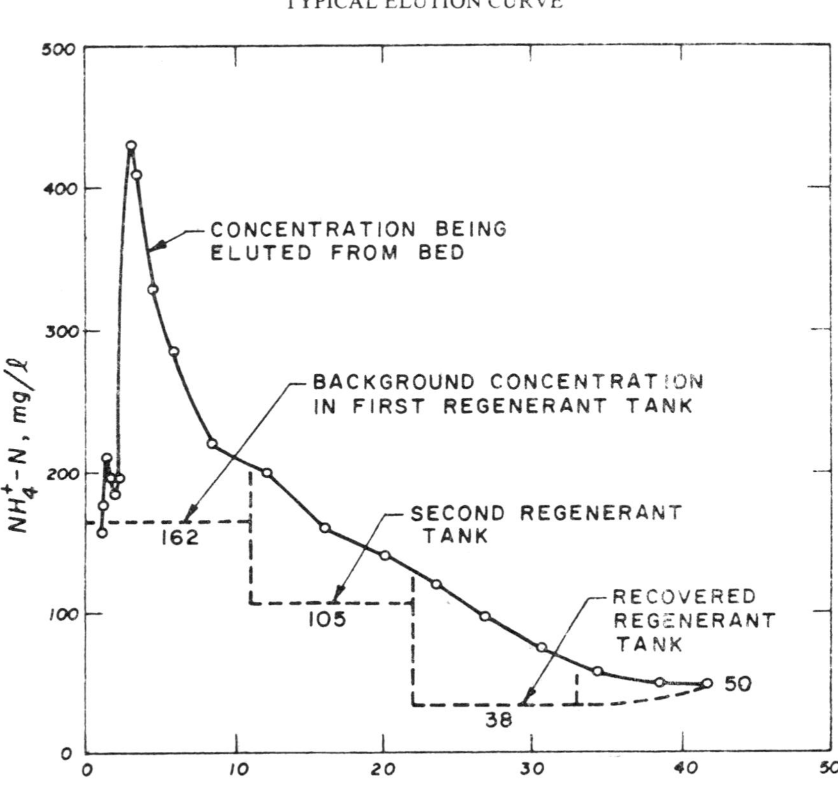

used as a fertilizer. A stripping tower depth of 24 feet (7.3 m) and a loading of 7 gpm/ sq ft (21 l/m^2/sec) are being used at Rosemount. Ceramic saddles are used rather than wooden slat packing because wood is not a suitable packing in a high pH- steam environment.

Heat exchangers are used to transfer heat from the stripped regenerant to the incoming, cold regenerant. Heat transfer to the incoming regenerant from the condenser used to condense the stripped regenerant may also be attractive. Provisions for scale control in the heat exchangers should be provided. The steam requirements have been estimated to be 15 pounds per 1,000 gallons (1.8 g/l).[9] Added information may be found in the Rosemount design example, Section 9.5.4.2.

7.3.3.4 Electrolytic Treatment of Neutral pH Regenerant

In this approach, ammonium in the regenerant solution is converted to nitrogen gas by reaction with chlorine which is generated electrolytically from the chlorides already present in the neutral pH regenerant solution. The regenerant solution is rich in NaCl and $CaCl_2$ which provide the chlorine produced at the anode of the electrolysis cell. A diagram of the regeneration system is presented in Figure 7-15. The regeneration of the clinoptilolite beds is accomplished with a two percent sodium chloride solution. The spent regenerant is collected in a large holding tank and then subjected to soda ash treatment for calcium removal. After the soda ash addition, the regenerant is clarified and transferred to another holding tank where the regenerant is recirculated through electrolysis cells for ammonia destruction.

FIGURE 7-15

SIMPLIFIED FLOW DIAGRAM OF ELECTROLYTIC REGENERANT TREATMENT SYSTEM

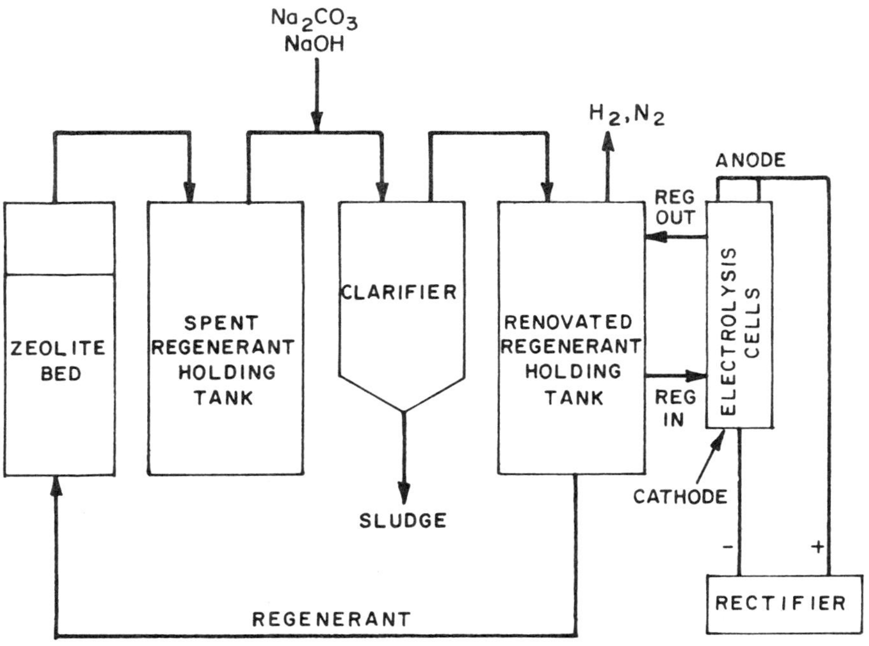

During the regeneration of the ion exchange bed, a large amount of calcium is eluted from the zeolite along with the ammonia. This calcium tends to scale the cathode of the electrolysis cell, greatly reducing its life. Calcium may be removed from the spent regenerant solution by a soda ash softening process prior to passing the spent regenerant through the electrolytic cells. High flow velocities through the electrolysis cells are required in addition to a low concentration of $MgCl_2$ to minimize scaling of the cathode by calcium hydroxide and calcium carbonate. The effects of flow rate are well illustrated by pilot test data.[5] Using the system shown in Figure 7-14, the flow rate through the cell was initially set at velocities of 0.13 to 0.16 ft per second (0.04-0.05 m/sec) and a thin buildup of scale was observed on the cathode at the bottom cell inlet end after 160 hours of operation. After 230 hours of operation, the flow velocity was reduced to 0.06 ft per second (0.018 m/sec) and very light scale buildup was observed depositing over the entire cathode area. Scale was removed from a one-square inch (6.45 cm^2) area of the cathode and the flow velocity through the cell was increased to 0.21 ft per second (0.064 m/sec) to determine the effect of scaling at higher cell velocities. At this increased flow which was maintained for most of the period of this study, no new scale was deposited on the cathode. Visually it appeared that from 25 to 50 percent of the previously deposited scale was removed. These observations suggest that scaling within the cell can be controlled by sufficient flow velocities. Acid flushing of the cells is necessary to remove this scale when the cell resistance becomes too high for economical operation.

In pilot tests of the electrolytic treatment of the regenerant at Blue Plains, about 50 watt-hours of power were required to destroy one gram of ammonia nitrogen.[2] When related to the treatment of water containing 25 mg/l NH_4^+–N, the energy consumed would be 4.7 kwh/1,000 gallons (1.2 watt-hrs/l). Tests at South Tahoe also indicated that a value of 50 watt-hours per gram is reasonable for design.[5]

The electrolytic process also results in about 56 cu ft (1586 l) of hydrogen gas being evolved per pound of ammonia nitrogen destroyed. Provisions must be made to vent, burn, or otherwise adequately control the hydrogen gas evolved in the electrolytic process.

The major disadvantage of the electrolytic approach is the substantial amount of electrical energy required. The electrical requirements of the air stripping (ARRP) system described in the preceding section are only about 10 percent of that required by the electrolytic process.

7.4 Considerations in Process Selection

The selective ion exchange process has the advantages of high efficiency, insensitivity to temperature fluctuations, and removal of ammonium with a minimal addition of dissolved solids. It may also be used with regenerant recovery systems which enable the recovery of the nitrogen removed from the wastewater in a reusable form. Its major disadvantage is its relatively complex operation. The process should be controlled by a system which will automatically initiate and program the regeneration cycle and return the ion exchangers to normal service.

The process is particularly attractive for those cases requiring year-round high level removal of nitrogen and where effluent TDS is of major concern. Although the effluent TDS is increased by the process (see Section 7.3.2.3), the overall increase is much less than for the breakpoint chlorination process. It must be recognized in the sizing of the upstream process capacities that there will be backwash wastes returned from the ion exchange process. The capacity of the clinoptilolite may be predicted accurately, based on the concentrations of ions present in the wastewater, minimizing the need for pilot tests for defining ion exchange capacity. Pilot tests of the overall ion exchange-regenerant recovery system may be useful, however, in evaluating physical and economic aspects of the proposed system design as applied to a specific wastewater.

7.5 References

1. Koon, J.H. and W.J. Kaufman, *Optimization of Ammonia Removal by Ion Exchange Using Clinoptilolite.* Environmental Protection Agency Water Pollution Control Research Series No. 17080 DAR 09/71.

2. Battelle Northwest and the South Tahoe Public Utility District, *Wastewater Ammonia Removal by Ion Exchange.* Environmental Protection Agency Water Pollution Control Research Series No. 17010 ECZ 02/71, February, 1971.

3. Battelle Northwest, *Ammonia Removal from Agricultural Runoff and Secondary Effluents by Selective Ion Exchange.* Robert A. Taft Water Research Center Report No. TWRC-5, March, 1969.

4. Mercer, B.W., *Clinoptilolite in Water Pollution Control.* The Ore Bin, published by Oregon Dept. of Geology and Mineral Industries, p. 209, November, 1969.

5. Prettyman, R., *et al., Ammonia Removal by Ion Exchange and Electrolytic Regeneration.* Unpublished report, CH2M/Hill Engineers, December, 1973.

6. Suhr, L.G., and L. Kepple, *Design of a Selective Ion Exchange System for Ammonia Removal.* Presented at the ASCE Environmental Engineering Division Conference, Pennsylvania State University, July, 1974.

7. *Physical-Chemical Plant Treats Sewage Near the Twin Cities.* Water and Sewage Works, p. 86, September, 1973.

8. Larkman, D., *Physical-Chemical Treatment.* Chemical Engineering, Deskbook Issue, p. 87, June 18, 1973.

9. Personal communication, B.W. Mercer, Battelle Northwest, December 14, 1973.

CHAPTER 8

AIR STRIPPING FOR NITROGEN REMOVAL

8.1 Chemistry and Engineering Principles

The ammonia stripping concept is based on very simple principles. Because of its simplicity, it offers a reliable means of ammonia removal when applied under appropriate conditions. The following section describes the basic concept.

8.1.1 Basic Concept

The equilibrium equation for ammonia in water is represented by:

$$NH_4^+ \rightleftharpoons NH_3\uparrow + H^+ \qquad (8\text{-}1)$$

At ambient temperatures and pH 7, the reaction is nearly complete to the left and only ammonium ions are present. As the pH is increased above 7, the reaction is driven to the right, and the fraction of dissolved ammonia gas increases until at pH values of 10.5-11.5, essentially all of the ammonium is converted to NH_3 gas (see Figure 6-2). The gaseous form may be removed by stripping.

The ammonia stripping process itself (Figure 8-1) consists of: (1) raising the pH of the water to values in the range of 10.8 to 11.5 generally with the lime used for phosphorus removal, (2) formation and reformation of water droplets in a stripping tower, and (3) providing air-water contact and droplet agitation by circulation of large quantities of air through the tower. The towers used for ammonia stripping of municipal wastewaters closely resemble conventional cooling towers.[1] Countercurrent towers, as opposed to cross-flow towers, appear best suited to ammonia stripping applications.

Detailed discussions of mass and enthalpy relationships and theoretical mathematical models of the stripping process are available in references 2, 3 and 4. However, these models are not normally used for stripping tower design, and an empirical design procedure is used.

Before addressing detailed design considerations, the general environmental impacts of the stripping process must be evaluated. It is obvious from Figure 8-1 that ammonia is being discharged into the atmosphere. Does the process solve a water pollution problem while creating an air pollution problem? What is the fate of the ammonia in the atmosphere? These questions must be satisfactorily addressed prior to the selection of air stripping for ammonia nitrogen removal.

8.2 Environmental Considerations

There are three major potential environmental impacts which must be evaluated if use of the

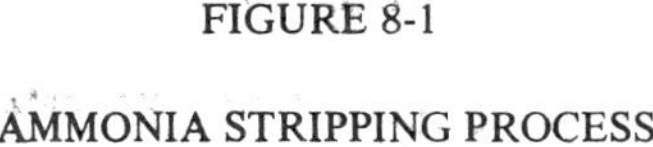

FIGURE 8-1

AMMONIA STRIPPING PROCESS

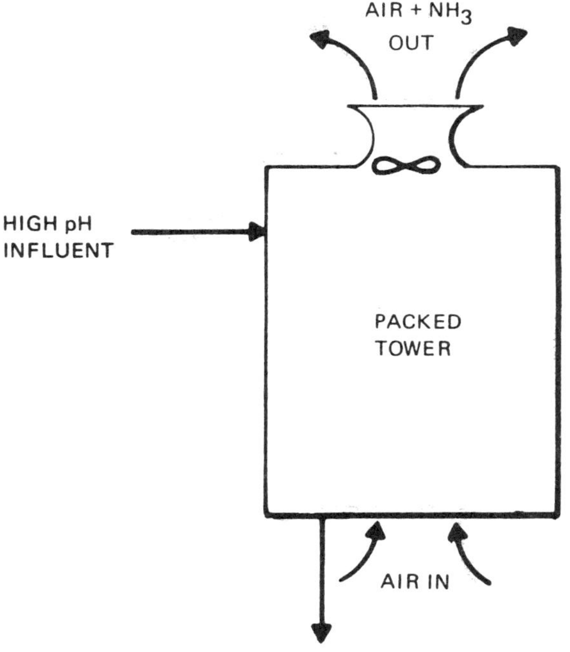

ammonia stripping process is proposed: air pollution, washout of ammonia from the atmosphere, and noise. If these three concerns cannot be favorably resolved for any given situation, then the potential process advantages of simplicity and low cost may become only academic.

8.2.1 Air Pollution

At an air flow of 500 cu ft per gallon (3.7 m^3/l) and at an ammonia concentration of 23 mg/l in the tower influent, the concentration of ammonia in the stripping tower discharge is about 6 mg/m^3. As the odor threshold of ammonia is 35 mg/m^3, the process does not present a pollution problem in this respect. Concentrations of 280-490 mg/m^3 have been reported to cause eye, nose, and throat irritation.[5] Concentrations of 700 mg/m^3 can have adverse effects on plants. Concentrations of 1,700-4,500 mg/m^3 must be reached before human or animal toxicities begin to occur. Ammonia discharged to the atmosphere is a stable material that is not oxidized to nitrogen oxides in the atmosphere.[6] Ammonia can react with sulfur dioxide and water to form an ammonium sulfate aerosol. However, for this consideration to be a limitation, the stripping tower would have to be located adjacent to a point source of sulfur dioxide.

The production and release of ammonia as part of the natural nitrogen cycle is about 50,000,000,000 tons per year. Roughly 99.9 percent of the atmosphere's ammonia concentration is produced by natural biological processes, primarily the bacterial breakdown of amino acids.[5,6] Although they are relatively insignificant sources, burning of coal and oil produces measurable quantities of ammonia.[6] The background levels of ammonia in the atmosphere have been observed to vary from .001 mg/m^3 to 0.02 mg/m^3 with a value of 0.006 mg/m^3 being typical.[6]

Available diffusion technology can be used to estimate the atmospheric concentration of ammonia at any point downwind of the stripping tower.[7] Calculations were made for the Orange County, California stripping tower for low mixing conditions (wind speed 1 m/sec). The resulting surface concentrations at the center of the downwind discharge zone including natural background levels were as follows:

Distance from Tower ft.	m	Surface Air Concentration of Ammonia, mg/m^3
300	91	5.2
1,000	305	1.6
1,600	488	0.6
3,200	975	0.2
16,000	4,877	0.0006

Background levels of ammonia are reached within 3 miles (4.8 km). No U.S. ammonia emission standards have been established by regulatory agencies because there are no known public health implications at concentrations normally encountered.

The American Conference of Governmental Industrial Hygenists recommended in 1967 an occupational threshold limit of 35 mg/m^3.[5] The permissible limit for ammonia in a submarine during a 60 day dive is 18 mg/m^3.[5] The Navy's Bureau of Medicine and Surgery has recommended an ammonia threshold limit for 1 hour of 280 mg/m^3. All of these values are above the 6 mg/m^3 which will typically occur at the tower discharge. As noted above, no ambient air quality standards for ammonia exist for the United States. However, such ambient air standards exist for Czechoslovakia, the U.S.S.R., and Ontario, Canada, as shown below:[5]

Location	Basic Standard mg/m^3	Basic Standard Averaging Time	Permissible mg/m^3	Permissible Averaging Time
Czechoslovakia	0.1	24 hr	0.3	30 min
U.S.S.R.	0.2	24 hr	0.2	20 min
Ontario, Canada	3.5	30 min		

8.2.2 Washout of Ammonia from the Atmosphere

There is a large turnover of ammonia in the atmosphere with the total ammonia content being displaced once per week on the average. Ammonia is returned to the earth through gaseous deposition (60 percent), aerosol deposition (22 percent), and precipitation (18 percent).[5] Although not the most significant mechanism for removal of ammonia from the atmosphere, precipitation does provide one pathway for the return of atmospheric ammonia to bodies of water and to soil. In rainfall, the natural background ranges from 0.01 to 1 mg/l with the most frequently reported values of 0.1 to 0.2 mg/l. The amount of ammonia in rainfall is directly related to the concentration of ammonia in the atmosphere. Thus, an increase in the ammonia in rainfall would occur only in that area where the stripping tower discharge increases the natural background ammonia concentration in the atmosphere. Calculations for the ammonia washout in the rainfall rate of 3 mm/hr (0.12 in./hr) have been made for the Orange County, California project with the following results:

Distance From Tower ft	Distance From Tower m	Peak Rainfall Ammonia Concentration, mg/l
300	91	60
1,000	305	18
1,600	488	11
3,200	975	5
16,000	4,877	0.5

The concentrations of ammonia in the rainfall would approach natural background levels within 16,000 ft (4.8 km) of the tower. The ultimate fate of the ammonia which is washed out by rainfall within this 16,000 ft (4.8 km) downwind distance depends on the nature of the surface upon which it falls. Most soils will retain the ammonia. That portion which lands on paved areas or directly on a stream will appear in the runoff from that area. Unless the stripping tower is located upwind in close proximity to a lake or reservoir, the direct return of ammonia to the aquatic environment by atmospheric washout should not make a significant contribution to the total ammonia discharged to the aquatic environment. However, this is a factor which must be carefully evaluated for each potential application.

8.2.3 Noise

There are three potentially significant noise sources in an ammonia stripping tower: (1) motors and fan drive equipment; (2) fans; and (3) water splashing. The following control measures are available:

- Motors – proper installation, maintenance and insulation
- Fans – reduction in tip speed and exhaust silencers
- Water – shielding of the tower packing and air inlet plenum

Based on sound level measurements from the tower at Lake Tahoe, the expected noise level at the tower is calculated to be about 64 decibels (dBA). This noise level can be reduced to 46 dBA at 600 ft (183 m) from the towers by control measures. The Orange County project (Sec. 9.5.5.2) includes several specific noise control measures. Before construction of the plant, ambient nighttime noise levels in the residential neighborhood around the Orange County plant were 40-45 dBA.

8.3 Stripping Tower System Design Considerations

The major factors affecting design and process performance include the tower configuration, pH, temperature, hydraulic loading, tower packing depth and spacing, air flow, and control of calcium carbonate scaling.

8.3.1 Type of Stripping Tower

There are two basic types of stripping towers now being used in full-scale applications: countercurrent towers and cross-flow towers (see Figure 8-2). Countercurrent towers (the entire airflow enters at the bottom of the tower while the water enters the top of the tower and falls to the bottom) have been found to be the most efficient. In the crossflow towers, the air is pulled into the tower through its sides throughout the height of the packing. This type of tower has been found to be more prone to scaling problems (see Section 8.3.7).

8.3.2 pH

The pH of the water has a major effect on the efficiency of the process. The pH must be raised to the point that all of the ammonium ion is converted to ammonia gas (see Section 8.1.1). If phosphorus removal is required, the use of lime as the coagulant will generally enable the necessary pH elevation to be achieved concurrent with phosphorus removal. If pH elevation does not occur in some upstream processes, then the economics of the stripping process are adversely affected since the costs of pH elevation must then be incurred solely for ammonia stripping.

FIGURE 8-2

TYPES OF STRIPPING TOWERS (REFERENCE 8)

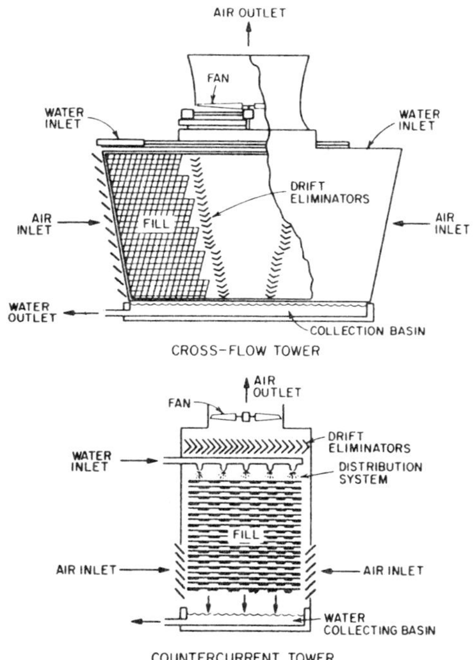

8.3.3 Temperature

A critical factor is the air temperature. The water temperature reaches equilibrium at a value near the air temperature in the top few inches of the stripping tower. As the water temperature decreases, the solubility of ammonia in water increases and it becomes more difficult to remove the ammonia by stripping. The amount of air per gallon must be increased to maintain a given degree of removal as temperature decreases. However, it is not practical to supply enough air to fully offset major temperature decreases. For example, at 20 C, 90-95 percent removal of ammonia is typically achieved. At 10 C, the maximum practical removal efficiency drops to about 75 percent. Data collected in pilot tests by EPA at the Blue Plains plant in Washington, D.C. well illustrate the temperature effects and are shown in Figure 8-3.[9] In warm weather tests at pH 11.5, with inlet air and water temperatures averaging 25.5 C and 26 C respectively, air stripping cooled the outlet water temperature by evaporation of the liquid within the tower to an average of 22.2 C. In a

FIGURE 8-3

EFFECT OF TEMPERATURE ON AMMONIA REMOVAL EFFICIENCY OBSERVED AT BLUE PLAINS PILOT PLANT (REFERENCE 9)

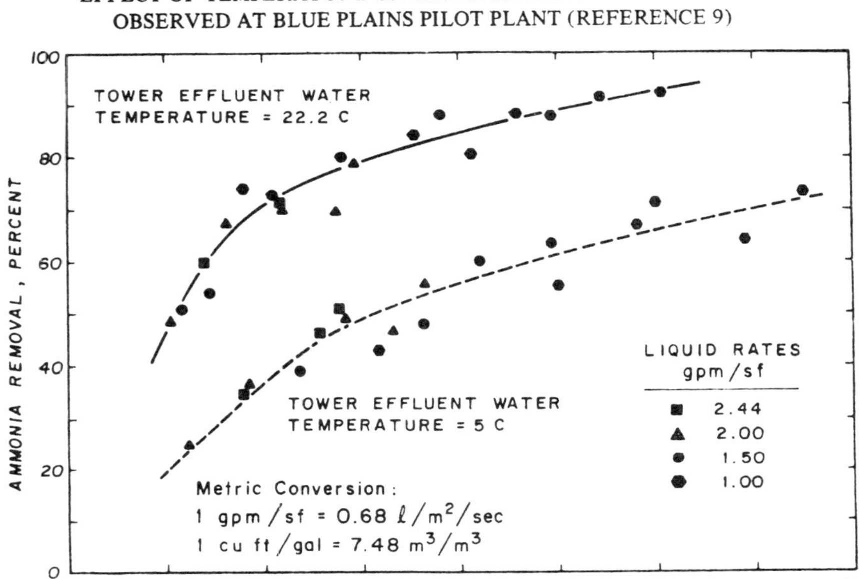

similar test with the inlet air temperature averaging 6 C and the inlet water temperature averaging 16 C, the air stripping cooled the outlet water to an average temperature of 5 C. Data from both the 22.2 C and 5 C conditions are shown in Figure 8-3. The decrease in efficiency from the warm to cold temperatures was approximately 30 percent over a wide range of air to water flows.

When air temperatures reach freezing (or when the wet bulb temperature of the air within the tower reaches 0 C), the tower operation must generally be shut down due to icing problems. The very large volumes of air required for the stripping process make it impractical to heat the air in cold climates. Waste heat from potential on-site sources such as sludge incinerators is typically only a small percentage of that needed.

8.3.4 Hydraulic Loading

The hydraulic loading rate of the tower is an important factor. This is typically expressed in terms of gallons/minute applied to each square foot (or $l/m^2/sec$) of the plan area of the tower packing. When the hydraulic loading rates become too high, the good droplet formation needed for efficient stripping is disrupted and the water begins to flow in sheets. If the rate is too low, the packing may not be properly wetted resulting in poor performance and scale accumulation. Data collected in pilot tests at South Tahoe illustrate this relationship and are shown in Figure 8-4.[10] In optimum summer conditions, the pilot data indicate that a flow rate of 2 gpm/sf (1.4 $l/m^2/sec$) is compatible with efficient tower operation at 20-24 ft (6.1-7.3 m) packing depths. Adequate flow distribution over the entire packing area is a critical factor. Full-scale towers at Orange County and Pretoria, South Africa are based on tower loadings of 1-1.13 gpm/sf (0.68-0.77 $l/m^2/sec$).

FIGURE 8-4

PERCENT AMMONIA REMOVAL VS. SURFACE LOADING RATE
FOR VARIOUS DEPTHS OF PACKING (REFERENCE 10)

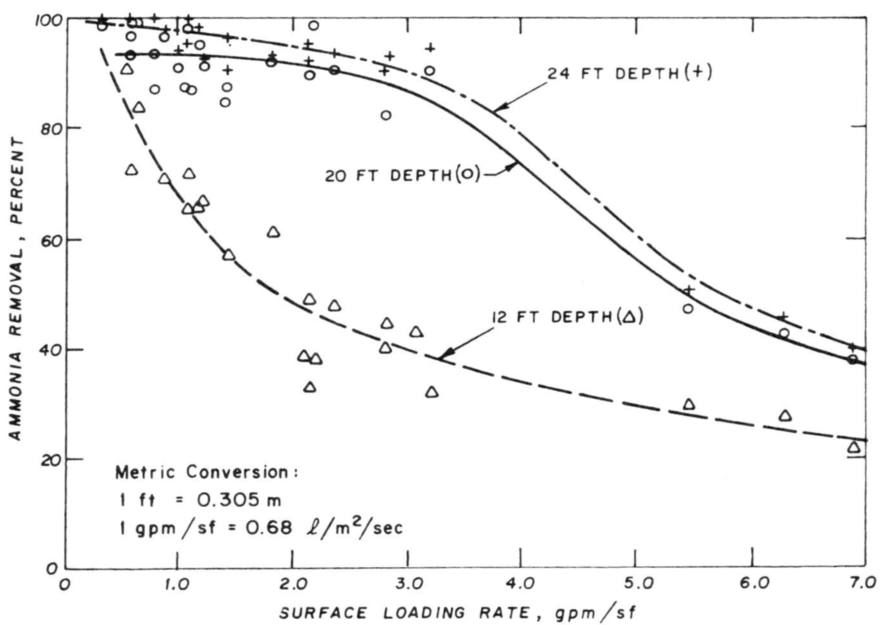

8.3.5 Tower Packing

8.3.5.1 Packing Depth

The depth of tower packing required for maximum ammonia removal will depend on the tower packing selected. Most stripping tower designs are based on the use of an open, cooling tower-type packing (horizontal packing members spaced about 2 in (5.1 cm) apart both horizontally and vertically) to minimize the power required to move adequate air quantities through the tower (see Sec. 8.3.6). If maximum removals are desired, tower packing depth should be at least 24 ft (7.3 m) with this type of packing, unless pilot plant data indicate that a lesser depth of a specific packing will accomplish the required removal. Packing with members spaced more than 2 inches (5.1 cm) apart may require greater depths, and pilot tests should be run to determine the required depth if greater spacings are proposed.

8.3.5.2 Packing Material and Shape

Both wood (Lake Tahoe) and plastic packings (Orange County) have been used in full-scale towers. The smooth plastic surfaces appear to be one factor accounting for reduced calcium carbonate scaling at the Orange County facility. Plastic packing has an advantage in that it does not suffer from the delignification that occurs with wood at elevated pH values.

Pilot studies at Orange County evaluated three different types of packing: ½ inch (1.27 cm) diameter PVC pipe, triangular shaped splash bars, and vertical film packing like that used in cooling towers.[11] With vertical packing the water moves in a thin film down vertical sheets of packing rather than moving as droplets as occurs in packing composed of horizontal splash bars. The film packing was found to provide only 50 percent or less ammonia removal and was eliminated from consideration early in the tests. Film packing fails to provide the repeated droplet formation and rupture needed for efficient stripping.

Since repeated splashing and droplet formation is a key parameter in ammonia stripping, a triangular shaped splash bar was tested. It was thought that it might provide two points of droplet formation compared to only one for a round splash bar. It was observed that droplet formation throughout the tower still occurred at only one point. The water flowed down the sides of the triangle and around the corners, where it collected on the base and dripped from a single point. Air flow and pressure drop measurements were made on both the circular and triangular packing. The static pressure drop (24 ft or 7.3 m packing depth) was 0.40-0.44 in (1-1.1 cm) of water when the triangular packing was used, compared to 0.36-0.40 in (0.9-1 cm) of water when the circular packing was used. No significant differences in ammonia removals were noted between the round and triangular shaped packings.

8.3.5.3 Packing Spacing and Configuration

Figure 8-5 is a sideview of a typical packing configuration using wood or plastic slats. In this example, the slats are spaced 2 in. (5.1 cm) apart (center to center) on the horizontal and 1.5 in. (3.8 cm) apart vertically. This spacing is referred to as 1.5 x 2 in. (3.8 x 5.1 cm). Figure 8-6 shows that the spacing of the tower packing is important in determining the air requirements for ammonia stripping. The 1.5 x 2 in. (3.8 x 5.1 cm) packing has 2.66 more slats for droplet formation and coalescing than does a 4 x 4 inch (10.2 x 10.2 cm) packing. Although spacing the packing members closer than 1.5 x 2 inches (3.8 x 5.1 cm) would improve performance, the increased pressure drop would greatly increase power costs (see Sec. 8.3.6).

Tests at Orange County indicate that packings in which alternate layers of packing are placed at right angles, rather than the parallel position shown in Figure 8-5, maintains better flow distribution and may be less susceptible to scale accumulation.

FIGURE 8-5

ILLUSTRATIVE PACKING CONFIGURATION

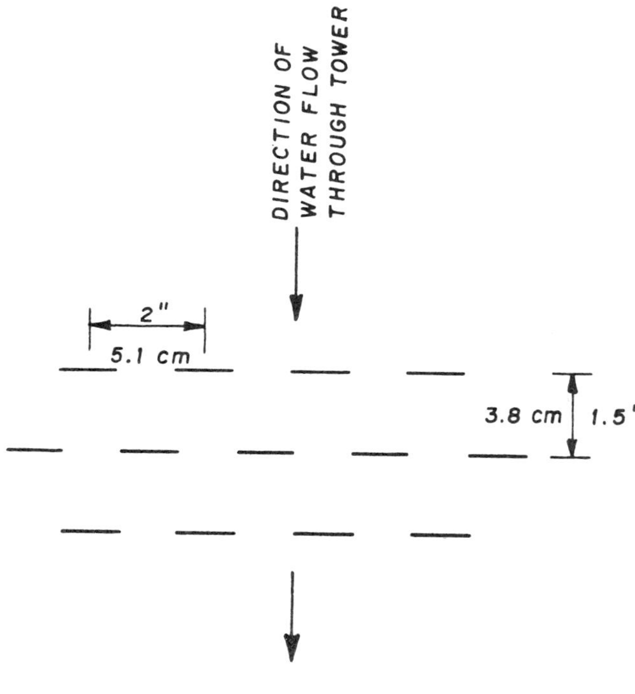

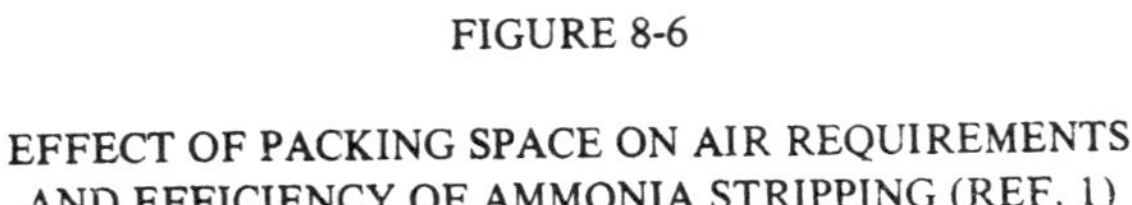

FIGURE 8-6

EFFECT OF PACKING SPACE ON AIR REQUIREMENTS AND EFFICIENCY OF AMMONIA STRIPPING (REF. 1)

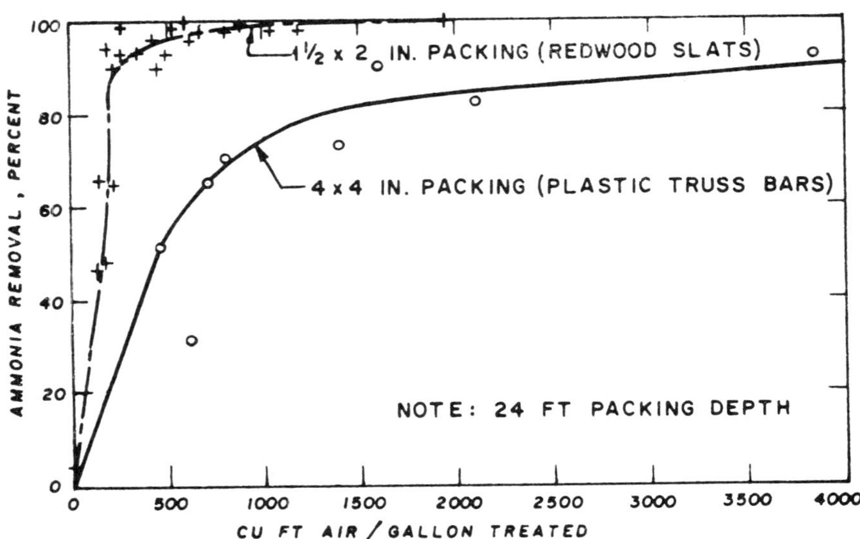

8.3.6 Air Flow

Gas transfer relationships indicate that an increase in ammonia removal can be achieved by increasing the air flow for a given tower height (see Figures 8-3 and 8-6). However, there is a practical limit on air flow rate due to the increase in air pressure drop with increasing flow rate. This results in higher capital investment for fans and increased power costs. The air pressure drop in a countercurrent tower is given as: [2]

$$P = f \cdot z \cdot Q_{air} \tag{8-2}$$

where:

P	=	Pressure drop, in. of water
f	=	Fanning friction factor
Q_{air}	=	Air flow rate, cu ft/min/sq ft ($m^3/min/m^2$)
z	=	Packing height, ft

Pressure drop increases exponentially with air flow rate. In general, air velocities of 550 cu ft/min/sq ft (1600 $m^3/min/m^2$) are considered to be the practical upper limit for countercurrent towers. The friction factor should be obtained from the packing manufacturer. General guides for wood grids are available in reference 3.

Figure 8-3 reflects the effects of the ratio of air to wastewater as observed at the Blue Plains pilot plant. These data are in general agreement with similar pilot data collected at South Lake Tahoe (Figure 8-6). For warm weather conditions, typical air requirements are about 300 cu ft/gal (2240 m^3/m^3) for 90 percent removal and 500 cu ft/gal 3740 m^3/m^3 for 95 percent removal. In cold weather conditions, the air requirements to achieve maximum tower efficiency increase substantially. Full-scale data at Tahoe indicate that, for their packing design, air flows of about 800 cu ft/gal (5980 m^3/m^3) would be needed to achieve 90 percent removal at an air temperature of 4 C. However, reliance solely on the stripping process in cold weather conditions is usually not practical, and most designs are based on moderate to warm weather conditions. Typical air design quantities for 90 percent removal are as follows: Orange County, California – 400 cf/gal (2990 m^3/m^3); South Lake Tahoe – 390 cf/gal (2920 m^3/m^3); Pretoria, South Africa – 338 cf/gal (2530 m^3/m^3).

The required air quantities are usually provided by a fan located on top of the tower. Two speed fan motors may be used to better match air supplied to the actual requirements. Because of the low pressure drops associated with the types of packings typically used (less than 1 inch or 2.5 cm water), the horsepower requirements for the fans are not great for these large quantities of air. For the 15 mgd (0.65 m^3/sec) Orange County plant, the total installed fan brake horsepower is 1380 HP.

8.3.7 Scale Control

A factor which may have an adverse effect on tower efficiency is scaling of the tower packing resulting from deposition of calcium carbonate from the unstable, high pH water flowing through the tower. Scaling potential can be minimized by maximizing the extent of completion of the calcium carbonate reaction in the lime treatment step. Using a high level of solids recycle in the clarification step will ensure more complete reaction. Another approach is to eliminate CO_2 from the air (see Section 8.4).

The original crossflow tower at the South Lake Tahoe plant has suffered a severe scaling problem. The severity of the scaling problem was not anticipated from the pilot studies in which a countercurrent tower was used. As a result, the full-scale cross-flow tower packing was not designed with access for scale removal in mind. Thus, portions of the tower packing are inaccessible for cleaning. Those portions which were accessible were readily cleaned by high pressure hosing.

The severity of the scaling problem has varied widely. Perhaps the most severe case is that reported at the Blue Plains pilot plant.[9] When operating at a pH of 11.5, a heavy scale of calcium carbonate formed on the crossflow tower packing (polypropylene grids). The scale was crystalline, hard, and could not be removed by a high pressure water hose. In contrast, several months of operation of the countercurrent pilot tower at Orange County at pH values above 11 resulted in only a thin coat of calcium carbonate scale on the 0.5 in. (1.3 cm) diameter PVC pipe packing. The thin coat of scale stabilized and did not continue to accumulate. The scale was friable and easily removed by water hosing. The Orange County pilot tower was later moved to South Tahoe where several months of operation indicated no scale buildup. The differences in operation between the pilot tower at Tahoe and the full-scale Tahoe tower were as follows: packing of plastic rather than wood, packing shape round rather than rectangular, countercurrent tower rather than parallel. The relative importance of these factors in eliminating the scaling noted in the full-scale Tahoe tower is uncertain. The experience with the full-scale (1 mgd) countercurrent tower at Pretoria, South Africa is similar to that at the full-scale Tahoe tower in that the scale formed can be readily flushed from the packing by a water jet.[12] The feeding of a scale-inhibiting polymer to the tower influent may also offer a means of scale control and such provisions are being made in the Orange County facility.

In light of the as-yet unpredictable nature of factors contributing to scaling of tower packing, it is prudent to conduct pilot tests for 3-6 months on the specific wastewater involved with the specific tower configuration proposed. The pilot vs. full-scale experience at Tahoe and the independent pilot tests at Orange County indicate that the use of countercurrent rather than crossflow towers will reduce scaling problems. Also, access should be provided to all of the tower packing for cleaning (see Sec. 9.5.5.2 for a discussion of the Orange County plant).

8.4 Ammonia Recovery or Removal From Off-Gases

As noted earlier, the calcium carbonate scaling problem can be minimized or eliminated by removing carbon dioxide (CO_2) from the stripping air. This section describes two approaches which accomplish this goal by removing ammonia and CO_2 from the off-gas from the tower and recycling the air.

8.4.1 Acid Systems

An approach to overcoming the limitations of the stripping process is currently being developed.[13] It appears that the process overcomes many limitations of the stripping process and has the advantage of recovery of ammonia as a byproduct.

The improved process is shown diagramatically on Figure 8-7. The process includes an ammonia stripping unit and an ammonia absorption unit. Both of these units are sealed from the outside air but are connected by appropriate ducting. The stripping gas, which initially is air, is maintained in a closed cycle. The stripping unit operates in the same manner as is now being or has been done in a number of systems with the exception that the gas stream is recycled rather than outside air being used in a single pass manner.

FIGURE 8-7

PROCESS FOR AMMONIA REMOVAL AND RECOVERY

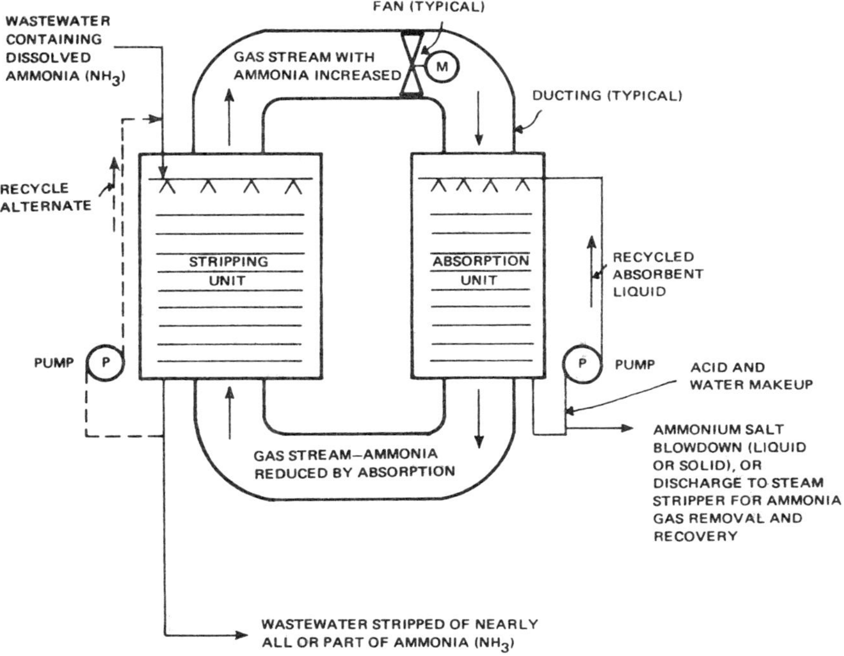

Most of the ammonia discharged to the gas stream from the stripping unit is removed in the absorption unit. Because of the favorable kinetics of the absorption reaction, the absorption unit may be reduced in size by about one third from that required for the stripping unit. The absorbing liquid is maintained at a low pH to convert absorbed and dissolved ammonia gas to ammonium ion. This effectively traps the ammonia and also has the effect of maintaining the full driving force for absorbing the ammonia since dissolved ammonia gas does not build up in the absorbent liquid. The absorption unit can be a slat tower or packed tower using sprays similar to the stripping unit, but will usually be smaller due to the kinetics of the absorption process.

The absorbent liquid initially consists of water with acid to obtain low pH (usually below 7). In the simplest case, as ammonia gas is dissolved in the absorbent and converted to ammonium ions, acid is added to maintain the desired pH. If sulfuric acid is added, as an example, an ammonium sulfate salt solution is formed. This salt solution continues to build up in concentration and the ammonia is finally discharged from the absorption device as a liquid or solid (precipitate) blowdown of the absorbent. With shortages of ammonia based fertilizers, a saleable byproduct may result. Ammonia sulfate concentrations of 50 percent are obtainable.

Mist eliminators are necessary between the absorber and stripper to prevent carryover of the ammonia laden moisture from the absorber to the stripper effluent. Because of the headloss in the mist eliminators and absorber packing, total headloss for the air approaches 2 inches (5.1 cm). It is believed that the usual scaling problem associated with ammonia stripping towers will be reduced by the improved process, since the carbon dioxide which normally reacts with the calcium and hydroxide ions in the water to form the calcium carbonate scale is eliminated from the stripping air during the first few passes. The freezing problem is eliminated due to the exclusion of nearly all outside air. The treatment system will normally operate at the temperature of the wastewater.

As discussed in Chapter 7, this approach is being used for stripping of ammonia from selective ion exchange process regenerant. Because of the higher power requirements (as compared to single-pass stripping), the use of this process may be limited to regeneration of the brines from the selective ion exchange process. Full-scale designs of this application are underway for a 22.5 mgd (1.0 m^3/sec) plant serving the Upper Occoquan Sewage Authority, Virginia and for a 6 mgd (0.26 m^3/sec) plant for the Tahoe-Truckee Sanitation Agency, California. Design criteria for this application are presented in Sec. 9.5.4.1 which describes the Upper Occoquan plant.

8.4.2 Nitrification-Denitrification

Another approach to achieving many of the same objectives of the acid system described above is also being developed.[14] The system, termed the Ammonia Elimination System (AES), is shown in Figure 8-8. Basic elements of the process are as follows:

1. Ammonia Stripping Tower. Ammonia is transferred in this tower from the liquid stream to the gas stream at high pH.

2. Ammonia Absorption-Oxidation Tower. Ammonia is transferred from the gas stream to the liquid stream at about pH 8. Ammonia is oxidized to nitrate by nitrifying bacteria in this tower. The following reaction summarizes the reactions in this tower:

$$NH_3 + HCO_3^- + 2O_2 \longrightarrow NO_3^- + H_2CO_3 + H_2O \qquad (8\text{-}3)$$

3. Denitrifying Reactor. Nitrate is reduced to nitrogen gas in this reactor (methanol or a waste carbon source is added to the reactor) as shown by the reaction:

$$NO_3^- + 1/6\ H_2CO_3 + 5/6\ CH_3(OH) \longrightarrow 1/2\ N_2\uparrow + 8/6\ H_2O + HCO_3^- \qquad (8\text{-}4)$$

4. Solids Separation. Denitrification organisms are settled from the process stream and are returned to the denitrification reactor or wasted.

5. The overall reaction in the AES system can be found by adding equations 8-3 and 8-4:

$$NH_3 + 2O_2 + 5/6\ CH_3(OH) \longrightarrow 1/2\ N_2\uparrow + 7/3\ H_2O + 5/6\ (H_2CO_3) \qquad (8\text{-}5)$$

FIGURE 8-8

AMMONIA ELIMINATION SYSTEM

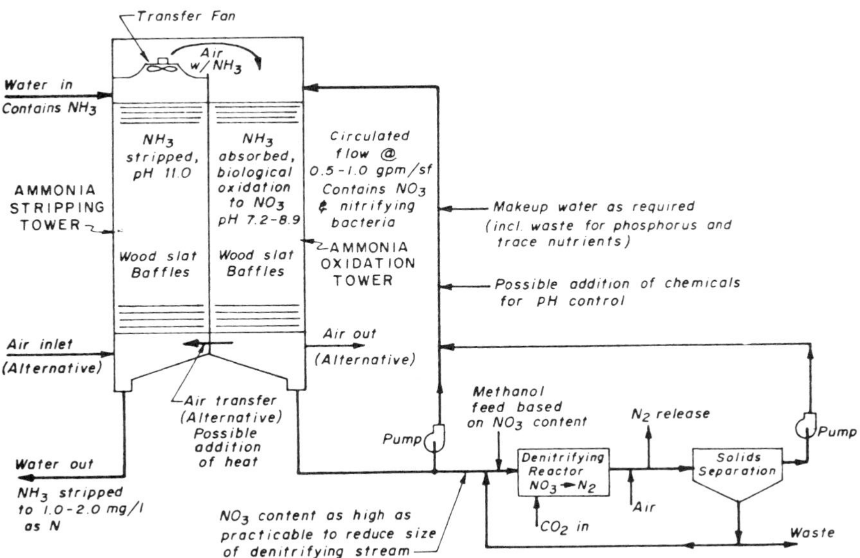

The process has two recycle streams, a gas recycle stream and a liquid recycle stream. The process configuration allows nitrogen to be treated separately from other waste contaminants; the AES can be inserted in a treatment scheme where nitrogen is in the ammonia form.

The AES has the following advantages in common with either ammonia stripping or the three-sludge biological system:

1. Removal of nitrogen from the main flow stream by a physical process (which has advantages over biological systems in terms of ease of process control).

2. Isolation of the nitrification stage from most agents in the plant influent which are toxic to nitrifying organisms.

3. Isolation of the ammonium oxidation stage from the carbon oxidation or removal stage.

4. Nitrogen is eliminated from the system as inoffensive nitrogen gas.

In addition to the above advantages, the AES has the following potential advantages not shared by either ammonia stripping or the three sludge system:

1. The insulation and heating of the liquid recycle streams, oxidation column, denitrification reactor, and clarifier(s) to increase process efficiency becomes economically feasible.

2. A waste carbon source may be used in place of methanol, since denitrification is accomplished on a side stream.

The AES shares the following advantages of the acid system described in Sec. 8.4.1:

1. Free ammonia gas is not discharged to the atmosphere.

2. Water vapor is not discharged to the atmosphere in large volumes.

3. The heating of air to prevent tower freezing and increase process efficiency becomes economically feasible due to gas recycle.

4. Gas recycle reduces scaling problems in the stripping tower as recycle gas will be very much lower in CO_2 content than atmospheric air.

From Equation 8-4, it can be seen that there is no net requirement for acid, as in the acid system described in Section 8.4.1.

8.5 Stripping Ponds

The South Tahoe system has been modified to reduce the impact of temperature and scaling limitations encountered at the plant.[15] Basically, the modified process consists of three steps (See Figure 8-9): (1) holding in high pH, surface agitated ponds, (2) stripping in a modified, crossflow forced draft tower through water sprays installed in the tower, and (3) breakpoint chlorination (see Sec. 9.5.5.1 for pilot plant results). This system was inspired by observations in Israel of ammonia nitrogen losses from high pH holding ponds.[16]

FIGURE 8-9

AMMONIA STRIPPING POND SYSTEM

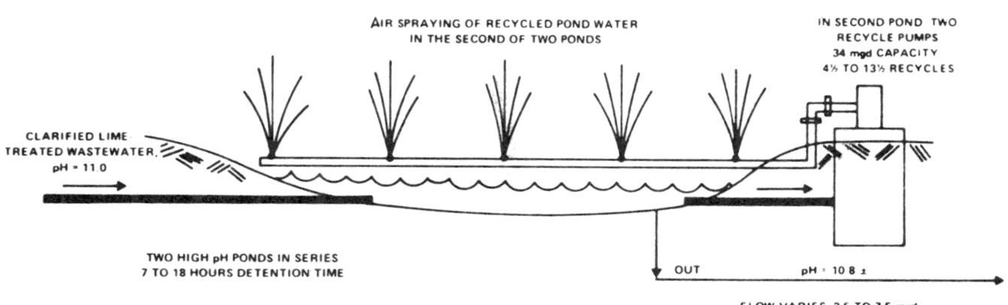

New High pH Flow Equalization Ponds

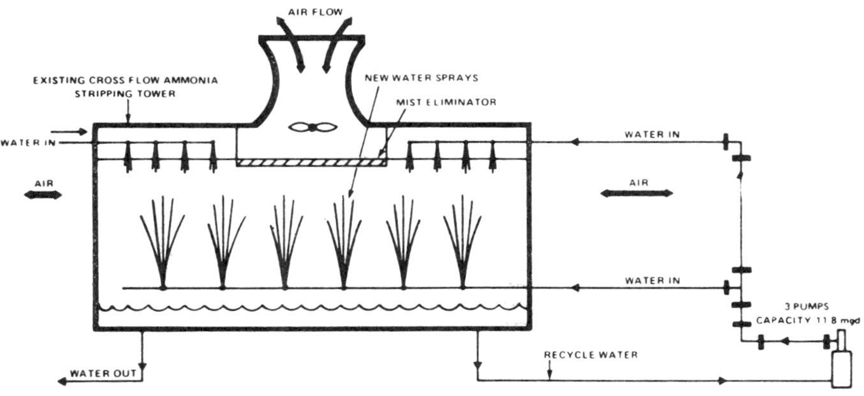

Stripping Tower Modified with New Sprays

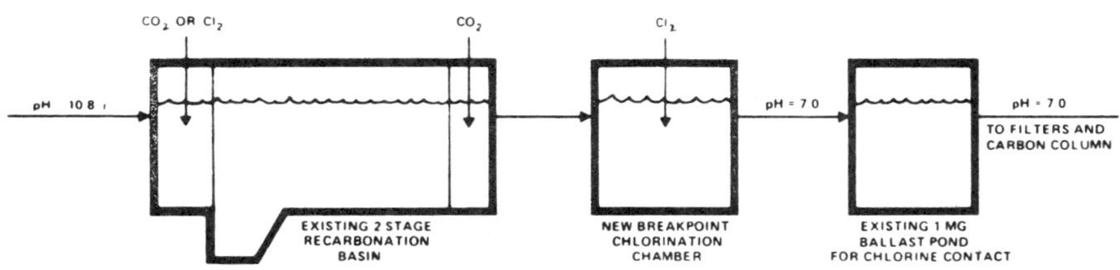

Breakpoint Chlorination

Pilot tests at South Tahoe indicated that the release of ammonia from high pH ponds could be accelerated by agitation of the pond surface. In the modified Tahoe system, the high pH effluent from the lime clarification process flows to holding ponds. Holding pond detention times of 7-18 hours are used in the modified South Tahoe plant. The pond contents are agitated and recycled 4-13 times by pumping the pond contents through vertical spray nozzles into the air above the ponds. The holding pond detention time and number of recycles vary with plant flow with the time and cycles decreasing as plant flow increases. At least 37 percent ammonia removal in the ponds is anticipated, even in cold weather conditions. The pond contents are then sprayed into the forced draft tower. The packing has been removed from the tower and the entire area of the tower is equipped with water sprays. At least 42 percent removal of the ammonia in the pond effluent is anticipated, based on pilot tests, from this added spray in cold weather that includes recycling of the pond effluent through the tower to achieve 2-5 spraying cycles. The ammonia escaping this process is removed by downstream breakpoint chlorination. It appears that stripping ponds offer an approach that takes advantage of the low cost and simplicity of the stripping process for removal of the bulk of the nitrogen, making breakpoint chlorination more attractive for complete removal of ammonia.

8.6 Considerations in Process Selection

One of the great advantages of air stripping is its extreme simplicity. Water is merely pumped to the top of the tower at a high pH, air is drawn through the fill, and the ammonia is stripped from the water droplets. The only control required is the proper pH in the influent water. This simplicity of operation enhances the reliability of the process. The process costs are also significantly less than any alternate method of nitrogen removal, assuming that the needed pH elevation occurs in conjunction with upstream phosphorus removal.

The major engineering limitations on application of the process result from its sensitivity to temperature variations and from potential scale accumulation on the tower packing. The first of these limitations is that of temperature as discussed in Section 8.3.3. Ammonia removals decrease as air and water temperatures decrease. Although increased air flows can offset temperature effects to some degree, it is not practical to supply enough air to offset major temperature drops. It has not been practical to operate stripping towers at ambient air temperatures below 0 C. Of course, this is not a limitation in climates where freezing temperatures do not occur or for plants where nitrogen removal is not required during cold or freezing weather. Modifications (Section 8.4) of the ammonia stripping process are being developed which may eliminate temperature limitations.

For applications in cold weather where a high degree of nitrogen removal is required, the stripping process itself will generally not be adequate. It is usually not practical to heat the large quantities of air required for the stripping process unless one is so fortunate as to have a large source of waste heat in proximity to the stripping tower (see Section 9.5.5.2 for a description of the use of waste heat from a desalting plant to heat stripping tower air while providing needed cooling of desalting plant water and wastes). The use of the stripping process supplemented as needed in cold weather by breakpoint chlorination is a process combination that may be attractive in some cases where efficient cold weather operation is needed and cold weather conditions do not persist for prolonged portions of the year.

A potentially serious problem is the formation and accumulation of calcium carbonate scale on the tower packing. Designs should anticipate this problem and provide for easy access for cleaning the packing as has been done in the Orange County tower (Sec. 9.5.5.2). It appears that scaling is not as severe in a countercurrent tower as in a crossflow tower and that scale does not adhere as tightly to the smooth, hard surface of plastic packing as it does to the rough, soft wood surface. The more open "criss-cross" type of packing developed at Orange County may also be more resistant to scale accumulation than the parallel packing arrangement used at Tahoe. Experience to date indicates that if adequate provisions are made in the design of stripping towers for access to packing, in many cases scale removal can be accomplished by water sprays without the use of chemicals or mechanical means. However, this is a factor that deserves special consideration and investigation at each location, since the scaling characteristics of different wastewaters may differ markedly. In light of the as-yet unpredictable nature of causes contributing to scaling of tower packing, it would be prudent to conduct pilot tests for 3-6 months on the specific wastewater involved with the specific tower configuration proposed.

8.7 References

1. Slechta, A. F., and G. L. Culp, *Water Reclamation Studies at the South Tahoe Public Utility District.* JWPCF, 39, No. 5, pp 787-814 (1967).

2. Roesler, J. F., Smith, R., and R. G. Eilers, *Mathematical Simulation of Ammonia Stripping Towers for Wastewater Treatment.* U.S. Department of Interior – FWPCA, Cincinnati, Ohio, January, 1970.

3. Roesler, J.F., Smith, R., and R.G. Eilers, *Simulation of Ammonia Stripping From Wastewater.* JSED, Proc. ASCE, 97, No. SA 3, pp 269-286 (1971).

4. Perry, J.H., *Chemical Engineering Handbook.* McGraw Hill Book Co., New York, NY.

5. Miner, S., *Preliminary Air Pollution Survey of Ammonia.* U.S. Public Health Service, Contract No. PH22-68-25, October, 1969.

6. *Nitrogenous Compounds in the Environment.* EPA Report SAB-73-001, December, 1973.

7. Pasquill, F., *Atmospheric Diffusion.* D. Nostrand Co., Ltd., London, 1962.

8. Culp, R.L., and G.L. Culp, *Advanced Wastewater Treatment.* Van Nostrand Reinhold, New York, 1971.

9. O'Farrell, T.P., Bishop, D. F., and A.F. Cassel, *Nitrogen Removal by Ammonia Stripping.* EPA Report 670/2-73-040, September, 1973.

10. South Tahoe Public Utility District, *Advanced Wastewater Treatment as Practiced at South Tahoe.* EPA Report 17010ELQ 08/71, August, 1971.

11. Wesner, G.M., and D.G. Argo, *Report on Pilot Waste Water Reclamation Study.* Orange County Water District Report, July, 1973.

12. van Vuuren, L.R., personal communication, May 30, 1972.

13. Kepple, L.G., *Ammonia Removal and Recovery Becomes Feasible.* Water and Sewage Works, p. 42, April, 1974.

14. Brown and Caldwell, *Application for a Research Contract for the Ammonia Elimination System.* Submitted to the Office of Research and Development, Water Quality Office, EPA, March, 1971.

15. Gonzales, J.G., and R.L. Culp. *New Developments in Ammonia Stripping.* Public Works, 104, No. 5, p 78 (1973) and No. 6, p 82 (1973).

16. Folkman, Y., and A.M. Wachs, *Nitrogen Removal Through Ammonia Release From Ponds.* Proceedings, 6th Annual International Water Pollution Research Conference, Tel Aviv, Israel, June 18-23, 1972.

17. Culp, G.L., *Physical-Chemical Techniques for Nitrogen Removal.* Prepared for the EPA Technology Transfer Program, July, 1974.

CHAPTER 9

TOTAL SYSTEM DESIGN

9.1 Introduction

With a defined set of effluent quality objectives, the environmental engineer must develop a cost-effective treatment system that is suited to the local situation. A variety of nitrogen removal options are presented in this manual that may form a part of the total treatment system. No one combination of processes is universally applicable.

The main thrust of this chapter is to present specific examples of treatment systems that have been selected and implemented. Design concepts that have evolved to suit local circumstances are emphasized.

9.2 Influence of Effluent Quality Objectives on Total System Design

The effluent quality obtainable from a treatment system has the most significant impact on its selection or rejection. When only ammonia removal (or conversion) is required, as opposed to total nitrogen removal, cost considerations would dictate selection of nitrification over any other method of ammonia removal in most cases. Exceptions would be low temperature (< 5 C) situations or where toxicants which cannot be effectively removed by a source control program are present in the wastewater. Other possible exceptions are cases wherein influent nitrogen levels are so low that they can be economically removed by breakpoint chlorination and where the carbon to nitrogen ratio is such that all the nitrogen can be assimilated into biomass during biological oxidation of organics.

In Section 2.5.5 the effectiveness of each type of process in removing the various nitrogen species was summarized. Tables 9-1, 9-2 and 9-3 contain representative nitrogen data for final effluents from systems incorporating nitrification-denitrification, ion exchange, and breakpoint chlorination. Ammonia stripping has not been included, since it usually must be backed up by breakpoint chlorination to achieve consistently low effluent nitrogen levels. These systems are capable of producing average effluent nitrogen levels of 2.0 to 3.0 mg/l, whether the nitrogen removing processes are biological or physical-chemical in nature.

Each nitrogen removal system should be assessed from the standpoint of its reliability in meeting effluent objectives. Factors resulting in the failure of a system to achieve nitrogen removal objectives are as follows:

1. Toxicity upsets

2. Overloading

3. Design deficiencies

4. Poor operation

5. Mechanical failures

6. Changes in influent quality

TABLE 9-1

EFFLUENT NITROGEN CONCENTRATIONS IN TREATMENT SYSTEMS INCORPORATING NITRIFICATION - DENITRIFICATION

Type and process sequence	Location	Ref.	Scale mgd (m^3/sec)	Period, days	Average effluent nitrogen, mg/l				
					Organic-N	NH_4^+-N	NO_3^--N	NO_2^--N	Total N
Lime treatment of raw sewage, nitrification, denitrification[b]	CCCSD,[a] Ca.	1	0.5 (0.022)	90	1.1	0.3	0.5	0.0	1.9
Primary treatment, high rate activated sludge, nitrification, denitrification, filtration[c]	Manassas, Va.	2	0.2 (0.0088)	120	0.8	0.0	0.7	0.0	1.5
.Primary treatment, roughing filters, nitrification, denitrification[d], filtration	El Lago, Texas	3	0.3 (0.0132)	55	0.8	0.9	0.6	0.0	2.3

[a]CCCSD = Central Contra Costa Sanitary District
[b]System 1, Fig. 9-1
[c]System 3B, Fig. 9-2
[d]Coarse media

TABLE 9-2

EFFLUENT NITROGEN CONCENTRATIONS IN TREATMENT SYSTEMS INCORPORATING ION EXCHANGE

Type and process sequence	Location	Ref.	Scale, mgd (m^3/sec)	Period days	Average effluent nitrogen, mg/l				
					Organic-N	NH_4^+-N	NO_3^--N	NO_2^--N	Total N
Lime treatment of raw wastewater, two-stage recarbonation, filtration, activated carbon, ion exchange[c]	Blue Plains, D.C.	4	0.05 (0.0022)	90	na[a]	3.6[b]	na	na	4.5[b,e]
Lime treatment of raw wastewater, recarbonation, filtration, activated carbon, ion exchange[c]	East Bay Municipal Utilities District, Ca.	5	Pilot scale	na	2.4	0.5	d	d	2.9[e]

[a]na = not available
[b]Intermittent operator attention only
[c]System 2A, Fig. 9-1
[d]Assumed negligible
[e]Estimated

TABLE 9-3

EFFLUENT NITROGEN CONCENTRATIONS IN TREATMENT SYSTEMS INCORPORATING BREAKPOINT CHLORINATION

Type and process sequence	Location	Ref.	Scale mgd (m^3/sec)	Period, days	Average effluent nitrogen, mg/l Organic-N	NH_4^+-N	NO_3^--N	NO_2^--N	Total N
Lime treatment of raw wastewater, two-stage recarbonation, filtration, breakpoint chlorination, activated carbon[b]	Blue Plains, D.C.	4	0.05 (0.0022)	120	na[a]	na	na	na	3.3
	Same, digital control	6	0.05 (0.0022)	9	na	na	na	na	1.6
Lime treatment of raw wastewater, filtration, activated carbon, breakpoint chlorination, and dechlorination by activated carbon[b]	Owosso, Michigan	7	0.02 (0.0009)	11	0.58	1.42	c	c	2.0[d]
Primary treatment, oxidation ponds, algae removal by alum-flotation, filtration, breakpoint chlorination	Sunnyvale, Ca.	8	0.01 (0.0004)	2	2.6	0.2	0.4	0.0	3.2

[a]na = not available from publication
[b]System 2A, Figure 9-1
[c]Assumed to be negligible
[d]Estimated

Toxicity upsets affect only biological nitrogen removal processes. A degree of protection against toxicity upset can be provided for the nitrification-denitrification system by providing pretreatment processes as described in Section 4.5.3. Several case histories are presented in this manual which show a very high stability for the biological processes, due in part to pretreatment. There are classes of toxicants, such as nonbiodegradable solvents, which are not effectively removed by pretreatment. Reliability under this circumstance is dependent on source control. Under some circumstances, the reliability of the source control program may be not effective enough to allow dependable nitrification.

The other factors listed affect both physical-chemical and biological systems to varying degrees. Overloading can be defined as operation which exceeds design conditions. Obviously both physical-chemical and biological processes can be expected to lose effectiveness when overloaded.

Theoretically sound processes can fail to meet objectives due to design deficiencies, mechanical breakdown or poor operation. This is true for both physical-chemical processes and biological processes.

Some of the nitrogen removal systems are more sensitive to the form of nitrogen in the influent than others. Generally, the physical-chemical processes are geared to a specific chemical form of nitrogen; for instance, urea cannot be removed by air stripping or breakpoint chlorination. Biological systems have the inherent capability to transform a multitude of nitrogenous compounds to ammonia for subsequent conversion to nitrate, a form suitable for denitrification. Thus, if changes in the distribution of influent nitrogen compounds occur with time, the biological processes may be more able to adapt to treatment of the new compounds than the physical-chemical processes.

In sum, the issue of relative reliability of the various approaches is mixed, and it cannot be claimed that some specific approach has a clear advantage over the others for general application. When stringent regulations require enhanced reliability, it is relatively simple to provide breakpoint chlorination for effluent polishing. Since breakpoint chlorination has no effect on nitrate or nitrite, it cannot make up for deficiencies in the denitrification process. However, in the nitrification-denitrification system, it is the nitrification step that is most susceptible to upset, and the breakpoint process provides full backup for it.

9.3 Other Considerations in Process Selection

Costs of the alternative nitrogen removal systems are specific to each situation and time-frame and generalizations about the alternatives are difficult to make. Long-run operating costs are of interest, but the long-term prices of chemicals and energy are particularly difficult to estimate.

Total dissolved solids (TDS) in the process effluent is sometimes a consideration. Biological nitrification-denitrification results in little change in TDS, whereas both ion exchange and breakpoint chlorination result in an increase in TDS.

Low liquid temperatures (< 10 C) often favor physical-chemical systems because the tankage requirements for biological nitrification-denitrification become very large. Biological nitrification-denitrification becomes less cost-effective below 5 C.

Receiving water standards or effluent requirements may dictate intermittent nitrogen removal. This requirement may favor breakpoint chlorination. An example is the nitrogen removal facility for the Sacramento Regional Treatment Plant, described in Section 9.5.3.1. In this case, breakpoint chlorination was chosen because its relatively low capital cost avoided the higher fixed costs of the other alternatives.

9.4 Interrelationships with Phosphorus Removal

Phosphorus removal is the subject of another publication of the EPA Office of Technology Transfer.[9] However, experience indicates that the majority of treatment plants being designed for nitrogen removal also have a requirement for phosphorus removal. As these two requirements have very different influences on treatment plant design, some consideration needs to be given to how nitrogen and phosphorus removal are interrelated in treatment system design.

9.4.1 Alternative Systems

Figures 9-1 and 9-2 are summaries of the five general approaches currently being considered or implemented for cases where high degrees of phosphorus and nitrogen removal are required. There are few exceptions to the listing; one exception concerns nitrogen removal from oxidation pond effluent, but these approaches tend to be very case specific and are difficult to generalize.

All flow diagrams in Figures 9-1 and 9-2 are capable of achieving low effluent levels of nitrogen and phosphorus. These levels are taken as averaging 2-3 mg/l of total nitrogen and 0.1 to 0.3 mg/l of total phosphorus. Multipoint chemical addition and filtration are shown to achieve low phosphorus levels in the final effluent. If lesser degrees of phosphorus removal are required, then some of these steps may be eliminated. Also, in each case, it is assumed that effluent BOD_5 objective is on the order of 10-15 mg/l. If further organic reduction is required, supplemental treatment is also required. Figures 9-1 and 9-2 also show variations within the five flowsheets where substitute processes are possible. The flowsheets are general process arrangements; for example, a block showing denitrification could mean either an attached growth process or a suspended growth process with a sedimentation tank.

System No. 1 is an integrated chemical-biological system using lime or other metal salt in the primary treatment stage to reduce phosphorus and organic loads followed by nitrification and denitrification stages and filtration. By moving lime treatment to the primary treatment stage, as opposed to tertiary applications, several advantages are gained. First, moving lime treatment to the primary stage causes enough organic reduction in the primary tanks to eliminate the need for a separate carbon removal step. Second, lime dose can be adjusted to elevate the pH in the nitrification step to the optimum range for nitrification as well as to compensate for any alkalinity depletion due to nitrification. Lastly, protection for the nitrifiers against most toxic heavy metals is provided.

Systems 2A and 2B are the independent physical-chemical treatment sequences incorporating physical-chemical nitrogen removal. A coagulant such as lime, or a metal salt and polymer, is used in the primary step for organics and phosphorus reduction. Activated carbon is provided for further organics reduction. In System 2A, either breakpoint chlorination or ion exchange is usual for nitrogen removal. In System 2B, ammonia stripping is used for nitrogen removal. Filtration is placed ahead of carbon adsorption in both variations of System 2; however, it may be placed after carbon adsorption in certain instances. For considerations in arrangement of the adsorption component of physical-chemical systems, the reader is referred to the *Process Design Manual for Carbon Adsorption,* a publication of the EPA Office of Technology Transfer.[10]

Another approach to integration of biological and physical-chemical treatment is provided by Systems 3A and 3B. System 3A takes advantage of the favorable effect of alkalinity depletion in nitrification on reducing lime dose in the chemical precipitation step as lime dose is directly affected by alkalinity.[11,12] System 3B shows a slightly different way of

FIGURE 9-1

ALTERNATE PROCESS SEQUENCING FOR SYSTEMS YIELDING COMBINED NITROGEN AND PHOSPHORUS REMOVAL - SYSTEMS WITH COAGULANT ADDITION TO PRIMARY SEDIMENTATION

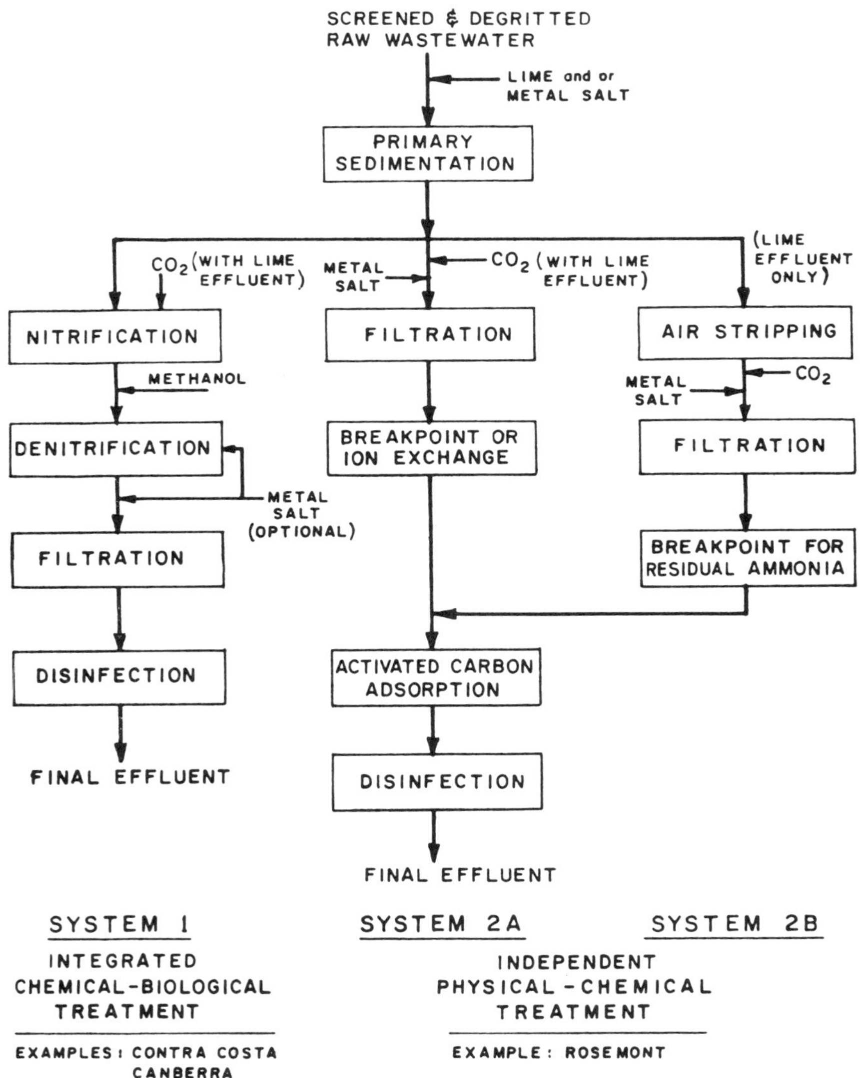

FIGURE 9-2

ALTERNATIVE PROCESS SEQUENCING FOR SYSTEMS YIELDING COMBINED NITROGEN AND PHOSPHORUS REMOVAL - SYSTEMS WITH COAGULANT ADDITION AFTER PRIMARY TREATMENT

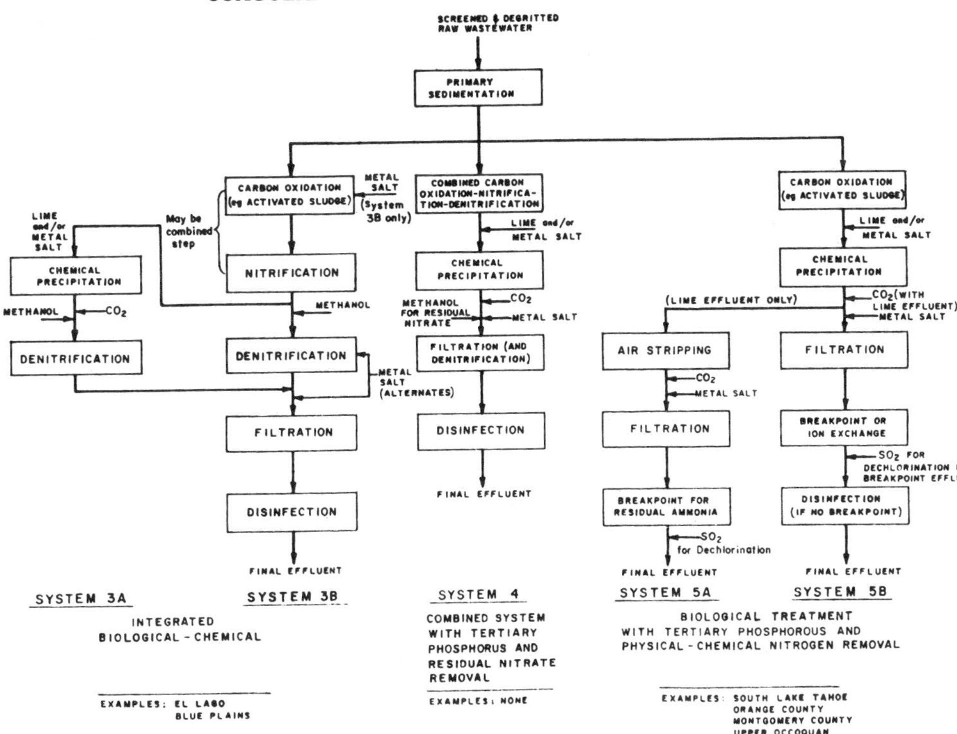

accomplishing the same objective. In this case metal salt is added to treatment units such as the carbon oxidation step and the filtration step, avoiding the need for the separate precipitation step of System 3A.

System 4 uses a combined carbon oxidation-nitrification-denitrification sequence. The organic carbon in the primary effluent serves as the carbon source for the removal of the bulk of the nitrogen in the influent wastewater. This effluent is polished with a tertiary phosphorus removal step. Residual nitrates are removed in the wastewater filter. System 4 has lower operating costs than System 1 or 3 because of the elimination of the bulk of the methanol costs incurred in the latter systems. On the other hand, the phosphorus removal step must be placed after the biological treatment step in System 4 for two reasons. First, precipitation of phosphorus in the primary step would reduce the organics in the primary effluent to the point where denitrification could not be supported. Second, metal salt addition to the biological step would add so many solids as to render unmanageable the simultaneous operation of carbon oxidation, nitrification and denitrification.

System 5 was the first system implemented in the U.S. and was used at South Lake Tahoe (in the System 5A configuration). It consists of conventional biological treatment followed by tertiary steps for phosphorus removal and physical-chemical nitrogen removal. System 5A employs air stripping with polishing by breakpoint chlorination while System 5B uses either breakpoint chlorination or ion exchange for nitrogen removal.

In conjunction with the system descriptions in Figures 9-1 and 9-2 are listed the case examples presented in this chapter which generally fit the system description. In most cases the examples do not precisely follow the system description because local requirements have dictated lesser or greater degrees of treatment. However, the case examples are close enough to be fitted into system categories.

9.4.2 Considerations in System Selection

Each of the systems outlined in Section 9.4.1 has its inherent advantages and disadvantages that need to be considered by the treatment plant designer in each individual situation. Some of these considerations are described in this section.

9.4.2.1 Phosphorus Removals Obtainable

Perhaps because of the long experience accumulated with System 5A (Figure 9-2) at South Lake Tahoe and the more recent development of alternative systems, System 5A or 5B were thought to have had a decided advantage in terms of low phosphorus residual over any of the integrated systems (System 1, 3A and 3B). It has been suggested that the integrated system, even with proper coagulant dosage, is limited to reduction of effluent phosphorus levels to 0.5 mg/l total phosphorus and that when lower phosphorus levels are required, System 5A or 5B should be employed.[12]

In actuality, the degree of phosphorus removal is not primarily affected by the system selected but by the pattern of chemical addition, the nature and doses of the chemicals used and the sophistication of process control. Regardless of the system selected, low effluent phosphorus levels are made possible by multiple phosphorus removal steps; this may be achieved by multipoint chemical addition or chemical addition in conjunction with other phosphorus removal methods such as tertiary filtration.

An example is provided by the well documented operation of the South Lake Tahoe plant.[13] A flowsheet for the plant as it existed in 1970 is presented in Figure 9-3. Multipoint chemical addition was practiced with lime treatment of the secondary effluent plus alum treatment at the tertiary filtration step. Phosphorus removal resulted at both these steps. In addition, phosphorus uptake occurred in the activated sludge step and possibly some removal occurred in the carbon adsorption step. The lime treatment alone reduces the phosphorus level to about 0.6 mg/l entering the filtration step. About 30 percent phosphorus removal occurred in the filter without alum addition. With alum addition, the effluent phosphorus level was reduced to 0.1 mg/l in a special one month test.[13] The phosphorus residuals for a one year period averaged 0.17 mg/l in the plant effluent as shown in Table 9-4. Alum dosage required to boost phosphorus removals by the filters from 30 to 90 percent was only 10 mg/l.[13]

From the Tahoe example it can be seen that the key to obtaining a low phosphorus residual is multiple removal steps. Multiple phosphorus removal steps have been included in all of the treatment systems portrayed in Figures 9-1 and 9-2. For instance, the option of metal salt addition to tertiary filters is available to *all* of the systems, not just System 5A or 5B, and comparable performance can be expected in each application. System performance is given in Table 9-4 for all systems except System 4. Favorable examples were chosen for each case in Table 9-4 and other cases for each system could be found with higher effluent phosphorus values. These examples are illustrative of what good design, operation, and control can produce.

9.4.2.2 Impacts on Sludge Handling

In Systems 3A, 4, 5A and 5B, (Figure 9-2) chemical precipitates can be kept separate from organic sludges. In Systems 1, 2A, 2B and 3B, (Figures 9-1 and 9-2) chemical sludges are combined with organic sludges. Separation of sludges allows the plant designers more options in sludge handling. For instance, chemical sludges can be subjected to coagulant recovery operations while processing of organic sludges can proceed without the hindering effects of the inert chemical sludges.

It has also been suggested that tertiary phosphorus removal (System 5A or 5B) may have an advantage in that lime recovery can be practiced.[12] However, System 1, 2, 3A and 4 all possess an advantage in common with 5A or 5B; namely, the ability to recover lime if it is the chosen coagulant.

FIGURE 9-3

SCHEMATIC FLOW DIAGRAM - SOUTH LAKE TAHOE, CALIFORNIA PLANT (1970)

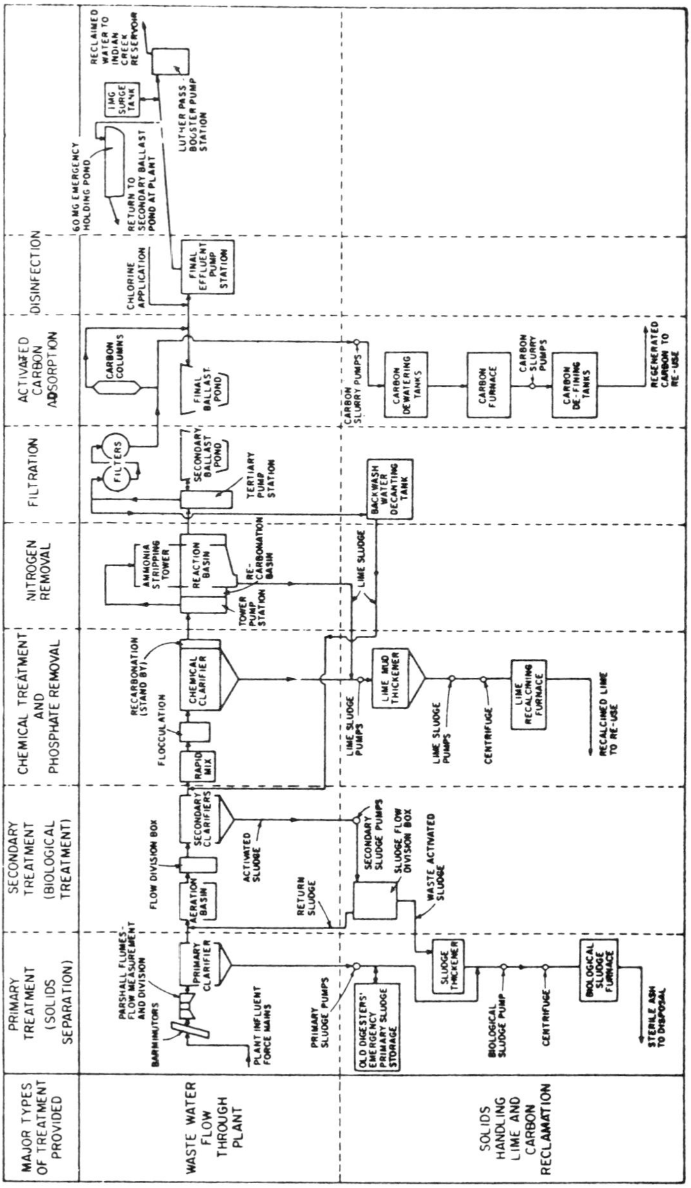

MAJOR TYPES OF TREATMENT PROVIDED
PRIMARY TREATMENT (SOLIDS SEPARATION)
SECONDARY TREATMENT (BIOLOGICAL TREATMENT)
CHEMICAL TREATMENT AND PHOSPHATE REMOVAL
NITROGEN REMOVAL
FILTRATION
ACTIVATED CARBON ADSORPTION
DISINFECTION

WASTE WATER FLOW THROUGH PLANT
PARSHALL FLUMES-FLOW MEASUREMENT AND DIVISION
BARMINUTORS
PRIMARY CLARIFIER
PLANT INFLUENT FORCE MAINS
FLOW DIVISION BOX
AERATION BASIN
SECONDARY CLARIFIERS
ACTIVATED SLUDGE
RETURN SLUDGE
PRIMARY SLUDGE PUMPS
FLOCCULATION
RECARBONATION (STAND BY)
RAPID MIX
CHEMICAL CLARIFIER
AMMONIA STRIPPING TOWER
REACTION BASIN
TOWER PUMP STATION
RE-CARBONATION BASIN
FILTERS
SECONDARY BALLAST POND
TERTIARY PUMP STATION
CARBON COLUMNS
FINAL BALLAST POND
CHLORINE APPLICATION
FINAL EFFLUENT PUMP STATION
60 MG EMERGENCY HOLDING POND
RETURN TO SECONDARY BALLAST POND AT PLANT
1MG SURGE TANK
LUTHER PASS BOOSTER PUMP STATION
RECLAIMED WATER TO INDIAN CREEK RESERVOIR

SOLIDS HANDLING LIME AND CARBON RECLAMATION
OLD DIGESTERS' EMERGENCY PRIMARY SLUDGE STORAGE
SLUDGE THICKENER
BIOLOGICAL SLUDGE PUMP
CENTRIFUGE
BIOLOGICAL SLUDGE FURNACE
STERILE ASH TO DISPOSAL
SECONDARY SLUDGE PUMPS
SLUDGE FLOW DIVISION BOX
WASTE ACTIVATED SLUDGE
LIME SLUDGE PUMPS
LIME MUD THICKENER
LIME SLUDGE PUMPS
CENTRIFUGE
LIME RECALCINING FURNACE
RECALCINED LIME TO RE-USE
LIME SLUDGE
BACKWASH WATER DECANTING TANK
CARBON SLURRY PUMPS
CARBON DEWATERING TANKS
CARBON FURNACE
CARBON SLURRY PUMPS
CARBON DE-FINING TANKS
REGENERATED CARBON TO RE-USE

TABLE 9-4

EFFLUENT PHOSPHOROUS CONCENTRATION FROM ALTERNATIVE SYSTEMS

Type and process sequence	Location	Ref.	Scale, mgd (m^3/sec)	Period, days	Average effluent phosphorus, mg/l		Average effluent total suspended solids, mg/l
					Total P	$PO_4^{\equiv}$-P	
System 1 Lime (with recalcination and recycle), nitrification, denitrification, without filtration	CCCSD, Ca.[a]	14	0.5 to (0.022)	10[b]	0.04	0.01	3.0
System 2A Lime with iron in second settling tank, filtration, ion exchange or break-point, filtration	Blue Plains, D.C.	4	0.05 (0.0022)	480	0.14	na	4.0
System 3B Carbon oxidation-nitrification-denitrification, filtration with alum addition to carbon oxidation and denitrification	Manassas, Va.	2	0.02 (0.0009)	120	0.3	na	0
System 5A Carbon oxidation, lime, ammonia stripping, recarbonation, filtration with alum addition, carbon adsorption	South Lake Tahoe, Ca.	13	2.4 (0.11)	365[c]	0.17	na	0

[a]CCCSD = Central Contra Costa Sanitary District

[b]August 1 to 10, 1973

[c]1970; representative year, the plant has been operational since 1968.

9.4.2.3 Reliability

Factors affecting the reliability of nitrogen removal processes have already been described in Section 9.2. Most of these same factors affect phosphorus removal and will not be repeated here.

The longest record of reliable experience in obtaining low phosphorus residuals is at the South Lake Tahoe Plant, operational since 1968 (Table 9-4). Low values have also been obtained consistently for long periods with physical-chemical systems (System 2A). In the latter case, iron has been used in a second stage settler after lime treatment and recarbonation, achieving further phosphorus removals. Less experience is available with Systems 1, 3A or 3B, but the limited testing to date indicates very low phosphorus residuals can be consistently obtained.

9.4.2.4 Flexibility of Operation in Multipurpose Treatment Units

The systems portrayed in Figures 9-1 and 9-2 incorporate varying degrees of integration of process function into the various treatment units. Systems 4, 5A and 5B represent extremes in terms of combining functions; in System 5A or 5B the tendency is for individual steps to perform a minimum of purposes while in System 4 many functions are accomplished in parallel in each step.

The argument can be made that the level of integration in a plant can affect its flexibility of operation in terms of adjustability of the system to meet varying loads or in terms of providing redundancy for possible process failures. The degree of integration possible is best studied by examination of pilot or full-test results. There have been full-scale tests that have shown that nothing has been lost in terms of flexibility or performance with a degree of process integration. Examples are provided by the Manassas and CCCSD experience (Systems 1 and 3B) described in Section 5.2.4. No long-term test results are available for System 4, which is unfortunate, since a high degree of integration of function is provided in the combined carbon oxidation-nitrification-denitrification step.

9.4.2.5 Cost

Cost is an essential factor in process selection. It is widely recognized that the integrated approaches hold a potential of cost savings over the biological-tertiary approach (System 5).[1,12,15] The reality of this cost saving will be determined in individual situations by local factors and must be specifically evaluated in each case by cost-effective analyses of the alternative systems.

9.5 Case Examples

Fourteen case examples of nitrogen control are presented, each showing how the various nitrogen removal systems described in this chapter have been applied. These include: four examples of nitrification for ammonia reduction, four examples of nitrification-denitrification for nitrogen removal, and two examples each of breakpoint chlorination, ion exchange and air stripping for nitrogen removal.

9.5.1 Case Examples of Nitrification for Ammonia Reduction

Four examples of how biological nitrification has been implemented are presented in the following discussion. The Jackson, Michigan plant design was oriented to reducing the nitrogenous oxygen demand (NOD) of the plant effluent in the receiving waters. The designs of the Valley Community Services District plant and the City of Livermore's plant were oriented to satisfying the very low coliform requirements set by the State of California. In these cases, ammonia reduction allowed efficient disinfection after breakpoint chlorination. In the design of another California plant, operated by the San Pablo Sanitary District, nitrification was included so that effluent toxicity requirements could be met.

Other case examples of nitrification were presented previously in Section 4.3.8. These included the Whittier Narrows Reclamation Plant in California which is oriented to groundwater recharge and the Flint, Michigan plant, designed for NOD removal.

9.5.1.1 Jackson, Michigan

The City of Jackson is operating a 17 mgd (0.74 m^3/sec) activated sludge plant that is designed to nitrify year-round. Nitrification is provided for removal of nitrogenous oxygen demand so that the receiving water, the Grand River, can be maintained at dissolved oxygen levels of about 4.0 mg/l. Since the implementation of full nitrification at this plant, this requirement has been consistently met in the zone of influence of the plant's discharge.

The City of Jackson, Michigan, with an equivalent population of 60,000 has been served by a conventional activated sludge treatment plant since 1936.[16] The current upgrading of the plant was completed in 1973 and resulted in the plant depicted in Figure 9-4 with the design data shown in Table 9-5. During this upgrading, the following facilities were added: (1) new primary effluent pumps, (2) a stormwater retention basin, (3) stormwater pumps for filling the retention basin, (4) additions to the aeration tanks, (5) additions to the secondary sedimentation tanks, (6) new return activated sludge pumps, (7) three new blowers and a blower building for the secondary treatment additions, and (8) a new plant electrical and control system.[17] This work was bid in December, 1970 and totaled $3,200,000 including legal, engineering and contingency costs. Operation and maintenance costs for the entire plant for fiscal year 1973/1974 totaled $464,159 for 5255 million gallons treated or $88/mil gal.[18]

Several features have been incorporated into this plant that have been stressed in this manual. First, the activated sludge system is operated in the conventional or plug flow manner to gain highest efficiency of nitrification even at the coldest temperature conditions (as low as 8 C). Coarse bubble aeration is utilized. Another feature of the plant is the

TABLE 9-5

DESIGN DATA
JACKSON, MICHIGAN WASTEWATER TREATMENT PLANT

Average dry weather flow (ADWF)	17 mgd (0.74 m^3/sec)
Peak dry weather flow (PDWF)	22 mgd (0.96 m^3/sec)
Peak wet weather flow (PWWF)	30 mgd (1.31 m^3/sec)
Raw wastewater quality at ADWF	
BOD_5	145 mg/l
SS	200 mg/l
Primary sedimentation tanks	
Number	6
Length	60 ft (18.3 m)
Width	24 ft (7.3 m)
Depth	11 ft (3.4 m)
Overflow rate at ADWF	1,970 gpd/sf (80.2 m^3/m^2/day)
Detention time at ADWF	1 hr
Retention basin	
Volume	12 mil gal (45,430 m^3)
Air blowers	
Number	6
Discharge pressure	6.5 psig (0.46 kgf/cm^2)
Capacity - total	33,000 cfm (940 m^3/min)
Aeration tanks	
Old tanks	
Number tanks	4
Passes per tank	1
Width	25.5 ft (7.8 m)
Length/pass	240 ft (73.2 m)
Depth	14.5 ft (4.4 m)
Volume (4 tanks)	355,000 cu ft (10,000 m^3)
New tanks	
Number of tanks	2
Passes per tank	2
Width/pass	25 ft (7.6 m)
Length/pass	150 ft (45.7 m)
Depth	14.5 ft (4.4 m)
Volume (2 tanks)	217,500 cu ft (6,159 m^3)
Detention time at ADWF[a]	6.0 hr
Final sedimentation tanks	
Old tanks	
Number	4
Diameter	70 ft (21.3 m)
Sidewater depth	11 ft (3.4 m)

TABLE 9-5

DESIGN DATA
JACKSON, MICHIGAN WASTEWATER TREATMENT PLANT
(CONTINUED)

New tanks	
Number	2
Diameter	80 ft (24.4 m)
Sidewater depth	12 ft (3.7 m)
Overflow rate at ADWF [a]	670 gpd/sf (27.3 m^3/m^2/day)
at PDWF [a]	865 gpd/sf (35.2 m^3/m^2/day)
Detention time at ADWF [a]	3.0 hr
Chlorine contact chamber	
Number of passes	4
Detention time at ADWF	47 min
Sludge digestion	
Number digesters	6
Volume	297,410 cu ft (8,422 m^3)
Sludge drying beds	
Surface area, sf	328,000 sf (30,489 m^2)

[a] Including both old and new tanks

FIGURE 9-4

JACKSON, MICHIGAN WASTEWATER TREATMENT PLANT FLOW DIAGRAM

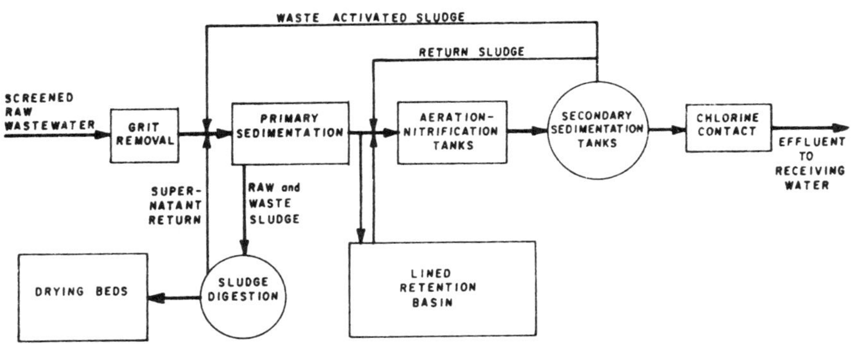

primary effluent retention basin. Rather than designing the secondary facilities for the full wet weather flows, flows above about 22 mgd are pumped to the retention basin. This flow is brought back by gravity to the secondary facilities when storm flows subside. It is also anticipated that the retention basin will serve as flow equalization storage during dry weather, once plant flows reach design capacity. This will be done to prevent ammonia bleedthrough during peak flow periods.[18]

The retention basin is lined, but has no provision for mixing. Actual operation indicates very satisfactory performance without odor development during storage. Except for completely draining the basin, cleaning is limited to once per year.[18]

Performance of the plant has been exceptionally stable and was previously summarized in Section 4.3.8.3. It should be noted that the plant is not yet being operated at its design flow.

9.5.1.2 Valley Community Services District, California

The Valley Community Services District (VCSD) Wastewater Treatment Plant at Dublin, California, is treating an average daily flow of 3.7 mgd (0.16 m^3/sec) from a largely residential service area. The original plant, consisting of raw wastewater screening, grit removal, primary sedimentation, activated sludge aeration, secondary sedimentation, digestion and sludge lagooning, was constructed in 1960 for approximately $1.5 million. In 1972, an additional primary sedimentation tank and appurtenances, an additional secondary clarifier, a digester, dual media filters, and chlorination and dechlorination facilities were constructed for about $3 million. These facilities raised the average dry weather flow capacity of the plant to 4 mgd (0.17 m^3/sec). The design data for the plant are shown in Table 9-6.[19]

Waste discharge requirements mandate effluent filtration, as the State of California requires that any effluent that may be used for water contact sports must be coagulated and filtered. Requirements also dictated that the median coliform content must not exceed an MPN of 2.2 per 100 ml. This plant incorporates nitrification for ammonia reduction so that residual ammonia can be economically breakpointed. This allows disinfection to proceed with a free chlorine residual so that the stringent bacteriological requirements may be met.

A flow diagram of the existing facilities (Fig. 9-5) shows a holding basin which is routinely used for flow equalization after primary treatment. Figure 9-6 is a photo of the holding basin. The holding basin is asphalt-lined and is equipped with a sprinkler system for washing out accumulated solids when the basin is drained. The sprinkler system was added by plant staff and was found to be very effective for odor control. The basin is emptied daily. A peak-to-average flow ratio of 3.4:1 is equalized in the holding basin to maintain stable biological treatment conditions. It was found prior to the 1972 additions to the plant that operation without flow equalization resulted in ammonia bleedthrough and eventual complete loss of nitrifying capability. The aeration tank is generally operated with the first

TABLE 9-6

VCSD WASTEWATER TREATMENT PLANT DESIGN DATA

Population	48,000
Average dry weather flow (ADWF)	4 mgd (0.17 m^3/sec)
Peak dry weather flow (PDWF)	8 mgd (0.33 m^3/sec)
Peak wet weather flow (PWWF)	12 mgd (0.50 m^3/sec)
Raw wastewater quality at ADWF	
BOD_5	330 mg/l
Suspended solids	330 mg/l
Primary treatment	
Preaeration and grit removal tanks	
Number	2
Detention time at ADWF	24 min
Primary sedimentation tanks	
New tanks	
Number	1
Length	100 ft (30.5 m)
Width	19 ft (5.8 m)
Old tanks	
Number	1
Length	110 ft (33.5 m)
Width	19 ft (5.8 m)
Average water depth	9 ft (2.74 m)
Detention time at ADWF[a]	1.7 hr
Overflow rate at ADWF[a]	960 gpd/sf (38.9 m^3/m^2/day)
Assumed removals	
BOD_5	40 percent
Suspended solids	70 percent
Activated sludge	
Aeration tanks	
Number	1
Passes per tank	2
Width each pass	30 ft (9.2 m)
Length each pass	210 ft (64.0 m)
Average water depth	15 ft (4.6 m)
Detention time, based on ADWF	8.5 hrs
BOD_5 loading	35 lb/1000 cf/day
Aeration blowers	
Number	4
Capacity per blower	2,900 cfm (82 m^3/min)
Discharge pressure	7.5 psig (0.53 kgf/cm^2)
Secondary sedimentation tank	
Old tanks	
Number	1
Diameter	65 ft (19.8 m)
Sidewater depth	9 ft (2.7 m)
New tanks	
Number	1
Diameter	90 ft (27.4 m)
Sidewater depth	14 ft (4.3 m)
Detention time, based on ADWF[a]	5.3 hrs
Overflow rate at ADWF[a]	410 gpd/sf (16.6 m^3/m^2/day)
Dual media filters	
Number	3
Area each filter	728 sf (67.66 m^2)
Filtration rate at PWWF	3.8 gpm/sf (1.36 l/m^2/sec)
Anthracite media	
Depth	36 inches (0.91 m)
Effective size	2.4 - 4.8 mm

TABLE 9-6

VCSD WASTEWATER TREATMENT PLANT DESIGN DATA (CONTINUED)

Sand media	
Depth	18 inches (0.45 m)
Effective size	0.8 - 1.0 mm
Pea gravel size	
Depth	8 inches (203 mm)
Backwash water rate (max.)	20 gpm/sf (13.6 $l/m^2/sec$)
Chlorine contact tanks	
Number	1
Volume	22,500 cu ft (637 m^3)
Detention time @ ADWF	1.0 hr
Sludge digestion	
Digester loading	
Primary solids	7,700 lb/day (3,500 kg/day)
Biological solids	3,800 lb/day (1,730 kg/day)
Total solids	11,500 lb/day (5,230 kg/day)
Digesters	
Number	2
Diameter	55 ft (16.8 m)
Sidewater depth	33 ft (10.1 m)
Total volume	157,000 cu ft (4,500 m^3)
Sludge disposal	
Sludge lagoons	
Number	2
Volume	350,000 cu ft (9,920 m^3)

[a]Including both old and new tanks

half of the first pass being used for reaeration of the return sludge, and with the primary effluent being step fed in increments to the remainder of the first pass. Fine bubble aeration is employed.

The VCSD plant has consistently nitrified the influent ammonia as shown by the performance data in Table 9-7.[20] The nitrogen figures are from once monthly 24-hour composite samples, while the BOD and suspended solids data is the average of daily composite samples. For the first ten months of 1974, the VCSD wastewater treatment plant averaged 98.6 percent BOD removal and 99.3 suspended solids removal. The ammonia nitrogen concentration in the effluent was typically less than 1 mg/l and has been at this level since August, 1973, apart from a notable process upset caused by an industrial spill in March, 1974. The nitrate-nitrogen concentration in the effluent has generally been around 24 mg/l and this is about 99 percent of the nitrogen in the effluent. For several months before August, 1973, the aeration capability was limited to two on-line blowers which were not enough to sustain complete nitrification. The data given in Table 9-8 shows the change in process performance after mechanical difficulties were overcome and a third blower was started up.[21] Before the aeration capacity was increased, the average ammonia nitrogen concentration in the effluent composite samples was 3.9 mg/l with the nitrate nitrogen level averaging 13.9 mg/l. A dissolved oxygen level of 2 to 4 mg/l is now maintained in the last portion of the aeration tank, whereas before August, 1973, the level was often less than 2 mg/l.

FIGURE 9-5

VALLEY COMMUNITY SERVICES DISTRICT (CALIF.)
WASTEWATER TREATMENT PLANT FLOW DIAGRAM

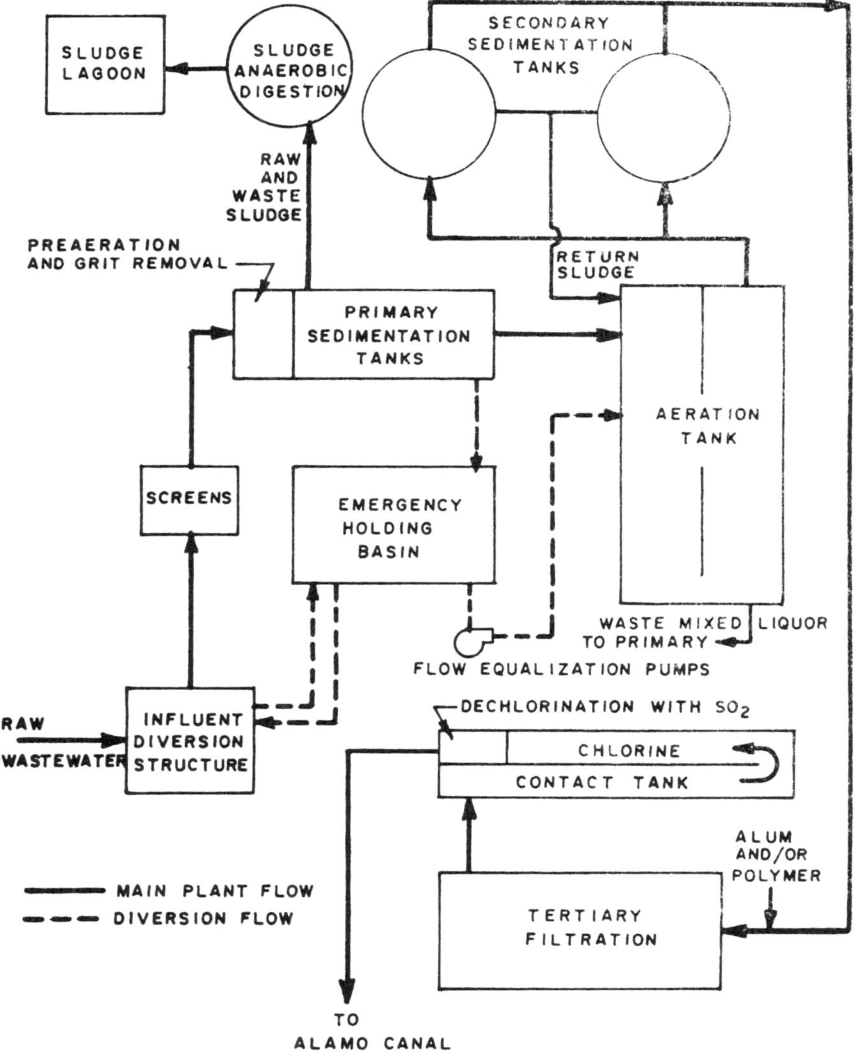

TABLE 9-7

NITRIFICATION PERFORMANCE AT VALLEY COMMUNITY SERVICES DISTRICT WASTEWATER TREATMENT PLANT, CALIFORNIA

Month 1974	Raw waste-water flow, mgd (m^3/sec)	Mixed liquor recycle ratio	Temp. C	MLSS, mg/l	Sludge volume index	Solids retention time, days	BOD_5, mg/l				Suspended solids, mg/l				Nitrogen,[a] mg/l	
							Primary effluent	Secondary effluent	Final effluent	Percent removal	Primary effluent	Secondary effluent	Final effluent	Percent removal	Effluent NH_4^+-N	Effluent NO_3^--N
January	4.09 (0.179)	0.49	na[b]	na	119	10.5	na	na	2.5	na	91	19.2	2.5	99.0	0.23	17.0
February	3.64 (0.159)	0.41	na	5,512	119	11.9	231	10.0	5.6	98.6	87	9.6	1.3	99.5	2.4	21.8
March	3.99 (0.175)	0.46	16	6,130	126	15.2	167	21.0	3.7	98.5	99	8.2	1.7	99.3	16	6.8
April	4.12 (0.181)	0.42	19	1,533	73	8.8	140	14.7	3.8	98.8	86	9.2	1.9	99.1	0.17	24.4
May	3.64 (0.159)	0.42	22	4,759	81	9.6	147	16.0	5.6	96.0	88	9.0	2.8	98.7	0.06	26.6
June	3.76 (0.165)	na	25	7,696	82	10.9	142	8.8	3.0	98.9	84	10.3	1.6	99.4	0.84	24.9
July	3.49 (0.153)	0.55	28	4,811	73	11	106	6.5	3.0	98.4	90	9.4	2.5	99.0	0.06	23.1
August	3.41 (0.149)	0.43	29	5,132	77	10.9	116	8.0	2.5	98.4	84	12.2	1.5	99.4	0.22	21.9
September	3.56 (0.156)	0.45	27	4,771	85	8.1	164	7.7	2.0	99.4	93	15.7	1.4	99.6	0.78	28.9
October	3.55 (0.156)	0.47	26	4,872	103	8.3	183	10.0	1.6	99.4	91.7	10.6	1.3	99.7	0.11	21.9

[a] Results of one 24-hour composite sample per month.

[b] na = not available.

FIGURE 9-6

HOLDING BASIN AT THE VALLEY COMMUNITY SERVICES DISTRICT (CALIFORNIA) WASTEWATER TREATMENT PLANT

TABLE 9-8

NITROGEN ANALYSES ON 24 HOUR COMPOSITE EFFLUENT SAMPLES AT THE VALLEY COMMUNITY SERVICES DISTRICT TREATMENT PLANT

Date sampled, 1973	Nitrate nitrogen as N, mg/l	Ammonia nitrogen as N, mg/l
June[a]	9.9	6.3
July[a]	16.5	2.0
August	9.5	4.7
September[b]	14.0	0.39
October[b]	26.7	<0.06
November[b]	26.0	0.30
December[b]	27.1	<0.06

[a]Two blowers on-line.

[b]Three blowers on-line.

The VCSD staff feels that this type of nitrification system is particularly subject to upset due to toxicants in the primary effluent. One recurring loss of nitrifying ability was ultimately traced to the periodic discharge of a solvent, trichlorethylene.[22] Once the industrial discharger was located and the spills ceased, the problem disappeared.

Operation and maintenance costs have averaged $350 per million gallons since the 1972 additions.

9.5.1.3 Livermore, California

Since 1967, the City of Livermore has operated the 5 mgd (0.22 m^3/sec) Water Reclamation Plant whose flow diagram is shown in Figure 9-7. Before 1967, the original plant included primary treatment, trickling filters, secondary sedimentation, polishing treatment with oxidation ponds and sludge digestion. During the 1967 enlargement, existing structures were rearranged in the flowsheet and additional facilities were added.[23] After the enlargement, the plant consisted of preliminary treatment and primary sedimentation, roughing filters, activated sludge for nitrification, and chlorination. The existing oxidation ponds were converted to emergency holding basins. Sludge disposal is by digestion with digested sludge being applied to sludge lagoons. Drying beds are used intermittently.

The plant layout was oriented to meeting discharge requirements mandating an effluent that contained no more than 20 mg/l of BOD_5, 20 mg/l of SS, and a five-day-median total coliform of 5 MPN per 100 ml. The low bacteriological requirements dictated that the plant

FIGURE 9-7

CITY OF LIVERMORE WATER RECLAMATION PLANT (CALIF.) FLOW DIAGRAM

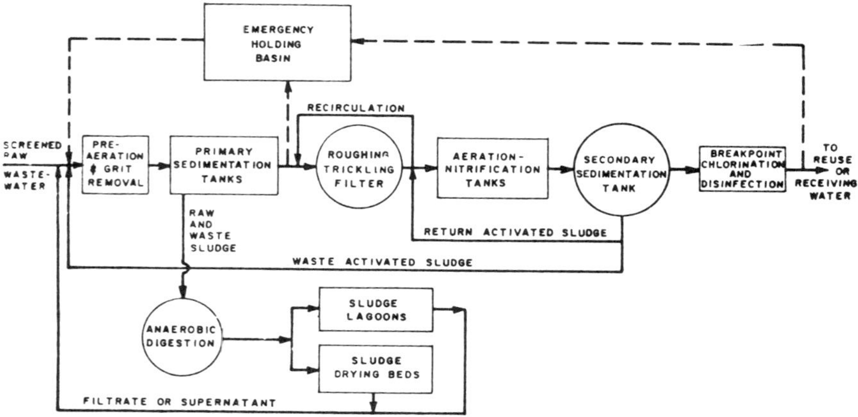

be designed to dependably nitrify on a year-round basis. A free chlorine residual was thought to be required for disinfecting to such low levels and that the ammonia level must be minimized in the secondary effluent so that disinfection through breakpoint chlorination could be economically practiced.

The trickling filters were retained in the plant layout to act as pretreatment for the nitrification step, to reduce organic loads and to moderate any shock loads that might upset nitrification. The sloughed solids from the roughing filter pass directly to the aeration tanks without any intermediate clarification.

Plant design data are shown in Table 9-9 and operating data for the year 1971 in Table 9-10.[23,24] The plant has consistently met effluent requirements and demonstrated stable year-round operation. Wastewater temperatures are favorable for nitrification, with a range of 15 to 24 C. Even when ammonia breakthrough has occurred in the secondary effluent, it has been effectively reduced by breakpoint chlorination. Recently, the plant has not had sufficient aeration capability to completely nitrify during peak nitrogen loan conditions, but this condition is being rectified in plant modifications currently underway. Coarse bubble aeration is used in the plant. A photograph of the aeration tank is shown in Figure 9-8. The activated sludge exhibits very good settling properties with sludge volume index (SVI) measurements consistently below 100 ml/g.

On the few occasions when plant upsets occur, making it likely that bacteriological requirements will not be met, the effluent is directed to the emergency holding basins. The chlorine contact tank has also served as a supplementary settling tank and yielded further reductions of BOD_5 and SS. The chlorine contact tank must be occasionally drained to remove accumulated solids.

Twenty to twenty-five percent of the effluent is used for irrigation purposes at a nearby golf course and on agricultural land. Effluent is also used at the golf course to fill several small lakes.

The initial plant had a construction cost of $900,000 (1957 dollars), which included land purchase, while the cost of the 1967 plant expansion was $1,300,000 (1968 dollars). Operational expenditures for 1971 were approximately $275,000 which, when expressed on a unit basis, is $224 per million gallons treated.

9.5.1.4 San Pablo Sanitary District, California

The San Pablo Sanitary District, California, operates a 12.5 mgd (0.55 m^3/sec) wastewater treatment plant designed for year-round complete nitrification. The original plant consisted of a primary treatment plant with effluent chlorination and digestion for solids processing. Additions completed in 1972 resulted in the plant flow diagram shown in Figure 9-9 and included additional treatment facilities, a new roughing trickling filter, new aeration-nitrification tanks, new secondary clarifiers, an additional chlorine contact tank, new

TABLE 9-9

DESIGN DATA - LIVERMORE WATER RECLAMATION PLANT

Population	62,500
Average dry weather flow (ADWF)	5 mgd (0.22 m^3/sec)
Peak dry weather flow (PDWF)	10 mgd (0.44 m^3/sec)
Peak wet weather flow (PWWF)	18 mgd (0.79 m^3/sec)
Raw wastewater loadings	
BOD_5	12,500 lb/day (5,670 kg/day)
Suspended solids	12,500 lb/day (5,670 kg/day)
Primary treatment	
Preaeration and grit removal tanks	
Number	2
Detention time at ADWF	36 min
Primary sedimentation tanks	
Number	2
Length	124 ft (37.8 m)
Width	19 ft (5.8 m)
Average water depth	9 ft (2.7 m)
Detention time at ADWF	1.5 hr
Overflow rate at ADWF	1,050 gpd/sf (42.8 m^3/m^2/day)
Emergency holding basin	
Volume	31 mil gal (117,000 m^3)
Roughing filters	
Number	2
Diameter	110 ft (33.5 m)
Depth of media	4.25 ft (1.3 m)
Total volume of media	80,152 cf (2,270 m^3)
Media type	Rock
Media size	2 to 4 in.
Recirculation ratio at ADWF	3.0 to 1.0
BOD_5 load	100 lb BOD_5/1,000 cf/day (1.62 kg/m^3/day)
Air blowers	
Number	3
Discharge pressure	7.5 psi (0.53 kgf/cm^2)
Capacity - total	6,000 cfm (170 m^3/min)
Aeration-nitrification tank	
Number	1
Passes per tank	2
Length/pass	160 ft (48.8 m)
Width/pass	30 ft (9.2 m)
Depth	15 ft (4.6 m)
Detention time at ADWF	5.2 hr
BOD_5 load	28 lb BOD_5/1,000 cf/day (0.45 kg/m^3/day)
Secondary sedimentation tank	
Number	1
Diameter	90 ft (27.4 m)
Sidewater depth	12 ft (3.7 m)
Overflow rate at ADWF	767 gpd/sf (31.2 m^3/m^2/day)
Chlorine contact tank (Breakpoint chlorination)	
Number	1
Passes per tank	2
Detention time at ADWF	1 hr
Anaerobic digestion	
Number	2
Volume	27,500 cf (779 m^3)
Loading, total solids	11,800 lb/day (5,350 kg/day)
Volatile matter, percent	75
Sludge disposal	
Digested sludge lagoons	
Number	2
Total volume	320,000 cf (9,060 m^3)
Sludge drying beds	
Number	4
Total area	22,400 sf (2,080 m^2)

dissolved air flotation thickeners, and two new digesters. This flowsheet is very similar to that used at the Livermore plant, described in Section 9.5.1.3, except that plastic media is used in the roughing filter in place of rock media. Design data for the plant are shown in Table 9-11.[25,26]

FIGURE 9-8

AERATION TANK AT THE LIVERMORE WATER RECLAMATION PLANT (CALIFORNIA) WITH ROUGHING TRICKLING FILTERS IN BACKGROUND

TABLE 9-10

NITRIFICATION PERFORMANCE AT
THE LIVERMORE WATER RECLAMATION PLANT

Month 1971	Flow, mgd (m^3/sec)	Recycle[a] ratio	MLSS, mg/l	SVI, ml/g	Θ_c, days	HT, hr	Air use MCF/day (l/sec)	Primary effluent		Roughing filter effluent			Secondary effluent					Final effluent				
								BOD_5, mg/l	SS, mg/l	BOD_5, mg/l	SS, mg/l	NH_4^+-N mg/l	BOD_5, mg/l	SS, mg/l	NH_4^+-N, mg/l	NO_2^--N, mg/l	NO_3^--N, mg/l	BOD_5, mg/l	SS, mg/l	NH_4^+-N, mg/l	Organic N, mg/l	Coliform MPN per 100 ml
January	3.3 (0.14)	0.35	1803	99	4.3	7.8	6.4 (2100)	140	80	129	137	33.5	8.9	20	0.86	0.07	17.8	9.3	19	<0.1	2.0	1.5
February	3.3 (0.14)	0.34	1756	66	4.7	7.8	6.4 (2100)	128	85	79	122	41.0	11.8	17	1.07	0.32	17.6	6.3	13	<0.1	3.1	0.6
March	3.2 (0.14)	0.35	1702	79	4.2	8.1	5.9 (1900)	156	84	121	128	41.9	16.9	24	6.73	0.78	19.3	5.2	16	1.3	1.9	3.4
April	3.2 (0.15)	0.34	1748	82	4.3	8.1	8.0 (2100)	125	74	107	96	33.3	8.7	15	0.94	0.02	18.8	3.1	8	<0.1	2.0	7.0
May	3.4 (0.15)	0.35	1743	91	5.0	7.6	6.3 (2100)	125	98	112	118	31.0	11.6	19	0.76	0.05	20.2	8.5	12	<0.1	<0.1	3.0
June	3.5 (0.15)	0.38	2112	79	5.4	7.4	6.8 (2200)	110	81	74	110	27.1	6.6	24	0.48	0.02	18.5	6.8	12	<0.1	7.8	3.8
July	3.3 (0.14)	0.44	2189	78	7.7	7.8	6.9 (2300)	118	71	81	102	23.3	9.1	19	0.88	0.04	18.3	7.3	9	<0.1	1.0	0.4
August	3.4 (0.15)	0.43	2160	79	7.2	7.6	7.2 (2400)	108	81	46	118	25.2	5.2	31	0.88	0.12	19.4	8.1	17	<0.1	1.3	0.5
September	3.6 (0.16)	0.44	2123	74	8.0	7.2	7.3 (2400)	114	60	82	115	22.5	5.4	19	1.75	0.17	18.2	6.1	11	<0.1	0.4	3.3
October	3.5 (0.15)	0.42	2157	64	9.6	7.4	7.3 (2400)	106	75	66	74	32.6	10.6	18	1.03	0.03	18.2	7.8	15	<0.1	1.7	1.9
November	3.4 (0.15)	0.42	2275	58	5.8	7.6	7.1 (2300)	146	63	115	89	35.6	11.3	31	0.46	0.10	18.9	12.4	12	<0.1	4.0	3.4
December	3.3 (0.14)	0.44	2316	52	5.0	7.8	6.5 (2100)	147	94	104	107	44.4	12.6	27	4.32	0.13	16.6	7.0	17	0.4	1.3	1.7

[a]Return activated sludge

Current discharge requirements are essentially those defined by EPA for municipal secondary treatment plants with additional requirements set on effluent toxicity.[26] In addition, the effluent pH is restricted to the range of 6.5 to 8.5. The acid production from nitrification and subsequent chlorination normally forces the effluent pH below 6.5. This has necessitated the addition of caustic to the final effluent to raise the pH to or above 6.5. Toxicity requirements state that fish bioassays must be run on the undiluted effluent and that 90 percent of a series of 10 consecutive tests must show 70 percent fish survival for 96 hr. Experience at this plant has indicated that the requirements cannot be met without removal of ammonia through nitrification.[26]

A primary design consideration in laying out the plant for nitrification was the presence of a significant volume fraction (11 to 13 percent) of potentially toxic industrial wastes in the influent wastewater. Tank truck washing residues and the waste from a manufacturer of organic peroxide and phenol formaldehyde are the major industrial waste sources. The roughing filter is used in the treatment plant to protect the nitrifying organisms from influent wastewater toxicity. Toxic dumps have caused severe sloughing and loss of growth on the media in the roughing filter, but nitrification remained unaffected.[26]

FIGURE 9-9

SAN PABLO SANITARY DISTRICT TREATMENT PLANT (CALIFORNIA) FLOW DIAGRAM

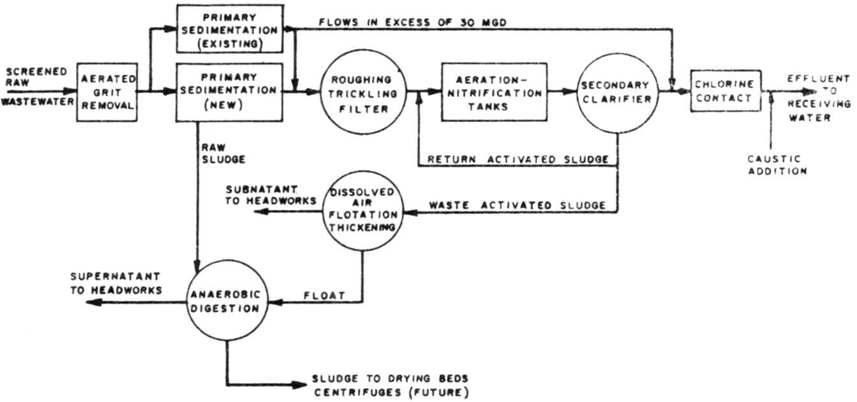

TABLE 9-11

DESIGN DATA, SAN PABLO SANITARY DISTRICT TREATMENT PLANT

Average dry weather flow (ADWF)	12.5 mgd (0.55 m^3/sec)
Peak wet weather flow (PWWF)	30 mgd (1.32 m^3/sec)
Raw wastewater quality	
BOD_5	340 mg/l
SS	300 mg/l
Primary sedimentation tanks	
New (Dry and wet weather use)	
Number	2
Diameter	70 ft (21.3 m)
Sidewater depth	10 ft (3.1 m)
Detention time (ADWF)	2 hr
Overflow rate (ADWF)	1624 gpd/sf(66/ m^3/ m^2/day)
Existing (Wet weather use only)	
Number	1
Diameter	100 ft (30.5 m)
Sidewater depth	10 ft (3.1 m)
Roughing filter	
Number	1
Diameter	52 ft (15.6 m)
Media type	Plastic corrugated sheet modules
Depth of media	18 ft (5.5 m)
Volume of media	38,200 cf (1080 m^3)
Media specific surface	29 sf/cu ft (95 m^2/m^3)
Recirculation ratio	2.4 to 1.0
BOD_5 load	350 lb/1000 cf/day (5.6 kg/m^3/day)
Air blowers	
Number	4
Discharge pressure	6.5 psig (0.46 kgf/cm^2)
Capacity - total	24,000 cfm (629 m^3/min)
Aeration-nitrification tanks	
Number	2
Passes/tank	1
Length/pass	252 ft (76.9 m)
Width/pass	50 ft (15.2 m)
Depth	15 ft (4.6 m)
Volume (total)	378,000 cu ft(10,700 m^3)
Detention time (ADWF)	5.4 hr
Secondary sedimentation tanks	
Number	2
Length	180 ft (54.9 m)
Width	60 ft (18.2 m)
Depth	8 ft (2.4 m)
Overflow rate at ADWF	580 gpd/sf(23.6 m^3/m^2/ day)
at PDWF	1390 gpd/sf(56.6 m^3/m^2/day)
Detention time at ADWF	2.5 hr
Chlorine contact chambers	
Number	2
Passes/tank	2
Length/pass	110 ft (33.5 m)
Width/pass	15 ft (4.6 m)
Depth	9 ft (2.7 m)
Detention ADWF	0.85 hr
Caustic addition (Na(OH))	
Dose	20 mg/l
Waste activated sludge thickening	
Number	1
Diameter	35 ft (10.7 m)
Sidewater depth	8 ft (2.4 m)
Solids loading	48 lb ds/sf/day (235 kg/m^2/day)
Anaerobic Digestion	
Number	4
Volume - total	367,000 cu ft (10,400 m^3)
Sludge drying bed	
Surface area	158,000 sf (14,700 m^2)

Consistent year-round nitrification is obtained in this plant as is shown in Table 9-12. While only once-monthly analyses of nitrogen are shown in Table 9-12, the consistency of nitrification in the plant is confirmed by daily ammonia nitrogen analyses of grab samples which normally show less than 0.2 mg/l ammonia nitrogen. Wastewater temperatures are favorable for nitrification, typically dropping to only 17 C, the average monthly temperature in January.

The treatment plant is currently operating at only one-half of its design capacity. To test the nitrification portion of the system at close to its design condition, one of the two aeration tanks was taken out of service during May, June and July of 1974. Both secondary sedimentation tanks remained in service during this period. Operating conditions and plant performance for this period are summarized in Tables 9-13 and 9-14 respectively.[27] Full nitrification was maintained throughout this special test period.

Examination of Table 9-14 provides some insight into the operation of the roughing filter. Total BOD_5 and total COD remained relatively unaffected by the roughing filter operation. Evidence of treatment, however, is provided by the soluble BOD and soluble COD tests and the total suspended solids (TSS) determinations. A reduction in soluble BOD_5 and soluble COD occurred coincidentally with an increase in TSS. This indicates organic removal with associated growth of biomass. In this plant's operation, the roughing filter converts influent organic matter to biological organisms. Subsequent treatment in the aeration-nitrification tank provides further oxidation and nitrification.

Construction cost of the added facilities totaled $4,900,000, including legal, engineering and contract administration. The contract was awarded in October 1970 and construction essentially completed by October, 1972. Treatment plant operating costs totaled $397,500 for fiscal year 1973/1974, during which time a total of 2,600 million gallons (9.8 million m^3) were processed through the secondary treatment facilities. On a unit basis, treatment plant O & M costs total $153/mil gal for fiscal year 1973/1974.[26]

9.5.2 Case Examples of Nitrification-Denitrification for Nitrogen Removal

Four examples of the incorporation of biological nitrification-denitrification in treatment plants for nitrogen removal are presented in this section. The Central Contra Costa Sanitary District's plant design is oriented towards reuse of the reclaimed water by nearby industries as well as meeting discharge requirements. The designs of the Canberra, Australia, Washington, D.C., and El Lago, Texas plants are laid out so that nitrogen and phosphorus are removed to protect the receiving waters.

9.5.2.1 Central Contra Costa Sanitary District, California

The Central Contra Costa Sanitary District (CCCSD) has under construction a new 30 mgd (1.31 m^3/sec) Water Reclamation Plant near Concord, California. Due to go on-line in 1976, the plant is designed to produce water for sale to the Contra Costa County Water District

TABLE 9-12

NITRIFICATION PERFORMANCE AT THE SAN PABLO SANITARY DISTRICT TREATMENT PLANT

Month	Year	Flow, mgd (m^3/sec)	Return sludge ratio	Temp. C	MLSS (% volatile)	SVI, ml/gram	Θc, days	HT, hr	Air use MCF/day (1/sec)	Roughing filter effluent			Secondary effluent						
										BOD_5[b] mg/l	COD[b] mg/l	SS[b], mg/l	BOD_5[c], mg/l	COD[c], mg/l	SS[c], mg/l	Organic-N[d], mg/l	NH_4^+-N[c], mg/l	NO_3^--N[d], mg/l	NO_2^--N[c], mg/l
7	1973	6.3 (0.28)	0.30	22.2	1403 (79)	73	12.3	10.8	na[a]	121	279	8.2	16	78	8	7.8	<0.2	18.6	0.05
8	1973	5.6 (0.25)	0.34	23.6	1497 (80)	68	13.9	12.1	na	125	306	97	4	56	7	3.8	<0.2	19.8	0.02
9	1973	5.7 (0.25)	0.25	24.4	1758 (76)	48	15.4	11.9	20.0	134	283	67	6	55	8	3.1	<0.2	18.2	0.02
10	1973	5.9 (0.26)	0.31	23.2	1765 (76)	66	11.8	11.5	19.4	131	281	95	6	62	6	3.4	<0.2	20.4	0.02
11	1973	9.7 (0.43)	0.24	20.3	1609 (72)	73	9.1	7.0	19.4	88	212	59	7	54	3	3.4	0.7	24.8	0.26
12	1973	8.4 (0.37)	0.29	18.5	1545 (73)	85	12.1	8.1	20.0	81	214	73	4	59	7	4.8	0.1	17.6	0.05
1	1974	9.1 (0.40)	0.22	17.0	1481 (74)	87	7.4	7.5	na	91	249	96	3	53	4	4.8	<0.2	17.0	0.22
2	1974	6.6 (0.29)	0.29	17.8	1415 (78)	96	9.2	10.3	na	92	240	79	4	53	4	2.2	<0.2	15.8	0.02
3	1974	8.3 (0.36)	0.19	18.2	1522 (74)	101	7.5	8.2	18.3	107	314	135	4	61	6	2.8	<0.2	11.8	0.09
4	1974	7.4 (0.32)	0.23	19.0	1581 (70)	69	6.8	9.2	na	111	247	95	5	54	7	4.8	<0.2	11.2	0.10
5	1974	6.2 (0.27)	0.44	22.0	e	82	e	e	na	140	332	130	3	50	4	5.1	<0.2	18.7	0.06
6	1974	6.2 (0.27)	0.44	22.0	2833 (78)	77	6.0	5.5	15.3	123	327	118	2	46	4	3.6	0.3	19.0	0.05

[a] na = not available

[b] grab sample at peak flow each week day

[c] composite sample, once per week

[d] composite sample, once per month

[e] on May 7, one aeration tank taken out of service

TABLE 9-13

AVERAGE PROCESS LOADING CONDITIONS AT THE SAN PABLO SANITARY DISTRICT TREATMENT PLANT DURING SPECIAL TEST, MAY 19TH TO JULY 8TH, 1974

Flow, mgd (m^3/sec)	6.30(0.28)
Temperature, C	23.0
Roughing filter	
BOD loading,	
lb BOD_5/1,000 cf/day (kg/m^3/day)	199 (3.19)
Hydraulic loading	
gpm/sf (m^3/m^2/min)	2.1 (0.086)
Aeration-nitrification tanks	
MLSS, mg/l	3070
Percent volatile	78.2
SVI, ml/gram	80.2
Average detention time, hr	5.4
BOD load	
lb BOD_5/1,000 cf/day (kg/m^3/day)	35.8 (0.57)
lb BOD_5/lb MLVSS/day (kg/kg/day)	0.24 (0.24)
Solids retention time, days	6.6
Secondary sedimentation tanks	
Average overflow rate, gpd/sf (m^3/m^2/day)	292 (11.89)
Average return ratio	0.48
Average solids load, lb/sf/day (kg/m^2/day)	11.0 (53.7)
Return activated sludge, mg/l	6835

TABLE 9-14

PERFORMANCE SUMMARY FOR THE SAN PABLO SANITARY DISTRICT TREATMENT PLANT DURING SPECIAL TESTING, MAY 19TH TO JULY 8TH, 1974

Characteristic	Raw wastewater[a]	Primary effluent	Roughing filter effluent[a]	Secondary effluent[a]
Total BOD_5, mg/l	220	145	129	3.3
Soluble BOD_5, mg/l	na[b]	97.5	52.8	na
Total COD, mg/l	na	322	334	47.5
Soluble COD, mg/l	na	191	137	40.7
TSS, mg/l	191	86.8	121	4.9
Ammonia-N	na	na	19.8	<0.2
Nitrate-N	na	na	na	19.0
Nitrite-N	na	na	na	0.04[c]

[a]Composite sample each weekday, except as indicated

[b]na = not available

[c]Grab sample at peak flow

which will resell the water to five large industries for cooling and process water. The reclamation contract calls for production of a water containing less than 10 mg/l of BOD_5, 1 mg/l total phosphorus and 5 mg/l total nitrogen.[1]

The liquid processing flowsheet for the CCCSD Water Reclamation Plant is shown in Figure 9-10 and design data are in Table 9-15.[29] Primary treatment follows lime addition and preaeration and is followed with a separate stage nitrification step. The use of lime in the primary treatment stage removes the bulk of the organic carbon before nitrification, resulting in a very stable oxidation of ammonia to nitrate. Addition of lime also enhances the removal of organic nitrogen, phosphorus, heavy metals and viruses. Biological denitrification follows nitrification, converting nitrate to nitrogen gas. Multimedia filtration will also be provided prior to distribution of reclaimed water to industry. Not shown in Figure 9-10 is a 140 million gallon (530,000 m^3) storage basin that can be used to store primary effluent to reduce peak wet weather loads on the nitrification and denitrification units. Also, the filtration facility has been provided with a 5 million gallon (18,900 m^3) storage basin for equalizing filter influent and a 30 million gallon (114,000 m^3) clear well for storage of filtered water before pumping it into the distribution system.

There is no intervening recarbonation stage between the primary clarification and nitrification stages. External carbon dioxide (CO_2) is added directly to the first pass of the aeration-nitrification tanks as needed. External requirements are minimal as the chief source of CO_2 is not the external source, but the CO_2 generated in the process. Carbon dioxide is derived from the oxidation of both organic carbon and ammonia. The in-process CO_2 generation capability is an example of lime clarification-nitrification compatibility. The nitrification pH is also kept in the 7.0 to 8.5 range, which is optimal for nitrification.

Based upon tests by the City of Milwaukee in the 1960's[30,31] and testing performed at the South Eastern Purification Plant in Melbourne, Australia,[32] the decision was made to use flat porous plates arranged uniformly over the bottom of the aeration-nitrification tanks. This method of fine bubble aeration gives an oxygen transfer efficiency of between 14 and 20 percent under standard conditions. The porous plates are 14 in. (36 cm) in diameter by 1¼ in. (3.2 cm) thick and are secured in polypropylene holders as shown in Figure 9-11. Thirty plates are arranged in a single precast panel; there are 40 panels per pass. Each panel has an inverted channel shape and is grouted to the tank. By using an inverted channel shape, the channel forms an air duct with the bottom slab of the aeration-nitrification tanks. Each channel has a manual drain so that each pass may be drained during start-up and shutdown operations. Four panels are fed by each downcomer pipe, thereby allowing throttling of the downcomer pipes and a tapered aeration operation. Air for the nitrification tanks, channel aeration, and preaeration is provided by two 60,000 scfm (1,698 m^3/min) steam turbine driven single-stage centrifugal blowers and is filtered in large compartment-type bag filters, to achieve a particulate concentration of less than 0.09 mg/1,000 cu ft (0.32 mg/100 m^3) of air.

FIGURE 9-10

LIQUID PROCESS FLOW SHEET - CCCSD WATER RECLAMATION PLANT (CALIF.)

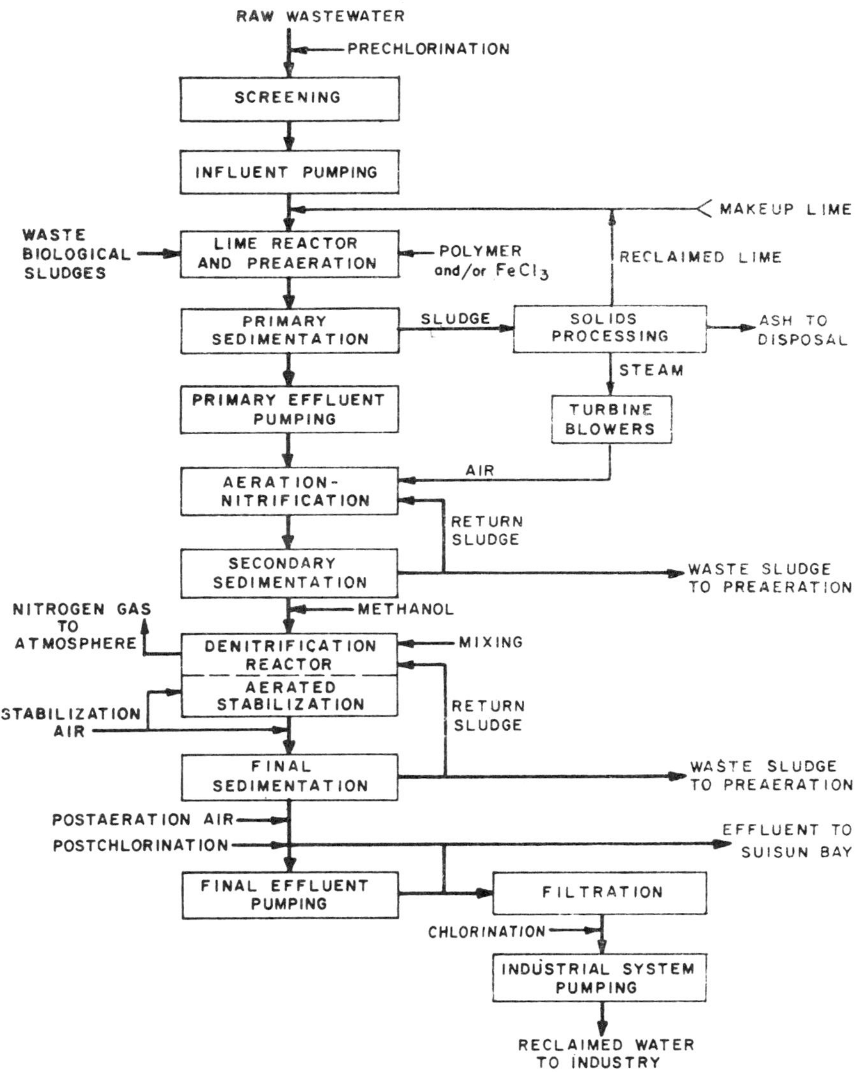

TABLE 9-15

CENTRAL CONTRA COSTA SANITARY DISTRICT
WATER RECLAMATION PLANT - DESIGN DATA

Population	310,000
Average dry weather flow (ADWF)	30 mgd (1.31 m^3/sec)
Peak dry weather flow (PDWF)	48 mgd (2.10 m^3/sec)
Peak wet weather flow (PWWF)	140 mgd (6.13 m^3/sec)
Raw wastewater quality at ADWF	
BOD_5	216 mg/l
SS	240 mg/l
TKN	30 mg/l
Total phosphorus as P	11 mg/l
Primary solids separation	
Chemical addition	
Lime dose as CaO	303 mg/l
Ferric chloride dose as $FeCl_3$	14 mg/l
pH	11.0
Preaeration, flocculation, grit removal tanks	
Number (one existing)	2
Volume, ea.	27,700 cu ft (785 m^3)
Detention time, ADWF	30 min
Primary sedimentation tanks	
Number (two existing)	4
Length	254 ft (77.4 m)
Width	38 ft (11.6 m)
Average depth	9.5 ft (2.9 m)
Overflow rate at ADWF	780 gpd/sf (31.8 m^3/m^2/day)
Detention time at ADWF	2.2 hr
Air blowers - steam turbine driven	
Number	2
Discharge pressure	8.0 psig (0.56 kgf/cm^2)
Capacity, ea.	60,000 cfm
Horsepower/ea. turbine	2,750
Aeration-nitrification tanks	
Number	2
Passes per tank	4
Length/pass	270 ft (82.3 m)
Width/pass	35 ft (10.7 m)
Depth	15 ft (4.6 m)
Detention time, ADWF	6.8 hr
Secondary sedimentation tanks	
Number	4
Diameter	115 ft (35.1 m)
Sidewater depth	16 ft (4.9 m)
Overflow rate, ADWF[a]	720 gpd/sf (29.3 m^3/m^2/day)
Denitrification tanks (anoxic contact) and aerobic stabilization)	
Number	2
Length/tank	315 ft (96.0 m)
Width/tank	30 ft (9.1 m)
Depth	15 ft (4.6 m)
Detention time, ADWF	102 min
Reactors/tank	9
Reactors used for stabilization (max)	4
Methanol to Nitrate - N ratio	3.0

TABLE 9-15

CENTRAL CONTRA COSTA SANITARY DISTRICT WATER RECLAMATION PLANT - DESIGN DATA (CONTINUED)

Final sedimentation tanks	
Number	4
Diameter	115 ft (35.1 m)
Sidewater depth	20 ft (6.1 m)
Overflow rate, ADWF[a]	720 gpd/sf (29.3 m^3/m^2/day)
Effluent filtration[b]	
Number	4
Total rated capacity	36 mgd (1.46 m^3/sec)
Total hydraulic capacity	54 mgd (2.20 m^3/sec)
Media depth	
Anthracite	2 ft (0.61 m)
Sand	1 ft (0.30 m)
Filtration rate at rated capacity	4.0 gpm/sf (2.72 l/m^2/sec)
Water backwash rate, max	25 gpm/sf (17.0 l/m^2/sec)
Surface wash rate	0.75 gpm/sf (0.51 l/m^2/sec)
Solids disposal and Lime reclamation	
Sludge thickening	
Primary sludge thickener (converted digester)	
Number	1
Diameter	62 ft (18.9 m)
Sludge to thickener	243,000 lb/day (110,500 kg/day)
Solids loading	80 lb DS/sf/day (390 kg/m^2/day)
Centrate thickener (converted digester)	
Number	1
Diameter	62 ft (18.9 m)
Solids loading	42 lb DS/sf/day (205 kg/m^2/day)
Centrifugation	
Number	4
Type	vertical, solid bowl
Max feed rate	260 gpm (0.016 m^3/sec)
Max g force, G	3,100
Cake solids conc., first stage	55 percent
Cake solids conc., second stage	14 percent
Furnaces	
Number	2
Diameter of hearth	22 ft (6.7 m)
Number of hearths	11
Rated capacity	
Sludge burning duty	70,000 lb DS/day (32,000 kg/day)
Recalcination duty	150,000 lb DS/day (68,000 kg/day)
Recycled lime fraction	60 - 65 percent

[a] Loading applied to sedimentation tanks at PWWF can be equalized by a primary effluent holding basin

[b] Data on filtration from reference 28

Nitrified mixed liquor flows to four circular sedimentation tanks. The tanks have a center-feed, peripheral discharge arrangement with sludge removal by vacuum-type sludge collectors. Sludge return rate is controlled by the blanket level in each secondary sedimentation tank. Solids retention time is controlled by use of waste activated sludge flow meters and return sludge concentration.

FIGURE 9-11

NITRIFICATION-DENITRIFICATION SYSTEM AT THE CCCSD WATER RECLAMATION PLANT (CALIFORNIA)

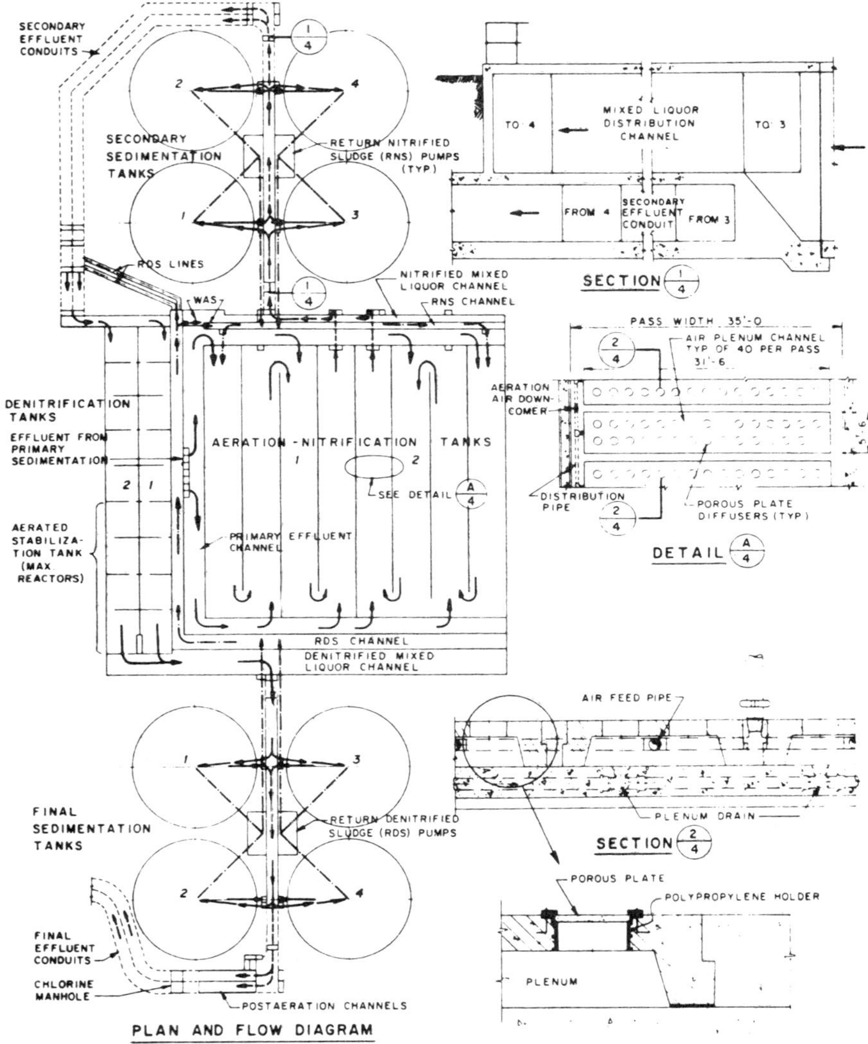

Suspended growth denitrification is used for nitrate removal. Uncovered reactors are employed, as initial testing showed them to be acceptable.[1] Two parallel denitrification tanks consist of nine completely mixed reactors in series. This arrangement allows approximation of a plug flow hydraulic regime. Each reactor is equipped with a low-shear turbine type mixer to keep the mixed liquor solids in suspension. The last four reactors in the series are equipped with spargers for aeration. With this arrangement, the first five cells in the series can be used for anoxic denitrification reactors and the last four cells can be used for denitrification or aerated stabilization depending on whether or not aeration is used. The volume devoted to each function can be varied to meet seasonal loads and temperatures. The anoxic reactors are designed to operate at 0.1 lb NO_3 –N rem./lb MLVSS/day at MLSS levels ranging from 3000 to 4000 mg/l. The gates between the last five reactors allow positive prevention of backmixing of oxygen between the reactors selected for denitrification and the reactors selected for aerated stabilization. The gates are open at the top to prevent trapping of floating solids in any reactor. The denitrified mixed liquor channel between the denitrification tanks and the final sedimentation tanks is aerated for further stabilization. Final sedimentation tanks are similar to the secondary sedimentation tanks.

All waste sludge produced in the CCCSD Water Reclamation Plant is eventually cycled into the primary sedimentation tanks and appears in the primary underflow. Sources of sludge include the suspended solids associated with the raw wastewater, the solids that are wasted to the primary sedimentation tanks from the subsequent biological treatment stages, and the inorganic sludges that are precipitated due to chemical action.

To maintain a pH of 11.0 in the primary sedimentation tank, large quantities of lime, approximately 400 mg/l as $Ca(OH)_2$, must be used. The need for such a large dosage predicates the economical recovery and reuse of the lime. The solids flowsheet is shown in Figure 9-12. The heart of the system is a two-stage centrifuge process using vertical solid-bowl centrifuges, where primary sludge is separated into two components, sludge cake rich in calcium carbonate ($CaCO_3$) and centrate containing most of the organic material, magnesium and phosphorus. The first stage centrifuge cake is approximately 70 percent $CaCO_3$. This cake is recalcined in a multiple hearth furnace, subsequently dry classified and the lime returned to storage. The lime recovery is expected to be approximately 60-65 percent of the lime used.[33]

The first stage centrate is thickened before being clarified in centrifuges identical to those used for primary sludge classification. The resulting cake is reduced in a multiple-hearth furnace (MHF) to a sterile ash which is used for landfill. The MHF is identical to that used for recalcining.

Energy in the hot off-gases from each MHF is reclaimed via waste heater boilers. Recovered steam is used to run the turbines which power the aeration blowers.

FIGURE 9-12

SOLIDS FLOW DIAGRAM AT THE CCCSD WATER RECLAMATION PLANT (CALIF.)

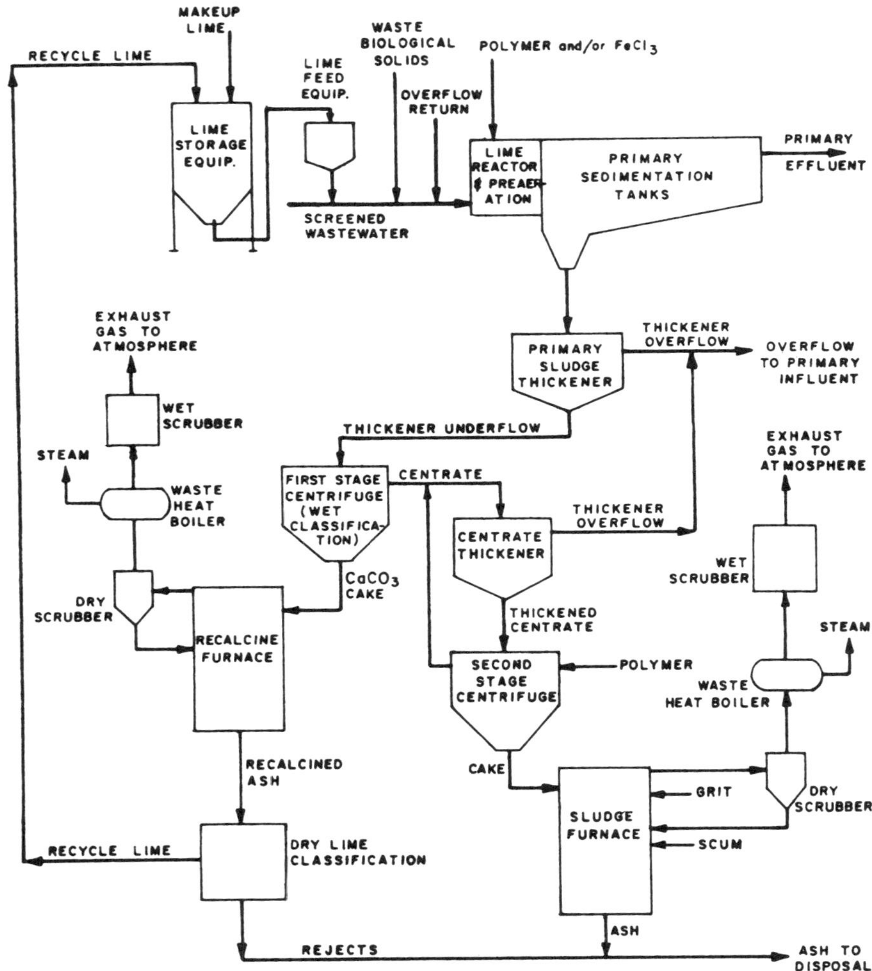

Operational control of the CCCSD Water Reclamation Plant will be divided into four areas, each of which will be manned by a senior operator, operator, and various maintenance men during each shift. The four areas are: (1) primary treatment, (2) biological treatment, (3) filtration, and (4) solids handling and conditioning. Because of the highly automated control system, cathode ray tubes (CRT) are provided in each operator control room, eliminating the need for lighted control panels. In addition to the main CRT's in the computer room, CRT's are located throughout the plant for data monitoring. Because of the very exacting standards required of industrial water, this plant is designed to prevent potential plant upsets by incorporating a direct digital control (DDC) dual-computer system to monitor and control all process functions.[34] Operation of the on-line computer is by a "management by exception" basis which means that as long as the process status is within normal limits, information is not printed, displayed, or needlessly alarmed.

While this plant will not go on-line until 1976, the process was rigorously monitored in a full-scale test at the existing CCCSD plant for a 23-month period. Portions of the test data are summarized in Section 5.2.4.2.

The construction contract for the first phase of work, excluding effluent filtration, totaled $47,000,000 and was let in mid-1972. The construction contract for the effluent filtration and appurtenances totaled approximately $14,000,000 and was let in the fall of 1974. Operation and maintenance costs were estimated for fiscal year 1976/1977 at $300/mil gal based on an average dry weather flow of 30 mgd.

9.5.2.2 Canberra, Australia

The Lower Molonglo Water Quality Control Centre (LMWQCC), an advanced wastewater treatment plant under construction, is designed to serve the City of Canberra, Australia's national capital. Present discharges from Canberra's existing plants are causing algal growth problems in the receiving water and in a downstream reservoir. To circumvent these problems, it was established that the LMWQCC should produce an effluent that contains:[35]

1. Substantially no settleable solids, turbidity, color or odor.

2. BOD_5 and suspended solids concentrations of less than 5.0 mg/l.

3. A median fecal coliform content less than 50 per 100 ml and a 90 percentile value of less than 400 per 100 ml.

4. Total nitrogen not exceeding 2.0 mg/l as N and total phosphorus not exceeding 0.15 mg/l as P.

5. Detergent concentrations less than 0.5 mg/l of MBAS.

6. No substances toxic to aquatic biota.

Unit processes employed at the LMWQCC include raw wastewater screening, lime addition, grit removal, flocculation, primary sedimentation, nitrification, secondary solids separation, denitrification, effluent filtration, effluent disinfection by chlorination, and dechlorination prior to discharge. Figure 9-13 is the process flow diagram for the treatment of the liquid fraction. Design data adopted and used for the LMWQCC is presented in Table 9-16.[36] The solids processing flowsheet is very similar to that shown in the case history for the Central Contra Costa Sanitary District's Water Reclamation Plant in Section 9.5.2.1 and will not be duplicated here.

Very steep site conditions and confinement of the plant to a limited area mandated an unusual arrangement of treatment structures. An example is the nitrification tanks which step down the hillside as shown in Figure 9-14. A total of 4 parallel tanks are provided. Each tank is subdivided into 8 compartments. A very close approach to plug flow is provided by these tanks since backmixing is prevented because the only way mixed liquor can pass along the tank is by overflowing weirs between compartments. The plug flow arrangement is the recommended process configuration for separate stage nitrification when very low residual ammonia nitrogen levels are required. Aeration air is provided to each compartment through porous plate diffusers spread across the tank floor. Diffuser arrangement is very similar to that described in Section 9.5.2.1 for the CCCSD plant. Carbon dioxide (in furnace stack gas) is added to the first two compartments on a continuous basis according to pH level in the nitrification tanks.

Due to site restrictions, an attached growth reactor system was chosen for the denitrification unit. The reactor chosen was specifically developed for this project and is the nitrogen gas filled denitrification column described in Section 5.3.2.1. Design details of the column are shown in Figure 5-7.

Bids for the LMWQCC were received in February, 1974 and the winning tender was \$A 27,000,000 (\$US 35,600,000). The plant is expected to be completed by late 1976. O&M costs calculated in February 1974, exclusive of amortization, were estimated at the equivalent of \$US 266/mil gal.[36]

9.5.2.3 Washington, D.C.

In 1969, regulatory agencies established stringent effluent standards for treatment plants that discharge into the Potomac River in the vicinity of Washington, D.C. These standards required upgrading the Washington, D.C. Blue Plains Plant to provide phosphorus and nitrogen removal as well as improved BOD and SS removals.[37,38,39,40] In early 1975 the EPA announced that the construction of the denitrification portion of the Blue Plains plant would be delayed for two years. This decision came after study of the energy and construction costs for the facility. During the postponement period, water quality improvement due to phosphorus removal and other treatment will be evaluated to determine if denitrification is necessary to achieve eutrophication control goals.[41] The

FIGURE 9-13

PROCESS FLOW DIAGRAM FOR THE LOWER MOLONGLO WATER QUALITY CONTROL CENTRE (CANBERRA, AUSTRALIA)

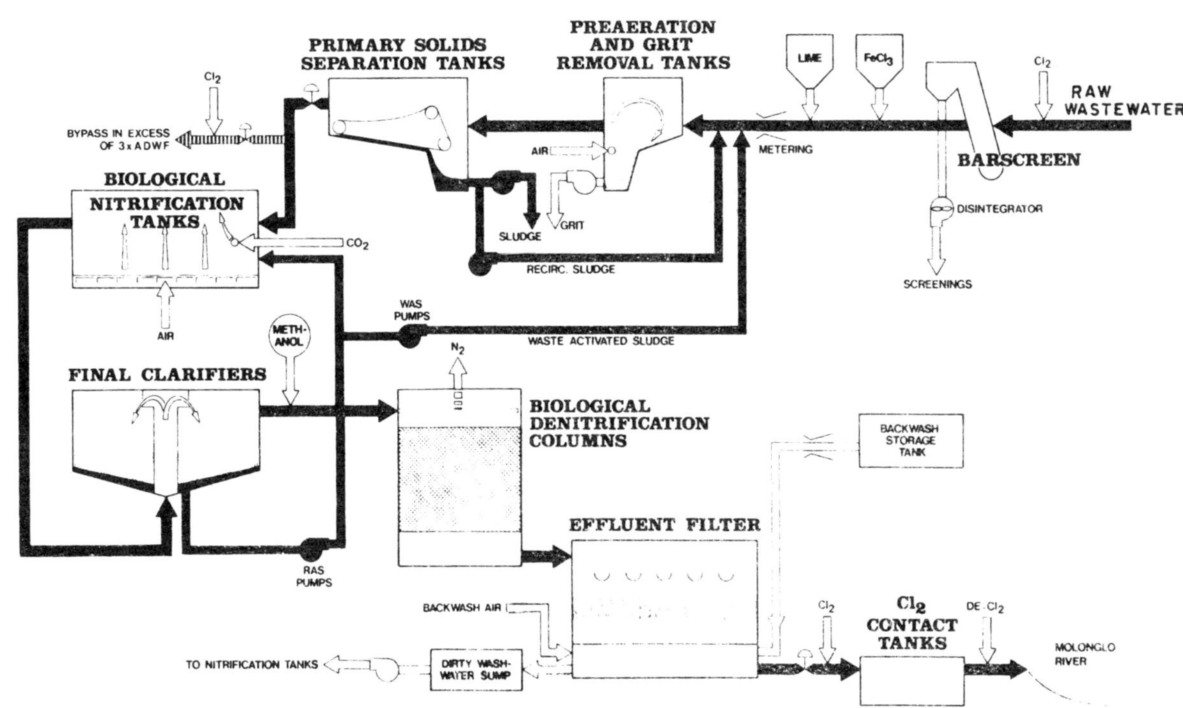

TABLE 9-16

LOWER MOLONGLO WATER QUALITY CONTROL CENTRE (AUSTRALIA), DESIGN DATA

Population	269,000
Average dry weather flow (ADWF)	28.8 mgd (1.26 m^3/sec)
Peak dry weather flow (PDWF)	43.3 mgd (1.90 m^3/sec)
Peak wet weather flow (PWWF)	144.0 mgd (6.30 m^3/sec)
Raw wastewater quality at ADWF	
BOD_5	221 mg/l
SS	242 mg/l
NH_4^+-N	35 mg/l
TKN	59 mg/l
Primary solids separation	
Chemical addition	
Lime dose as CaO	230 mg/l
Ferric chloride dose	20 mg/l
pH	11.0
Flocculation and grit removal tanks	
Number	2
Volume (each)	27,700 cu ft (785 m^3)
Detention time (ADWF)	20 min
Primary sedimentation tanks	
Number	4
Length	212 ft (64.7 m)
Width	38.4 ft (11.7 m)
Depth	8.9 ft (2.7 m)
Overflow rate at ADWF	880 gpd/sf (35.7 m^3/m^2/day)
Detention time at ADWF	1.8 hr
Air blowers	
Number	3
Discharge pressure	7.5 psig (0.53 kgf/cm^2)
Capacity - total	90,000 cfm (2,550 m^3/min)
Biological nitrification tanks	
Number of tanks	4
Compartments per tank	8
Width	35 ft (10.5 m)
Length/compartment	39.4 ft (12 m)
Depth	16.2 ft (4.95 m)
Volume (4 tanks)	565 cu ft (19,960 m^3)
CO_2 required	3,600 lb/day (13,600 kg/day)
Detention time at ADWF	4.4 hr
BOD_5 load	0.18 lb BOD_5/lb MLVSS/day
Final Sedimentation tanks	
Number	5
Diameter	120 ft (36.6 m)
Sidewater depth	20.6 ft (3.28 m)
Overflow rate at 3 x ADWF	1,635 gpd/sf (66.6 m^3/m^2/day)
Biological denitrification columns	
Number of cells	8
Width/cell	32.2 ft (9.8 m)
Length/cell	40.2 ft (12.3 m)
Media depth	20 ft (5.8 m)
Total media volume	196,750 cu ft (5,572 m^3)
Application rate	146 gal/cu ft/day (19.0 m^3/m^3/day)
Methanol to nitrate - N ratio[a]	2.8
Effluent filtration	
Number	4
Width	23.75 ft (7.24 m)
Length	105 ft (32.00 m)
Media depth	
Anthracite	3.5 ft (1.06 m)
Sand	1.5 ft (0.45 m)
Filtration rate at 3 x ADWF	6.0 gpm/sf (4.1 l/m^2/sec)
Water backwash rate	18.6 gpm/sf (12.65 l/m^2/sec)

TABLE 9-16

LOWER MOLONGLO WATER QUALITY CONTROL CENTRE (AUSTRALIA), DESIGN DATA (CONTINUED)

Effluent chlorination-dechlorination	
Number of tanks	1
Contact time at ADWF	50 min
Solids disposal and lime reclamation	
Primary underflow solids	198,450 lb/day (90,000 kg/day)
First stage centrifugation (classification)	
Number of centrifuges	1
Calcium carbonate recovery	90 percent
Cake solids contraction	50-60 percent
Total solids capture	60 percent
Second stage centrifugation (clarification)	
Number of centrifuges	3
Cake solids concentration	12-18 percent
Total solids capture, minimum	70 percent
Furnaces	
Number	2
Diameter of hearth	22 ft (6.7 m)
Number of hearths	9
Rated capacity	
Sludge burning duty	70,400 lb DS/day (32,000 kg/day)
Recalcination duty	237,600 lb DS/day (108,000 kg/day)
Reclaimed lime to storage	47,960 lb/day (21,800 kg/day)
Recycled lime fraction	68 percent

[a]Assumed influent nitrate nitrogen: 28 mg/l

discussion which follows includes the design of the denitrification facilities for this plant as it provides an example of how denitrification can be incorporated in a large plant. Most of the discussion in this section is drawn from the *Process Design Manual for Upgrading Wastewater Treatment Plants*, an EPA Technology Transfer publication.[37]

In 1969, very little performance data were available on the alternative phosphorus and nitrogen removal methods that might be used in this situation. Through the cooperation of the Joint EPA-DC Pilot Plant, it was possible to pilot and evaluate several alternative nutrient removal treatment sequences. Based on these studies, two-point addition of a metal salt was selected for phosphorus removal, and it was determined that nitrogen removal would be best achieved through biological nitrification and denitrification processes. The pilot studies also indicated that to consistently meet the established effluent standards, multimedia filtration was required. Anticipated performance data for the upgraded plant are presented in Table 9-17. Figures 9-15, 9-16 and 9-17 show, respectively, the flow diagrams for the primary and secondary systems, nitrification and denitrification systems and filtration and disinfection systems of the plant.

The existing secondary system consists of four aeration tanks and 12 sedimentation units. To handle the anticipated increase in plant design flow from 240 mgd to 309 mgd, (10.5 m^3/sec to 13.6 m^3/sec) the existing secondary system will be enlarged with two additional

FIGURE 9-14

SECTION THROUGH NITRIFICATION TANKS AT THE LMWQCC, CANBERRA, AUSTRALIA

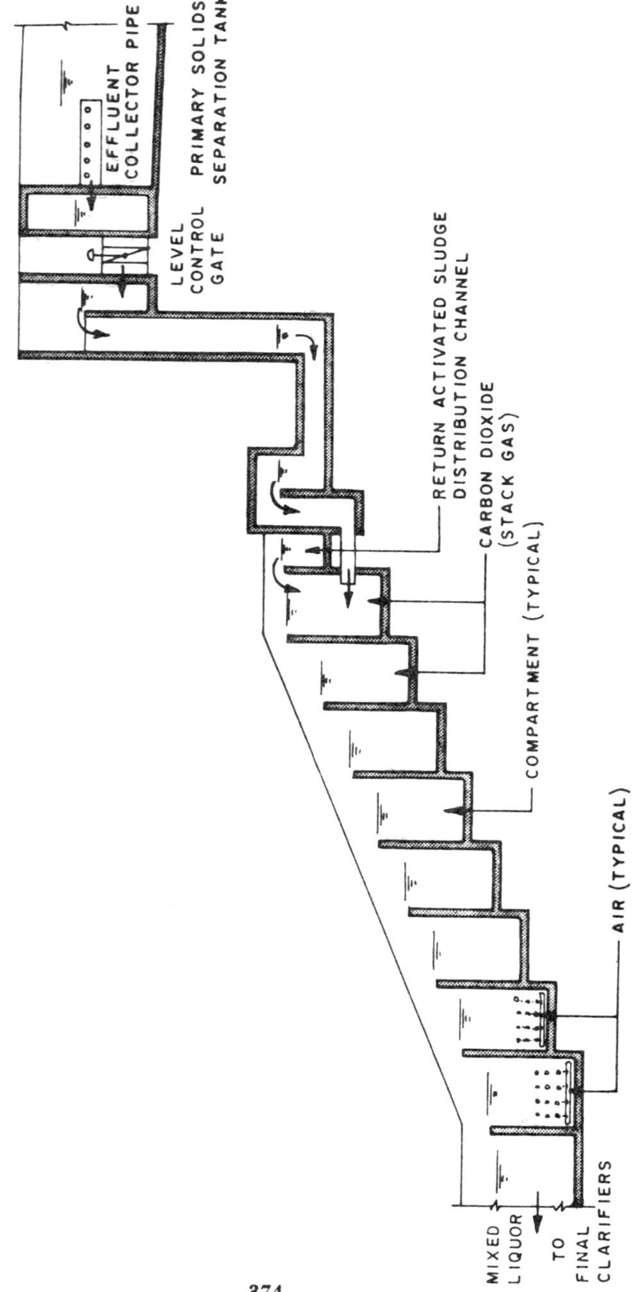

TABLE 9-17

ANTICIPATED PERFORMANCE DATA AND EFFLUENT STANDARDS - BLUE PLAINS PLANT (REFERENCE 39)

	Influent	Secondary Effluent	Nitrification Effluent	Denitrification Effluent	Filtration Effluent	Effluent Standard
BOD, mg/l	206	35	10	6	4	5
Total phosphorus, mg/l	8.4	2.0	1.0	0.5	0.2	0.22
Nitrogen:						
Organic-N, mg/l	8.6	3.0	1.0	1.0	0.5	-
NH_4^+-N, mg/l	13.7	14.8	1.5	1.0	1.0	-
NO_2^- + NO_3^--N, mg/l	0	0.2	11.1	1.0	0.5	-
Total N, mg/l	22.3	18.0	13.6	3.0	2.0	2.4

aeration tanks and 12 additional final sedimentation tanks. The aeration tanks are designed for a volumetric loading of 120 lb BOD_5/1,000 cu ft/day (1.92 kg/m^3/day), an organic loading of 2.4 lb BOD/lb MLSS/day and a MLSS concentration of 1,300 mg/l. Since the increased design loadings require more air per unit volume than the existing aeration system can deliver, the existing aeration capability will be increased. This system will be modified from a coarse-bubble, spiral-roll system to a coarse-bubble, spread-pattern system to improve oxygen transfer efficiency. The secondary system air capacity has been designed to provide 0.54 lb O_2/lb BOD_5 removed.

Alum or ferric chloride will be added to the mixed liquor of the secondary system and is expected to remove approximately 70 percent of the phosphorus contained in the plant influent. The addition of metallic salts to the secondary system is also expected to improve the BOD removal in the secondary system from 75 percent to 85 percent. This will ensure a secondary effluent BOD concentration of less than 40 mg/l, which was found during pilot testing to be desirable for successful nitrification in the second stage. To remove most of the remaining phosphorus, metallic salts will also be added to the nitrogen release tanks.

Biological nitrification facilities are designed for oxidation of 0.066 lb NH_4^+-N/lb MLVSS/day at minimum wastewater temperatures and a MLVSS concentration of 1,700 mg/l. At the stoichiometric oxygen requirement of 4.6 lb O_2/lb NH_4^+-N oxidized, one hundred and twenty - 75 hp turbine aerators are required. Maximum air supply to the turbines will be 88,000 cfm (2500 m^3/min). The turbines were selected in this instance because, due to the limitations of the site, the nitrification tanks are designed to have depths of 30 feet (9.15 m) to obtain the required volume. The turbines will provide adequate mixing to this depth and are capable of supplying a range of oxygen to the system as required by varying ammonia-nitrogen influent loads and varying wastewater temperatures. Lime will be added to the nitrification reactor to maintain a minimum wintertime pH of 7.5.

FIGURE 9-15

WASHINGTON, D.C. BLUE PLAINS TREATMENT PLANT
FLOW DIAGRAM OF PRIMARY AND SECONDARY SYSTEMS

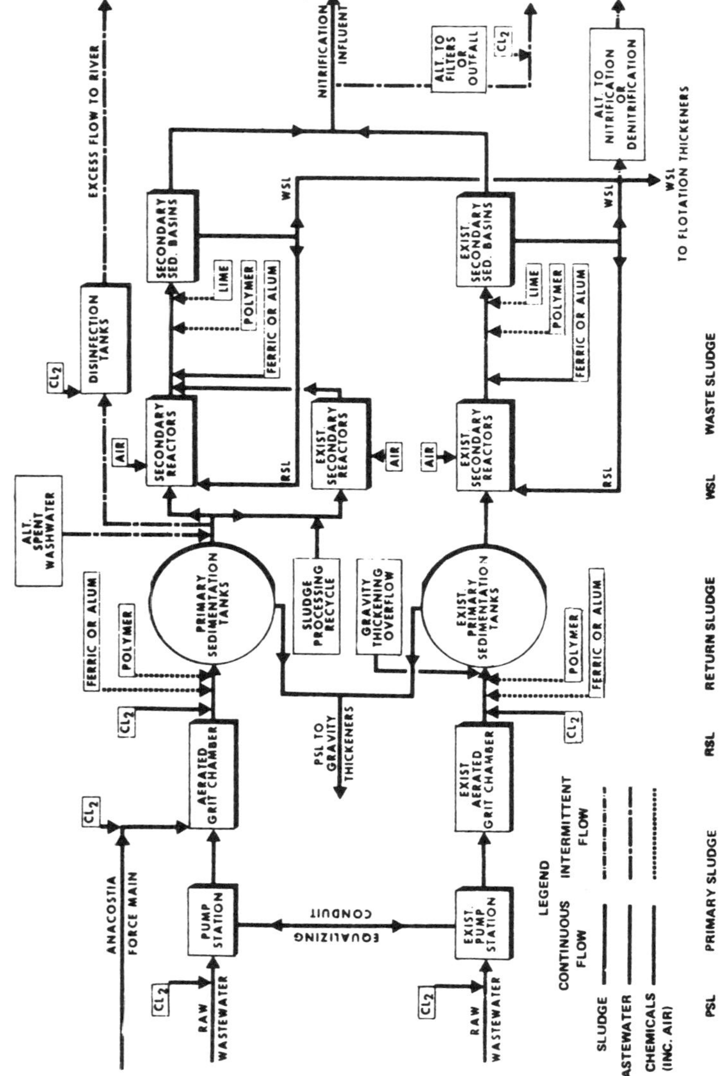

FIGURE 9-16

WASHINGTON, D.C. BLUE PLAINS TREATMENT PLANT
FLOW DIAGRAM OF NITRIFICATION AND DENITRIFICATION SYSTEMS

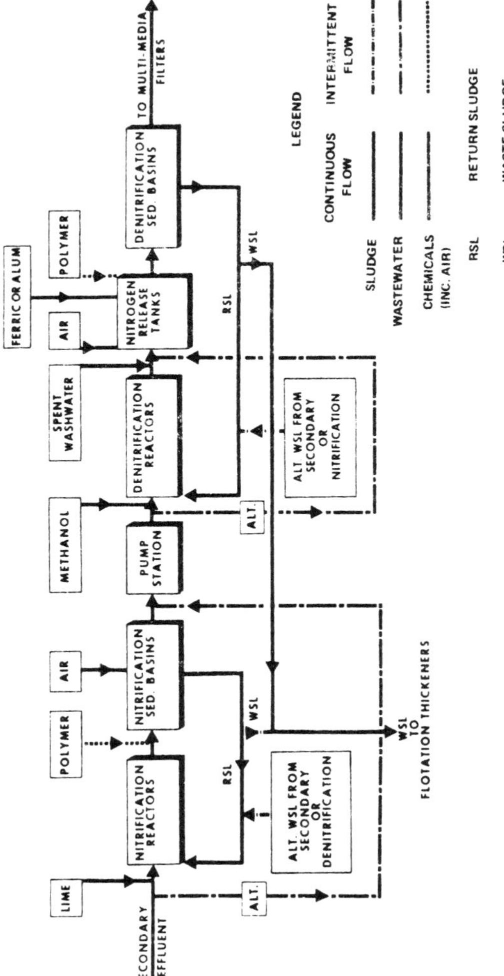

FIGURE 9-17

WASHINGTON, D.C. BLUE PLAINS TREATMENT PLANT
FLOW DIAGRAM OF FILTRATION AND DISINFECTION SYSTEMS

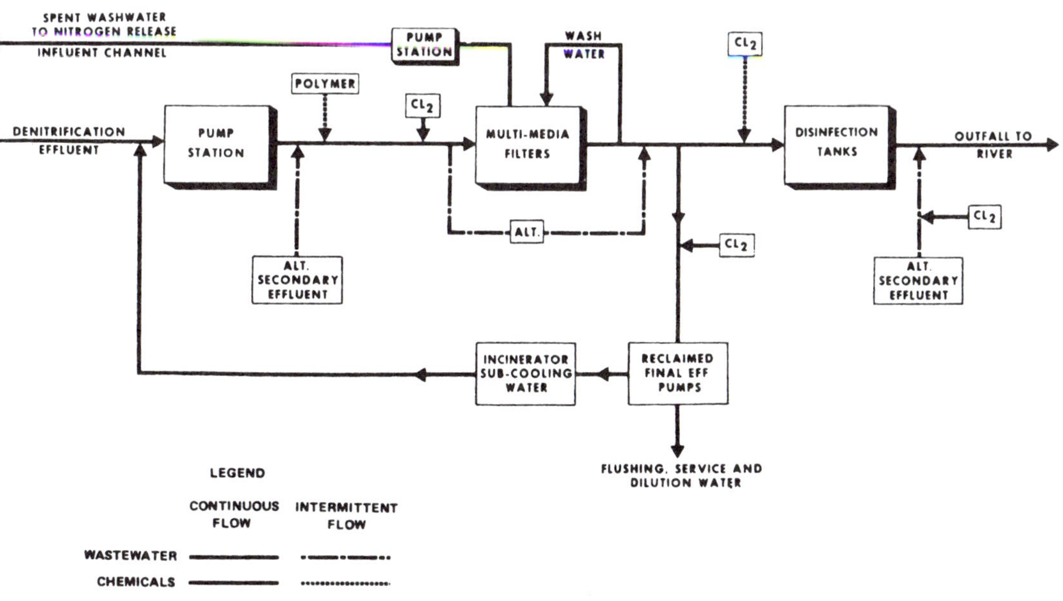

The nitrification sedimentation tanks are designed for average and peak hydraulic and solids loadings of 580 and 1,210 gpd/sq ft (23.6 m^3/m^2/day and 49.3 m^3/m^2/day) and 17.4 and 36.6 lb/sq ft/day (85 and 177 kg/m^2/day), respectively. The sludge return system is designed to provide return of 40 percent of peak flow. However, the system will normally be operated to return 30 percent. of the average flow. Continuous monitoring of the DO content of the nitrification effluent will be provided to ensure that the influent DO to the denitrification system is minimized.

The biological denitrification system is laid out to include reactors, nitrogen release tanks and sedimentation tanks. The reactors have been designed for removal of 0.0425 lb NO_3^- –N/day/lb MLVSS at a design MLVSS concentration of 2,100 mg/l, with up to 4.5 lb methanol added/lb NO_3^- –N applied. The reactors will be 44 ft (13.4 m) deep and be equipped with forty-eight - 75 hp mixers, and will be covered but not airtight.

The nitrogen release tanks were designed to serve three functions: (1) to strip supersaturated nitrogen gas, (2) to provide mixing for second-stage metal salt addition for residual phosphorus removal and (3) to provide an aerobic zone for removal of excess methanol. These tanks will furnish a 20-minute detention period at average flow and 12 minutes at peak flow.[40]

The denitrification sedimentation tanks are designed for hydraulic loadings of 670 gpd/sq ft (27.3 m^3/m^2/day) at average flow, and 1,410 gpd/sq ft (57.4 m^3/m^2/day) at peak flow. Solids loadings are 25.6 and 54.0 lb/sq ft/day (125 and 264 kg/m^2/day) at average and peak flows, respectively.

The 36 multimedia filters are designed for filtration rates of 3.0 gpm/sq ft (2 l/m^2/sec) at average flow and 6.2 gpm/sq ft (4.2 l/m^2/sec) at peak flow. Backwashing will be done at intervals of 24 hours at a rate of 25 gpm/sq ft (17 l/m^2/sec). The backwash water will be equalized in conduits and may be returned upstream of either the secondary reactors or the nitrogen release tanks.

Provision has been made to chlorinate either upstream or downstream of the filters with 24 minutes detention provided in contact tanks following the filters.

Sludge processing facilities will include gravity thickening of primary sludge, flotation thickening of secondary and advanced treatment sludges, vacuum filtration and sludge incineration.

9.5.2.4 El Lago, Texas

El Lago, Texas, is a small suburban community of 3,000 persons located near the Lyndon B. Johnson Space Center. The operating agency for wastewater treatment is the Harris County Water Control and Improvement District #50. This district currently operates a 0.3 mgd (0.013 m^3/sec) treatment plant. In 1969, the District received an order from the Texas

Water Quality Board that mandated protection of Clear Lake from excessive eutrophication. Two means were available for compliance with this order at that time; export of wastewater or providing nutrient removal prior to discharge to Clear Lake. The second option was chosen and the District obtained a grant from the EPA to demonstrate full-scale nitrogen and phosphorus removal.[3]

The original plant consisted of a rock trickling filter plant with anaerobic sludge digestion for solids processing. The modified flowsheet incorporating nutrient removal is shown in Figure 9-18. Added facilities are identified by asterisks and include new aeration-nitrification tanks, new denitrification columns, new tertiary filtration, and facilities for metal salt, polymer and methanol addition. All existing structures were incorporated into the upgraded plant. Design criteria for the modified plant are shown in Table 9-18.[3] Two separate types of denitrification columns were provided so that alternative designs could be compared. One set of columns was of the submerged high porosity media type described in Section 5.3.2.2 and was filled with Koch Flexirings (coarse media column in Table 9-18). The other type was the submerged low porosity media type described in Section 5.3.2.3 and as supplied by the Dravo Corp. (Fine media column in Table 9-18.) Both column types are shown on Figure 9-19.

Tables 9-19 and 9-20 are tabular summaries of the initial performance with the fine media and coarse media respectively.[3] Phosphorus removal during the fine media evaluation was erratic due to the cessation of iron addition during storm conditions. The fine media columns produced an effluent containing 17 mg/l of suspended solids because the columns are backwashed with nitrified effluent rather than clear tertiary filter effluent. Since tertiary filtration is also provided, this has not been a problem at El Lago.

Table 9-21 reproduces a month of operating data for the fine media denitrification columns. This data shows the improvements in all parameters of effluent quality obtained after one year of experience in plant operation.[42]

The coarse media denitrification columns also performed well during the investigation. While the fine media columns required at least daily backwashing to prevent excessive headlosses, the high void volume of the coarse media allowed operation without frequent backwash. The routine procedure was to backwash every 4 weeks.[3] This proved to be an important difference to the plant operators, who found that the coarse media units required much less attention than the fine media units.[43]

The capital costs for the modifications to the El Lago Facility were incurred over a two-year period (1971-1973) and totaled $312,365 including change orders. This cost includes the provision of dual denitrification facilities; had only one type of denitrification system been included it is estimated that the total would be about $75,000 less. The only increase in operating costs has been for chemicals and power and this has totaled $96/mil gal.[3] It was found that the existing plant operators could adapt to advanced waste treatment processes and no increase in staff was required.

FIGURE 9-18

EL LAGO, TEXAS WASTEWATER TREATMENT PLANT, FLOW DIAGRAM

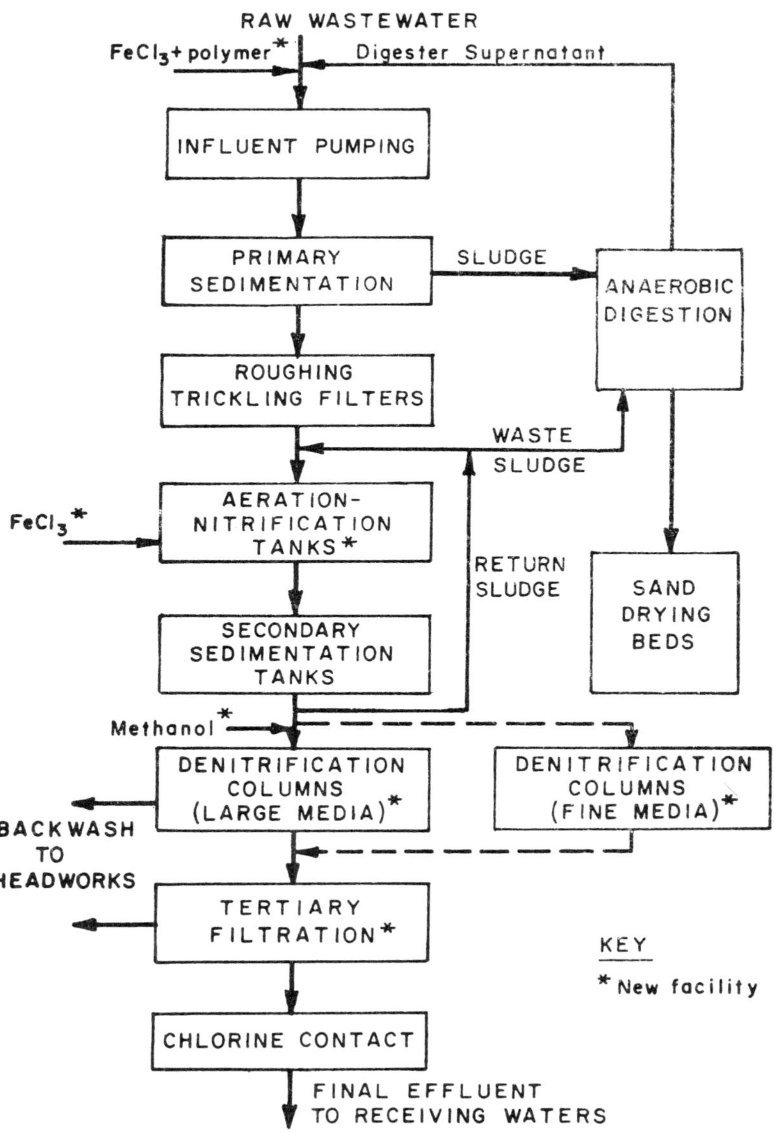

TABLE 9-18

DESIGN DATA, EL LAGO, TEXAS WASTEWATER TREATMENT PLANT

Population	3,000
Average dry weather flow (ADWF)	0.3 mgd (0.013 m^3/sec)
Peak dry weather flow (PDWF)	0.5 mgd (0.022 m^3/sec)
Peak wet weather flow (PWWF)	1.0 mgd (0.044 m^3/sec)
Raw wastewater quality	
BOD_5	161 mg/l
SS	195 mg/l
Total Kjeldahl nitrogen	37.5 mg/l
Primary sedimentation tanks (existing)	
Number	2
Detention time (ADWF)	1.6 hr
Overflow rate (ADWF)	440 gpd/sf (30 m^3/m^2/day)
Roughing trickling filter (rock, existing)	
Number	2
Depth	6.5 ft (2.0 m)
Rock size	4 in (100 mm)
Volume of media	20,900 cu ft (590 m^3)
Organic load (ADWF)	12 lb BOD_5/1,000 cf/day (0.192 kg/m^3/day)
Recirculation rate	0.3 mgd (0.013 m^3/sec)
Air blowers (new)	
Number	2
Discharge pressure	6.5 psig (0.46 kgf/cm^2)
Capacity - total	900 cfm (25.2 m^3/min)
Aeration-nitrification tanks (new)	
Number	2
Arrangement	Series
Volume - Total	10,100 cu ft (302 m^3)
Detention time (ADWF)	6.1 hr
MLVSS	1,000 mg/l
Solids retention time	10 days
Secondary sedimentation tanks (existing)	
Number	2
Detention time (ADWF)	5.4 hr
Overflow rate at ADWF	320 gpd/sf (13 m^3/m^2/day)
Overflow rate at PWWF	1,060 gpd/sf (42 m^3/m^2/day)
Denitrification columns	
Coarse media	
Number (series)	2
Type	50 psig (3.5 kgf/cm^2) pressure vessel
Diameter	10 ft (3 m)
Media depth	10 ft (3 m)
Media type	Koch Flexirings
Specific surface	105 sf/cu ft (346 m^2/m^3)
Voids	92 percent
Empty bed contact time	1 hr
Surface application rate at ADWF	2.5 gpm/sf (1.6 l/m^2/sec)
Air backwash rate	10 cfm/sf (3.1 m^3/m^2/min)
Water backwash rate	20 gpm/sf (13.5 l/m^2/sec)
Fine media	
Number (series)	2
Type	50 psig (3.5 kgf/cm^2) pressure vessel
Diameter	6 ft (1.8 m)
Media height	6.5 ft (2.0 m)
Media type	3 to 4 mm uniform sand
Specific surface	250 sf/cu ft (825 m^2/m^3)
Voids	40 percent
Empty bed contact time	0.25 hr
Surface application rated at ADWF	7.4 gpm/sf (5 l/m^2/sec)
Air backwash rate	8 cfm/sf (2.4 m^3/m^2/min)
Water backwash rate	20 gpm/sf (13.5 l/m^2/sec)

TABLE 9-18

DESIGN DATA, EL LAGO, TEXAS WASTEWATER TREATMENT PLANT (CONTINUED)

Tertiary filtration	
Number (parallel)	2
Type	30 psig (2.1 kgf/cm^2) pressure vessel
Height	8 ft (2.4 m)
Diameter	3.5 ft (1.15 m)
Media height and type	3 ft (1.0 m) of 0.3 to 0.8 mm sand
Media support base	0.5 ft (0.15 m) gravel
Surface application rate at ADWF	2.2 gpm/sf (1.5 l/m^2/sec)
Water backwash rate	15 gpm/sf (10 l/m^2/sec)
Chlorine contact tank	
Detention time (ADWF)	1 hr
Anaerobic digestion	
Volume	8,830 cf (2,472 m^3)
Sand drying bed	
Area	6,300 sf (580 m^2)

FIGURE 9-19

EL LAGO, TEXAS DENITRIFICATION COLUMNS, COARSE MEDIA TYPE ON RIGHT AND FINE MEDIA TYPE ON LEFT

TABLE 9-19

INITIAL PERFORMANCE OF FINE MEDIA DENITRIFICATION COLUMNS AT EL LAGO, TEXAS - JUNE 4 TO JULY 6, 1973

Constituent	Mean value, mg/l at indicated sample location					
	Raw wastewater[a,b]	Primary influent	Primary effluent	Nitrified effluent	Denitrified effluent[c,d]	**Final effluent**
Total P	12.8	15.4	8.4	7.3	6.6	**4.8**
Soluble P	10.3	4.7	4.1	3.4	5.5	**3.6**
SS	113	289	72	37	17	**3**
NH_4^+-N	18.7	21.7	21.5	0.9	0.8	**0.6**
TKN	42.6	38.6	30.2	3.7	2.4	**3.3**
NO_3^--N[e]	-	-	-	15.2	2.6	**2.3**
BOD_5	175	222	-	65[f]	9	**9**
COD	297	488	181	121[f]	72	**51**
Temperature[g]	26.5	-	-	-	-	**-**

[a] Average flow to plant: 0.307 mgd (0.013 m^3/sec)
[b] Peak daily flow to plant: 1.0 mgd (0.044 m^3/sec)
[c] Average flow to denitrification columns: 0.254 mgd (0.011 m^3/sec)
[d] Peak daily flow to denitrification columns: 0.420 mgd (0.018 m^3/sec)
[e] Nitrite - N always less than 0.2 mg/l
[f] Includes methanol
[g] Degrees C

TABLE 9-20

INITIAL PERFORMANCE OF COARSE MEDIA DENITRIFICATION COLUMNS AT EL LAGO, TEXAS - JULY 8 TO AUGUST 31, 1973

Constituent	Mean value, mg/l at indicated sample location					
	Raw wastewater[a,b]	Primary influent	Primary effluent	Nitrified effluent	Denitrified effluent[c,d]	**Final effluent**
Total P	12.3	13.1	6.7	-	-	**2.8**
Soluble P	10.3	3.1	2.4	-	-	**2.3**
SS	102	231	63	43	19	**4.5**
NH_4^+-N	16.3	14.6	14.4	0.9	1.2	**0.9**
TKN	29.7	31.8	26.7	2.6	2.5	**1.7**
NO_3^--N[e]	-	-	-	13.6	0.9	**0.6**
BOD_5	143	156	87	43[f]	15	**8**
COD	248	336	167	107[f]	52	**38**
Temperature[g]	27.2	-	-	-	-	**-**

[a] Average flow to plant: 0.320 mgd (0.014 m^3/sec)
[b] Peak daily flow to plant: 0.900 mgd (0.039 m^3/sec)
[c] Average flow to denitrification columns: 0.315 mgd (0.014 m^3/sec)
[d] Peak daily flow to denitrification columns: 0.632 mgd (0.028 m^3/sec)
[e] Nitrite - N always less than 0.2 mg/l
[f] Includes methanol
[g] Degrees C

TABLE 9-21

SUBSEQUENT PERFORMANCE OF FINE MEDIA DENITRIFICATION COLUMNS AT EL LAGO, TEXAS - OCTOBER 1 THROUGH OCTOBER 31, 1974

	Mean value, mg/l at indicated sample location					
Constituent	Primary influent[a,b]	Primary effluent	Nitrified effluent	Denitrified effluent[c,d] Column No. 1	Denitrified effluent[c,d] Column No. 2	Final effluent
Total P[g]	12	3.6	3.1	-	-	0.41
Soluble P	1.8	1.0	0.41	-	-	0.40
SS	295	51	81	51	44	1
NH_4^+-N	-	18	0.4	-	0.4	0.3
TKN	-	24	2.6	-	1.5	0.9
NO_3^--N	-	-	15	1.9	0.9	0.6
BOD_5	-	-	62[e]	16	12	3
COD	-	-	113[e]	44	36	19
Temperature[f]	-	-	21	-	-	-

[a] Average daily flow to plant: 0.301 mgd (0.013 m^3/sec)
[b] Peak daily flow to plant 0.47 mgd (0.021 m^3/sec)
[c] Average flow to denitrification columns: 0.282 mgd (0.012 m^3/sec)
[d] Peak daily flow to denitrification columns: 0.470 mgd (0.021 m^3/sec)
[e] Includes methanol
[f] Degrees C
[g] Reduction of P due to ferric chloride addition to primary and nitrification step (37 mg/l as Fe). Also polymer, DOW A-23, added to primary at 0.23 mg/l and to tertiary filter at 0.17 mg/l.

9.5.3 Case Examples of Breakpoint Chlorination for Nitrogen Removal

Two examples of the use of breakpoint chlorination for nitrogen removal are presented in this section. The Sacramento Regional County Sanitation District's plant will incorporate breakpoint chlorination of approximately one-half of the plant effluent to achieve the partial nitrogen removal dictated by plant effluent requirements. The Montgomery County, Maryland facility is designed for breakpoint chlorination of the entire flow to meet rigid limitations on total nitrogen set on the plant's effluent.

9.5.3.1 Sacramento, California

The proposed Sacramento Regional Wastewater Treatment Plant will be owned and operated by the Sacramento Regional County Sanitation District to serve the City and County of Sacramento. The Regional Plant is designed for 125 mgd (5.43 m^3/sec) average seasonal dry weather flow. It consolidates 23 existing plants presently discharging to the Sacramento and American Rivers above Sacramento and provides for discharge to the Sacramento River, downstream.[44] At minimum river flows maintained by upstream dam development, the 125 mgd average daily flow will be about 2.7 percent of the total river flow at the plant discharge site.[45]

Effluent from the plant will comply with waste discharge requirements adopted by the California Regional Water Quality Control Board on October 25, 1974. Effluent quality requirements require BOD and suspended solids to average less than 30 mg/l on a monthly basis. Total nitrogen is limited to 15 mg/l when Sacramento River flow is below 12,000 cfs (340 m^3/sec) at a specified gauging station. The 15 mg/l total nitrogen requirement under low flow conditions in the receiving water is to reduce algae blooms. Based on long-term rainfall and river flow data, it is anticipated that nitrogen removal will be required for an average of 67 days per year.[45,46] One consideration in developing design criteria for the regional plant was to find the most cost-effective solution for intermittent nitrogen removal.

In the regional facility mixed municipal and cannery waste will undergo treatment steps including prechlorination, preaeration, grit removal, primary sedimentation, oxygen activated sludge treatment, secondary sedimentation, post aeration to strip carbon dioxide and raise pH, chlorine disinfection, and sulfur dioxide dechlorination. For intermittent nitrogen removal at Sacramento, breakpoint chlorination was most cost-effective. In comparison to biological nitrification-denitrification, breakpoint chlorination was cost-effective if the total nitrogen limitation did not exceed a duration of about 300 days per year. Breakpoint via on-site production of hypochlorite solution from direct mixing of liquid chlorine and caustic was cost-effective in comparison to electrolytic generation of hypochlorite if the duration of the nitrogen limitation did not exceed about 200 days per year.[46]

Most significant in the cost analysis is the capital expenditure to meet the maximum chlorination capacity. Calculations indicate the maximum required chlorination capacity for breakpoint will be 170 tons per day (154,000 kg/day). This figure represents treating the entire plant flow. The average required chlorination capacity for breakpoint is 78 tons of chlorine per day (70,700 kg/day). One alternative for meeting maximum required capacity would be to provide 42 standard 8,000 lb/day (3,600 kg/day) vacuum chlorinators. Alternatively, generation of hypochlorite using the electrolytic process would require 51 units of 3.3 ton/day (2,990 kg/day) capacity as well as brine and salt storage. Another alternative was reviewed, that employed by the Los Angeles County system which utilizes liquid chlorine and base mixed in a water stream.[47] This system appeared to provide the required maximum operating flexibility with minimum investment cost. Thus, it was decided to feed liquid chlorine and caustic soda into a recirculated effluent water stream directly forming hypochlorite.

Liquid chlorine is available in rail tank cars from San Francisco Bay area producers as well as Washington state. The rail facilities allow flexibility of plant deliveries and eliminate the need for permanent on-site chlorine storage tanks.

Design criteria for the breakpoint facilities at Sacramento are summarized in Table 9-22. The breakpoint system is designed as two complete and separate units each capable of chlorinating half of the plant flow. This allows breakpoint chlorination of only the portion of the plant flow required to meet the 15 mg/l total nitrogen limitation. It is estimated that

TABLE 9-22

DESIGN CRITERIA FOR HYPOCHLORITE PRODUCTION FACILITY SACRAMENTO REGIONAL WASTEWATER TREATMENT PLANT

Plant Loading	
Flows	
Maximum hourly	150 mgd (6.57 m^3/s)
Average dry weather	110 mgd (4.82 m^3/s)
Average seasonal dry weather	125 mgd (5.48 m^3/s)
Total nitrogen (influent)	
Maximum hourly	37 mg/l
Average dry weather	33 mg/l
Chemical requirements	
Chemical ratios	
Ammonia nitrogen to total nitrogen	0.75 : 1.00
Chlorine to ammonia nitrogen ratio	
Maximum	10.0 : 1.0
Average	9.0 : 1.0
Caustic to chlorine	
Maximum	1.3 : 1.0
Average	1.0 : 1.0
Chemical feed rates	
Chlorine	
Maximum	170 tons/day (154,195 kg/day)
Average[a]	78 tons/day (70,750 kg/day)
Caustic	
Maximum	221 tons/day (200,450 kg/day)
Average[a]	78 tons/day (70,750 kg/day)
Chemical storage and delivery	
Chlorine storage	
Number single unit railroad tank car spots	12
Maximum capacity, each unit	90 tons (81,630 kg)
On-line storage capacity	5 days
In-plant storage capacity	7 days
Sodium hydroxide storage	
Number tanks	3
Size, dia. x height	40 ft x 20 ft (12.2m x 6.1 m)
Capacity, each tank (25% conc.)	188,000 gal (711,580 l)
Number single unit railroad tank car spots	2
Maximum capacity, each unit (50% conc.)	90 tons (81,630 kg)
Sodium hydroxide feed pumps	
Number	3
Capacity, each unit (25% conc.)	75 gpm (4.73 l/s)
Padding air compressors	
Number	2
Capacity, each unit	130 scfm (61.3 m^3/s)
Discharge pressure	220 psig (15.2 bar)
Sodium Hypochlorite Generators	
Number	2
Maximum capacity, each unit	85 tons/day (77,100 kg/day)
Channel mixers	
Number	?
Capacity	40 hp (29.8 kW)
Ammonia analyzers	2
Chlorine analyzers, total	2
Chlorine analyzers, free	2
Water supply pumps	
Number	2
Capacity, each unit	800-2100 gpm (50.5-132.5 l/s)

[a]Represents breakpoint chlorination of three-quarters of average plant flow.

both units will be required to be operational at certain times to maintain effluent requirements. Figure 9-20 is a simplified schematic of the breakpoint chlorination system and its control instrumentation. The process consists of producing sodium hypochlorite by inline mixing of liquid chlorine, water, and caustic at a chlorine concentration of 7,000-8,500 ppm and a pH of 7.5. Maintenance of proper pressures in the chlorine feed system is the key to successful operation of the hypochlorite generation system. Experience at Los Angeles County White Point plant indicates a necessary pressure of 40 psig (2.8 kgf/cm^2) in the mixing tee and greater than 110 psig (7.73 kgf/cm^2) at the chlorine feed valve. These pressures require pressurizing ("padding") the chlorine railcars with 175 psig (12.30 kgf/cm^2) dry air.

A chlorine reserve tank is used to provide chlorine during periods when empty chlorine tank cars are being replaced with full cars and can provide about a one-hour supply of chlorine.

The feed rate of liquid chlorine is measured with a steel tube rotameter and controlled with an automatic valve. Pressure drop from the chlorine feed line to the mixing tee and initial mixing of chlorine, caustic and water is partially accomplished using a plug injector. To improve mixing immediately downstream of the mixing tee, an inline powered mechanical mixer is provided. The hypochlorite solution line then includes three taps for pH metering and a back pressure valve to maintain 40 psig at the mixing tee. The pH meters are duplicated for each unit. The duplication is for improved reliability of the pH system and the multiple taps are for additional flexibility in selecting the point which is monitored for control purposes.

The related feeds of caustic and water are provided by pumping from the chemical storage area and the effluent channel. Caustic is fed from storage tanks through centrifugal pumps, automatic control valves, and magnetic flow meters. The control signal is proportional to the chlorine feed with feedback control provided by the pH meters. Manual caustic feed is provided for breakpoint start-up and for pH adjustment of final effluent.

Plant effluent is used for sodium hypochlorite solution water and is provided by two variable speed pumps located as shown in Figures 9-20 and 9-21. The control of the solution flow (via pump speed) is proportional to the chlorine feed. The solution water flow is measured by magnetic flow meters just ahead of the mixing tee.

The plan view of Figure 9-21 also shows the application points for the breakpoint chlorination located at the end of each battery of secondary sedimentation tanks. To insure rapid mixing, two spargers are installed with orifices that insure a minimum exit velocity of about 10 fps (3.1 m/sec). Automatic valving provides one sparger for lower feed rates and two when flows cause back pressure to reach the control limit of the hypochlorite back pressure valve. Immediately downstream of the breakpoint spargers is a mechanical mixer followed by a submerged overflow weir. The mixing channel and the post-aeration channel following are covered. The exhaust can be passed through mobile activated carbon filters to remove odors, such as nitrogen trichloride, if necessary. The activated carbon filters are

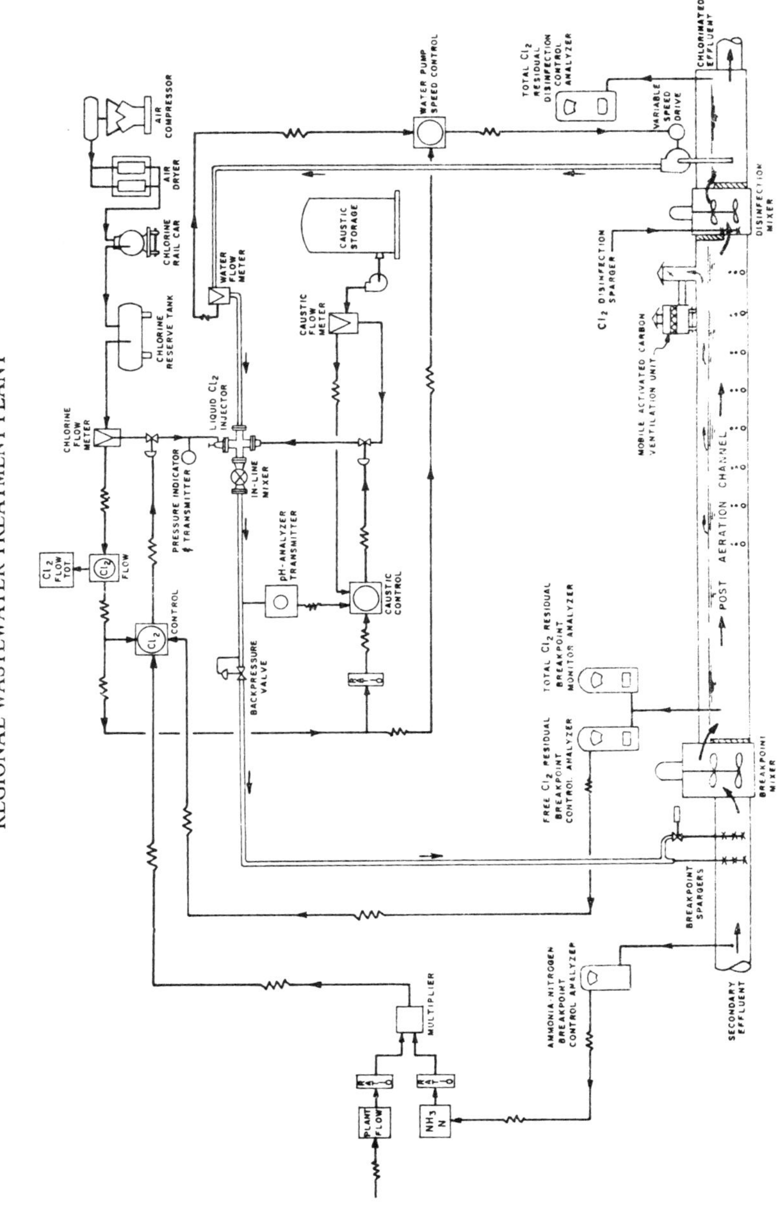

FIGURE 9-20

HYPOCHLORITE GENERATION SCHEMATIC - SACRAMENTO REGIONAL WASTEWATER TREATMENT PLANT

FIGURE 9-21

PLAN AND SECTION OF THE BREAKPOINT FACILITY AT THE SACRAMENTO REGIONAL WASTEWATER TREATMENT PLANT

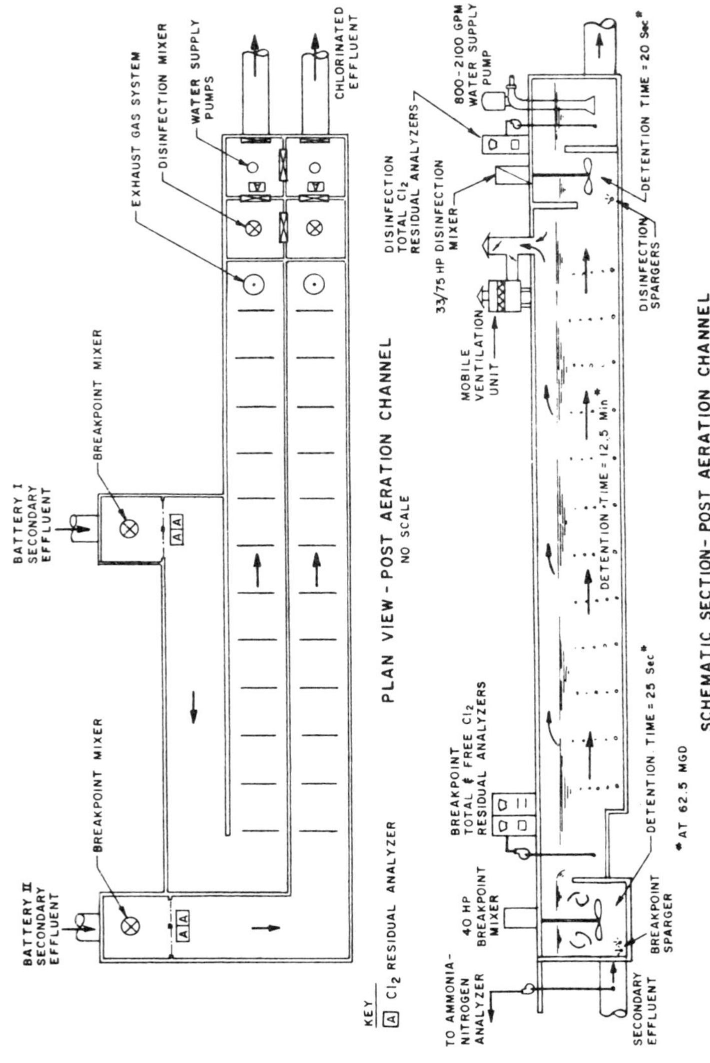

mobile units that are identical to those used throughout the plant in various applications. When the carbon is exhausted, the mobile units are replaced with duplicate units and taken to a central carbon handling facility for bed replacement. Breakpoint chlorination control is closed loop. Feed forward control includes ammonia nitrogen concentration, effluent flow, and a ratio proportioner of chlorine to ammonia nitrogen. Feedback control is achieved by free chlorine residual measured after the breakpoint reaction. Ammonia nitrogen samples are pumped continuously from each battery upstream of the hypochlorite sparger to ammonia nitrogen analyzers. Chlorine residual analyzers are of the amperometric type and are in duplicate. One analyzer measures free chlorine residual and is used for control; the other analyzer measures total chlorine residual and is used as a monitor. Separate provision is made for feeding chlorine for disinfection when breakpointing is not being practiced. This was done because different magnitudes of chemicals are involved when dealing with breakpoint than with disinfection. Estimates for capital and operating costs for the breakpoint chlorination facility are in Tables 9-23 and 9-24.

TABLE 9-23

CAPITAL COST BREAKDOWN FOR BREAKPOINT CHLORINATION AT THE SACRAMENTO REGIONAL WASTEWATER TREATMENT PLANT

Item	Estimated cost[a]
Breakpoint generation equipment	$ 95,000
Outside piping	240,000
Chlorine unloading facilities	110,000
Caustic storage and pumping	162,000
Railroad	244,000
Air padding facilities	55,000
Subtotal	$ 906,000
Engineering and contingencies	272,000
Total	$1,178,000

[a]Cost basis: October, 1974

TABLE 9-24

TOTAL ANNUAL COST BREAKDOWN FOR BREAKPOINT CHLORINATION AT THE SACRAMENTO REGIONAL WASTEWATER TREATMENT PLANT

	1,000/yr	$/mil gal treated	$/mil gal - annual average
Chemical cost[a,b]	1,265	172.00	31.50
Labor cost	30	4.00	3.00
Subtotal	1,295	176.00	34.50
Amortization of capital	101	13.50	2.50
Total annual cost[c]	1,396	189.50	37.00

[a]Costs are based on chlorine @ $144/ton and caustic soda @ $168/ton, October 1974 prices; average plant flow of 110 mgd; and breakpoint chlorination required 67 days/aver. yr. at an average of three quarters of the plant flow.

[b]Power costs negligible

[c]Capital recovery at 7 percent and 25 years

9.5.3.2 Montgomery County, Maryland

This new 60 mgd (2.6 m^3/sec) facility is now being designed.[48] It will be owned and operated by the Washington Suburban Sanitary Commission to serve the Maryland suburbs of Washington, D.C. The plant will discharge to the Potomac River above the raw water intakes for the Washington, D.C. water treatment plants.[49] At minimum river flows, the effluent will make up about 15 percent of the water volume at the water plant intakes. Because of the critical nature of the downstream water use, the following effluent goals have been established:

Parameter	Value
BOD_5, mg/l	1.0
Suspended solids, mg/l	0
Total nitrogen (as N), mg/l	2.0
Total phosphorus (as P), mg/l	0.1
Chloride, mg/l	200-350
Total dissolved solids, mg/l	850-1,120
Coliform bacteria – MPN/100ml	2.2
Fecal coliform bacteria – MPN/100ml	2.2

Figure 9-22 is a flow diagram of the treatment process. The final effluent will be stored in a reservoir with a 10 day holding capacity before discharge to the Potomac. The lime sludges will be recalcined and reused and the granular carbon will be regenerated on-site for reuse.

The breakpoint process was selected over alternate approaches for several reasons.[50] It was felt that the results of the first scale-up of the selective ion exchange process underway at the nearby Upper Occoquan (see Sec. 9.5.4.1) plant in Virginia should be available before a 60 mgd (2.6 m^3/sec) selective ion exchange facility was attempted.

The prolonged cold weather periods made ammonia stripping inadequate for substantial portions of the year. The biological approach seemed to offer no significant cost savings and was prone to occasional inefficiencies which would be particularly significant because of the effluent quality goals. The effluent TDS additions from the breakpoint process were not a limitation in this case because the effluent does not enter a reservoir or confined watershed where the solids would be recycled and continue to build up but enters the Potomac River shortly before it flows into the Potomac Estuary.

Concern over the hazards of transporting and storing large quantities of chlorine gas led to the decision to use sodium hypochlorite for the breakpoint process. An investigation of the availability and cost of sodium hypochlorite led to a decision to install an on-site hypochlorite generation facility of the membrane cell type.[51,52] For this electrolytic process of sodium hypochlorite generation, the raw materials required are electrical power, salt, and water. The membrane cells and the overall system are shown schematically in

FIGURE 9-22

FLOW DIAGRAM OF THE MONTGOMERY COUNTY, MARYLAND PLANT

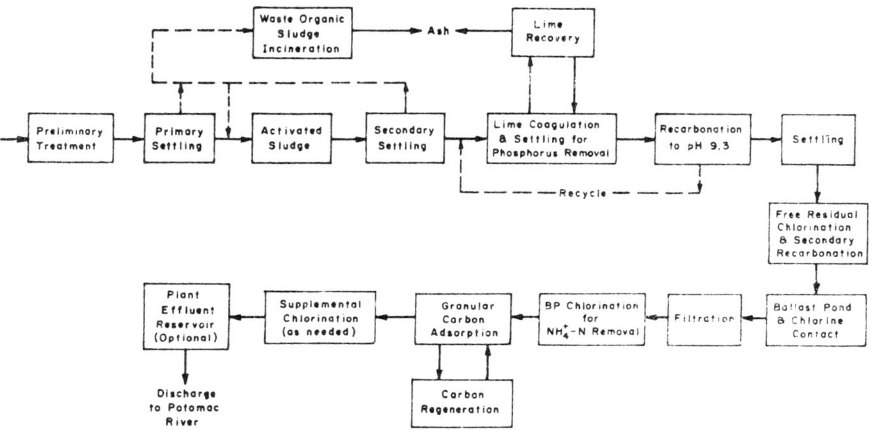

Figures 9-23 and 9-24. Saturated brine is fed to the anode compartment of the cells. At the anode surface, chlorine gas is generated. The effluent from the cathode compartment is sent to a gas-liquid separator in which the hydrogen is removed from the caustic solution. Chlorine and caustic are fed to the reactor, with the caustic in slight excess, to produce sodium hypochlorite. The reactor is water cooled to avoid decomposition of hypochlorite and formation of oxygen. Insoluble gases, mainly oxygen, are removed in a gas-liquid separator before the hypochlorite is sent to product storage. Brine feed for the system comes from a salt storage tank which is used for salt storage and production of saturated brine. The cell typically produces hypochlorite with 8 percent available chlorine at 1.7 kwh/lb (0.77 kwh/gm) available chlorine.

The design criteria for the hypochlorite production facility at Montgomery County are summarized in Table 9-25. Concentrated brine is created by combining softened water (softened to minimize potential for scaling of the membrane with calcium) and solid salt to achieve a saturated brine solution (approximately 26 percent salt by weight). Both the solid salt storage and salt solution are contained in lixators which are fitted with a brine collection system and a brine level control system. Delivery of the solar-type salt to the lixators is by 22 ton (19.95 metric tons) dump trucks. The brine is pumped from the lixators to a brine treatment system, consisting of successive addition and mixing of caustic soda (NaOH) and soda ash (Na_2CO_3) followed by settling to precipitate calcium and magnesium. The brine is then filtered through a rapid sand filter and a cartridge filter to remove suspended solids and pumped to storage. Storage for a one-day supply of treated

FIGURE 9-23

MEMBRANE CELL USED FOR HYPOCHLORITE PRODUCTION

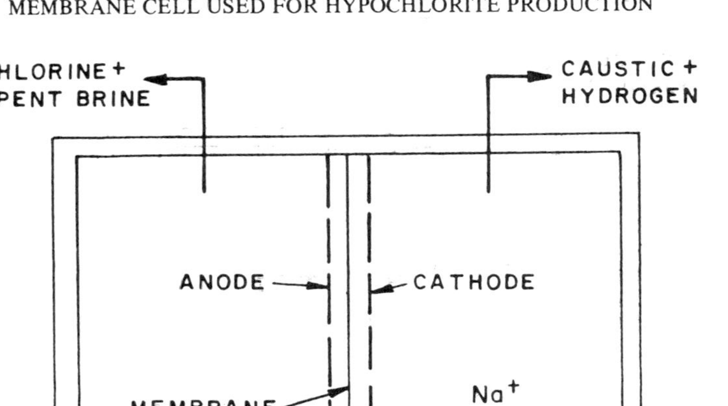

brine is provided. The direct current source required by the membrane cells is provided by a solid state rectifier, with side-mounted controls. Safety features in the control panel include automatic shutdown in case of high reactor temperature, high DC voltages, blower failure, or high or low DC current. The hydrogen from the electrolytic cells can be either (1) diluted with air as it is formed to maintain a hydrogen concentration of less than 0.25 percent by volume, the explosive limit, or (2) compressed and piped to the solids processing building for use as a fuel. Sodium hypochlorite leaves the reactors by gravity flow to pumps from which it is sent to storage.

The breakpoint reaction is accomplished by adding a sodium hypochlorite solution to the wastewater at a dosage slightly in excess of the stoichiometric requirement for oxidation of the ammonia nitrogen to gaseous nitrogen. Figure 9-25 illustrates the design of the system and Table 9-26 presents the design criteria. Filter effluent flows by gravity to the breakpoint chlorination reactors.

FIGURE 9-24

OVERALL SYSTEM USING MEMBRANE CELLS FOR HYPOCHLORITE PRODUCTION

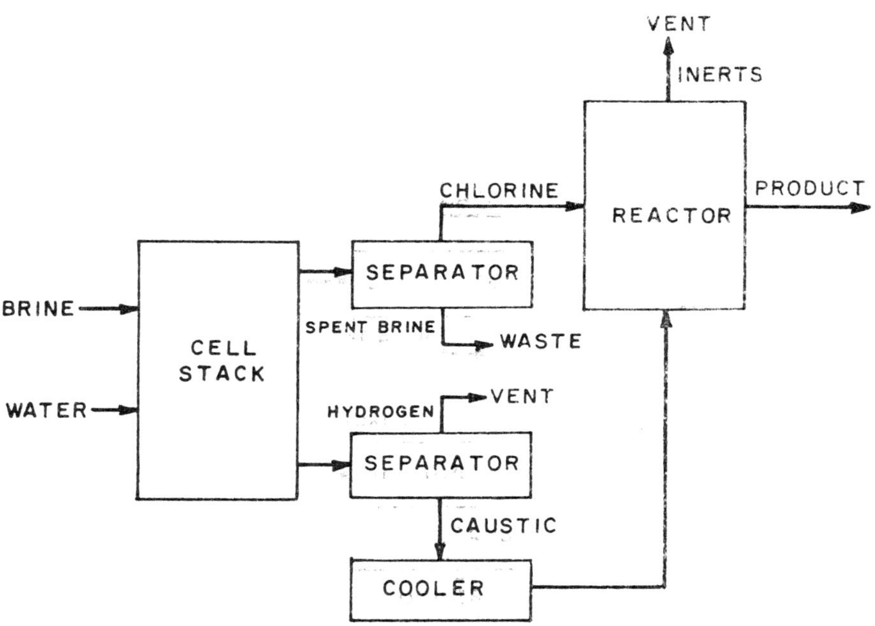

The wastewater first passes through two in-line mechanical mixers in series. The sodium hypochlorite is added to the first mixer along with sodium hydroxide if needed for pH adjustment. The second mixer is provided for protection if the first mixer malfunctions. The second mixer is normally operated to provide thorough mixing. The mixer was sized so that a single mixer would provide violent mixing (G = 1,000 sec^{-1}) to insure instant and complete mixing of the hypochlorite and the wastewater.

Wastewater then flows to breakpoint reactors (closed concrete tanks) where it is air mixed to complete the chemical reaction and where 30 minutes contact time for disinfection is provided. Flow is distributed over the first half of the length of each basin through a multiport header. Distribution of flow in this manner minimizes the decrease in pH caused by the reaction of sodium hypochlorite with ammonia nitrogen (by avoiding a single point of injection into the basin of the wastewater-hypochlorite mixture) and thereby minimizes the formation of nitrogen trichloride. Air is diffused into the wastewater over the bottom of

TABLE 9-25

DESIGN CRITERIA FOR HYPOCHLORITE PRODUCTION FACILITY AT THE MONTGOMERY COUNTY FACILITY

Average dry weather flow	60 mgd (2.63 m^3/sec)
Sodium chloride storage and treatment	
1. NaCl required @ 60 ton/day Cl_2 production[a]	105 tons/day (95.2 metric tons/day)
2. Salt storage	
Liquid level	4 @ 22 ft x 33 ft x 18 ft (6.7 m x 10.1 m x 5.5 m)
Storage capacity	1,470 tons (1333 metric tons)
Maximum saturated brine production	60 gpm (3.8 l/sec) each (min.)
3. Type of salt used	Solar
4. Brine treatment system	
A. Soda ash requirements	450 lb/day (204 Kg/day)
3 minute rapid mix	3 ft x 3 ft x 3 ft swd (0.91 m x 0.91 m x 0.91 m)
B. NaOH requirements	360 lb/day (163 Kg/day)
3 minute rapid mix, 3 hour settling	3 ft x 3 ft x 3 ft swd (0.91 m x 0.91 m x 0.91 m)
C. Rapid sand filters and cartridge filtration follows settling	
5. Treated brine storage, fiberglass tanks[b]	6 @ 12 ft x 12 ft (3.66 m x 3.66 m)
	16,000 gal each (60,560 l)
6. Truck delivery	22 tons/load (19.95 metric tons)
7 trucks/day/5 day week	
Sodium hypochlorite generators	
1. Generating capacity	60 tons/day (54.4 metric tons)
2. Modules required @ 3.3 tons each	20
On line	18
Redundant	2
3. Rectifiers required	7
4. Power required	10,000 KVA
5. Building required	87 ft x 96 ft x 20 ft (26.5 m x 39.2 m x 6.1 m)
Builder provides for housing, modules, and rectifiers.	
Monorail with 5 ton capacity provided for cell stack maintenance.	
Rectifiers separated from modules by a glass wall to prevent corrosion.	
6. Generator module dimensions	4 ft x 16 ft x 8 ft (1.2 m x 4.9 m x 2.4 m)
7. Rectifier dimensions	4 ft x 6 ft x 8 ft (1.2 m x 1.8 m x 2.4 m)
Sodium hypochlorite storage	
1. Storage sized to provide 1 day storage for power outage plus storage for 7 days @ 90 mgd (3.94 m^3/sec)	
2. Storage capacity	700,000 gal (2650 m^3)
	230 tons equivalent (209 metric tons)
	6-117,000 gal ground level tanks (443 m^3)
	6-30 ft x 22 ft tanks (9.1 m x 6.7 m)

[a]Assumes 90 percent utilization

[b]Each tank will provide 4 hours storage at full production

the second half of the basin to strip any nitrogen trichloride from solution. Additional air may be introduced in the space between the liquid surface and the basin cover to further dilute any nitrogen trichloride that might be present to prevent the development of an explosive concentration that occurs at 0.5 percent by volume. The diffusion of air into the breakpoint reactor contents also strips gaseous nitrogen and carbon dioxide from solution. The latter will result in a desirable increase in pH.

Exhaust gases from the reactors are recirculated to provide mixing with some of the gas bled off to the recalcining furnace for thermal decomposition of the nitrogen trichloride. The alkaline environment in the recalcining furnace will avoid discharge of hydrochloric acid (HCl) to the atmosphere that might otherwise occur.

FIGURE 9-25

SCHEMATIC OF MONTGOMERY COUNTY, MARYLAND
BREAKPOINT CHLORINATION PROCESS

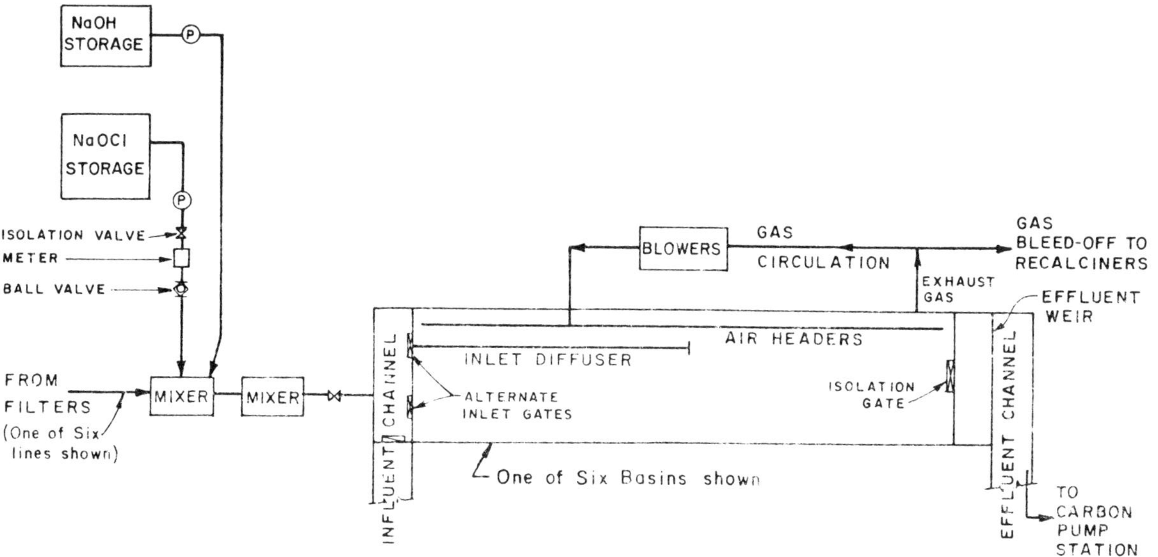

TABLE 9-26

BREAKPOINT CHLORINATION DESIGN CRITERIA FOR THE MONTGOMERY COUNTY FACILITY

Average dry weather flow	60 mgd (2.63 m^3/sec)
In-line mixers	
1. Lines normally in service	5, 3 ft dia. (91.4 cm) influent lines
2. Lines normally on standby	1
3. In-line mixers	
Size	36 in. (91.4 cm)
Mixers per line	2 (in series)
HP	3 HP/Mixer (2.23 Kw), 6 HP/Line (4.46 Kw)
G	1000 sec^{-1}
4. Flow rate through mixer with 5 in service	
Average	12.81 mgd (0.56 m^3/sec)
Maximum	19.21 mgd (0.84 m^3/sec)
Peak instant	21.34 mgd (0.93 m^3/sec)
Chemical feed	
1. NaOCl	
Feed rate @ 8% available Cl_2 and with 5 units in service	210,000 gpd @ 90 mgd (552 l/min @ 3.94 m^3/sec)
Total feed rate per basin	32.5 gpm @ 100 mgd (123 l/min @ 4.38 m^3/sec)
	30.0 gpm @ 90 mgd (113 l/min @ 3.94 m^3/sec)
	20.0 gpm @ 60 mgd (75.7 l/min @ 2.63 m^3/sec)
2. NaOH[a]	
Feed @ 20% NaOH with 5 units in service (total)	15 mg/l average
	52 mg/l max. w/no alkalinity
	5500 gpd @ 90 mgd @ avg. (14.5 l/min @ 3.94 m^3/sec)
	21,000 gpd @ 100 mgd @ maximum (55 l/min @ 4.38 m^3/sec)
3. Feed rates based on:	
Influent NH_4^+-N	18.7 mg/l
Cl_2 : NH_4^+-N	10:1 by weight
Reaction basins and air mixing	
1. Basins normally in service[b]	5
2. Basins normally on standby	1
3. Basin dimensions	20 ft x 120 ft x 15 ft swd (6.1 m x 36.6 m x 4.6 m)
Volume	36,000 gal (136 m^3)
4. Theoretical mix time	5 basins in service
	30 min @ average flow
	20 min @ max. flow
5. Diffuser air requirements	30 scfm/1,000 cu ft (30.3 l/m^3)
Total	5,400 scfm (150 m^3/min) - 5 basins in service
Per basin	1,080 scfm (30 m^3/min)
6. Air headers and diffusers	
Main headers	2 (one at each end) 8 in. dia. (20.3 cm)
	540 scfm normal (15 m^3/min)
	1,080 scfm max. (30 m^3/min)
Cross headers	24, 2.5 in. dia. (6.4 cm) 455 scfm (12.6 m^3/min) normal
	90 scfm max. (2.5 m^3/min)
Diffusers	11/header, 264/basin, 2 ft (0.61 m) O.C. on cross headers,
	4.1 scfm (0.11 m^3/min) normal, 8.2 scfm max. (0.22 m^3/min)
7. Blowers	
Normally in service	2
Normally on standby	1
Capacity	3,000 cfm (85 m^3/min) @ 8 psi (0.60 kgf/cm^2) ea.
HP	150 HP each (112 Kw)
8. Inlet diffuser	
Diameter	48 in. (121.9 cm)
Material	FRP
Length	42 ft 6 in. (12.95 m)
Inlet ports	8 @ 3 ft 0 in. (0.9 m) O.C.

[a]NaOH needed only when alkalinity is 150 mg/l and no air stripping.

[b]Basins covered to contain product gases and lined with corrosion protection membrane to 1 ft (30.5 cm) below water surface.

The breakpoint process is controlled by pacing the sodium hypochlorite feed rate to the influent flow and influent ammonia nitrogen concentration. The pH is also monitored and controls the addition, when necessary, of sodium hydroxide to maintain an optimum pH for the breakpoint reaction.

Ammonia nitrogen in the breakpoint effluent is monitored to determine efficiency of the process. Free and combined chlorine residuals and pH are also continuously monitored in breakpoint reactor effluent.

The estimated costs of the Montgomery County facility are shown in Table 9-27. The hypochlorite generation facility will also provide hypochlorite for uses other than breakpoint although its entire cost has been shown for the breakpoint process. As noted in Table 9-27, the hydrogen liberated in the breakpoint process has a potential value of $10.00/mil gal ($0.0026/m^3) if it is collected and used as a fuel.

TABLE 9-27

ESTIMATED COSTS OF BREAKPOINT CHLORINATION AT THE MONTGOMERY COUNTY PLANT

Capital[a,b]	
Breakpoint reaction basins	$2,300,000
Operations building (including off-gas treatment and blowers)	565,000
Hypochlorite plant	
Storage	817,000
Generation	3,780,000
Salt dissolution	705,000
Total	$8,167,000
Operation and maintenance[b]	
Power	
Hypochlorite production	$ 26.20/mil gal
Mixing, stripping	18.00
Salt	25.20
Labor	12.60
Subtotal	$ 82.00 ($0.022/m^3)
Amortized capital, $8,167,000, 20 years @ 7% @ 60 mgd	35.20 ($0.009/m^3)
	$ 177.20 ($0.031/m^3)

[a]Not including contingencies and engineering

[b]December, 1974 cost levels

9.5.4 Case Examples of Selective Ion Exchange for Nitrogen Removal

Two examples of the incorporation of ion exchange into wastewater treatment plant layouts are presented in this section. In the case example for Upper Occoquan Sewage Authority, ion exchange is used in a tertiary treatment step following biological treatment. The objective of this plant is to meet an effluent limitation of 1 mg/l total nitrogen. In the Rosemont case, ion exchange is used in a physical-chemical flowsheet to meet an effluent limitation of 1 mg/l ammonia nitrogen.

9.5.4.1 Upper Occoquan Sewage Authority, Va.

This new regional plant now under construction will replace 11 small secondary plants which discharge into tributaries of a water supply reservoir which serves as the raw water source for water treatment plants serving about 500,000 people in the Virginia suburbs of Washington, D.C. Information about the water reuse aspects of the project is available in reference 49. The effluent eventually reaches a water supply reservoir in which nitrogen is believed to be one of the principal eutrophication factors. The effluent standards are shown below:

Parameter	Value
BOD_5, mg/l	1.0
COD, mg/l	10.0
Suspended solids	Unmeasurable
Phosphorous, mg/l	0.1
Methylene Blue Active Substances (MBAS), mg/l	0.1
Turbidity, JTU	0.4
Coliforms, total/100 ml	2
Nitrogen, total	1.0 mg/l

The main processes which are included in this plant are shown in Figure 9-26.

The initial plant capacity will be 22.5 mgd (0.99 m^3/sec), although an initial daily capacity of 10.9 mgd (0.48 m^3/sec) is all that will be initially certified by the State of Virginia so as to provide 100 percent complete backup facilities in the initial operation. Provisions are included in the Virginia State Water Control Board regulatory policy to increase the rated capacity to 15.0 mgd (0.66 m^3/sec) after a one year satisfactory demonstration period. Backup facilities would then constitute approximately 50 percent of the rated capacity.

The methods of nitrogen removal of biological, ammonia stripping, and breakpoint chlorination were evaluated before selection of the selective ion exchange process. This process was selected primarily because of its inherent reliability and efficiency and the

FIGURE 9-26

FLOW DIAGRAM - UPPER OCCOQUAN SEWAGE AUTHORITY PLANT (VIRGINIA)

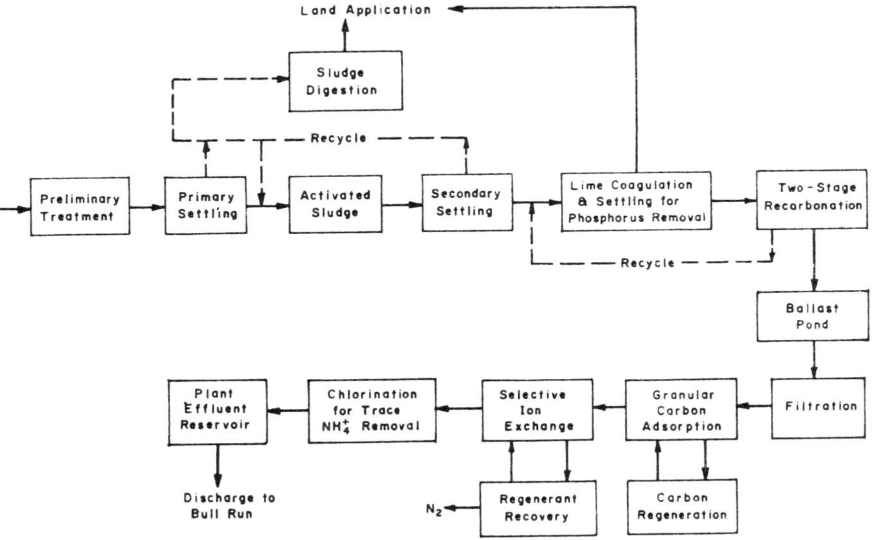

minimal effect on total dissolved solids (TDS). The selective ion exchange process was located after the carbon columns to take advantage of the available head remaining after pumping through the pressure carbon columns. Also, as an incidental benefit of this process sequence, the clinoptilolite will serve as a final polishing filter to remove the small amount of carbon fines or other suspended solids which may be present in the carbon column effluent.

Carbon fines and other solids trapped in the clinoptilolite bed are removed by backwashing before the regeneration cycle. These backwash wastes will be returned to the treatment process, typically to the chemical coagulation process where these solids will be removed in the precipitation process and be trapped in the chemical sludge.

The selective ion exchange regenerant recovery was initially planned to be accomplished by breakpoint chlorination of the ammonium using electrolytic cells (see Chapter 7). A delay in the project of about one year occurred following final design while awaiting the project funding to develop. During this period, the ammonia removal and recovery process (ARRP) described in Sections 7.3.3.2 and 8.4.1 was developed by the design engineers. Based on pilot plant results, it was concluded that the annual operating cost for the plant could be reduced by \$375,000 at a flow of 15 mgd (0.66 m^3/sec) (\$0.07/1,000 gal or \$0.18/m^3). In addition, the electrical energy requirement would be only 10 percent of the electrolytic cell breakpoint chlorination process needs. Also, a byproduct would be obtained in the form of ammonium sulfate, a common chemical fertilizer.[53] Based on this information, the Authority authorized a redesign of the regeneration facilities to incorporate the ARRP process. The following paragraphs describe the full-scale ion exchange and regenerant recovery facilities.

Eight ion exchange beds operate in parallel and are separated into two independent trains, each with four beds and a common manifold. Each bed is a horizontal steel pressure vessel, 10 ft (3.05 m) in diameter by 50 ft (15.24 m) long containing a four ft (1.2 m) deep bed of clinoptilolite (see Figures 9-27 and 9-28).[54] Each parallel train is completely independent, including piping, instrumentation and control, and electrical supply. In addition, certain backup facilities are available in each train such as key instrumentation and control. Such measures were necessary to conform to the design policy for reliability established by the Virginia State Water Control Board.

Table 9-28 is a summary of the design criteria for the ion exchange process at the future anticipated rating of 15 mgd (0.66 m^3/sec). The system will be entirely automated using automatic valves in a manner similar to most larger water treatment plant filtration facilities. Regeneration will be initiated either on a run time basis, volume throughput basis, or manually. Backwashing will be done before each regeneration. Backwash water will be returned to the wastewater process, typically to the chemical coagulation process, or to the plant headworks.

TABLE 9-28

DESIGN CRITERIA SELECTIVE ION EXCHANGE PROCESS FOR AMMONIUM REMOVAL AT THE UPPER OCCOQUAN PLANT (VIRGINIA)

Flow rate	15 mgd (0.66 m^3/sec)
Beds in service	4
Beds in regeneration	2
Beds - backup capacity	2
Flow per bed	3.75 mgd (0.16 m^3/sec)
Bed loading rate	10.82 bed volumes/hr 5.25 gpm/sf (3.6 l/m^2/sec)
Backwash rate	8 gpm/sf (5.4 l/m^2/sec)
Bed volumes to exhaustion	145
Average ammonia removal efficiency	95%
Average influent ammonia nitrogen concentration	20 mg/l
Average effluent ammonia nitrogen concentration	1 mg/l
Normal concentration of ammonia nitrogen at initiation of regeneration	2.5 mg/l
Clinoptilolite size	20 x 50 mesh
Clinoptilolite depth	4 ft (1.22 m)

FIGURE 9-27

PLAN AND SECTION OF ION EXCHANGE BEDS AT UPPER OCCOQUAN PLANT

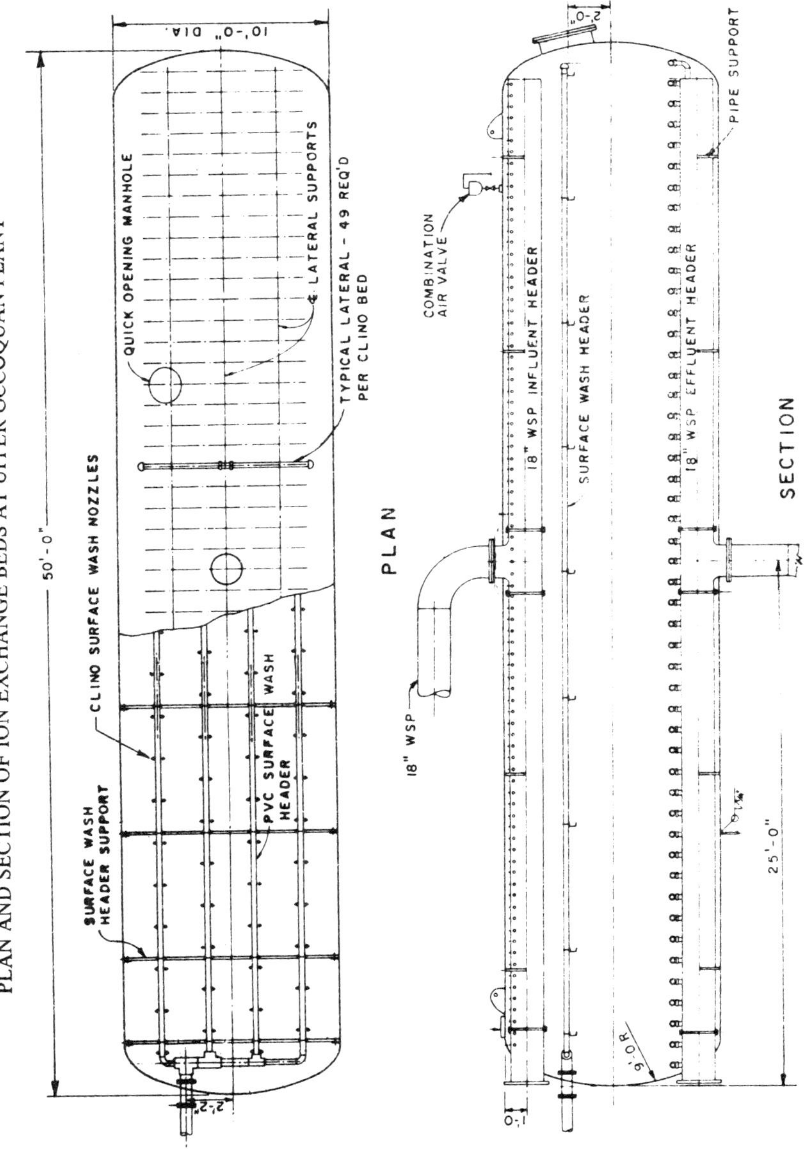

FIGURE 9-28

ADDED DETAILS - ION EXCHANGE BEDS AT UPPER OCCOQUAN PLANT

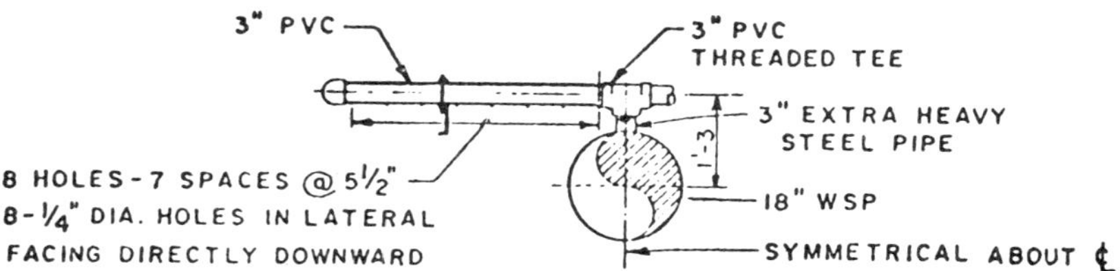

TYPICAL UNDERDRAIN LATERAL

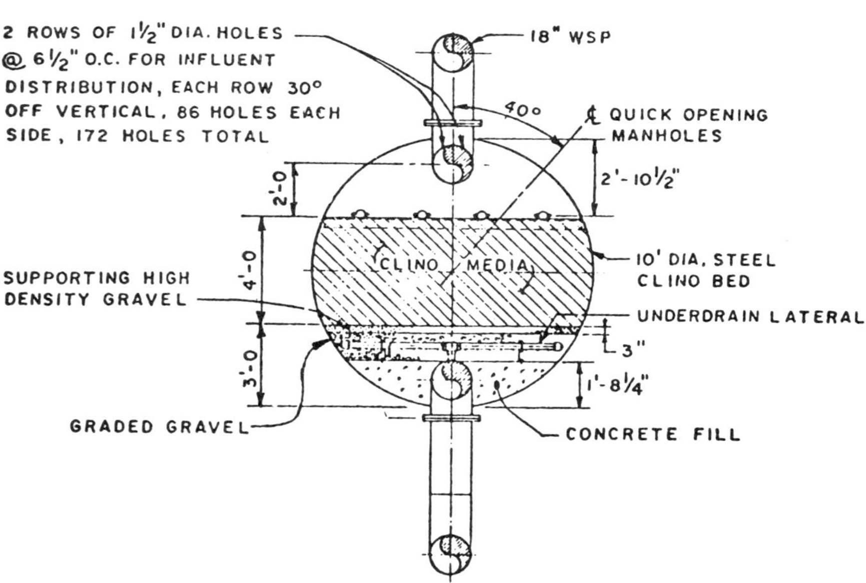

The beds will be regenerated with a 2 percent sodium chloride solution. The regeneration process will be that shown in Figure 7-13 and as described in Section 7.3.3.2. The regenerant recovery system consists of four 375,000 gallon (1420 m^3) tanks, a regenerant pumping system and associated automatic valves, two 35-ft (10.7 m) diameter clarifiers for magnesium hydroxide removal, and 18 ARRP modules. Figures 9-29 and 9-30 illustrate the ARRP module design. The ARRP units will be shop fabricated. The basic tower units are 12-ft (3.66 m) diameter fiberglass tanks. All materials will be fiberglass or PVC. Each tower has a 25 HP (18.6 kw) fan. The air rate is approximately 34,000 cfm/tower (952 m^3/min) at an air to liquid ratio of 566 ft^3/gal (4.15 m^3/l). Tower air velocities are 300 fpm (91.4 m/min). Knitted mesh mist eliminators prevent moisture carryover from tower to tower. The total system head loss is about 2.5-3.0 inches (6.4 cm - 7.6 cm) of which about 1.5 inches (3.8 cm) is in the media. The media is 2-inch (5.1 cm) diameter polypropylene plastic packing (Tellerette). A summary of the regeneration system design criteria is shown in Table 9-29.

FIGURE 9-29

PLAN VIEW OF ARRP MODULE - UPPER OCCOQUAN PLANT

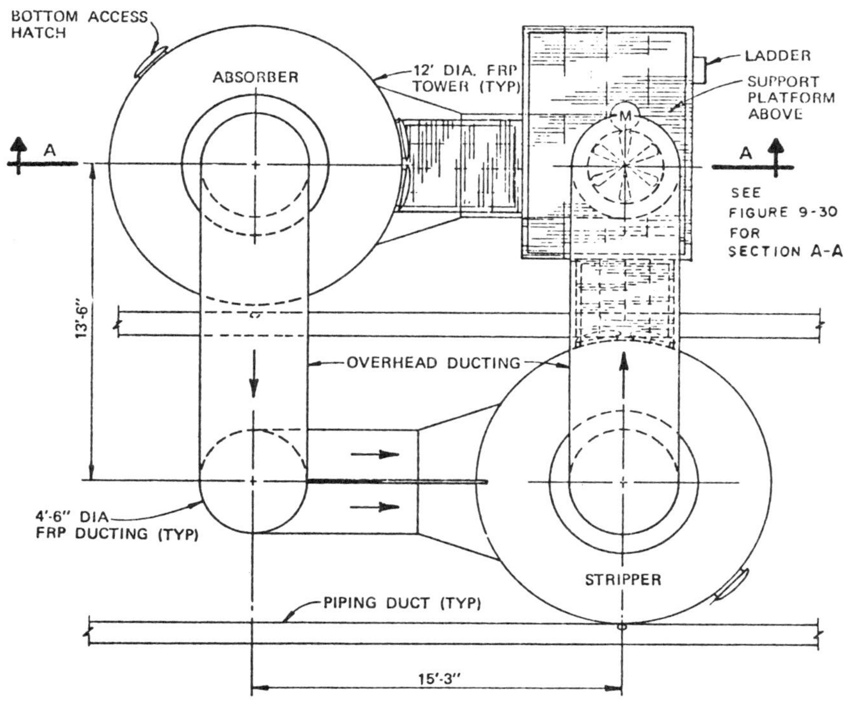

FIGURE 9-30

SECTION OF ARRP MODULE - UPPER OCCOQUAN PLANT

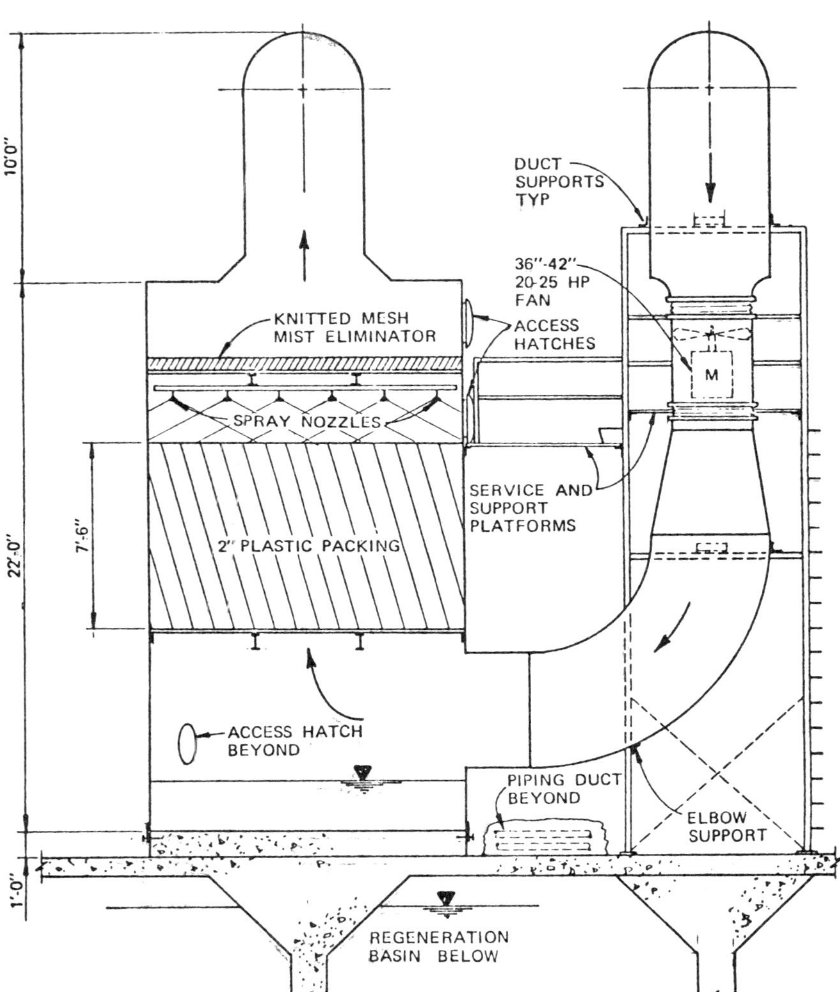

TABLE 9-29

REGENERATION AND REGENERANT RECOVERY SYSTEM DESIGN CRITERIA AT THE UPPER OCCOQUAN PLANT (VIRGINIA)

Regeneration system[a]	
Number of regenerant tanks	4
Size of each tank	375,000 gals (1420 m^3)
Number of beds regenerated at once	2
Number of regeneration cycles per day	3.58
Regeneration bed volumes	39-44
Regenerant recovery system[a]	
Recovery system flow rate	1,080 gpm (68 l/sec)
Operation time per day	16 hr
Clarifiers	
Number of units	2
Diameter	35 ft (10.66 m)
Overflow rate	800 gpd/sf (32.6 m^3/m^2/day)
Ammonia removal and recovery process[a]	
Number of ARRP modules	18
Liquid loading rate	760 gpd/sf (31.1 m^3/m^2/day)
Air to liquid loading rate	566 cf/gal (4234 m^3/m^3)
Media height	7.5 ft (2.29 m)
Removal efficiency at 10 C	90%
at 20 C	95%

[a]At 15 mgd flow rate

The average ammonia-nitrogen concentration from ion exchange beds will be about one mg/l. The organic nitrogen is expected to be 0.5-0.8 mg/l and the nitrate nitrogen is expected to be from 0.1-0.2 mg/l. Thus, the total nitrogen leaving the ion exchange process will be about 1.6-2.0 mg/l. Since the discharge standard is 1.0 mg/l, additional nitrogen removal is necessary. This will be accomplished by breakpoint chlorination of the ion exchange effluent. A dosage of approximately 8-10 mg/l will result in nearly complete removal of ammonia nitrogen. The final effluent is then expected to have a total nitrogen concentration of less than 1 mg/l.

The estimated costs are shown in Table 9-30. Since the initial constructed capacity of 22.5 mgd (0.99 m^3/sec) may be operated at no more than 15 mgd (0.66 m^3/sec) because of state requirements for backup capacity, costs are shown on the basis of operation of two-thirds of the constructed capacity and at the full constructed capacity, which would be the more generally applicable circumstance. The income from sale of ammonium sulfate is based on the lowest wholesale price in effect at the time of this writing, $43/ton ($39/metric ton). In some areas, the wholesale price is as high as $65/ton ($59/metric ton).

9.5.4.2 Rosemount, Minnesota

This new 0.6 mgd (2271 cu.m/day) plant operated by the Metropolitan Sewer Board of the Twin Cities area (Minneapolis-St. Paul) provides independent physical-chemical treatment of a municipal wastewater.[55,56] This plant is the first full-scale physical-chemical plant to be

TABLE 9-30

ESTIMATED COSTS OF SELECTIVE ION EXCHANGE AT THE UPPER OCCOQUAN PLANT (VIRGINIA)

Item	Estimated costs, $/mil gal	
	at 15 mgd ($0.66\ m^3/sec$)	at 22.5 mgd ($0.99\ m^3/sec$)
Operating and maintenance[a]		
Chemicals		
NaOH	$ 26.80	$ 26.80
NaCl	7.10	7.10
H_2SO_4	9.80	9.80
	$ 43.70	$ 43.70
Income from sale of $(NH_4)_2\ SO_4$ @ $43/ton	$ (12.60)	$ (12.60)
Net chemical cost	31.10	31.10
Power, 18 HP/mil gal @ $0.0192/kwh	6.90	6.90
Labor	17.70	17.70
Total, O & M	$ 55.70	$ 55.70
Capital[a]		
$4,470,000, 20 years @ 7%	$ 77.22	$ 51.59
Total annual cost[a]	$132.92 ($0.035/m^3)	$107.29 ($0.027/m^3)

[a]August, 1974 costs

placed in operation in the U.S. A schematic of the process is shown in Figure 9-31. The entire plant is enclosed in a single 14,500 sq ft (1347 m^2) steel building. Final effluent standards are as follows:

Parameter	Value
BOD_5, mg/l	10
Suspended solids, mg/l	10
COD, mg/l	10
Ammonia – Nitrogen, mg/l	1
Phosphorus, total (P), mg/l	1
pH	8.5

FIGURE 9-31

SCHEMATIC OF ROSEMOUNT, MINNESOTA PLANT

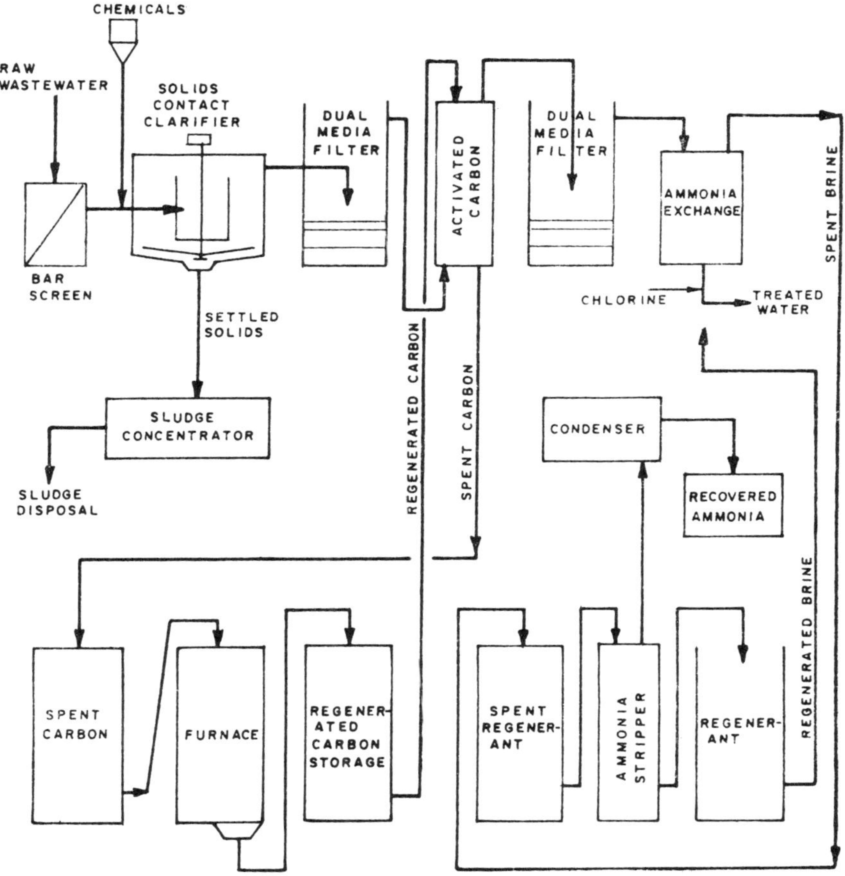

Selective ion exchange is accomplished by clinoptilolite in 2 of 3 downflow columns in series, each containing a 6-ft (1.82 m) depth of clinoptilolite. When the ammonia-nitrogen reaches 1 mg/l in the effluent from the polishing column, the lead column is removed from service and regenerated (upflow) and the third column placed on line. The steam process discussed in Section 7.3.3.3 is used for clinoptilolite regeneration and ammonia recovery. Brine is stored at 77 C and is cooled to 27 C while passing through a heat exchanger on the way to the column. The waste brine leaving the column is passed through the other side of the heat exchanger to elevate its temperature to 71 C before entering the stripping process. Brine temperature in the storage tanks is controlled by steam supplied to internally mounted coils. Waste brine is collected in a mixed storage tank before being stripped. Soda ash is added to the storage tank to elevate the pH to 12, which results in the precipitation of calcium carbonate and magnesium hydroxide. Mixing is discontinued 20 minutes after the pH has reached 12 to allow these precipitates to settle. The sludge is then pumped from the bottom of the waste brine storage tank. The waste brine is then pumped to the steam stripper at a rate of 53 gpm (3.34 l/sec) and a steam flow of 3000 lb/hr (1498 kg/hr). The stripper operates at an equilibrium temperature of 104 C and the condensor at 38 C. The time required for a stripping/reclaiming operation is about 5 hr. Design criteria are summarized in Table 9-31. The plant was in the start-up phase at the time of this writing and no operating data were available.

TABLE 9-31

ROSEMOUNT (MINNESOTA) ION EXCHANGE DESIGN CRITERIA

Ammonium exchange columns (Two trains of 3)	
Loading rate[a]	4.2 gpm/sf (2.85 l/m²/sec), 5.6 BV/hr
Clinoptilolite capacity	
per unit volume	0.3 lb/cu ft (4.86 kg/m³)
per column	90 lb (40.8 kg)
Ammonia nitrogen loading rate	50 lb/day (22.7 kg/day)
Ammonia removal	95%
Clinoptilolite depth per column	6 ft (1.83 m)
Clinoptilolite size	20 x 50 mesh
Normal operation	2 columns in series, 250 BV/cycle
Backwash rate	8 gpm/sf (5.43 l/m²/sec)
Regeneration system	
Brine solution to columns[b]	
Hydraulic application rate	2.0 gpm/sf (1.36 l/m²/sec)
Volume	4.5 BV
Strength	6% NaCl
Temperature	71 C
pH[c]	11
Brine solution regeneration	
Regeneration cycle length	5 hr
Hydraulic loading rate to steam stripping tower	7 gpm/sf (4.75 l/m²/sec)
Tower depth	24 ft (7.3 m)
Caustic soda added	3 lb/lb NH_4^+-N
Bed rinse[a]	
Rinse rate	300 gpm (18.9 l/sec)
Time	70 min
Ammonia recovery	
Aqueous ammonia strength	1%
Aqueous ammonia volume	1,000 gpd (3,785 lpd)
Ammonia stripper	
Steam @ 10 psig	3,300 lb/hr (1,498 kg/hr)
Throughput	53 gpm (3.34 l/sec)
Size - diameter	3 ft (0.91 m)
height	18 ft (5.5 m)

[a]Downflow

[b]Upflow

[c]Elevated with caustic soda

9.5.5 Case Examples of Air Stripping for Nitrogen Removal

Two plants which utilize air stripping for nitrogen removal are described in this section. Both of these, South Lake Tahoe and Orange County, use the biological-tertiary approach in which ammonia stripping is used after biological treatment. At Tahoe, nitrogen removal is incorporated on an experimental basis, as no nitrogen removal requirement exists. At the Orange County Water District plant, nitrogen is removed to permit recharge of groundwater.

9.5.5.1 South Lake Tahoe, California

At South Tahoe an experimental full-scale ammonia stripping tower built to handle half of the total plant design flow of 7.5 mgd (0.33 m^3/sec) has been operated on an intermittent basis since 1969.[13,57,58,59] The plant flow sheet is shown in Figure 9-3. The stripping process was installed at South Lake Tahoe as an EPA research and demonstration installation and not as the result of any requirement to remove nitrogen from the plant effluent. In the absence of any need for full time operation, the purposes of this stripping tower have included: (1) demonstration of full-scale tower efficiencies as compared to pilot plant test results; (2) determination of cold weather operating limitations and other operating problems with investigation of solutions to these problems and (3) collection of data for design purposes for future expansion of this process to full plant design capacity as well as for use in planning similar facilities at other locations. The large tower capability to remove ammonia almost exactly duplicates the results of pilot plant operation, reaching 95 percent removal in warm weather. The cold weather operating limitations and recommended tower design improvements have been determined as discussed in Chapter 8. The new design criteria obtained from operating the tower have already been used to prepare plans for stripping towers to treat wastewaters at Orange County, California[60] and the operating experience has also provided the direction and the basis for expansion of the Tahoe nitrogen removal facilities to meet anticipated future requirements for nitrogen removal. As discussed later in this section and in Section 8.5, the original packed tower system has recently been modified to provide year round, full-scale nitrogen removal.

The design data for the original packed tower are given in Table 9-32. It is a crossflow cooling type tower modified for ammonia stripping. The overall dimensions of the tower are 32 ft (9.75 m) x 64 ft (19.5 m) x 47 ft (14.3 m) high. Water, at pH 11, was pumped to the top of the tower by either or both of two constant-speed pumps. These pumps were backflushed two or three times daily to minimize buildup of calcium carbonate scale in the pump units. When the plant inflow was less than the rate at which the pumps were delivering water to the tower, some water was recycled from the tower effluent back to the pump suction well. This avoided the need for variable speed pump control, and at the same time provided some recirculation through the tower, which improved ammonia removal. At the top of the tower, the influent water entered a covered distribution box and overflowed to a distribution basin. The distribution basin is a flat deck with a series of holes fitted with plastic nozzles. Further distribution of the inflow was provided by diffusion decks immediately below the distribution basin. Three other diffusion decks were provided at 6 ft (1.82 m) vertical intervals in the fill. The tower fill provided, theoretically, 215 successive droplet formations as the water passed down through the tower. The tower effluent fell into a concrete collection basin which also formed the base for the tower structure. From the collection basin, the tower effluent passed through a Parshall measuring flume into the first-stage recarbonation chamber, where excess pumpage returned through a flap gate into the tower pump sump to recirculate through the tower.

TABLE 9-32

DESIGN DATA, AMMONIA STRIPPING TOWER AT SOUTH LAKE TAHOE, CALIFORNIA

Nominal capacity:	3.75 mgd (0.164 m^3/sec)			
Type:	Cross flow with central air plenum and vertical air discharge through fan cylinder at top of tower			
Fill:	Plan area, 900 sq ft (83.6 m^2) Height, 24 ft (7.3 m) Splash bars: Material, rough sawn treated hemlock, size, 3/8 in. x 1-1/2 in. (0.95 cm x 3.8 cm) spacing, vertical 1.33 in. (3.37 cm) horizontal 2 in. (5.08 cm)			
Air flow:	Fan, two-speed, reversible, 24-ft diameter, horizontal			
	Water rate		Air rate	
	gpm (l/sec)	gpm/sf (l/m^2/sec)	cfm (m^3/min)	cf/gal (m^3/m^3)
	1,350	1.0	750,000	550
	(85)	(0.68)	(21,000)	(4,115)
	1,800	2.0	700,000	390
	(114)	(1.36)	(19,600)	(2,918)
	2,700	3.0	625,000	230
	(170)	(2.04)	(17,500)	(1,720)
Tower structure:	redwood			
Tower enclosure:	corrugated cement asbestos			
Air pressure drop:	1/2 in. of water at 1 gpm/sf (1.27 cm @ 0.68 l/m^2/sec)			

Air entered the tower through side louvers, passed horizontally through the tower fill and drift eliminators (or air flow equalizers), and entered a central plenum. At the top center of the plenum is a 24 ft (7.3 m) diameter, six-bladed, horizontal fan. Fan blades and fan cylinder are both made of glass-reinforced polyester. The fan takes suction from the plenum and discharges to the atmosphere through the fan cylinder. The fan has a maximum capacity of about 750,000 cfm (21,000 m^3/min). It is equipped with a two-speed reversible 75 HP (56 kw) motor.

The limitations of the packed tower at Tahoe are caused by the cold winter temperatures and the scaling of the tower packing. Because the scaling problem was not anticipated from the pilot studies, the access to the tower packing needed to remove the scale with water jets was not provided. As a result, portions of the original hemlock packing became hopelessly fouled with calcium carbonate scale, thereby decreasing effective tower packing area.

The costs of the operation of the packed tower at Tahoe at a 2 gpm/sq ft (1.36 l/m^2/sec) rate as estimated for a 7.5 mgd (0.33 m^3/sec) scale for continuous operation are listed in Table 9-33. Operating labor included backflushing of tower pumps and cleaning the distribution deck to remove $CaCO_3$ precipitation, process inspection, lubrication, and daily determination of tower ammonia removal efficiency. Maintenance costs reflect the cost of cleaning tower fill.

Because of the cold weather limitations of packed tower inherent to the Tahoe location and the reduced efficiency of the existing tower due to scaling of the tower packing, a research program was initiated at the South Lake Tahoe plant to develop alternate, low-cost techniques which could be applied to the full-scale plant while using as much of the existing facilities as possible. Although there are no current regulatory agency requirements for nitrogen removal for exported wastes, future requirements for effluent reuse or disposal in the Lake Tahoe basin will probably include nitrogen removal. At the current time, the States of California and Nevada require that all wastewaters, regardless of the degree of treatment be exported from the Lake Tahoe basin. The South Tahoe effluent is exported to Alpine County, California where it is used to form the 1 billion gallon (3.785 billion liters) Indian Creek Reservoir.[59] An excellent trout fishery has been established in this recreation reservoir and it is necessary to control pH and ammonia concentrations to prevent fish toxicity. Thus, anticipated future regulatory agency requirements and current effluent reuse practices required that a system for full-scale nitrogen removal be developed for the South Lake Tahoe plant.

Inspired by observations of ammonia release from holding ponds in Israel,[61] research was undertaken at Tahoe to improve the release of ammonia from the high pH effluent. The results led to the following three steps being applied to a full-scale modification of the Tahoe system:[62]

1. Holding in high pH ponds (with surface agitation in one pond).

2. Stripping in a modified crossflow forced draft tower through air sprays installed in the tower.

3. Breakpoint chlorination.

Design data are given in Table 9-34 and the entire system is depicted in Figure 8-9. Because ammonia removals by stripping vary so much with temperature, and since low temperatures, high ammonia concentrations, and high flows never occur simultaneously at this location, design data are presented for two sets of conditions which are expected to occur: low flow (2.5 mgd or 0.11 m^3/sec), low water temperature (3 C), and low ammonia content; and high flow (7.5 mgd or 0.33 m^3/sec), high water temperature (22 C), and high ammonia content.

TABLE 9-33

OPERATING COSTS FOR AMMONIA STRIPPING FOR CONTINUOUS OPERATION OF TAHOE AIR STRIPPING TOWER AT 7.5 MGD

Operating cost per day[a]	$/Day
Electricity[b]	60.78
Operating labor	4.63
Maintenance labor	5.17
Repair material	.78
Instrument maintenance	.94
Total operating cost	72.30
Total cost per mil gal[a]	$/mil gal
Operating	9.64
Capital	8.00
Total	17.64

[a]1970 dollars

[b]Average cost per day at 7.5 mgd from 10 months of continuous operation, i.e., May, 1969, through September, 1970.

TABLE 9-34

DESIGN DATA AND ESTIMATED NITROGEN REMOVALS
FOR ALL-WEATHER AMMONIA STRIPPING AT SOUTH TAHOE, CALIFORNIA

Description	Flow 2.5 mgd ($0.11\ m^3/sec$) Water Temp. 3 C NH_3-N		Flow 7.5 mgd ($0.33\ m^3/sec$) Water Temp. 22 C NH_3-N	
	Estimated reduction, percent	Remaining, mg/l	Estimated reduction, percent	Remaining, mg/l
Influent, pH = 11.0		15		35
Holding ponds				
Detention = 7 hr			15	30
Detention = 18 hr	10	13.5		
Air spraying in second pond				
Turnovers = $4\frac{1}{2}$			28	
Turnovers = $13\frac{1}{2}$	30	9.5		21.5
Stripping tower, air spraying with forced draft				
Recycle turnovers = 1.6			23	16
Recycle turnovers = 5	42	5.4		
Overall removal	64		40	
NH_3-N remaining to be removed by by breakpoint chlorination, mg/l		5.4		16
Chlorine required for breakpoint chlorination, lb/day		1,130		10,000

It is difficult to predict accurately the performance of the full-scale ponds and natural wind-vented sprays from the results of laboratory and pilot plant data, because the efficiency of ammonia removal by these methods depends to a great extent on the immediate removal of the released ammonia gas from the vicinity of the air-water interface. The efficiency of air sweeping in removing ammonia declines as the area of the ponds increases. In the absence of full-scale plant data, the removal efficiencies obtained in the laboratory and pilot plant[62] were heavily discounted in estimating the performance of the full-scale plant shown in Table 9-34. The following paragraphs describe the modified Tahoe stripping process.

The two high pH ponds also serve to equalize flow to the modified stripping tower and breakpoint facilities. This reduces design capacity requirements for these two processes and improves their operating characteristics. The total surface area of these ponds is about 64,000 sq ft (5946 m^2). The water depth varies from 3.0 to 7.25 ft (0.9 m - 2.2 m). Detention time ranges from 7 hr, at 7.5 mgd (0.33 m^3/sec) to 18 hr at 2.5 mgd (0.11 m^3/sec). The second pond in the series is provided with surface agitation in order to increase the ammonia removal. The system of surface agitation consists of sprinkling with 34 mgd (1.5 m^3/sec) of recycled pond water through 2 pumps and a spray system consisting of about 75 vertical nozzles each delivering about 320 gpm (20.1 l/sec). At the 2.5 mgd (0.11 m^3/sec) plant flow, the spray system recycles the pond water 13 times, and at the 7.5 mgd (0.33 m^3/sec) flow, 4 recycles occur. The spray nozzles are 4 in. x 2½ in. (10.2 cm x 6.35 cm) female pipe reducers each fitted with a 2½ in. x 1½ in. (6.35 cm x 3.8 cm) bushing. Each spray orofice is about 1-7/8 in. (4.8 cm) in diameter. Nozzles having interval vanes or other obstructions introduce air containing carbon dioxide to the water in the nozzle. This causes deposition and rapid buildup of calcium carbonate within the nozzle. Such nozzles are unsatisfactory because of the resulting plugging and flow restriction problems. Only nozzles with unobstructed clear opening, and without internal vanes, should be used.

A major modification has been made to the existing stripping tower. The existing packing has been removed, and the entire area of the tower equipped with water sprays. The existing trays at the top of the tower distribute part of the flow, and 4 nozzle equipped headers in the bottom of the tower spray water upward into the tower. The pump capacity to the tower is 11.8 mgd (0.52 m^3/sec). At cold weather plant flow rates of 2.5 mgd (0.11 m^3/sec), this flow will provide a recycle rate of 5 through the sprays in the tower. The capacity and type of nozzle used in the tower is similar to the nozzles used in the ponds. Based on plant scale tests of this spray system, with the induced draft fan operating at high speed, it is anticipated that at least 42 percent of the ammonia in the pond effluent will be removed in the tower under cold weather operating conditions.

Chlorine may be added at two points in the process (Figure 8-9). The first point of application is in the primary recarbonation chamber at a pH of 11.0. Only enough chlorine is added to reduce the pH to about 9.6, thus eliminating the need for addition of carbon dioxide (CO_2) at this point. About 65 mg/l is required to reduce the pH from 11.0 to 9.6. The balance of the chlorine needed to reach the breakpoint for complete ammonia removal is added in a chamber immediately downstream from the secondary recarbonation chamber. At times sufficient chlorine is added at this point to reduce the pH to 7.0 or less, so that no CO_2 will be required. A dose of approximately 160 mg/l of chlorine is required to reduce the pH from 11.0 to 7.0. When the breakpoint is reached with a lesser dose of chlorine, then some CO_2 must be added in the secondary stage of recarbonation to produce a pH of 7.0. About 10 mg/l of chlorine is required for each mg/l of NH_4^+ –N to reach breakpoint. After breakpoint chlorination treatment, water from the ballast pond is pumped to the existing filters and carbon columns. The carbon columns remove any excess chlorine. The modified Tahoe process was being placed in operation at the time of this writing and full-scale data are not yet available.

9.5.5.2 Orange County Water District, California

The Orange County Water District (OCWD) at Santa Ana, California has under construction a 15 mgd (0.66 m^3/sec) wastewater reclamation plant and a 3 mgd (0.13 m^3/sec) seawater desalting plant at the same site.[60] The water from the two plants will be blended and pumped into a line of injection wells which are located on the land side of a line of seawater pumping wells to form a barrier against seawater intrusion into the fresh water aquifer.[63] The injection system also serves to replenish the supply of groundwater available for use. The OCWD water reclamation plant will take 15 mgd (0.66 m^3/sec) of trickling filter effluent from the secondary treatment plant of the Orange County Sanitation District. It will be subjected to high lime treatment at pH of 11.0 and clarification in a basin equipped with settling tubes. The clarified high pH water will be pumped to two countercurrent ammonia stripping towers. In the climate at this location, freezing temperatures are not experienced, and waste heat from the desalting plant operation will be used to heat the inlet air to the stripping towers for increased ammonia removal efficiency. The design also fully realizes scaling problems encountered elsewhere and incorporates provisions for scale control even though scaling was not a problem in pilot tests at this site.

The plant is designed with two ammonia stripping/cooling towers each equipped with six 18 ft (5.49 m) diameter fans. An end section view of a tower is shown in Figure 9-32 and an overall view is shown in Figure 9-33. The stripping sections are designed for a hydraulic loading of 1.0 gpm/ft^2 (0.68 l/m^2/sec) and an air flow of 400 cu ft/gal (2,990 m^3/m^3). The cooling sections are designed to cool the desalting process water and brines from 46-49 C to 27-29 C and will raise the air temperature to the stripping section to 31-33 C. Splash-bar packing will be used for ammonia stripping and a film packing located in the air inlet plenum will be used for cooling.

A prime design criterion was that the ammonia stripping packing be accessible and removable for cleaning because scaling of the packing might reduce air flow and ammonia removal efficiency. Provisions have been made to feed a scale inhibiting polymer, if needed, to the tower influent. The warm saturated air exhausted from the cooling sections theoretically would permit the tower packing to be less than the design height of 25 ft (7.6 m). Because during the first few years of operations the desalting plant will operate on an intermittent basis, it was not considered prudent to reduce the packing depth.

The tower fill or packing is made from ½ in. (1.27 cm) diameter Schedule 80 PVC pipe at 3 in. (7.6 cm) centers horizontally and with alternate layers placed at right angles and at 2 in. (5.1 cm) centers vertically (see Figure 9-34). The fill was factory prefabricated in modules which are about 6 ft (1.82 m) by 6 ft (1.82 m) by 4 ft (1.22 m) high. Each module is supported within its own steel or fiberglass frame, so that it is easily removable if necessary for cleaning. An overhead hoist and moveable dolly are provided to assist in packing removal for cleaning. However, the access corridors and the removable air baffle panels in the tower should make it possible to reach all of the fill in its normal operating position within the tower for hosing down to remove any excessive calcium carbonate scale which may form.

Based on the results of extensive pilot plant tests at the OCWD, it is expected that the towers will remove more than 90 percent of the ammonia from the wastewater. The ammonia remaining in the tower effluent will be removed by breakpoint chlorination using about 10 mg/l of chlorine for each mg/l of ammonia nitrogen still present. No difficulty is anticipated in meeting the limits set by the regulatory agency of 1.0 mg/l of ammonia-nitrogen in the injection water. The Orange County design offers an example of a design with adequate scale control provisons incorporated.

FIGURE 9-32

ORANGE CO. AMMONIA STRIPPING/COOLING TOWER SECTION

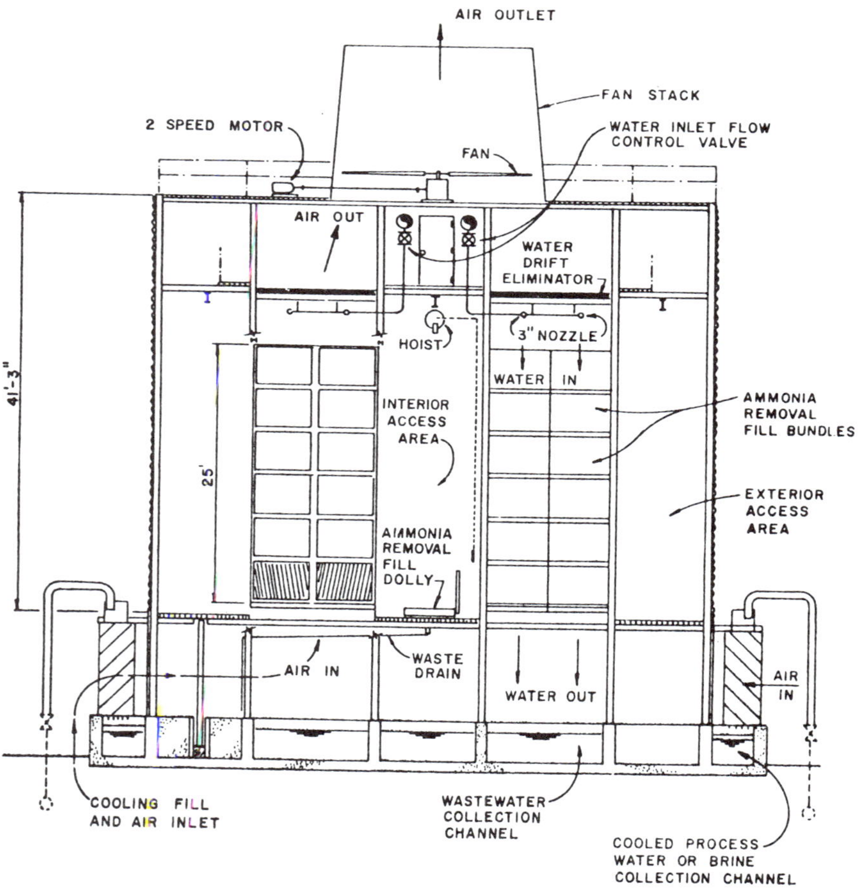

The estimated costs (1974 costs) for the Orange County ammonia stripping towers are as follows:

Capital	\$42/mil gal
Operating	\$29/mil gal
Total	\$71/mil gal

These costs are substantially higher than those reported for the Tahoe facility. The difference results from the following factors: Orange County towers were designed at 1 gpm/ft^2 (0.68 l/m^2/sec) loading rate while Tahoe towers were designed at 2 gpm/ft^2 (1.36 l/m^2/sec); extensive provisions for tower cleaning were included at Orange County; heat exchange facilities were included in the Orange County design but no credit taken in the tower design; Orange County costs were for 1974 vs the 1970 basis for the Tahoe costs.

FIGURE 9-33

OVERALL VIEW OF THE ORANGE COUNTY WATER DISTRICT (CALIF.) (DESALTING PLANT IS IN RIGHT BACKGROUND, CHEMICAL CLARIFIER IN RIGHT FOREGROUND. NOTE WALLS AT BASE OF TOWERS TO SHIELD NOISE AT AIR INLET.)

FIGURE 9-34

STRIPPING TOWER PACKING MODULE AT THE ORANGE COUNTY WATER DISTRICT PLANT (CALIF.)

9.6 References

1. Horstkotte, G.A., Niles, D.G., Parker, D.S., and D.H. Caldwell, *Full-Scale Testing of a Water Reclamation System.* JWPCF, 46, No. 1, pp 181-197 (1974).

2. Mulbarger, M.C., *The Three Sludge System for Nitrogen and Phosphorus Removal.* Presented at the 44th Annual Conference of the Water Pollution Control Federation, San Francisco, California, October, 1971.

3. *Description of the El Lago, Texas, Advanced Wastewater Treatment Plant.* Harris County Water Control and Improvement District Number 50, Seabrook, Texas, March, 1974.

4. Bishop, D.F., O'Farrell, T.P., Cassel, A.F., and A.P. Pinto, *Physical-Chemical Treatment of Raw Municipal Wastewater.* Report prepared for the Environmental Protection Agency, Contract Number 14-12-818, February, 1973.

5. Koon, J.H., and W.J. Kaufman, *Optimization of Ammonia Removal by Ion Exchange Using Clinoptilolite.* Environmental Protection Agency Water Pollution Control Research Series No. 17080 09/71, 1971.

6. Bishop, D.F., Schuk, W., Yarrington, R., Bowers, J.F., Fein, E.D., and H.W. Treupel, *Physical-Chemical Wastewater Treatment Under Direct Digital Control.* Presented at the International Workshop: Instrumentation, Control and Automation for Wastewater Treatment Plants, London, September 15-20, 1973.

7. Barnes, R.A., Atkins, P.F., Jr., and D.A. Scherger, *Ammonia Removal in a Physical-Chemical Wastewater Treatment Process.* Prepared for the EPA, Report No. R2-72-123, November, 1972.

8. Brown and Caldwell, *Report on Tertiary Treatment Pilot Plant Studies.* Prepared for the City of Sunnyvale, California, February, 1975.

9. *Process Design Manual for Phosphorus Removal.* U.S. EPA, Office of Technology Transfer, Washington, D.C. (1971).

10. *Process Design Manual for Carbon Adsorption.* U.S. EPA, Office of Technology Transfer, Washington, D.C. (1973).

11. Brown and Caldwell, Consulting Engineers, *Lime Use in Wastewater Treatment.* Report submitted to the U.S. Environmental Protection Agency, 1975.

12. Culp, G.L., Wesner, G.M., and R.L. Culp, *1974 Lake Tahoe Advanced Wastewater Treatment Seminar Manual.* Clean Water Consultants, El Dorado Hills, California, 1974.

13. South Tahoe Public Utility District, *Advanced Waste Treatment as Practiced at South Tahoe.* EPA WPCRS 17010ELQ8/71, August, 1971.

14. Brown and Caldwell, *Lime Sludge Recycling Study.* Report to the Central Contra Costa Sanitary District, June, 1974.

15. Battelle Memorial Institute, Pacific Northwest Laboratories, *Evaluation of Municipal Sewage Treatment Alternatives.* Prepared for the Council on Environmental Quality and the Environmental Protection Agency, Contract EQC 316, February, 1974.

16. Greene, R.A., *Complete Nitrification by Single Stage Activated Sludge.* Presented at the 46th Annual Conference of the Water Pollution Control Federation, Cleveland, Ohio, October, 1973.

17. McNamee, Porter and Seeley, Consulting Engineers, *Plans for Sewage Treatment Plant Additions, City of Jackson, Michigan.* Submitted to the Michigan Water Resources Commission, October, 1970.

18. Greene, R.A., Personal Communication to Parker, D.S., City of Jackson, Michigan, September, 1974.

19. Brown and Caldwell, *Project Report for the Wastewater Treatment Plant Stage 3 Improvements.* Valley Community Services District, September, 1972.

20. Valley Community Services District, *Wastewater Treatment Plant Operating Reports.*

21. Johnson, James L., Personal Communication to G.A. Carthew. Valley Community Services District, November, 1974.

22. Environmental Quality Analysts, Inc., Letter Report to Valley Community Services District, March, 1974.

23. Brown and Caldwell, *Contract for Treatment Plant Enlargement – 1965.* Prepared for the City of Livermore, December, 1965.

24. City of Livermore, California, *Water Reclamation Plant Operating Reports,* 1971.

25. John Carollo Engineers, *Contract Documents for San Pablo Sanitary District Wastewater Treatment Plants – 1970 Additions.* 1970.

26. Kennedy, Bill, personal communication with D.S. Parker. San Pablo Sanitary District, November, 1974.

27. San Pablo Sanitary District, California, *Wastewater Treatment Plant Operating Reports.* June, 1973 to July, 1974.

28. Weddle, C.L., Personal communication to D.S. Parker, Bechtel Corp. October, 1974.

29. Eisenhauer, D.L., Sieger, R.B. and D.S. Parker, *Design of an Integrated Approach to Nutrient Removal.* Presented at the EED-ASCE Specialty Conference, Penn. State University, Pa., July, 1974.

30. Leary, R.D., *et al, Effect of Oxygen-Transfer Capabilities on Wastewater Treatment Plant Performance.* Journal of the Water Pollution Control Federation, 40, p 1978 (1968).

31. Leary, R.D., *et al. Full Scale Oxygen Transfer Studies of Seven Diffuser Systems.* Journal of the Water Pollution Control Federation, 41, p 459 (1969).

32. Aberley, R.C., Rattray, G.B. and P.P. Dougas, *Air Diffusion Unit.* JWPCF, 46, No. 5, pp 895-910 (1974).

33. Parker, D.S., Carthew, G.A. and G.A. Horstkotte, *Lime Recovery and Reuse in Primary Treatment.* EED of the ASCE, in press.

34. Flanagan, M.J., *Direct Digital Control of Central Contra Costa Sanitary District Water Reclamation Plant.* Presented at IAWPR Specialized Conference, London, England, September, 1973.

35. Caldwell Connell Engineers, *Design Report, Lower Molonglo Water Quality Control Centre.* Report to the National Capital Development Commission, April, 1971.

36. Caldwell Connell Engineers, *Revisions to Design Report, Lower Molonglo Water Quality Control Centre.* Report to the National Capital Development Commission, May, 1972.

37. *Process Design Manual for Upgrading Existing Wastewater Treatment Plants.* U.S. EPA, Office of Technology Transfer, Washington, D.C. (1974).

38. Schwinn, D.E., *Design Features of the District of Columbia's Water Pollution Control Plant.* Presented at the Sanitary Engineering Specialty Conference, ASCE, Sanitary Engineering Division, Rochester, New York, June, 1972.

39. Schwinn, D.E., and G.K. Tozer, *Largest Advanced Waste Treatment Plant in the U.S. and in the World.* Environmental Science and Technology, 8, No. 10, (1974).

40. Sawyer, C.N., *Supplementary Comments on Nitrification and Denitrification Facilities.* Prepared for the EPA Technology Transfer Seminar, Denver, Colorado, November 13, 1974.

41. *EPA Postpones Denitrification.* Reporter, Interstate Commerce Commission on the Potomac River Basin, 32, No. 2, February, 1975.

42. Harris County Water Control and Improvement District No. 50, *Monthly Progress Report for October 1974, on Project 11010 GNM,* prepared for the EPA.

43. Barth, E.F., EPA Technology Transfer Design Seminar on Nitrogen Control Technology. Presented at Denver, Colorado, November 13, 1974.

44. Brown and Caldwell/Dewante and Stowell, *Feasibility Study for the Northeast-Central Sewage Service Area.* Prepared for the County of Sacramento, November, 1974.

45. Sacramento Area Consultants, *Outfall Project and Environmental Impact Report.* Prepared for the Sacramento Regional County Sanitation District, December, 1974.

46. Sacramento Area Consultants, *Chemical Report, Sacramento Regional Wastewater Treatment Plant.* January, 1975.

47. White, G.C., Personal Communication to E. Appel, June, 1974.

48. CH2M/Hill, *Design of Montgomery County, Maryland Plant for the Washington Suburban Sanitary Commission,* 1975.

49. Culp, G.L., Culp, R.L. and C.L. Hamann, *Water Resource Preservation by Planned Recycling of Treated Wastewater.* JAWWA, 65, No. 10, pp 641-647 (1973).

50. CH2M/Hill, *Wastewater Treatment Study, Montgomery County, Maryland.* November, 1972.

51. *Cloromat.* Ionics, Inc. Brochure

52. Michalek, S.A., and F.B. Leitz, *On-Site Generation of Hypochlorite.* JWPCF, 44, No. 9, pp 1697-1712 (1972).

53. Suhr, L.G., and L. Kepple, *Design of a Selective Ion Exchange System for Ammonia Removal.* Presented at the ASCE Environmental Engineering Division Conference, Pennsylvania State University, July 1974.

54. CH2M/Hill, *Design of the Upper Occoquan Sewage Authority.* (1974).

55. *Physical-Chemical Plant Treats Sewage Near the Twin Cities.* Water and Sewage Works, p. 86, September, 1973.

56. Larkman, D., *Physical-Chemical Treatment.* Chemical Engineering, Deskbook Issue, p. 87, June 18, 1973.

57. Culp, R.L., *Nitrogen Removal by Air Stripping.* Proceedings, Wastewater Reclamation and Reuse Workshop, SERL, Univ. of California, Lake Tahoe, Ca., June 25-27, p. 128, 1970.

58. Culp, R.L. and G.L. Culp, *Advanced Wastewater Treatment.* Van Nostrand Reinhold Co., New York, p. 51, 1971.

59. Culp, R.L., and H.E. Moyer, *Wastewater Reclamation and Export at South Tahoe.* Civil Engineering, p. 38, June, 1969.

60. Wesner, G.M., and R.L. Culp, *Wastewater Reclamation and Seawater Desalination.* JWPCF, 44, No. 10, pp 1932-1939 (1972).

61. Folkman, Y., and A.M. Wachs, *Nitrogen Removal Through Ammonia Release From Holding Ponds.* Proceedings, 6th Annual International Water Pollution Research Conference, Tel Aviv, Israel, June 18-23, 1972.

62. Gonzales, John G., and R.L. Culp, *New Developments in Ammonia Stripping,* Public Works, 104, No. 5, p. 78 (1973) and No. 6, p. 82 (1973).

63. Wesner, G.M., and D.C. Baier, *Injection of Reclaimed Wastewater Into Confined Aquifers.* JAWWA, 62, No. 3, pp 203-210 (1970).

APPENDIX A

GLOSSARY OF SYMBOLS

Symbol	*Definition*
A	air flow per unit of tank volume, standard cu ft per min per 1000 cu ft; dimensionless number in contact stabilization calculations
ADWF	average dry weather flow
(a)	activity of an ion
B	fraction of sludge in stabilization tank; oxygen required for carbonaceous oxidation, mg/l
BHP	brake horsepower
C	fraction of total sludge in contact tank
C_l	process dissolved oxygen level, mg/l
C_b	denitrifier biomass production, mg/l
C_m	required methanol concentration, mg/l
C_s	oxygen saturation in water at temperature T, mg/l
D	concentration of nitrate nitrogen, mg/l; axial dispersion coefficient, sq ft/hr
D_o	influent NO_3^--N, mg/l
D_l	effluent NO_3^--N, mg/l
$\bar{D}_o$	mass average influent NO_3^--N level over 24 hr, mg/l

Symbol	*Definition*
$\bar{D}_1$	mass average effluent NO_3^--N level over 24 hr, mg/l
DO	dissolved oxygen, mg/l
e_o	aerator rated oxygen transfer efficiency at standard conditions, percent
F/M	food to microorganism ratio
f	nitrifier fraction of the mixed liquor solids; fanning friction factor
HT	hydraulic detention time
I	inventory of VSS under aeration, lb
K_B^A	selectivity coefficient for ion exchange equilibria.
K_D	half saturation constant for nitrate, mg/l NO_3^--N
K_d	"decay" coefficient, day^{-1}
K_M	half saturation constant for methanol, mg/l of methanol
K_N	half saturation constant for oxidation of ammonia nitrogen, mg/l
K_{O_2}	half saturation constant for oxygen, mg/l
K_s	half saturation constant = substrate concentration, mg/l at half the maximum growth rate
L	tank length, ft
M	methanol concentration, mg/l
M_c	mass of heterotrophs grown through oxidation of organic carbon

Symbol	*Definition*
M_N	mass of nitrifiers grown through oxidation of ammonia
MLSS	mixed liquor suspended solids, mg/l
MLVSS	mixed liquor volatile suspended solids, mg/l
MPN	most probable number
N	NH_4^+-N concentration, mg/l
N_o	TKN in the influent, mg/l
N_1	NH_4^+-N in the effluent, mg/l
$\bar{N}_o$	24 hr-average influent TKN, mg/l
$\bar{N}_1$	24 hr-average effluent NH_4^+-N, mg/l
$(NO_3^-)_c$	NO_3^--N level in the contact tank, mg/l
$(NO_3^-)_s$	NO_3^--N level in the stabilization tank, mg/l
NOD	nitrogenous oxygen demand, mg/l
NT	ammonia nitrogen oxidized, lb/day
PDWF	peak dry weather flow
PWWF	peak wet weather flow
Q	influent flow rate, mgd
$\bar{Q}$	mean flow rate (ADWF), mgd
Q_A	air flow, cfm
q_b	rate of substrate removal, lb BOD (or COD) removed/lb VSS/day

Symbol	*Definition*
q_D	nitrate removal rate, lb NO_3^--N rem./lb VSS/day
$\hat{q}_D$	peak nitrate removal rate, lb NO_3^--N rem./lb VSS/day
q_N	ammonia oxidation rate, lb NH_4^+-N oxidized/lb VSS/day
$\hat{q}_N$	peak ammonia oxidation rate, lb NH_4^+-N oxidized/lb VSS/day
r_N	nitrification rate, lb NH_4^+-N oxidized/lb MLVSS/day
$\hat{r}_N$	peak nitrification rate, lb NH_4^+-N oxidized/lb MLVSS/day
R	recycle flow rate, mgd
S	growth limiting substrate concentration, mg/l; total sludge wasted in lb/day
SCFM	standard cubic ft per minute
S_o	influent total BOD (or COD), mg/l
S_1	effluent soluble BOD (or COD), mg/l
SF	safety factor
SS	suspended solids, mg/l
SVI	sludge volume index, ml/g
T	temperature, degrees C
TKN	Total Kjeldahl Nitrogen
Δt	time increment

Symbol	*Definition*
u	mean displacement velocity, ft per hr
V	volume of aeration tank or reactor, mil gal
V_c	volume of contact tank, mil gal
V_s	volume of stabilization tank, mil gal
VSS	volatile suspended solids
W	waste sludge flow rate, mgd; tank width, ft; total oxygen demand, mg/l; oxygen transferred under process conditions, lb/day
W_O	oxygen transferred under standard conditions, lb/day
X	a coefficient in oxygen transfer calculations
X_1	MLVSS, mg/l
X_2	effluent VSS, mg/l
X_c	contact tank MLVSS, mg/l
X_s	stabilization tank MLVSS, mg/l
X_w	waste sludge VSS, mg/l
Y_b	heterotrophic yield coefficient, lb VSS grown per lb of substrate removed
$\bar{Y}_b$	net yield of VSS of heterotrophs per unit of carbon (BOD_5 or COD) removed
Y_D	denitrifier gross yield, lb VSS grown/lb NO_3^--N rem.
$\bar{Y}_D$	denitrifier net yield, lb VSS grown/lb NO_3^--N rem.

Appendix A—Glossary of Symbols

Symbol	*Definition*
Y_N	organism yield coefficient, lb *Nitrosomonas* grown (VSS) per lb NH_4^+-N removed
α_B^A	selectivity coefficient for ion exchange equilibria
θ_c	solids retention time, days
θ_c^d	solids retention time of design, days
θ_c^m	minimum solids retention time, days, for nitrification at given pH, T and DO
μ	growth rate of microorganism, day^{-1}
$\hat{\mu}$	maximum growth rate of microorganism, day^{-1}
μ_b	net growth rate of heterotrophic population
μ_c	growth rate of nitrifiers in contact tank, day^{-1}
μ_D	denitrifier growth rate, day^{-1}
$\hat{\mu}_D$	maximum denitrifier growth rate, day^{-1}
$\bar{\mu}_D$	design denitrifier growth rate, day^{-1}
μ_N	*Nitrosomonas* growth rate, day^{-1}
$\hat{\mu}_N$	peak *Nitrosomonas* growth rate, day^{-1}
$\dot{\mu}_N$	maximum possible nitrifier growth rate under environmental conditions of T, pH and DO, and $N \gg K_N$
μ_s	growth rate of nitrifiers in stabilization tank, day^{-1}

APPENDIX B

METRIC CONVERSION CHART

Multiply	By	To Get
Inches	2.54	Centimeters
Feet	0.3048	Meters
Square Feet	0.0929	Square Meters
Cubic Feet	0.0283	Cubic Meters
Pounds	0.454	Kilograms
Gallons	3.79	Liters
Gallons/Minute	5.458	Cubic Meters/Day
Feet/Second	0.305	Meters/Second

PART II

PHOSPHORUS REMOVAL

ACKNOWLEDGMENTS

The original edition of this Design Manual (October 1971) was prepared for Technology Transfer by Black & Veatch, Consulting Engineers. This first revision to the basic text was prepared for Technology Transfer by the firm of Shimek, Roming, Jacobs and Finklea. Major EPA contributors were E. F. Barth, J. M. Smith, C. A. Brunner, and J. B. Farrell of the U.S. EPA Office of Research and Development, Cincinnati, Ohio. Major reviewers were D. J. Lussier, J. C. Dyer, and R. M. Madancy of Technology Transfer.

Revisions to Chapters 4, 5, 10 and 11 have been made by J. Laughlin of Shimek, Roming, Jacobs and Finklea in order to update the information in these chapters. Appendix A has been added to describe an EPA computer model for evaluating phosphorus removal strategies.

NOTICE

Chapter 1
INTRODUCTION

1.1 Purpose

The technology for removal of phosphorus from wastewater has developed rapidly in the last few years. The need for practical phosphorus removal procedures is a result of the over-fertilization and eutrophication of the Country's surface waters. This further resulted in the establishment of state water quality standards that limit the concentration of phosphorus in receiving waters. As an example, states that have streams tributary to the Great Lakes have established wastewater effluent phosphorus limits or require a fixed percentage reduction of phosphorus. Several other states have also established phosphorus standards.

The Environmental Protection Agency has sponsored many research and demonstration studies at several cities in the past few years to advance the knowledge of phosphorus removal. Local and state governments and private industries have also contributed to this work. This manual is intended to summarize process design information for the best developed removal methods that have resulted from this governmental and private effort.

1.2 Scope

This manual discusses a number of phosphorus removal methods that involve chemical precipitation. Phosphorus removal obtainable by biological activity alone is not included. Treatment methods in which phosphorus removal occurs, but is not a principal objective, are also omitted. The latter group of processes includes ion exchange, reverse osmosis and other demineralization treatments which at present are more closely associated with wastewater renovation and reuse than with pollution control. These will be included in updated versions of the manual when appropriate. One application of a precipitation method that has been omitted and which is of growing interest is complete lime treatment of raw sewage. This application will be included when the manual is updated.

The information included on chemical methods of phosphorus removal was obtained from the available literature, progress reports of demonstration studies, and private communications with investigators actively working in the field. Design guidelines have been developed from these sources.

The information contained in this manual is oriented toward design methods and operating procedures for removal of phosphorus from wastewater to comply with state water quality standards. It is recognized that present state standards may eventually be superseded with more stringent requirements for phosphorus removal. Such revisions in state standards must be considered in design of treatment facilities to minimize future modifications to treatment plant structures.

Engineers are often faced with circumstances that require the design of treatment facilities with a limited amount of data, testing, and time. Such circumstances require a more conservative design to allow for contingencies. Since most phosphorus removal methods are of recent development, they fall into this category. Conservative values are given, therefore, when stating design parameters.

Chapter 2
BASIC DESIGN CONSIDERATIONS

2.1 Sources of Phosphorus

Domestic wastewater normally has a substantial concentration of phosphorus with the primary sources being a result of man's activities in the home. Human wastes such as feces, urine, and waste food disposal account for about 30 to 50% of the phosphorus in domestic wastewater (1). Detergents containing phosphate builders and used principally for laundering of clothes account for the remainder of phosphorus, or about 50 to 70%. Other sources of phosphorus may cause deviation from these percentages. For example, where sodium hexametaphosphate or other phosphorus compounds are used as corrosion and scale control chemicals in water supplies, the phosphorus added will be present in the same concentration, although not necessarily in the same form, as in the water supply. This source can account for 2 to 20% of the total phosphorus present in the wastewater.

Industrial wastewater discharges to the sanitary sewer system may either dilute or increase the concentration of phosphorus present in the combined wastewater. For example, wastewaters from pulp and paper mills may be phosphorus deficient and therefore dilution will occur. On the other hand, wastewaters discharged from some industries, such as potato processing plants, may contain high concentrations of phosphorus and consequently may increase the phosphorus concentration of the combined wastewater.

The quantity of phosphorus resulting from human excretions reportedly ranges from 0.5 to 2.3 lb per capita per year (2). The mean annual excretion is estimated to be about 1.2 lb per capita. The mean annual contribution of phosphorus from synthetic detergents with phosphate builders is estimated to be about 2.3 lb per capita, at present. Thus, exclusive of industrial wastes and other phosphorus sources, such as water softening or sequestering agents, the domestic phosphorus contribution to wastewater is about 3.5 lb per capita per year.

On this basis, the wastewater phosphorus concentration may be estimated for a community where there are no phosphorus data available. Assuming a population of 5,000, the total phosphorus discharged equals 5,000 x 3.5 = 17,500 lb/year, or 48 lb/day. If a wastewater flow of 100 gal. per capita per day is assumed, the total flow is 0.5 million gal. per day (mgd) and the phosphorus concentration in the wastewater is as follows:

$$\frac{48}{0.5 \times 8.34} = 11.5 \text{ mg/l (as P)}$$

The average total phosphorus concentration in domestic raw wastewater is found to be about 10 mg/l expressed as elemental phosphorus (P).

The above information is provided to serve as a rough guide to engineers and not as a basis of design. Sampling of the wastewater and analyses for phosphorus are recommended in all cases. It may also be necessary to survey the industries that will be discharging wastewater to the sanitary sewer system, to determine their influence on the concentration of phosphorus in the combined wastewater.

Among industrial sources of phosphorus may be potato processing wastes (3), fertilizer manufacturing wastes (4), animal feedlot wastes (5), certain metal finishing wastes (1, 6), flour processing wastes (7), dairy wastes (8, 9), commerical laundry wastes (10), and slaughterhouse wastes (11). The amount of phosphorus may be quite variable depending on the specific industrial plant.

2.2 State Regulatory Agencies

Each state has at least one organization responsible for water pollution prevention, control, and abatement activities. In about half of the 50 states this responsibility is divided between two state agencies. Usually one of these is the state health department and the other an organization charged specifically by statute with the conduct of the state's water pollution control program. Although there has been an intensification of Federal activities in water pollution control over the past twenty years, much of the responsibility for program operations rests with the states. Invariably the appropriate state agencies should be consulted in connection with the planning and conduct of water pollution control programs at cities and other communities. The Environmental Protection Agency regional office personnel work closely with state agencies and can give advice regarding the proper state agency contacts.

2.3 State Standards

As of June 1971, sixteen states had adopted wastewater effluent phosphorus standards. In general, these standards have taken the form of an effluent concentration limit or a requirement for a specified percentage reduction in the phosphorus concentration in the raw wastewater. In most cases, effluent concentration limits range from 0.1 to 2.0 mg/l as P, with many established at 1.0 mg/l. Percentage reduction requirements range from 80 to 95%.

It should be noted that neither an effluent nor a percentage-reduction standard actually limits the phosphorus load in terms of pounds of phosphorus discharged per day. Load (lb/day) is a direct function of both effluent P concentration and daily effluent volume. Assuming the effluent concentration standard is being met, an increase in the daily flow of effluent will produce a proportionate increase in the mass of phosphorus discharged daily to the receiving waters. If the phosphorus load then becomes intolerable, adjustment downward of the effluent concentration standard will be necessary. Similarly, adjustment upward of a percentage-reduction standard will be required if the phosphorus load becomes objectionable as a result of increased waste-water flow or increased P concentration in the raw wastewater or both.

Effluent concentration standards can be derived from receiving water standards as indicated in the following example.

Receiving Water
 Phosphorus concentration upstream 0.01 mg/l

Receiving Water - (continued)

Maximum allowable phosphorus concentration	0.05 mg/l
Low flow	140 ft^3/sec (90 mgd)

Wastewater

Average daily flow	4.8 mgd
Effluent concentration standard	X

Mass Balance

$$(0.01)(8.34)(90) + (X)(8.34)(4.8) = (0.05)(8.34)(90+4.8)$$

$$X = 0.8 \text{ mg/l P}$$

Implicit in the above calculation is the assumption that the effluent is mixed thoroughly with the waters of the receiving stream. The low flow selected is not necessarily the low flow of record. It is usually the low flow averaged over a period of 10 days or less and having a return period of once in 10 years.

2.4 Provision for Future Plant Expansion and More Stringent Effluent Requirements

The plant site is an important consideration in the design of treatment works. It is highly desirable that sufficient site area be obtained to permit initial construction of the required system without crowding and to provide space for future expansion. Treatment systems are likely to become larger and more complex in the future rather than less so, and failure to obtain ample site area initially is almost certain to create costly problems later.

Reduction of the allowable effluent phosphorus concentration may be necessary in some states as increases in wastewater flow or phosphorus concentration of the raw wastewater occur. In some instances, merely increasing the chemical dosage may be the only necessary change. On the other hand, it may be necessary to add an additional treatment sequence such as filtration to the existing facilities. In nearly every case, planning for plant expansion or more stringent effluent requirements will result in economy. Such planning should include selection of a plant layout that allows space for addition of facilities without removing existing processes from service.

Other provisions that should be considered are space within chemical buildings for additional coagulant storage and feeding equipment, space for polymer storage and feed equipment, and adequate space for handling the additional solids generated by phosphorus removal processes.

2.5 Flow Equalization

The cyclic nature of wastewater flows, in terms of volume and strength, is well established. While the concept of flow equalization has been employed in the field of water supply, and in the treatment of some industrial wastes, it has not been widely accepted in the municipal pollution control field. Anticipated problems with solids settling, odor, and septicity can be cited as the major factors limiting its use.

The advent of stricter stream standards, which sometimes require removal of contaminants during peak flows, the elimination of plant bypassing, and the increased removal efficiencies that are possible when biological or chemical treatment processes are operated at near steady-state conditions, have all favored the increased use of flow-smoothing, and flow equalization devices ahead of or integral with the design of major treatment works.

There are two major objectives in the design of flow equalization basins. The first of these is simply to dampen the diurnal flow variation that normally exists in typical municipal wastewater collection systems to achieve a constant or nearly constant flow rate through the downstream treatment processes. In this type of system, little consideration is given to controlling the concentration changes that take place during storage. The major design factors are supplying enough air to keep the basin aerobic and to provide adequate mixing to prevent solids deposition.

The second objective of flow equalization is to provide the capacity to distribute shock loads of toxic or process inhibiting substances over a reasonable period of time so as to prevent system failure and to minimize the periodic discharge of harmful contaminants to the receiving stream or surface impoundment. Time-dependent concentration profiles and flow-through curves are normally used to analyze the flow characteristics of these systems and to determine the effect of tank geometry, placement of effluent weirs and mixing regime on changes in contaminant concentrations through the basin.

In all cases the added costs of flow equalization must be measured against the possible reduction in downstream process costs and the increased efficiencies that can be achieved by operating these processes under constant loading conditions.

2.6 References

1. "Phosphates in Detergents and the Eutrophication of America's Waters", Hearings before a Subcommittee of the Committee on Government Operations, House of Representatives, 91st Congress, December 15-16, 1969. U.S. Government Printing Office, Washington (1970).

2. Task Group Report, "Sources of Nitrogen and Phosphorus in Water Supplies", *JAWWA*, 59:3, p 344 (1967).

3. Mackenthun, K. M., "The Phosphorus Problem", *JAWWA*, 60:9, p 1047 (1968).

4. Nemerow, N. L., *Theories and Practices of Industrial Waste Treatment*, Addison-Wesley, Reading, Mass. (1963).

5. *Relationship of Agriculture to Soil and Water Pollution*, Cornell University Conference on Agricultural Waste Management, Rochester, N.Y., Jan. 19-21, 1970.

6. Anderson, J. S., and Iobst, E. H., Jr., "Case History of Wastewater Treatment in a General Electric Appliance Plant", *JWPCF*, 40:10, p 1786 (1968).

7. Dickerson, B. W., and Farrell, P. J., "Laboratory and Pilot Plant Studies on Phosphate Removal from Industrial Wastewater", *JWPCF*, 41:1, p 56 (1969).

8. McKee, F. J., "Dairy Waste Disposal by Spray Irrigation", *Sew. & Ind. Wastes*, 29:2, p 157 (1957).

9. Lawton, G. W., et al., "Spray Irrigation of Dairy Wastes", *Sew. & Ind. Wastes*, 31:8, p 923 (1959).

10. Flynn, J. M., and Andres, B., "Launderette Waste Treatment Processes", *JWPCF*, 35:6, p 783 (1963).

11. Wymore, A. H., and White, J. E., "Treatment of Slaughterhouse Waste Using Anaerobic and Aerated Lagoons", *Wat. & Sew. Works*, 115:10, p 492 (1968).

Chapter 3

THEORY OF PHOSPHORUS REMOVAL BY CHEMICAL PRECIPITATION

3.1 Forms and Measurement of Phosphorus

Phosphorus is found in wastewater in three principal forms: orthophosphate ion, polyphosphates or condensed phosphates, and organic phosphorus compounds. Orthophosphate was in the past often considered to be PO_4^{3-} and was reported in this way. Actually, there are a number of forms of orthophosphate in equilibrium, with the predominant form changing as pH changes. At the usual pH of municipal wastewater, the predominant form is HPO_4^{2-}. With the present practice of reporting phosphorus as P, one does not usually have to be concerned with the actual form. It is well to remember, however, that the predominant form does change with pH. The removal of phosphorus with lime results primarily from pH increase. Polyphosphates can be looked upon as polymers of phosphoric acid from which water has been removed. Materials of various molecular weight are in widespread use. Complete hydrolysis results in formation of orthophosphate. The chemistry of organic phosphorus compounds is complicated. Their decomposition also leads to orthophosphate.

In raw sewage there are substantial amounts of all three principal phosphorus forms. During biological treatment, significant changes take place. As organic materials are decomposed, their phosphorus content is converted to orthophosphate. On the other hand, inorganic phosphates are utilized in forming biological floc. The polyphosphates are for the most part converted to orthophosphate. The result is that in a well treated secondary effluent, a large fraction of the phosphorus is present as orthophosphate. From the standpoint of removal by precipitation, this is fortunate since the orthophosphate is the easiest form to precipitate.

Because phosphorus will be present in a variety of forms, both organic and inorganic, the only satisfactory measure of treatment plant removal efficiency must be based on total phosphorus entering the plant in the raw wastewater and total phosphorus discharged in the plant effluent. Inasmuch as all colorimetric tests for phosphorus depend upon the formation of an orthophosphate color complex, it is essential in the determination of total phosphorus, to convert any polyphosphates (condensed phosphates) and organic phosphorus compounds present in the sample to the orthophosphate form. This conversion is accomplished by acid digestion of the sample in the presence of a strong oxidizing agent. *Standard Methods,* 13th edition (1), describes several procedures for determining total phosphorus as well as the various forms of phosphorus in wastewater. A suitable procedure for measuring total phosphorus, including the preliminary digestion stage, may be selected from this reference. An automated method is described in *FWPCA Methods for Chemical Analysis of Water and Wastes (2).*

3.2 Chemistry of Removal by Precipitation

The materials found practical for phosphorus precipitation include the ionic forms of aluminum, iron, and calcium. The calcium is added as lime and the hydroxyl ions also have a role in phosphorus removal. The chemistry of phosphorus removal by interaction with metallic ions is complex. In the interests of brevity and simplicity, it is assumed in the following discussion that phosphorus reacts as the orthophosphate form, PO_4^{3-}, and that the reaction products are as indicated. The polyphosphates and organic phosphorus are also removed, probably by a combination of more complex reactions and sorption on floc particles. The reactions presented and the associated computations are principally for illustrative purposes and do not purport to represent the true reactants, products and reaction mechanisms. Calculated requirements for phosphorus precipitants must be viewed as no more than crude estimates subject to large variations in actual practice because the forms of phosphorus and the precipitates may be substantially different from those assumed.

3.2.1 Aluminum Compounds as Phosphorus Precipitants

Aluminum ions can combine with phosphate ions to form aluminum phosphate as follows:

$$Al^{3+} + PO_4^{3-} \longrightarrow \underline{AlPO_4}$$

The above equation indicates that the mole ratio for $Al:PO_4$ is 1:1. Inasmuch as the mole ratio of $P:PO_4$ is also 1:1, the mole ratio for Al:P is 1:1 or $\frac{Al}{P} = 1$ when both aluminum and phosphorus are expressed in terms of gram-moles or lb-moles. On a weight, rather than a mole basis, this means that 27 lb of Al will react with 95 lb of PO_4 to form 122 lb of $AlPO_4$. Since each 95 lb (1 lb-mole) of PO_4 contains 31 lb (1 lb-mole) of P, the weight relationship between Al and P is 27 lb of Al to 31 lb of P or 0.87 for this reaction.

The principal source of aluminum for use in phosphorus precipitation is "alum", a hydrated aluminum sulfate, having the approximate formula $Al_2(SO_4)_3 \cdot 14H_2O$ (molecular weight of 594). The chemical, which is also known as "filter alum" or "papermaker's alum", averages about 17% soluble aluminum expressed as Al_2O_3 or about 9.1% expressed as Al. Its reaction with PO_4^{3-} may be written as follows:

$$Al_2(SO_4)_3 \cdot 14H_2O + 2PO_4^{3-} \longrightarrow \underline{2AlPO_4} + 3SO_4^{2-} + 14H_2O$$

The sulfate remains in solution as SO_4^{2-}. The above reaction indicates that 1 lb-mole of alum (594 lb) will react with 2 lb-moles (190 lb) of PO_4^{3-} containing 62 lb phosphorus to form 2 lb-moles (244 lb) of $AlPO_4$. The weight ratio of alum to phosphorus is, therefore, 9.6:1. The alum requirement per pound of phosphorus may also be derived from the Al:P mole ratio as follows:

Mole ratio Al:P = 1:1
Therefore weight ratio Al:P = 27:31 = 0.87:1
Alum contains 9.1% Al
Therefore, alum required per lb of P = $\frac{0.87}{0.091}$ = 9.6 lb

Stumm and Morgan (3), state that the solubility of $AlPO_4$ is pH dependent and varies as follows:

pH	Approximate Solubility (mg/l)
5	0.03
6	0.01 (minimum solubility)
7	0.3

The optimum pH for removal of phosphorus probably lies in the range of 5.5 to 6.5, although some removal occurs above pH 6.5.

Addition of alum will lower the pH of wastewater because of neutralization of alkalinity and release of carbon dioxide. The extent of pH reduction will depend principally on the alkalinity of the wastewater. The higher the alkalinity, the less is the reduction in pH for a given alum dosage. Most wastewaters contain sufficient alkalinity so that even large alum dosages will not lower the pH below about 6.0 to 6.5. In exceptional cases of low wastewater alkalinity, pH reduction may be so great that addition of an alkaline substance, such as sodium hydroxide, soda ash, or lime will be required.

Adjustment of the pH downward can be accomplished by the addition of sulfuric acid but this complicates the treatment process and it may be preferable simply to use a higher alum dosage.

Bench, pilot, and full-scale studies have shown that considerably higher than stoichiometric quantities of alum usually are necessary to meet phosphorus removal objectives. A competing reaction, responsible for the pH reduction mentioned above, at least partially accounts for the excess alum requirement. It occurs as follows:

$$Al_2(SO_4)_3 \cdot 14H_2O + 6HCO_3^- \longrightarrow \underline{2Al(OH)_3} + 6CO_2 + 14H_2O + 3SO_4^{2-}$$

The following ratios of alum (9.1%Al) to phosphorus are believed reasonably representative for alum treatment of municipal wastewater (4).

P Reduction Required	Al:P Mole Ratio	Al:P Weight Ratio	Alum:P Weight Ratio
75%	1.38:1	1.2:1	13:1
85%	1.72:1	1.5:1	16:1
95%	2.3:1	2.0:1	22:1

To achieve 85% P removal from a wastewater containing 11 mg/l of P the alum dosage needed would be (16)(11) = 176 mg/l or 1470 lb/10^6 gal.

Sodium aluminate will also serve as a source of aluminum for the precipitation of phosphorus. The chemical formula for sodium aluminate is $Na_2Al_2O_4$ or $NaAlO_2$.

One commercial form is the granular trihydrate, which may be written $Na_2O \cdot Al_2O_3 \cdot 3H_2O$ and which contains about 46% Al_2O_3 or 24% Al.

The sodium aluminate - phosphate reaction may be represented as follows:

$$Na_2O \cdot Al_2O_3 + 2PO_4^{3-} + 4H_2O \longrightarrow \underline{2AlPO_4} + 2NaOH + 6OH^-$$

In contrast to alum, which reduces pH, a rise in pH may be expected upon addition of sodium aluminate to wastewater.

The above reaction indicates an Al:P mole ratio of 1:1, an Al:P weight ratio of 0.87:1, and a sodium aluminate to P weight ratio of about 3.6:1.

As with the use of alum, mole and weight ratios somewhat in excess of the theoretical must be anticipated in practice.

3.2.2 Iron Compounds as Phosphorus Precipitants

Both ferrous (Fe^{2+}) and ferric (Fe^{3+}) ions can be used in the precipitation of phosphorus. With Fe^{3+} a reaction can be written similar to that shown earlier for precipitation of aluminum phosphate. A 1:1 mole ratio of Fe:PO_4 results. Since the molecular weight of iron is 55.85, the weight ratio of Fe:P is 1.8:1. Just as in the case of aluminum, a larger amount of iron is required in actual situations than the chemistry of the reaction predicts. With Fe^{2+} the situation is more complicated and not fully understood. Ferrous phosphate can be formed. The mole ratio of Fe:PO_4 would be 3:2. Experimental results indicate, however, that when Fe^{2+} is used, the mole ratio of Fe:P will be essentially the same as when Fe^{3+} is used.

A number of iron salts are available for use in phosphorus precipitation. These include ferrous sulfate, ferric sulfate, ferric chloride, and pickle liquor. Additional discussion of these is given later. All will lower the pH of wastewater because of neutralization of alkalinity. Pickle liquor contains substantial amounts of free sulfuric acid or hydrochloric acid.

Iron salts are most effective for phosphorus removal at certain pH values. For Fe^{3+} the optimum pH range is 4.5 to 5.0. This is an unrealistically low pH, not attained in most wastewaters. Significant removal of phosphorus can be attained at higher pH. For Fe^{2+} the optimum pH is about 8 and good phosphorus removal can be obtained between 7 and 8. The acidity of ferrous salts, and especially pickle liquor, necessitates addition of lime or sodium hydroxide for good results. Where the water is aerated following Fe^{2+} addition, the use of a base may not be necessary.

3.2.3 Lime as a Phosphorus Precipitant

Calcium ion reacts with phosphate ion in the presence of hydroxyl ion to form hydroxyapatite. This material has a variable composition, but an approximate equation for its formation can be written as follows, assuming in this case that the phosphate present is HPO_4^{2-}:

$$3\ HPO_4^{2-} + 5Ca^{2+} + 4OH^- \longrightarrow \underline{Ca_5(OH)(PO_4)_3} + 3\ H_2O$$

The reaction is pH dependent. The solubility of hydroxyapatite is so low, however, that even at a pH as low as 9.0, a large fraction of the phosphorus can be removed. In lime treatment of wastewater, the operating pH may be predicated on the ability to obtain good suspended solids removal rather than on phosphorus removal.

Although it is possible to calculate an approximate lime dose for phosphorus removal, this is generally not necessary. In contrast to iron and aluminum salts, the lime dose is largely determined by other reactions that take place when the pH of wastewater is raised. Some of these reactions are discussed later. Only in waters of very low bicarbonate alkalinity would the phosphate precipitation reaction consume a large fraction of the lime added.

3.3 References

1. *Standard Methods for the Examination of Water and Wastewater,* 13th Edition, American Public Health Assoc., New York (1971).
2. *FWPCA Methods for Chemical Analysis of Water and Wastes,* Federal Water Pollution Control Administration, U.S. Department of the Interior (Nov. 1969).
3. Stumm, W. and Morgan, J. J., *Aquatic Chemistry,* p 521 Wiley-Interscience, John Wiley & Sons, Inc., New York (1970).
4. Allied Chemical Corporation, Industrial Chemicals Division, "Use of Aluminum Sulfate for Phosphorus Reduction in Wastewaters", Unpublished paper.

Chapter 4

PHOSPHORUS REMOVAL BY MINERAL ADDITION BEFORE THE PRIMARY SETTLER

4.1 General Considerations

Removal of phosphorus from wastewater can be accomplished by modifying the conventional primary treatment process to include chemical precipitation with aluminum or iron salts (1,2). The advantages of phosphorus removal during primary treatment include flexibility in chemical feeding, adequate reaction times and mixing conditions, flocculation and removal, of more suspended solids and suspended BOD in the primary settler, and reduced loading of suspended solids and BOD to secondary treatment processes. Significant increases in removals of both suspended solids and BOD can occur concurrently with phosphorus removal. Phosphorus removal by chemical addition is amenable to automatic control using process instrumentation to measure waste flow and/or phosphorus concentration.

The disadvantage of addition of metal salts during primary treatment alone is that a significant amount of the phosphorus may not be in the ortho form and may not be easily precipitated. Higher phosphorus removals can be obtained with chemical addition beyond primary treatment and with chemical additions at several points.

The available precipitation-flocculation processes (3,4,5,6,7,8) include addition of ferrous or ferric chloride, ferrous or ferric sulfate, aluminum sulfate, or sodium aluminate, followed by flocculation using an anionic polymer to enhance solids separation. A strong base is often added between additions of ferrous iron and polymer to counteract the depression of pH by the ferrous compounds.

In order to provide effective chemical additions to primary treatment units, the following sequence is recommended:

1. Salt addition to raw sewage with thorough mixing. If a base is to be used, it should be added at least 10 seconds later.
2. Reaction for at least five minutes.
3. Addition of anionic organic polymer followed by flash mixing for 20 to 60 seconds.
4. Mechanical or air flocculation for 1 to 5 minutes.
5. Gentle delivery of flocculated wastewater to sedimentation units.

Mixing may be carried out in specially designed basins or advantage may be taken of existing high turbulence areas such as in a Parshall flume. Flocculation may similarly be carried out in specially constructed equipment or may be allowed to take place within some part of an existing system. Schematic diagrams of 2 plants, which demonstrate how existing equipment can be used with mineral addition, are shown in Figures 4-1 and 4-2.

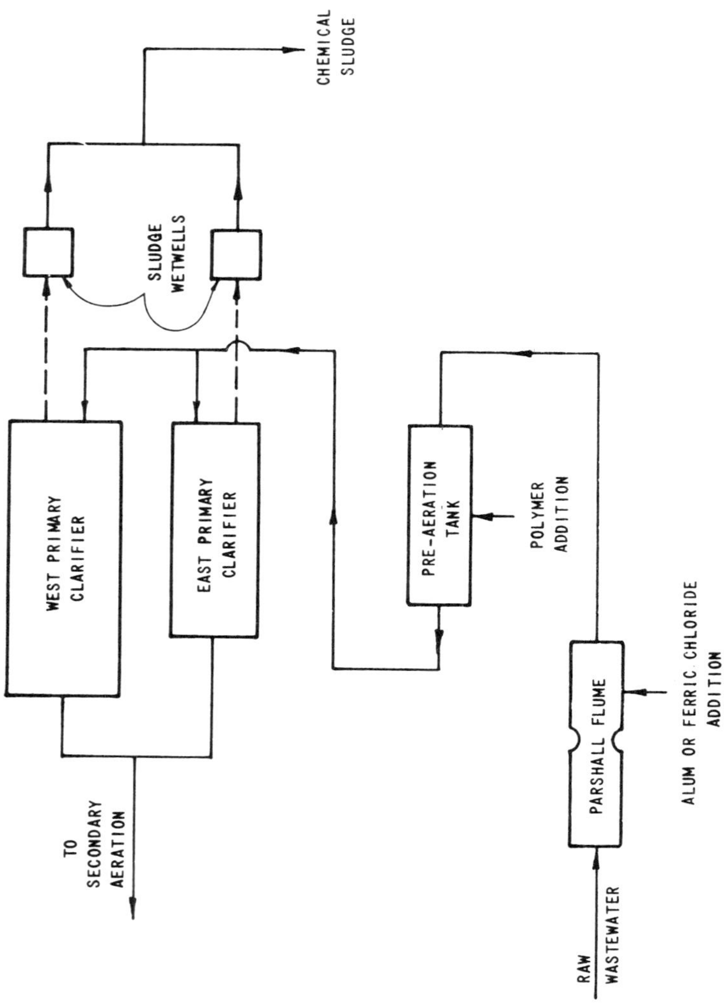

FIGURE 4-1 LEBANON, OHIO PRIMARY TREATMENT AND CHEMICAL FEEDING SCHEMATIC

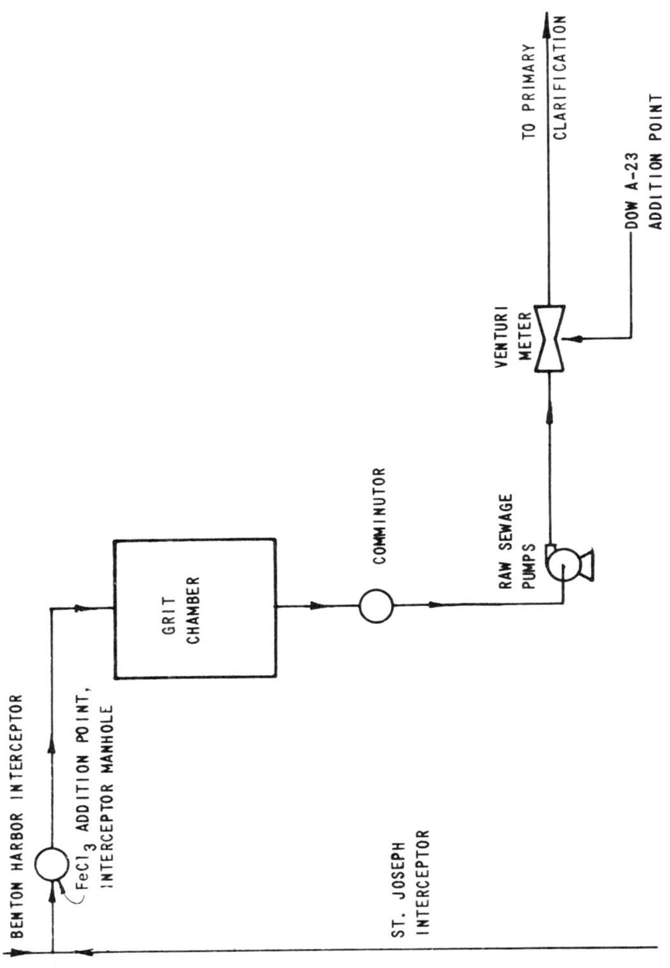

FIGURE 4-2 SCHEMATIC OF PRETREATMENT OF PLANT FLOW AT BENTON HARBOR - ST. JOSEPH WASTEWATER DISPOSAL PLANT

The points of chemical addition will vary depending on the particular situation. It is important, however, that addition points be downstream of return streams, such as digester supernatant, if a high degree of phosphorus removal is required.

Optimum conditions for phosphorus removal processes are best determined in two steps: a laboratory feasibility study followed preferably by full-scale plant trials if the plant exists, or by pilot trials. The feasibility study should be a short-term laboratory evaluation of various chemical systems to obtain preliminary estimates of chemicals required to meet established phosphorus removal standards. During the plant trials, chemicals and equipment should be supplied for full-scale operation over a period of 30 days or longer, to obtain detailed design data.

Chemical removal of phosphorus from wastewaters during primary treatment is a practical and controllable process which can result in low discharges of phosphorus from existing primary treatment plants and secondary treatment plants of either the trickling filter or activated sludge type. Additional costs for chemicals and attendant mixing, dispersing, and monitoring equipment are moderate.

Data and correlations given in this chapter will provide guidance and permit making design estimates. However, it would be risky to eliminate on-site experimental studies because of variations in equipment design and operation, and in wastewater characteristics.

4.2 Phosphorus Removal Data for Alum Addition

Alum has been used as a source of aluminum for phosphorus removal in raw wastewater. Sodium aluminate can also be used if the natural alkalinity of the raw wastewater is too low to prevent excessive reduction of pH by alum. However, no investigations using aluminate have been reported. Tests have been conducted on laboratory, pilot, and plant scales to determine the feasibility of alum addition to raw wastewater.

Bench-scale tests of alum addition were conducted at Springfield, Ohio (9) and Two Rivers, Wisconsin (10). All bench testing was conducted in the following manner. Samples of wastewater were stirred on a six-place stirrer for approximately 2 minutes at 100 rpm, followed by 18 minutes of gentle stirring (40 to 50 rpm). Alum additions were made during the rapid mix period. When polymers were used, they were added approximately 2 minutes after the alum addition. Treated samples were allowed to settle for 30 minutes before total phosphorus analyses were made.

The respective states have required phosphorus removals of 80% at Springfield and 85% at Two Rivers. At Springfield, an average Al:P weight ratio of 1.9:1 (molar ratio of 2.2:1) was required to reduce the average influent total phosphorus of 5 mg/l by 80%. Raw wastewater at Two Rivers required an average Al:P weight ratio of 0.93:1 (molar ratio of 1.07:1) to reduce the average influent phosphorus of 20.1 mg/l by 85%. These test data indicate that as phosphorus concentration increases, the Al:P ratio necessary to obtain a specific per cent removal decreases. Individual tests at Two Rivers, for a broad range of influent phosphorus concentrations, also showed the same variation. Influent total phosphorus ranged from 7.6 to 28.5 mg/l, and the Al:P weight ratios necessary to remove 85% ranged from 2.0:1 to 0.58:1, respectively.

A study of alum addition for phosphorus removal was conducted at Lebanon, Ohio on both pilot and plant scales (11) to determine whether chemical treatment of raw wastewater could be used successfully to precede activated carbon adsorption for organic removal.

A 15 gpm pilot plant was utilized to determine the most effective dosages of alum to be studied in the full-scale tests. Data from the pilot plant indicated that liquid alum dosages of 150, 250, and 350 mg/l, as $Al_2(SO_4)_3 \cdot 14H_2O$, and anionic polymer dosages of 0.25 and 0.5 mg/l should be evaluated. Only the Parshall flume, preaeration basin, and primary clarifiers of the 1.5 mgd Lebanon plant were used in this study. Alum was fed to the raw wastewater at the Parshall flume to assure complete mixing. The preaeration basin was used for flocculation, with polymer being added toward the end of the basin. The primary basins served as settlers for the precipitated phosphorus. Figure 4-1 shows the chemical addition arrangement at Lebanon which was used for both alum and iron studies. The iron experiments will be covered later in this chapter.

Each dosage was evaluated for five days. A polymer dose of 0.5 mg/l was found to be satisfactory for improving floc formation and settling. An alum dose of 250 mg/l, Al:P weight ratio of 3.4:1 (molar ratio of 3.9:1), removed 93% of the phosphorus in this high alkalinity (400 mg/l as $CaCO_3$) wastewater. Effluent phosphorus concentration did not exceed 0.7 mg/l and averaged 0.5 mg/l. The alum dose of 150 mg/l, Al:P weight ratio of 1.5:1 (molar ratio of 1.7:1), removed only 74% of the phosphorus; while the alum dose of 350 mg/l, Al:P weight ratio of 3.9:1 (molar ratio of 4.5:1), reduced the phosphorus by 94%. The alum dose needed to meet the State removal requirement of 80% would fall between the 150 and 250 mg/l dosages studied.

4.3 Phosphorus Removal Data for Iron Addition

Operating data from several plants utilizing mineral addition before the primary clarifier for phosphorus removal are shown in Table 4-1.

With primary treatment alone, phosphorus removals varied from 60 to 91%. In secondary plants, iron addition before the primary resulted in phosphorus removals from 75 to 93% over the entire plant. Iron doses varied between 10 and 90 mg/l, as Fe. The Lebanon study showed that doubling the iron dose (from 45 to 90 mg/l) with the same polymer addition raised phosphorus removal from 68 to 91%. Polymer doses in all of the iron studies ranged from 0.26 to 0.7 mg/l.

Table 4-1

PHOSPHORUS REMOVAL OPERATING DATA FOR MINERAL ADDITION BEFORE PRIMARY CLARIFIER

Location	Flow mgd	Metal	Dose mg/l or M:P (molar)	Polymer Used	Total Phosphorus: Plant Influent mg/l	Total Phosphorus: Primary Effluent mg/l	Phosphorus Removal in Primary percent	Reference
Lebanon, Ohio	1.01	$FeCl_3$	45 or 0.98:1	- - -	- - -	- - -	68	11, 12
	0.86	$FeCl_3$	90 or 1.9:1	Yes	- - -	- - -	91	
Grayling, Michigan	0.3	$FeCl_2$	15-25	Yes	10-20	3-6	60-80	8, 13
Lake Odessa, Michigan	- - -	$FeCl_2$	15-25	Yes	- - -	- - -	75*-93*	14, 15
Benton Harbor, Michigan	5.5** (activated sludge return to primary)	$FeCl_3$	21 or 1.6:1	Yes	7.6	0.8*	90*	16
	5.5** (no sludge return)	$FeCl_3$	21 or 1.6:1	Yes	7.6	0.8*	65 91*	
Warren, Michigan	36	$FeCl_3$	15	- - -	- - -	- - -	80*	17
Texas City, Texas	0.7	$FeCl_2$	1.3 to 1.6:1	No	- - -	- - -	80*-90*	18
	0.7	$Al_2(SO_4)_3$	1.6 to 1.8:1	No	- - -	- - -	80*	
	0.7	$Al_2(SO_4)_3$	2.5:1	Yes	- - -	- - -	>90*	
Trenton, Michigan	4.5	$FeCl_2$	20-40	- - -	11.7	1.5	87	19
Waukegan, Illinois	15	$FeCl_3$	2.5:1	Yes	8.2	0.7	91	20
	15	$FeCl_3$	1.6:1	Yes	9.6	0.9	91	
Pontiac, Michigan	10	$FeCl_3$	15	Yes	7.0	0.9	85-90	21
	10	$FeCl_3$	16 (split)	Yes	7.2	0.5	93	
	10	$FeCl_3$	10	Yes	7.5	1.6	80	

See footnotes at end of table.

Table 4-1 (Continued)

Location	Flow mgd	Metal	Dose mg/l or M:P (molar)	Polymer Used	Total Phosphorus: Plant Influent mg/l	Total Phosphorus: Primary Effluent mg/l	Phosphorus Removal in Primary percent	Reference
Michigan City, Indiana	9	$FeCl_3$	14.4	...	10.0	1.7	83	22
	9	$Al_2(SO_4)_3$	7.2	...	9.0	1.0	89	
Babbitt, Minnesota	0.4	$Na_2Al_2O_4$	1.1:1	No	11.0	0.8	93	23
Dogue Creek STP, Virginia	3	$FeCl_3$	1.6:1	Yes	9.0	1.0	89	24
Little Hunting Creek STP, Virginia	4	$FeCl_3$	1.9:1	Yes	10.0	1.0	90	24
Essexville, Michigan	0.5	$FeCl_3$	17 or 2:1	Yes	4.0	0.5*	88*	25
Mentor, Ohio	4	$FeCl_2$	3.1:1	...	...	...	80	26,27

NOTE: *Values are for entire plant.
**Design Capacities: 8 mgd (primary), 4 mgd (secondary).

Iron carry-over occurred at Mentor, where $FeCl_2$ addition alone resulted in an effluent Fe concentration of 42.5 mg/l. Lime addition reduced the iron leakage to an effluent Fe concentration of 11.6 mg/l. Polymers were ineffective in reducing iron leakage.

Increased suspended solids and BOD removals were also obtained with iron additions. At Grayling, Michigan, suspended solids removal was increased by 27% to an average of 78%. At Benton Harbor, Michigan, where waste activated sludge is returned to the primary, iron addition gave a 3.0% increase in solids removal over the entire treatment plant. An example of increased BOD removal is seen in the Grayling, Michigan study. Removal of BOD in the primary was increased by 45% after iron addition. This resulted in a total BOD removal of 58% in the primary.

Typical values for primary settler detention times were 2.3 hours at Mentor and 2.25 hours at Benton Harbor. At Mentor, air at a rate of 42.5 to 85 ft^3/minute was used for flocculation in the mixing area of the primary tank.

Although lab test procedures should be adjusted to best approximate individual plant conditions, outlines of procedures used at two locations may be useful as guidelines. Jar tests were conducted at Mentor in the following manner. Iron, lime, and polymer were added one minute apart to 1,000 ml of raw wastewater at a stirrer speed of 100 rpm. The chemicals and raw wastewater were then mixed for five minutes at 20 rpm, and finally the mixture was allowed to settle quiescently for 1/2 hour before a portion of the supernatant was pipetted off for analysis.

4.4 Overall Performance of Treatment Plants with Mineral Addition to the Primary Settler

4.4.1 Primary Treatment Alone

The relative efficiencies of primary and secondary treatment with and without primary phosphorus removal are summarized in Table 4-2. Phosphorus removal without chemical precipitation in the primary clarifiers ranges from only 10 to 20% through a conventional secondary treatment plant. At least one half or more of this removal is due to sedimentation of the insoluble phosphorus fraction in the raw wastewater during primary clarification. The addition of the precipitation-flocculation process in primary clarification will result in 70 to 90% removal of total phosphorus in the primary. These removal levels can be expected only in well designed primary clarifiers that are not hydraulically overloaded. Suspended solids removals in the primaries will be increased from 40 to 70% to 60 to 75%. This increase occurs despite the additional solids generated by phosphorus removal. BOD removals in the primaries will be increased from 25 to 40% to 40 to 50%. The additional removal is primarily due to the suspended BOD fraction.

4.4.2 Effect on Secondary Treatment

Additional phosphorus removal through either trickling filter or activated sludge units following chemical precipitation in the primaries can amount to 10 to 15%. This

Table 4-2

EFFECTIVENESS OF PRIMARY AND SECONDARY TREATMENT WITH AND WITHOUT MINERAL ADDITION FOR PHOSPHORUS REMOVAL

	Phosphorus Removal (%)		Suspended Solids Removal (%)		BOD Removal (%)	
	Without	With	Without	With	Without	With
Primary Treatment	5-10	70-90	40-70	60-75	25-40	40-50
Secondary Treatment						
Trickling Filter	10-20	80-95	70-92	85-95	80-90	80-95
Activated Sludge	10-20	80-95	85-95	85-95	85-95	85-95

removal is attributed to biological flocculation. Split addition of metal in both the primary and secondary portions of the plant should be considered if exceptionally high levels (>90%) of total phosphorus removal are desired. Addition points for the secondary metal feed at or near the effluent end of the biological treatment unit should be included in the design.

Increased capture of suspended solids in the primaries results in reduced loading to the secondary units. Although the solids entering the secondary are reduced, the liquid volumes remain the same and, therefore, normal detention times should be used. Essentially the same amount of soluble BOD must be insolubilized and removed. The additional removal of suspended solids and BOD across the primaries will reduce the amount of waste activated sludge to be handled. Savings in plant operating costs may result from reduced air requirements for aeration, decreased chlorine demand, and improved sludge filterability (28).

4.5 Dosage Selection and Correlation

The material presented in Chapters 6 and 7 on choice of chemicals and dosage in trickling filter plants and activated sludge plants generally applies also to mineral addition in primary plants. Therefore, the reader is referred to these chapters for discussions of requirements for influent phosphorus data, procedures for obtaining dosage design data, calculations of daily doses, and scheduling of dose rate changes to match phosphorus variation. The reader is also referred to Chapter 10 for general information on various chemicals. Material presented here will be limited to comments and/or data on dosage selection, dosage correlation, and influent phosphorus variations.

It is generally desirable to plot or correlate test results on a simple basis to make them more useful. A reasonably good correlation can be obtained by plotting the log of the fraction of soluble phosphorus remaining as a function of the weight ratio of metal to soluble phosphorus. Typical of such plots is Figure 4-3, where both the fraction of soluble phosphorus remaining and the percent removed are plotted on the log scale. The data in Figure 4-3 are from several ferric chloride phosphorus removal studies conducted in the Great Lakes region.

Since there is usually considerable scatter in the data, such correlations should be used very conservatively as a guide for predicting doses. On-site tests under actual plant conditions are always preferred as the basis for such a plot. The required ratio of metal ion to influent soluble phosphorus for a specified phosphorus removal can be obtained from such a plot. This ratio is multiplied by the soluble influent phosphorus concentration in mg/l to determine the required chemical dose in mg/l.

As will be discussed in Chapter 6, the dose rate should be varied during the day to match more closely the varying influent phosphorus concentration. The need for doing this is illustrated by Figure 4-4 where some typical data on the diurnal variation of ortho phosphorus in plant influent are presented.

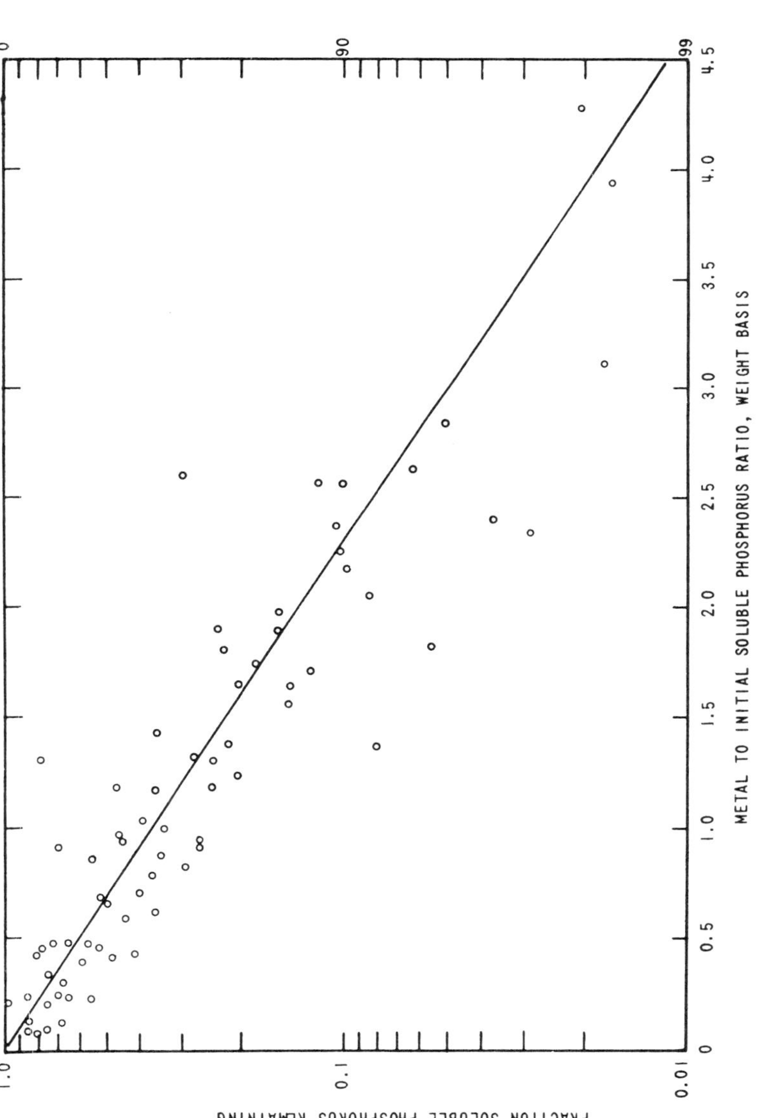

FIGURE 4-3 SOLUBLE PHOSPHORUS REMOVAL BY FERRIC CHLORIDE ADDITION

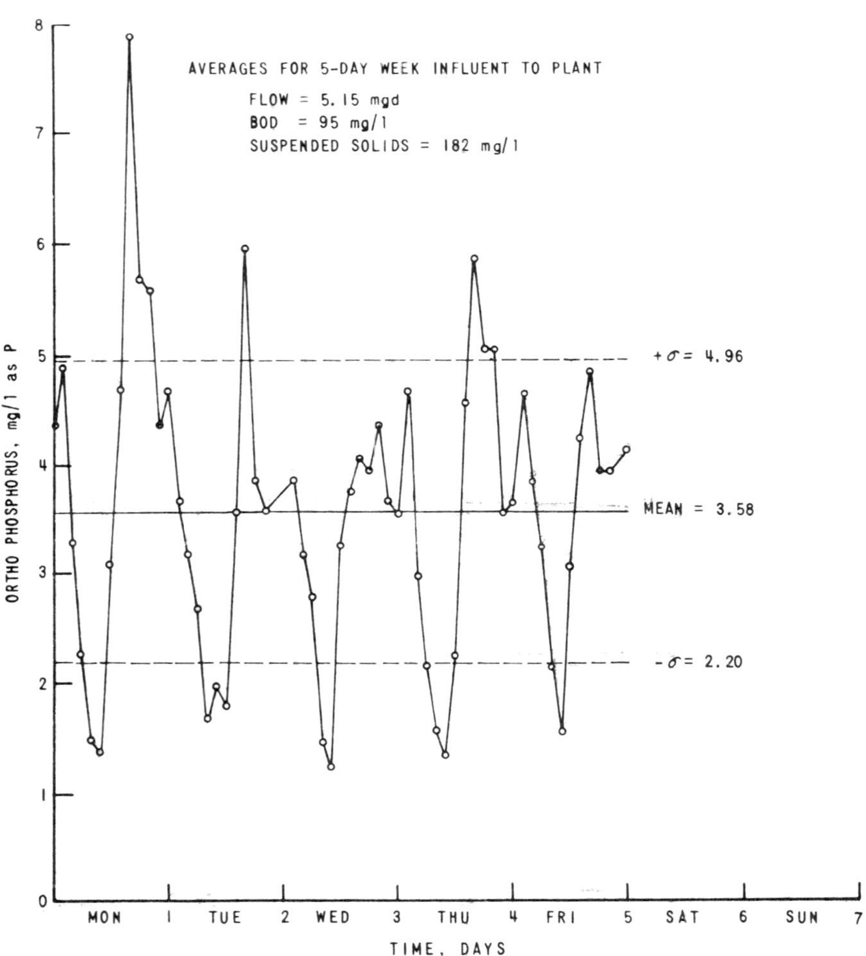

FIGURE 4-4 DIURNAL VARIATION OF ORTHO PHOSPHORUS

4.6 Case Histories

This section presents information on design and operation of three treatment plants which remove phosphorus by adding metals before primary settling and by adding polymers as required to supplement these metals.

The plant at Pontiac, Michigan employs commercial grade ferric chloride as a metal source; Trenton, Michigan uses waste pickle liquor; Waukegan, Illinois has successfully used three liquid forms of metals: ferric chloride, alum, and waste pickle liquor. Operations have been effective at all three plants, and details are presented here to illustrate the variety of approaches to phosphorus removal which are presently being applied successfully.

4.6.1 Pontiac, Michigan

The Auburn Plant in Pontiac, Michigan is a conventional activated sludge treatment plant with an average design flow of 10 mgd. The facility was constructed in 1963 according to design parameters presented in Table 4-3. Very simple chemical feed facilities were added to permit addition of ferric chloride and polymers to the primary section of the plant as shown in Figure 4-5.

A test program in 1969 (21) showed that feeding ferric chloride and polymer ahead of the primary clarifier was an effective procedure to remove 80 to 90 percent of total phosphorus in the wastewater. Long-term analysis of concentrations and forms of phosphorus in plant influent indicate soluble compounds represent 60 to 80 percent of the total phosphorus present, and that orthophosphates represent 35 to 45 percent of total phosphate. There appeared to be minimal iron leakage in the effluent, and overall plant performance, in terms of BOD and SS removals was improved by this technique.

The best point for addition of ferric chloride was just before the aerated grit chamber at the head of the plant. Theoretical displacement through the grit chamber was approximately 10 minutes, but observations indicated actual detention was more on the order of 2 to 5 minutes at average flow rates. This was sufficient time, however, for mixing and initial flocculation of the metal salt. Passage through the grit unit provided adequate lag time before polymer addition.

Since 1972, the plant has been operating approximately as shown by the performance data in Table 4-4 (29).

These data, mostly derived from 1973 experience, reflect a high-quality effluent and well-stabilized sludge. When feeding chemicals, the primary section of the plant removes 40 to 60 percent of incoming BOD, 50 to 75 percent of incoming SS, and 50 to 75 percent of incoming total phosphorus.

Table 4-3

DESIGN DATA FOR PONTIAC, MICHIGAN

Description	
Design Flow, mgd	
Average Dry Weather	10
Peak Dry Weather	20
Maximum	36
Primary Clarifiers	
Detention Time,[1] hr	1.5
Overflow Rate,[1] gpd/sq ft	1,000
Sludge Solids Concentration, percent	6.5
Sludge Volume, percent of total flow	0.08
Aeration Tanks	
Detention Time,[1] hr	6.5
MLSS, mg/l	2,000-3,000
Return MLSS, percent	40
Final Clarifiers	
Detention Time,[1] hr	1.9
Overflow Rate,[1] gpd/sq ft	790
Sludge Digesters	
Total Active Volume, cu ft	195,000
Vacuum Filters	
Total Area, sq ft	375
Nominal Capacity, psf/hr	4

[1]Based on Average Flow

There has been no evidence of serious problems related to escape of colloidal iron from the treatment system. Some 2 to 3 mg/l of soluble or colloidal iron is found both in plant influent and effluent.

Both ferric chloride and polymer are needed in this operation for good overall performance. If iron dosage is reduced below the indicated mole ratio of 1.7, or if polymer addition is discontinued, effluent quality begins to degrade in a matter of a few hours.

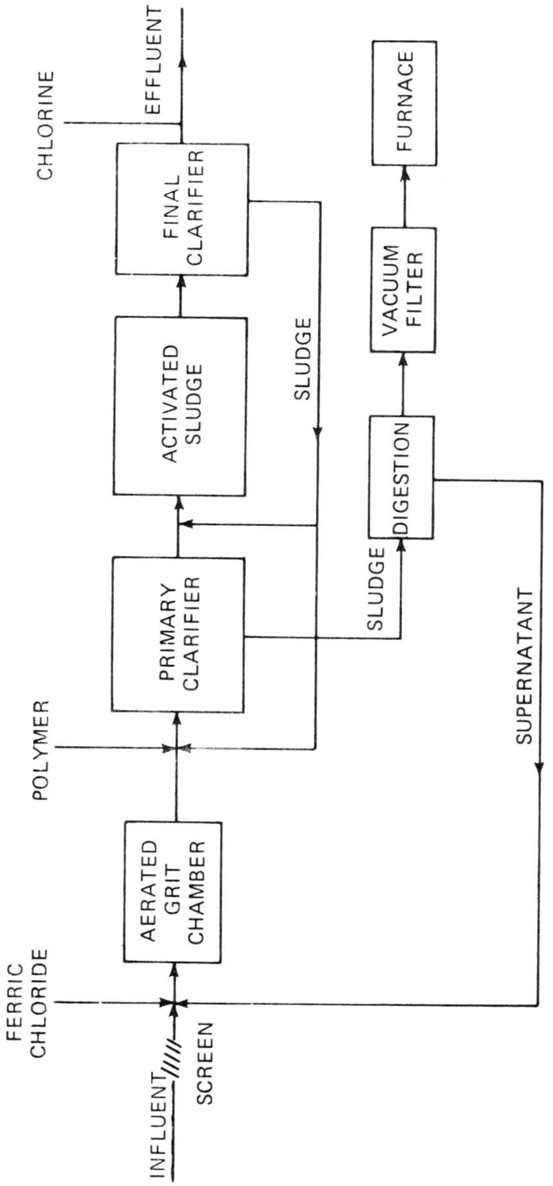

FIGURE 4-5 TREATMENT FACILITIES AT PONTIAC, MICHIGAN

INFLUENT
SCREEN
FERRIC CHLORIDE
AERATED GRIT CHAMBER
POLYMER
PRIMARY CLARIFIER
SLUDGE
ACTIVATED SLUDGE
SLUDGE
FINAL CLARIFIER
CHLORINE
EFFLUENT
DIGESTION
SUPERNATANT
VACUUM FILTER
FURNACE

Table 4-4

OPERATING DATA FOR PONTIAC, MICHIGAN

Description	
Flow, mgd	10
Influent Characteristics	
BOD, mg/l	100
SS, mg/l	140
Total P, mg/l	5
Chemical Addition	
Iron Dose, mg/l	15
Iron Dose, Fe/P	1.7
Polymer Dose, mg/l	0.3
Overall Plant Performance	
Effluent BOD, mg/l	10
Effluent SS, mg/l	20
Effluent Total P, mg/l	1
BOD Removal, percent	90
SS Removal, percent	86
Total P Removal, percent	80
Digesters	
Temperature, deg F	90
Solids Loading, lb VSS/cu ft/day	0.09
Digested Sludge Concentration, percent	7.2
Digested Sludge Volatile Fraction, percent	45
Vacuum Filters	
Loading, psf/hr	3-4
Cake Solids Concentration, percent	30-35

Volume of sludge produced by chemical addition is approximately the same as generated during conventional treatment. The chemical sludge is more concentrated but there have been no problems with digestion or dewatering of sludges produced in this location. Digested sludge is conditioned with ferric chloride and lime (at about the same dosage rates as previously required for conventional sludge treatment) prior to vacuum filtration. Cake forms readily on the filters but at modest yield rates. Both the vacuum filters and the multiple hearth furnace which follows are operated continuously for periods of 60 to 75 hr during midweek. The units are shut down, cleaned, and held on standby the rest of the time.

Cost of treatment chemicals in 1973 was 2¢/1,000 gal of wastewater treated. Of this cost, 85 percent was for ferric chloride. Exact cost records are not available on capital equipment installed in this system. That modest equipment outlay consisted of fiberglass tanks and chemical metering pumps. Although quite workable, this temporary chemical feed arrangement is to be replaced by permanent improvements in a 1974 plant expansion.

4.6.2 Trenton, Michigan

Since December 1970, Trenton, Michigan has operated a 5.5 mgd activated sludge treatment plant with addition of pickle liquor and polymer for phosphorus removal and effluent polishing (19,30). Design data for the Trenton plant are listed in Table 4-5.

Pickle liquor, the only metal salt used on a long-term basis at Trenton, is stored in two rubber-lined steel tanks having a capacity of 25,000 gal each. These tanks have hot water heating coils to keep coagulant temperature high enough to prevent crystallization of solids during cold weather. The storage tanks are filled by tank trucks bringing pickle liquor from a nearby steel plant; the trucks discharge into the storage tanks by applying low pressure compressed air. The liquid metal salts are fed, by gravity, in proportion to the rate of wastewater flow through the plant at a preselected concentration of metal. The concentration can and is changed from time to time to meet changing needs.

Two polymer tanks, each holding 500 gal, are outfitted with variable flow metering pumps which can be manually set to predetermined rates. These inject polymer between the activated sludge aeration tankage and the final clarifiers. Mechanical mixing equipment is provided in a small box between these two operations, and medium energy mixing is provided there to blend polymers into the main flow.

Predesign pilot scale testing and plant scale startup operations indicated the system performed best in the mode illustrated by Figure 4-6. Pickle liquor was added just ahead of the aerated grit chamber, and the polymer was added to the activated sludge effluent en route to the final clarifier.

Both the polymer addition and tube settlers installed in the final clarifier were intended to improve collection and capture of fine floc observed in primary effluent during studies in the late 1960's.

Table 4-6 shows performance data typically observed at this treatment facility since it has been in full scale service.

Subsequent to these operating data, an additional inflow of industrial waste amounting to 1 mgd has been added to the plant flow. The treatment system is presently accepting this

Table 4-5

DESIGN DATA FOR TRENTON, MICHIGAN

Description	
Design Flow, mgd	
Average Dry Weather	5.5
Peak Dry Weather	6.5
Maximum Rate	16.5
Primary Clarifiers	
Detention Time,[1] hr	1.3
Overflow Rate,[1] gpd/sq ft	1,440
Weir Rate,[1] gpd/ft	23,000
Aeration Tanks	
Detention Time,[2] hr	5.5
Return Flow, percent of average	30
Final Clarifiers (Tube Module Type)	
Detention Time,[1] hr	3.5
Tube Overflow Rate, gpd/sq ft	1,200
Total Surface Overflow Rate, gpd/sq ft	500
Sludge Production	
Primary (dry), lb/day	4,000
Secondary (dry), lb/day	5,000
Chemical (dry), lb/day	4,000
Primary plus Chemical, percent	5
Secondary, percent	1.0-1.5
Gravity Thickening	
Solids Loading, psf/day	6.2
Vacuum Filtration	
Area, sq ft	360
Nominal Capacity, psf/hr	2-3

[1]Based on Average Flow
[2]Including Return Flow

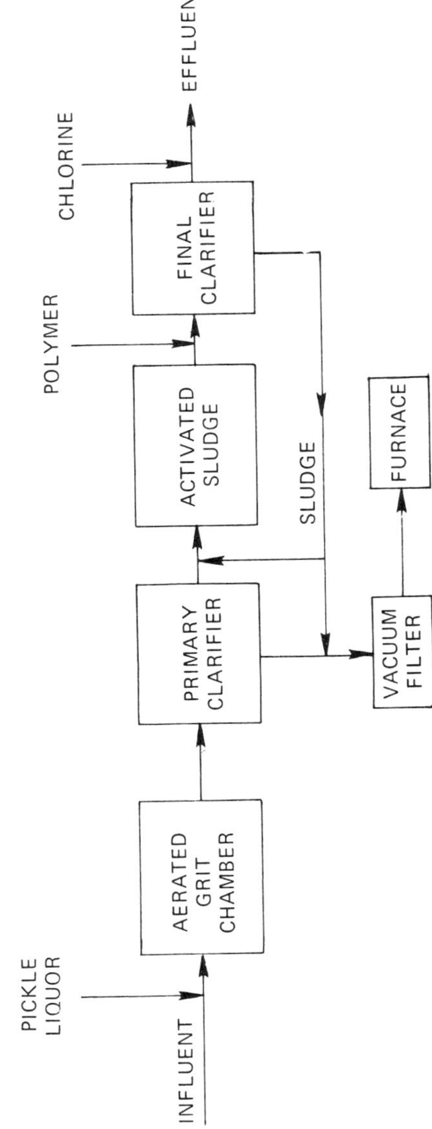

FIGURE 4-6
TREATMENT FACILITIES AT TRENTON, MICHIGAN

additional waste, plus other increased domestic loads, so that average daily flow is now 7.5 mgd. This additional waste includes an increase in BOD loading to the plant of about 4,000 lb/day. Plant performance at this increased load is reported as essentially the same as in Table 4-6. Preliminary operating data from early 1974 indicate removal of 90 percent of the BOD and SS, and 80 percent of total phosphorus. Sludge processing has continued on a stable basis as indicated in Table 4-6.

Total cost of the 1974 facilities was $4.8 million. In this major construction project, it was not possible to separate the relatively minor expenses related to the chemical treatment facilities themselves. Something on the order of $5,000 could be identified for expenditures

Table 4-6

OPERATING DATA FOR TRENTON, MICHIGAN

Description	
Flow, mgd	5.5
Influent Characteristics	
BOD, mg/l	220
SS, mg/l	300
Total P, mg/l	9.5
Ortho P, mg/l	5.1
Chemical Addition	
Iron Dose, mg/l	20
Iron Dose, Fe/P	1.2
Polymer Dose, mg/l	0.4
Overall Plant Performance	
Effluent BOD, mg/l	10
Effluent SS, mg/l	10
Effluent Total P, mg/l	1
Effluent Ortho P, mg/l	0.5
BOD Removal, percent	95
SS Removal, percent	97
Total P Removal, percent	89
Ortho P Removal, percent	90
Gravity Thickeners	
Thickened Sludge Concentration, percent	5.1
Thickened Sludge Mass, lb/day	16,500
Cake Solids Concentration, percent	15-20

for specific hardware, but concrete, electrical service, and other such features could not be realistically identified as if they were an independent capital improvement. At any rate, facility costs would be very low when capitalized over a period of years.

In 1973, the total operating costs of this facility was approximately 20¢ to 22¢/1,000 gal of wastewater treated; this cost included all phases of wastewater plant operation. Operating cost for chemical addition was quite low, on the order of 1¢/1,000 gal, consisting of about 0.5¢/1,000 gal for polymer, and slightly less than that for the very economical source of iron represented by pickle liquor. Iron in pickle liquor analyzed 0.75 lb of iron/gal and the delivered cost was 2¢/gal. Therefore, delivered cost of iron was 2.7¢/lb.

4.7 Costs

All cost estimates for the removal of phosphorus by chemical precipitation-flocculation during primary treatment are dependent to some extent on raw wastewater flow rates. The cost per 1,000 gal. for chemicals may be less in larger installations because of large volume purchasing at lower unit prices. Total capital costs for tankage, chemical feed pumps, and instrumentation will be less for smaller plants. Placed on a unit volume of waste treated basis, however, capital costs for small plants will be higher than for large plants.

4.7.1 Chemical Costs

The chemical costs for phosphorus removal in the primary settler have been computed for both liquid alum and liquid $FeCl_3$. The respective unit costs of these chemicals were estimated at \$60/ton for alum (on a 17.1% Al_2O_3 basis) and \$90/ton, as $FeCl_3$, for liquid $FeCl_3$. These figures represent average delivered prices.

As a basis for estimating costs in terms of ¢/1,000 gal., an influent P concentration of 10 mg/l was assumed. For removal of 80% of the total phosphorus in primary treatment, equal molar ratios of Al:P and Fe:P of 1.5:1 were assumed. Polymer costs were included, based on a dosage rate of 0.5 mg/l and a chemical cost of \$1.50/lb. The chemical costs of phosphorus removal treatment are estimated to be 3.6 ¢/1,000 gal. for liquid $FeCl_3$ and 4.2 ¢/1,000 gal. for liquid alum, respectively. As mentioned earlier, these costs may vary depending on the size of treatment plant and cost of freight. Actual costs should be determined prior to detail design.

The chemical cost for phosphorus removal at plants where waste pickle liquor is available is more difficult to predict accurately because of varying iron concentrations, larger storage requirements, freight charges, and the additional alkalinity requirement.

4.7.2 Capital Costs

Equipment costs for phosphorus removal with aluminum and iron are small relative to the chemical costs. Costs for facilities will be presented in Chapter 10 for single point addition of several different chemicals.

4.8 References

1. Duren, J. W., "Detergent Builders and the Environment", Presented to the Chemical Marketing Research Association, Chicago, Illinois (February 24, 1971).

2. Schuessler, R. G., "Phosphorus Removal, From Domestic Wastewater or From Detergents?", Presented at the 57th Annual Meeting of the Chemical Specialties Manufacturers Association, New York, N.Y. (December 15, 1970).

3. Leckie, J., and Stumm, W., "Phosphate Precipitation", in *Water Quality Improvement by Physical and Chemical Processes*, Edited by Gloyna, E. F., and Eckenfelder, W. W., Jr., University of Texas Press, Austin, Texas, p 237 (1970).

4. Daniels, S. L., "Phosphorus Removal from Wastewater by Chemical Precipitation and Flocculation", Presented at the American Oil Chemists' Society, 1971 Short Course, "Update on Detergents and Raw Materials", Lake Placid, New York (June 16, 1971).

5. Johnson, E. L., Beeghly, J. H., and Wukasch, R. F., "Phosphorus Removal with Iron and Polyelectrolytes", *Public Works*, 100:11, p 66 (1969).

6. Schuessler, R. G., "Phosphorus Removal – A Controllable Process", *Chem. Eng. Prog. Symp. Series*, 67:107, p 536 (1971).

7. Wukasch, R. F., "The Dow Process for Phosphorus Removal", Presented at the FWPCA Phosphorus Removal Symposium, Chicago, Illinois (June 19, 1968).

8. Wukasch, R. F., "New Phosphate Removal Process", *Wat. and Wastes Eng.*, 5:9, p 58 (1968).

9. Harriger, R. D., and Hoffman, F. L., "Phosphorus Removal Tests – Springfield, Ohio for Black and Veatch Consulting Engineers", Technical Service Department, Allied Chemical Corporation, Industrial Chemicals Division, Syracuse, New York (1971).

10. Harriger, R. D., and Hoffman, F. L., "Phosphorus Removal Tests – Two Rivers, Wisconsin for Black and Veatch Consulting Engineers", Technical Service Department, Allied Chemical Corporation, Industrial Chemicals Division, Syracuse, New York (1970).

11. Feige, W. A., "Full Scale Mineral Addition at Lebanon, Ohio", Inhouse Report, Environmental Protection Agency, Water Quality Office, Advanced Waste Treatment Research Laboratory, Cincinnati, Ohio (1971).

12. Hathaway, S. W., and Smith, J. E., Jr., "Fundamental Design Concepts for the Lime Stabilization of Lebanon Raw Sludge", Environmental Protection Agency, Water Quality Office, Advanced Waste Treatment Research Laboratory, Cincinnati, Ohio, Unpublished paper (1971).

13. Green, O., Eyer, F., and Pierce, D., "Studies on Removal of Phosphates and Related Removal of Suspended Matter and Biochemical Oxygen Demand at Grayling, Michigan, March – September 1967", Division of Engineering, Michigan Department of Health, Lansing, Michigan, and The Dow Chemical Company, Midland, Michigan (1967).

14. Leary, R. D. and Ernest, L. A., "Municipal Utilization of an Industrial Waste for Phosphorus Removal": Presented at the 32nd Porcelain Enamel Institute Technical Forum at the University of Illinois (October 8, 1970).

15. ________________, "Studies on Removal of Phosphates and Related Removal of Suspended Matter and Biochemical Oxygen Demand at Lake Odessa, Michigan, May-October, 1967," Wastewater Section, Division of Engineering, Michigan Department of Public Health, Lansing, Michigan, and The Dow Chemical Company, Midland, Michigan (1967).

16. Johnson, E. L., Beeghly, J. H., and Wukasch, R. F., "Phosphorus Removal with Iron and Polyelectrolytes," *Public Works,* 100:11, p 66 (1969).

17. Boggia, C., and Herriman, G. L., "Pilot Plant Operation at Warren, Michigan," *Proceedings of the 43rd Annual Conference of the Michigan Pollution Control Association* (1968).

18. Connell, C. H., "Phosphorus Removal and Disposal From Municipal Wastewater," U.S. EPA Report No. 17010 DYE (February, 1971).

19. Hennessey, T. L., "Pilot Studies, Design and Operation of a Modified Activated Sludge Plant Using Pickle Liquor and Tube Settlers," Annual Meeting WPCF, Atlanta, Georgia (October, 1972).

20. Wilson, Thomas E., Greeley and Hansen Engineers, Chicago, Personal Communication (1973).

21. Hennessey, John, Jelinski, Robert, Beeghly, Joel H., and Pawlak, Timothy J., "Phosphorus Removal at Pontiac, Michigan," The Dow Chemical Company, Midland, Michigan.

22. Harriger, R. D., and Zuern, H. E., "Michigan City, Indiana, Wastewater Treatment Plant, August 15-October 22, 1971, Chemical Precipitation of Phosphorus," Allied Chemical Corporation.

23. Cole, Don and Tveite, Paul, "Sodium Aluminate for Removal of Phosphate and BOD," *Public Works,* p. 86 (October, 1972).

24. Robson, C. Michael, Nickerson, Gary L., and Morrison, R., "Chemical Addition at Two Trickling Filter Plants to Enhance Treatment Efficiency," 45th Annual Conference, WPCF, Atlanta, Georgia (October, 1972).

25. Berthume, Don, City of Essexville, Michigan, Personal Communication (1973).

26. Gaughan, D. M. and Alvord, E. T., "Phosphorus Removal by Ferrous Iron and Lime," Final Report by Rand Development Corp. for the County of Lake, Ohio and EPA's Water Quality Office on Federal Grant No. 172-01-68, Prepublication Copy (1971).

27. Progress Reports, Mentor, Lake County, Ohio, FWPCA Grant WRPD 172-01-68, May 15, 1968 through May 31, 1969 and September 1, 1969 through July 31, 1970.

28. Voshel, D., and Sak, J. G., "Effect of Primary Effluent Suspended Solids and BOD on Activated Sludge Production," *JWPCF*, 40:5, Part 2, p R203 (1968).

29. Hennessey, J., City of Pontiac, MI, direct communication.

30. Hennessey, T. L., Consulting Engineer, Trenton, MI, direct communication.

Chapter 5

PHOSPHORUS REMOVAL BY LIME ADDITION BEFORE THE PRIMARY SETTLER

5.1 Introduction

Use of lime to treat raw wastewater dates back to the mid-1800's. Observers recognized that calcium formed a precipitate both with hardness and with phosphates. Under one patented technique both phosphates and lime (CaO) were added at the treatment plant to increase the amount of precipitate formed and to aid removal of SS (1).

More recent experimental efforts involved lime treatment of raw wastewater to remove phosphorus. These are typified by an investigation in New York, in which feeding 300 mg/l of slaked lime to a typical domestic wastewater removed 94 percent of the phosphorus to an effluent concentration of 0.6 mg/l, reduced SS by 95 percent to 4 mg/l, and reduced BOD by 71 percent to an average of 101 mg/l (2).

Other studies have suggested that lime addition to primary treatment is an efficient technique (3, 4). These studies were at bench-, pilot- and plant-scale and total phosphorus removals obtained were 80 percent or more with concurrent reduction in BOD of 60 to 70 percent. In this work, sufficient lime was added to bring primary effluent to a pH of 9.5 or 10. This pH level was tolerated without problems in the activated sludge process which followed; carbon dioxide generated by biological activity served to correct pH to near neutral levels. Recirculation of lime sludge was very beneficial. With sludge reseeding, as much as 60 to 75 percent reduction in overall lime requirements were obtained compared to tertiary lime treatment. Reaction time required to achieve settleable lime-phosphorus compounds was reduced from 1.0 hr without recirculation, to 0.25 hr with reseeding.

A number of other studies, from bench tests through plant-scale trials, helped to establish lime addition to primary treatment as a viable alternate for phosphorus removal operations (5, 6, 7).

The low-lime system described in Section 5.2.1 can provide 80 percent phosphorus removal on a consistent basis. By adding tertiary filtration and facilities for addition of metal coagulants and polymer, the technique can reduce total effluent phosphorus to near 1.0 mg/l.

The high-lime system described in Section 5.2.2 offers an even higher level of efficiency. With flocculant aids and filtration, it can consistently produce final phosphorus levels below 1.0 mg/l.

5.2 System Layout and Operation

Two basic systems are considered in this section. One involves relatively low lime dosages to primary influent, with biological treatment serving both to recarbonate the wastewater and remove additional phosphorus for metabolic needs. The other approach entails high-lime treatment in the primary section, followed by pH adjustment with supplemental carbon dioxide and biological treatment. The two systems are discussed separately in this section, after the comments below which apply to both approaches.

Sludge disposal during lime addition will be more complex than in a conventional biological plant. Lime sludge may not be compatible with existing disposal processes. Total sludge weight may be doubled or tripled over conventional operation. Recovery of lime by recalcining may be less attractive in this process because treatment in primary units yields a sludge which is high in concentration of organic and extraneous inorganic materials, and relatively low in recoverable lime compounds.

Lime treatment appeals most where incoming hardness and alkalinity are low because less lime is needed and less sludge is produced. However, turbid effluent may develop and settling aids may be required to supplement bioflocculation to insure a clear final effluent (8). Recirculation of lime sludge has proven beneficial to the basic reaction taking place. It promotes a more complete reaction, reduces scaling, and lowers lime demand.

Removal of phosphorus by lime addition in primary treatment is accompanied by increased BOD and SS removal at that point. This reduces organic load on secondary facilities and could aid in establishing nitrification.

In evaluating lime treatment systems, the effect of reduced carbon loads on their activated sludge unit should be considered. Since lower organic loadings to the secondary treatment units result in less cellular synthesis, a reduction in sludge wasting may be required. However, if the sludge is retained in the aeration system (in an attempt to keep MLSS up to some prescribed level) sludge age may extend to the point that settleability suffers and phosphorus escapes by cell lysis.

Also, lime precipitation in the primary clarifier will increase the capture of silts or other similar inorganic solids. These weighting agents will no longer be carried into the activated sludge unit for sorption onto mixed liquor solids, and this may result in a fluffier biological floc.

5.2.1 Low-Lime Treatment

The basic appeal of low lime addition in primary treatment stems from two fundamentals: the chemical law of mass action and the need for phosphorus in biological systems.

In the first of these, lime precipitates more phosphorus during initial stages of their mutual reaction than it does when pH has been elevated and phosphorus concentrations are quite low. This effective performance by lime usually continues through the first 75 percent of phosphorus reduction, and that occurs before pH 10 when lime demands for softening reactions increase. Overall lime demands become much higher when the process is operated to reduce phosphorus to very low levels.

Following initial phosphorus removal by lime, in the primary facilities, biological treatment removes more phosphorus as bacteria utilize it to support cell synthesis. Generally, phosphorus is not a limiting nutrient in domestic biological treatment systems as long as primary effluent contains at least 0.2 mg/l of soluble phosphorus. Microorganisms demand phosphorus to synthesize cells in direct response to the carbon concentration of the incoming waste; in municipal wastewater, carbon is the limiting element, not phosphorus. The result is lime precipitation can proceed at economical levels, then biological treatment can further reduce phosphorus residuals to an acceptable effluent concentration.

The basic low-lime treatment system is shown schematically in Figure 5-1. One advantage of this process is the opportunity to use existing primary clarifiers for phosphorus removal. Capital expenditure, therefore, could be relatively small. The process requires equipment for feeding and flash mixing lime. Flocculation usually occurs in the inlet zone of the clarifier but separate facilities may be preferred for this process. The elevated pH of primary effluent is reduced by recarbonation due to carbon dioxide produced in biological metabolism. The low-lime system may not be appropriate at trickling filter plants unless high effluent pH is tolerable, since recarbonation would be minimal in trickling filter units.

5.2.2 High-Lime Treatment

For the purposes of this manual, high-lime treatment is defined as the addition of sufficient lime in primary facilities to achieve pH 11. The basic system layout is generally as shown in Figure 5-1, for low-lime treatment. Recarbonation with stored carbon dioxide may be necessary and facilities for this would normally be provided. Recovery of lime by recalcination, an alternate feature discussed later in this Chapter, would frequently be included in high-lime systems.

High-lime treatment is applicable when effluent quality requirements include special provisions such as softening (for reuse), low levels of soluble compounds of metals, improved virus removal, or reliable and consistent reduction of phosphorus below 1.0 mg/l. Some form of tertiary solids removal, such as filtration, would usually be employed in these plants.

High-lime treatment would normally be combined with other biological, physical or chemical processes to provide an overall system of advanced waste treatment. If biological treatment is included, the primary clarifier phosphorus residual should be high enough to meet metabolic requirements.

High-lime treatment may result in a net increase in plant effluent levels of total dissolved solids and alkalinity. See Section 5.3.2.

FIGURE 5-1
TYPICAL FACILITIES FOR LOW-LIME ADDITION IN PRIMARY TREATMENT

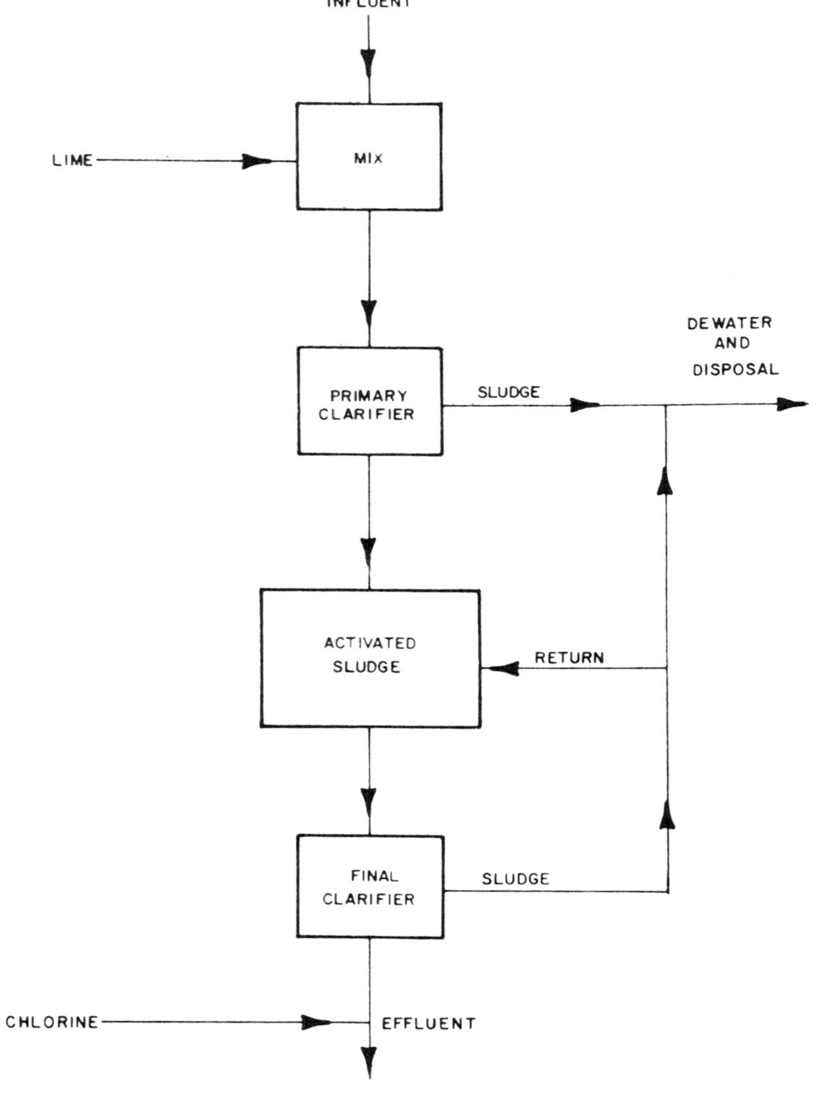

5.3 Case Histories

Two case histories are included in this section. One involves low-lime addition in primary treatment, while the other is a high-lime system. Both installations are further cited in Section 5.5, and it is intended that information in this section will give better perspective to those later comments on sludges involved.

5.3.1 Newmarket, Ontario—Low-Lime Process

Lime fed into the primary section of an existing 2.4 mgd conventional activated sludge treatment plant in Newmarket was the first permanent installation for phosphorus removal in Ontario, Canada (5, 9). Phosphorus removal facilities have been operational in Newmarket since March, 1971.

The original treatment facility (constructed in 1963) included screening and grit removal in an aerated tank with 5 to 10 minutes detention. The original design data and subsequent operating data are shown in Table 5-1. A schematic of the Newmarket treatment facility is shown in Figure 5-2.

Table 5-1

DESIGN AND OPERATING DATA FOR NEWMARKET, ONTARIO

Operation	Design Data	Operating Data
Flow, mgd	2.4	1.9
Primary Clarifiers		
Detention Time, hr	1.6	2.0
Overflow Rate, gpd/sq ft	1,330	1,050
Aeration Basins		
Detention Time, hr	8	10
F/M, lb BOD/day/lb MLSS	1.0	0.8[1]
Final Clarifiers		
Detention Time, hr	2.4	3.0
Overflow Rate, gpd/sq ft	1,000	800
Chlorine Contact		
Detention Time, hr	0.42	0.53

[1]When lime addition was established, efficient primary clarification reduced F/M to 0.2 lb BOD/day/lb MLSS; concurrently, one aeration basin was shut down so effective aeration time was reduced to 6.7 hr.

FIGURE 5-2

TREATMENT PLANT AT NEWMARKET, ONTARIO

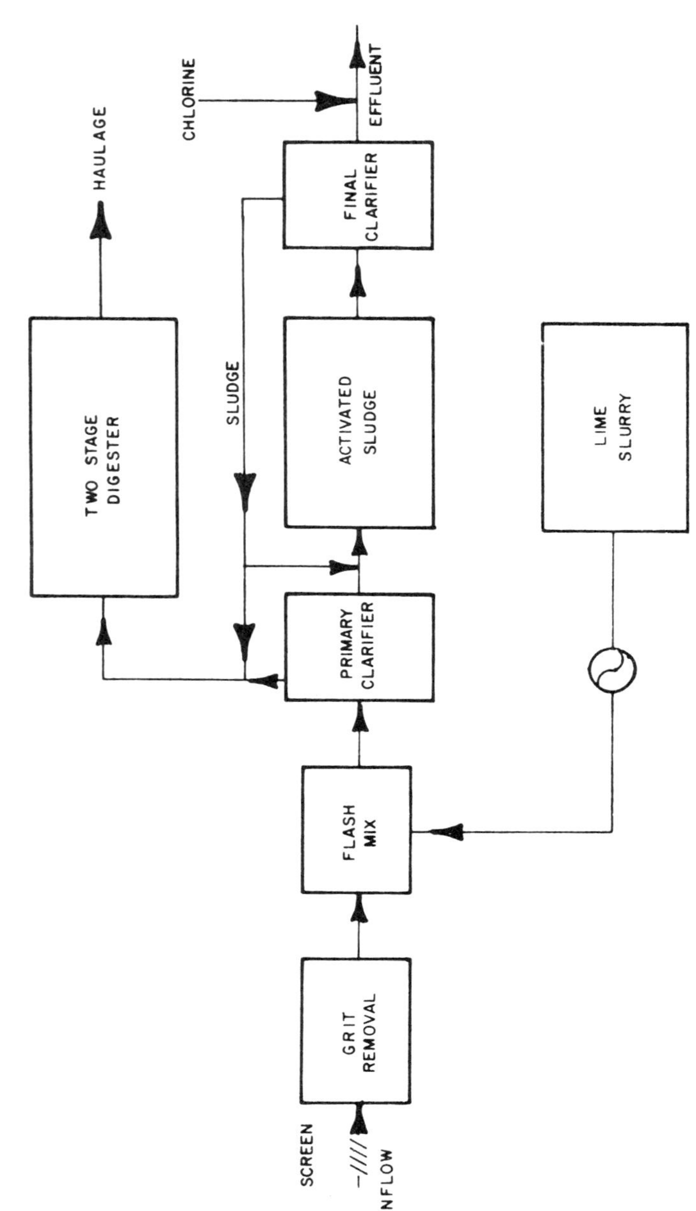

Sludge treatment includes anaerobic digestion of combined primary and waste activated sludges. The heated primary digester (40 ft. in diameter and 21.3 ft. deep) has a capacity of 2,000,000 gal and gas mixing is employed to distribute the daily solids loading of 0.1 lb/cu ft. A secondary digester, of similar size, completes the digestion process. Ultimate disposal is by wet spreading on farmland.

Lime storage is in a 25 ft diameter by 17 ft deep tank (mostly underground) which holds 30 tons of hydrated lime made up to near a 30 percent slurry by weight. To put lime into storage, bulk calcium hydroxide is discharged directly to the storage tank along with slurry water. The new slurry, after thorough blending, is sampled and titrated to determine actual lime concentration. This type of storage also allows use of waste carbide slurry from a local acetylene plant. Continuous slurry agitation is provided by either of two centrifugal slurry pumps mounted beside two positive displacement lime feed pumps. Slurry pumps require double mechanical seals for this severe service. Lime feed lines require flushing connections.

Lime feed is based on preselected dose rates with automatic interlock to rate of wastewater inflow. Metered slurry is discharged into a new flash mix box, where inflow is mechanically agitated for about 8 minutes.

Lime operating data reported in Table 5-2 is based on a 200 mg/l slaked lime dosage. Overall phosphorus removal at this dosage was greater than 80 percent.

Performance was slightly less effective at a slaked lime dose of 175 mg/l. Also, waste carbide lime, at rates equivalent to 150 mg/l of slaked lime, were reasonably effective in removing phosphorus. However, the acetylene waste proved highly abrasive to pumps and also formed heavy encrustations enroute to the biological unit. Carbonate scale problems were fewer using slaked lime, but occasional hosing down of primary launders was still necessary. Encrustation did not occur in aeration basins.

Discharge of relatively alkaline (average pH 9.6 with peaks to pH 11.5) primary effluent to the activated sludge system does not appear to distress the biological process.

Waste activated sludge is returned to the primary clarifiers and the resulting sludge mixture is drawn to anaerobic digestion. During conventional operation about 5 gal total sludge/1,000 gal plant flow was produced. Primary underflow was about 4 percent solids, of which 65 percent were volatile.

During lime addition, waste activated sludge production is reduced 50 percent due to the lower organic loads on secondary treatment. Additional sludge from lime reactions keep primary underflow at 5 gal/1000 gal plant flow. In this mode, however, the solids content averages 9 percent or more, of which 36 percent are volatile. The pH ranges from 8 to 10 in this sludge. Concentrations up to 15 percent solids were obtained when operators thickened in situ by allowing sludge depth to build to 2.5 ft on the clarifier bottom. No problems developed with mechanical scrapers, and sludge lines remained unplugged.

Table 5-2

OPERATING RESULTS FROM NEWMARKET, ONTARIO

	Conventional Treatment				Lime Treatment			
	Influent	Primary	Final	Percent Removal	Influent	Primary	Final	Percent Removal
BOD	249	172	21	92	227	84	9	96
SS	341	126	17	95	321	80	7	98
Total Phosphorus	14.2	12.5	8.7	39	10.3	2.7	2.0	81
Orthophosphate	8.1	8.3	7.6	6	6.7	1.2	1.8	73
Hardness	225	229	235	–	239	240	277	--
Alkalinity	386	377	227		391	374	247	-
pH	7.8	7.8	7.7		8.0	9.6	8.0	

Note: All data are mg/l except pH: hardness and alkalinity are expressed as calcium carbonate.

Chemical-biological sludge is pumped to digestion, as in prior conventional operation. The anaerobic system has adapted to this type sludge and complete digestion is obtained. Operation over a period of months has indicated that digester stability was enhanced when sludge was thickened by holding a sludge blanket nearly 2.0 ft deep in the primary clarifier.

This brought sludge pH down to 9.5 or less on a consistent basis, and thickened the mass to 11 percent solids. Digested sludge has a pH near 7.2. Digestion produces normal amounts of methane gas, but free carbon dioxide is reduced by consumption in recarbonation reactions. Digester supernatant contains less than 5 mg/l of soluble phosphorus.

Capital costs, in 1971 dollars, for the lime treatment additions are shown in Table 5-3. Operating costs are almost entirely limited to the 200 mg/l lime consumption. In this instance, there is no increase in the cost of sludge handling because the volume of digested sludge has remained unchanged.

5.3.2 Central Contra Costa Sanitary District (CCCSD), California–High-Lime Process

Recent plant-scale tests of lime addition before primary treatment have been conducted in Contra Costa County, California (10,11,12). A complete advanced treatment test facility has been in operation since November 1971. Full-scale units provide tests of all elements of a 30 mgd water reclamation system currently under construction and scheduled for start-up in 1976. Since the purpose of the CCCSD water reclamation system is to reclaim wastewater for industrial reuse effluent requirements are quite strict and are listed in Table 5-4.

Table 5-3

CONSTRUCTION COSTS AT NEWMARKET, ONTARIO

Structures	$23,000
Mechanical Work	20,000
Electrical Work	8,000
Equipment	5,000
Engineering	9,000
Miscellaneous	4,000
Total	$69,000

Table 5-4

EFFLUENT REQUIREMENTS FOR CCCSD TREATMENT FACILITY

Constituent	Limit (mg/l)
Total Phosphorus	1.0
BOD	10.0
Total Nitrogen	5.0
Hardness (as $CaCO_3$)	300
Alkalinity (as $CaCO_3$)	255
TDS (maximum increment above water supply)	375

In order to demonstrate effectiveness of the planned 30 mgd treatment system, investigators converted existing treatment facilities to various plant-scale forms of advanced waste treatment. The final layout of the overall test facility is shown in Figure 5-3. Table 5-5 gives pertinent data.

After screening, raw sewage passed a chemical addition point where lime was added and flash mixed with recycled primary sludge. Ferric chloride and polymer addition were also possible at the same point. Wastewater then flowed to the preaeration basin. Heavy grit

FIGURE 5-3
SYSTEM LAYOUT AT CONTRA COSTA COUNTY

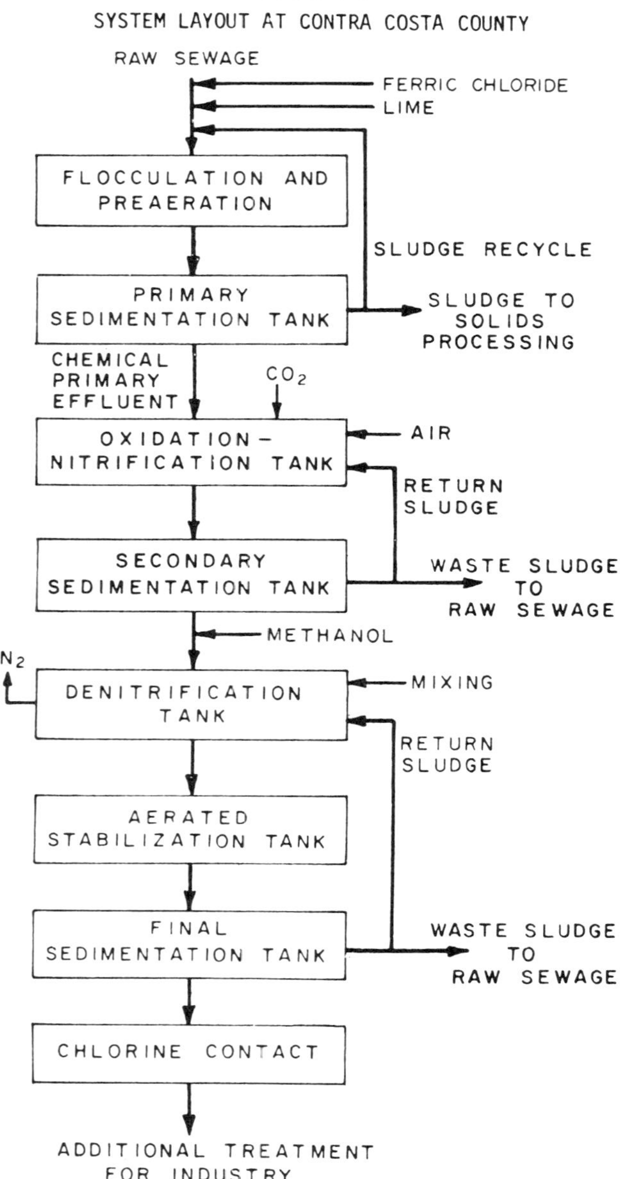

Table 5-5

BASIC DATA FOR CCCSD

Operation	
Preaeration and Grit Removal	
Average Water Depth, ft	11
Detention Time, hr	0.53
Design Flow/Tank, mgd	2.5
Air Supplied, cu ft/gal	0.10
Primary Clarification	
Average Water Depth, ft	10
Detention Time, hr	1.6
Overflow Rate, gpd/sq ft	1,200
Design Flow/Tank, mgd	2.5

particles such as sand and unreacted lime settled out there. The gentle air agitation following lime addition promoted the growth of large floc particles which were readily settleable. Long-term operation of this air mixing facility indicated it was effective, and this approach eliminated problems associated with mechanical flocculators in raw wastewater.

Waste activated sludge was returned to the preaeration section; there it mixed and settled with the chemical and primary sludges. The system had no separate facilities for treatment of waste activated sludge.

In the preaeration basin it appeared the carbon dioxide content of agitation air might, by the process of recarbonation, work against the lime reactions required in the preaeration unit. However, if all carbon dioxide in the air were transferred into solution, less than 0.5 mg/l dissolved carbon dioxide would be introduced. This was considered insignificant.

However, carbon dioxide in the preaeration air did cause scaling inside the orifices of the coarse bubble diffusers. Initially the tank had to be drained every week or two to clean plugged orifices. Installation of high-pressure air blowoff facilities extended that period to two months. In the 30 mgd plant, swing diffusers will be provided so tanks will not have to be taken out of service for diffuser maintenance.

The lime-treated effluent was then settled in the rectangular primary tanks under operating parameters described previously in Table 5-5.

During the course of the studies lime dosage was varied, and supplemental coagulants were utilized to modify effluent qualities. Data for various modes of operation are shown in Tables 5-6 and 5-7.

Operation at pH 11 was not as effective in phosphorus removal as operation at 11.5 (see Table 5-5). The phosphorus remaining in the primary effluent at pH 11 was 2.3 mg/l compared with less than 1 mg/l at the higher dose.

Ferric chloride was tested as a supplemental coagulant to enhance capture of colloidal phosphorus precipitates. At pH 11, a ferric chloride dose of 14 mg/l produced primary effluent phosphorus of less than 1 mg/l (see Table 5-6). Even the relatively low lime dose for pH 10.2 was possible if compensated by ferric chloride dosage of 24 mg/l. In this operation primary effluent total phosphorus averaged 0.68 mg/l.

Removals of BOD, SS, and other parameters were comparable in each mode of operation. However, increase in hardness across the primary varied considerably. Operation at pH 11 with ferric chloride yielded the lowest effluent hardness.

The data in Tables 5-6 and 5-7 represent results when primary clarifier underflow solids were recycled to the chemical addition point. This was done to improve precipitation reactions involving calcium, magnesium, and phosphorus. Total solids in the lime reactor were maintained in the range of 1,000 to 2,500 mg/l. Later with solids recycle discontinued, the major performance change was in effluent hardness. During operation at pH 11 with ferric chloride dose of 10 to 16 mg/l, hardness across the primary increased an average of 32 mg/l. Under similar chemical feed rates, hardness increase across the primary was 6 mg/l when flocculator solids were in the range of 900 to 1,500 mg/l TSS. When flocculator solids were increased to 1,700 to 3,900 mg/l, hardness increase across the primary was 16 mg/l (12).

Considerable sludge thickening occurred in the primary clarifiers. Concentration of underflow sludge varied with a number of factors, including the nature of the raw wastewater. When a nearby cannery was in operation, underflow solids concentration did not exceed 4.5 percent when operating at pH 11. This seemed due to anaerobic action and gasification of peach and tomato solids in the raw sludge. Otherwise, 5 percent total solids in the underflow was maintained. Occasionally underflow at total solids levels of 9 percent

Table 5-6

LIME TREATMENT RESULTS AT CCCSD (12)

	Flow: 1.30 mgd, pH: 11.5, $Ca(OH)_2$: 500 mg/l					Flow: 1.12 mgd, pH: 11.0, $Ca(OH)_2$: 400 mg/l				
	Influent	Control Primary		Chemical Primary		Influent	Control Primary		Chemical Primary	
	mg/l	mg/l	% Removal	mg/l	% Removal	mg/l	mg/l	% Removal	mg/l	% Removal
Total Phosphorus	9.4	-	--	0.96	90	9.2			2.3	75
BOD	190	103	46	50	74	192	121	37	60	69
SS	199	57	71	41	79	195	57	71	47	76
TOC	107	59	45	37	65	118	68	42	48	59
Turbidity (JTU)	–	35	--	16	-	–	40	-	26	
Calcium Hardness (as $CaCO_3$)	76	–	–	168	-	76		–	156	
Magnesium Hardness (as $CaCO_3$)	96	–	–	40	–	103	--	--	59	

Table 5-7

LIME AND METAL TREATMENT RESULTS AT CCCSD (12)

	Flow: 1.19 mgd, pH: 11.0, $Ca(OH)_2$: 400 mg/l, $FeCl_3$: 14 mg/l					Flow: 1.20 mgd, pH: 10.2, $Ca(OH)_2$: 289 mg/l, $FeCl_3$: 24 mg/l				
	Influent	Control Primary		Chemical Primary		Influent	Control Primary		Chemical Primary	
	mg/l	mg/l	% Removal	mg/l	% Removal	mg/l	mg/l	% Removal	mg/l	% Removal
Total Phosphorus	9.5	–	–	0.85	91	9.4	-	-	0.68	93
BOD	210	109	48	53	75	178	106	40	59	66
SS	305	69	78	27	91	235	59	75	31	87
TOC	130	72	45	37	72	117	68	42	43	63
Turbidity (JTU)	--	45	-	14	--	–	41	-	15	–
Calcium Hardness (as $CaCO_3$)	76	-	--	139	–	7,2	--	–	148	
Magnesium Hardness (as $CaCO_3$)	90	–	--	32	–	90	-		70	

were obtained. At about 7 percent total solids, coning and bridging occurred in hoppers of these particular tanks and septic sludge zones developed.

Contrary to experience with high-lime operation, underflow solids at pH 10.2 could not be thickened to more than 4.5 percent. Use of an anionic polymer at 0.25 mg/l allowed thickening to 6 percent. An alternate anionic polymer did not significantly improve solids underflow and neither polymer enhanced the effluent quality.

The pattern of deposition of solids in the primary clarifier was studied. Most of the sludge was found in the first three-eighths of tank length. The heavier solids settled out rapidly in this zone and the rest of the tank served to settle finally divided solids. Continuously operating scaper flights kept the bottom relatively clean. The rectangular primary clarifier, with preaerated grit removal, seemed suitable for the lime precipitation operation.

Primary clarifier effluent flowed to the activated sludge nitrification-oxidation tanks directly with no continuous intervening recarbonation stage. External CO_2 was added to mixed liquor only when needed by vaporization form liquid storage. Normally, CO_2 was only necessary when lime addition operations were at pH 11.5 or higher. In-process CO_2 generation was sufficient for complete recarbonation with primary operation at pH 11. Supplemental CO_2 was injected to the first bay of the activated sludge tanks to avoid scaling in the transfer channel. All CO_2 sources contributed to lower pH to 7.5 in a short time.

CO_2 generated in the biological process from the oxidation of carbon provided most of the recarbonation requirements in all operating modes. The amount of CO_2 generated from oxidation was calculated from a total carbon balance around the oxidation-nitrification stage and typically amounted to 55 mg/l of CO_2 for recarbonation.

The amount of CO_2 generated from the hydrogen ions produced during nitrification was calculated from alkalinity destruction, and appeared to amount to some 115 mg/l under typical conditions. The denitrification operation, occurring in a subsequent unit, was not involved in recarbonation of primary effluent.

Daily supplemental CO_2 requirements averaged about 25 mg/l after the process was stabilized at pH 11.5. CO_2 demand varied diurnally according to the strength of the wastewater.

The CCCSD treatment facility was equipped with a high-pH alarm to warn operators of the need for exterior CO_2. The 30 mgd water reclamation plant will use a fully automated pH control system.

It was recognized that, in this particular plug-flow activated sludge system, the highest pH levels would occur in the first part of the aeration tankage. With operation in the conventional mode, primary influent and return activated sludge blended at the influent end. Thus a large part of the oxygen demand also occurred in the first part of the tank.

Enough oxygen to match the load in the head end of this tank would maximize CO_2 generation there. In the 30 mgd reclamation plant, DO probes in each pass will allow varying the diffused air input according to load.

An unusual feature of this plant was application of two-stage wet classification of raw sludge by centrifuges. The first-stage centrifuge received primary clarifier underflow (thickened in situ) at a high rate. It captured two-thirds of the solids present: the resulting cake was 50 percent solids by weight and contained 85 percent of the calcium carbonate in the feed sludge. Feed sludge was not conditioned with chemicals. This first-stage separation provided sludge for recalcination. The centrate from the first centrifuge proceeded to a second centrifuge designed for efficient capture of solids remaining in the feed. When treated with polymer a drier cake was delivered. This cake which contained most of the organics and phosphorus precipitated in the primary clarifier, was incinerated. Further information on sludge handling is presented in Section 7.5.

5.4 Other Experience

This section contains information on lime treatment plants under construction, on full-scale studies, and on related information which may be of benefit to designers of phosphorus removal facilities. This is not intended to be a complete listing, but should be representative.

5.4.1 Rochester, N.Y.–Low-Lime Process

Bench scale treatability studies for phosphorus removal preceded design of a 100 mgd treatment plant at Rochester, New York (13, 14). Activated sludge treatment alone removed about 20 percent of the influent phosphorus (8 mg/l as P). This agreed reasonably well with phosphorus required for biological metabolism.

Jar tests of chemical treatment were performed. These tests examined alternated chemical procedures to achieve required 80 percent phosphorus removal. Lime, and ferric chloride with anionic polymers, were tried on raw wastewater. Either 100 mg/l of lime as CaO (pH 9.5) or the combination of 40 mg/l $FeCl_3$ and 0.5 mg/l anionic polymer reduced phosphorus 70 percent. This removal plus biological removal raised overall reduction to 80 percent or more. BOD and SS reductions in jar tests were about 50 percent and 85 percent respectively. In this case, lime was selected due to: 1) overall economy, and 2) ease of sludge handling in the plant configuration to be described.

Existing treatment plant facilities at Rochester include primary clarifiers, Imhoff tanks, sludge holding tanks, coil filters, and sludge incinerators. Proposed improvements include modification primary clarifiers to add flocculation sections, addition of new primary clarifiers incorporating flocculation sections, conversion of Imhoff tanks to complete-mix activated sludge units, and addition of chlorination facilities. Plant startup is scheduled for late 1975. Detention in flocculation sections of existing clarifiers will be 20 minutes. In the new primary clarifiers, flocculation time will range up to 40 minutes. Existing and new primary clarifiers will have overflow rates of 1,000 gpd/sq ft and detention times of 2.3 and 2 hr, respectively, at average design flow. A schematic diagram of the expanded plant is shown on Figure 5-4. Plant design is based on the criteria shown in Table 5-8.

Lime will be added just after flow passes the comminutors. The pH will be monitored ahead of the flocculation basins and lime addition will be automatically regulated to maintain pH 9.5.

Anionic polymer will be fed just ahead of the flocculation basins when such addition is necessary to improve settling. Provisions for returning solids from the primary clarifiers to the flocculation basins are included.

Although lime treatment and biological treatment should remove 80 percent of incoming phosphorus, there will be additional facilities for adding alum to aeration tanks. This subsystem may be used alone or may supplement lime treatment for greater phosphorus removal.

Complete-mix activated sludge is expected to disperse the primary effluent rapidly, permitting recarbonation by bacterial CO_2. F/M can be controlled over a broad range in the configuration provided for aeration tanks.

Since 80 to 90 percent SS removal is expected in the primary clarifiers, the weight of primary sludge (based only on increased SS removal) will be 1.5 to 2.0 times conventional primary sludge production. Sludge produced by chemical precipitation must be added to

Table 5-8

DESIGN DATA FOR ROCHESTER, N. Y.

Average design flow, mgd	100
Maximum flow, mgd	200
Minimum flow, mgd	50
BOD loading, lb/day	250,000
BOD concentration, mg/l	300
Suspended solids concentration, mg/l	300
Volatile suspended solids concentration, mg/l	240
Total hardness, mg/l	200
Phosphorus, mg/l as P	8

FIGURE 5-4

TREATMENT SYSTEM AT ROCHESTER, N.Y.

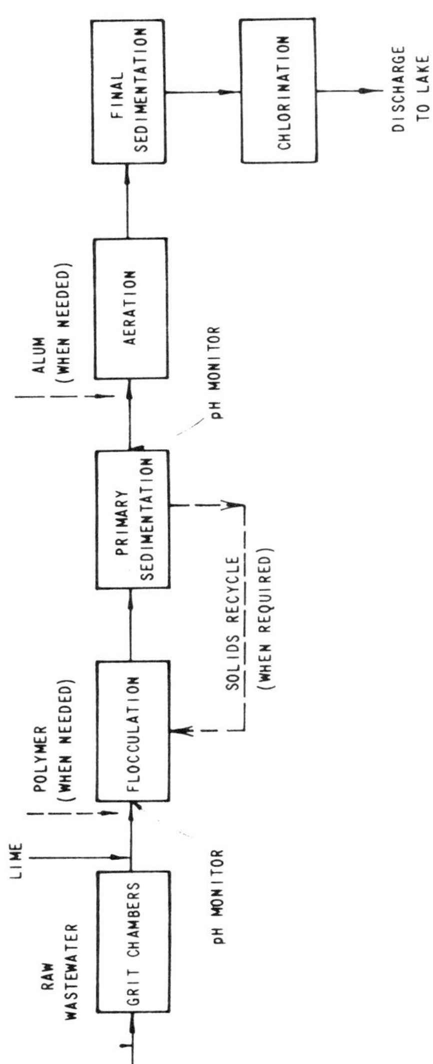

this, raising the total to about 3.0 times the normal amount. Primary sludge quantities at design flow with lime treatment are estimated at 175 tons/day. Since organic load on the activated sludge section will be reduced about 50 percent, activated sludge will be reduced substantially. At design loads with lime treatment, waste activated sludge is estimated at 30 tons/day, compared to about 75 tons/day if lime treatment were not used. Solids handling facilities have been designed for a capacity of 270 tons/day.

Sludges will be combined and gravity thickened to a concentration of 5 percent. Thickeners are designed for a solids loading of 13 lb/day/sq ft. After thickening, sludge will be dewatered on vacuum filters and incinerated in multiple hearth furnaces. Incinerator ash will be sluiced and pumped to two 0.5-acre lagoons, each having a water depth of 11 ft, operating in series. Effluent from lagoons will be pumped back to the head of the plant. Lagoons will probably have to be emptied every 9 to 12 months.

5.4.2 Lower Allen Township, Pa.–Low-Lime Process

Recalcination of sludge generated by lime addition in the primary is one feature of a new 6.0 mgd plant at Lower Allen Township, Pennsylvania (15). This plant, where startup began in February 1974, is a complete-mix activated sludge system with low-lime addition in the primary section.

Rough screening, degritting, and comminution precede lime injection. Flash mixing blends both recalcined lime and recycled seed sludge form the primary clarifier. Lime addition is controlled automatically within the range of pH 9.4 to pH 9.8. Mixed flow is divided between two flocculating clarifiers, each 85 ft in diameter and 12 ft in depth. These have a design average overflow rate of 700 gpd/sq ft.

Primary effluent is treated biologically in four activated sludge basins, each 45 ft. x 90 ft x 15 ft. Recarbonation reduces pH rapidly to between 7.5 and 8.5. Biological uptake is expected to complement lime precipitation in order that overall phosphorus reduction will be 80 percent. Phosphorus concentration in influent wastewater is 9.0 mg/l. Two final clarifiers are each 85 ft in diameter and 12 ft in depth.

Combined primary and waste activated sludges are blended and gravity thickened. After storage, sludge is conditioned before being fed to two parallel solid-bowl centrifuges. Dewatered cake from the centrifuges is fed to a multi-hearth dual purpose furnace. In the furnace, organic fractions are volatized while waste inert inorganic sludge is dried. The furnace is also designed to recalcine and recover quicklime for reuse at the head of the plant.

5.4.3 Manhattan, Kansas–Low-Lime Process

In Manhattan, Kansas, research into phosphorus removal using low lime and biological treatment began with laboratory scale investigations (16). It was found that both orthophosphates and polyphosphates could be precipitated readily. Complex phosphates present in wastewater seemed to inhibit calcium carbonate precipitation, to the extent that recalcination for lime recovery seemed impracticable. In these studies, 150 mg/l of slaked lime raised pH to 9.5. Total phosphorus removal before biological treatment was 60 percent, and SS were reduced 90 percent. Lime-primary sludge production was about twice that obtained in conventional operation and overall sludge production, including waste activated, was about 1.5 times that found with conventional secondary treatment.

The low lime process was operated at a constant pH. Performance of the downstream complete-mix activated sludge system was not disturbed. Microbial CO_2 production in the activated sludge unit was sufficient to maintain pH near neutral. A mixture of lime-precipitated primary sludge and waste activated sludge was dewatered and vacuum filtered with small additions of anionic polymers.

A pilot scale demonstration of the technique followed (17). The wastewater was typical domestic wastewater generated at the rate of 80 gal/capita/day. Untreated wastewater had a pH near 7, and other characteristics were BOD and SS near 200 mg/l, total phosphorus of 10 mg/l, alkalinity of 235 mg/l, hardness of 150 mg/l, calcium of 110 mg/l, and magnesium of 40 mg/l. The pilot facilities included a 15,000 gpd conventional treatment process, with lime addition to the primary. Influent wastewater was supplied to the system on a diurnal pattern from a variable speed pump and lime slurry was injected into the discharge line of that pump. Flow then entered a mixing and reaction tank where detention was near 2 minutes. Lime feed was made proportional to flow initially, but was later converted to pH control.

Primary clarification followed with a surface loading rate of 225 gpd/sq ft. Primary sludge was extracted hourly. Primary effluent flowed into plug flow aeration tankage where it was detained about 5 hr. The final clarifier had an overflow rate similar to the primary clarifier but sludge withdrawal was continuous.

Primary and secondary sludges were detained separately in holding tanks large enough to store the sludge from an entire day's operation. This permitted more effective sampling than could be obtained by compositing smaller portions. The two sludges were then blended into a third tank for dewatering studies. The biological system operated to give sludge age of 12 to 15 days when MLSS was 2,000 mg/l.

During the pilot study, a number of the operational features were changed deliberately or through circumstance. It appeared the most important operational variable was pH, following addition of lime to the primary effluent. If that pH was held between 9.5 and 10, phosphorus removals of 75 percent or more were generally obtained in the primary unit (see Figure 5-5). Consistent phosphorus reduction was not obtained unless the pH was kept at 9.8. When pH was increased to 11.5, 90 percent phosphorus removal resulted, but a considerably greater lime dosage was required.

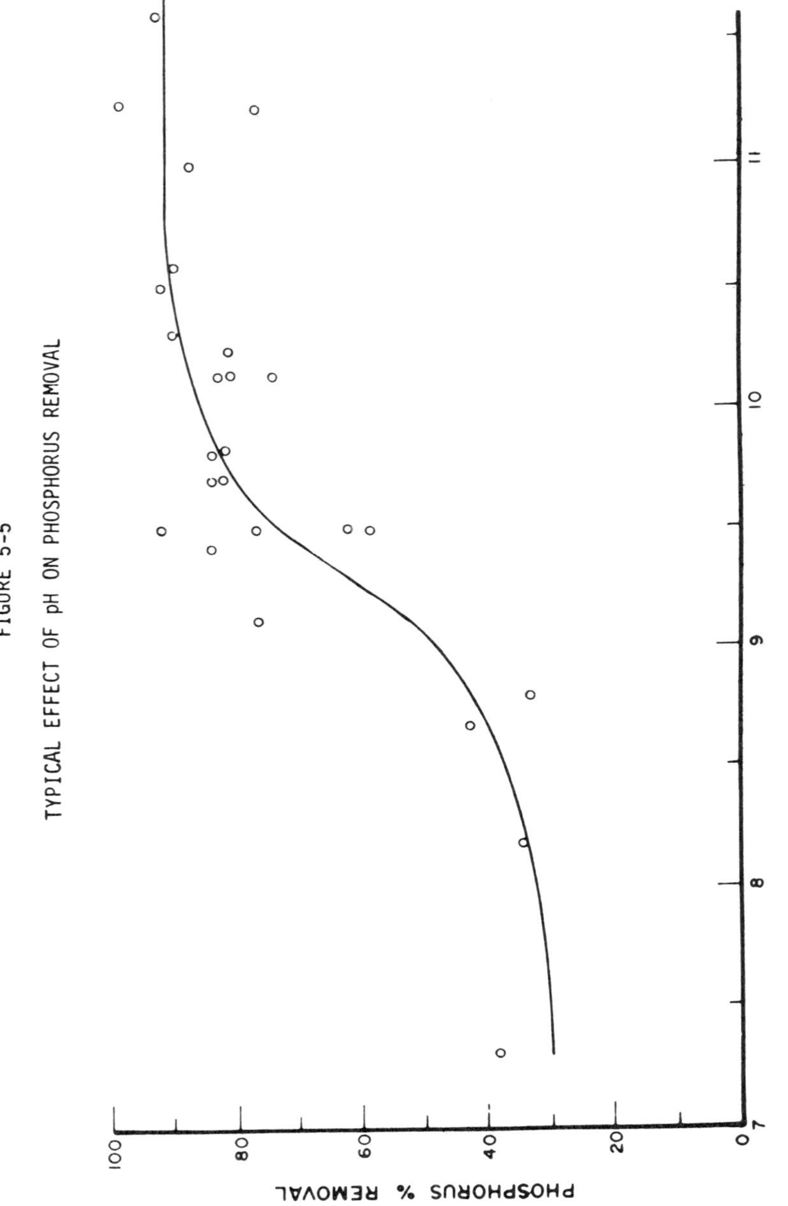

FIGURE 5-5

TYPICAL EFFECT OF pH ON PHOSPHORUS REMOVAL

It appeared that fine floc escaping the primary clarifier contained, in particulate form, most of the 25 percent total phosphorus residual which entered the biological section. Reduction of the flow rate (which was already quite low in terms of surface loading of the primary clarifier) did not materially improve capture of precipitated phosphorus. In this instance a higher degree of performance was not obtained by the use of polymer.

Phosphorus removal in the biological section was 5 percent or less of the total phosphorus coming into the treatment plant.

Lime treatment to pH 9.5 to 10 resulted in primary BOD reductions near 60 percent, a considerable improvement over the usual 30 or 40 percent obtained in plain clarification. SS concentrations were not materially reduced, as already mentioned in connection with lime floc which was escaping the primary unit. CO_2 production by bacteria in the activated sludge unit neutralized final effluent to levels near pH 8.

Addition of lime to the primary clarifier resulted in physical and biological changes in the activated sludge. Mixed liquor turned gold or light brown instead of the usual chocolate brown characteristic of this operation. Bacteria in the culture tended to shift into filamentous species. Among the scavenger organisms, population of free swimming ciliates declined, while rotifers increased to dominance.

The technique produced two to three times more total weight of sludge than generated during conventional treatment. However, with properly administered polymer, the sludge could be concentrated. The most attractive sludge treatment systems involved polymer conditioning, dewatering by vacuum filter or centrifuge, and incineration or land fill. Primary sludge had a volatile fraction of 50 percent during lime addition, as opposed to 65 percent in conventional systems. Chemical sludge would compact to a 3 percent density in 0.5 hr. This concentration could be considerably improved with an anionic polymer. Biological solids were produced at the rate of about 50 mg/l as the flow passed through the aeration tank. Biological sludge represented 7 percent by weight of the total sludge produced when using the lime technique.

Total sludge resulting from lime treatment was estimated at 3,750 lb/million gal of flow. This compared with 1,500 lb/million gal, of which some one-third would be waste activated sludge. Sludge conditioning with anionic polymer (2.5 lb/ton dry solids) yielded a filtration rate of 5 lb/sq ft/hr.

Several problems developed during the course of the study, and these should be considered by designers of such plants. Milk of lime solution had a marked tendency to scale everything it contacted prior to injection into the wastewater. Following injection, lime depositing out of treated wastewater caused scaling problems on the membrane of a pH probe, and on wetted parts of an automatic phosphorus analyzer. These and other surfaces also tended to collect a biological slime. Lime scale on pH probes was removed every other day by submersing them into diluted acid.

This pilot plant used 60 deg. hoppers for sludge collection in the primary clarifier. They did not work well with lime sludge which had a tendency to cling even to these sharply sloped bottoms. In due time denitrification occurred within that trapped layer and floating solids resulted. Observers noted that sludge also adhered to hopper walls in the final clarifier and became anaerobic prior to withdrawal. It appeared that this resulted in release of cellular phosphorus into the effluent.

5.4.4 Richmond Hill, Ontario—Low-Lime Process

Following laboratory and pilot scale studies, a three-month plant scale investigation of lime treatment for phosphorus removal was conducted using a 0.2 mgd section of a 2 mgd conventional activated sludge plant in Richmond Hill, Ontario (18). Lime was added before the primary clarifier on a continuous basis and no special mixing or flocculating equipment was provided.

Raw wastewater had alkalinity of 381 mg/l and total hardness of 401 mg/l. Average total influent phosphorus was 10.6 mg/l (as P). Normal plant operation reduced total phosphorus by 39 percent. BOD and SS removals in the primary clarifier were 21 and 37 percent, respectively.

Lime dosage of 175 mg/l produced a pH of 9.3. During lime addition, BOD and SS removals in the primary clarifier averaged 72 and 78 percent, respectively. Total phosphorus removal averaged 83 percent in primary treatment, and removal through the entire plant was 92 percent, resulting in a total phosphorus concentration in the secondary effluent of 0.9 mg/l.

Throughout the study, soft calcium carbonate scale accumulated on effluent weirs and troughs of the primary clarifier but was easily sluiced away using a water hose. There was no appreciable buildup on submerged surfaces within the primary clarifier.

Lime treatment did not upset the biological process in the test section. The pH of 9.3 in primary effluent was promptly reduced below 8 by CO_2 generated in the aeration tank. MLSS equilibrium was reached and maintained throughout the study. Due to improved primary performance, organic loading to the conventional (activated sludge system) was reduced and its behavior shifted to operation as an extended aeration process with no secondary sludge wastage.

Sludge precipitated in the primary clarifier was a mixture of calcium phosphate, calcium hydroxide, calcium carbonate, and coagulated organics and grit. This sludge was free flowing and considerably different from the gelatinous matter previously produced in bench scale studies. Sludge removed from the primary unit had a solids content of 6 to 8 percent. It gravity thickened readily to 10 to 12 percent solids, but was difficult to dewater without proper chemical conditioning. Biological activity in the lime-sludge was inhibited by the high lime content. The sludge proved free from noxious odors even after prolonged holding under anaerobic conditions. Supernatant from the chemical sludge, after 5 weeks of storage, contained no appreciable amount of phosphorus, either total or soluble. It was concluded that lime treatment could be readily controlled either in response to flow or on the basis of pH in a flash mixer or primary effluent.

5.4.5 Seneca Falls, New York

A combined biological-chemical precipitation process (proprietary name "PhoStrip") for phosphorus removal was applied in a short-term (30 day) study at the existing Seneca Falls, N.Y. treatment plant in 1973. This process invoives modification of the conventional activated sludge process. A schematic of the Seneca Falls treatment scheme is shown in Figure 5-6.

The major modification to the activated sludge process is the incorporation of the phosphate "stripping" tank in the return sludge line. This tank permits the return sludge to become anaerobic which, apparently, induces the sludge organisms to secrete phosphate. The phosphorus enriched supernatant from the stripper is then dosed with lime and returned to the primary clarifier for sedimentation of phosphorus. Data obtained during the 30-day trial are included in Table 5-9.

Reported chemical costs for the Seneca Falls trial, based on a lime dosage of 24.1 mg/l (as CaO), were \$2.01 per million gallons of raw wastewater. Additional costs would include the phosphate stripper tank, the lime feeding, storage and mixing facilities, and associated pumping facilities. At Seneca Falls, a spare clarifier was used as the phosphate stripper tank. It is claimed that total costs for the project due to chemical cost savings may result in significant savings compared to conventional mineral addition.

5.5 Sludge

Detailed information in Chapter 11 concerns amounts of sludge generated during lime addition, and offers design parameters required to select and size sludge treatment and disposal facilities. Additional information specifically related to sludge produced by lime addition in primary treatment operations is discussed in this section.

Plant scale centrifuge trials on underflow from lime treated primary units at Newmarket, Ontario, were considered successful (19). Centrifugation captured some 99 percent SS in the feed, and cake solids were near 30 percent by dry weight. The performance of the solid bowl centrifuge was such that centrate contained 700 mg/l SS and about 20 mg/l total phosphorus. Centrate was returned to the flash mixer preceding the primary clarifier. Polymer requirements were less than 1 lb/dry ton of solids. Overdosing of polymer caused foaming in the centrate.

Sludge production during low lime addition at Manhattan, Kansas (17) was 250 percent of the weight of sludge produced by conventional treatment. However, being primarily a grainy chemical sludge, this mixture of primary and waste activated sludge was much easier to handle, dewater, and dispose of than sludge produced during normal plant operation.

FIGURE 5-6
PHOSPHORUS REMOVAL PROCESS AT SENECA FALLS, N. Y.

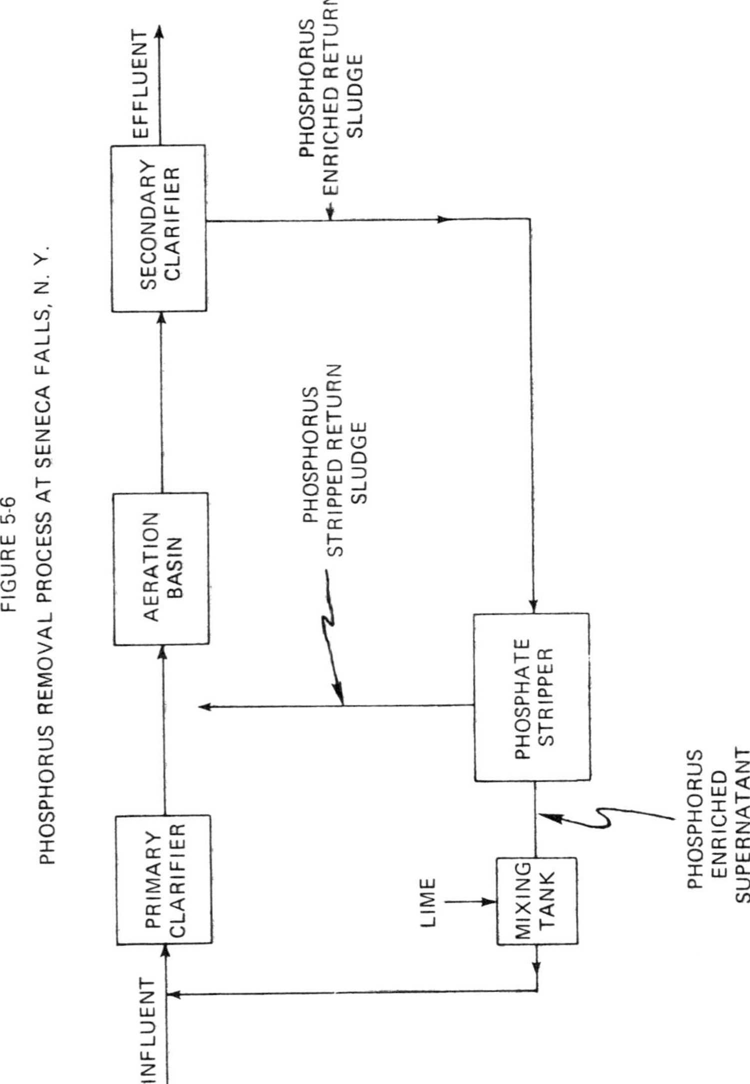

Table 5-9

RESULTS OF PHOSPHORUS REMOVAL TRIAL AT SENECA FALLS, N. Y.

Plant Flow, mgd	
Design	1.75
Average	0.9
Influent	
BOD, mg/l	158
P, mg/l	6.3
Detention Times (based on average flow), hours	
Primary Clarifier	3.9
Aerator	5.5
Secondary Clarifier	3.2
Phosphate Stripper	18
Return Flows, percent of raw flow	
Sludge to Phosphate Stripper	24
Sludge to Aeration Tank	10
Supernatant to Primary Clarifier	14
Lime Dose, mg/l as CaO	24.1
SS, mg/l	
Mixed Liquor	1,440
Return Sludge to Stripper	7,840
Return Sludge to Aerator	15,910
Effluent	
BOD, mg/l	3.5
P, mg/l	0.55
Plant Performance	
BOD Removal, percent	98
P Removal, percent	91

Canadian studies (5), taken as a group, indicate total sludge production at a conventional plant will increase from 1800 lb/million gallons to 4600 lb/million gallons with addition of some 200 mg/l lime to primary clarifiers. Solids concentration increased from 4.5 to 9 percent, so the volumetric increase was 25 percent.

In a primary plant, sludge production increased from 800 lb/million gallons to 2400 lb/million gallons with solids concentration rising from 7 to 15 percent (a net 50 percent increase in sludge volume).

Plant-scale testing of two-stage centrifugation of combined lime-treated primary and waste activated sludge has been applied at Central Costa County Sanitary District (CCCSD), California (11). This permitted a degree of sludge classification vital to the integrated system of lime recovery and sludge disposal to be utilized in the 30 mgd treatment plant under construction at CCCSD. Figure 5-7 depicts the solids handling and disposal facilities which will be incorporated in the 30 mgd plant; Figure 5-8 depicts the plant-scale sludge handling system used at CCCSD to develop design data for the full-scale system.

The purpose of the first of the two centrifuges is to maximize recovery of calcium carbonate while rejecting other sludge components, such as organic material, phosphorus and magnesium compounds, and other constituents coprecipitated with calcium carbonate. The first-stage cake represents phase, all of which have a slower settling rate than calcium carbonate, are deliberately carried out with the process centrate and passed on to a second-stage centrifuge operated to obtain high solids recoveries. Second-stage solids are incinerated (breaking down organic compounds in the process) and the total ash, including fine inert materials, can be disposed of without recycling back into the wastewater treatment process.

Optimum first-stage performance in the test system was obtained after varying critical centrifuge variables: feed rate, centrifugal force, pool depth, and conveyor speed. Observations were made on individual sludge constituent recovery throughout a broad range of these process variables. Figure 5-9 shows the generalized results. The definition of performance allowed setting the centrifuge variables to capture a high percentage of calcium carbonate but a low percentage of all other sludge constituents.

Certain trends were noted in the CCCSD studies which may have application in other locations. In general, results followed past practice in centrifugation of calcium carbonate and, almost with exception, followed basic centrifuge fundamentals.

When wastewater was lime treated to a pH of 11 or 11.5, the highest percentage of calcium carbonate was recovered in the first unit and the greatest reject of extraneous solids occurred. Sludge produced when flocculating at pH 10.2 was not as easily classified. Separation was most effective when pool depth in the bowl was at very shallow settings. Dewatering of the calcium carbonate cake was materially improved by raising centrifugal force to the range of 2,100 g's. Polymer preconditioning of sludge was not required.

FIGURE 5-7 PROPOSED SLUDGE FACILITY AT CONTRA COSTA COUNTY

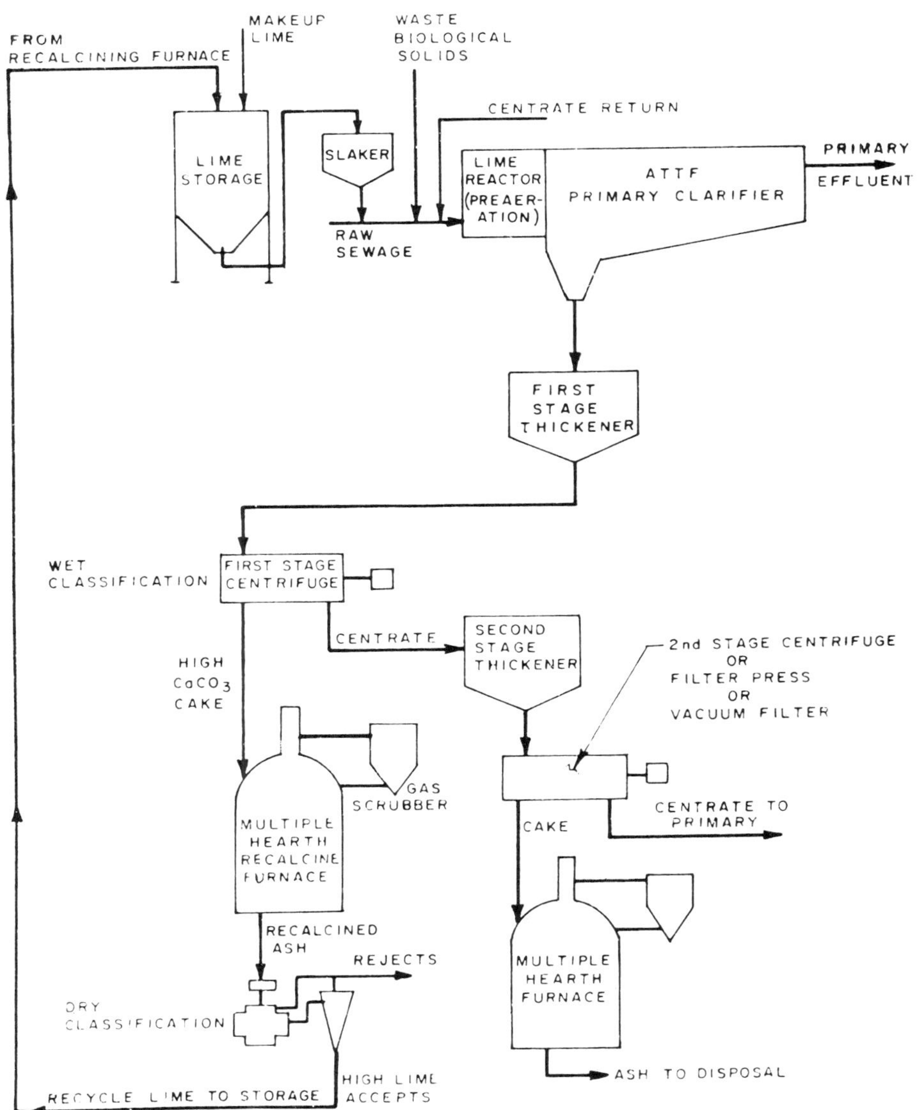

FIGURE 5-8 PLANT SCALE SOLIDS TESTING SYSTEM AT CONTRA COSTA

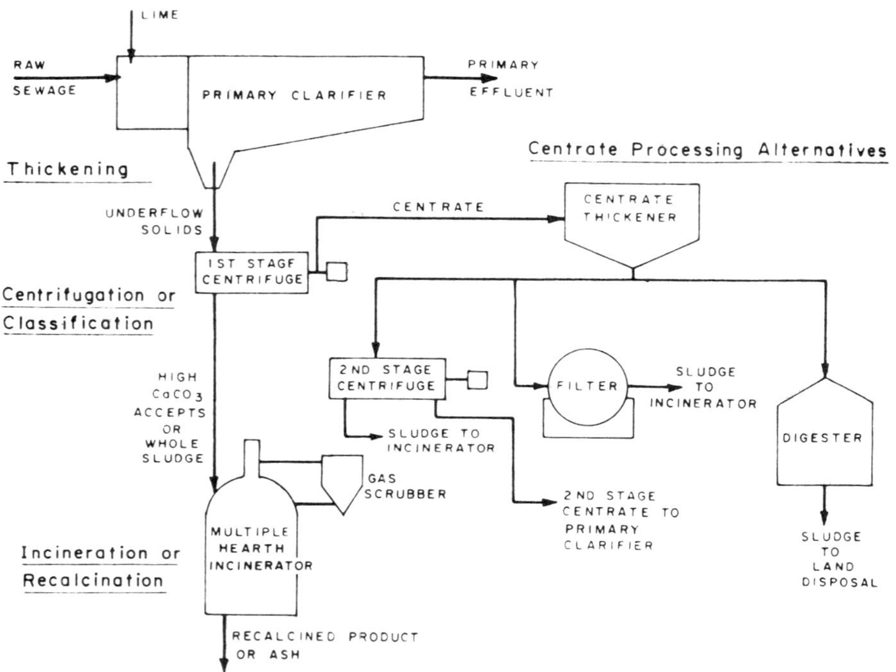

First-stage centrate passed to the second centrifuge for further treatment. Several polymers were tried and some proved effective in increasing solids capture when fed at the rate of 2 lb/ton of solids. Solids recovery of 80 percent and resulting cake dryness of 15 percent solids were typical operating levels obtained after experimentation. Setting the machine for maximum pool depth and operating near 2,100 g's gave satisfactory results in this particular installation.

FIGURE 5-9 SLUDGE CLASSIFICATION CURVES AT CONTRA COSTA

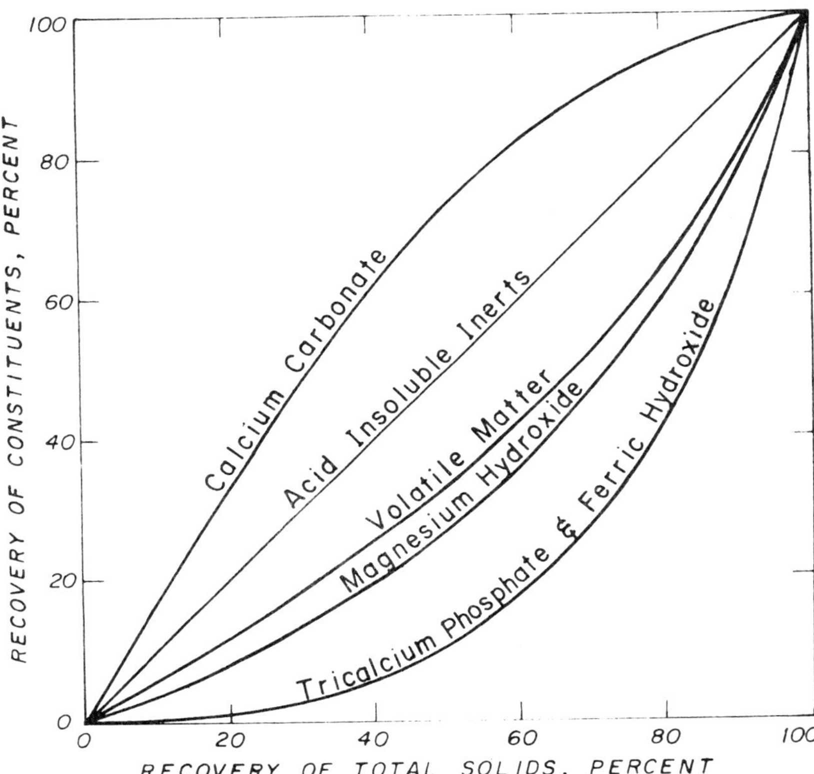

The extensive investigations at this installation (11) have only been briefly summarized here. The conclusion is that, in this instance, two-stage classification by centrifuges was applicable. Results of the operation were generally predictable by centrifuge laws. Other locations considering this type of operation should make comparable predesign studies.

It is also worthwhile to note that sludge production at the CCCSD test installation was predicted by theoretical procedures similar to those described in Chapter 9. Actual sludge production was then measured under four different operating modes, and operating experience agreed within about 10 percent with predicted values. These predictions and the actual production rates both included biological sludge from the nitrification-denitrification system, as well as settleable solids normally removed in primary treatment. In this case, the organic fraction of total sludge was one-quarter to one-third of the combined weight involved.

5.6 Costs

The capital cost of adding rapid mix and flocculation facilities to existing plants has been estimated for 1, 10, and 100 mgd plants. These should serve as a general guideline where phosphorus is to be removed by lime addition to the primary clarifier. Estimated costs for rapid mix facilities include concrete basins (with detention of one minute at design flow), and mixers with shafts and impellers of stainless steel. Flocculation facilities consist of baffle walls and mixers to provide a flocculation chamber with 30 minute detention within existing circular clarifiers. Estimated costs of these are given in Table 5-10 and include allowance for Contractor's installation, overhead, and profit, plus 20 percent for engineering and contingencies. Figures given relate to an EPA-STP Cost Index of 185.

Table 5-10

TYPICAL CAPITAL COSTS FOR LIME TREATMENT IN PRIMARY FACILITIES

	1 mgd	10 mgd	100 mgd
Rapid Mix	$ 2,200	$ 13,000	$ 82,000
Flocculation	64,000	264,000	983,000
Total	$66,200	$277,000	$1,065,000

Investment figures for lime storage and feed facilities are given in Chapter 10.

The cost/1000 gal for adding single-stage lime treatment to an existing plant was determined by amortizing capital cost at 6 percent over 25 years and adding the cost of chemicals. Operational expenses such as labor and power are not included.

Lime costs/1000 gal are based on a chemical cost of $22/ton with 90 percent available CaO at a dosage of 150 mg/l as CaO. The equipment and chemical costs (¢/1000 gal) are given in Table 5-11.

Table 5-11

TYPICAL TOTAL COSTS FOR LIME TREATMENT IN PRIMARY FACILITIES

	1 mgd	10 mgd	100 mgd
Equipment Amortization			
Rapid mix and flocculator	1.4	0.6	0.2
Chemical storage and feed system	0.4	0.2	0.1
Chemical Cost	1.5	1.5	1.5
Total Cost (¢/1,000 gal)	3.3	2.3	1.8

Costs do not include allowance for equipment for sludge handling and disposal and corresponding operating costs. Since a considerable quantity of additional sludge is produced, this expense is likely to be substantial.

5.7 References

1. Pearse, et al., "Chemical Treatment of Sewage," *Sewage Works Journal*, 7:6, p. 997 (1935).

2. Karanik, J. M., and Nemerow, N. L., "Removal of Algal Nutrients," *Water and Sewage Works*

3. Albertson, O. W., and Sherwood, R. J., "Removing Phosphates from Wastewater," *Industrial Water Engineering*, p. 30 (November, 1967).

4. Albertson, O. E., and Sherwood, R. J. "Phosphate Extraction Process," *JWPCF*, 41:8, p. 1467 (1969).

5. Boyko, B. I., and Rupke, J. W. G., "Technical Implementation of Ontario's Phosphorus Removal Programme," *Proc.*, 28th Purdue Industrial Waste Conference, Lafayette, Indiana (May, 1973).

6. Spohr, Guenter, and Talts, Andres, "Phosphate Removal by pH Controlled Lime Dosage," *Public Works*, p. 63 (July, 1970).

7. Tofflemire, T. J., and Hetling, L. J., "Treatment of a Combined Wastewater by the Low-Lime Process," *Journal Water Poll. Control Fed.*, 45, p. 210 (1973).

8. Bishop, Dolloff F., O'Farrell, Thomas P., and Stamberg, John B., "Physical-Chemical Treatment of Municipal Wastewater," *Journal WPCF*, 44:3, p. 361 (March, 1972).

9. Black, S. A., "Lime Treatment for Phosphorus Removal at the Newmarket/East Gwillimbury WPCP," Research Paper No. 2032, Research Branch, Ministry of the Environment, Ontario (May 1972).

10. Parker, D., Brown & Caldwell Engineers, Private Communication (June, 1973).

11. Parker, D. S., Zadick, F. J., and Train, K. E., "Sludge Processing for Combined Physical-Chemical-Biological Sludges," Grant #R801445, Office of Research and Monitoring, EPA (March, 1973).

12. Horstkotte, G. A., et al., "Full-Scale Testing of a Water Reclamation System," *Jour WPCF*, 46:1, p. 181-197 (January, 1974).

13. Wahbeh, V. N., "Design and Modification of the Rochester, New York Wastewater Treatment Plant for Phosphorus Removal," Presented at the Technical Seminar/Workshop on Advanced Waste Treatment, Chapel Hill, North Carolina (February, 1971).

14. Weller, L., Black and Veatch, Kansas City, Mo., direct communication (1974).

15. Paul, P., Gannett, Fleming, Corddry and Carpenter, Inc., Harrisburg, Pa., Direct Communication (1973).

16. Schmid, L. A., and McKinney, R. E., "Phosphate Removal by a Lime-Biological Scheme," *JWPCF*, 41:7, p. 1259 (1969).

17. Schmid, Lawrence L., "Pilot Plant Demonstration of a Lime-Biological Treatment Phosphorus Removal Method," Water Quality Office, Environmental Protection Agency, Project 17050 DCC (September, 1972).

18. ________________ and Lewandowski, W., "Phosphorus Removal by Lime Addition to a Conventional Activated Sludge Plant," Div. of Res. Publ. No. 36, Ontario Water Resources Commission (November, 1969).

19. Smith, A. G., "Centrifuge Dewatering of Lime Treated Sewage Sludge," Research Paper No. W2030, Res. Br., Ministry of the Environment, Ontario (May, 1972).

Chapter 6

PHOSPHORUS REMOVAL IN TRICKLING FILTERS BY MINERAL ADDITION

A logical approach to removal of phosphorus in trickling filter plants would be the use of precipitating chemicals somewhere in the system. Iron and aluminum can both be used for this application. Because of their similar action, iron and aluminum salts are commonly referred to collectively as minerals in wastewater treatment and are considered separately from calcium (lime). Required physical facilities for adding iron and aluminum are relatively simple. They are covered, along with special considerations for specific chemicals, elsewhere in this manual. This discussion will emphasize the merits of alternative addition points, review methods for determining dosage, and report studies where chemicals have been dosed directly to a trickling filter and to trickling filter effluent. Broader aspects of mineral addition for phosphorus removal are covered in other sections of this manual and in several recent reports (1, 2, 3, 4).

6.1 Pre-Design Decisions

State quality standards on phosphorus usually form the major basis for considering mineral addition for a given trickling filter plant. However, phosphorus concentration in the effluent is only one factor involved; effluent BOD and suspended solids will also be markedly reduced by this form of chemical treatment.

Modification of existing trickling filter plants for phosphorus removal is often a simple operation, and inclusion of chemical treatment facilities in new plants is a relatively minor addition. In either case, capital costs for chemical treatment equipment are relatively low. Also, the required degree of phosphorus removal has little effect on capital cost. Facilities for 80% phosphorus removal are identical to those for 95% removal.

On the other hand, operating costs are substantially higher to obtain successful treatment with chemical addition. More constant surveillance and more laboratory support are needed than with a conventional plant. Added operating costs are on the order of 4¢ to 5¢ per 1,000 gal. for reducing phosphorus (as P) to 2 mg/l with an accompanying reduction of BOD and solids to 15 mg/l. For an additional 1¢ to 2¢ per 1,000 gal., phosphorus can be reduced to 1 mg/l or less and BOD and solids to 10 or 12 mg/l. This simultaneous reduction of phosphorus, suspended solids and BOD may be the deciding factor for adopting chemical treatment in a trickling filter plant.

6.2 Process Options

There are three locations within a trickling filter plant where chemical precipitation can be incorporated: the primary clarifier, the trickling filter, and the final clarifier. The preferred approach is to provide facilities to serve both the first and last of these but not to add chemicals directly to the trickling filter.

Chemical addition to primary clarifiers will produce the greatest sludge yield of any one arrangement. Since treatment at this point will reduce influent BOD about 50%, it may be an appealing approach in an existing plant which is organically overloaded. Unfortunately, a significant amount of the phosphorus present may not be in the ortho form at this stage and may not be effectively precipitated and removed. Even after chemical treatment, primary effluent may remain fairly turbid, but will clear up during passage through the rest of the plant.

Mineral addition in the trickling filter does not seem to cause any functional problems although the filter may blotch and slough more than usual. A modest inorganic film often forms over the surface of the rocks but does not seem to interfere with biological activity or cause ponding. As data shown later will indicate, however, high degrees of phosphorus removal are not attained when chemicals are added only to the filter. Poor removal may result partly from short circuiting in the filter. This approach is not recommended.

Treatment in the final clarifier alone is a risky approach. If the operation is not done effectively, poorly treated effluent can escape to the receiving water. In spite of these reservations, experience has shown that precipitation in the final clarifier can be both effective and controllable. Orthophosphate predominates at this point and it precipitates readily. Also, bio-degradable detergents which can interfere with precipitation are largely absent. If underflow solids from a dosed final clarifier are returned to the primary clarifier, it will stimulate unusually effective clarification there.

In order to allow for incorporating the advantages of chemical treatment in both the primary and secondary clarifiers, it is recommended that chemical addition and mixing facilities be provided at both locations. Facilities required are not expensive and provision for mineral addition to both primary and secondary clarifier gives considerable flexibility of operation. Chemical treatment in both clarifiers can then be used if experience shows it to be the most effective technique at the particular plant involved.

6.3 Performance Data Using Aluminum Salts

6.3.1 Richardson, Texas

A plant-scale study was begun at Richardson, Texas, in 1970 to evaluate the potential of chemical addition to remove phosphorus from and improve overall performance of the City's trickling filter plant (5). Major objectives were to reduce the effluent phosphorus concentration to 1.0 mg/l (as P), or less, and to reduce effluent BOD and SS residuals to 15 mg/l or less.

The existing plant was a low-rate trickling filter facility with combination primary clarifier/digester units (clarigesters) as shown on Figure 6-1. Operating data for a control period before the addition of chemicals are shown in Table 6-1.

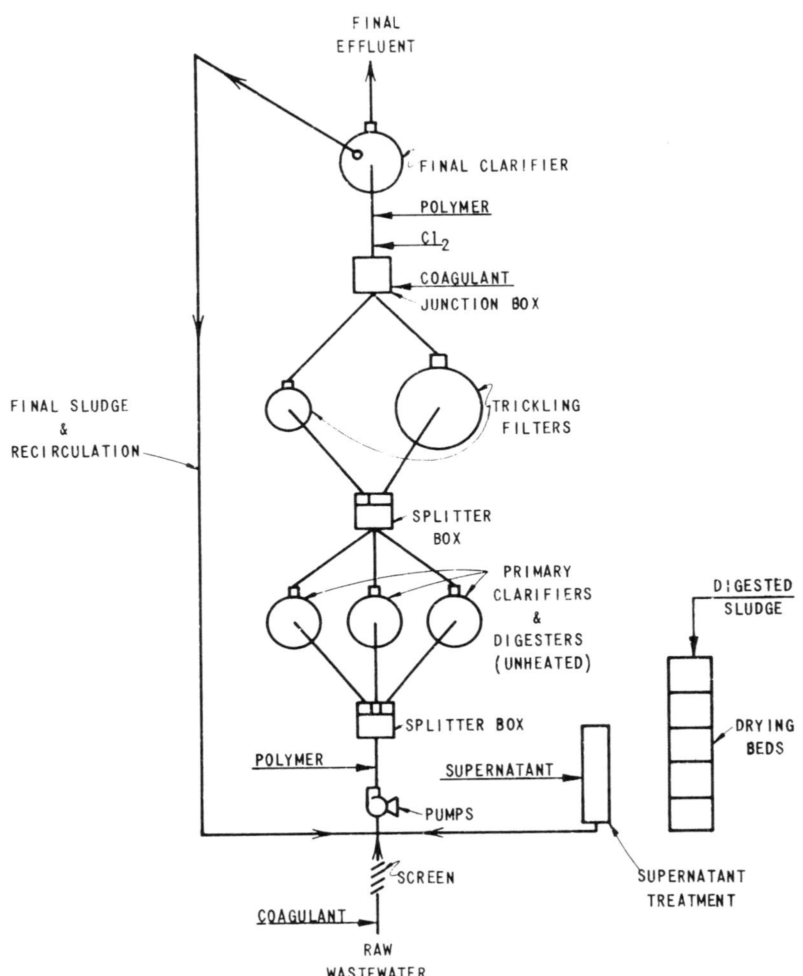

FIGURE 6-1 RICHARDSON, TEXAS
TREATMENT PLANT FACILITIES

Table 6-1

OPERATING DATA FOR RICHARDSON, TEXAS

Description	1970 Control Period	1971-72 Extended Alum Run (11½ Months)
Flow, mgd	1.5	1.6
Influent BOD, mg/l	166	170
Influent SS, mg/l	155	155
Influent P, mg/l	11	11.4
Primary Clarifier		
Overflow Rate, gpd/sq ft	400	425
Primary Effluent BOD, mg/l	*	115
Primary Effluent SS, mg/l	*	110
Primary Effluent P, mg/l	*	8.6
BOD Removal, percent	*	32
SS Removal, percent	*	29
P Removal, percent	*	25
Trickling Filter		
Depth, ft	6.5	6.5
Recirculation Ratio	0	0
Hydraulic Loading, mgd/acre	3.9	4.1
Organic Loading, lb BOD/1,000 cu ft/day	*	14.0
Secondary Clarifier		
Alum Dosage Rate, Average Al:P mole ratio		1.6**
Overflow Rate, gpd/sq ft	390	415
Secondary BOD Removal, percent	*	96
Secondary SS Removal, percent	*	94
Secondary P Removal, percent	*	94
Overall Plant Performance		
Effluent BOD, mg/l	20	5
Effluent SS, mg/l	15	7
Effluent P, mg/l	8	0.5
BOD Removal, percent	88	97
SS Removal, percent	90	95
P Removal, percent	27	96

*Not reported.
**Based on Influent P

Based on operational trials, alum addition ahead of the final clarifier was selected as the most promising chemical additive system for an extended 1½-month run. The flow diagram for the upgraded system is shown on Figure 6-1.

It was found that a mole ratio (Al : Influent P) of 1.5 to 1.7 consistently yielded effluent concentrations of 5 mg/l of BOD, 7 mg/l of SS and 0.5 mg/l of total phosphorus (as P). The corresponding effluent values for this plant prior to chemical addition were 20, 15 and 8 mg/l, respectively. The improved plant performance obtained with this upgrading technique was attributed in part to the low final clarifier overflow rate, careful management of final clarifier sludge withdrawal to prevent disruption to and loss of the alum floc blanket and frequent manual adjustment of the chemical feed pump rate to match alum dosage to mass inflow of phosphorus. Alum treatment doubled the volume of anaerobically digested sludge produced. However, the digested alum/biological sludge exhibited superior drying characteristics on sand beds and could be removed in about one-half the normal time. Operating data for the upgraded plant are shown in Table 6-1.

Chemical costs were $0.05/1,000 gal of plant flow or $0.36/lb of phosphorus removed, with phosphorus removal at the 96 percent level. The 1970 capital costs associated with plant modifications for chemical addition were $65,000, allocated as follows:

New laboratory building	$21,000
Laboratory equipment and furniture	6,000
Chemical storage and feed equipment, plant piping and metering modifications	38,000
Total	$65,000

6.3.2 Chapel Hill, North Carolina

Promising results were obtained in removing phosphorus and in generally improving plant performance at Richardson, Texas, by the addition of alum to trickling filter effluent as discussed in Subsection 4.5.4.1. In view of these results, the University of North Carolina Wastewater Research Center initiated a follow-up study to further explore and confirm this process at the Chapel Hill, North Carolina Wastewater Treatment Plant (6). Conducting a similar chemical addition project at Chapel Hill was considered worthwhile because it is a typical high-rate trickling filter plant using recirculation, whereas the Richardson filters were low-rate units. Furthermore, parallel and identical lines of treatment units were available at Chapel Hill, allowing direct comparison of results with and without alum addition. Comparison of parallel results was not possible at Richardson.

The wastewater treatment plant at Chapel Hill is a conventional high-rate installation treating predominately domestic wastewater. Incoming wastewater passes through pretreatment facilities for the removal of large solids and grit. The flow is then divided into equal portions for diversion to two identical lines of treatment, each consisting of a primary clarifier, trickling filter and final clarifier as shown on Figure 6-2. On one side of the plant, a feed pump system was installed to add liquid alum to the trickling filter effluent just prior to the final clarifier.

Operating data for the two sides of the plant are shown in Table 6-2. Significant improvements in BOD, SS and phosphorus removals were achieved with an alum dosage rate that ranged between an Al:Influent P mole ratio of 1.5 and 2.2. Within this range and for the loading rates shown, the final clarifier hydraulic loading appeared to be the significant factor affecting process efficiency. Subsequent phases of the study indicated that phosphorus removals of over 90 percent and BOD removals of about 95 percent could be consistently obtained by lowering the final clarifier overflow rate to 500 gpd/sq ft. Recirculation of the final clarifier settled alum/humus sludge to the primary clarifier for thickening had a beneficial effect on primary treatment efficiency and decreased the organic loading on the alum train trickling filter.

FIGURE 6-2

CHAPEL HILL, N.C. TREATMENT PLANT FACILITIES

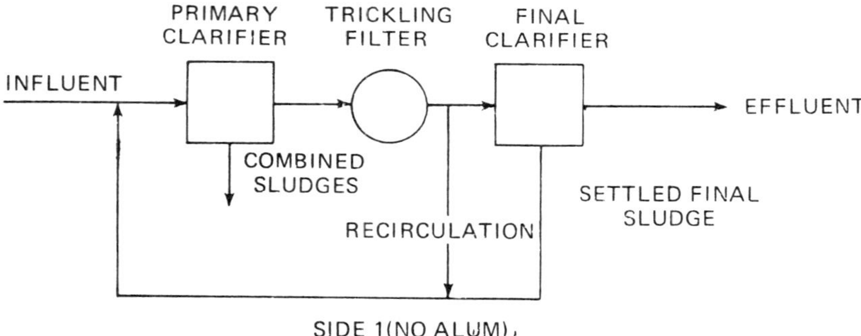

SIDE 1(NO ALUM),

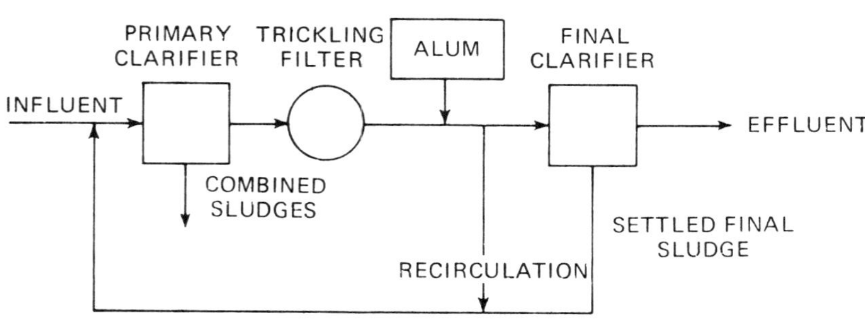

SIDE 2(WITH ALUM)

Table 6-2

OPERATING DATA FOR CHAPEL HILL, NORTH CAROLINA

Description	Side 1 (No Alum)	Side 2 (With Alum)
Flow, mgd	1.41	1.41
Influent BOD, mg/l	168	168
Influent SS, mg/l	229	229
Influent P, mg/l	11.3	11.3
Primary Clarifier		
Overflow Rate, gpd/sq ft	1,100	1,100
Primary Effluent BOD, mg/l	77	63
Primary Effluent SS, mg/l	89	68
Primary Effluent P, mg/l	9.7	6.6
BOD Removal, percent	54	63
SS Removal, percent	61	70
P Removal, percent	14	42
Trickling Filter		
Depth, ft	4.25	4.25
Recirculation Ratio	2.0	2.0
Hydraulic Loading, mgd/acre[1]	16.3	16.3
Organic Loading, lb BOD/1,000 cu ft/day	18.8	15.4
Secondary Clarifier		
Alum Dosage Rate, Average Al:P mole ratio		1.72[2]
Overflow Rate, gpd/sq ft	885	885
Secondary BOD Removal, percent	43	76
Secondary SS Removal, percent	29	53
Secondary P Removal, percent	5	73
Overall Plant Performance		
Effluent BOD, mg/l	44	15
Effluent SS, mg/l	63	32
Effluent P, mg/l	9.2	1.8
BOD Removal, percent	74	91
SS Removal, percent	72	86
P Removal, percent	19	84

[1] Includes recirculation.

[2] Based on Influent P.

6.4 Performance Data Using Iron Salts

Addition of iron to trickling filters has been studied at Detroit and Wyoming, Michigan.

The Detroit Metro Water Department (7) compared the performance of pilot activated sludge and plastic-media trickling filter systems in order to decide on the design basis for a full-scale plant. Chemicals were added before the primary settler for both the trickling filter and activated sludge studies, and to the primary effluent for the activated sludge study. The plastic-media trickling filter test program consisted of a series of 16 experimental runs. Application rates ranged from 1.0 to about 3.5 gpm/ft^2. The raw wastewater inorganic phosphate averaged 7.9 mg/l P with a range of 1.9 mg/l to 26.6 mg/l. Without the addition of chemicals, total phosphate removals ranged from 18 to 57%. With the addition of 15 mg/l of iron (pickle liquor) removals were 34 to 77%. With 15 mg/l iron and 0.3 mg/l Dow Purifloc A23 polymer, removals ranged from 47 to 80%. With 15 mg/l iron, 0.3 mg/l A23 polymer, and 30 mg/l NaOH, removals were in the range of 67 to 78%. It was concluded that control of conditions for removal of phosphorus using the plastic-media trickling filters would be more difficult than in the activated sludge process. The activated sludge process achieved more consistent and higher phosphorus removals with iron addition and was recommended for the full-scale plant.

Dow Chemical Company performed plant scale iron addition tests at the Wyoming, Michigan trickling filter plant (8, 9, 10). The 1969 study was designed for ferric chloride addition to the raw wastewater to take advantage of the additional BOD and suspended solids removal in the primary. However, due to the recirculation of the trickling filter sludge with its high phosphorus content, it was necessary to split the $FeCl_3$ addition between the raw wastewater and the trickling filter effluent. The chemical feed and phosphorus removal data are presented in Table 6-3.

To achieve the 80% phosphorus removal required by the State, iron addition of 15 mg/l to the raw wastewater and 3 to 4 mg/l Fe^{3+} to the trickling filter effluent along with 0.3 mg/l Purifloc A23 were necessary. Removal of BOD during chemical treatment ranged from 60% to 87% across the plant. Suspended solids removal in the primary averaged 75% with an overall removal of 87%. Iron addition to the Wyoming trickling filter plant permitted greater than 80% phosphorus removal only with the split chemical addition.

6.5 Choice of Chemical Addition

Since in-plant chemical precipitation is a new technique, pilot-scale data should be obtained prior to design of new facilities. However, when the plant already exists, full-scale trials may not be any more expensive than pilot-scale tests and they will be more reliable.

Table 6-3

IRON ADDITION FOR PHOSPHORUS REMOVAL AT WYOMING, MICHIGAN

	Test No.			
	1	2	3	4
Chemical Feed				
Fe^{3+} to raw wastewater, mg/l	15	15	15-20	15
Fe^{3+} to trickling filter effluent, mg/l	5-7	3-4	0	7
Purifloc A23, mg/l, before primary sedimentation	0.4	0.3	0.2-0.3	0
Total Phosphorus				
Raw, mg/l	10.4	11.1	14.0	17.5
Primary effluent, mg/l	1.68	2.10	3.4	6.5
Removal, %	83.9	81.0	75.7	62.9
Final effluent, mg/l	0.88	1.46	2.9	5.7
Removal, %	91.7	87.6	79.2	67.4

The jar test is a most important tool for both designer and operator when using chemical addition. Information can be obtained inexpensively and in a short period of time, however, certain precautions should be taken. An auxiliary flash mix is highly desirable if the jar test apparatus does not have capability for true high energy mixing. After intense dispersion of trial chemicals, flocculation can be simulated by careful programming of time and mixing energy levels so they approximate conditions in the plant. The most effective way to match jar test and plant conditions is to adjust the jar test mechanism until it appears to duplicate the hydraulic regime observed directly in the plant. Duration of each mixing interval should approximate plant conditions, assuming plug flow. Finally, the machine should be left turning very slowly to approximate settling conditions in the plant. A slight motion in the test jars is a better representation of clarifier conditions than if stirring paddles are turned off. No matter how carefully the rotation speed and timing intervals are selected to approximate plant conditions, the operator should devote considerable practice to use of the apparatus before his results are accepted and used.

A wide variety of other bench scale and laboratory tests (11, 12) hold considerable promise in wastewater technology. At this time, however, none of them have proved as effective as the jar test.

6.6 Nature and Role of Chemicals Involved

Compounds mentioned here are those most likely to be selected for chemical treatment in a trickling filter plant. Of the commercially available metal salts (13), liquid forms are more convenient to handle, especially at small plants. They may have the lowest overall cost and are generally more effective than dry forms. However, it may be necessary to use dry salts because transportation of pre-mixed solutions, containing both water and salt, may be costly.

Metal salt forms include liquid ferric chloride, pickle liquor, liquid alum, and sodium aluminate. Information on properties and handling these chemicals is given in Chapter 10.

Polymers are available in a variety of forms, mostly dry powders. Since the amount of polymer required is far less than the amount of metal salt, the problems of using dry polymer compounds are not severe. Of the three categories of polymers (cationic, anionic and nonionic) there is no universal choice regarding the most effective type.

Whenever chemicals are chosen, they must perform two main functions: precipitate phosphorus, and remove all types of colloids in the water. Phosphorus precipitation is caused solely by metal salts and involves conversion of soluble orthophosphate to insoluble phosphate colloids. Following that, both metal salts and polymers serve to coagulate (destabilize) colloids. Flocculation of destabilized colloids is also brought about by both metal and polymers.

6.7 Dosage Selection and Control

Proper dosage selection and control are major keys to the success of chemical treatment because this is where both cost and performance are determined. The technique selected must be effective but also should be simple and easily interpreted by the operator.

In the metal salt system, the key parameters are the mole ratio fed and the effluent phosphorus concentration. As a basis for dosage determination, complete composite hourly and daily phosphorus analyses should be performed initially. These data will provide a guide to both total daily dose required and to daily variations in dosing rate.

Once accomplished, this task need not be repeated because subsequent refinements can be based on phosphorus concentration in the final effluent. Phosphorus observations should be made at both the primary clarifier and the final sedimentation tank if direct dosing of the filter is discounted.

A suggested procedure for setting the dose is as follows. First, express incoming phosphorus in pounds per day, then convert to lb-mole/day by dividing by the atomic weight (lb/lb-mole). Atomic weights are 31 for phosphorus, 27 for aluminum and 56 for iron. Next set the mole ratio (metal:P) at the proper value, between 1.5:1 and 2:1, and compute the daily coagulant dose. An example calculation is shown below:

If incoming phosphorus (as P) is 310 lb/day,

$$310/31 = 10 \text{ lb-moles/day}$$

If desired mole ratio (M:P) is 2:1,

$$\left(\frac{2}{1}\right)(10) = 20 \text{ lb-moles metal required}$$

Using liquid alum having 4.37% Al,

(20) (27) = 540 lb Al required

(540)/(11.1 lb/gal.) (0.0437)

= 1,100 gal. liquid alum required/day

Then take the total volume to be fed per day and convert it into dosage rates with three to five changes of rate per day. These different feed rates should be selected to match incoming phosphorus at the point of injection. Dose must be matched carefully to phosphorus demand. Feeding twice the amount required part of the time and half the amount the remainder of the time will waste chemicals and lead to poor performance. In fact, overfeeding chemicals can give as poor results as underfeeding; this stems from the tendency of high metal concentrations to drive phosphorus colloids into a highly stabilized state. Dilution of sewage during rainy weather makes little difference in the amount of metal salts required since total phosphorus, in terms of lb/day, remains about the same.

Changes in feed pump settings can be made manually according to a schedule established and posted beforetime. Several types of cam regulated feed controls are also available and are effective (14).

After establishing the initial feed program, effluent phosphorus should be monitored hourly or continuously. The plant should be kept on a given feed schedule for at least three to five days, to allow the biological reactions to adjust to any effects of chemical dosing. The effluent phosphorus data will show peak concentrations resulting from dosing improperly for a period of time. The feed schedule can then be adjusted to remove these peaks. Finally, studies of different mole ratios fed can be made and the results plotted as shown on Figure 6-3. These curves can then be used to choose the mole ratio required to meet effluent phosphorus standards.

Continuing operation of the plant should be monitored carefully and adjustments should be made in the chemical feed schedule when indicated by consistently high effluent phosphorus. Chemical feeding can be automated.

Polymer may be needed for good solids removal. Control of polymer feeding is relatively straightforward. Polymer feed rates are usually kept constant, but a reduced feed rate may be used during periods of low flow. Use of polymers may reduce metal salt demand and make metal addition more attractive. Polymers which prove effective in jar tests should be carefully evaluated in plant scale operation. Typical dosage is 1 mg/l or less. One direct way to judge polymer effectiveness is to observe effluent turbidity. Polymer feeding can also be automated.

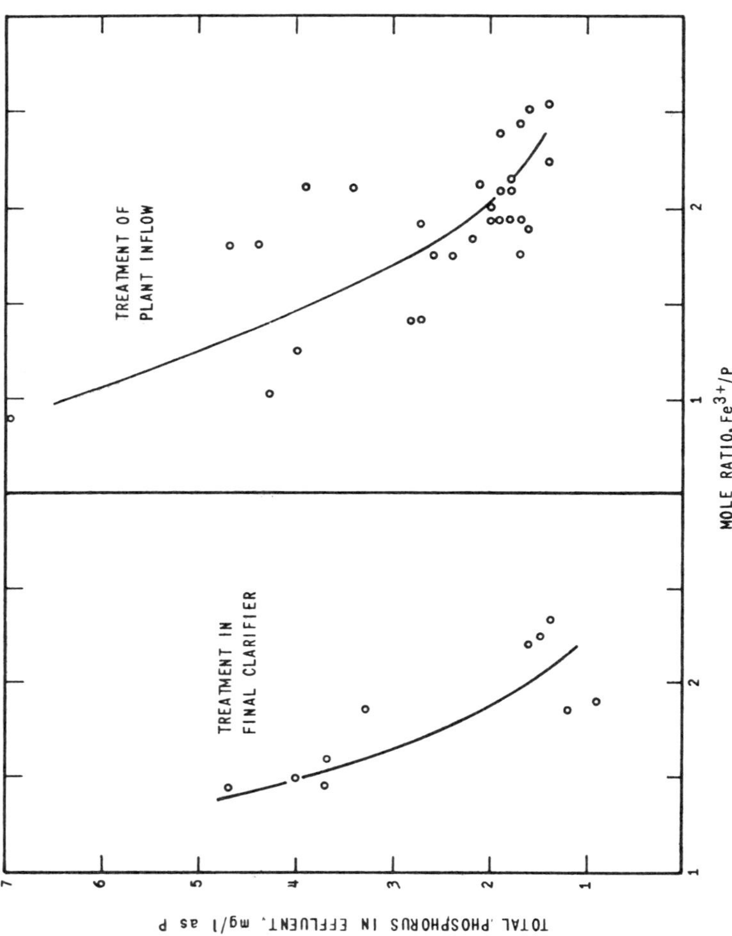

FIGURE 6-3 PHOSPHORUS REMOVAL BY IRON ADDITION (RICHARDSON, TEXAS)

6.8 Iron Leakage

Iron leakage may be a problem in trickling filter plants using iron precipitation. Figure 6-4 shows some typical data from Richardson, Texas (5). The problem appeared worse when iron was being fed in the final clarifier. It is evident that polymer treatment will reduce the escape of iron colloids.

6.9 Sampling and Analysis

The use of chemicals to remove phosphorus in trickling filter plants will require greater laboratory space, facilities, and reagents than normal treatment. Table 6-4 shows a typical matrix of data which is desirable at a conventional plant, as well as typical analyses required to support chemical treatment.

In the latter matrix, notice that the analyses include both anion and cation of the chemical being fed. Monitoring of alkalinity is optional unless the effluent concentration drops below 50 mg/l as $CaCO_3$. Some alkalinity is required to provide buffering and prevent a lowering of pH which would have an adverse effect on both chemical precipitation and biological activity.

Observation of turbidity in the final effluent is a good test. There is strong evidence that phosphorus and turbidity correlate in final effluents and the turbidity test provides a simple and responsive on-site control tool. Turbidity may be measured in the laboratory with a bench top meter or in the plant with a submersible disk lowered in the final clarifier. Observation of the height of the floc blanket in the clarifier is also a good means of control.

Effluent phosphorus concentration should be determined either on a semicontinuous basis with an automatic analyzer or on an hourly basis using automatic or manual sampling. In addition, a daily composite is also advisable. Sample portions should be flow-weighted rather than time-weighted. In plants enjoying stable effective performance, the schedule may be relaxed.

Signals from direct reading sensors and automatic analyzers can be used as a basis for automatic control of plant operation.

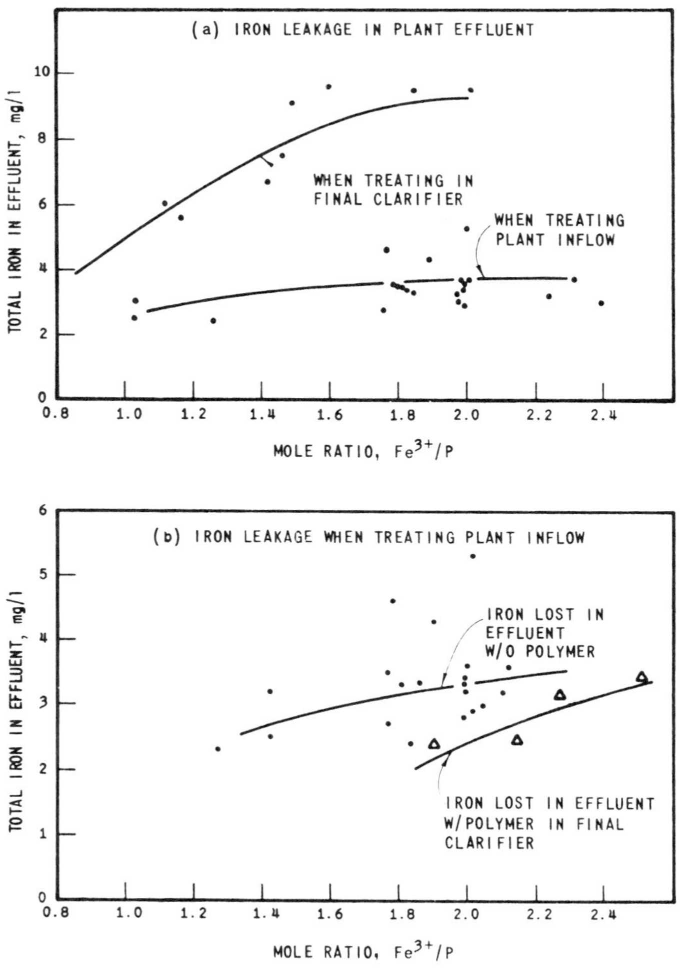

FIGURE 6-4 IRON LEAKAGE DURING PHOSPHORUS REMOVAL

TABLE 6-4 DESIRABLE LABORATORY ANALYSES

ANALYSES FOR CONVENTIONAL TREATMENT

	RAW	PRIM EFF	FILT EFF	FINAL EFF	RECIRC	SLUDGE	SUPERNAT
FLOW	X				X	X	X
TOT SOL	X	X	X	X	X	X	X
TOT VOL SOL	X	X	X	X	X	X	X
SUS SOL	X	X	X	X	X		X
SUS VOL SOL	X	X	X	X	X		X
SET SOL	X	X	X	X			
BOD	X	X		X			X
DO	X			X			
COD	X	X	X	X	X		X
pH	X	X	X	X		X	X
TEMP	X	X	X	X		X	X

ADDITIONAL ANALYSES FOR CHEMICAL TREATMENT

	RAW	PRIM EFF	FILT EFF	FINAL EFF	RECIRC	SLUDGE	SUPERNAT
PHOS	X	X		X	X	X	X
ALK	X	X		X		X	X
FE	X	X		X	X	X	X
AL	X	X		X	X	X	X
SO_4	X	X	X	X		X	X
CL	X	X		X			
TURB	X	X		X			

6.10 References

1. Nesbitt, J. B., "Phosphorus Removal The State-of-the-Art", *JWPCF*, 41:5, p 701 (1969).

2. Hall, M. W., and Engelbrecht, R. S., "Phosphorus Removal Past, Present, and Future", *Wat. and Wastes Eng.*, 6:8, p 50 (1969).

3. Scalf, M. R., et al, "Phosphate Removal: Summary of Papers", *Jour. SED, ASCE*, 95:SA5, p 817 (1969).

4. Stephan, D. G., and Schaffer, R. B., "Wastewater Treatment and Renovation Status of Process Development", *JWPCF*, 42:3, p 399 (1970).

5. Derrington, R. E., et al., "Enhancing Trickling Filter Performance by Chemical Precipitation," U.S. EPA Project No. 11010 EGL.

6. Preliminary Report, *Methods for Improvement of Trickling Filter Plant Performance*, Part I, University of North Carolina. U.S. EPA Contract 14-12-505.

7. Detroit Metro Water Department, "Development of Phosphate Removal Processes", Detroit, Michigan (July, 1970) (Program 17010 FAH Grant WPRD 51-01-67 Advanced Waste Treatment Laboratory, Cincinnati, Ohio), Pre-publication copy.

8. Condon, W. R., "Design of the Wyoming, Michigan Wastewater Treatment Plant Improvements", presented at the Technical Seminar/Workshop on Advanced Waste Treatment, Chapel Hill, North Carolina (Feb. 9-10, 1971).

9. "Report on Wastewater Treatment Plant Improvements, Wyoming, Michigan", Black and Veatch of Michigan, Consulting Engineers, Kansas City, Missouri (1970).

10. Stonebrook, W. J., Dykhuizen, V., Beeghly, J. H., and Pawlak, T. J., "Phosphorus Removal at a Trickling Filter Plant, Wyoming, Michigan", Unpublished (undated).

11. TeKippe, R. J. and Ham, R. K., "Coagulation Testing: A Comparison of Techniques – Part 1", *JAWWA*, 62:9, p 594 (1970).

12. TeKippe, R. J. and Ham, R. K., "Coagulation Testing: A Comparison of Techniques – Part 2", *JAWWA*, 62:10, p 620 (1970).

13. "Coagulants for Waste Water Treatment", *Chem. Eng. Prog.*, 66:1, p 36 (1970).

14. McAchran, G. E. and Hogue, R. D., "Phosphate Removal from Municipal Sewage", *Wat. & Sew. Works*, 118:2, p 36 (1971).

Chapter 7

PHOSPHORUS REMOVAL IN ACTIVATED SLUDGE PLANTS BY MINERAL ADDITION

7.1 Description of Process

The mechanism of phosphorus removal by mineral salt addition, usually iron or aluminum compounds, in a biological system is through a combination of precipitation, adsorption, exchange, and agglomeration as influenced by the pH and ionic composition of the water. The phosphorus removal technique is operationally simple and is accomplished by direct mineral addition to an aeration tank. Treatment costs are largely a function of the required effluent phosphate residual. Through optimization techniques, any degree of phosphorus removal may be provided. The main liability is the introduction of dissolved solids.

Phosphorus removal by biological systems alone, regardless of the internal mechanisms of phosphorus incorporation into the sludge solids, is limited, and removal is primarily controlled by the magnitude of solids wasting. The net amount of solids wasting is dictated by the quantity and type of substrate received and removed, system design and operation, and solids capture in the final effluent. A high rate system, with its greater solids production, would undoubtedly have the greatest background phosphorus removal potential by pure synthesis mechanisms. The magnitude of this background removal can be found by plant measurement with existing facilities or estimated by knowledge of the net solids production and phosphorus concentration in the waste activated sludge. Figure 7-1 shows the phosphorus removal capability for a variety of activated sludge systems at typically reported values for domestic wastewater treatment.

Historically, the benefits derived from mineral additions to activated sludge plants have been known since 1938 (1). Thomas (2) was among the first to point out the increased phosphorus removals. Wirts (3) also observed the increased phosphorus removals derived from a metal bearing industrial waste at a municipal treatment facility in Cleveland, Ohio. Early investigators of mineral addition for upgrading and modifying activated sludge systems for phosphorus removal include Tenny and Stumm (4), Barth and Ettinger (5), and Eberhardt and Nesbitt (6). Barth, et al., conducted pilot plant investigations with a three sludge system for combined chemical-biological nitrogen and phosphorus removal (7). Large scale, long term investigations have been recently completed at Manassas, Virginia (8), Pennsylvania State University (9), and Texas City, Texas (10). Recent publications have examined the use of aluminum (11) and iron (12) salts and investigated the kinetics and mechanisms of phosphorus removal using these minerals (13).

In the treatment of domestic wastewaters several investigators, most recently Rickert and Hunter (14), have shown that in a single sludge system a stable organic residual is found after aeration times of 30 to 60 minutes. Unfortunately, if treatment is stopped

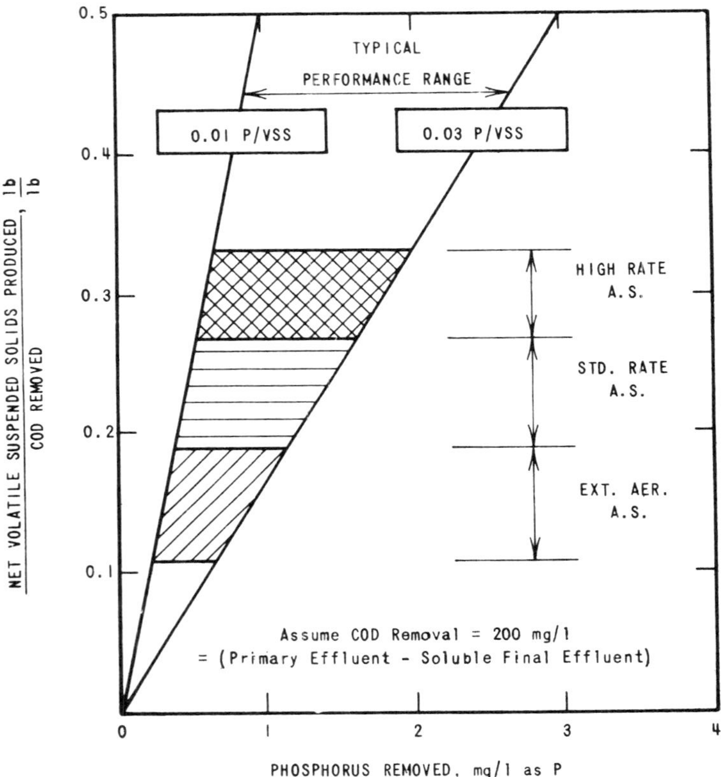

FIGURE 7-1 PHOSPHORUS REMOVAL CAPABILITY OF ACTIVATED SLUDGE SYSTEMS RECEIVING SETTLED DOMESTIC WASTEWATER

at this point, bio-flocculation is almost nonexistent and poor solids densities, settling characteristics and effluent quality result. However, when incorporating mineral addition, these liabilities are overcome and there exists an opportunity to derive capital savings from reduced tankage requirements to compensate for the increased operating expenditures associated with mineral addition. Indeed, in many cases, the use of a short residence time activated sludge system with mineral addition, settling, and filtration may result in a higher quality effluent for equal or lower costs than a purely physical-chemical approach of coagulation, settling, filtration and carbon adsorption (15). From a liability standpoint, however, the designer should be cognizant of the greater mass of generated waste activated sludge which must be successfully handled and treated for ultimate disposal.

Mineral addition to activated sludge for phosphorus removal offers the following advantages:

a. Ease of operation (direct chemical addition to existing unit processes).

b. Relatively small additional solids production (causing increases in sludge density and dewaterability).

c. Flexibility to changing conditions (treatment costs are largely a function of the influent and required final effluent phosphorus concentrations).

The main disadvantage is the introduction of dissolved solids which in some circumstances can be considered pollutants in their own right.

7.2 Mineral Selection and Addition

The selection of the mineral to use for phosphorus removal involves a multitude of considerations. Any commercial chemical, or industrial waste stream (for example, pickle liquor or alum sludge from a water treatment plant) bearing available aluminum or iron has potential application. Ideally, the best approach requires comparative full-scale plant tests of several months duration. This, of course, is usually impossible, and the next best approach is to run comparative jar tests, the results of which may not be directly applicable. However, they are at least indicative of performance. As an illustrative example, the procedures and results from the Manassas studies (8, 16) are utilized in the subsequent paragraphs.

Figure 7-2 shows the total soluble phosphorus as a function of the metal ion dose for the Manassas raw wastewater, for mixed liquor from a high rate activated sludge system, and for biologically stabilized final effluent. Since both the alum and ferric chloride can cause an alkalinity depletion and potential pH drop, pH values are also recorded. Superior performance for alum was found at all mass dosages with an apparent limiting residual of 1.0 mg/l as P for the raw wastewater and final effluent, and less than 0.1 mg/l as P for the aerator mixed liquor. A yellow color in the product water was noted with the higher iron dosages.

Replotting the aluminum data from Figure 7-2 in Figure 7-3 shows the influence of

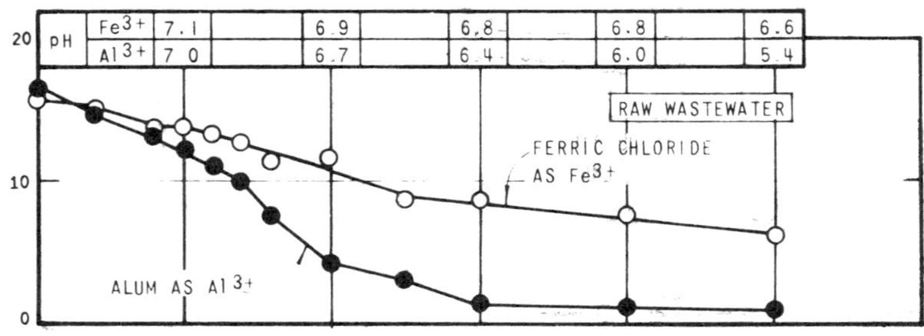

TOTAL SOLUBLE PHOSPHORUS, mg/l as P

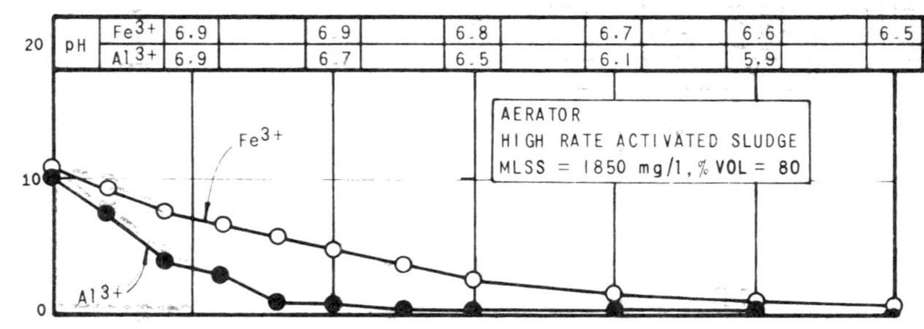

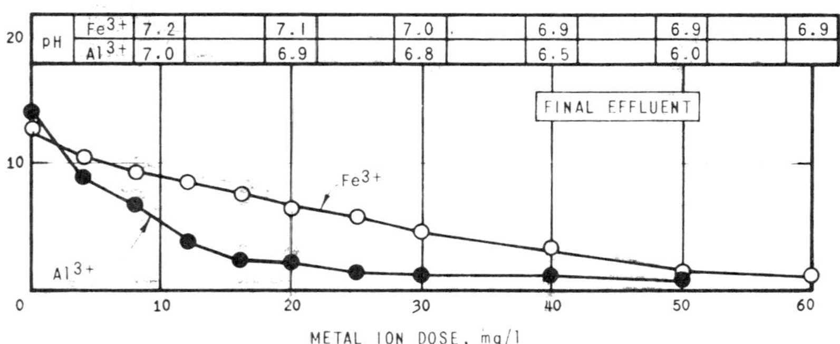

METAL ION DOSE, mg/l

FIGURE 7-2 TOTAL SOLUBLE PHOSPHORUS RESIDUAL VERSUS METAL ION DOSE

(Reprinted with permission from the American Institute of Chemical Engineers. New York, N.Y. 10017)

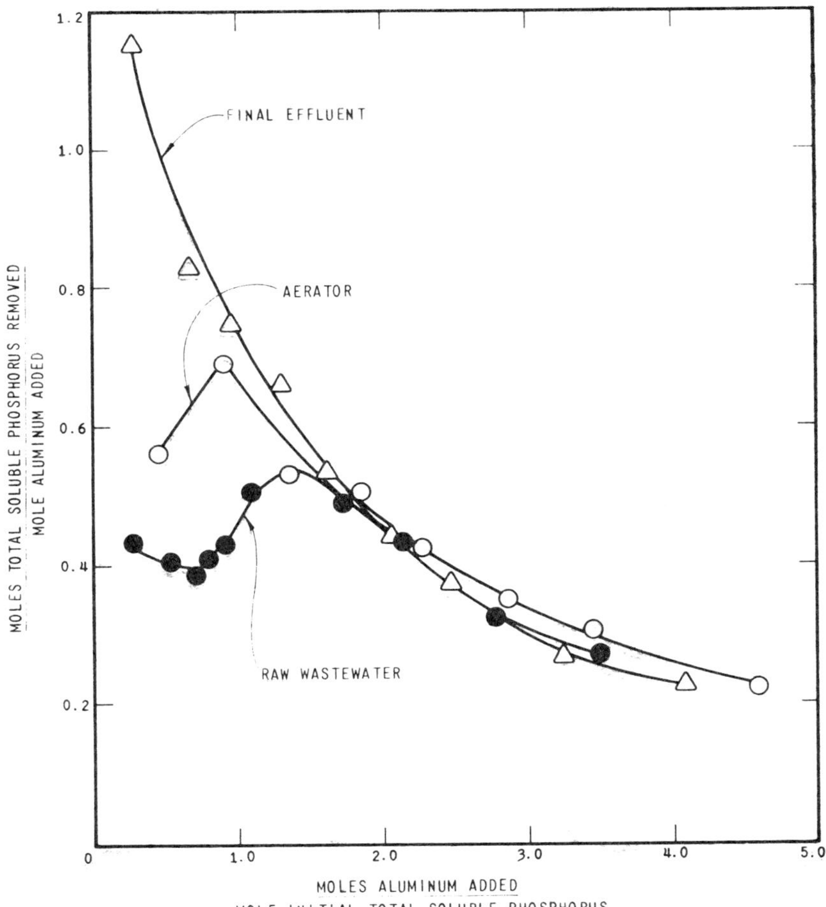

FIGURE 7-3 INFLUENCE OF POINT OF ADDITION ON PHOSPHORUS REMOVAL WITH ALUMINUM

(Reprinted with permission from the American Institute of Chemical Engineers, New York, N.Y. 10017)

the point of addition on the molar effectiveness of the added aluminum. Similar but lower molar curves are found with the iron versus phosphorus residual data. The lag in phosphorus removal effectiveness encountered with the raw wastewater and aerator mixed liquor is thought to be due to the presence of unhydrolyzed phosphorus and the occurrence of competing side reactions for the metal ion. Aluminum and iron are known to form insoluble heavy salts with certain hydrophilic colloids such as detergents and soluble proteins and their degradation products (17), all of which would be present in raw and partially stabilized wastewater. These curves indicate that at molar ratios of less than 1:1, there is an obvious advantage in phosphorus removal derived by adding the chemical to a well stabilized liquid, whereas at molar ratios above 1.5:1, enough metal ion has been added to satisfy the competing side reactions and all points of addition are equally effective. The data indicate that very low phosphorus residuals are possible in a combined chemical-biological system and that the ideal point of addition would be just before final solids-liquid separation to obtain maximum biological stabilization of the soluble liquid phase.

Figure 7-4 shows the influence of wastewater and process pH upon phosphorus removal effectiveness with and without mineral additions. In the upper graph of Figure 7-4, above pH 8, phosphorus removal is occurring naturally through the formation of an insoluble calcium-phosphate complex via the natural hardness of the wastewater. The lower graph of Figure 7-4 shows that phosphorus removal at any molar dosage of metal ion is actually controlled by pH dependent reactions, optimally around 6 and 5 for aluminum and iron, respectively. Since the optimum range for aerobic oxidation of carbonaceous compounds is about pH 6 to 8, it can be seen that there is an opportunity for the simultaneous optimization of carbon oxidation and phosphorus removal by aluminum addition. However, if nitrification is a treatment goal there exists a conflict and, in turn, a compromise in pH goals. Diminished nitrification performance has been reported when alum additions have depleted the wastewater's alkalinity to a point where the pH was near optimum for phosphorus removal (9).

Table 7-1 provides a further comparison of representative sources of aluminum and iron. It should be noted that dissolved solids or extraneous ions are introduced and in some circumstances these ions may be undesirable (18). Alum and $FeCl_3$ cause an alkalinity loss, and without buffering capacity in the wastewater, a substantial pH reduction may occur. This can be compensated for by adding a source of alkalinity, such as lime, or by using aluminate. Treatment performance can deteriorate dramatically if the system pH is allowed to fall much below optimum (5). The designer should be aware that periods of heavy storm water infiltration will substantially dilute the normal alkalinity of the wastewater, and if operation is carried out at a pH of 6, only 20 to 40 mg/l of residual $CaCO_3$ buffering alkalinity may remain (8). The bottom part of Table 7-1 shows a comparison of the metal salts on a mole per mole basis, assuming a mole of metal ion precipitates one mole of phosphorus. In practice this never occurs. On this basis the possible cost advantage and alkalinity advantage with iron is lost and the aluminum salts show a potential for less additional solids production.

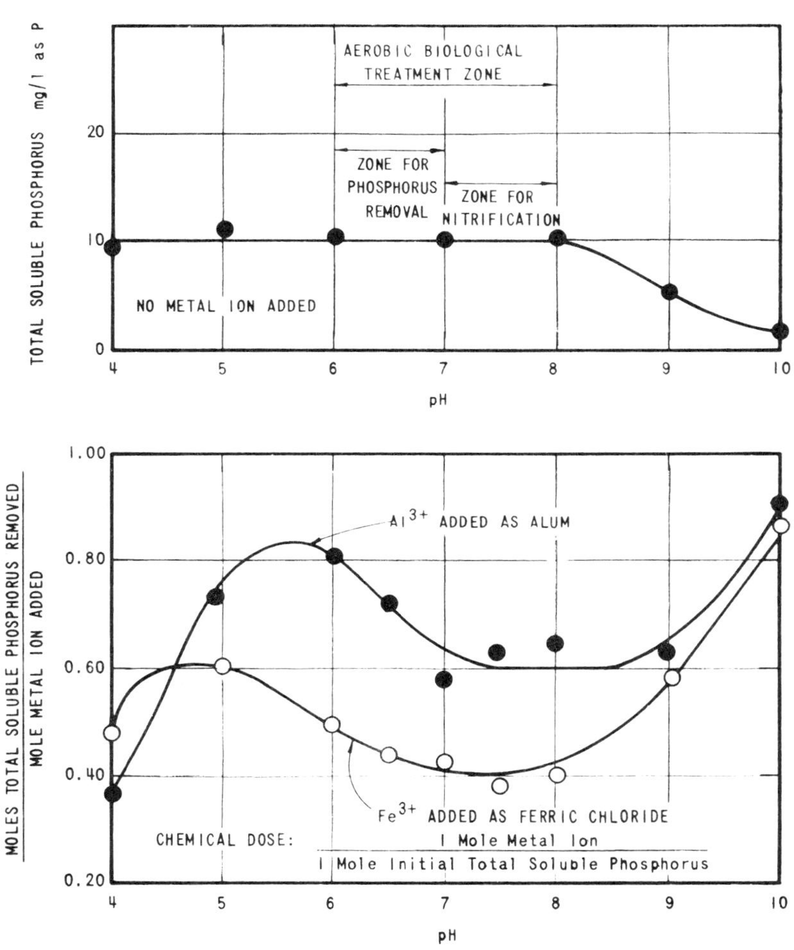

FIGURE 7-4 EFFECT OF pH ON PHOSPHORUS REMOVAL FROM A FINAL EFFLUENT

(Reprinted with permission from the American Institute of Chemical Engineers, New York, N.Y. 10017)

Table 7-1

METAL SALT COMPARISON

	Metal Salt		
	Liquid Sodium Aluminate (Al^{3+})	Liquid Alum (Al^{3+})	Ferric Chloride (Fe^{3+})
	(per lb of Metal Ion)		
Bulk Delivered Price at Manassas, $	0.69	0.32	0.23
Extraneous Ion	Na^+	SO_4^{2-}	Cl^-
Introduced, lb	0.85	5.4	1.9
Maximum Alkalinity Loss as $CaCO_3$, lb	None	5.5	2.7
	(Per lb of Phosphorus at a Mole Dosage of 1:1)		
Bulk Delivered Price at Manassas, $	0.60	0.28	0.41
Extraneous Ion	Na^+	SO_4^{2-}	Cl^-
Introduced, lb	0.74	4.7	3.4
Maximum Alkalinity Loss as $CaCO_3$, lb	None	4.8	4.8
Phosphorus Removal of Metal Ion, lb	0.87	0.87	1.8
Additional Solids Production, lb	$AlPO_4$ 3.9	$AlPO_4$ 3.9	$FePO_4$ 4.9

Several other points should be considered in mineral selection. First, in the process effluent the soluble Al, less than 0.5 mg/l (16), is apparently much lower than found with iron additions, 6 mg/l Fe reported at Cleveland (19). Second, alum adds a divalent ion (SO_4^{2-}) which has a proven coagulation benefit (20). Third, both Mulbarger (8) and Long, et al. (9), report improved product water clarity when using alum in comparison to aluminate.

Chemicals can be added with confidence in the activated sludge process just before solids-liquid separation, assuming adequate dispersal. Studies by Allied Chemical Company (21) in a 25,000 gpd pilot plant indicated that the addition of alum at the seven-eights point in the aeration basin, along with anionic polymer in the transfer line, provided the optimum phosphorus removal, coagulation and settling. Polymer dosages were found to be optimum in the 0.2 to 0.4 mg/l range. An average Al:P weight ratio of 2.1:1 was required to reduce the average influent total phosphorus of 3.6 mg/l to 0.45 mg/l in the effluent.

It should be noted that utilization of activated sludge aeration tanks for chemical mixing and flocculation is a compromise. Typically, a diffused aeration system will have a velocity gradient of over 100 ft/second/ft and the velocity gradient of a mechanical aeration system will be much higher. Water treatment experience has shown that velocity gradients greater than 75 ft/second/ft will result in the onset of chemical floc disintegration (22). Undoubtedly, some chemical floc as well as biological floc disintegration does occur in activated sludge tanks and, because of solids recycle, is unavoidable. Floc disintegration and mineral overdose both lead to product water quality degradation. In new design situations it may be advantageous to provide a brief period of gentle solids agitation prior to solids separation to promote reflocculation of the chemical-biological solids. This can be done by gentle air agitation, a flocculation tank, or a flocculator-clarifier. The designer should also provide the capability of adding organic polymer, if needed. Nalco 676 and Atlas 2A2, slightly anionic organic polyelectrolytes, have been successfully used at dosages between 0.3 and 0.4 mg/l (8).

7.3 Performance and Optimization

7.3.1 Phosphorus Removal

At molar dosages of 1 to 2, metal ion to phosphorus, soluble phosphorus residuals of 0.3 to 3.0 mg/l as P are frequently encountered in the literature. Interpretation of some of the data is complicated by failures to establish biological equilibrium [90% equilibrium values are reached at approximately 2.2 times the operating cell retention time (16)] prior to changing variables or initiating an investigative run, and failure to consider and report pH.

Detailed data from iron additions to activated sludge systems are limited in the literature and consequently the more extensively available data on aluminum addition provide a better source of definitive design information. However, a few investigations of iron additions to secondary treatment processes have been conducted on the pilot and full plant scale.

Pilot plant studies by the Detroit Metro Water Department, in which several types of activated sludge systems were investigated, indicated that the most effective total phosphate reduction was provided by the conventional activated sludge process with iron ($FeCl_2$ derived from pickle liquor) fed to the primary effluent (23). The majority of the test runs exhibited removals varying from 75 to 85% for inorganic phosphate concentrations ranging from 1.9 to 26.6 mg/l as P in the raw wastewater. Phosphate removals by the activated sludge process alone (no iron fed) fell principally in the 55 to 70% range. It was found that, in general, the addition of iron to the conventional activated sludge process provided a treatment system capable of consistently high phosphorus reduction. The step-feed, activated sludge process, with the addition of iron or iron, polymer and caustic soda, also provided consistent total phosphorus reduction. It was concluded that the full-scale plant should be designed for the activated sludge process, arranged for both conventional and step-feed configurations, and that provision be made for phosphorus removal, utilizing injection of steel pickle liquor ($FeCl_2$).

A pilot plant (5,760 gpd) at Warren, Michigan (24) was designed to simulate the existing 36 mgd activated sludge plant, with the addition of one sand filter and one multimedia filter operated in parallel after the final clarifiers. Ferric chloride was fed to the effluent of the aeration basins to determine whether or not this method would meet the State-required 80% phosphorus removal. Ferric chloride addition at a ratio of 1.5 mg/l of Fe to 1.0 mg/l of soluble PO_4^{3-} reduced the average total phosphorus concentration from 7.0 mg/l to 1.2 mg/l in the effluent. This amounted to a reduction of 83%.

Leary, et al., of the Milwaukee Sewerage Commission (25), conducted a one-year plant-scale study of the use of waste pickle liquor ($FeSO_4$) to enhance phosphorus removal. The Milwaukee Jones Island Plant consists of a single primary facility followed by two parallel activated sludge plants. Pickle liquor was employed for phosphorus removal at the 115 mgd East plant during the test period. The 85 mgd West plant served as a control. Pickle liquor feed was equivalent to an average Fe^{2+} dosage of 9.4 mg/l. The pickle liquor was added before the aeration tank about 55 feet downstream from the point of addition of return sludge. Based on a 1970 average total phosphorus concentration of 8.2 mg/l P in the screened sewage, the East Side plant, with iron addition, accomplished 91.3% phosphorus removal while the control West plant removed 83.1%. Effluent phosphorus averaged 0.7 mg/l at the East plant and 1.4 mg/l at the control plant. The return sludge phosphorous concentration was 2.29% P in the control plant and 2.61% P in the East (test) plant. The iron content of the return sludge was increased from 1.86 to 5.08%.

The pickle liquor addition increased the East plant effluent SO_4^{2-} concentration from 123 to 145 mg/l and decreased the alkalinity from 213 to 169 mg/l as $CaCO_3$. Yearly average pH values were 7.0 and 7.1, respectively, for the East and West plants. The free acid in the pickle liquor used (two sources) ranged from 2.1 to 9.3% H_2SO_4.

Comparison of the BOD, COD, and suspended solids removal efficiencies of the West and East plants indicated that the addition of unneutralized pickle liquor did not adversely affect purification, and it did not cause problems with the plant physical facilities. The results of this one-year study indicated that the use of waste sulfuric acid pickle liquor as a source of iron for phosphorus precipitation and removal was practical and effective in maintaining low plant effluent phosphorus residuals.

In studies at the University of Missouri at Rolla (26), alum and aluminate at various times were fed to laboratory activated sludge units and compared to a control unit. The alum was more effective, as shown on Figure 7-5, than aluminate, based on the Al:P mole ratio. In the activated sludge units which were dosed with aluminum, the volatile mixed liquor suspended solids dropped to 50 to 60% of the total mixed liquor suspended solids, compared to a value of 80% in the control. The aluminate units had higher pH, but all pH values were in the range of 7.7 to 8.2.

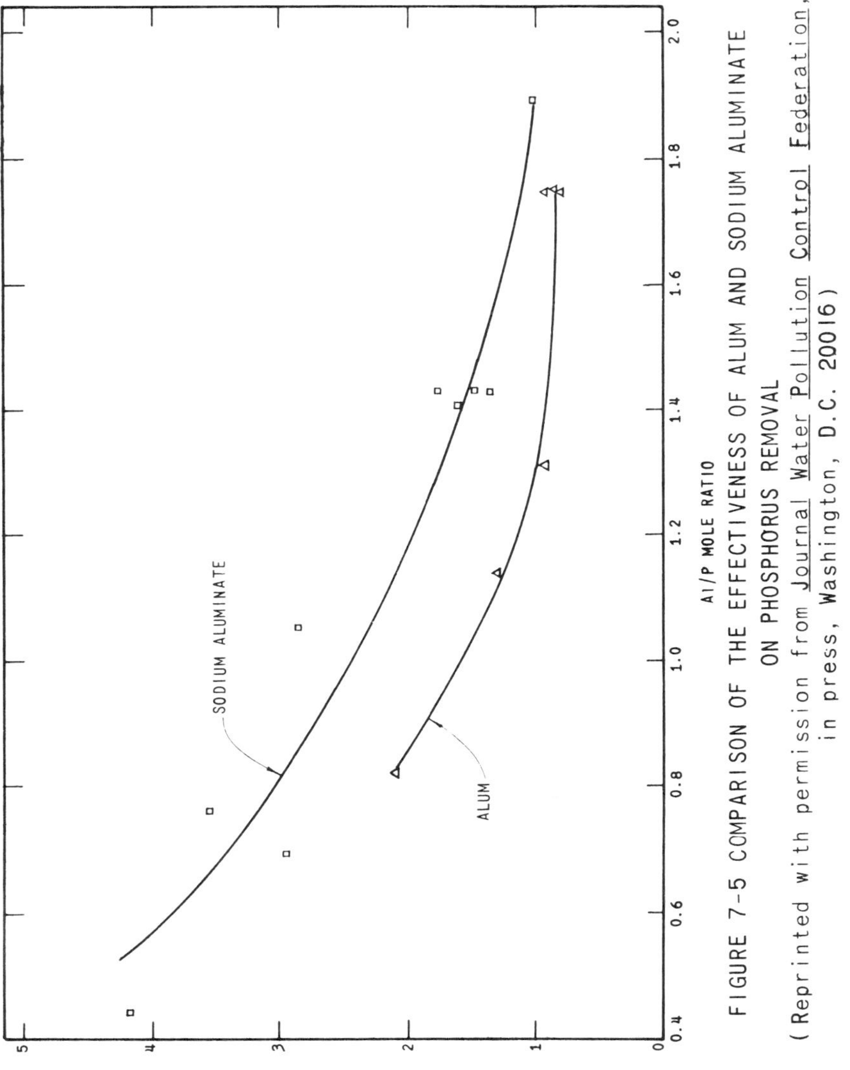

FIGURE 7-5 COMPARISON OF THE EFFECTIVENESS OF ALUM AND SODIUM ALUMINATE ON PHOSPHORUS REMOVAL

(Reprinted with permission from *Journal Water Pollution Control Federation*, in press, Washington, D.C. 20016)

Full-scale studies were conducted at Pennsylvania State University on the removal of phosphorus by addition of aluminum to a 2.0 mgd activated sludge plant. These studies began in 1969 and continued through 1970 (9). In the first phase of the work, alum and sodium aluminate were compared with respect to their ability to precipitate phosphorus. To prevent biological upset due to mineral solids displacing biological solids in the mixed liquor, the volatile suspended solids concentration was held at the same level. Since the volatile fraction of the total mixed liquor suspended solids decreased due to mineral solids buildup, this required the maintenance of a higher total suspended solids concentration in the mixed liquor.

The data shown by Table 7-2 indicate that alum was slightly superior to sodium aluminate in removing phosphorus from the moderately alkaline wastewater. The best point of addition for alum was found to be at the effluent channel of the aeration basin which carries mixed liquor to the final settling basin. Addition of alum both at the inlet end of the aerator and at a point about two-thirds of the way along the basin from the inlet resulted in lower removal efficiencies. Alum addition ahead of aeration produced a cloudy effluent.

Table 7-2

COMPARISON OF USE OF ALUM AND SODIUM ALUMINATE AT PENNSYLVANIA STATE UNIVERSITY

Parameter	Chemical*			
	Alum		Sodium Aluminate	
	Influent	Effluent	Influent	Effluent
Suspended Solids, mg/l	134	19	94	25
BOD_5, mg/l	87	6	50	7
COD, mg/l	244	31	160	36
Phosphorus as P, mg/l				
Filtered Ortho	6.4	0.28	5.4	0.51
Filtered Total	7.6	0.27	6.9	0.57
Unfiltered Ortho	10.4	1.17	6.2	1.14
Unfiltered Total	12.4	1.48	8.6	1.53
Alkalinity as $CaCO_3$, mg/l			156	103
pH	7.4	7.1	7.15	7.25
Alum Dosage as $Al_2(SO_4)_3 \cdot 14H_2O$, mg/l	160			
$Na_2Al_2O_4$ dosage, mg/l			44	
Al:P Ratio **, weight	2.04:1		2.04:1	
SO_4^{2-}, mg/l	24	131		

*Chemicals added at the effluent end of the aeration tank.
**Al:P ratio based on expression of phosphorus in the filtered ortho form.

Aluminate addition was found to be most effective at a point in the aeration basin near the outlet. Aluminate addition either before or after aeration produced a more turbid effluent.

Sludge production in the studies comparing alum and aluminate showed that the volume of sludge produced per million gallons of wastewater treated was about the same for both coagulants. However, the weight of solids produced was considerably less for aluminate than alum.

Because alum was found to be superior to aluminate in removing phosphorus, further study of alum was performed for a period of one year. One-half of the activated sludge plant was used in the alum study and the other half was operated without chemical addition for comparison. Table 7-3 shows average results for operation at rates equal to or less than the plant design capacity.

Table 7-3

COMPARISON OF ALUM ADDITION TO THE AERATOR AND CONVENTIONAL ACTIVATED SLUDGE AT PENNSYLVANIA STATE UNIVERSITY FOR FLOWS NOT EXCEEDING THE DESIGN CAPACITY OF THE PLANT

Parameter	Influent	Effluent with Alum Addition*	Effluent without Alum Addition
Suspended Solids, mg/l	110	22	26
BOD_5, mg/l	71	9	13
COD, mg/l	172	55	68
Phosphorus as P, mg/l			
Filtered Ortho	6.7	0.28	6.7
Filtered Total	6.3	0.36	6.7
Unfiltered Ortho	8.9	1.15	7.2
Unfiltered Total	10.0	1.41	7.3
Alkalinity as $CaCO_3$, mg/l	168	80	120
pH	7.6	6.75	7.35
Al:P Ratio**, Weight		2.97:1	

*Alum added to effluent of aeration tank. Values shown in this column are means for the period of study.

**The Al:P ratio is based on expression of phosphorus in the filtered ortho form.

Problems were experienced with loss of solids from the final basin during peak flow periods where alum treatment was being practiced. It was believed, however, that proper allowance for peak flows in design of such facilities would eliminate excessive solids loss. The effluent insoluble phosphorus concentration was found to correlate well with the effluent suspended solids concentration.

A 2,000 gpd pilot operation at the main wastewater treatment plant of the District of

Columbia was operated at a constant alum dose rate (27). The pilot plant is a four-stage step aeration process followed by secondary clarification. Alum was added to the wastewater of moderate alkalinity (100 to 150 mg/l as $CaCO_3$) at the fourth stage of aeration using an Al:P weight ratio of 2:1 (molar ratio 2.3:1), based on an influent phosphorus concentration of 8.15 mg/l as P. Mixed liquor suspended solids averaged 6,300 mg/l with a volatile content of 60% and a cell retention time of about 5 days. Total phosphorus residuals averaged 1.9 mg/l as P (78% removal) in the clarifier effluent.

Figure 7-6 shows the chemical phosphorus removal found for the Manassas high rate activated sludge plant with aluminate (16) and alum (8) additions at slightly alkaline pH values. Phosphorus concentrations for Figure 7-6 were corrected for biological removal. The Manassas Pilot Plant, shown in Figure 7-7, is a 0.2 mgd three-stage (nitrification-denitrification) activated sludge system operated at the Greater Manassas Sanitary District of Prince William County, Virginia. Table 7-4 shows the system performance for 200 determinations with alum addition. The reader is referred to the original publication for more detailed information.

Table 7-4

COMBINED CHEMICAL–HIGH RATE ACTIVATED SLUDGE SYSTEM PERFORMANCE

Item	Min.	10	20	50	80	90	Max.
			Percent Less Than				
Primary Effluent							
COD, mg/l	27	64	86	152	227	250	324
P, mg/l	2.2	3.8	5.0	8.4	11.4	12.8	18.5
BOD_5:COD*	0.26			0.43			0.56
High Rate Activated Sludge Effluent							
Soluble Phase							
COD, mg/l	13	23	26	33	41	48	79
P, mg/l	0.2	0.8	1.0	2.0	3.0	3.6	6.8
BOD_5:COD*	0.14			0.28			0.36
Solid Phase							
Suspended Solids, mg/l	0	10	14	21	36	44	196
BOD_5:SS*	0.10			0.23			0.31
COD:SS*	0.43			0.58			0.80
P:SS*	0.039			0.051			0.063
Suspended Solids After Polymer Addition,** mg/l	0	0	1	2	8	14	18
Suspended Solids After Mixed Media Filtration**, mg/l	0	0	0	0	2	5	12

*Ratio of minimum, median, and maximum monthly averages for each parameter
**94 determinations

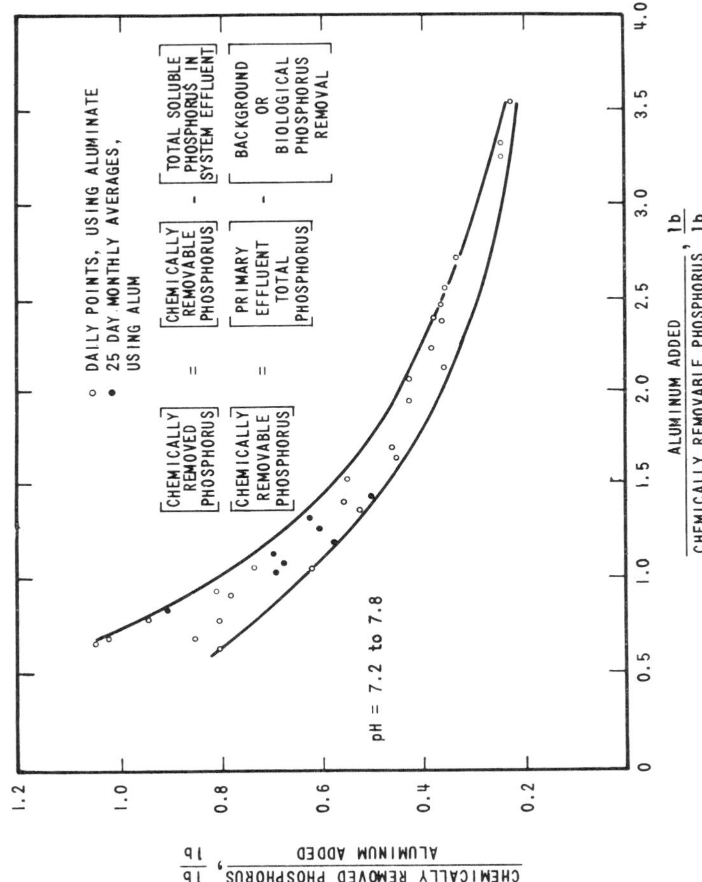

FIGURE 7-6 CHEMICAL PHOSPHORUS REMOVAL ENVELOPE: HIGH RATE ACTIVATED SLUDGE SYSTEM

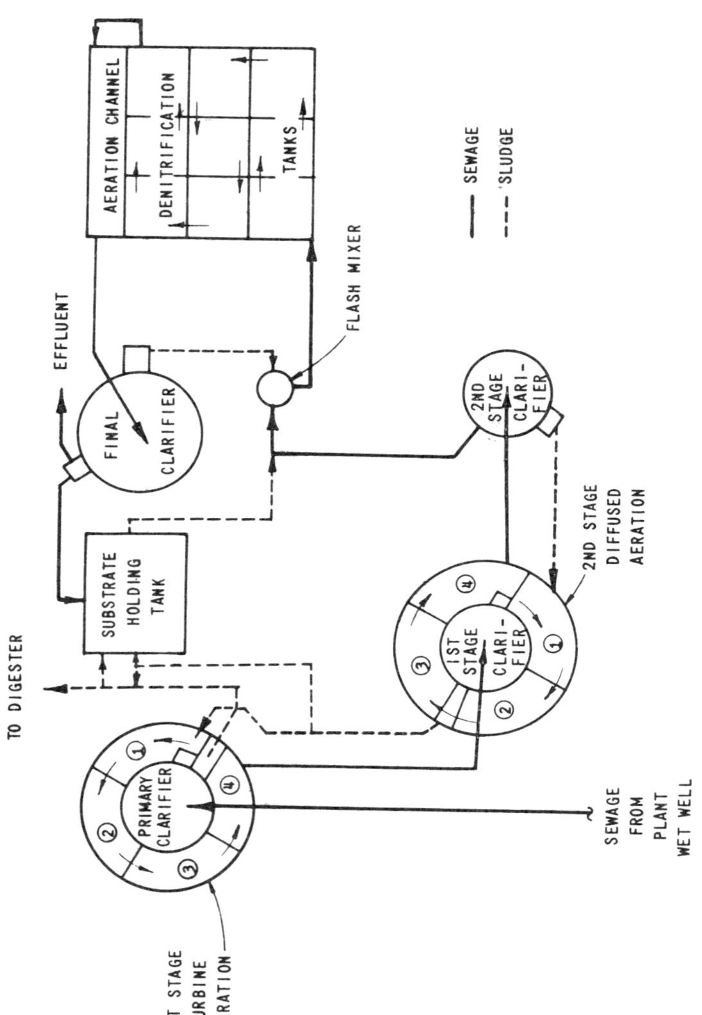

FIGURE 7-7 0.2 mgd BIOLOGICAL NITRIFICATION - DENITRIFICATION PILOT PLANT
Greater Manassas Sanitary District
Prince William County, Virginia

Optimization of phosphorus removals with mineral additions can be done two ways: by pH adjustment to the point of minimum solubility (pH 6 to 6.5, residual alkalinity of 20 to 40 mg/l as $CaCO_3$) and by splitting the chemical dose to separate parts of the treatment system (8). Optimization maximizes the efficient use of chemical, minimizes additional solids production, and produces soluble phosphorus residuals of less than 0.5 mg/l. Hais, et al. (27), indicated, using step aeration with alum addition, that above pH 6.6 a rapid increase in the effluent phosphorus residual occurred with increasing pH. However, operation with a mixed liquor pH of less than 6.2 caused an upset in the activated sludge process. It was concluded that optimum performance (0.6 mg/l effluent phosphorus residual) was achieved by maintaining a final stage pH between 6.3 and 6.6 at an Al:P weight ratio of 2:1. Mulbarger (8) showed the capability of pH adjustment with a denitrification system and demonstrated the advantages of split chemical treatment in a multi-sludge system. Results from the pH adjusted system are shown in Figure 7-8. Long, et al. (9), found a natural pH adjustment with high doses of alum in a poorly buffered wastewater and report phosphorus residuals below 0.5 mg/l at Al:P weight ratios of from 2:1 to 3:1.

Organism activity and type (9) as well as the completeness of a chemical reaction affect phosphorus removal and are influenced by the temperature. No temperature effect has been reported, however, for the rate of chemical reaction (13) operating between 10 and 20° C.

Mulbarger (8, 16) notes that the presence or absence of nonsettleable effluent suspended solids from a combined chemical-biological system is more dependent upon the ratio of net volatile solids produced to aluminum added than upon pH, exceeding an aluminum to phosphorus ratio, or exceeding a given aluminum dosage. It is recommended that this ratio not drop below 3 to 5. It is believed that the biological volatile solids provide exchange and/or sorption sites of limited capacity for the precipitated aluminum-hydroxy-phosphate complex, and that these natural polymers aid in clarification. Thus, the more biological solids produced from the system the greater the aluminum dosage that can be used without effluent suspended solids problems.

When designing the phosphorus removal system, the designer should be cognizant of the phosphorus levels in the effluent suspended solids. There was a nominal 1.0 mg/l of phosphorus in the effluent suspended solids from the high rate system at Manassas (8). After pH adjustment and addition of 0.3 to 0.4 mg/l of anionic polymer (Nalco 676), effluent suspended solids were reduced to 2 mg/l. Consequently, the effluent total phosphorus was also reduced. Polymer performance was also reported to be influenced by the pH (8). If total phosphorus residuals of less than 0.5 mg/l as P are required, a multimedia filtration system is recommended. Dual and tri-media filtration both provide effective tertiary solids separation. Hais, et al. (27) showed that tri-media filtration consistently removed between 5 and 10% more suspended material than did the dual-media filter. Filter runs between 24 and 32 hours were regularly maintained. Tri-media filtration performance was reported to be superior to dual-media filtration at Manassas (8).

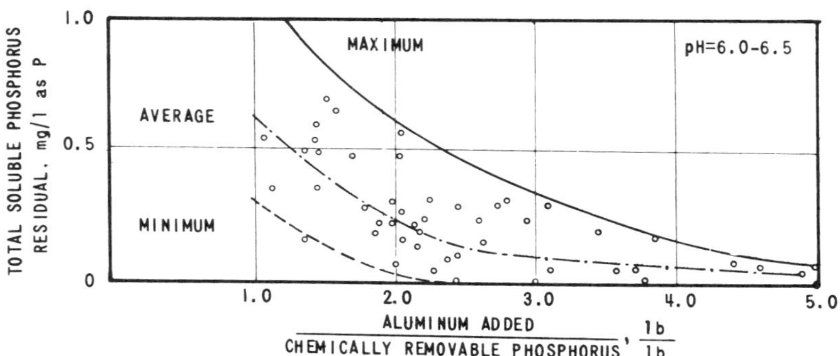

FIGURE 7-8 EFFECT OF ALUMINUM ADDITION WITH pH ADJUSTMENT ON TOTAL SOLUBLE PHOSPHORUS RESIDUAL

7.3.2 Costs

Power costs for combined chemical-biological treatment are negligible. Maintenance costs will increase somewhat due to the additional equipment. Capital expenditures for storage facilities, instrumentation, and equipment will be very low at larger facilities unless multi-media filtration is incorporated. Chemical costs will vary as a function of the incoming phosphorus concentration and the required phosphorus residual. Additional expenditures for pH control (alkalinity dependent) and solids control (polymer additions) may be necessary. See Chapter 10 for further information.

7.4 Process Design Examples

Four design example problems were selected for an arbitrary wastewater. Each illustrates specific points in combined chemical-biological wastewater treatment with alum additions for phosphorus removal. Results from the Manassas work are used for some of the calculations.

Assumptions:

a. Net biological solids production = 0.28 lb of volatile solids produced per lb of COD removed; the volatile solids contain 2% phosphorus; and the total solids average 80% volatile.

b. The biological system is to be designed with a minimum hydraulic contact time of 1 hour at the maximum 4 hour hydraulic peak. The ratio of maximum flow rate to average flow rate = 1.5. A cell retention time (CRT) of two days is selected.

c. Primary Effluent Characteristics:

$$COD = 200 \text{ mg/l}$$
$$P = 10 \text{ mg/l}$$
$$\text{Alkalinity} = 150 \text{ mg/l as } CaCO_3$$

Find:

1. Effluent COD, BOD_5 and phosphorus
2. Total solids production
3. Alkalinity loss
4. Dissolved solids addition
5. Size of final clarifier, return and waste sludge pumping systems.
6. Chemical costs with the following delivered prices:

Alum	= \$0.32/lb Al^{3+}
Polymer	= \$1.50/lb polymer
Acid	= \$0.02/lb H_2SO_4
Lime	= \$0.01/lb $Ca(OH)_2$

Example No. 1: High rate system for most efficient chemical addition with pH above 7 and no optimization by pH adjustment.

1. **Effluent COD, BOD_5, and Phosphorus**
(See Table 7-4. Note: Soluble organics from a biological system vary as a function of the wastewater and dilution of the refractory fraction. Manassas values are used for illustration only.)

$$\text{Effluent soluble COD} = \frac{200\text{-}152}{227\text{-}152}(41\text{-}33) + 33 \quad \text{[Interpolated from data of Table 7-4]}$$
$$= 5 + 33 = 38 \text{ mg/l}$$

$$\text{Assume effluent suspended solids} = 20 \text{ mg/l}$$
$$\text{Total effluent COD} = 20\,(0.58) + 38$$
$$= 12 + 38$$
$$= \underline{50 \text{ mg/l}}$$

Effluent BOD_5 = 38 (0.28) + 20 (0.23)

= 10.6 + 4.6 [Using BOD_5 ratios from Table 7-4]

= 15 mg/l

Background phosphorus removal per lb COD removed

= (biological solids production ratio) (phosphorus content)

= (0.28) (0.02)

$= 0.0056 \frac{\text{lb phosphorus removed}}{\text{lb COD removed}}$

[The background phosphorus removal rate at Manassas was found to be approximately $0.01 \frac{\text{lb P Removed}}{\text{lb COD removed}}$ (8)]

Background phosphorus removal expressed as concentration

= (COD removal) (phosphorus removal per lb COD removed)

= (200-38) (0.0056)

= 0.9 mg/l

Chemically removable phosphorus = 10.0 - 0.9 = 9.1 mg/l

Use Al^{3+}/chemically removable phosphorus dose = 1.25 lb/lb

From Figure 7-6 find 0.6 lb of phosphorus removed per lb of aluminum added at the dosing ratio of 1.25

Aluminum dose = 1.25 (9.1) = 11.3 mg/l Al^{3+}

Chemical phosphorus removal = 0.6 (11.3) = 6.8 mg/l P

Soluble phosphorus residual = 9.1 - 6.8 = 2.3 mg/l

Total effluent phosphorus = 2.3 + 20 (0.05) [P/SS ratio from Table 7-4]

= 3.3 mg/l P

The preceding calculations are useful from the standpoint that they show the pollutant fractions in both the soluble and solid phases. Often a remarkable treatment improvement can result from effluent solids control and minimization.

2. **Alkalinity Loss**

Maximum possible alkalinity loss due to aluminum hydrolysis

= 5.4 (5.5) = 30 mg/l as $CaCO_3$

which gives $\frac{30}{11.3}$ = 2.7 lb $CaCO_3$ loss/lb Al^{3+}

At an aluminum added to chemically removable phosphorus ratio of 1.16, an average alkalinity loss of 2.3 lb $CaCO_3$/lb Al^{3+} was measured at Manassas (8). Like solids production values, alkalinity losses are a function of the proximity of aluminum dose to stoichiometric phosphorus requirements and the reaction pH.

3. **Dissolved Solids Addition**

$$SO_4^{2-} = (11.3 \text{ mg } Al^{3+}/l)\left(5.4 \frac{\text{mg } SO_4^{2-}}{\text{mg } Al^{3+}}\right) = \underline{61 \text{ mg}/l}$$

Although dissolved solids are added to the system, the designer should remember that other soluble wastewater components (i.e., organics, phosphorus, alkalinity, etc.) are being removed. Prior to the addition of sulfuric acid for pH adjustment at Manassas, no increase in dissolved solids was measured on a primary to final effluent basis because of a compensating loss of soluble wastewater components (8).

4. **Final Clarifier and Pumping System**

Organic solids = 57 mg/l

Inorganic solids = 43 mg/l

Total = 100 mg/l

Solids lost in effluent = 20 mg/l

Solids accumulation = 80 mg/l

CRT = 2 days, and aeration tank detention at average flow

= 1.5 (1) = 1.5 hours

$$\text{MLSS in aerator} = \frac{(24)(2)}{1.5}(80) = 2{,}560 \text{ mg/l}$$

Since this is the final clarifier a conservative overflow rate should be utilized. From Figure 7-9 the minimum settling rate of 1,150 gpd/ft^2 was observed at a MLSS of 2,600 mg/l. To provide for peak flow, the clarifier should be designed for an overflow rate of 770 gpd/ft^2, or less. It should have a 12 ft effective water depth, and might be provided with a surface skimmer.

Maximum SVI at 2,600 mg/l MLSS = 82 ml/gm (Figure 7-9)

Approximate return sludge concentration at worst condition

$$= \frac{10^6}{SVI} = \frac{10^6}{82} = 12{,}200 \text{ mg/l}$$

Solids balance equation where R is sludge return rate

$$(Q_{max} + R)(MLSS) \leq (R)(\text{return sludge conc.})$$

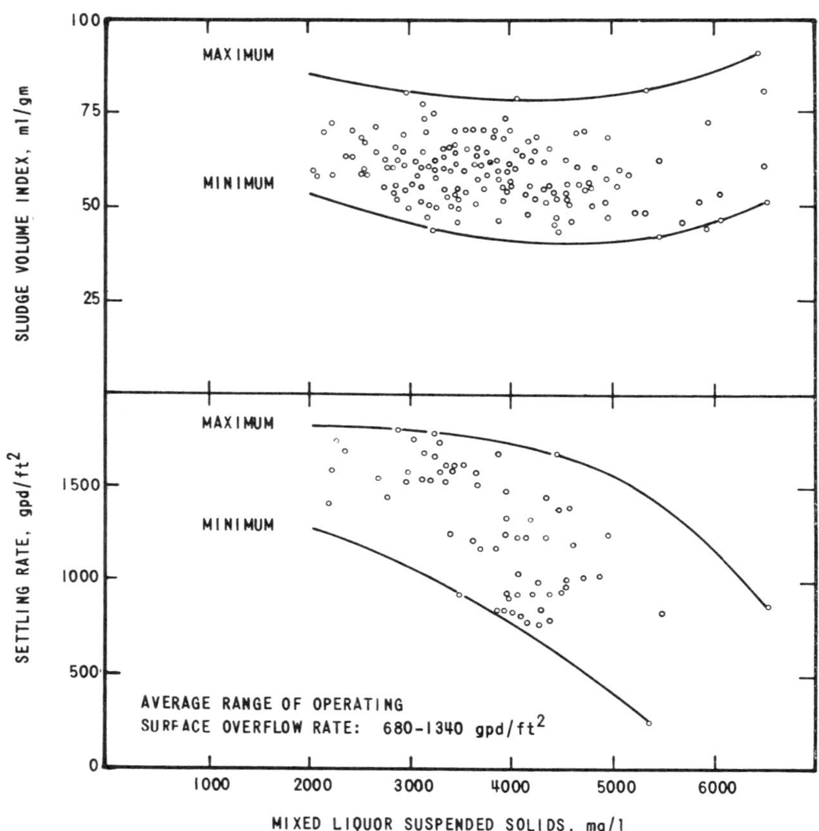

FIGURE 7-9 PHYSICAL CHARACTERISTICS OF HIGH RATE ACTIVATED SLUDGE WITH MINERAL ADDITION (8,16)

or considering just the case of equality,

$$R = 1.5\ Q_{ave}\ \frac{2{,}600}{12{,}200 - 2{,}600}$$

$$R = 0.4\ Q_{ave}$$

(Actually, effective design would indicate a ±50% capability; i.e., 0.2 to 0.6 Q_{ave} capability)

Solids to be wasted per day = 80 mg/l at a minimum sludge solids concentration = 12,200 mg/l. Assume continuous wastage with separate pump.

$$W = Q_{ave}\ \frac{80}{12{,}200} = \underline{0.0065\ Q_{ave}}$$

(Again, effective design would indicate a ±50% capability; i.e., 0.003 to 0.01 Q_{ave} capability).

5. **Alum Cost**

$$= (11.3)\,(0.32)\,(8.34)\,(10^{-3}) = \underline{3.1¢/1{,}000\ \text{gal.}}$$

Additional Comments

No mention has been made of the effect of sidestream phosphorus contributions. With conventional solids handling and treatment processes, gravity and flotation thickening, filtration and centrifugation, aerobic and anaerobic digestion, and incineration, phosphorus recycle should not be a problem and any that occurs will largely be associated with the solids fraction of the minor process streams. However, processes that cause biological solids liquefaction and hydrolysis, such as heat treatment processes, will result in a substantial recycle of soluble phosphorus with a corresponding increase of metal ion requirements.

Example No. 2: Same system as Example No. 1 but with sulfuric acid addition for pH adjustment (residual alkalinity = 30 mg/l as $CaCO_3$) and polymer addition for solids control.

Detailed calculations are not shown for the problem since the techniques are essentially the same as in Example No. 1. Results are summarized, along with results from the other examples, in Table 7-5. Highlights from this problem are as follows:

1. COD, BOD_5, and suspended phosphorus concentrations in the effluent are substantially reduced by an assumed polymer dose of 0.3 mg/l. The chemically removed phosphorus was calculated for this case by multiplying the 0.6 lb P removed/lb Al^{3+} added, from Figure 7-6, by 1.33. The 1.33 was obtained from Figure 7-4, assuming an adjusted pH of about 6. The result was a phosphorus residual of 0.1 mg/l with an aluminum dose of 11.3 mg/l. But

Figure 7-8 showed from experimental data a nominal 0.5 mg/l residual phosphorus at this Al:P ratio and, therefore, the conservative residual of 0.5 mg/l soluble phosphorus was chosen.

Table 7-5

SUMMARY OF DESIGN CALCULATIONS

Item	Example 1	Example 2	Example 3	Example 4
pH Adjustment	No	Yes	Yes	Yes
Polymer Addition	No	Yes	Yes	Yes
Split Treatment*	No	No	No	Yes
Mixed Media Filtration	No	No	Yes	Yes
Total Aluminum Dose, mg/l	11.3	11.3	27.3	18.2
Effluent Quality, mg/l				
Suspended Solids	20	2	0	0
COD	50	39	38	–
BOD_5	15	11	11	–
Total Phosphorus	3.3	0.6	0.2	0.2
Total Soluble Phosphorus	2.3	0.5	0.2	0.2
Solids Production, mg/l				
Biological	57	57	57	57+
Chemical	43	45	92	66
Alkalinity Loss Due to Al^{3+} Addition mg/l as $CaCO_3$	30	21	109	59
SO_4^{2-} Increase Due to Al^{3+} Addition, mg/l	61	61	147	98
Alum Cost, ¢ / 1,000 gal.	3.1	3.1	7.4	4.9
Polymer Cost, ¢ / 1,000 gal.	0	0.3	0.3	0.3
Acid Cost, ¢ / 1,000 gal.	0	1.7	0	1.0
Total Cost, ¢ / 1,000 gal.	3.1	5.1	7.7	6.2
Clarifier Overflow Rate, gpd/ft^2	770	670	360	
Return Sludge Rate	0.4Q	0.6Q	0.9Q	
Waste Sludge Rate	0.0065Q	0.008Q	0.012Q	

*Split treatment here incorporates mineral addition as in Example No. 1 without pH adjustment, followed by mineral addition with pH adjustment at a later point.

2. The alkalinity loss due to hydrolysis was also reduced due to increased $AlPO_4$ formation. Using 96% strength sulfuric acid and assuming a residual alkalinity of 30 mg/l as $CaCO_3$, a calculated sulfuric acid requirement of 101 mg/l was found. The sequence of chemical addition is acid, alum, and polymer, with about 5 minutes of mixing at peak flows between each application.

3. Dissolved solids additions totaled 156 mg/l SO_4^{2-}; 61 mg/l due to the alum and 95 mg/l due to the acid.

4. Because of improved solids capture, the operating MLSS in the aerator increased to 3,200 mg/l necessitating a lower surface overflow rate in the clarifier, and greater return and waste sludge handling capability.

5. Chemical costs increased to 5.1¢/1,000 gal. with benefits in phosphorus, suspended solids, BOD_5 and COD removals.

Example 3: Assuming mixed media filtration provides complete solids control, what is the change in system calculations if a residual effluent phosphorus of 0.2 mg/l as P is to be provided?

Highlights from this problem are as follows:

1. Essentially no differences from Example No. 2 are found in effluent COD and BOD. However, the mixed media filter does provide positive effluent solids control in the event of system upset. An Al:P weight ratio of 3 (Figure 7-8) was used.

2. The addition of alum caused an alkalinity consumption of 109 mg/l, which would deplete the background alkalinity of the wastewater to 41 mg/l as $CaCO_3$. If a substantially higher alum dosage had been required, all of the alkalinity might have been consumed, and addition of lime might have been necessary.

3. Dissolved solids addition consisted of 147 mg/l SO_4^{2-} from the alum.

4. Additional solids productions resulted in an operating solids level of 4,800 mg/l in the aerator, necessitating a lower surface overflow rate and higher return and waste sludge capabilities than in earlier examples.

5. Total chemical treatment costs are raised to 7.7¢/1,000 gal.

Example No. 4: Split treatment (chemical addition at more than one point) is utilized for a required effluent phosphorus concentration of 0.2 mg/l. Assume the first point of addition is as in Example No. 1. Further mineral addition will be made to the effluent in Example No. 1 either before filtration or during a second stage of biological treatment.

1. Changes in effluent COD and BOD are not calculated since this is dependent upon whether split treatment is practiced with just filtration, or whether another biological system is added. In a real situation it would be better to reduce the aluminum dose below that in Example No. 1 in the first stage system for more efficient chemical use. This was not done here for design simplicity. Another possibility is addition of part of the chemical dose in the primary settler.

   The Al^{3+} dose of 6.9 mg/l for the second addition was arrived at using Figure 7-8 for a pH adjusted system.

2. The alkalinity depletion due to aluminum addition was calculated as 59 mg/l as $CaCO_3$. This would leave a residual alkalinity of 91 mg/l. Sixty mg/l as H_2SO_4 was added to reduce the alkalinity to 30 mg/l. Again, in a real situation acid might not have to be added to this wastewater.

3. Dissolved solids additions were calculated as 98 mg/l SO_4^{2-} due to alum addition and 59 mg/l SO_4^{2-} due to acid addition.

4. Clarifiers were not sized in this example. If two biological processes were used in series, the final clarifier of the first process could be designed with an overflow rate from the middle of the performance curve in Figure 7-9.

5. The cost of acid is 1.0¢ / 1,000 gal. Total chemical costs are 6.2¢ / 1,000 gal.

Comment: This final example, in comparison to example No. 3, shows the benefit derived from split chemical treatment for low phosphorus residuals.

7.5 References

1. Guggenheim Process Described by: Kiker, J. E., "Waste Disposal for Dairy Plants", *Sanitarian* (Los Angeles) 16, p 11 to 17 (1953). Florida Engng. Exp. Station 7, Leaflet Series No. 51 (1953). Moore, R. B., "Biochemical Treatment for Anderson, Ind.", *Wat. Works and Sew.*, 85:11 (1938).

2. Thomas, E. A., "Phosphat-Elimination in der Belebtechlemannlage vonMannedorf und Phosphat-Fixation in See-und Klarschlamm", *Vierteljahrsschrift der Naturforschenden Gesellschaft in Zurich*, 110, Schlussheft, S., p 419 (1965).

3. Wirts, J. J., "Communications", *Buckeye Bulletin* (Summer, 1966).

4. Tenny, M. W., and Stumm, W. J., "Chemical Flocculation of Mirco-organisms in Biological Waste Treatment", *JWPCF*, 37:10, p 1370 (1965).

5. Barth, E. F., and Ettinger, M. B., "Mineral Controlled Phosphorus Removal in the Activated Sludge Process", *JWPCF*, 39:8, p 1361 (1967).

6. Eberhardt, W. A., and Nesbitt, J. B., "Chemical Precipitation of Phosphorus in a High Rate Activated Sludge System", *JWPCF*, 40:7, p 1239 (1968).

7. Barth, E. F., Brenner, R. C., and Lewis, R. F., "Chemical-Biological Control of Nitrogen and Phosphorus in Wastewater Effluent", *JWPCF*, 40:12, p 2040 (1968).

8. Mulbarger, M. C., "The Three Sludge System for Nitrogen and Phosphorus Removal", To be presented at the 44th Annual Conference of the Water Pollution Control Federation, San Francisco, California (October, 1971).

9. Long, D. A., Nesbitt, J. B., and Kountz, R. R., "Soluble Phosphate Removal in the Activated Sludge Process – A Two Year Plant Scale Study", Presented at the 26th Annual Purdue Industrial Waste Conference, Purdue University, LaFayette, Indiana (May, 1971).

10. Connell, C. H., and Frey, S. M., Private Communications (January, 1971).

11. Humenick, M. J., and Kaufman, W. J., "An Integrated Biological-Chemical Process for Municipal Wastewater Treatment", Presented at the 5th International Water Pollution Conference, San Francisco, California (July, 1970).

12. Schmidt, F., and Ewing, L., "Phosphate Removal System for Small Activated Sludge Plants", Presented at the Pennsylvania Water Pollution Control Association Meeting, State College, Pennsylvania (August, 1970).

13. Recht, H. L., and Ghassemi, M., "Kinetics and Mechanism of Precipitation and Nature of the Precipitate Obtained in Phosphate Removal from Wastewater Using Aluminum (III) and Iron (III) Salts", Water Pollution Control Research Series, 17010 EKI, Contract 14-12-158, USDI, FWQA (April, 1970).

14. Rickert, D. A., and Hunter, J. V., "Effects of Aeration Time on Soluble Organics During Activated Sludge Treatment", *JWPCF*, 43:1, p 134 (1971).

15. Culp, R. L., and Culp, G. L., *Advanced Wastewater Treatment*, Van Nostrand Reinhold Company, New York (1971).

16. Mulbarger, M. C., and Shifflett, D. G., "Combined Biological and Chemical Treatment for Phosphorus Removal", *Chem. Eng. Prog. Symp. Series*, 67:107, p 107 (1970).

17. Sawyer, C. N., *Chemistry for Sanitary Engineers*, McGraw-Hill, New York (1960).

18. McKee, J. E., and Wolf, H. W., *Water Quality Criteria*, Publication No. 3-A, California Water Quality Board, Sacramento, California (1963).

19. Cherry, A. L., and Schuessler, R. G., "Private Company Improves Municipal Waste Facility", *Wat. and Wastes Eng.* 8:3, p 32 (1971).

20. *Water Quality and Treatment*, 2nd Ed., American Water Works Association, New York (1951).

21. Ockershausen, R. W., Harriger, R. D., and Zuern, H. E., "Phosphorus Removal Tests with Alum" for Buffalo Sewer Authority, Buffalo, New York, by Technical Service

Department, Allied Chemical Corporation, Industrial Chemicals Division, Morristown, New Jersey (1971).

22. Fair, G. M., and Geyer, J. C., *Water Supply and Wastewater Disposal*, Wiley and Sons, New York (1961).

23. Detroit Metro Water Department, "Development of Phosphate Removal Processes", Detroit, Michigan (Program 17010 FAH Grant WPRD 51-01-67 Advanced Waste Treatment Laboratory, Cincinnati, Ohio), Prepublication Copy (July, 1970).

24. Boggia, C., and Herriman, G. L., "Pilot Plant Operation at Warren, Michigan", *Proceedings of the 43rd Annual Conference of the Michigan Pollution Control Association* (1968).

25. Leary, R. D., Ernest, L. A., Powell, R. S., and Manthe, R. M., "Phosphorus Removal with Pickle Liquor in a 115 MGD Activated Sludge Plant", Sewerage Commission of the City of Milwaukee, Wisconsin, Grant No. 11010FLQ, Water Quality Office, Environmental Protection Agency, Advanced Waste Treatment Research Laboratory, Cincinnati, Ohio, Prepublication Copy, (1971).

26. Grigoropoulos, S. G., Vedder, R. C., and Max, D. W., "Fate of Aluminum-Precipitated Phosphorus in Activated Sludge and Anaerobic Digestion", Presented at the 43rd Annual Conference of the Water Pollution Control Federation, Boston, Massachusetts (1970).

27. Hais, A. B., Stamberg, J. B., and Bishop, D. F., "Alum Addition to Activated Sludge with Tertiary Solids Removal", Presented at the 68th National Meeting of the AIChE, Houston, Texas (March 1971).

28. Dean, R. B., "Sludge Handling", Presented at the Advanced Waste Treatment and Water Reuse Symposium, 4th Session, Sponsored by the Environmental Protection Agency, Dallas, Texas (Jan. 12-14, 1971).

29. Smith, J. E. Farrell, J. B., and Dean, R. B., Unpublished Data from the Advanced Waste Treatment Laboratory, Office of Research and Monitoring, Environmental Protection Agency, Cincinnati, Ohio (Jan., 1971)

30. Dotson, G. K., Unpublished Data from the Advanced Waste Treatment Laboratory, Office of Research and Monitoring, Environmental Protection Agency, Cincinnati, Ohio (July, 1971).

Chapter 8

PHOSPHORUS REMOVAL BY LIME TREATMENT OF SECONDARY EFFLUENT

8.1 Description of Process

8.1.1 Theory

Lime treatment of wastewater is essentially the same process as the familiar lime softening of drinking water supplies. The objectives, however, are quite different. While softening may occur, the primary objective is to remove phosphorus by precipitation as hydroxyapatite. This reaction was described in Chapter 3.

During phosphorus precipitation other important reactions occur. The reaction of lime with alkalinity, that results in calcium removal when carrying out softening, not only takes place when treating wastewater, but may have a very important effect on the general efficiency of the process. This reaction can be considered to take place in the two following ways:

$$Ca(HCO_3)_2 + Ca(OH)_2 \rightarrow 2\ \underline{CaCO_3} + 2\ H_2O$$

$$NaHCO_3 + Ca(OH)_2 \rightarrow \underline{CaCO_3} + NaOH + H_2O$$

The first equation is that for softening. Some wastewaters do not contain enough calcium, however, for that equation to be satisfied. Calcium carbonate precipitation may still occur, but by the second reaction. The reactions to form $CaCO_3$ are important for two reasons: the lime consumption determines to a considerable extent the lime dose required for operating the process, and the resulting $CaCO_3$ acts as a weighting agent to aid in settling of sludge.

Another reaction that may be important is precipitation of $Mg(OH)_2$ as follows:

$$Mg^{2+} + 2\ OH^- \rightarrow \underline{Mg(OH)_2}$$

This reaction does not approach completion until the pH is raised to 11. Magnesium hydroxide is a gelatinous precipitate which aids colloid removal, but which hinders sludge thickening and dewatering. Where the pH must be raised to 11 or above to meet a phosphorus removal requirement or some other treatment objective, $Mg(OH)_2$ formation must be considered.

In the two-stage lime treatment process which will be described below, it is necessary to include recarbonation after the first stage to reduce the pH and precipitate the excess lime as $CaCO_3$ in the second stage. The following reaction occurs:

$$Ca^{2+} + CO_2 + 2\ OH^- \rightarrow \underline{CaCO_3} + H_2O$$

Carbon dioxide may also be used to lower pH after lime treatment. The important reaction in this case would be the conversion of Co_3^{2-} to HCO_3^-.

A final reaction that is of major concern, where lime is to be recovered by sludge recalcination, is as follows:

$$CaCO_3 \xrightarrow{\Delta} CaO + CO_2$$

The CaO produced would then be slaked to form $Ca(OH)_2$ before use.

8.1.2 Treatment Systems

Two lime treatment systems may be used with wastewater, single-stage and two-stage. Figure 8-1 shows a single-stage system. In single-stage treatment lime is mixed with feed water to raise the pH to a desired value. Although the pH will depend upon the required phosphorus removal, it is likely to be substantially less than 11, and may be less than 10. Precipitation of phosphate and other materials, as indicated by reactions discussed above, takes place. Time is allowed in the appropriate equipment for precipitate particles to flocculate to sufficient size for good settling. The clarified water from the settler may be discharged directly or may be filtered to improve solids removal. Adjustment of pH with CO_2 may be necessary before discharge and will almost certainly be required before filtration to prevent post-precipitation of $CaCO_3$ from the unstable water. The settled lime sludge may be disposed of as landfill or may be recalcined for recovery of lime. In the latter case, the sludge is thickened, dewatered by centrifuge or vacuum filter, and calcined. The calcined product is then slaked and reused. To avoid buildup of inerts in the lime, some of the sludge or recalcined lime must be wasted or the inerts must be separated from the sludge before calcination.

Two-stage treatment is somewhat more complicated than single-stage treatment. In typical two-stage treatment, shown in Figure 8-2, enough lime is added to the water in the first stage to raise the pH above 11. Precipitation of hydroxyapatite, $CaCO_3$, and $Mg(OH)_2$ occurs. Consideration of the solubility product for $CaCO_3$ and the equilibrium between CO_3^{2-} and HCO_3^- shows that the minimum solubility for Ca^{2+} occurs at a pH of about 10. At a pH of 11 or above there is a considerable concentration of Ca^{2+} present in the water. In two-stage treatment CO_2 is added after the first-stage settler to bring the pH down to about 10 where $CaCO_3$ precipitation results. The $CaCO_3$ is settled out and the clarified water is either discharged or sent to filtration. As in the case of single-stage treatment, pH reduction may be necessary before discharge and probably would be required before filtration.

8.2 Typical Performance Data

A number of full-scale and pilot-scale tertiary lime treatment plants are in operation and more full-scale plants are beginning operation. Only one plant, that at South Lake Tahoe, California, has operated for a significant period with recalcination of sludge for lime recovery. Most aspects of this 7.5 mgd plant have been discussed in detail by Culp and Culp (1). Although the data on handling of wastewater sludges for recalcination and the recalcination process itself are limited mainly to this plant,

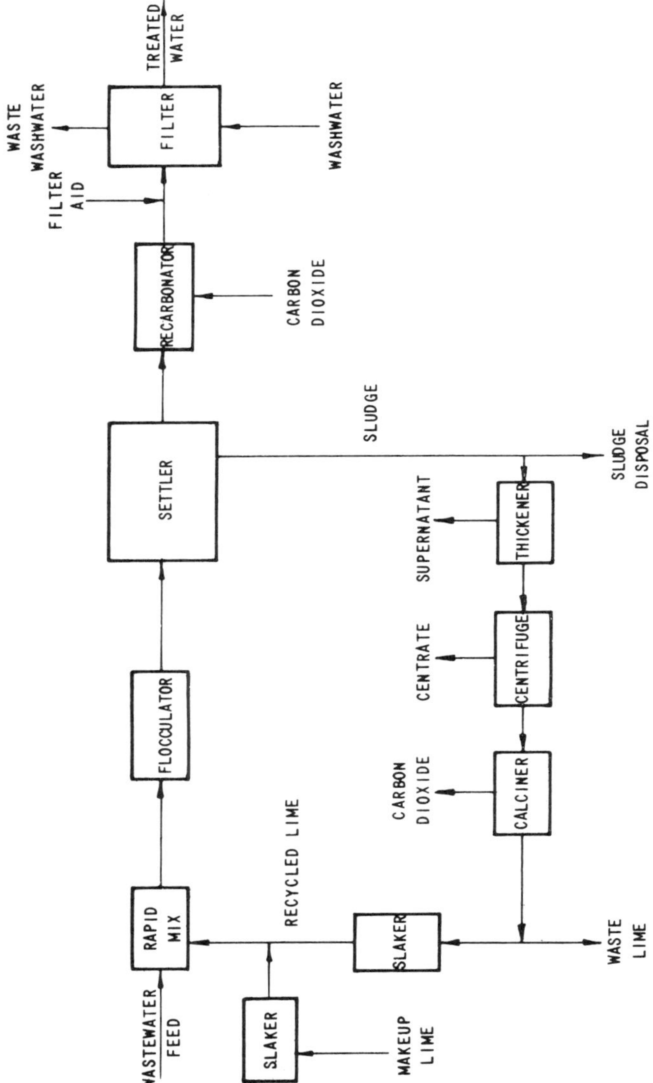

FIGURE 8-1 SINGLE STAGE LIME TREATMENT SYSTEM

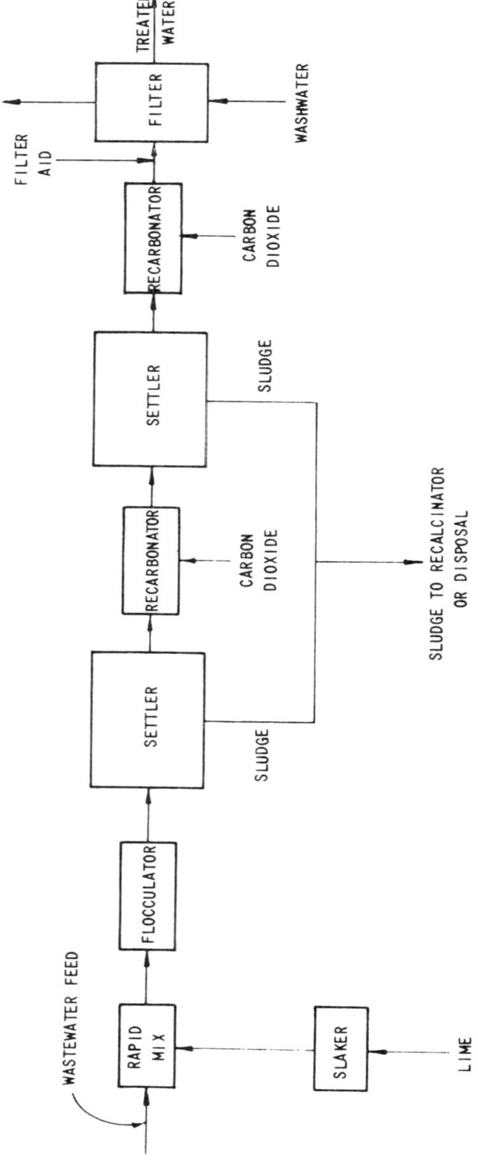

FIGURE 8-2 TWO STAGE LIME TREATMENT SYSTEM

performance data on other parts of a lime treatment system are available from more than one location. Extensive data have been reported from pilot plants located in Washington, D. C., (2) and Lebanon, Ohio (3).

The lime treatment system at Tahoe is a two-stage clarification system which is followed by pressure filters of the multimedia type. These contain 3 ft. of a mixture of coal, sand, and garnet. The water passes through two filters in series. The first stage clarifier is operated at a pH of 11. To reach that pH requires about 300 mg/l CaO. A small amount of polymer is used to improve flocculation. The pH is reduced to 9.6 by recarbonation before the second-stage settler. Before filtration, the pH is further reduced to 7.5. A small amount of alum (1 to 20 mg/l) or a combination of alum and polymer is used as filter aid. Filter run length has varied between 4 and 60 hours.

Phosphorus removal at this plant always has been good and has improved as operating experience has increased. By returning plant waste streams containing precipitated phosphorus to the first stage flocculator, it is now possible to obtain routinely an effluent with less than 0.1 mg/l P. Before the filters the phosphorus concentration is about 0.4 mg/l.

In addition to phosphorus removal, there is significant removal of organic materials. During a period of intensive study of the first stage of clarification (unpublished data) 77% removal of BOD and 61% removal of COD were obtained. Although there was not a large removal of suspended solids, there was a significant change in the character of the solids from organic to largely inorganic. Further removal of organic materials and suspended solids occurs over the remainder of the system. Culp and Culp report typical filter effluents with a BOD of 3 mg/l, COD of 25 mg/l, and turbidity of 0.3 JTU.

Sludges from the first stage settler and the settler following recarbonation are sent to a gravity thickener. Solids concentration increases during thickening from about 1% to from 8 to 20%. Thickened sludge is then centrifuged to form a cake with from 30% to more than 40% solids. The cake is next calcined in a multiple hearth furnace and the recovered lime slaked for reuse. Initially it was planned to operate the centrifuge for high solids recovery in the cake. This necessitates discarding a significant amount of the recovered lime to prevent buildup of precipitated phosphate and other inerts. Since operation began, it has been found that much of the phosphate and some other inert materials can be separated from the $CaCO_3$ in the sludge by operating with a rather high solids concentration in the centrate. Approximately 90% of the phosphorus can be removed in the centrate, for example, when 25% of the solids entering the centrifuge are allowed to remain in that stream. This classification procedure results in a loss of about 15% of the recoverable lime. A greater degree of utilization of recalcined lime compensates for the loss, however, and the load on the calcining furnace is reduced. Over three years of operation, mostly without classification in the centrifuge, the average concentration of CaO in the recalcined product was 66%. Recalcined lime made up 72% of the total used at the plant.

Considerable operating data have been obtained by O'Farrell and Bishop (2) on a two-stage system operated at the EPAWQO-DC Pilot Plant in Washington, D. C. This 50,000 gpd plant treated secondary effluent from a pilot activated sludge system at the municipal treatment plant. The first stage pH was maintained at about 11.7, using a lime dose of about 400 mg/l as CaO, and the pH after recarbonation was maintained at about 10.3. Ferric iron was added at a rate of 5 mg/l in the second stage to improve flocculation. Water from the second-stage settler was filtered without further pH adjustment through gravity-flow dual-media filters consisting of anthracite coal over sand. The average filter run length was more than 50 hours.

Phosphorus removal for the system was similar to that obtained at Tahoe; 0.09 mg/l as P was the average concentration remaining. Phosphorus concentration before filtration averaged 0.13 mg/l.

In addition there was significant organic and suspended solids removal. The average BOD was reduced from 15 mg/l to 2.1 mg/l before and 1.5 mg/l after filtration. Suspended solids were reduced from 33 mg/l to 17 mg/l before filtration and 3.8 mg/l after filtration.

Sludge from the second stage settler was returned to the first stage and all sludge removed from the system was from the settler of the first stage. This sludge contained about 5% solids and constituted about 1.5% of the feed volume. Although there was sludge thickening and calcining equipment available, only preliminary study was made of lime reuse. Limited results showed improved thickening of the sludge when recalcined lime was recycled to the system.

A 75 gpm single-stage pilot system was operated at the Lebanon, Ohio Sewage Treatment Plant. Feed water was activated sludge effluent. This system, consisting of a flocculator-clarifier followed by dual-media filters of anthracite coal and sand, is described by Berg, et al. (3). Sludge was gravity thickened and discharged to sand drying beds. The system has been operated over a pH range from about 9 to 11 with most data taken at a pH of 9.5. Before filtration the pH was reduced to about 8.8 with sulfuric acid to prevent precipitation on the filter media. Filter run length averaged about 90 hours.

Effluent phosphorus concentrations for the system are shown in Figure 8-3. The effect of pH is clearly indicated. Since the average phosphorus concentration of the secondary effluent was about 10 mg/l as P, approximately 95% removal was obtained at a pH as low as 9.5. Before filtration the average phosphorus concentration was 0.75 mg/l at a pH of 9.5. Removal improved only slightly at higher pH because of the presence of precipitated phosphorus in the effluent.

Average suspended solids concentration of the secondary effluent was 43.5 mg/l. This was reduced to 16.5 mg/l in the clarifier. Much of the suspended matter in the clarifier effluent consisted of inorganic precipitates. After filtration the water was of high clarity. During a six month period when the biological treatment plant operated well, turbidity averaged 0.2 JTU. Even when the suspended solids load from the activated

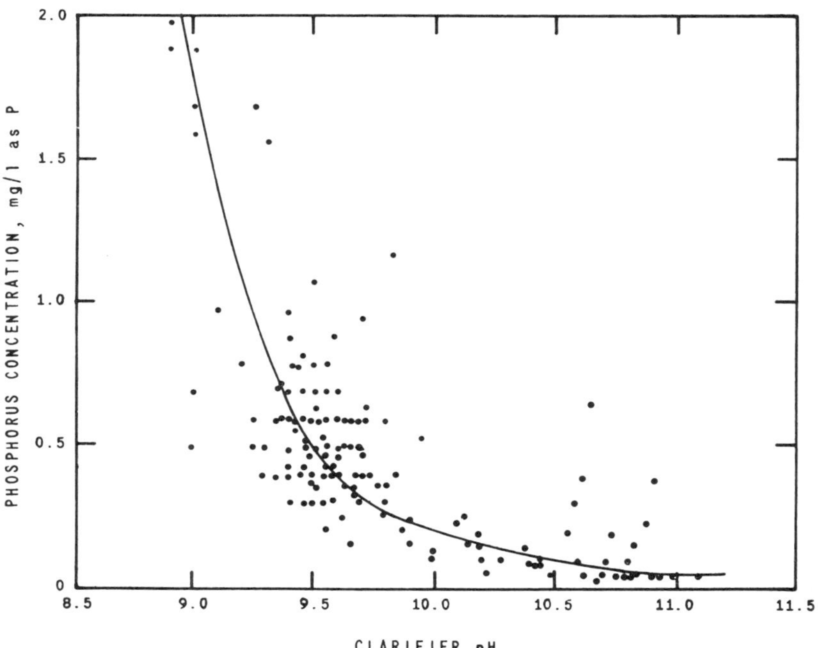

FIGURE 8-3 EFFECT OF pH ON PHOSPHORUS CONCENTRATION OF EFFLUENT FROM FILTERS FOLLOWING LIME CLARIFIER

sludge plant was high, the lime treatment system produced water with a daily average turbidity not exceeding 2.5 JTU.

During a two month period in which extensive measurement of organic removal was made, total organic carbon was reduced by 63% over the clarifier and 68% over the clarifier and filters. Organic carbon measurements taken at other times indicated removals of from 55 to 74%. Occasional BOD and COD samples indicated removals of 86% and 62%.

Sludge was usually removed from the settler at 1.5 gpm giving a sludge concentration of 2.8% solids. This was thickened to about 10% before being pumped to drying beds. The sludge dewatered quickly to about 50% solids. Equipment was not available for recalcining the sludge.

The lime requirement is an important consideration in lime treatment. This can vary over a wide range depending on operating pH and water composition. From reactions given earlier it was seen that alkalinity has an important effect on the lime dose. Buzzell and Sawyer (4) have shown, for example, that for several wastewaters the lime dose required to reach a pH of 11 correlated approximately with alkalinity. Examination of the earlier equations would show also that calcium hardness can affect lime dose. One part by weight of CaO can react with from 0.89 to 1.79 parts of bicarbonate alkalinity expressed as $CaCO_3$, the lower value applying to very soft waters and the higher value to very hard waters. In addition to the reaction of lime with hardness, other competing reactions occur in lime treatment of wastewater. Also, there may be incomplete reaction of the lime. All of these complications make calculation of lime dose difficult. The result is that, at present, determination of lime dose is largely empirical. Approximate values have already been given for the plants at Tahoe and Washington, D.C. Some approximate values for the Lebanon work are shown in Table 8-1. It appears from the data already available that the lime dose will usually be in the range of 300 to 400 mg/l as CaO for two-stage treatment, and from 150 to 200 mg/l where single-stage treatment is satisfactory.

Table 8-1

LIME REQUIREMENTS

Feed Water Alkalinity (mg/l as $CaCO_3$)	Clarifier pH	Approximate Lime Dose (mg/l of CaO)
300	9.5	185
300	10.5	270
400	9.5	230
400	10.5	380

8.3 Criteria for Selection of Process

Lime treatment of secondary effluent represents a significantly higher capital cost and somewhat higher operating cost for phosphorus removal than mineral addition to a conventional treatment plant. Lime treatment has, however, several advantages over mineral addition. It can be considered more dependable since it adds additional flocculation and sedimentation steps to the system. Upsets in the conventional plant, which would reduce the efficiency of phosphorus removal by mineral addition, would have less effect on tertiary lime treatment. Because tertiary lime treatment is separate from the conventional treatment plant, it adds flexibility to operation of the system. For very high degrees of phosphorus removal, lime treatment would be the method of choice, not only because of the inherent greater dependability mentioned above, but because of the ability to produce an effluent slightly lower in phosphorus content. Lime treatment decreases the total dissolved solids content of the water by removal of hardness and alkalinity while mineral addition adds to the total dissolved solids. In the case of alum addition, for example, 5.3 parts by weight of sulfate are added for each part of aluminum. Lime treatment has the capability of removing turbidity to very low levels. Where there are plans for recreational reuse or certain industrial reuses of the treated water, low turbidity along with low phosphorus content of the lime treatment effluent is very desirable. Lime treatment offers the opportunity for recovery of the treatment chemical. At the present time there are no acceptable methods for recovery of aluminum or iron salts.

The choice of single-stage or two-stage lime treatment depends partly upon the degree of phosphorus removal required, but more importantly, on the alkalinity of the water. Unless a high treatment pH is used, waters with low alkalinity, in the range of 150 mg/l as $CaCO_3$ or less, form a poorly settleable floc because of the low fraction of dense $CaCO_3$. A pH above 10 is needed even to obtain measurable $CaCO_3$ production. A pH of 11 or above is then needed to precipitate $Mg(OH)_2$ which aids in settling of fine particles. Since neither discharge nor reuse of the high-pH water is likely to be acceptable, addition of a second treatment stage after recarbonation to a pH of about 10 becomes generally necessary. It would be possible to recarbonate to a much lower pH and avoid the second treatment stage, but this would result in a high calcium effluent and would eliminate the chance to produce high $CaCO_3$ sludge in situations where lime recovery was contemplated. Although most of the phosphorus removal occurs in the high pH first stage, in accordance with the pH dependence of phosphorus solubility as shown earlier by Figure 8-3, some removal does also occur in the second stage. Another important reason for including the second stage is to assure better control of clarification. With low alkalinity waters there is sometimes difficulty in settling the sludge in the first stage even at high pH. The second stage settler with its high $CaCO_3$ sludge, prevents solids carryover when the first stage settler does not operate properly.

With high alkalinity waters, a well settling floc is formed at pH values as low as 9.5. There is no need for two-stage treatment unless the required degree of phosphorus removal necessitates a high pH. In addition to obtaining a small degree of phosphorus removal, the second stage would be used in these cases to lower calcium content in the effluent and to obtain high $CaCO_3$ sludge for lime recovery. There is not yet available operating experience at enough plants to state positively the alkalinity at which single-stage treatment will perform satisfactorily, from the standpoint of floc settleability. At present, experience indicates that at an alkalinity of 150 mg/l as $CaCO_3$, single-stage treatment probably cannot be used. Between 150 mg/l and 200 mg/l settleability will probably depend on the amount of organic floc present. Above 200 mg/l, settleability is likely to be satisfactory.

8.4 Description of and Criteria for Choice of Equipment

8.4.1 Single-Stage System and First Stage of a Two-Stage System

Diagrams for single and two-stage lime treatment systems are shown in Figures 8-1 and 8-2. The first part of each system is made up of clarification equipment essentially the same as that found in water treatment plants. It includes a chemical feeding system (see Chapter 10), a rapid mix tank, a flocculator, a settler, and a means for sludge removal. Mixing, flocculation, and settling may be carried out in separate vessels or tanks, or they may be combined in one integrated unit. The system consisting of separate tanks has been used for many years in water treatment and was the system adopted for the first stage of the Tahoe lime treatment plant. The integrated type of equipment is sometimes referred to as an upflow clarifier or a sludge blanket clarifier. The latter term can be misleading, however, since not all such units operate with a sludge blanket. A diagram of a unit that does not operate with a sludge blanket is shown in Figure. 8-4. Culp and Culp (1) recommend the system of separate tanks because that system allows separate control of each part of the process. They point out that such a system has greater flexibility in points of addition for chemicals. They also point out that in the integrated units with a sludge blanket there may be difficulty in controlling the blanket height and the blanket may become anaerobic, leading to poor phosphorus and solids removal. Pilot studies at other locations have shown that excellent solids removal can be obtained with sludge blanket equipment, but that blanket instability could be a problem. On the other hand, pilot upflow units designed to operate without a sludge blanket have proved to be very effective at Washington, D.C. (2). Results available at this time indicate that both the system of separate tanks and the integrated units without sludge blankets should be considered for design of new plants.

The Tahoe plant is an excellent example of a system with the first stage consisting of separate tanks for each unit process. At the present time the plant must be relied upon heavily as a guide to design of such systems. Description of the equipment and some design parameters are given by Culp and Culp. The rapid mix tank which is located in one corner of the flocculator has about a 30 second residence time at design flow. (The plant usually runs with a daily average flow of about one third of design flow.) Mixing is accomplished with a vertical shaft mixer. The flocculator is a

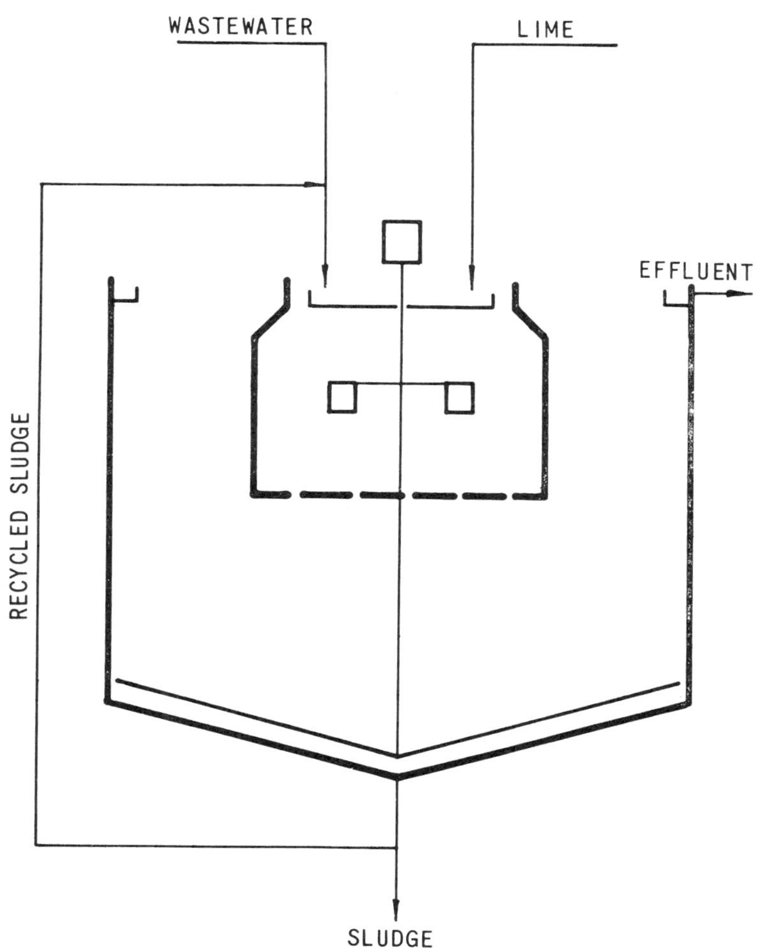

FIGURE 8-4 TYPICAL UPFLOW CLARIFIER

square tank with a depth of 8 ft. At design flow the residence time is 4.5 minutes. The flocculator is provided with air agitation, but operating experience has shown that this agitation is not necessary. The settler is circular with a center inlet. The depth is 10 ft and the design overflow rate is 950 gpd/ft^2. There are two sludge pumps, one a variable-speed centrifugal, and the other a positive displacement Moyno. Provision has been made to return part of the sludge to the rapid mix tank. The concept of returning sludge to the section of the clarification equipment where precipitation is taking place is called solids contact. The objective is to hasten precipitation and to obtain larger precipitate particles which will settle well.

The ease with which flocculation occurs at Tahoe would suggest that a flocculator for this service presents no design problems. Experience is, however, limited. Pilot testing would be very desirable to determine the flocculating characteristics of a particular wastewater. Jar tests may be of some value. Unfortunately, the effect of solids contact, which experience at Tahoe and elsewhere indicates is beneficial, is difficult to duplicate in a jar test. If jar tests indicated good flocculating characteristics without benefit of solids contact, however, flocculation should be as good or better with solids contact. When pilot studies are carried out, provision should be made for solids contact.

Comments concerning the effect of solids contact in jar tests also apply to settling rates determined from these tests. Rapid settling without benefit of solids contact would be a strong indication that settling with solids contact would also be rapid.

Clarification equipment of the integrated type for use in municipal water treatment is manufactured by a large number of companies. Units of this type used for treatment of wastewater have been essentially of the same design as used for municipal water supplies. The basins are usually circular with the rapid mix and flocculating sections located at the center of the settler. Agitation for mixing and flocculating as well as power for sludge scrapers is provided at the center of the settler. Solids contact is usually provided, and may be carried out with best control by external circulation of sludge to the mixing section. In some designs, however, internal recirculation is used. There is considerable variation in the geometry and complexity of the mixing and flocculating sections depending upon the manufacturer. The unit shown in Figure 8-4 is a relatively simply type. Units from three different manufacturers were used in pilot work at Lebanon, Ohio and Washington, D.C. These and others at additional locations have performed effectively. However, in some cases with operation of a sludge blanket, difficulties have been encountered. A sludge blanket can be eliminated by avoiding designs in which the wall separating the center compartments from the concentric settler reaches close to the bottom of the settler. The design in Figure 8-4 prevents sludge blanket formation.

There has been little effort to modify this type of equipment from use in water treatment to use with wastewater. Experience at Tahoe would suggest, for example, that less flocculation time, and probably less agitation, is required than is generally provided. A unit such as that shown by Figure 8-4 may have about 40 minutes flocculation time at design capacity for water treatment, nearly nine times that at

Tahoe. Whether modifications in such equipment would be worthwhile remains to be seen.

If a designer chooses to use equipment that is on the market, he does not have complete flexibility in specifying the design. Fortunately, equipment satisfactory for water supply treatment has proved satisfactory for wastewater, except that a lower settler overflow rate is necessary. The single-stage treatment system at Lebanon, Ohio was operated at a constant overflow rate of 1,440 gpd/ft^2. This proved satisfactory even with a sludge blanket. At Washington, D. C., tests were made with settler rates as high as 1,950 gpd/ft^2 although the average rate was 1,120 gpd/ft^2. For design, a peak, dry weather overflow rate of 1,200 to 1,400 gpd/ft^2, with the average somewhat lower, appears reasonable.

8.4.2 Recarbonation

In a two-stage system recarbonation follows the first stage settler. In a single-stage system and following the second stage of a two-stage system, pH reduction by recarbonation may be necessary to make the effluent suitable for filtering or for discharge. An excellent discussion of all aspects of wastewater recarbonation is given by Culp and Culp (1). The reader is referred to that publication, as well as one by Haney and Hamann (5), especially for information on sources of carbon dioxide. Sources include stack gas from sludge incinerators and lime recalciners, CO_2 generators, and commercial liquid CO_2.

The equipment used for contacting the CO_2 and water may be simply a tank with the gas being bubbled through the water. This is the system used at the Tahoe plant where the height of water over the CO_2 source is 8 ft. Pilot studies at Washington, D. C., with a tank depth of 11 ft and a turbine mixer to reduce bubble size and distribute bubbles, showed almost 100% absorption of CO_2.

When considering residence time, the recarbonation tank following the first stage of a two-stage system must be differentiated from the recarbonation tank used just to reduce effluent pH. The residence time of the recarbonator between the first and second stages is not particularly important. At Tahoe the recarbonation tank is only large enough to give a 5 minute residence time at design flow. In the Washington, D. C. studies, the residence time was 15 minutes. A problem that may arise when recarbonating wastewater is foam formation. This is most pronounced when the flow of bubbles is concentrated in parts of the recarbonation tank. The CO_2 distributing system should cover as well as possible the whole lateral area of the tank bottom. This would be especially true if the residence time were as low as 5 minutes.

In the recarbonation tank used just for effluent pH adjustment, sufficient residence time must be allowed for completion of reactions taking place. Culp and Culp recommend 15 minutes. In the Tahoe plant only 4 minutes is provided at design flow, but the water flows to storage ponds with a much longer residence time before the water is filtered. Just as in the case of recarbonation between stages of a two-stage system, good bubble distribution is important for prevention of excessive foaming.

Since one reason for using a two-stage system is to prevent a high calcium concentration in the effluent, the pH to which the water should be recarbonated between stages is that for maximum conversion of the calcium to $CaCO_3$. Where lime recovery is practiced, maximum formation of $CaCO_3$ is also desirable because it results in maximum CaO production. The optimum pH is about 10. At Tahoe, for example, 10.3 was selected because tests showed that pH to give maximum $CaCO_3$ production.

The pH to be selected when the objective of recarbonation is just pH reduction, will depend on a number of factors. An effluent standard may determine this pH. If stabilizing the water to prevent post precipitation is the objective, the pH may be determined by consideration of the Langelier Index. If, as in the case of Tahoe, a treatment step such as activated carbon adsorption is to follow lime treatment, this will determine pH. In the latter case a pH of about 7.5 has usually been selected.

The CO_2 dose requirements can be calculated from the chemical reactions taking place and a knowledge of the concentrations of the various forms of alkalinity in the water. For design of a new plant, alkalinity data must be obtained from samples of liquor taken from jar tests run at the same pH values as planned for the plant. Results must be considered very approximate, but they do help in sizing CO_2 feeding equipment. Culp and Culp discuss the calculations in detail. It was found in pilot plant work at Washington, D.C. that the CO_2 dose for recarbonation between the stages of a two-stage treatment system can be calculated reasonably well by considering the reduction in calcium concentration that occurs. This reduction is due to $CaCO_3$ formation with the CO_3^{2-} coming from the CO_2. The calcium content before recarbonation can be determined satisfactorily from jar tests run at the desired operating pH. The calcium content after recarbonation can also be determined approximately by jar test. Results from several pilot plants indicate the soluble calcium content at the pH of minimum solubility should be about 40 mg/l as Ca. This figure is probably just as good for calculations as a figure obtained from a jar test. The CO_2 dose in mg/l is then equal to:

$$(\text{Ca reduction in mg/l})\left(\frac{44}{40}\right).$$

A safety factor of about 20% should be added to the calculated dose to compensate for inefficiency in absorption.

8.4.3 Second Stage Flocculation and Sedimentation

In two-stage treatment, recarbonation is followed by the second stage flocculation and settling. Experience with equipment for the second stage of treatment with wastewater is limited. Two very different systems have been tested. The simplest is the system at Tahoe. In the Tahoe plant the equipment used is just a longitudinal settler with a 30 minute detention time and a 2,400 gpd/ft^2 overflow rate at design flow. No flocculation equipment is provided. Culp and Culp (1) refer to the settler as a reaction and settling basin. It has been found at Tahoe that a significant part of the recarbonation reaction actually occurs in the settler. This is shown by the pH decrease of about 0.7 unit that occurs between the recarbonation tank and the settler outlet. The clarification that results at Tahoe is unusually good for the simplicity of the equipment.

Jar test and pilot plant work at Washington, D. C. indicated that flocculation after addition of a flocculating aid was required to obtain satisfactory clarification of that water. The pilot tests were run in upflow equipment of the integrated type discussed earlier. Without a flocculating aid, fine $CaCO_3$ precipitate escaped the settler, even at a settler overflow rate as low as 1,100 gpd/ft^2. Using 5 mg/l of Fe^{3+}, good operation was observed at overflow rates as high as 1,950 gpd/ft^2. Hydraulic limitations prevented testing at higher rates. Apparently, the good contact between solids and water during flocculation was beneficial in reducing $CaCO_3$ supersaturation in the effluent. Precipitation in the filters following the second stage did not occur, and no pH reduction was required.

In this work the sludge from the second stage was returned to the mixing section of the first stage to serve as a weighting agent. In waters of low alkalinity or expected high loads of biological solids, this procedure should be considered.

It is difficult to recommend the amount of flocculation to provide and the overflow rate to use for second-stage treatment. Jar tests would be of little value because of the complication of first raising the pH with lime and then lowering with CO_2. Pilot testing may be desirable. Such a pilot system would have to include the first stage equipment and recarbonation equipment, although these could be of crude design. In the absence of pilot tests to indicate otherwise, flocculation should be provided in the second-stage system. If filtration is included, a peak dry weather overflow rate of up to 2,000 gpd/ft^2 in the settler may be used. If filtration is not provided, a somewhat lower overflow rate should be chosen to assure reasonable effluent quality during periods when an upstream part of the system is not operating properly.

8.4.4 Filtration

The last step in a complete lime treatment system is filtration. Although there are a variety of filter types that could be used, essentially all the recent test work has been done with downflow filters of multimedia type. These are essentially the same in design as a rapid sand filter, except that the media are graded from coarse at the filter surface to fine at the filter outlet. This is accomplished by using media of different densities with the largest particles being composed of the least dense material. The result is a filter bed which has far more capacity for removing suspended solids than an ordinary sand filter, without impairing effluent clarity. Both gravity and pressure filters have been used. Each has certain advantages. Generally, however, pressure filters are more appropriate at smaller plants.

Two media combinations have been tested extensively. These are dual-media of coal and sand, and tri-media of coal, sand, and garnet. Filters used at Lebanon, Ohio following a single-stage system (3) and filters at Washington, D.C. following a two-stage system (2) were of the dual-media type. Both contained 18 in. of anthracite coal over 6 in. of sand. Average particle sizes of the media at Lebanon were 0.75 mm and 0.46 mm; at Washington, D. C., 0.9 mm and 0.45 mm. At 2 gpm/ft^2 the average run length at Lebanon was about 90 hours using 8 ft of water as the terminating pressure loss. At Washington, D. C., filter run length averaged about 50 hours for an average rate of 3.4 gpm/ft^2 and a terminating pressure loss of 7 ft of water. High clarity waters were obtained in each case. Backwash in both cases was carried out at 20 gpm/ft^2. Backwash time was 5 minutes at Lebanon and 10 minutes at Washington, D. C. In the latter case, a surface wash was included.

These run length figures give a rough idea of the results that may be expected at other locations. For maximum run length at a given product quality, the depths of media and their particle sizes should be optimized. Although this was not done for the above-mentioned filters, pressure loss distribution in the filters at Lebanon as reported by Berg, et al. (3) indicates that good solids storage in the anthracite was being obtained. Small pilot filters could be used for optimization of the media.

A reasonable design rate for gravity-flow dual-media filters appears to be about 3 gpm/ft^2. Higher rates could result in inconveniently short runs during periods of high solids - load. Rates much lower than 3 gpm/ft^2 result in excessive filter costs, not justified by the longer filter runs. Very long filter runs could result in backwash problems from biological activity in the filters. Recommendation of a 3 gpm/ft^2 rate is based upon having good operation of the second-stage settler. Frequent settler upsets must be avoided.

Tri-media filters are used at the Tahoe plant. These are pressure filters which at design flow operate at 5 gpm/ft^2. Each bed holds 3 ft of mixed coal, sand, and garnet as supplied by Neptune Microfloc. Two filters are used in series, and are backwashed in series, usually when the head loss reaches 16 ft of water. Run length has varied from 4 hours during heavy solids load to about 60 hours under good conditions. Backwash is carried out at 15 gpm/ft^2. The backwash water is reprocessed through the lime treatment system. More details are given by Culp and Culp (1).

From the available data, it is difficult to give precise criteria for choosing between dual-media or tri-media filters following lime treatment. The use of garnet allows for a smaller particle size at the bottom of the filter than is possible with sand. In case of floc weakness, the tri-media filter offers, therefore, more protection from turbidity breakthrough. The cost, however, is slightly higher. Where the tri-media fill is not used, filters can still be backwashed at the onset of turbidity breakthrough. Experience with dual-media filters on lime treated water has not shown sudden breakthrough to be an important problem. Where water of the highest clarity is required, tri-media filters may be of most value.

8.4.5 Sludge Handling

The sludge from a lime treatment system may be handled in two general ways. It may be thickened, dewatered, and disposed of or it may be thickened, dewatered, and recalcined to recover lime for reuse. For small plants, recalcination will not be economically competitive with disposal and should not be considered unless there are restrictions on disposal which make that alternative difficult. It can be assumed generally that for plants over 10 mgd, recalcination of lime sludge will be practical. Recalcination may also be practical at plants somewhat smaller than 10 mgd, depending on the cost of purchased lime and other local conditions.

Data given earlier indicate that the volume of sludge from lime treatment will vary from 1.5% to several percent of feed volume. Sludge concentration will probably be in the range of 1% to 5%. The actual weight of sludge will vary with the chemical composition and amount of suspended solids in the feed water. Values have been observed in the range of 4 to 7 lb/1,000 gal. Sludge production from the first stage of a two-stage system or from a single-stage system can be estimated approximately by weighing the dried sludge from jar tests run at the planned operating pH. Additional sludge produced in the second stage can be assumed to be the $CaCO_3$ formed from calcium concentration reduction during recarbonation. Calcium content before recarbonation can be obtained from the above-mentioned jar tests and after recarbonation can be assumed to be 40 mg/l as Ca.

Design information for the Tahoe sludge handling system is reported by Culp and Culp. (1) The reader is referred to that publication for further information. The Tahoe plant is the only one which includes recalcination of sludge from the lime treatment of wastewater, and that has been operated for a long period of time. Paramenters from that system can be used as a rough guide for design. The gravity thickener has a bottom scraper mechanism and is 8 ft deep. Overflow from the thickener is returned to either the primary clarifier or the first stage of lime treatment. The design solids loading is 200 lb/day/ft^2 and the design overflow rate is 1,000 gpd/ft^2. Some preliminary data from pilot work at Washington, D.C. suggest these loadings are high. The Tahoe equipment was sized, however, to handle a volume of sludge equal to about 9% of the plant design flow. The actual volume of sludge should usually be less than one third of that volume. At Washington, D.C., recalcined lime was found to produce a sludge which thickened significantly faster than sludge from virgin lime.

If lime is not to be recovered from the sludge, the underflow from the thickener can be placed directly on drying beds for final dewatering, or it can be dewatered by centrifuge or vacuum filter.

If sludge recalcination is planned, the thickened sludge would be dewatered, by centrifuging or filtering and fed to the calcining furnace. The centrifuge may have advantages over filters for this purpose, such as the ability to separate phosphate sludge from $CaCO_3$. Lime from the furnace is stored for reuse. It, along with any new lime, must be slaked before use. Sludge can be pumped from the thickener to the centrifuge. Cake from the centrifuge must be transported by conveyor. At Tahoe the centrate is sent to the primary clarifiers. It has been found at Tahoe that, by operating the centrifuge with less than maximum solids capture, much of the precipitated phosphorus can be retained in the centrate. This results in a higher quality lime. This mode of operation requires a second centrifuge to remove the phosphorus rich solids from the centrate. These solids may have value for use in fertilizer.

The Tahoe dewatering and recalcining system is described by Culp and Culp. The centrifuge is a 24 by 60 in. concurrent flow type. The calciner is a 14 ft-3 in. diameter, 6 hearth furnace operated at a top temperature of 1,850^{o} F. Other types of furnaces have been used in water treatment plants and these should also be applicable to the sludge from wastewater treatment. The reader is referred to equipment manufacturers for further information about centrifuges and furnaces.

8.4.6 Control of Lime and Carbon Dioxide Feed

The feeding of lime to a lime softening system can be controlled in a number of ways. For the wastewater application, the most appropriate appears to be control of pH, with the pH value being selected for good suspended solids removal or to meet a phosphorus removal requirement. Where flow equalization is employed, pH sensing alone should be sufficient. Where there is substantial diurnal variation in flow, better control can be maintained by flow proportional-pH control. Systems are available for this type of control.

Carbon dioxide dose is also controlled by pH. As in the case of lime feed, flow proportional-pH control would be preferred where there is diurnal variation in flow.

There is a tendency for pH electrodes to become coated with precipitates and lose sensitivity. The precipitate can be dissolved with acid. In some instances, however, it has been found satisfactory to use manual control because of difficulties with the pH control system.

8.4.7 Scale Formation on Equipment

Because the water in a lime treatment system is supersaturated to some degree with $CaCO_3$ and other precipitating substances, there is a tendency for scale to form on equipment and pipe surfaces. The problem is particularly serious with the lime slurry from the slaker. This can quickly plug the slurry line to the rapid mix tank. There should be easy access to all parts of this line for cleaning. At one small plant a flexible hose was used to feed lime slurry. By periodically flattening or flexing the hose, the scale was removed. The mechanical mixer in the rapid mix tank will also become scaled and must be cleaned.

Scale can also form in sludge lines and the effluent line from the first-stage clarifier. It is recommended that open troughs be used wherever possible. Provision should be made for cleaning lines when open conduits are not possible.

Possible scaling of filters has already been mentioned. Recarbonation before the filters will minimize scale formation.

8.5 Capital and Operating Costs

Because of the short history of lime treatment of wastewater, there is a scarcity of capital and operating cost information. Costs are available for the Tahoe plant and are reported by Culp and Culp (1). Capital cost for the 7.5 mgd lime treating facility was $1,115,000 and cost of the filters was an additional $705,000. For 1969 the estimated operating costs exclusive of equipment amortization were 7.3¢/1,000 gal. for lime treatment without filtration and 2.8¢/1,000 gal. for filtration. Amortization of equipment, based on costs adjusted to the 1969 national average, interest of 5% for 25 years, and the assumption of the plant operating at full capacity, would add 2.7¢/1,000 gal. for treatment without filtration and 1.8¢/1,000 gal. for filtration. For the plant operating at full capacity the total cost of operation would then be 10.0¢/1,000 gal. without filtration and 14.6¢/1,000 gal. including filtration. The fraction of the cost resulting from amortization is a substantial 31% even with the assumption of full capacity. There is an obvious need to keep equipment size to a minimum. Flow equalization deserves strong consideration when planning for lime treatment of secondary effluent to minimize the initial cost of new equipment.

Additional costs for lime treatment based on information from Tahoe and other sources have been reported by Smith and McMichael (6). Tables 8-2 and 8-3 show

Table 8-2

CAPITAL COST OF LIME TREATING FACILITIES

Treatment	Cost ($) Plant Size (mgd) 1.0	10	100
Single-Stage without Filtration	100,000	1,200,000	5,500,000
Two-Stage without Filtration	160,000	1,500,000	7,900,000
Dual-Media Filtration	110,000	510,000	2,300,000

Table 8-3

TOTAL COST FOR LIME TREATMENT OF WASTEWATER

Treatment	Cost (¢/1,000 gal.) Plant Size (mgd) 1.0	10	100
Single-Stage without Filtration	13	7	4
Two-Stage without Filtration	16	9	6
Dual-Media Filtration	8	3	1.4

costs based largely on their results. Capital costs have been updated to December, 1970. Amortization is at 6% for 25 years. For 1 mgd plants, recalcination equipment is not included. All lime would be purchased. The applicability of recalcination was discussed earlier. Component costs making up the total cost figures represent an average for the whole country. Local conditions may cause significant deviation from these values.

8.6 References

1. Culp, R. L., and Culp, G. L., *Advanced Wastewater Treatment,* Van Nostrand Reinhold Company, New York (1971).

2. O'Farrell, T. P., and Bishop, D. F., "Lime Precipitation in Raw, Primary, and Secondary Wastewater", Presented at the 68th National Meeting of AIChE, Houston, Texas (March 1971).

3. Berg, E. L., Brunner, C. A., and Williams, R. T., "Single-Stage Lime Clarification of Secondary Effluent", *Wat. and Wastes Eng.*, 7:3, p 42 (1970).

4. Buzzell, J. C., and Sawyer, C. N., "Removal of Algal Nutrients from Raw Wastewater with Lime", *JWPCF,* 39:10, Part 2, p R16 (1967).

5. Haney, P. D., and Hamann, C. L., "Recarbonation and Liquid Carbon Dioxide", *JAWWA,* 61:10, p 512 (1969).

6. Smith, R., and McMichael, W. F., "Cost and Performance Estimates for Tertiary Wastewater Treating Processes", Robert A. Taft Water Research Center, Report No. TWRC-9 (June, 1969).

Chapter 9

PHOSPHORUS REMOVAL BY MINERAL ADDITION TO SECONDARY EFFLUENT

9.1 Description of Process

Alum and to a lesser extent iron salts have been evaluated as precipitants for phosphorus in effluents from conventional treatment plants. Both batch studies and continuous-flow systems up to 2.5 mgd have successfully reduced phosphorus to very low levels.

Equipment typically consists of mixers, flocculators, settlers, and filters. Some work has been done, however, with the chemicals added directly to filters. Specific studies are discussed later and provide additional information on equipment commonly used.

9.2 Summary of Design Information

An alum dosage of about 200 mg/l is required for phosphate removal from typical municipal wastewater while dosages of 50 to 100 mg/l are sufficient for effluent clarification. Iron salts, while successful in precipitating phosphate, have found little application because of residual iron remaining in the treated water. An Al:P molar ratio of 1:1 to 2:1 is required. The optimum pH for alum treatment is near 6.0 while for iron the optimum pH is near 5.0 (1). The pH of high alkalinity waters may be reduced either by using high dosages of alum or by adding supplementary dosages of sulfuric acid. Anionic polyelectrolytes are useful with alum in improving settleability of floc. Settling alone will reduce residual P to about 1 mg/l.

If higher removals are required, filtration must be employed. With proper operation, residual P may be reduced to less than 0.1 mg/l. Multimedia filters are preferred over sand filters because of the extended length of run. Microstrainers have not been successful in removing alum floc.

Surface overflow rates for clarifiers have ranged from 580 to 1,440 gpd/ft^2, but for consistent removal of phosphate, the lower end of the range is preferred. Filtration rates of 2 to 5 gpm/ft^2 are used.

No process has yet been developed for successfully recovering alum from sludge for reuse although both alkaline and acid regeneration were investigated (2, 3). The sludge can be dewatered by conventional methods. Because dewatering may be difficult, design of such facilities should be conservative.

Costs for tertiary treatment for phosphate removal range from 6 to 15¢/1,000 gal. depending on plant size and degree of treatment required.

9.3 Laboratory and Pilot Studies

9.3.1 South Lake Tahoe (3, 4, 5, 6)

Liquid alum was used to precipitate phosphate in a 2.5 mgd tertiary plant at South Lake Tahoe, California. The treated wastewater was applied to multimedia filters for solids removal without prior sedimentation.

Chemical feed was proportioned automatically to flow, and flash mixing was provided by an in-line mixer in the filter influent line. The filter media consisted of layers of anthracite coal, graphite, fine garnet and coarse garnet overlaying a supporting bed of gravel. Two filters were operated in series at 5 gpm/ft^2. Backwash with secondary plant effluent was initiated either by headloss or by a rise in turbidity of the effluent. Polyelectrolyte was used as a filter aid to control the depth of penetration of floc into the filter beds. With an alum dosage of 200 mg/l, 8 to 9 mg/l P in the secondary plant effluent was reduced to 0.03 to 0.3 mg/l P in the filter effluent. Turbidities were correspondingly reduced from 30 to 70 JTU to 0.2 to 3.0 JTU.

Efforts to regenerate alum from the sludge by both acid and alkaline treatment proved to be uneconomical and a lime treatment process was developed as a substitute for the alum process. Lime could be economically recovered for reuse and had the added advantage of producing a high pH which facilitated removal of nitrogen by ammonia stripping.

Total costs for phosphate removal only, as estimated in 1966, are shown in Table 9-1.

Table 9-1

1966 ESTIMATED COSTS FOR TERTIARY PHOSPHATE REMOVAL WITH ALUM

Plant Capacity (mgd)	Total Cost (¢/1,000 gal.)
2.5	11.7
10	8.5
50	7.1
100	6.6
200	6.3

9.3.2 Nassau County (7, 8)

Nassau County, New York, is evaluating treatment methods for its wastewater to provide a product water which is suitable for ground injection as a barrier to salt water intrusion. A pilot plant has been operated at 400 gpm producing a water which in most respects exceeds USPHS drinking water standards. Effluent from high rate activated sludge is pumped into a 40 ft diameter by 14 ft deep clarifier where alum and polyelectrolyte are added. Sludge is recirculated into the coagulation zone to improve efficiency of floc formation. The treated wastewater then passes downward through a flocculation zone and upward through a clarification zone to orifice type collectors. The clarified wastewater passes to 2 dual-media filters in parallel operated at 3.0 gpm/ft^2. Each filter consists of a 36 in. bed of No. 1-1/2 anthracite (effective size of 0.90 mm) above a 12 in. layer of sand (effective size of 0.40 mm). Filter backwash includes air scour, surface wash and high and low-rate backwashing. The filter effluent passes through 4 granular carbon contactors operating in series to remove organics.

Typical operating results with about 200 mg/l alum are shown in Table 9-2.

Table 9-2

NASSAU COUNTY PERFORMANCE DATA

	Influent	Effluent
Turbidity, JTU	20 - 25	0.2 - 1.5
Color, Pt-Co Units	20 - 40	none
COD, mg/l	78	13
P, mg/l	9	0.1 - 1.0

Costs have been estimated, for separate tertiary treatment only, at plant sizes of 1, 10, and 100 mgd as shown in Table 9-3.

Table 9-3

NASSAU COUNTY COST ESTIMATES

	Plant Capacity (mgd)		
	(1)	(10)	(100)
	(Costs, ¢/1,000 gal)		
Process			
Coagulation	4.9	3.5	3.2
Filtration	1.8	1.1	1.0
Carbon adsorption	6.3	4.5	4.0
	13.0	9.1	8.2
Operating labor	28.0	5.6	1.8
Total Cost	41.0	14.7	10.0

9.3.3 Chicago, Hanover Park (9)

Alum was used to coagulate solids and precipitate phosphate from secondary effluent at Chicago's Hanover Park tertiary treatment plant. Treatment consisted of coagulation-sedimentation in circular clarifiers followed by either rapid sand filtration or microstraining for polishing. Average annual flow was 1.5 mgd.

Combined coagulation plus sand filtration removed 34, 58, and 82% P at dosages of 38, 60, and 140 mg/l of alum, respectively. Combined coagulation plus microstraining effected a lower phosphate removal of 23% at an alum dosage of 38 mg/l. In general, alum coagulation under the test conditions gave little or no improvement in suspended solids removal over that obtained by sand filtration of uncoagulated secondary effluent. The microstrainer did not significantly remove alum-coagulated solids.

9.3.4 Dallas (10)

Dallas, Texas, is operating a demonstration project designed to provide necessary design data for upgrading of the Dallas-White Rock trickling filter plant. Phosphorus removal facilities include a solids contact upflow clarifier unit and two multimedia filters. The clarifier has a rise rate of 0.5 gpm/ft^2 at the design flow of 100 gpm and a detention time of 4 hours. Both filters contain anthracite over sand media. One is a standard gravity filter. The other is a biflow type, with feed water entering at both the top and bottom of the bed. Filtered water is removed by a manifold-lateral collector located at the mid-depth of the bed. The gravity filter is designed for 50 gpm at a filtration rate of 4 gpm/ft^2. The biflow filter is designed for 100 gpm at a filtration rate of 8 gpm/ft^2, or 4 gpm/ft^2 for both the bottom and top of the bed.

The pilot plant process has not yet been optimized for phosphorus removal, however, limited data from two short test periods on trickling filter and activated sludge effluents are shown in Table 9-4.

Table 9-4

REMOVAL OF TOTAL ORTHOPHOSPHATE BY COAGULATION-CLARIFICATION IN THE DALLAS DEMONSTRATION PLANT

	Trickling Filter Effluent	Activated Sludge Effluent
Lime Dose, mg/l CaO	128	133
$FeCl_3$ Dose, mg/l Fe^{3+}	25	56
Soluble Phosphate In, mg/l P	6.5	7.0
Soluble Phosphate Out, mg/l P	0.43	0.21
% Soluble P Removed	93.5	97.0
Test Period, days	12	2

9.3.5 Dayton (11)

Dayton, Ohio, used alum in a 310 gpm pilot plant to remove phosphate and suspended solids from trickling filter effluent. The rectangular sedimentation basin was designed for 105 minutes detention and an overflow rate of 1,014 gpd/ft^2. Two filters designed for loadings up to 8 gpm/ft^2 were used for polishing effluent. One contained 30 in. of sand while the other used 20 in. of anthracite overlaying 10 in. of sand. The sand had an effective size of 0.40 to 0.45 mm with a uniformity coefficient of 1.35 to 1.70. The anthracite had an effective size of 0.85 to 1.20 mm with a uniformity coefficient of 1.35 to 1.80. Both air and water were used in backwashing.

Alum with activated silica or anionic polyelectrolyte as coagulant aids achieved phosphate removals at a dosage rate of 5 lb alum/lb PO_4^{3-} down to residuals of 4 mg/l PO_4^{3-} in filter effluents (89% removal of phosphate). Further removals down to 1 to 2 mg/l PO_4^{3-} required in excess of 200 mg/l alum. The clarifier was operated at overflow rates of 655 to 1,035 gpd/ft^2.

The filters removed 30 to 70% of the applied phosphate load. The dual media filter produced an effluent comparable to that from the sand filter with filter runs 2 to 3 times the length of the sand filter runs.

Costs to provide 90% phosphate removal with alum for the present 50 mgd flow in a 75 mgd plant were estimated at 6.5 to 8.2¢/1,000 gal.

9.3.6 Oxidation Pond Effluents

At Lancaster, California, a research project was initiated to renovate wastewater for use in recreational lakes (12). Existing treatment consisted of primary sedimentation followed by 45 to 60 days detention in oxidation ponds. For coagulation, laboratory jar tests showed alum to be more effective on a cost basis than either lime or ferric sulfate. Phosphate residuals were reduced to 1.1, 0.2, and 0.01 mg/l P with alum dosages of 150, 200, and 300 mg/l, respectively, at an optimum pH near 6.0. Pilot studies showed that both sedimentation and air flotation would provide satisfactory separation of most of the precipitated phosphates, however, filtration was necessary to obtain the required low residuals of less than 0.15 mg/l P.

Final specifications for a full scale process included coagulation with about 300 mg/l alum and 20 minutes of flocculation with a conventional paddle flocculator, sedimentation in a horizontal clarifier with an overflow rate of 400 gpd/ft^2, and filtration through a dual media gravity filter consisting of 18 in. of 0.5 mm anthracite over 8 in. of No. 20 sand.

Capital costs of a 0.5 mgd facility were estimated at $150,000 with operating costs of $184/$10^6$gal. The total cost of capital, operation, and maintenance for a 3 mgd facility was estimated at $150/$10^6$ gal.

Alum was recently evaluated by Shindala and Stewart (13) for phosphorus removal and polishing of waste stabilization pond effluents in Mississippi. Jar tests were used in the evaluation. Using a minimum of 90% removal of phosphorus and 70% removal of COD as criteria, the optimum alum dosage was 85 mg/l at pH 5.5. With higher doses, up to 97% phosphorus removal and 85% COD removal were observed. Ferric chloride and ferric sulfate were also tried but left a residual iron color in the supernatant.

9.3.7 Moving Bed Filter (14)

Alum and polyelectrolyte were used to treat trickling filter effluent in a proprietary moving bed filter at Liberty Corner, New Jersey. The device, a product of Johns-Manville Products Corp., used a bed of sand which is periodically moved upward in an inclined filter vessel. Clean sand is fed in at the bottom of the filter and floc-laden sand is removed mechanically at the top as required by pressure drop, leaving a fresh filtration surface. The wastewater being filtered passes down through the bed and flows through screens at the sides of the filter body.

A major factor controlling phosphate removal is the molar ratio of Al:P. At an alum dosage of 200 mg/l and influent total phosphorus concentrations on the order of 25 to 28 mg/l as P (Al:P ratio = 0.6 to 0.7), the total phosphorus removal efficiency averaged 90%. With lower total phosphorus concentrations (Al:P ratios = 1.2 to 2.6) removal efficiencies averaged 95% and reached as high as 99%. The alum treatment also removed about 70% of suspended solids leaving an average of 15 mg/l in the effluent.

Capital costs for a 1 mgd moving bed filter are estimated at $264,000. Total operating costs for such an installation using 200 mg/l alum are estimated at 12¢/1,000 gal. treated.

9.4 References

1. Recht, H. L., and Ghassemi, M., "Kinetics and Mechanism of Precipitation and Nature of the Precipitate Obtained in Phosphate Removal from Wastewater Using Aluminum (III) and Iron (III) Salts", Water Pollution Control Research Series 17010EKI, Contract 14-12-158, USDI, FWQA (April, 1970).

2. Farrell, J. B., Salotto, B. V., Dean, R. B., and Tolliver, W. E., "Removal of Phosphate from Wastewater by Aluminum Salts with Subsequent Aluminum Recovery", *Chem. Eng. Prog. Symp. Series,* 64:90, p 232 (1968).

3. Slechta, A. F., and Culp, G. L., "Water Reclamation Studies at the South Tahoe Public Utility District," *JWPCF,* 39:5, p 787 (1967).

4. Culp, R. L., "Wastewater Reclamation by Tertiary Treatment", *JWPCF,* 35:6, p 799 (1963).

5. Culp, R. L., and Roderick, R. E., "The Lake Tahoe Water Reclamation Plant", *JWPCF,* 38:2, p 147 (1966).

6. Culp, R. L., "Wastewater Reclamation at South Tahoe Public Utilities District", *JAWWA,* 60:1, p 84 (1968).

7. Rose, J. L., "Injection of Treated Waste Water into Aquifers", *Wat. & Wastes Eng.,* 5:10, p 40 (1968).

8. Rose, J. L., "Advanced Waste Treatment in Nassau County, N. Y.", *Wat. & Wastes Eng.,* 7:2, p 38 (1970).

9. Lynam, B., Ettelt, G., and McAloon, T., "Tertiary Treatment at Metro Chicago by Means of Rapid Sand Filtration and Microstrainers", *JWPCF,* 41:2 (Part 1), p 247 (1969).

10. Graeser, H. J., and Haney, P. D., "Dallas Builds Center to Study Wastewater Reclamation", *Wat. & Wastes Eng.,* 5:12, p 34 (1968).

11. Tossey, D. F., Fleming, P. J., and Scott, R. F., "Tertiary Treatment by Flocculation and Filtration", *Jour. SED, ASCE,* 96:SAI, p 75 (1970).

12. Dryden, F. D., and Stern, G., "Renovated Waste Water Creates Recreational Lake", *Environmental Science & Tech.,* 2:4, p 268 (1968).

13. Shindala, A., and Stewart, J. W., "Chemical Coagulation of Effluents from Municipal Waste Stabilization Ponds", *Wat. & Sew. Works,* 118:4, p 100 (1971).

14. "Phosphorus Removal Using Chemical Coagulation and a Continuous Countercurrent Filtration Process", Water Pollution Control Research Series 17010EDO USDI, FWQA (June 1970).

CHAPTER 10

STORAGE AND FEEDING OF CHEMICALS

10.1 Aluminum Compounds

The principal aluminum compounds that are commercially available and suitable for phosphorus precipitation are alum and sodium aluminate (1). Both are available in either liquid or dry forms. Alum is acidic in nature while sodium aluminate is alkaline and this may be an important factor in choosing between them. Aluminum chloride should be considered, and is discussed herein.

10.1.1 Dry Alum

10.1.1.1 Properties and Availability

The commercial dry alum most used in wastewater treatment is known as "filter alum" and has the approximate chemical formula $Al_2(SO_4)_3 \ 14H_2O$ and a molecular weight of 594. Alum is white to cream in color. The pH varies between 3.0 and 3.5 in aqueous alum solutions having concentrations between 1% and 10%. Commercially available grades and their corresponding bulk densities and angles of repose are given in Table 10-1:

TABLE 10-1

AVAILABLE GRADES OF DRY ALUM

Grade	Bulk Density lb/cu ft	Angle of Repose
Lump	62 to 68	varies
Ground	60 to 71	43°
Rice	57 to 61	38°
Powdered	38 to 45	65°

Each grade has a minimum aluminum content of 17%, expressed as Al_2O_3. This corresponds to a 9% concentration as aluminum. Viscosity and solution crystallization temperatures are included in the subsequent section on liquid alum.

Since dry alum is only partially hydrated, it is slightly hygroscopic. However, it is relatively stable when stored under the extremes of temperature and humidity encountered in the contiguous United States.

The solubility of commercial dry alum at various temperatures is listed in Table 10-2:

TABLE 10-2

SOLUBILITY OF ALUM AT VARIOUS TEMPERATURES

Temperature °F	Solubility lb/gal
32	6.03
50	6.56
68	7.28
86	8.45
104	10.16

Dry alum is not corrosive unless it absorbs moisture from the air, such as during prolonged exposure to humid atmospheres. Therefore, precautions should be taken to ensure storage space is free of moisture.

Alum is shipped in 100 lb bags, drums, or in bulk (minimum of 40,000 lb) by truck or rail. Bag shipments may be ordered on wood pallets. In 1973, the price range for dry alum in bulk quantities is $60 to $65 per ton, F.O.B. the point of manufacture. Freight costs must be added to this. Bagging adds about $5 per ton to the base cost for bulk alum. Names and locations of major producers are in a subsequent section.

10.1.1.2 General Design Considerations

Water utilities usually use ground and rice alum because of their superior flow characteristics. These grades have less tendency to lump or arch in storage and therefore provide more consistent feeding qualities. Hopper agitation is seldom required with these grades, and in fact may be determined to feeding due to packing the contents of the bin.

Alum dust is present in the ground grade and will cause minor irritation of eyes and respiratory tract. A respirator will protect against alum dust. Alum dust should be thoroughly flushed from the eyes immediately and washed from the skin with water. Gloves should be worn to protect the hands. Because of minor irritation in handling and the possibility of alum dust causing rusting of adjacent machinery, dust removal equipment is desirable.

10.1.1.3 Storage

A typical storage and feeding system for dry alum is shown in Figure 10-1. Bulk alum can be stored in mild steel or concrete bins with dust collector vents located in, above, or adjacent to the equipment room. Recommended storage capacity is about 30 days. Dry alum in bulk can be transferred with screw conveyors, pneumatic conveyors, or bucket elevators made of mild steel. Pneumatic conveyor elbows should have a reinforced backing to withstand abrasion.

Bags and drums of alum should be stored in a dry location. Bag or drum loaded hoppers should have storage capacity for eight hours at the nominal maximum feed rate so personnel are not required to charge the hopper more than once per shift. Converging hopper sections should have a minimum wall slope of 60° to prevent arching.

Bulk Storage hoppers should have a discharge bin gate so feeding equipment may be isolated for servicing. The bin gate should be followed by a flexible connection, and transition hopper chute or hopper which acts as a conditioning chamber over the feeder.

10.1.1.4 Feeding Equipment

The feed system includes all components required for preparation of the chemical solution. Capacities and assemblies should be selected to fulfill individual system requirements. Three basic types of chemical feed equipment are used: volumetric, belt gravimetric, and loss-in-weight gravimetric. Volumetric feeders are usually used where low initial cost and lower delivery capacities are the basis of selection. Volumetric feeder mechanisms are usually exposed to corrosive vapors from the dissolving chamber. Manufacturers usually control this problem by use of an electric heater to keep the feeder housing dry or by using plastic components in the exposed areas.

Volumetric dry feeders are generally of the screw type. Two designs of screw feed mechanism are available. Both allow even withdrawal across the bottom of the feeder hopper to prevent dead zones. One screw design is the variable pitch type with the pitch expanding evenly to the discharge point. The second screw design is the constant pitch-reciprocating type. This type has each half of the screw turned in opposite directions so the turning and reciprocating motion alternately fills one half of the screw while the other half is discharging. The variable pitch screw has one point of discharge, while the constant pitch-reciprocating screw has two points of discharge, one at each end of the

FIGURE 10-1

TYPICAL DRY FEED SYSTEM

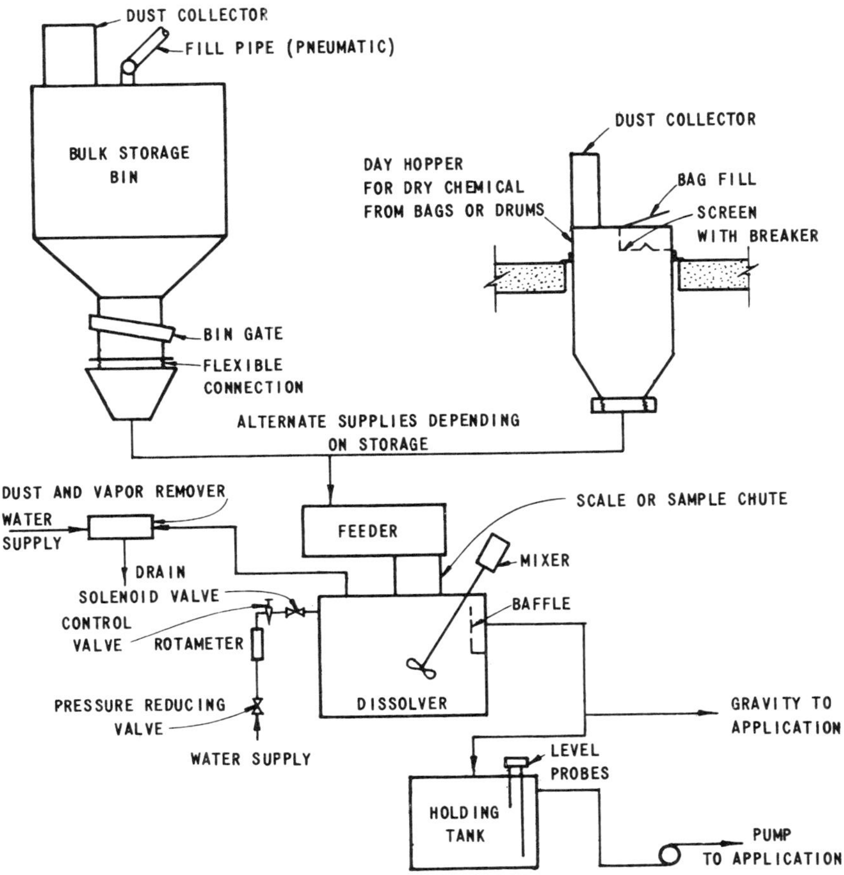

screw. The accuracy of volumetric feeders is influenced by the character of material being fed and ranges between 1% for free-flowing materials to 7% for cohesive materials. This accuracy is volumetric and is not related to accuracy by weight (gravimetric).

Where the greatest accuracy and more economical use of chemical is desired, the loss-in-weight type feeder should be selected. This feeder is suited to low and medium feed rates with a maximum of approximately 4,000 lb/hour. The unit consists of a material hopper and feeding mechanism mounted on enclosed scales. The feed rate controller retracts scale poise weight to deliver the dry chemical at the desired rate. The feeding mechanism must feed at this rate to maintain the balance of the scale. Any unbalance of the scale beam causes a corrective change in output. Continuous comparison of actual hopper weight with set hopper weight prevents cumulative errors. Accuracy of the loss-in-weight feeder is near 1% by weight of the set rate.

Belt-type gravimetric feeders span the capacity ranges of volumetric and loss-in-weight feeders and can be sized for most applications in wastewater treatment. Initial expense falls between that for the volumetric feeder and the loss-in-weight feeder. Belt-type gravimetric feeders consist of a basic belt feeder incorporating a weighing and control system. Feed rates can be varied by changing either the weight per foot of belt, belt speed, or both. Controllers in general use are mechanical, pneumatic, electric and mechanical-vibrating. Accuracy specified for belt-type gravimetric feeders should be within 1% of set rate. This equipment normally includes mild steel hoppers, stainless steel mechanism components, and rubber surfaced feed belts.

Because alum solution is corrosive, dissolving or solution chambers should be constructed of type 316 stainless steel, fiberglass reinforced plastic (FRP), and plastics. Dissolvers should be sized for preparation of the desired solution strength. The most dilute solution strength usually recommended is 0.5 lb of alum to 1 gallon of water, or a 6% solution. The dissolving chamber is designed for a minimum detention time of 5 minutes at the maximum feed rate. Because excessive dilution may be detrimental to coagulation, eductors, or float valves that would ordinarily be used ahead of centrifugal pumps, are not recommended. Dissolvers should be equipped with water meters and mechanical mixers so the water to alum ratio may be properly established and controlled.

10.1.1.5 Piping and Accessories

Pipe made from FRP or plastics (polyvinyl chloride, polyethylene, polypropylene, and similar materials) is recommended for alum solution. Care must be taken to provide adequate support for these piping systems, with close attention given to spans between supports so objectionable deflection will not occur. Lined steel pipe is generally tougher and more rigid, but the cost of providing a near perfect lining may detract from its suitability.

Solution flow by gravity to the point of discharge is desirable. When gravity flow is not possible, transfer components should require little or no dilution. When metering pumps or proportioning weir tanks are used, return of excess flow to a holding tank should be considered. Metering pumps are discussed in the section on liquid alum.

Valves used in solution lines should be plastic, type 316 stainless steel or rubber-lined iron or steel.

10.1.1.6 Pacing and Control

Volumetric and gravimetric feeders are usually adaptable to operation from any standard instrument control and pacing signals.

When solution must be pumped, consideration should be given to use of holding tanks between the dry feed system and feed pumps; solution water supply should be controlled to prevent excessive dilution. The dry feeders may be started and stopped from tank level probes. Variable control metering pumps can transfer alum stock solution to the point of application without further dilution.

Means should be provided for calibration of chemical feeders. Volumetric feeders may be mounted on platform scales. Belt feeders should include a sample chute and box to check actual delivery with set delivery.

Gravimetric feeders are usually furnished with totalizers only. Remote instrumentation is frequently used with gravimetric equipment, but seldom used with volumetric equipment.

10.1.2 Liquid Alum

10.1.2.1 Properties and Availability

Liquid alum is shipped in insulated tank cars or trucks. During the winter it is heated prior to shipment so crystallization will not occur during transit. Liquid alum is shipped at a solution strength of about 4.37% as aluminum or about 8.3% as Al_2O_3 or 49% as $Al_2(SO_4)_3 \cdot 14H_2O$. This solution has a density of 11.1 lb/gal at 60°F and contains about 5.4 lb dry alum (17% Al_2O_3) per gallon of liquid. This solution will begin to crystallize at 30°F and crystallizes at 18°F.

Crystallization temperatures of other solution strengths are as follows (1) in Table 10-3:

TABLE 10-3

CRYSTALLIZATION TEMPERATURES OF LIQUID ALUM

Al_2O_3 percent	Temperature of Crystallization °F
5.19	26
6.42	21
6.67	19
6.91	17
7.16	15
7.40	13
7.66	12
7.92	14
8.19	17
8.46	20
8.74	28

The viscosity of various alum solutions is given in Figure 10-2.

Tank truck lots of 3,000 to 5,000 gallons are available. Tank car lots are available in quantities of 7,000 to 18,000 gallons. The 1974 price of liquid alum on an equivalent dry alum (17% Al_2O_3) basis is about $60 per ton, F.O.B. the point of manufacture. Since liquid alum is an intermediate compound in production of dry alum, the liquid form costs less. However, liquid alum costs more to transport since it is nearly half water, by weight. Therefore, a cost tradeoff point can be established when exact chemical costs and local freight rates are determined. Liquid alum will generally be more economical than dry alum if the point of use is within 100 miles of the manufacturing plant; ease of handling, storing and feeding liquid alum extend its practical transport limit to 200 miles or more.

FIGURE 10-2

VISCOSITY OF ALUM SOLUTIONS

(COURTESY OF ALLIED CHEMICAL CO.)

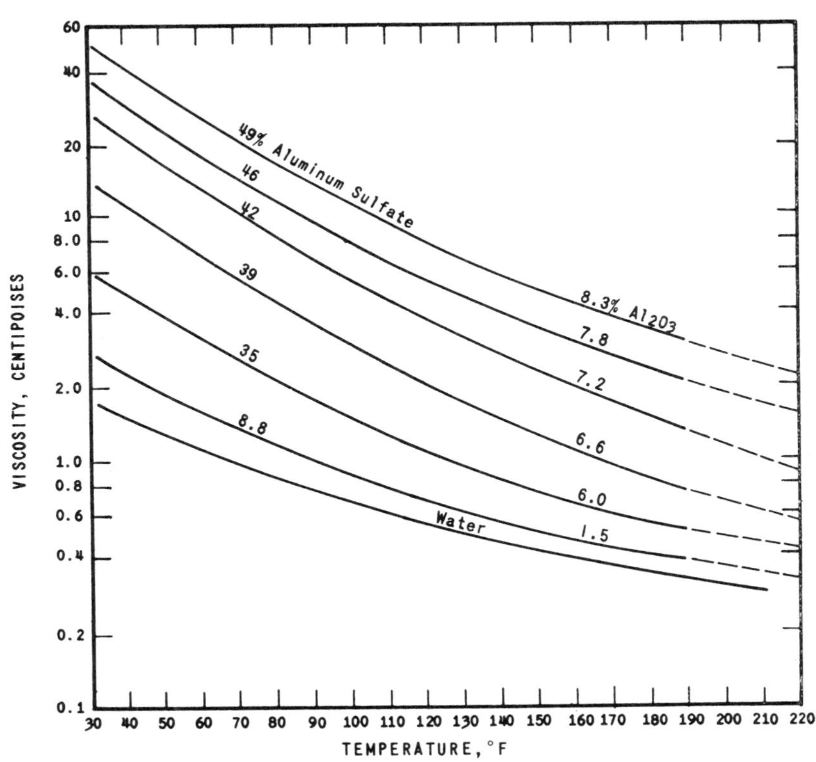

10.1.2.2 General Design Considerations

Bulk unloading facilities usually must be provided at the treatment plant. Rail cars are constructed for top unloading and therefore require an air supply system and flexible connectors to pneumatically displace the alum from the car. U.S. Department of Transportation regulations concerning chemical tank car unloading should be observed. Tank truck unloading is usually accomplished by gravity or by a truck mounted pump.

Established practice in the treatment field has been to dilute liquid alum prior to application. However, recent studies have shown that feeding undiluted liquid alum results in better coagulation and settling. This is reportedly due to prevention of hydrolysis of the alum.

No particular industrial hazards are encountered in handling liquid alum. However, a face shield and gloves should be worn around leaking equipment. The eyes or skin should be flushed and washed upon contact with liquid alum. Liquid alum becomes very slick upon evaporation and therefore spillage should be avoided.

10.1.2.3 Storage

Liquid alum is stored without dilution at the shipping concentration. Storage tanks may be open if indoors but must be closed and vented if outdoors. Outdoor tanks should also be heated, if necessary, to keep the temperature above 25°F to prevent crystallization. Storage tanks should be constructed of type 316 stainless steel; FRP, steel lined with rubber, polyvinyl chloride, or lead; see subsequent section for further details. Liquid alum can be stored indefinitely without deterioration.

Storage tanks should be sized according to maximum feed rate, shipping time required, and quantity of shipment. Tanks should generally be sized for 1-1/2 times the quantity of shipments. A 10-day to 2-week supply should be provided to allow for unforeseen shipping delays.

10.1.2.4 Feeding Equipment

Various types of gravity or pressure feeding and metering units are available. Figures 10-3 and 10-4 illustrate commonly used feed systems. The rotodip-type feeder or rotameter is often used for gravity feed and the metering pump for pressure feed systems.

The pressure or head available at the point of application frequently determines the feeding system to be used. The rotodip feeder can be supplied from overhead storage by gravity with the use of an internal level control valve, as shown by Figure 10-3. It may also

FIGURE 10-3

ALTERNATIVE LIQUID FEED SYSTEMS

FOR OVERHEAD STORAGE

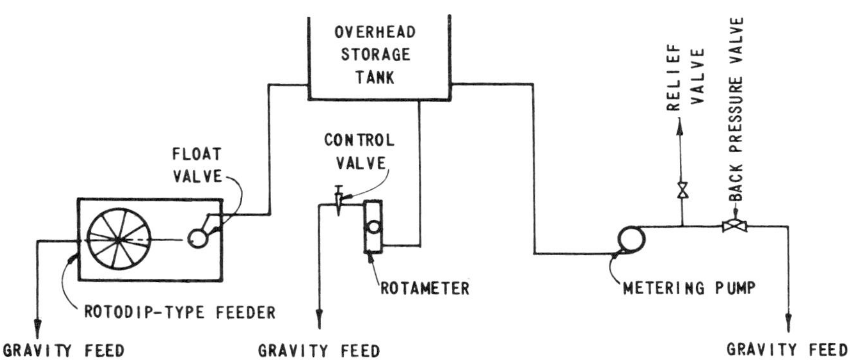

FIGURE 10-4

ALTERNATIVE LIQUID FEED SYSTEMS

FOR GROUND STORAGE

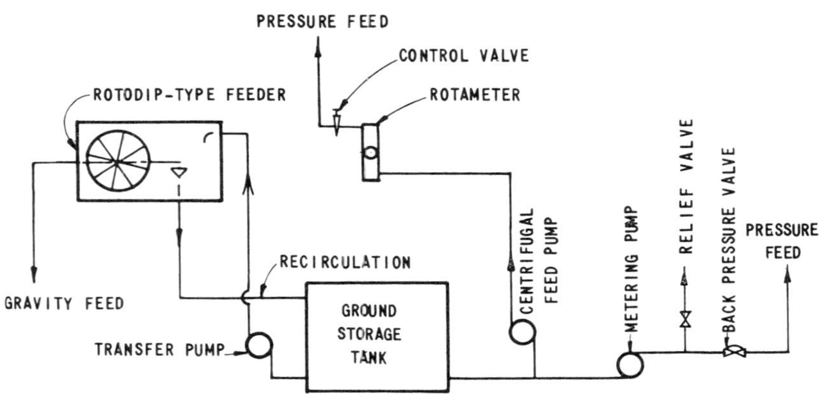

be supplied by a centrifugal pump. The latter arrangement requires an excess flow return line to the storage tank, as shown by Figure 10-4. Centrifugal pumps should be direct-connected but not close-coupled because of possible leakage into the motor, and should be constructed of type 316 stainless steel, FRP, and plastics.

Metering pumps, currently available, allow a wide range of capacity to compare with the rotodip and rotameter systems. Hydraulic diaphragm type pumps are preferable to other type pumps and should be protected with an internal or external relief valve. A back pressure valve is usually required in the pump discharge to provide efficient check valve action. Materials of construction for feeding equipment should be as recommended by the manufacturer for the service, but depending on the type of system, will generally include type 316 stainless steel, FRP, plastics, and rubber.

10.1.2.5 Piping and Accessories

Piping systems for alum should be FRP, plastics (subject to temperature limits), type 316 stainless steel, or lead. Piping and valves used for alum solutions are also discussed in the preceding section on dry alum.

10.1.2.6 Pacing and Control

The feeding systems described above are volumetric, and the feeders generally available can be adapted to receive standard instrument pacing signals. The signals can be used to vary motor speed, variable speed transmission setting, stroke speed and stroke length where applicable. A totalizer is usually furnished with a rotodip-type feeder, and remote instruments are available. Instrumentation is rarely used with rotameters and metering pumps.

10.1.3 Dry Sodium Aluminate

10.1.3.1 Properties and Availability

Dry sodium aluminate, $Na_2Al_2O_4$, is available from manufacturers listed in a subsequent section. Dry sodium aluminate is shipped in 50 lb bags and has a bulk density ranging from 40 to 50 lb/ft^3. The Al_2O_3 content ranges from 41 to 46%. Dry sodium aluminate is noncorrosive and the pH of a 1% solution is about 11.9. Manufacturers should be consulted for more precise specifications of their product.

The 1974 price of dry sodium aluminate in 40,000 lb shipments ranges from $0.13 to $0.15 per lb, F.O.B. the point of manufacture. Prices increase substantially for smaller shipments.

10.1.3.2 General Design Considerations

Requirements for dry sodium aluminate feed systems are generally similar to those for dry aluminum sulfate. Dry sodium aluminate is not available in bulk quantities. Therefore, the small, day-type hoppers with manual filling arrangements as shown by Figure 10-1 are used. Precautionary measures for handling sodium aluminate are similar to those for strong alkalies, such as caustic soda. Contact with skin, eyes, and clothing should be avoided. Aluminate dust or solution spray should not be breathed.

10.1.3.3 Storage

Dry sodium aluminate is stored as received, in bags, and at optimum conditions of 60 to 90°F; the recommended storage limit is 6 months. Hopper material of mild steel is completely adequate. This chemical may or may not be free flowing, depending on the manufacturer and grade used. Therefore, hopper agitation may be required. Sodium aluminate deteriorates on exposure to the atmosphere and care should be taken to avoid tearing of bags.

10.1.3.4 Feeding Equipment

Materials of construction for dissolving chambers may be mild steel or stainless steel and selection may be influenced by conformity with adjacent equipment. Equipment similar to that shown by Figure 10-1 is applicable. Standard practice for the free-flowing grade of sodium aluminate calls for dissolvers sized for 0.5 lb per gallon or 6% solution strength with a dissolver detention time of 5 minutes at the maximum feed rate.

After dissolving dry sodium aluminate in the preparation of batch solutions, agitation should be minimized or eliminated to prevent deterioration of the solution. Air agitation is not recommended, and solution tanks should be covered to prevent carbonation of the solution.

10.1.3.5 Piping and Accessories

Materials for piping and transporting dry sodium aluminate solution may be mild steel, iron, type 304 stainless steel, concrete, or plastics. The use of copper, copper alloys, and rubber should be avoided.

10.1.3.6 Pacing and Control

Pacing and control fundamentals are similar to those described for dry alum. The amount of dilution is not a consideration in the use of sodium aluminate. Therefore, the use of float valves to satisfy centrifugal pump suction, and the use of eductors, is permissible.

10.1.4 Liquid Sodium Aluminate

10.1.4.1 Properties and Availability

Liquid sodium aluminate is available from the manufacturers listed in a subsequent section. There is considerable variety in the composition of liquid sodium aluminate from the manufacturers listed. The Al_2O_3 content varies from 4.9 to 26.7%. The lower solution strengths are usually more expensive because of the cost of freighting the solution water. Because of the variety of solution strengths available, the manufacturers should be contacted for more specific information on density, viscosity, and cost.

Liquid sodium aluminate is available in 30-gallon drums (380 lb), tank truck, and tank car quantities. The approximate 1974 price of liquid sodium aluminate in 40,000 lb quantities is about $0.05 per lb ($Al_2O_3$ content, 26.4% or 19.9%) F.O. B. point of manufacture.

10.1.4.2 General Design Considerations

Because of the alkaline nature of sodium aluminate, it should not be used in contact with brass, copper, aluminum or rubber. Liquid sodium aluminate is a strong alkali and the same precautions should be exercised in handling it as in handling caustic soda.

10.1.4.3 Storage

Liquid sodium aluminate is usually stored at the shipping concentration, either in the shipping drums or in mild steel tanks. Storage tanks may be located indoors or outdoors; however, outdoor tanks should be provided with facilities for indirect heating. The maximum recommended length of storage is two to three months. Bulk shipments can be unloaded by gravity, pumping, or air pressure. However, if air is used, it should first be passed through lime-caustic soda breathers to remove carbon dioxide. Steam injection facilities are required at the unloading site.

10.1.4.4 Feeding Equipment

Feeding equipment and systems as described for liquid alum generally apply to liquid sodium aluminate except with changes of requirements regarding dilution and materials of construction as described above.

Liquid sodium aluminate may be fed at shipping strength or diluted to a stable 5 to 10% solution. Stable solutions are prepared by direct addition of low hardness water and mild agitation. Air agitation is not recommended.

10.1.4.5 Piping and Accessories

Piping requirements are the same as previously indicated for solutions of dry aluminate.

10.1.4.6 Pacing and Control

System pacing and control requirements are the same as described for liquid alum.

10.1.5 Aluminum Chloride

Aluminum chloride is not as readily available as other compounds previously discussed, and in some cases it does not enjoy a competitive price structure; nevertheless, it should be considered as a source of aluminum by designers undertaking phosphorus precipitation by metal addition.

The majority of the aluminum chloride production facilities in this country are located near petroleum refining centers and petrochemical complexes. This is because those industries consume three-fourths of the total production. Miscellaneous users of the balance of the overall production include makers of antiperspirants and a variety of other unrelated manufacturing operations. Therefore, a reliable and economical source of aluminum chloride may, in fact, be near the proposed wastewater treatment project.

Much of the aluminum chloride produced is the form of semi-pure anhydrous crystals with an off-white color. This product is derived from direct chlorination of scrap aluminum, and has a molecular weight of 133.3 and purities near 99 percent. In some instances purities may drop to 96 percent which is satisfactory for wastewater treatment if analysis shows other compounds present are acceptable.

Another solid form is produced by crystallization from hydrochloric acid in the form of a hexahydrate, in which six molecules of water attach to each molecule of aluminum chloride.

The most usual commercial form of liquid aluminum chloride is in a 32-degree Baume solution; it contains 28 percent aluminum chloride by weight. This particular liquor is not universally produced, however, so designers should be prepared to consider other concentrations with a fairly wide variety of other chemical compounds present.

Containers for shipping and storing should be similar to those used for dry and liquid alum. Handling and feeding requirements are also similar to those used with alum. Designers should check with local producers for further details.

Designers may contact manufacturers operating in at least seventeen plants in the U.S. shown in Table 10-4:

TABLE 10-4

SUPPLIERS OF ALUMINUM CHLORIDE

Allied Chemical Corp.:	El Segundo, Calif.; Hegewisch, Ill.; Elberta, N.Y.
Clinton Chemical Co.:	Houston, Texas
Dow Chemical Co.:	Freeport, Texas
E. I. du Pont de Nemours & Co., Inc.:	Grasselli, N.J.; East Chicago, Ind.
FMC Corporation:	Nitro, W. Va.
Hercules Alchlor Chemical Co., Inc.:	Ravenna, Ohio
Monsanto Co.:	Everett, Mass.; Texas City, Texas
Pearsall Chemical Corp.:	Phillipsburg, N.J.; La Porte, Texas
Reheis Co., Inc.:	Berkeley Heights, N.J.
Stauffer Chemical Co.:	Dominguez, Calif.; Baton Rouge, La.; Elkton, Md.

10.1.6 Estimated Initial Costs of Adding Liquid Alum Feed Facilities to Existing Wastewater Treatment Plants

Cost estimates of chemical storage and feeding equipment for 1, 10, and 100 mgd plants have been prepared based on the use of liquid alum (8.3% Al_2O_3) for phosphorus precipitation. A mole ratio of Al:P of 2:1 was assumed to treat an influent phosphorus concentration of 10 mg/l as P. This corresponds to an alum dose of 191 mg/l (17.4 mg/l as aluminum).

Chemical feed equipment was sized for a peak feed rate of twice that calculated from the mole ratio. Storage was provided for at least 15 days at the average feed rate. This storage time is arbitrary and will vary at each installation depending on the distance to and the reliability of the source of chemical supply. Piping and buildings to house the feeding equipment are not included in the estimates. At many locations the chemical feeding equipment can be installed in existing buildings. In all estimates, the costs include an allowance for the contractor's installation, overhead, and profit, and an allowance of 20% of the construction cost for engineering and contingencies. The estimated costs for each size plant are shown in Table 10-5. These figures correspond to an EPA Construction Cost Index of 185:

TABLE 10-5

COST OF LIQUID ALUM FEED FACILITIES

Plant Size (mgd)	Estimated 1973 Cost ($)
1	15,000
10	50,000
100	600,000

For the 1.0 mgd plant, feeding equipment for liquid alum includes two 25 gph hydraulic diaphragm pumps (one operating and one standby) with the necessary accessories and equipment to pace the feed rate with plant flow. Two 3,000-gallon FRP tanks with accessories are provided for storage of liquid alum. The total capacity of 6,000 gallons allows some flexibility in ordering and receiving 4,000-gallon tank truck shipments.

For the 10 mgd plant, liquid alum is fed by three 125 gph rotodip-type feeders (two operating and one standby) with the necessary accessories and control panel for proportioning chemical feed to flow. Alum is stored in four 11,500-gallon FRP tanks.

For the 100 mgd plant, liquid alum is fed with seven 415 gph rotodip-type feeders (six operating and one standby) with the necessary control equipment. Storage costs are based on ten 50,000-gallon underground concrete tanks for the rotodip feeders but eliminates the need for heating the tanks. Three 500-gallon FRP day tanks are included in the estimate; one for each pair of feeders. Four alum transfer pumps (three operating and one standby), with a capacity of 50 gpm at 50 ft of head are provided.

The cost of feeding and storage facilities for the 100 mgd plant is greater than 10 times the cost of facilities for the 10 mgd plant. This is because underground storage facilities were the basis of design for the 100 mgd plant in contrast to FRP tanks located in an existing building for the 10 mgd plant. The cost of a building for the 1.0 and 10.0 mgd plants would raise the cost substantially. FRP tanks could be used for the larger plant; however, structural design may necessitate special construction requirements for underground FRP tanks. FRP tanks located above ground and on concrete foundations would require special insulation and heating.

10.2 Iron Compounds

10.2.1 Liquid Ferric Chloride

10.2.1.1 Properties and Availability

Liquid ferric chloride (2) is a corrosive, dark brown oily-appearing solution having a typical unit weight between 11.2 and 12.4 lb/gal (35 to 45% $FeCl_3$). The ferric chloride content of these solutions is 3.95 to 5.58 lb/gal. Shipping concentrations vary from summer to winter due to the relatively high crystallization temperature of the more concentrated solutions as shown by Figure 10-5. The pH of a 1% solution is 2.0.

The molecular weight of ferric chloride is 162.22. Viscosities of ferric chloride solutions at various temperatures are presented in Figure 10-6.

Liquid ferric chloride is shipped in 3,000 to 4,000 gallon bulk truckload lots, in 4,000 to 10,000 gallon bulk carload lots, and in 5 and 13 gallon carboys. Liquid ferric chloride is produced at the locations listed in a subsequent section. The 1974 price of liquid ferric chloride in bulk quantities is $0.045 per lb (as ferric chloride), F.O.B. point of manufacture.

FIGURE 10-5

FREEZING POINT CURVE FOR COMMERCIAL

FERRIC CHLORIDE SOLUTIONS

(ADAPTED FROM DOW CHEMICAL CO.)

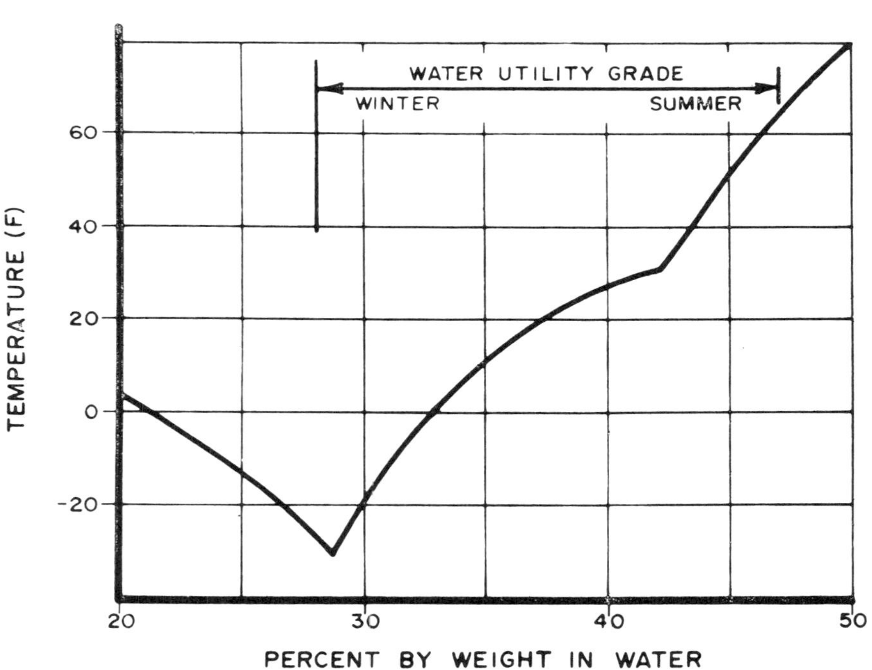

FIGURE 10-6

VISCOSITY VS COMPOSITION OF FERRIC CHLORIDE SOLUTIONS AT VARIOUS TEMPERATURES

(COURTESY OF DOW CHEMICAL CO.)

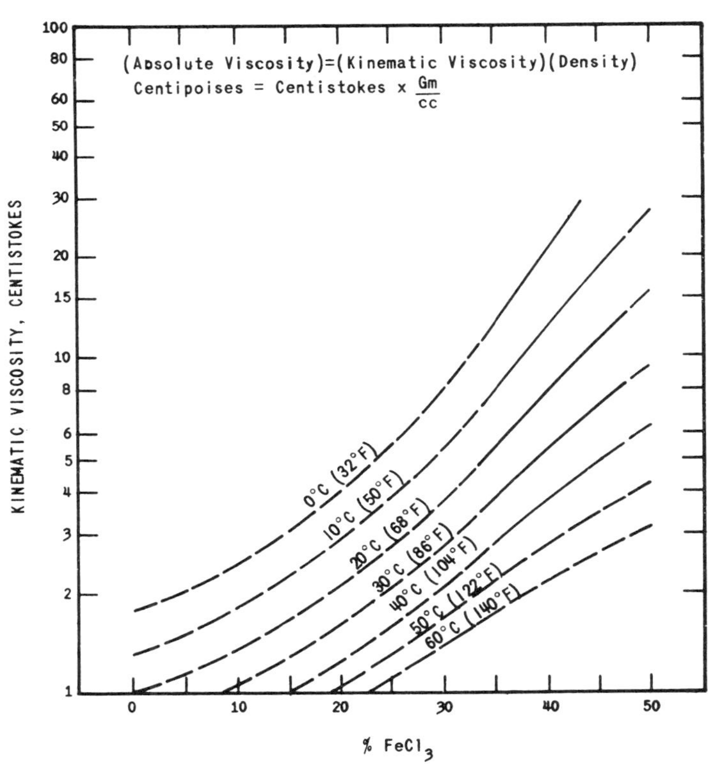

Tank trucks and cars are normally unloaded pneumatically, and operating procedures must be closely followed to avoid spills and accidents. The safety vent cap and assembly (painted red) should be removed prior to opening the unloading connection to depressurize the tank car or truck, prior to unloading.

10.2.1.2 General Design Considerations

Ferric chloride solutions are corrosive to many common materials and cause stains which are difficult to remove. Areas which are subject to staining should be protected with resistant paint or rubber mats.

Normal precautions should be employed when cleaning ferric chloride handling equipment. Workmen should wear rubber gloves, rubber apron, and goggles or a face shield. If ferric chloride comes in contact with the eyes or skin, flush with copious quantities of running water and call a physician. If ferric chloride is ingested, induce vomiting and call a physician.

10.2.1.3 Storage

Ferric chloride solution can be stored as shipped. Storage tanks should have a free vent or vacuum relief valve. Tanks may be constructed of FRP, rubber lined steel, or plastic lined steel. Resin impregnated carbon or graphite are also suitable materials for storage containers.

It may be necessary in most instances to house liquid ferric chloride tanks in heated areas or provide tank heaters or insulation to prevent crystallization. Ferric chloride can be stored for long periods of time without deterioration. The total storage capacity should be 1-1/2 times the largest anticipated shipment, and should provide at least a 10-day to 2-week supply of the chemical at the design average dose.

10.2.1.4 Feeding Equipment

Feeding equipment and systems described for liquid alum generally apply to ferric chloride except for materials of construction, and the use of glass tube rotameters.

It may not be desirable to dilute the ferric chloride solution from its shipping concentration to a weaker feed solution because of possible hydrolysis. Ferric chloride solutions may be transferred from underground storage to day tanks with impervious graphite or rubber lined self-priming centrifugal pumps having teflon rotary and stationary seals. Because of the tendency for liquid ferric chloride to stain or deposit, glass-tube rotameters should not be used for metering this solution. Rotodip feeders and diaphragm metering pumps are often used for ferric chloride, and should be constructed of materials such as rubber lined steel and plastics.

10.2.1.5 Piping and Accessories

Materials for piping and transporting ferric chloride should be rubber or Saran lined steel, hard rubber, FRP, or plastics. Valving should consist of rubber or resin lined diaphragm valves, Saran lined valves with teflon diaphragms, rubber sleeved pinch type valves, or plastic ball valves. Gasket material for large openings such as manholes in storage tanks should be soft rubber; all other gaskets should be graphite-impregnated blue asbestos, teflon, or vinyl.

10.2.1.6 Pacing and Control

System pacing and control requirements are similar to those discussed previously for liquid alum.

10.2.2 Ferrous Chloride (Waste Pickle Liquor)

10.2.2.1 Properties and Availability

Ferrous chloride, $FeCl_2$, as a liquid is available in the form of waste pickle liquor from steel processing. The liquor weighs between 9.9 and 10.4 lb/gal and contains 20 to 25% ferrous chloride or about 10% available iron. A 22% solution of ferrous chloride will crystallize at a temperature of -4°F. The molecular weight of ferrous chloride is 126.76. Free acid in waste pickle liquor can vary from 1 to 10% and usually averages about 1.5 to 1.0%. Ferrous chloride is slightly less corrosive than ferric chloride.

Waste pickle liquor is available in 4,000-gallon truckload lots and a variety of carload lots. In most instances the availability of waste pickle liquor will depend on the proximity to steel processing plants. Dow Chemical Company produces a waste pickle liquor, having a ferrous chloride content of about 22% at a 1974 price of $0.045 per lb of ferrous chloride in bulk car or truckload quantities, F.O.B. Midland, Michigan.

10.2.2.2 General Design Considerations

Since ferrous chloride or waste pickle liquor may not be available on a continuous basis, storage and feeding equipment should be suitable for handling ferric chloride. Therefore, the ferric chloride section should be referred to for storage and handling details.

10.2.3 Ferric Sulfate

10.2.3.1 Properties and Availability

Ferric sulfate, $Fe_2(SO_4)_3 \cdot X\ H_2O$, is marketed as a dry, partially hydrated product with seven water molecules. Typical properties are presented in Table 10-6 below (3):

TABLE 10-6

PROPERTIES OF FERRIC SULFATE

Molecular Weight	526
Bulk Density, lb/cu ft	56-60
Water Soluble Iron Expressed as Fe	21.5%
Water Soluble Fe^{+3}	19.5%
Insolubles Total	2.0%
Free Acid	2.5%
Moisture @ 105°C	2.0%

Ferric sulfate is shipped in car and truckload lots of 50 lb and 100 lb moistureproof paper bags and 200 lb and 400 lb fiber drums. Bulk carload shipments in box and closed hopper cars are available. The major producer is Cities Service Company, with a plant located at Copper Hill, Tennessee.

The 1973 price of bulk ferric sulfate (21.8% Fe) is about $39 per ton, F.O.B. Copper Hill, Tennessee. Bagging adds some $6 to $11 per ton over the bulk rate.

General precautions should be observed when handling ferric sulfate, such as wearing goggles and dust masks, and areas of the body that come in contact with the dust or vapor should be washed promptly.

10.2.3.2 General Design Considerations

Aeration of ferric sulfate should be held to a minimum because of the hygroscopic nature of the material, particularly in damp atmospheres. Mixing of ferric sulfate and quicklime in conveying and dust vent systems should be avoided as caking and excessive heating can result. The presence of ferric sulfate and lime in combination has been known to destroy cloth bags in pneumatic unloading devices (4). Because ferric sulfate in the presence of moisture will stain, precautions similar to those discussed for ferric chloride should be observed.

10.2.3.3 Storage

Ferric sulfate is usually stored in the dry state either in the shipping bags or in bulk in concrete or steel bins. Bulk storage bins should be as tight as possible to avoid moisture absorption, but dust collector vents are permissible and desirable. Hopper walls should slope 36% or more.

Bins may be located inside or outside and the material transferred by bucket elevator, screw or air conveyors. Ferric sulfate stored in bins usually absorbs some moisture and forms a thin protective crust which retards further absorption until the crust is broken.

10.2.3.4 Feeding Equipment

Feed solutions are usually made up at a water to chemical ratio of 2:1 to 8:1 (on a weight basis) with the usual ratio being 4:1 with a 20-minute detention time. Care must be taken not to dilute ferric sulfate solutions to less than 1% to prevent hydrolysis and deposition of ferric hydroxide. Ferric sulfate is actively corrosive in solution, and dissolving and transporting equipment should be fabricated of type 316 stainless steel, rubber, plastics, ceramics or lead.

Dry feeding requirements are similar to those for dry alum except that belt type feeders are rarely used because of their open type of construction. Closed construction, as found in the volumetric and loss-in-weight type feeders, generally exposes a minimum of operating components to the vapor, and thereby minimizes maintenance. A water jet vapor remover should be provided at the dissolver to protect both the machinery and operator.

10.2.3.5 Piping and Accessories

Piping systems for ferric sulfate should be FRP, plastics, type 316 stainless steel, rubber or glass.

10.2.3.6 Pacing and Control

System pacing and control are the same as discussed for dry alum.

10.2.4 Ferrous Sulfate

10.2.4.1 Properties and Availability

Ferrous sulfate or copperas is a byproduct of pickling steel and is produced as granules, crystals, powder, and lumps. The most common commercial form of ferrous sulfate is $FeSO_4 \cdot 7H_2O$, with a molecular weight of 278, and containing 55 to 58% $FeSO_4$ and 20 to 21% Fe. The product has a bulk density of 62 to 66 lb/ft^3. When dissolved, ferrous sulfate is acidic. The composition of ferrous sulfate may be quite variable and should be established by consulting the nearest manufacturers.

Bulk, drum (400-lb) and bag (50 and 100 lb) shipments are available from producers at locations listed in a subsequent section. The 1974 bulk price of ferrous sulfate in bulk carload and truckload quantities is about \$28 per ton (21% Fe), with an additional \$8 per ton for bagging.

Ferrous sulfate is also available in a wet state in bulk form from some plants. This form is likely to be difficult to handle and the manufacturer should be consulted for specific information and instructions.

Dry ferrous sulfate cakes at storage temperatures above 68°F, is efflorescent in dry air, and oxidizes and hydrates further in moist air.

General precautions similar to those for ferric sulfate with respect to dust and handling acidic solutions, should be observed when working with ferrous sulfate. Mixing quicklime and ferrous sulfate produces high temperatures and the possibility of fire.

10.2.4.2 General Design Considerations

The granular form of ferrous sulfate has the best feeding characteristics and gravimetric or volumetric feeding equipment may be used.

The optimum chemical to water ratio for continuous dissolving is 0.5 lb/gal or 6% with a detention time of 5 minutes in the dissolvers. Mechanical agitation should be provided in the dissolver to assure complete solution. Lead, rubber, iron, plastics, and type 304 stainless steel can be used as construction materials for handling solutions of ferrous sulfate.

Storage, feeding and transporting systems probably should be suitable for handling ferric sulfate as an alternative to ferrous sulfate.

10.2.5 Estimated Initial Costs of Adding Liquid Ferric Chloride Feed Systems to Existing Wastewater Treatment Plants

Costs of chemical storage and feeding equipment for 1, 10, and 100 mgd plants have been estimated based on the use of liquid ferric chloride (basis: 35% ferric chloride) for phosphorus precipitation. A mole ratio of Fe:P of 2.0:1 was assumed to treat an influent phosphorus concentration of 10 mg/l as P. This corresponds to a ferric chloride dose of 104 mg/l (36 mg/l as Fe^{3+}).

Chemical feed equipment was sized for a peak feed rate of twice that calculated from the mole ratio. Storage was provided for at least 15 days at the average feed rate. This storage time is arbitrary and will vary at each installation depending on the distance to and the reliability of the souce of chemical supply. Piping and buildings to house the feeding equipment are not included. At many locations the chemical feeding equipment can be installed in existing buildings. All estimated costs include an allowance for the contractor's installation, overhead, and profit plus an allowance of 20% of the construction cost for engineering and contingencies. The estimated costs for each size plant are shown in Table 10-7. These figures correspond to an EPA Construction Cost Index of 185:

TABLE 10-7

COSTS OF LIQUID FERRIC CHLORIDE FEED FACILITIES

Plant Size (mgd)	Estimated 1973 Cost ($)
1	15,000
10	50,000
100	500,000

For the 1 mgd plant, the feeding equipment for liquid ferric chloride includes two 18 gph, hydraulic diaphragm pumps (one operating and one standby) with the necessary accessories and equipment to pace the feed rate to the plant flow. Liquid ferric chloride is stored in two 3,000-gallon FRP tanks with accessories. The 6,000-gallon total capacity allows use of liquid ferric chloride in 4,000-gallon tank truck quantities.

For the 10 mgd plant, liquid ferric chloride is fed by three 90 gph rotodip-type feeders (two operating and one standby) with the necessary accessories and control panel for proportioning chemical feed to flow.

For the 100 mgd plant, seven 305 gph rotodip-type feeders (six operating and one standby) are provided along with the necessary control equipment. Storage is provided in eight 50,000-gallon rubber-lined concrete tanks located underground. The underground storage necessitates transfer pumps and day tanks for the rotodip feeders, but eliminates the need for heating the tanks. Three 500-gallon FRP day tanks are included along with four ferric chloride transfer pumps sized for 35 gpm each at 50 ft of head.

With regard to the cost of the feed system for the 100 mgd plant being out of proportion to that for a 10 mgd plant, the reader is referred to Section 10.1.5 of this chapter for the discussion of underground concrete storage vessels as opposed to FRP tanks.

10.3 Lime

10.3.1 Quicklime

10.3.1.1 Properties and Availability

Quicklime, CaO, has a density range of approximately 55 to 75 lb/ft^3, and a molecular weight of 56.08. A slurry for feeding, called milk of lime, can be prepared with up to 45% solids. Lime is only slightly soluble, and both lime dust and slurries are caustic in nature. A saturated solution of lime has a pH of about 12.4.

Lime can be purchased in bulk in both car and truckload lots. It is also shipped in 80 and 100 lb multiwall "moistureproof" paper bags. Lime is produced at the locations listed in a subsequent section (5).

1974 prices for bulk pebble quicklime range from $18 per ton to $22 per ton with the higher prices generally in the far west, and higher than average in the north. Bagging adds approximately $5 per ton to the cost.

The CaO content of commercially available quicklime can vary quite widely over an approximately range of 70 to 96%. Content below 88% is generally considered below standard in the municipal use field (6). Purchase contracts are often based on 90% CaO content with provisions for payment of a bonus for each 1% over and a penalty for each 1% under the standard. A CaO content less than 75% probably should be rejected because of excessive grit and difficulties in slaking.

Workmen should wear protective clothing and goggles to protect the skin and eyes, as lime dust and hot slurry can cause severe burns. Areas contacted by lime should be washed immediately. Lime should not be mixed with chemicals which have water of hydration. The lime will be slaked by the water of hydration causing excessive temperature rise and possibly explosive conditions. Conveyors and bins used for more than one chemical should be thoroughly cleaned before switching chemicals.

10.3.1.2 General Design Considerations

Pebble quicklime, all passing a 3/4-in screen and not more than 5% passing a No. 100 screen, is normally specified because of easier handling and less dust. Hopper agitation is generally not required with the pebble form. Published slaker capacity ratings require "soft or normally burned" limes which provide fast slaking and temperature rise, but poorer grades of limes may also be satisfactorily slaked by selection of the appropriate slaker retention time and capacity.

10.3.1.3 Storage

Storage of bagged lime should be in a dry place, and preferably elevated on pallets to avoid absorption of moisture. System capacities often make the use of bagged quicklime impractical. Maximum storage period is about 60 days.

Bulk lime is stored in air-tight concrete or steel bins having a 60° slope on the bin outlet. Bulk lime can be conveyed by conventional bucket elevators and screw, belt, apron, drag-chain, and bulk conveyors of mild steel construction. Pneumatic conveyors subject the lime to air-slaking and particle sizes may be reduced by attrition. Dust collectors should be provided on manually and pneumatically filled bins.

10.3.1.4 Feeding Equipment

A typical lime storage and feed system is illustrated in Figure 10-7. Quicklime feeders are usually limited to the belt or loss-in-weight gravimetric types because of the wide variation of the bulk density. Feed equipment should have an adjustable feed range of at least 20:1 to match the operating range of the associated slaker. The feeders should have an over-under feed rate alarm to immediately warn of operation beyond set limits of control. The feeder drive should be instrumented to be interrupted in the event of excessive temperature in the slaker compartment.

Lime slakers for wastewater treatment should be of the continuous type, and the major components should include one or more slaking compartments, a dilution compartment, a grit separation compartment and a continuous grit remover. Commercial designs vary in regard to the combination of water to lime, slaking temperature, and slaking time, in obtaining the "milk of lime" suspensions.

The "paste" type slaker admits water as required to maintain a desired mixing viscosity. This viscosity therefore sets the operating retention time of the slaker. The paste slaker usually operates with a low water to lime ratio (approximately 2:1 by weight), elevated temperature, and five minute slaking time at maximum capacity.

The "detention" type slaker admits water to maintain a desired ratio with the lime, and therefore the lime feed rate sets the retention time of the slaker. The detention slaker operates with a wide range of water to lime ratios (2.5:1 and 6:1), moderate temperature, and a 10-minute slaking time at maximum capacity. A water to lime ratio of from 3.5:1 to 4:1 is most often used. The operating temperature in lime slakers is a function of the water to lime ratio, lime quality, heat transfer, and water temperature. Lime slaking

FIGURE 10-7

TYPICAL LIME FEED SYSTEM

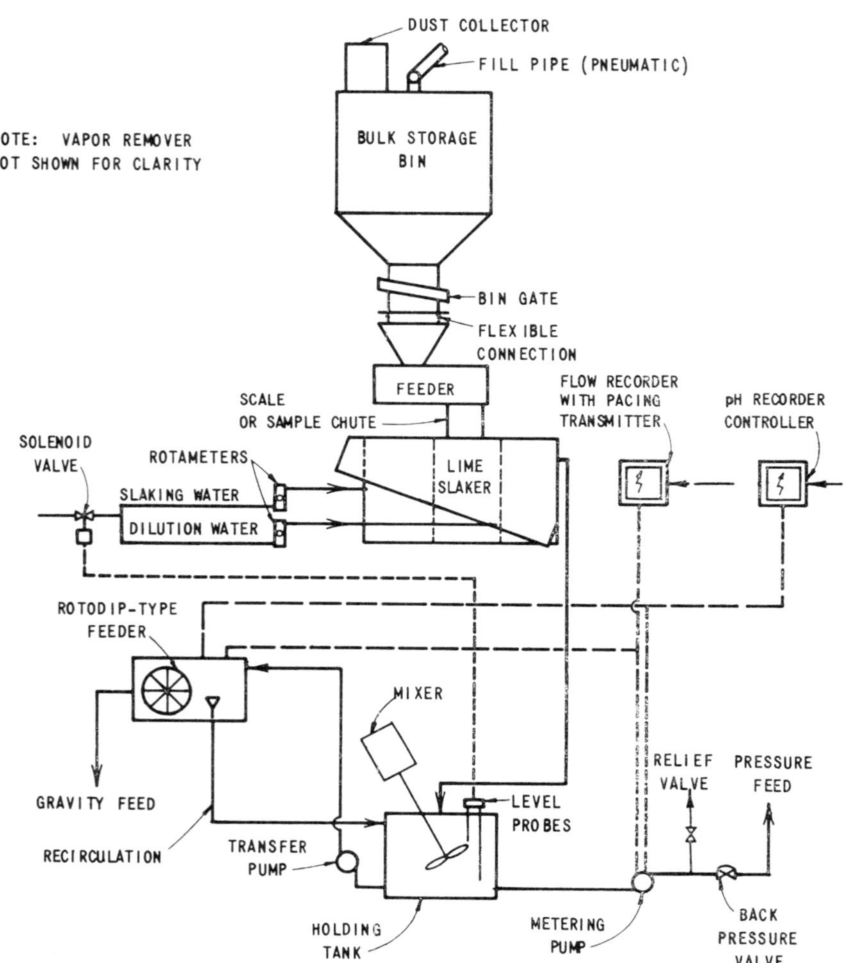

evolves heat in hydrating the CaO to $Ca(OH)_2$ and, therefore, vapor removers are required for feeder protection.

10.3.1.5 Piping and Accessories

Lime slurry should be transported by gravity in open channels wherever possible. Piping channels and accessories may be rubber, iron, steel, concrete, and plastics. Glass tubing, such as that in rotameters, will cloud rapidly and therefore should not be used. Any abrupt directional changes in piping should include plugged tees or crosses to allow rodding-out of deposits. Long sweep elbows should be provided to allow the piping to be cleaned by the use of a cleaning "pig". Daily cleaning is desirable.

Milk of lime transfer pumps should be of the open impeller centrifugal type. Pumps having an iron body and impeller with bronze trim are suitable for this purpose. Rubber lined pumps with rubber covered impellers are also frequently used. Make-up tanks are usually provided ahead of centrifugal pumps to ensure a flooded suction at all times. "Plating-out" of lime is minimized by the use of soft water in the make-up tank. Turbine pumps and eductors should be avoided in transferring milk of lime because of scaling problems.

10.3.1.6 Pacing and Control

Lime slaker water proportioning is integrally controlled or paced from the feeder. Therefore, the feeder-slaker system will follow pacing controls applied to the feeder only. As discussed previously, gravimetric feeders are adaptable to receive most standard instrumentation pacing signals. Systems can be instrumented to allow remote pacing with telemetering of temperature and feed rate to a central panel for control purposes.

The lime feeding system may be controlled by an instrumentation system integrating both plant flow and pH of the wastewater after lime addition. However, it should be recognized that pH probes require daily maintenance in this application to monitor the pH accurately. Deposits tend to build up on the probe and necessitate frequent maintenance. The low pH lime treatment systems (pH 9.5 to 10.0) can be more readily adapted to this method of control than high lime treatment systems (pH 11.0 or greater) because less maintenance of the pH equipment is required. In a closed loop pH-flow control system, milk of lime is prepared on a batch basis and transferred to a holding tank with variable output feeders set by the flow and pH meters to proportion the feed rate. Figure 10-7 illustrates such a control system.

10.3.2 Hydrated Lime

10.3.2.1 Properties and Availability

Hydrated lime, $Ca(OH)_2$, has a working density of 20 to 50 lb/ft^3, tends to flood the feeder, and will arch in storage bins if packed. The molecular weight is 74.08. The dust and slurry of hydrated lime are caustic in nature. The 1973 cost of bulk hydrated lime varies from $20 to $22 per ton. Bagged lime is available but increases the cost about $5 per ton. The availability of hydrated lime may be determined by contracting manufacturers listed in a subsequent section. The pH of a saturated hydrated lime solution is the same as that given for quicklime.

10.3.2.2 General Design Considerations

Hydrated lime is slaked lime and needs only enough water added to form milk of lime. Wetting or dissolving chambers are usually designed to provide 5 minutes detention with a ratio of 0.5 lb per gallon of water or 6% slurry at the maximum feed rate. Hydrated lime is usually used where maximum feed rates do not exceed 250 lb/hour. Hydrated lime and milk of lime will irritate the eyes, nose, and respiratory system and will dry the skin. Affected areas should be washed with water.

10.3.2.3 Storage

Information given for quicklime also applies to hydrated lime except that bin agitation must be provided. Bulk bin outlets should be provided with non-flooding rotary feeders and with hopper slopes of 65%.

10.3.2.4 Feeding Equipment

Volumetric or gravimetric feeders may be used, but volumetric feeders are usually selected only for installations where comparatively low feed rates are required. Dilution does not appear to be important; therefore, control of the amount of water used in the feeding operation is not considered necessary. Inexpensive hydraulic jet agitation may be furnished in the wetting chamber of the feeder as an alternative to mechanical agitation. The jets should be sized for the available water supply pressure to obtain proper mixing.

10.3.2.5 Piping and Accessories

Piping and accessories as described for quicklime are also appropriate for hydrated lime.

10.3.2.6 Pacing and Control

Controls as listed for dry alum apply to hydrated lime. Hydraulic jets should operate continuously and only shut off when the feeder is taken out of service. Control of the feed rate with pH as well as pacing with the plant flow may be used with hydrated lime as well as quicklime.

10.3.3 Estimated Initial Costs of Adding Lime Feed Facilities to Existing Wastewater Treatment Plants

Cost estimates of chemical storage and feeding equipment have been prepared based on the use of hydrated lime for a 1 mgd plant, and pebble quicklime for 10 and 100 mgd plants. Lime feed rates are based on a dosage of 150 mg/l and allow for peak rates of twice this capacity. Storage was provided for at least 15 days at the average rate although it may be desirable to provide 30 days for dry chemicals. This storage time is arbitrary and will vary at each installation depending on the distance to and the reliability of the source of chemical supply. Piping and buildings to house the feeding equipment are not included in the estimates. The estimated costs of steel bins with dust collector vents and filling accessories are included for all three plants. The chemical feeding equipment can be installed in existing buildings at many locations. In all estimates, the costs include an allowance for the contractor's installation, overhead, and profit, and an allowance of 20% of the construction cost of engineering and contingencies. The estimated costs for each size plant are shown in Table 10-8. These figures correspond to an EPA Construction Cost Index of 185.

TABLE 10-8

COST OF LIME FEED FACILITIES

Plant Size (mgd)	Estimated Base Bid ($)	Additional Cost of Flow and pH Controls ($)	Total Cost ($)
1	19,000	5,000	24,000
10	95,000	15,000	110,000
100	365,000	100,000	475,000

Equipment for the 1.0 mgd plant includes two volumetric feeder systems with manually loaded bins, dissolving chambers and accessories. Equipment for the 10.0 mgd plant includes two gravimetric feeder-slaker systems. Steel bins are estimated to hold 1-1/2 truckloads of quicklime each, and it is assumed that the truck will be equipped with a pneumatic blower. Equipment for the 100.00 mgd plant includes four gravimetric feeder-slaker systems. The base estimate includes the bin gate and feeder-slaker accessories and controls.

Costs for the flow and pH controls include holding tanks with mixers, and slurry feeders with controls. Flow and pH metering instruments are not included.

10.4 Compounds for pH Adjustment

10.4.1 Soda Ash

10.4.1.1 Properties and Availability

Soda ash, Na_2CO_3, is available in two forms. Light soda ash has a bulk density range of 35 to 50 lb/ft^3 and a working density of 41 lb/ft^3. Dense soda ash has a density range of 60 to 76 lb/ft^3 and a working density of 63 lb/ft^3. The pH of a 1% solution of soda ash is 11.2.

The molecular weight of soda ash is 106. Commercial purity ranges from 98 to 99%. The viscosities of sodium carbonate solutions are given in Figure 10-8. Soda ash by itself is not particularly corrosive, but in the presence of lime and water, caustic soda is formed which is quite corrosive.

Soda ash is available in bulk by truck, box car and hopper car, and in 100 lb bags from locations listed in a subsequent section. The 1974 price for soda ash ranges from $50 to $60 per ton, F.O.B. the point of manufacture; however, prices vary substantially between manufacturers and should be obtained from the closest manufacturers or local distributors.

10.4.1.2 General Design Considerations

Dense soda ash is generally used in municipal applications because of superior handling characteristics. It has little dust, good flow characteristics, and will not arch in the bin or flood the feeder. It is relatively hard to dissolve and ample dissolver capacity must be provided. Normal practice calls for 0.5 lb of dense soda ash per gallon of water or a 6% solution retained for 20 minutes in the dissolver.

The dust and solution are irritating to the eyes, nose, lungs and skin and therefore general precautions should be observed and the affected areas should be washed promptly with water.

10.4.1.3 Storage

Soda ash is usually stored in steel bins and where pneumatic filling equipment is used, bins should be provided with dust collectors. Bulk and bagged soda ash tend to absorb atmospheric carbon dioxide and water forming the less active sodium bicarbonate ($NaHCO_3$). Material recommended for unloading facilities is steel.

FIGURE 10-8

VISCOSITY OF SODA ASH SOLUTIONS

(COURTESY PPG INDUSTRIES INC., CHEMICAL DIV.)

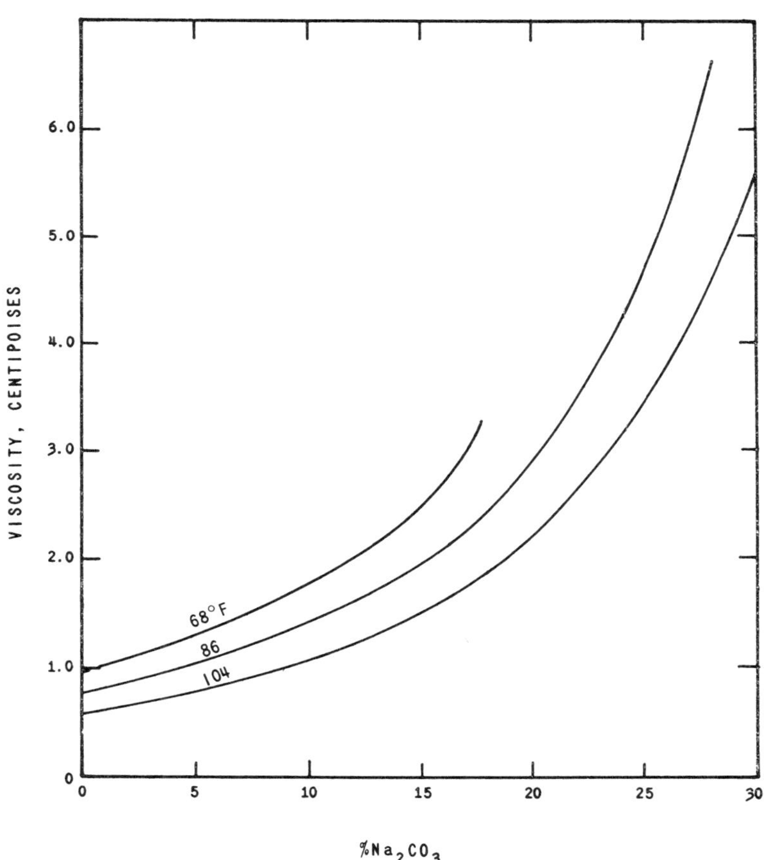

10.4.1.4 Feeding Equipment

Feed equipment as described for dry alum is suitable for soda ash. Dissolving of soda ash may be hastened by the use of warm dissolving water. Mechanical or hydraulic jet mixing should be provided in the dissolver.

10.4.1.5 Piping and Accessories

Materials of construction for piping and accessories should be iron, steel, rubber, and plastics.

10.4.1.6 Pacing and Control

Controls as discussed for dry alum apply also to soda ash equipment.

10.4.2 Liquid Caustic Soda

Anhydrous caustic soda (NaOH) is available but its use is generally not considered practical in water and wastewater treatment applications. Consequently, only liquid caustic soda is discussed below.

10.4.2.1 Properties and Availability

Liquid caustic soda is shipped at two concentrations, 50% and 73% NaOH. The densities of the solutions as shipped are 12.76 lb/gal for the 50% solution and 14.18 lb/ga for the 73% solution. These solutions contain 6.38 lb/gal NaOH and 10.34 lb/gal NaOH, respectively. The crystallization temperature is 53°F for the 50% solution and 165°F for the 73% solution. The molecular weight of NaOH is 40. Viscosities of various caustic soda solutions are presented in Figure 10-9. The pH of a 1% solution of caustic soda is 12.9.

Truckload lots of 1,000 to 4,000 gallons are available in the 50% concentration only. Both shipping concentrations can be obtained in 8,000, 10,000, and 16,000 gallon carload lots. Tank cars can be unloaded through the dome eduction pipe using air pressure or through the bottom valve by gravity or by using air pressure or a pump. Trucks are usually unloaded by gravity or with air pressure or a truck mounted pump.

Major producers of caustic soda and their respective plant locations are listed in a subsequent section. The current price for liquid caustic soda ranges from about $75 to $80 per ton (NaOH), F.O.B. the point of manufacture.

FIGURE 10-9

VISCOSITY OF CAUSTIC SODA SOLUTIONS

(COURTESY OF HOOKER CHEMICAL CO.)

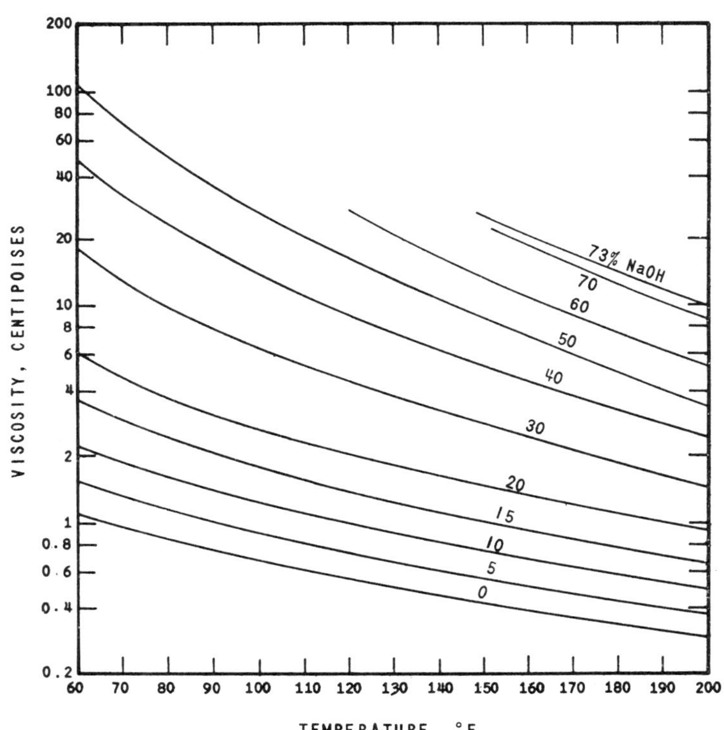

10.4.2.2 General Design Considerations

Liquid caustic soda is received in bulk shipments, transferred to storage and diluted as necessary for feeding to the points of application. Caustic soda is poisonous and is dangerous to handle. U.S. Department of Transportation Regulations for "White Label" materials must be observed. However, if handled properly caustic soda poses no particular industrial hazard. To avoid accidental spills, all pumps, valves, and lines should be checked regularly for leaks. Workmen should be thoroughly instructed in the precautions related to the handling of caustic soda. The eyes should be protected by goggles at all times when exposure to mist or splashing is possible. Other parts of the body should be protected as necessary to prevent alkali burns. Areas exposed to caustic soda should be washed with copious amounts of water for 15 minutes to 2 hours. A physician should be called when exposure is severe. Caustic soda taken internally should be diluted with water or milk and then neutralized with dilute vinegar or fruit juice. Vomiting may occur spontaneously but should not be induced except on the advice of a physician.

10.4.2.3 Storage

Liquid caustic soda may be stored at the 50% concentration. However, at this solution strength, it crystallizes at $53^{o}F$. Therefore, storage tanks must be located indoors or provided with heating and suitable insulation if outdoors. Because of its relatively high crystallization temperature, liquid caustic soda is often diluted to a concentration of about 20% NaOH for storage. A 20% solution of NaOH has a crystallization temperature of about $-20^{o}F$. Recommendations for dilution of both 73% and 50% solutions should be obtained from the manufacturer, because special considerations are necessary.

Storage tanks for liquid caustic soda should be provided with an air vent for gravity flow. The storage capacity should be equal to 1-1/2 times the largest expected delivery, with an allowance for dilution water, if used, or 2 weeks supply at the anticipated feed rate, whichever is greater. Tanks for storing 50% solution at a temperature between $75^{o}F$ and $140^{o}F$ may be constructed of mild steel. Storage temperatures above $140^{o}F$ require more elaborate materials selection and are not recommended. Caustic soda will tend to pick up iron when stored in steel vessels for extended periods. Subject to temperature and solution strength limitations, rubber, 316 stainless steel, nickel, nickel alloys, or plastics may be used when iron contamination must be avoided.

10.4.2.4 Feeding Equipment

Further dilution of liquid caustic soda below the storage strength may be desirable for feeding by volumetric feeders. Feeding systems as described for liquid alum generally apply to caustic soda with appropriate selection of materials of construction. A typical system schematic is shown in Figure 10-10. Feeders will usually include materials such as ductile iron, stainless steels, rubber, and plastics.

FIGURE 10-10

TYPICAL CAUSTIC SODA FEED SYSTEM

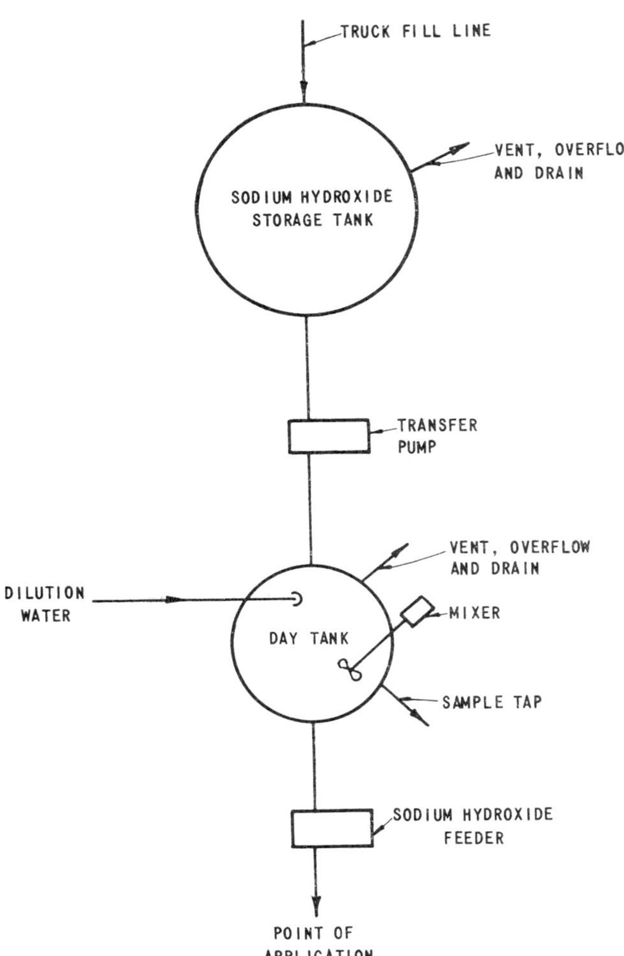

10.4.2.5 Piping and Accessories

Transfer lines from the shipping unit to the storage tank should be spiral-wire-bound neoprene or rubber hose, solid steel pipe with swivel joints, or steel hose. Because caustic soda attacks glass, use of glass materials should be avoided. Other miscellaneous materials for use with liquid caustic soda feeding and handling equipment are listed below (7) in Table 10-9:

TABLE 10-9

MATERIALS SUITABLE FOR CAUSTIC SODA SERVICE

Components	Recommended Materials for Use With 50% NaOH Up to 140°F
Rigid Pipe	Standard Weight Black Iron
Flexible Connections	Rigid Pipe with Ells or Swing Joints, Stainless Steel or Rubber Hose
Diluting Tees	Type 304 Stainless Steel
Fittings	Steel
Permanent Joints	Welded or Screwed Fittings
Unions	Screwed Steel
Valves - Non-leaking (Plug)	
Body	Steel
Plug	Type 304 Stainless Steel
Pumps (Centrifugal)	
Body	Steel
Impeller	Ni-Resist
Packing	Blue Asbestos
Storage Tanks	Steel

10.4.2.6 Pacing and Control

Controls as listed for liquid alum also apply to liquid caustic soda equipment.

10.5 Carbon Dioxide

10.5.1 Properties and Availability

Carbon dioxide, CO_2, is available for use in wastewater treatment plants in gas and liquid form. The molecular weight of carbon dioxide is 44. Dry carbon dioxide is not chemically active at normal temperatures and is a non-toxic safe chemical; however, the gas displaces oxygen and adequate ventilation of closed areas should be provided. Solutions of carbon dioxide in water are very reactive chemically and form carbonic acid. Saturated solutions of carbon dioxide have a pH of 4.0 at 68°F.

The gas form may be produced on the treatment plant site by scrubbing and compressing the combustion product of lime recalcining furnaces, sludge furnaces, or generators used principally for the production of carbon dioxide gas only. These generators are usually fired with combustible gases, fuel oil, or coke and have carbon dioxide yields as shown in Table 10-10 (8).

TABLE 10-10

CARBON DIOXIDE YIELDS OF COMMON FUELS

Fuel	Quantity	Yield lb
Natural Gas	1,000 ft^3	115
Coke	1 lb	3
Kerosene	1 gal	20
Fuel Oil (No. 2)	1 gal	23
Propane	1,000 ft^3	352
Butane	1,000 ft^3	454

The gas forms, as generated at the plant site, usually have a carbon dioxide content of between 6% and 18% depending on the source and efficiency of the producing system.

The liquid form is available from commercial suppliers in 20 to 50 lb cylinders, 10 to 20 ton trucks and 30 to 50 ton rail cars. The commercial liquid form has a minimum carbon dioxide content of 99.5%.

1974 prices range from \$33-\$36 per ton for 3,000 tons per year and over, to \$71-\$74 per ton for a quantity of 150 tons per year. These prices include an allowance for freight within a 100 mile radius of the point of manufacture. Another \$6 or \$7 per ton may be added for each additional 100 miles to the point of destination. Major producers of commercial carbon dioxide are listed in a subsequent section.

10.5.2 General Design Considerations

Recovery of carbon dioxide from recalcining furnaces or incinerators is the least expensive source, but maintenance of stack gas systems is likely to be extensive because of the corrosive nature of the wet gas and the presence of particulate matter. Scrubber systems are required to clean the stack gas and especially designed gas compressors are necessary to provide the process injection pressure.

Pressure generators and submerged burners require less maintenance because the system pressure is established by compressors or blowers handling dry air or gas. On-site generating units have a limited range of carbon dioxide production as compared with the liquid storage and feed system, and therefore may require multiple units.

The liquid carbon dioxide storage and feed system generally includes a temperature-pressure controlled, bulk storage tank, an evaporation unit, and a gas feeder to meter the gas. Solution feeders, similar in construction to chlorinators, may also be used to feed carbon dioxide.

10.5.3 Storage

This section applies only to use of commercial liquid carbon dioxide. Liquid system capacities encountered in wastewater treatment usually usually require on-site bulk storage units. Standard pre-packaged units are available, ranging in size from 3/8 to 50 tons capacity, and are furnished with temperature-pressure controls to maintain approximately 300 lb/in^2 at 0^oF conditions. The typical package unit contains refrigeration, vaporization, safety and control equipment. The units are well insulated and protected for outdoor location. The gas from the evaporation unit usually passes through two stages of pressure reduction before entering the gas feeder to prevent the formation of dry ice.

10.5.4 Feeding Equipment

Feeding systems for the stack gas source of carbon dioxide consist of simple valving arrangements, for admitting varying quantities of make-up air to the suction side of the constant volume compressors, or for venting excess gas on the compressor discharge. A typical system is described by Culp and Culp (9).

Pressure generators and submerged burners are regulated by valving arrangements on the fuel and air supply. Generation of carbon dioxide by combustion is usually difficult to control, requires frequent operator attention and demands considerable maintenance over the life of the equipment, when compared to liquid CO_2 systems.

Commercial liquid carbon dioxide is becoming more generally used because of its high purity, the simplicity and range of feeding equipment, ease of control, and smaller, less expensive piping systems. After vaporization, carbon dioxide with suitable metering and pressure reduction may be fed directly to the point of application as a gas. Metering of direct fed pressurized gas is difficult due to high adiabatic expansion characteristics of the gas. Also, direct feed requires extremely fine bubbles to insure the gas goes into solution; this, in turn, can lead to scaling problems. However, vacuum operated, solution type gas feeders are preferred. Such feeders generally include safety devices and operating controls in a compact panel housing, with materials of construction suitable for carbon dioxide service. Absorption of carbon dioxide in the injector water supply approaches 100% when a ratio of 1.0 lb of gas to 60 gal of water is maintained.

10.5.5 Piping and Accessories

Mild steel piping and accessories are suitable for use with cool, dry, carbon dioxide. Hot, moist gases, however, require the use of type 316 stainless steel or plastic materials. Plastics or FRP pipe are generally used for solution piping and diffusers. Diffusers should be submerged at least 8 ft, and preferably deeper, to assure complete absorption of the gas.

10.5.6 Pacing and Control

Standard instrument signals and control components can be used to pace or control carbon dioxide feed systems.

Using stack gas as the source of carbon dioxide, the feed rate can be controlled by proper selection and operation of compressors, by manual control of vent or bleed valves, or by automatic control of such valves by a pH meter-controller system.

In commercial carbon dioxide feed systems, solution feeders may function as controllers and can be paced by instrument signals from pH monitors and plant flow meters.

In feeding commercial carbon dioxide directly to the point of application as a gas, a differential pressure transmitter and a control valve may function as the primary elements of a control system. Standard instrument signals may be used to pace or control the rate of carbon dioxide feed.

Carbon dioxide generators are difficult to pace or control other than by manual or automatic operation of vent or bleed valves that waste a portion of the produced gas according to the plant requirements.

10.6 Polymers

10.6.1 Dry Polymers

10.6.1.1 Properties and Availability

Types of polymers vary widely in characteristics. Manufacturers should be consulted for properties, availability, and cost of the polymer being considered. References are available that indicate the types and characteristics of polymers available (10, 11). Bulk shipments are generally not desirable. Polymers are available in a variety of container or package sizes.

10.6.1.2 General Design Considerations

Dry polymer and water must be blended and mixed to obtain a recommended solution for efficient action. Solution concentrations vary from fractions of a percent up. Preparation of the stock solution involves wetting of the dry material and usually an aging period prior to application. Solutions can be very viscous, and close attention should be paid to piping size and length and pump selections. Metered solution is usually diluted just prior to injection to the process to obtain better dispersion at the point of application.

10.6.1.3 Storage

General practice for storage of bagged dry chemicals should be observed. The bags should be stored in a dry, cool, low humidity area and used in proper rotation, i.e., first in, first out.

Solutions are generally stored in type 316 stainless steel, FRP, or plastic lined tanks.

10.6.1.4 Feeding Equipment

Two types of systems are frequently combined to feed polymers. The solution preparation system includes a manual or automatic blending system with the polymer dispensed by hand or by a dry feeder to a wetting jet and then to a mixing-aging tank at a controlled ratio. The aged polymer is transported to a holding tank where metering pumps or rotodip feeders dispense the polymer to the process. A schematic of such a system is shown by Figure 10-11. It is generally advisable to keep the holding or storage time of polymer solutions to a minimum, 1 to 3 days or less, to prevent deterioration of the product.

10.6.1.5 Piping and Accessories

Selection must be made after determination of the polymer; however, type 316 stainless steel or plastics are generally used.

FIGURE 10-11

MANUAL DRY POLYMER FEED SYSTEM

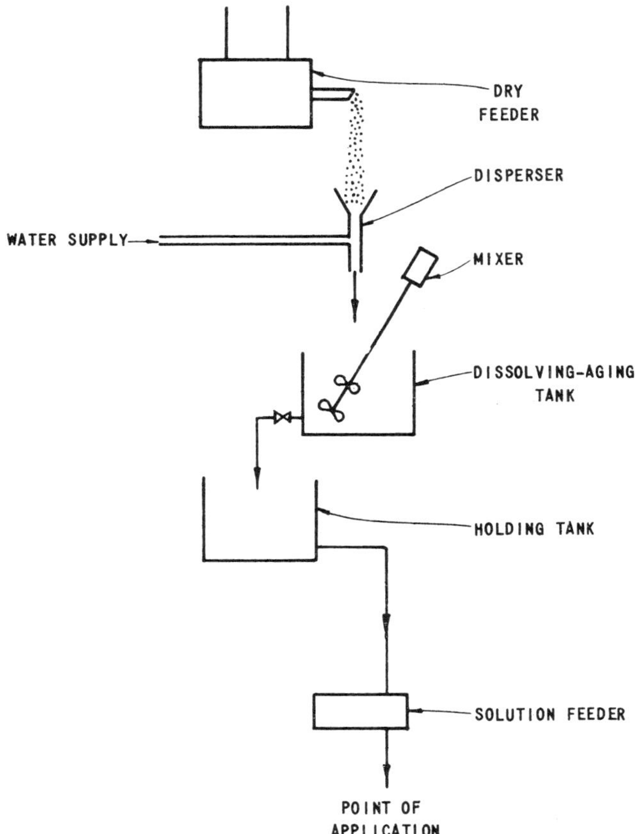

10.6.1.6 Pacing and Control

Controls as listed for liquid alum apply to the control of liquid dispersing feeders.

The solution preparation system may be an automatic batching system, as shown by the schematic on Figure 10-12, that fills the holding tank with aged polymer as required by level probes. Such a system is usually provided only at large plants. Unitized solution preparation units are available, but have a limited capacity.

10.6.2 Liquid Polymers

10.6.2.1 Properties and Availability

As with dry polymers, there is a wide variety of products, and manufacturers should be consulted for specific information.

10.6.2.2 General Design Considerations

Liquid systems differ from the dry systems only in the equipment to blend the polymer with water to prepare the solution. Liquid solution preparation is usually a hand-batching operation with manual filling of a mixing-aging tank with water and polymer.

10.6.2.3 Feeding Equipment

Liquid polymers need no aging and simple dilution is the only requirement for feeding. The dosage of liquid polymers may be accurately controlled by metering pumps or rotodip feeders.

The balance of the process is generally the same as described for dry polymers.

10.6.3 Estimated Initial Costs of Adding Polymer Feed Facilities to Existing Wastewater Treatment Plants

Cost estimates of solution preparation and feeder equipment have been prepared, based on the use of dry polymer, for 1, 10, and 100 mgd plants. Chemical feed equipment was sized to have a capacity sufficient to prepare and feed a 0.25% stock solution at a dosage of 1 mg/l. Piping and buildings to house the feeding equipment and store the bags are not included. At many locations, the chemical feeding equipment can be installed in existing buildings. All estimated costs include an allowance for the contractor's installation, overhead, and profit plus an allowance of 20% of the construction cost for engineering and contingencies. The estimated costs for each size plant are shown in Table 10-11. These figures correspond to an EPA Construction Cost Index of 185:

FIGURE 10-12

AUTOMATIC DRY POLYMER FEED SYSTEM

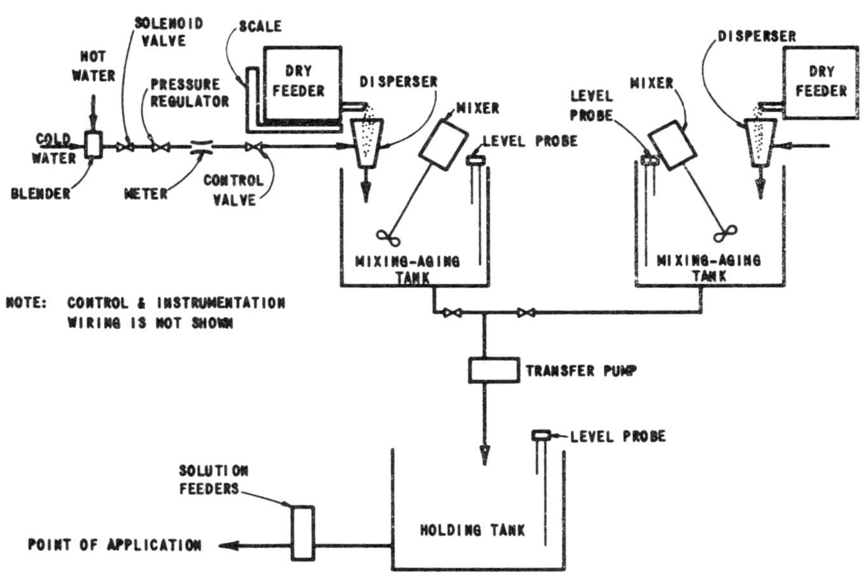

TABLE 10-11

COST OF POLYMER FEED FACILITIES

Plant Size (mgd)	Estimated Cost ($)
1	7,000
10	31,000
100	225,000

The system for the 1 mgd plant was based on the polymer being manually fed to the mixing tank, with mixing and solution feed equipment arranged for manual control. Two independent systems of tanks and feeders are included. The system for the 10 mgd plant includes two volumetric dry feeders discharging to two mixing tanks, arranged for batch control. The mixing tanks discharge to a single holding tank from which two solution feeders take their supply for application to the process. The system for the 100 mgd plant includes four volumetric dry feeders complete with steel day hoppers, dust collectors, bin gates and flexible connectors. The system operation is automatic for the preparation and transfer of aged polymer solution from four mixing tanks to two holding tanks. Ten solution feeders meter the polymer to the treatment process.

10.7 Chemical Suppliers

10.7.1 Metal Salts and Lime

The first and largest list of suppliers concerns metal salts and lime. It is divided geographically into states, which are tabulated alphabetically. Then, under each state, production locations are listed for the following alphabetized chemicals:

alum (dry and/or liquid)

ferric chloride

ferric sulfate

ferrous sulfate

lime (high calcium and/or dolomitic)

sodium aluminate (dry and/or liquid)

Table 10-12 summarizes extent and location of presently available sources of metal salts and lime:

TABLE 10-12

GEOGRAPHICAL SUMMARY OF SUPPLIERS OF METAL SALTS AND LIME

Location	Chemical
25 States	Alum
Michigan	Ferric chloride
Tennessee	Ferric sulfate
Ga., Md., Mo., N.J.	Ferrous sulfate
32 states	Lime
Ark., Ill., La., Ga., Mo.	Sodium aluminate

Table 10-13 offers a detailed listing of metal salt and lime suppliers. Persons interested in contacting chemical suppliers may direct inquiries to either the point of production or to a nearby sales office for the chemical company involved. This list does not include potential or established suppliers of waste pickle liquor. Nor are pH adjusting chemicals included here; these are normally available through most major chemical supply firms.

TABLE 10-13

SUPPLIERS OF METAL SALTS AND LIME

Location	Manufacturer	Chemical
ALABAMA		
Coosa Pines	American Cyanamid	Alum, Liquid
Demopolis	American Cyanamid	Alum, Liquid
Mobile	American Cyanamid	Alum, Liquid and Dry
Naheola	Stauffer	Alum, Liquid
Allgood	Cheney Lime & Cement Co.	Lime, High Calcium
Keystone	Martin Marietta Cement, Southern Division	Lime, High Calcium
Landmark	Cheney Lime & Cement Co.	Lime, High Calcium
Montevallo	U.S. Gypsum Co.	Lime, High Calcium
Roberta	Southern Cement Co., Div. Martin Marietta Corp.	Lime, High Calcium
Saginaw	Longview Lime Co., Div. Woodward Co., Div. Mead Corp.	Lime, High Calcium
Siluria	Alabaster Lime Co.	Lime, High Calcium
ARIZONA		
Douglas	Paul Lime Plant, Inc.	Lime, High Calcium
Globe	Hoopes & Co.	Lime, High Calcium
Nelson	U.S. Lime Div., The Flintkote Co.	Lime, High Calcium
ARKANSAS		
Pine Bluff	Allied	Alum, Liquid
Batesville	Batesville White Lime Co., Div. Rangaire Corp.	Lime, High Calcium
Bauxite	Reynolds Chemicals	Sodium Aluminate, Dry

SUPPLIERS OF METAL SALTS AND LIME

Location	Manufacturer	Chemical
CALIFORNIA		
Bay Point	Allied	Alum, Liquid and Dry
Antioch	Imperial West Chemicals	Ferric Chloride
El Segundo	Allied	Alum, Liquid
Richmond	Stauffer	Alum, Liquid
Vernon	Stauffer	Alum, Liquid
City of Industry	U.S. Lime Div., The Flintkote Co.	Lime, High Calcium
Diamond Springs	Diamond Springs Lime Co.	Lime, High Calcium
Lucerne Valley	Pfizer, Inc., Minerals, Pigments & Metals Div.	Lime, High Calcium
Richmond	U.S. Lime Div., The Flintkote Co.	Lime, High Calcium
Salinas	Kaiser Aluminum & Chemical Corp.	Lime, Dolomitic
Westend	Stauffer Chemical Co.	Lime, High Calcium
COLORDAO		
Denver	Allied	Alum, Liquid and Dry
Ft. Morgan	Great Western Sugar Co.	Lime, High Calcium
CONNECTICUT		
Canaan	Pfizer, Inc., Minerals Pigments & Metals Div.	Lime, Dolomitic
FLORIDA		
Fernandina Beach	Tennessee Corp.	Alum, Liquid
Jacksonville	Allied	Alum, Liquid
Port St. Joe	Allied	Alum, Liquid

SUPPLIERS OF METAL SALTS AND LIME

Location	Manufacturer	Chemical
FLORIDA (cont.)		
Brooksville	Chemical Lime Co.	Lime, High Calcium
Sumterville	Dixie Lime and Stone Co.	Lime, High Calcium
GEORGIA		
Atlanta (2)	Burris; Allied	Alum, Liquid and Dry
Augusta	Tennessee Corp.	Alum, Liquid and Dry
Cedar Springs	Tennessee Corp.	Alum, Liquid
Macon	Allied	Alum, Liquid
Savannah	Allied	Alum, Liquid
Savannah	American Cyanamid Co.	Ferrous Sulfate
Vinings	Vinings Chemical Co.	Sodium Aluminate, Liquid
ILLINOIS		
E. St. Louis	Allied	Alum, Liquid and Dry
Joliet	American Cyanamid	Alum, Liquid and Dry
Marblehead	Marblehead Lime Co.	Lime, High Calcium
McCook	Vulcan Materials Co.	Lime, Dolomitic
Quincy	Marblehead Lime Co.	Lime, High Calcium
So. Chicago	Marblehead Lime Co.	Lime, High Calcium
Thornton	Marblehead Lime Co.	Lime, Dolomitic
Chicago	Nalco Chemical Company	Sodium Aluminum, Liquid and Dry
INDIANA		
Buffington	Marblehead Lime Co.	Lime, High Calcium

Searsport	Northern	Alum, Liquid and Dry
MARYLAND		
Baltimore	Olin	Alum, Dry
Baltimore (3)	Byproducts Processing Co., Inc.;	Ferrous Sulfate
	Glidden Co.;	Ferrous Sulfate
	Cosmin Corp.;	Ferrous Sulfate
Woodsboro	S.W. Barrick & Sons, Inc.	Lime, High Calcium

SUPPLIERS OF METAL SALTS AND LIME

Location	Manufacturer	Chemical
MASSACHUSETTS		
Adams	Holland	Alum, Liquid
Salem	Hamblet & Hayes	Alum, Liquid and Dry
Adams	Pfizer, Inc., Minerals Pigments & Metals Div.	Lime, High Calcium
Lee	Lee Lime Corp.	Lime, Dolomitic
MICHIGAN		
Detroit	Allied	Alum, Liquid
Escanaba	American Cyanamid	Alum, Liquid
Kalamazoo (2)	Allied; American Cyanamid	Alum, Liquid
Midland	Dow Chemical Co.	Ferric Chloride
Wyandotte	Pennwalt Corp.	Ferric Chloride
Detroit	Detroit Lime Co.	Lime, High Calcium
Ludington	Dow Chemical Co.	Lime, High Calcium
River Rouge	Marblehead Lime Co.	Lime, High Calcium
Cloquet	American Cyanamid	Alum, Liquid
Pine Bend	North Star	Alum, Liquid and Dry
Duluth	Cutler Magner Co.	Lime, High Calcium
MISSISSIPPI		
Monticello	American Cyanamid	Alum, Liquid
Vicksburg	Allied	Alum, Liquid

SUPPLIERS OF METAL SALTS AND LIME

Location	Manufacturer	Chemical
MISSOURI		
St. Louis	National Lead	Ferrous Sulfate
Bonne Terre	Valley Mineral Products Co.	Lime, Dolomitic
Hannibal	Marblehead Lime Co.	Lime, High Calcium
Ste. Genevieve	Mississippi Lime Co.	Lime, High Calcium
Springfield	Ash Grove Cement Co.	Lime, High Calcium
Kansas City	Conservation Chemical Co.	Sodium Aluminate, Liquid
NEVADA		
Apex	U.S. Lime Div., The Flintkote Co.	Lime, High Calcium
Henderson	U.S. Lime Div., The Flintkote Co.	Lime, Dolomitic and High Calcium
McGill	Morrison-Weatherly Corp.	Lime, High Calcium
Sloan	U.S. Lime Div., The Flintkote Co.	Lime, Dolomitic and High Calcium
NEW JERSEY		
Newark	Essex	Alum, Liquid and Dry
Warners	American Cyanamid	Alum, Liquid and Dry
Sayreville	National Lead	Ferrous Sulfate
Newton	Limestone Products Corp. of America	Lime, High Calcium, Hydrate Only

SUPPLIERS OF METAL SALTS AND LIME

Location	Manufacturer	Chemical
NORTH CAROLINA		
Acme	Wright	Alum, Liquid
Plymouth	American Cyanamid	Alum, Liquid
OHIO		
Chillicothe	Allied	Alum, Liquid
Cleveland	Allied	Alum, Liquid and Dry
Hamilton	American Cyanamid	Alum, Liquid and Dry
Middletown	Allied	Alum, Liquid
Ashtabula	Union Carbide Olefins Co.	Lime, High Calcium
Carey	National Lime & Stone Co.	Lime, Dolomitic
Cleveland	Cuyahoga Lime Co.	Lime, High Calcium
Delaware	MCQ Co.	Lime, High Calcium
Genoa	U.S. Gypsum Co.	Lime, Dolomitic
Gibsonburg (2)	Pfizer, Inc., Minerals, Pigments & Metal Div.; National Gypsum Co.	Lime, Dolomitic
Huron	Huron Lime Co.	Lime, High Calcium
Marble Cliff	MCQ Co.	Lime, High Calcium
Millersville	J.E. Baker Co.	Lime, Dolomitic
Woodville (2)	Ohio Lime Co.; Standard Lime & Refractories Div., Martin Marietta Corp.	Lime, Dolomitic
OKLAHOMA		
Marble City	St. Clair Lime Co.	Lime, High Calcium
Sallisaw	St. Clair Lime Co.	Lime, High Calcium

SUPPLIERS OF METAL SALTS AND LIME

Location	Manufacturer	Chemical
OREGON		
North Portland	Stauffer	Alum, Liquid and Dry
Baker	Chemical Lime Co. of Oregon	Lime, High Calcium
Portland	Ash Grove Cement Co.	Lime, High Calcium
PENNSYLVANIA		
Johnsonburg	Allied	Alum, Liquid
Marcus Hook	Allied	Alum, Liquid and Dry
Newell	Allied	Alum, Liquid
Annville	Bethlehem Mines Corp.	Lime, High Calcium
Bellefonte (2)	National Gypsum Co.; Warner Co.	Lime, High Calcium
Branchton	Mercer Lime & Stone Co.	Lime, High Calcium
Devault	Warner Co.	Lime, Dolomitic
Pleasant Gap	Chemical Lime Inc.	Lime, High Calcium
Plymouth Meeting	G.&W.H. Corson, Inc.	Lime, Dolomitic
SOUTH CAROLINA		
Catawba	Burris	Alum, Liquid
Georgetown	American Cyanamid	Alum, Liquid and Dry
SOUTH DAKOTA		
Rapid City	Pete Lien & Sons, Inc.	Lime, High Calcium

SUPPLIERS OF METAL SALTS AND LIME

Location	Manufacturer	Chemical
TENNESSEE		
Chattanooga	American Cyanamid	Alum, Liquid and Dry
Counce	Stauffer	Alum, Liquid
Springfield	Burris	Alum, Liquid
Copper Hill	Cities Service Co.	Ferric Sulfate
Knoxville (2)	Foote Mineral Co.; Williams Lime Mfg. Co.	Lime, High Calcium
TEXAS		
Houston (2)	Stauffer; Ethyl	Alum, Liquid and Dry
Texas City	Gulf Chemical & Metallurgical	Ferrous Sulfate, Ferric Sulfate
Blum	Round Rock Lime Companies	Lime, High Calcium
Cleburne	Texas Lime Co., Div. Rangaire Corp.	Lime, High Calcium
Clifton	Chemical Lime Inc.	Lime, High Calcium
Houston	U.S. Gypsum Co.	Lime, High Calcium
McNeil	Austin White Lime Co.	Lime, High Calcium
New Braunfels	U.S. Gypsum Co.	Lime, High Calcium
Round Rock	Round Rock Lime Companies	Lime, High Calcium
San Antonio	McDonough Bros., Inc.	Lime, High Calcium
UTAH		
Grantsville	U.S. Lime Div., The Flintkote Co.	Lime, Dolomitic and High Calcium
Lehi	Rollins Mining Supplies Co.	Lime, High Calcium

SUPPLIERS OF METAL SALTS AND LIME

Location	Manufacturer	Chemical
VERMONT		
Winooski	Vermont Assoc. Lime Industries, Inc.	Lime, High Calcium
VIRGINIA		
Covington	Allied	Alum, Liquid
Hopewell	Allied	Alum, Liquid
Norfolk	Howerton Gowen	Alum, Liquid
Clearbrook	W.S. Frey Co., Inc.	Lime, High Calcium
Kimballton (2)	Foote Mineral Co.; National Gypsum Co.	Lime, High Calcium
Stephens City	M.J. Grove Lime Co., Div. The Flintkote Co.	Lime, High Calcium
Strasburg	Chemstone Corp.	Lime, High Calcium
WASHINGTON		
Kennewick	Allied	Alum, Liquid
Tacoma (2)	Allied; Stauffer	Alum, Liquid
Vancouver	Allied	Alum, Liquid and Dry
Tacoma	Domtar Chemicals Inc.	Lime, High Calcium
WEST VIRGINIA		
Riverton	Germany Valley Limestone Div., Greer Steel Co.	Lime, High Calcium

SUPPLIERS OF METAL SALTS AND LIME

Location	Manufacturer	Chemical
WISCONSIN		
Manasha	Allied	Alum, Liquid
Wisconsin Rapids	Allied	Alum, Liquid
Eden	Western Lime & Cement Co.	Lime, Dolomitic
Green Bay	Western Lime & Cement Co.	Lime, High Calcium
Knowles	Western Lime & Cement Co.	Lime, Dolomitic
Manitowoc	Rockwell Lime Co.	Lime, Dolomitic
Superior	Cutler-LaLiberte-McDougall Corp.	Lime, High Calcium

10.7.2 Polymer Suppliers

Table 10-14 lists names and addresses of firms offering polymer which may prove effective in treatment approaches described herein.

TABLE 10-14

POLYMER SOURCES AND TRADE NAMES

Source	Trade Name(s)
Allied Colloids, Inc. One Robinson Lane Ridgewood, NJ 07450	Percol
Allstate Chemical Co. Post Office Box 3040 Euclid, OH 44117	Allstate

POLYMER SOURCES AND TRADE NAMES

Sources	Trade Name(s)
Allyn Chemical Co. 2224 Fairhill Road Cleveland, OH 44106	Claron
American Cyanimid Co. Berdan Avenue Wayne, NJ 07470	Superfloc Magnifloc
Atlad Chemical Ind., Inc. Wilmington, DE 19899	Sarbo Atlasep
Berdell Industries 28-01 Thomson Avenue Long Island City, NY 11101	Berdell
Betz Laboratories, Inc. Somerton Road Trevose, PA 19047	Betz Polyfloc
Bond Chemicals, Inc. 1500 Brookpark Road Cleveland, OH 44109	Bondfloc
Brennan Chemical Co. 704 N. First Street St. Louis, MO 631J2	Brenco
The Burtonite Company Nutley, NJ 07110	Burtonite
Calgon Corporation Post Office Box 1346 Pittsburgh, PA 15222	Cat-Floc

POLYMER SOURCES AND TRADE NAMES

Source	Trade Name(s)
Commercial Chemical 11 Patterson Avenue Midland Park, NJ 07432	Speedifloc
Dearborn Chemical Div. W. R. Grace & Co. Merchandise Mart Plaza Chicago, IL 60654	Aquafloc
Dow Chemical USA Barstow Building 2020 Dow Center Midland, MI 48640	Dowell PEI Purifloc Separan XD
Drew Chemical Corp. 701 Jefferson Road Parsippany, NJ 07054	Drewfloc Amerfloc
DuBois Chemicals Div. W. R. Grace & Co. 3630 E. Kemper Road Sharonville, OH 45241	Flocculite
E. I. DuPont de Nemours & Co. Eastern Laboratory Gibbstown, NJ 08027	DuPont
Environmental Pollution Investigation & Control, Inc. 9221 Bond Street Overland Park, KS 66214	Dynafloc

POLYMER SOURCES AND TRADE NAMES

Source	Trade Name(s)
Fabcon International 1275 Columbus Avenue San Francisco, CA 94133	Zulcar Fabcon
Henry W. Fink & Co. 6900 Silverton Avenue Cincinnati, OH 45236	Kleer-Floc
Gamlen Sybron Corp. 321 Victory Avenue S. San Francisco, CA 94080	Gamafloc Gamlose Gamlen
Garrett-Callahan 111 Rollins Road Millbrae, CA	Garrett-Callahan
General Mills Chemicals 4620 N. 77th Street Minneapolis, MN 55435	Supercol Guartec
Hercules, Inc. 910 Market Street Wilmington, DE 19899	Hercofloc
Frank Herzl Corp. 299 Madison Avenue New York, NY 10017	Perfectamyl
ICI America, Inc. Wilmington, DE 19899	Atlasep
Illinois Water Treatment Co. 840 Cedar Street Rockford, IL 61102	Illco

POLYMER SOURCES AND TRADE NAMES

Source	Trade Name(s)
Kelco Company 8225 Aero Drive San Diego, CA 92123	Kelgin Kelcosol
Key Chemicals 4346 Tacony Philadelphia, PA 19124	Key-Floc
Metalene Chemical Co. Bedford, OH 44014	Metalene
The Mogul Corporation 20600 Chagrin Boulevard Cleveland, OH 44122	Mogul
Nalco Chemical Co. 6216 W. 66th Pl. Chicago, IL 60638	Nalcolyte
Narvon Mining & Chemical Co. Keller Avenue & Fruitville Pike Lancaster, PA 17604	Sink-Floc Zeta-Floc
National Starch & Chemical Corp. 1700 W. Front Street Plainfield, NJ 07063	Floc-Aid Natron
O'Brein Industries, Inc. 95 Dorsa Avenue Livingston, NJ 07039	O'B Floc
Oxford Chemical Div. Consolidated Foods Corp. Post Office Box 80202 Atlanta, GA 30341	Oxford-Hydro-Floc

POLYMER SOURCES AND TRADE NAMES

Source	Trade Name(s)
Reichhold Chemicals, Inc. RCI Building White Plains, NY 10602	Aquarid
Standard Brands Chem. Ind., Inc. Post Office Drawer K Dover, DE 19901	Tychem
A. E. Staley Mfg. Co. Post Office Box 151 Decatur, IL 62525	Hamaco
Stein, Hall & Co., Inc. 605 Third Avenue New York, NY 10016	Hallmark Jaguar Polyhall
Swift & Company Oakbrook, IL 60521	Swift
James Varley & Sons, Inc. 1200 Switzen Avenue St. Louis, MO 63147	Varco-Floc
W. E. Zimmie, Inc. 810 Sharon Drive Westlake, OH 44145	Zimmite

10.7.3 Other Chemicals

A large number of manufacturers generate sodium hydroxide; it is produced concurrently with chlorine, and sometimes elemental sodium. The compound is produced at nearly 100 plant sites in this country, and bulk storage is available in most major cities. Designers are referred to chemical suppliers in their geographical area for this commodity.

Carbon dioxide is generated at some 75 locations around the country, and in addition the liquefied gas is stored in bulk in nearly all major cities. Some twenty manufacturers are involved in these operations, and designers are referred to local chemical suppliers for further information on carbon dioxide.

About ten manufacturers produce soda ash in the United States. The 15 plant sites involved are widely scattered geographically, as follows in Table 10-15:

TABLE 10-15

SODA ASH MANUFACTURERS

Ammonia-Soda Plants

Allied Chemical Corp., Syracuse, N.Y.; Detroit, Mich.; Baton Rouge, La.

Diamond Alkali Co., Painesville, Ohio

Olin Mathieson Chemical Corp., Saltville, Va.; Lake Charles, La.

Pittsburgh Plate Glass Co., Barberton, Ohio; Corpus Christi, Texas

Wyandotte Chemicals, Corp., Wyandotte, Mich.

Natural Soda Plants

American Potash and Chemical Corp., Trona, Calif.

FMC Corp., Green River, Wyoming

Pittsburgh Plate Glass Co., Bartlett, Calif.

Stauffer Chemical Co., Westend, Calif.; Green River, Wyoming

Carbonation of Caustic Liquors

Dow Chemical Co., Freeport, Texas

The preceding list is from a reference written to supply such information on these and many other related compounds:

Faith, W. L.; Keyes, D. B.; & Clark, R. L., Industrial Chemicals, 3rd Ed., John Wiley & Son, New York (1965).

10.8 Chemical Feeders

Chemical feed systems must be flexibly designed to provide for a high degree of reliability in light of the many contingencies which may affect their operation. Thorough waste characterization in terms of flow extremes and chemical requirements should precede the design of the chemical feed system. The design of the chemical feed system must take into account the form of each chemical desired for feeding, the particular physical and chemical characteristics of the chemical, maximum waste flows and the reliability of the feeding devices.

In suspended and colloidal solids removal from wastewaters the chemicals employed are generally in liquid or solid form. Those in solid form are generally converted to solution or slurry form prior to introduction to the wastewater stream; however, some chemicals are fed in a dry form. In any case, some type of solids feeder is usually required. This type of feeder has numerous different forms due to wide ranges in chemical characteristics, feed rates and degree of accuracy required. Liquid feeding is somewhat more restrictive, depending mainly on liquid volume and viscosity.

The capacity of a chemical feed system is an important consideration in both storage and feeding. Storage capacity design must take into account the advantage of quantity purchase versus the disadvantage of construction cost and chemical deterioration with time. Potential delivery delays and chemical use rates are necessary factors in the total picture. Storage tanks or bins for solid chemicals must be designed with proper consideration of the angle of repose of the chemical and its necessary environmental requirements, such as temperature and humidity. Size and slope of feeding lines are important along with their materials of construction with respect to the corrosiveness of the chemicals.

Chemical feeders must accommodate the minimum and maximum feeding rates required. Baker (12) indicates that manually controlled feeders have a common range of 20:1, but this range can be increased to about 100:1 with dual control systems. Chemical feeder control can be manual, automatically proportioned to flow, dependent on some form of process feedback, or a combination of any two of these. More sophisticated control systems are feasible if proper sensors are available. If manual control systems are specified with the possibility of future automation, the feeders selected should be amenable to this conversion with a minimum of expense. An example would be a feeder with an external motor which could easily be replaced with a variable speed motor or drive when automation is installed (12). Standby or backup units should be included for each type of feeder used. Reliability calculations will be necessary in larger plants with a greater multiplicity of these units. Points of chemical addition and piping to them should be capable of handling all possible changes in dosing patterns in order to have proper flexibility of operation. Designed flexibility in hoppers, tanks, chemical feeders and solution lines is the key to maximum benefits at least cost (12).

Liquid feeders are generally in the form of metering pumps or orifices. Usually these metering pumps are of the positive-displacement variety, plunger or diaphragm type. The choice of liquid feeder is highly dependent on the viscosity, corrosivity, solubility, suction and discharge heads, and internal pressure-relief requirements (13). Some examples are shown in Figure 10-13 and Figure 10-14. In some cases control valves and rotameters may be all that is required. In other cases, such as lime slurry feeding, centrifugal pumps with open impellers are used with appropriate controls (14). More complete descriptions of liquid feeder requirements can be found in the literature and elsewhere (13).

Solids characteristics vary to a great degree and the choice of feeder must be considered carefully, particularly in the smaller-sized facility where a single feeder may be used for more than one chemical. Generally, provisions should be made to keep all chemicals cool and dry. Dryness is very important, as hygroscopic (water absorbing) chemicals may become lumpy, viscous or even rock hard; other chemicals with less affinity for water may become sticky from moisture on the particulate surfaces, causing increased arching in hoppers. In either case, moisture will affect the density of the chemical and may result in under-feed. Dust removal equipment should be used at shoveling locations, bucket elevators, hoppers and feeders for neatness, corrosion prevention and safety reasons. Collected chemical dust may often be used.

The simplest method for feeding solid chemicals is by hand. Chemicals may be preweighed or simply shoveled or poured by the bagful into a dissolving tank. This method is of economic necessity limited to very small operations, or to chemicals used in very weak solutions.

Because of the many factors, such as moisture content, different grades and compressibility, which can affect chemical density (weight to volume ratio), volumetric feeding of solids is normally restricted to smaller plants, specific types of chemicals which are reliably constant in composition and low rates of feed. Within these restrictions, several volumetric types are available. Accuracy of feed is usually limited to plus or minus 2% by weight but may be as high as plus or minus 5%.

FIGURE 10-13

PLUNGER TYPE METERING PUMP

(COURTESY OF WALLACE & TIERNAN)

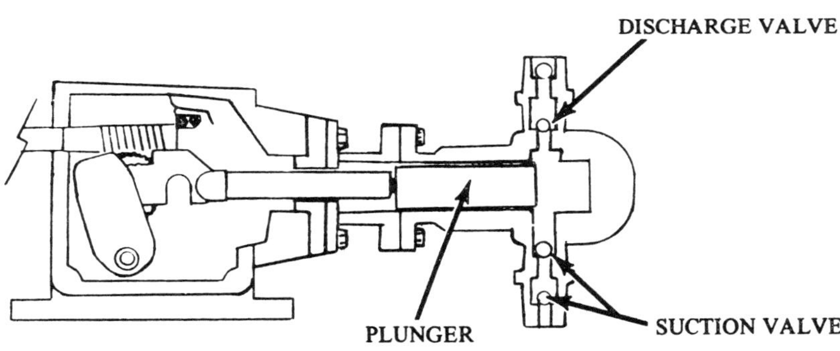

FIGURE 10-14

DIAPHRAGM TYPE METERING PUMP

(COURTESY OF WALLACE & TIERNAN)

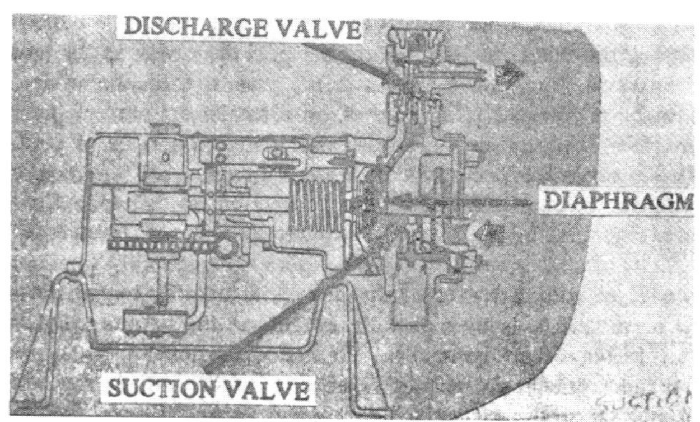

One type of volumetric dry feeder uses a continuous belt of specific width moving from under the hopper to the dissolving tank. A mechanical gate mechanism regulates the depth of material on the belt, and the rate of feed is governed by the speed of the belt and/or the height of the gate opening. The hopper normally is equipped with a vibratory mechanism to reduce arching. This type of feeder is not suited for easily fluidized materials.

Another type employs a screw or helix from the bottom of the hopper through a tube opening slightly larger than the diameter of the screw or helix. Rate of feed is governed by the speed of screw or helix rotation. Some screw-type designs are self-cleaning, while others are subject to clogging. Figure 10-15 shows a typical screw-feeder.

Most remaining types of volumetric feeders generally fall into the positive-displacement category. All designs incorporate some form of moving cavity of a specific or variable size. In operation, the chemical falls by gravity into the cavity and is more or less fully enclosed and separated from the hopper's feed. The size of the cavity, and the rate at which the cavity moves and is discharged, governs the amount of material fed. The positive control of the chemical may place a low limit on rates of feed. One unique design is the progressive cavity metering pump, a non-reciprocating type. Positive-displacement feeders often utilize air injection to enhance flowability of the material. Some examples of positive-displacement units are illustrated in Figure 10-16.

The basic drawback of volumetric feeder design, i.e., its inability to compensate for changes in the density of materials, is overcome by modifying the volumetric design to include a gravimetric or loss-in-weight controller. This modification allows for weighing of the material as it is fed. The beam balance types measure the actual mass of material. This is considerably more accurate, particularly over a long period of time, than the less common spring-loaded gravimetric designs. Gravimetric feeders are used where feed accuracy of about plus or minus 1% is required for economy, as in large-scale operations and for materials which are used in small, precise quantities. It should be noted, however, that even gravimetric feeders cannot compensate for weight added to the chemical by excess moisture. Many volumetric feeders may be converted to loss-in-weight function by placing the entire feeder on a platform scale which is tared to neutralize the weight of the feeder.

Good housekeeping and need for accurate feed rates dictate that the gravimetric feeder be shut down and thoroughly cleaned on a regular basis. Although many of these feeders have automatic or semi-automatic devices which compensate to some degree for accumulated solids on the weighing mechanism, accuracy is affected, particularly on humid days when hygroscopic materials are fed. In some cases, built-up chemicals can actually jam the equipment.

FIGURE 10-15

SCREW FEEDER

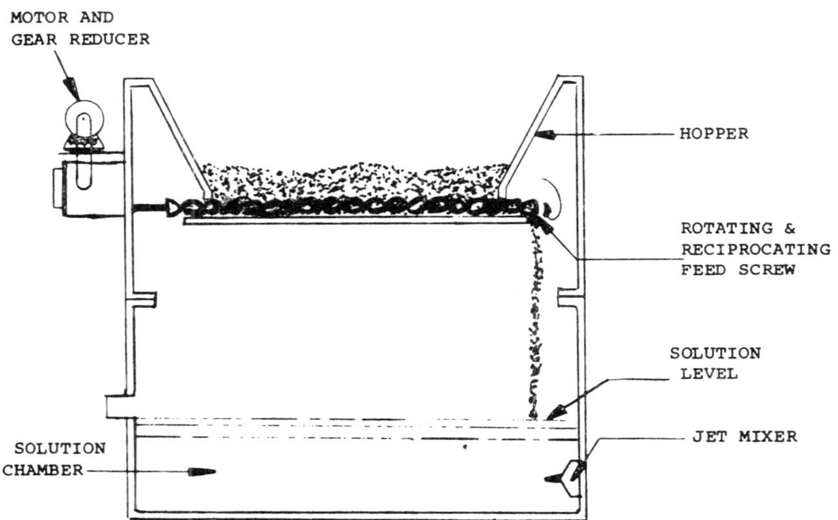

FIGURE 10-16

POSITIVE DISPLACEMENT SOLID FEEDER-ROTARY

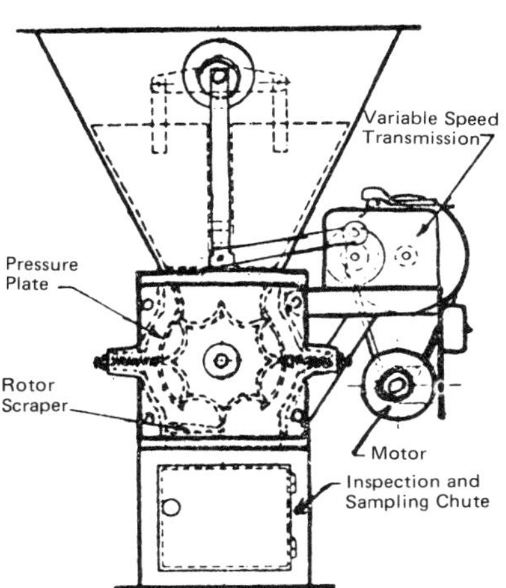

No discussion of feeders is complete without at least passing reference to dissolvers, as any metered material must be mixed with water to provide a chemical solution of desired strength. Most feeders, regardless of type, discharge their material to a small dissolving tank which is equipped with a nozzle system and/or mechanical agitator depending on the solubility of the chemical being fed. Solid materials, such as polyelectrolytes, may be carefully spread into a vortex spray or washdown jet of water immediately before entering the dissolver. It is essential that the surface of each particle become thoroughly wetted before entering the feed tank to ensure accurate dispersal and to avoid clumping, settling or floating.

A dissolver for a dry chemical feeder is unlike a chemical feeding mechanism, which by simple adjustment and change of speed can vary its output tenfold. The dissolver must be designed for the job to be done. A dissolver suitable for a rate of 10 lb/hr may not be suitable for dissolving at a rate of 100 lb/hr. As a general rule, dissolvers may b oversized, but dissolvers for commercial ferric sulfate or lime slakers do not perform well if greatly oversized.

It is essential that specifications for dry chemical feeders include specifications on dissolver capacity. A number of factors need to be considered in designing dissolvers of proper capacity. These include detention times and water requirements, as well as other factors specific to individual chemicals.

The capacity of a dissolver is based on detention time, which is directly related to the wettability of rate of solution of the chemical. Therefore, the dissolver must be large enough to provide the necessary detention for both the chemical and the water at the maximum rate of feed. At lower rates of feed, the strength of solution or suspension leaving the dissolver will be less, but the detention time will be approximately the same unless the water supply to the dissolver is reduced. When the water supply to any dissolver is controlled for the purpose of forming a constant strength solution, mixing within the dissolver must be accomplished by mechanical means, because sufficient power will not be available from the mixing jets at low rates of flow. Hot water dissolvers are also available in order to minimize the required tankage.

The foregoing descriptions give some indication of the wide variety of materials which may be handled. Because of this variety, a modern facility may contain any number of a variety of feeders with combined or multiple materials capability. Ancillary equipment to the feeder also varied according to the material to be handled. Liquid feeders encompass a limited number of design principles which account for density and viscosity ranges. Solids feeders, relatively speaking, vary considerably due to the wide range of physical and chemical characteristics, feed rates and the degree of precision and repeatability required.

TABLE 10-16

TYPES OF CHEMICAL FEEDERS

Type of feeder	Use	Limitations	
		Capacity (cubic feet per hr)	Range
Dry feeder:			
Volumetric:			
Oscillating plate	Any material, granules or powder.	0.01 to 35	40 to 1
Oscillating throat (universal)	Any material, any particle size.	0.002 to 100	40 to 1
Rotating disc	Most materials including NaF, granules or powder.	0.01 to 1.0	20 to 1
Rotating cylinder (star)	Any material, granules or powder.	8 to 2,000 or 7.2 to 300	10 to 1 or 100 to 1
Screw	Dry, free flowing material, powder or granular.	0.05 to 18	20 to 1
Ribbon	Dry, free flowing material, powder, granular, or lumps.	0.002 to 0.16	10 to 1
Belt	Dry, free flowing material up to 1½-inch size, powder or granular.	0.1 to 3,000	10 to 1 to 100 to 1
Gravimetric:			
Continuous–belt and scale	Dry, free flowing, granular material, or floodable material.	0.02 to 2	100 to 1
Loss in weight	Most materials, powder granular or lumps.	0.02 to 80	100 to 1
Solution feeder:			
Nonpositive displacement:			
Decanter (lowering pipe)	Most solutions or light slurries.	0.01 to 10	100 to 1
Orifice	Most solutions	0.16 to 5	10 to 1
Rotameter (calibrated valve)	Clear solutions	0.005 to 0.16 or 0.01 to 20	10 to 1
Loss in weight (tank with control valve).	Most solutions	0.002 to 0.20	30 to 1
Positive displacement:			
Rotating dipper	Most solutions or slurries	0.1 to 30	100 to 1
Preportioning pump:			
Diaphragm	Most solutions. Special unit for 5% slurries.[1]	0.004 to 0.15	100 to 1
Piston	Most solutions, light slurries.	0.01 to 170	20 to 1
Gas feeders:			
Solution feed	Chlorine	8,000 lb/day max	20 to 1
	Carbon dioxide	6,000 lb/day max	20 to 1
Direct feed	Chlorine	300 lb/day max	10 to 1

[1] Use special heads and valves for slurries.

Table 10-16 describes several types of chemical feeders commonly used in wastewater treatment.

10.9 Additional Reading

10.9.1 General

1. *Water Quality and Treatment,* Third Edition, American Water Works Association, Inc., McGraw-Hill, NY (1971).

2. *Water Treatment Plant Design,* American Society of Civil Engineers, American Water Works Association, and Conference of State Sanitary Engineers, NY (1969).

3. Ockershausen, R. W., "Safe Handling of Water Works Chemicals," *JAWWA,* 63:6, p. 336 (1971).

4. "Chemicals Used in Treatment of Water and Wastewater," BIF Reference No. 1.21-15, BIF Industries, Providence, RI (1970).

10.9.2 Aluminum Compounds

5. Truitt, V., "We Converted to Liquid Alum Treatment," *Public Works,* 99:11, p. 88 (1968).

6. "Aluminum Sulfate," Allied Chemical Corporation, Morristown, NJ (1968).

7. "Alum," American Cyanamid Company, Wayne, NJ (1964).

8. "Aluminum Sulfate," Stauffer Chemical Company, NY (1967).

10.9.3 Iron Compounds

9. "Handling Ferric Chloride," Dow Chemical Company, Midland, MI (1965).

10. "Pennsalt Ferric Chloride," Pennsalt Chemicals Corporation, Philadelphia, PA (1965).

10.9.4 Lime

11. "Chemical Lime Facts," National Lime Association, Washington, D.C. (1964).

12. Carr, R. L., "Limestone to Lime to Slaked Lime," *Water and Sewage Works,* Reference No. 113:R.N., p. R-61 (1966).

13. "Chemicals Used in Water and Wastewater Treatment - Calcium Hydroxide (Hydrated Lime)," *Water and Wastes Engineering,* 7:1, p. 53 (1970).

14. "Chemicals Used in Water and Wastewater Treatment - Quicklime," *Water and Wastes Engineering,* 7:5, p. 52 (1970).

15. "Lime Handling, Application & Storage in Treatment Processes," National Lime Association, Washington, D. C. (1971).

16. "Chemistry and Technology of Lime and Limestone," Boynton, R. S., John Wiley & Sons, New York, NY (1966).

10.9.5 Soda Ash

17. "Soda Ash," Allied Chemical Corporation, NY (1966).

18. "What's Different About FMC Soda Ash?", FMC Corporation, NY (1964).

19. "Stauffer 'Natural' Soda Ash," Stauffer Chemical Company, NY (1964).

20. "Wyandotte Soda Ash," Wyandotte Chemicals Corporation, Wyandotte, Michigan (1955).

10.9.6 Caustic Soda

21. "Caustic Soda," PPG Industries, Inc., Chemical Division, Pittsburgh, PA (1969).

22. "Unloading Liquid Caustic from Tank Cars," Manufacturing Chemists Association, Washington, D.C. (1968).

23. "Properties and Essential Information for Safe Handling and Use of Caustic Soda," Manufacturing Chemists Association, Washington, D.C. (1968).

24. "Caustic Soda," Allied Chemical Corporation, NY (1963).

25. "Caustic Soda Handbook," Diamond Alkali Company, Cleveland, OH (1967).

26. "Dow Caustic Soda Handbook," Dow Chemical Company, Midland, MI (Undated).

27. "Hooker Caustic Soda," Hooker Chemical Corporation, Niagara Falls, NY (1966).

28. "Olin Caustic Soda," Olin Mathieson Chemical Corporation, NY (1961).

29. "Pennsalt Caustic Soda," Pennsalt Chemicals Corporation, Philadelphia, PA (1964).

30. "Caustic Soda," Stauffer Chemical Company, NY (1965).

31. "Caustic Soda by Wyandotte," Wyandotte Chemicals Corporation, Wyandotte, MI (1961).

10.10 References

1. "Chemicals Used in Water and Wastewater Treatment-Aluminum Sulfate," *Water and Wastes Engineering,* Dec. 6:12, p. 46 (1969).

2. "Chemicals Used in Water and Wastewater Treatment-Ferric Chloride," *Water and Wastes Engineering,* 7:3, p. 65 (1970).

3. "Ferri-Floc for Water and Wastewater Treatment," Cities Service Co., Atlanta, GA (1972).

4. Schworm, W. B., "Iron Salts for Water and Waste Treatment," *Public Works,* 94:10, p. 118 (1963).

5. "Lime for Water and Wastewater Treatment," BIF Reference No. 1.21-24, BIF Industries, Providence, RI (June, 1969).

6. National Lime Association, direct communication (May, 1971).

7. "Caustic Soda," PPG Industries, Inc., Chemical Division, Pittsburgh, PA (1969).

8. Haney, P. D. and Hamann, C. L., "Recarbonation and Liquid Carbon Dioxide," *JAWWA*, 61:10, p. 512 (1969).

9. Culp, R. L. and Culp, G. L., *Advanced Wastewater Treatment*, Van Nostrand-Reinhold Company, NY (1971).

10. Carr, R. L., "Polyelectrolyte Coagulant Aids - Dry and Liquid Handling and Application," *Water and Sewage Works*, Reference No. 114:R.N., p. 4-64 (1967).

11. Russo, F. and Carr, R. L., "Polyelectrolyte Coagulant Aids and Flocculants: Dry and Liquid, Handling and Application," *Water and Sewage Works*, Reference No. 117:R.N., p. R-72 (1970).

12. Baker, R. J., "Chemical Feed Systems Determine Plant Efficiency and Reliability," *Water and Sewage Works*, Reference No. 116:R-21 (November, 1969).

13. Russo, F., and Carr, R. L., Jr., "Polyelectrolyte Coagulant Aids and Flocculants: Dry and Liquid, Handling and Application," *Water and Sewage Works*, Reference No. 117:R-72 (November, 1970).

14. Culp, R. L., and Culp, G. L., *Advanced Wastewatewater Treatment*, Van Nostrand Reinhold Company, NY (1971).

15. R. P. Lowe, "Chemical Feed Systems," 10th Annual Water Conf. of Eng. Soc. of W. PA (October 17-19, 1949).

CHAPTER 11

SLUDGE HANDLING AND DISPOSAL

11.1 Introduction

A high proportion of wastewater treatment capital and operating cost is due to sludge handling and disposal. Consequently, it is important that design engineers have reliable information on quantities and dewatering properties of sludges so that solids handling facilities can be properly designed.

Information on sludges produced in conventional wastewater treatment is widely scattered, and results from different locations are not related well to each other. There is even less information available on sludges produced by the newer processes for phosphorus removal.

There has recently been a concerted effort to provide this necessary information. One group involved is the staff of the Ultimate Disposal Research Program operating as part of the Advanced Waste Treatment Research Laboratory of the National Environmental Research Center in Cincinnati.

Material in this chapter comes from information compiled and organized by that group. Their basic data were obtained in several ways:

a. Grants and contracts aimed at obtaining comparative information on phosphorus-removal sludges have been funded.

b. Reports have been prepared which summarize information obtained on sludges in EPA Research and Development Grants where useful sludge data have been collected.

c. Field crews have visited certain plants where promising processes are being implemented and have obtained comparative data on sludges involved.

d. Visits have been made to municipalities and sludge data have been obtained from plant staffs.

e. Published literature has been surveyed and information of value on sludges has been extracted.

f. Memoranda have been prepared to organize the information available and to respond to specific needs.

g. Experimental investigations of sludge dewatering are being conducted at several locations by Advanced Waste Treatment Laboratory staff.

All of these routes are still being actively pursued, although the collection of information is far from complete. However, there was such urgent need for this information (e.g., inclusion in this manual) that an interim report was prepared, and this chapter was developed from it.

Progress on sludge handling and disposal technology will be dealt with in detail in a forthcoming design manual on the subject (1).

11.2 Generalized Assessment of Filtration Rates

At least two groups (2) (3) have developed cost information computer programs which can be used to predict the cost of alternative processing schemes for wastewater treatment and for phosphorus removal processes. These programs have included in their inherent logic (not as input data) typical values for dewatering rates, normal sludge mass, and additional sludge produced by phosphorus removal processes.

Aside from the output from these cost estimated computer programs, reliable estimates of the dewaterability of various sludges can be of substantial value to those concerned with process design. That type of information is presented in this chapter.

When attempting to decide on proper values for sludge dewatering properties, there is a statistical problem of within-plant variations and between-plant variations. The between-plant variations are sometimes extreme. This is illustrated in the following hypothetical compilation:

TABLE 11-1

HYPOTHETICAL COMPILATION OF DEWATERING RATES

	Vacuum Filter Yield (lb/hr-sq ft)	
	Avg. (40 Plants)	Std. Deviation
Digested Mixed Sludge	4.0	2.0
Raw Mixed Sludge	5.0	3.0

The designer wishing to use data of this type to predict within-plant differences between raw and digested sludges will find tabulated values are of little value because standard deviations are so great. Within-plant variations between raw and digested sludge are less than indicated in the table. Unfortunately, within-plant comparisons are virtually unavailable because usually only one type of sludge at a given plant is dewatered.

Information relating various within-plant sludges to each other should eventually be collected. In this chapter there has been an attempt to estimate such relationships based on experience and information from the sources previously cited. Relative values are suggested here. These generalized figures should be used only when there is no other information available.

The information in this section relates only to vacuum filtration. The important area of centrifugation has been introduced in other chapters, when such information was available. Section 11.4 also describes some case studies in which centrifugation was used as a dewatering means, but data are insufficient for development of generalized design parameters.

There has been no consideration given to lime sludges in this introductory section. Sludges produced by lime addition, even when added to the primary clarifier, contain large quantities of calcium carbonate and filter better than conventional or alum and iron sludges. In addition, a different approach is needed than for alum, iron, and conventional sludges. Section 11.4 describes case studies where lime sludge is dewatered by various means, and Section 11.3.2 presents fundamentals involved.

11.2.1 Conventional Sludges

The digested total sludge from a conventional primary plus activated sludge plant is probably the commonest sludge encountered in wastewater treatment. Consequently, it has been selected as the standard against which other sludges are compared.

Table 11-2 shows estimated relative filtration yields and conditioning costs of various wastewater sludges compared to a standard digested primary plus waste activated sludge. These values are for belt-type vacuum filters operated at optimum conditions, preceded by a reasonable processing sequence including thickening if clarifier sludge is dilute. Filter cloth selection was appropriate for the situation to get satisfactory solids capture.

TABLE 11-2

DESIGN FACTORS FOR FILTRATION OF CONVENTIONAL SLUDGES

Sludge Type	Relative* Yield	Relative Cost of Conditioning**	Cake Solids Content
Raw Primary Plus WAS	1.4	0.7	28
Digested Primary Plus WAS	1.0	1.0	26
Raw Primary Plus TF	1.8	0.6	32
Digested Primary Plus TF	1.3	0.9	30
Digested Primary	2.3	0.7	33
Raw Primary	3.2	0.5	35

*Avg. yield for digested primary plus WAS is estimated at 4.0 lb/hr-sq ft

**Average cost of conditioning is $10/dry ton, 1974 costs

An average value of 4.0 lb/hr-sq ft is suggested for the yield when filtering digested primary plus waste activated sludge. If a good estimate of the yield for this or any other sludge is available, it should be utilized and estimates of the yields for the other sludges obtained by means of the factors in Table 11-2.

Mass of sludge produced in wastewater treatment can be estimated by means of well-established procedures discussed in Section 11.3. Reduction in mass by digestion can be estimated by volatile solids reduction. Since digestion reduces mass approximately 50 percent, required filter area will be less for digested sludge than for raw sludge despite slightly higher yields for raw sludges.

Some shortcomings of the tabulated numbers are evident. No influence of solids content is indicated, although it is well known that, up to about 8% solids, filter yield is proportional to solids content. It must be presumed that appropriately designed thickeners are used. It becomes clear that when a filter yield is chosen, there are implicit assumptions about the operations upstream.

Effect of elutriation is not included in Table 11-2. Elutriation would reduce lime demand substantially. Polymer required for conditioning would be reduced as well; elutriation is virtually a necessity for some digested sludges (4).

No indication of range in solids content of filtered sludge cake is given. This is of importance when the sludge is to be incinerated. Some data (5) indicate that digested primary and waste activated sludge would have a solids content of about 28%. For raw primary and waste activated sludge, it would be about 30%. Slightly lower values are suggested in Table 11-2.

11.2.2 Iron and Alum Sludges

Sludges produced when adding ferric or aluminum salts have a substantially different character than conventional sludges. Quantities are greater and handling and dewatering properties are different. Changing the addition point in the processing scheme also produces changes in the character of the sludge.

Table 11-3 compares estimated vacuum filtration yields and conditioning costs of the various metal-effected sludges relative to conventional digested primary plus waste activated sludge.

TABLE 11-3

DESIGN FACTORS FOR FILTRATION OF CONVENTIONAL PLUS METAL SLUDGES

Sludge Type	Relative Yield*	Relative Cost of Conditioning*
Aluminum to Primary		
Raw Primary Plus WAS	1.2	1.3
Digested Primary Plus WAS	0.85	1.4
Raw Primary Plus TF	1.35	1.2
Digested Primary Plus TF	1.0	1.3
Digested Primary	1.05	1.0
Raw Primary	1.5	0.9
Aluminum to Aerator		
Raw Primary Plus WAS	1.2	1.3
Digested Primary Plus WAS	0.85	1.4

Iron to Primary		
Raw Primary Plus WAS	1.3	1.1
Digested Primary Plus WAS	0.95	1.2
Raw Primary Plus TF	1.5	1.0
Digested Primary Plus TF	1.1	1.1
Digested Primary	1.2	0.8
Raw Primary	1.6	0.7
Iron to Aerator		
Raw Primary Plus WAS	1.3	1.1
Digested Primary Plus WAS	0.95	1.2

*Yields and costs are related to yield and cost obtained with digested primary plus WAS when no chemicals are added to the wastewater. See Table 11-2.

Estimates will be more accurate if absolute yields for one or more sludges can be determined for a particular situation. The factors in the table can then be used to predict other yields. If no information is available, filter yield for digested conventional primary plus waste activated sludge may be estimated to be 4.0 lb/hr-sq ft.

11.3 Quantity of Phosphorus Removal Sludges

11.3.1 Iron and Alum Sludges

When ferric iron or trivalent aluminum are added to wastewater containing phosphorus, a metal phosphate precipitates and excess metal forms an insoluble hydroxide. The hydroxide acts as a flocculant precipitate which carries with it other suspended solids when it settles. In addition, soluble organic material apparently is adsorbed on the precipitate and removed, as discussed later.

The amount and nature of additional sludge formed is a function of the process addition point. For example, addition to the primary clarifier causes removal of much more suspended solids than occurs in conventional sedimentation. Also, soluble organic material is removed there. However, the biological stage which follows receives a lower BOD wastewater; consequently, the amount of biological sludge is reduced correspondingly.

There are available tabulations which present average observed values of the amount of sludge typically produced when iron and aluminum are added to wastewater (6). However, there is still need for a more precise method of predicting sludge production so designers can more accurately size sludge processing equipment.

A detailed procedure is presented here for the case where these metals are added to the conventional processing sequence of primary treatment plus either activated sludge or trickling filter. In this procedure the chemicals may be added in primary or secondary treatment, or as a tertiary process. The method requires that the following information be available: complete processing sequence, wastewater characteristics (e.g., SS, BOD, soluble TOC), and efficiency of solids removal in clarifiers. Methods are outlined for predicting the following: chemical sludge produced, extra sludge from improved clarifier efficiency, extra sludge from removal of soluble materials by the chemicals, and reduced amount of biological sludge when biological plant load is reduced. In this manner, total sludge production, related to processing conditions and wastewater quality, can be determined. The method is given in Sections 11.3.1.1 through 11.3.1.5.

11.3.1.1 Theoretical Production of Chemical Sludge

Chemical sludge production may be calculated by assuming the metal added reacts first with phosphorus compounds which are thereby removed, and that any excess metal forms hydroxide. The calculation is illustrated in Table 11-4.

TABLE 11-4

CALCULATION OF CHEMICAL SLUDGE PRODUCTION

<u>Given</u>: P_{in} = 10 mg/l

P_{out} = 1 mg/l

Al Case: dose of 19 mg/l Al
(Al/P atomic ratio = 2.18)

Fe Case: dose of 40 mg/l Fe
(Fe/P atomic ratio = 2.22)

<u>Aluminum</u>

Stoichiometry:

$$Al + PO_4 = Al\ PO_4$$

$$Al + 3\ OH = Al\ (OH)_3$$

(9 mg/l P)/(31) = 0.29 millimoles/1 Al PO_4
(M.W. = 122)

(19 mg/l Al)/27 = 0.70 mmol/1 Al

0.70 - 0.29 = 0.41 mmol/1 Al excess, goes to Al $(OH)_3$
(MW = 78)

(0.29 mmol/1 Al PO_4)(122) = 35.4 mg/l Al PO_4

(0.41 mmol/1 Al $(OH)_3$)(78) = 32.0 mg/l Al $(OH)_3$

Total = 35.4 + 32.0 = 67.4 mg/l

Sludge Produced:

67.4 mg chemical sludge per liter of wastewater treated
or 67.4/19 = 3.55 mg chemical sludge per mg Al
or 67.4/9 = 7.48 mg chemical sludge per mg P removed

Iron

Stoichiometry:

Fe + PO_4 = Fe PO_4

Fe + 3 OH = Fe $(OH)_3$

(9 mg/l P)/(31) = 0.29 mmol/1 Fe PO_4
(M.W. = 150.8)

(40 mg/l Fe)/(55.84) = 0.72 mmol/1 Fe

0.72 - 0.29 = 0.43 mmol/1 Fe excess, goes to $Fe(OH)_3$
(M.W. = 106.8)

(0.29 mmol/1 Fe PO_4)(150.8) = 43.7 mg/l Fe PO_4

(0.43 mmol/1 Fe $(OH)_3$)(106.8) = 45.9 mg/l Fe $(OH)_2$

Total = 43.7 + 45.9 = 89.6 mg/l

Sludge Produced:

89.6 mg of chemical sludge per liter of wastewater treated
or 89.6/40 = 2.13 mg chemical sludge per mg Fe
or 89.6/9 = 9.96 mg chemical sludge per mg P removed

The illustration in Table 11-4 shows that at approximately the same atomic ratio of metal to phosphorus, 48 percent more chemical sludge is predicted for iron addition than for aluminum addition.

Sludge production is related in Table 11-4 to mass of phosphorus removed per liter of sewage treated. This is the most rational procedure when sludge mass is calculated. However, it is frequently convenient to relate sludge mass to the mass of metal added. The following are estimated values for the ratio of sludge mass to metal dose as the metal/P atomic ratio increases from zero to infinity:

Al:	4.52 to 2.89 mg sludge/mg Al
Fe:	2.70 to 1.92 mg sludge/mg Fe

Information to verify these calculated values is scarce. Swedish experience (7) indicates the additional weight of sludge may be calculated from the following relationships:

sludge (mg/l of wastewater) = (4) (Al dose, mg/l)

sludge (mg/l of wastewater) = (2.5) (Fe dose, mg/l)

That reference does not elaborate on the source of information although it was drawn from Swedish practice at the time. It is not stated whether additional mass from improved suspended and dissolved solids removal is included. The great majority of plants in Sweden were adding chemicals as a post-treatment after biological treatment. Consequently, it may be presumed that these figures indicate total increase in sludge mass from Al or Fe added to the final effluent from a biological treatment plant.

Other studies (8) estimated additional sludge produced by alum addition to the activated sludge process. They report 4 to 3 mg addition solids per mg of Al as dose level increased from 10 to 20 mg Al/1.

Another report (9) has estimated chemical sludge production when alum was added to secondary effluents. The following relationship was observed:

chemical sludge (mg/l of wastewater) = (5.2)(Al dose, mg/l)

Al/P atomic ratio range: 2.0 to 2.8

This is about 20 percent higher than the Swedish figure and about 45 percent higher than indicated in Table 11-4 for the stoichiometric relationship.

Data available, then, indicate chemical sludge mass higher than the values represented by the simple stoichiometry indicated in Table 11-4. There is reason to conclude that chemical sludge production could be higher than stoichiometric values because other cations (such as heavy metals) and anions can be incorporated into the complex ligand structures of aluminum and ferric hydroxides. Consequently, it is recommended that with either Al or Fe, sludge mass be estimated to be 35 percent higher than the simple stoichiometric values.

11.3.1.2 Theoretical Production of Biological Sludge

Biological sludge is produced and wasted in both the activated sludge process and the trickling filter process.

One source (10) reports the following equation for excess sludge production in the conventional activated sludge process:

$$W_V/M = (0.79\ F/M) - (0.0714)$$

where W_V = lb volatile suspended solids produced per day

M = lb mixed liquor volatile suspended solids in the aerator

F = lb 5-day BOD removed per day

Rearranging this equation,

$$W_V/F = (0.79) - (0.0714/F/M)$$

The equation now gives the mass of volatile solids produced per unit mass of BOD removed (W_V/F) as a function of the food to mass ratio.

A recent report (11) states that 50 percent of the sum of the BOD and suspended solids fed to a trickling filter is converted to waste sludge.

If suspended solids and BOD concentrations in the wastewater as it enters the biological treatment step are known, sludge production can be estimated by means of the above relationships.

11.3.1.3 Increase in Production of Sludge From Sewage Solids

Suspended solids and dissolved solids in wastewater are removed by chemical treatment. Sludge mass is increased by the direct contribution of the mass of these removed solids.

One report (12) has summarized data on suspended solids removal from well designed physical-chemical pilot plants. For an average sewage and effective clarifiers, suspended solids removal of 85 percent can be achieved. If the sewage is high in suspended solids, removals will be better. This figure is substantially higher than the 50-75% removal range suggested previously (13). However, those earlier figures included plants where chemicals were added to an existing facility which might have been overloaded or were functioning poorly for other reasons.

The mass of sludge produced from suspended solids removed is calculated directly by multiplying the change in suspended solids concentration by the volume of sewage.

There is also evidence (12) which indicates an appreciable mass of dissolved materials (i.e., filtrable through a 0.45 micron membrane filter) is removed by chemical treatment. The following table presents this information along with data collected from other sources.

TABLE 11-5

REMOVAL OF DISSOLVED MATERIALS DURING CHEMICAL ADDITION

Source	Percent Removals			
	Fe		Al	
	SCOD	STOC	SCOD	STOC
Eimco				
pilot plant (14)			40	30
laboratory (14)				ca. 30
Friedman et al (15)		30		
Farrell (16)				27
Westrick and Cohen (17)	45	42		

Note: SCOD is soluble COD
STOC is soluble TOC

Some of that date (16) represent the average removal obtained in seven batch experiments in which raw sewage was treated with alum in a small clarifier (150 liter batches). Al/P atomic ratio ranged from 1.6 to 2.8 and removals from 20 to 32 percent.

Based on the above data, it is concluded that soluble TOC removals will average 30 percent. Data are scant for COD and none appear available for BOD. It is assumed, however, that removals will be the same for STOC.

The mass of sludge associated with the soluble TOC removed is more than the carbon content involved. It is estimated that, on the average, the soluble compounds removed by Al and Fe treatment will average 40 percent carbon on a volatile solids basis. This figure is in accord with the carbon content of compounds tested in developing the TOC test for wastewaters (18). Thus, to obtain an estimate of the mass of sludge (volatile basis) produced, TOC removed should be multiplied by 1/0.40 or 2.5.

If only soluble COD removed or soluble BOD removed is known, the problem is more difficult, primarily because the relationship between mass and COD or BOD changes as the degree of oxidation of the sewage changes. For a secondary effluent, volatile mass/COD is generally considered to be about 1.5 (19). Assuming the BOD is about 70 percent of the COD, volatile mass/BOD is 2.15. For raw sewage, which has a low degree of oxidation, the COD is expected to be higher for a given mass. Volatile mass/COD for a raw sewage is estimated to be 1.1, approximately 10 percent lower than lowest values previously found (19) for secondary effluent. Volatile mass/BOD would be about 1.6.

In the above discussion, mass relates to mass of volatile solids. To obtain total mass, non-volatile solids must be included. It is estimated that the ash content of these soluble solids is about 15 percent. Mass of volatile solids should be multiplied by 100/(100-15) or 1.18 to get total solids.

Calculation of sludge mass resulting from removal of soluble material will require information on the absolute value of soluble TOC, COD or BOD. These values are rarely measured. An examination of previous data (14), where both SCOD and STOC were measured, showed that ratios of soluble to total COD and TOC in the influent sewage were 0.30 and 0.38 respectively. Other data (16) showed an STOC/TOC ratio of 0.64. If data on soluble values of TOC or COD are not available, soluble TOC or COD may be estimated to be 50 percent of the total TOC or COD.

The correlations for predicting sludge production require information on total BOD removal. Some data (12) indicate total BOD, COD and TOC reductions are about the same and range from 50 to 85 percent depending on clarifier efficiency. Removal of

65 percent is thereby chosen as a representative figure and will be used in example calculations.

11.3.1.4 Summary of Calculation Procedures

The calculation procedure is described for various chemical addition points and is illustrated with a specific example in Section 11.3.1.5. Factors involved in that example are summarized here.

Chemicals Added to Primary

Chemical sludge is calculated on the basis of the phosphorus level and the chemical dose in the manner illustrated in Table 4. Estimated mass should be increased by about 35 percent (see Section 11.3.1.1).

Additional suspended solids removals in primary treatment are estimated using quantities derived per Section 11.3.1.3. The additional suspended solids mass removed should be added directly to total sludge mass.

Mass resulting from dissolved solids removal has to be estimated indirectly from soluble TOC, COD or BOD removal. Factors to be used in predicting mass have been discussed in Section 11.3.1.3. Equations relating soluble TOC, COD and BOD input levels to mass of soluble solids removed are summarized below:

$$\text{(soluble solids mg/l)} = STOC_{in}\text{(mg/l)}x0.30x2.5x1.18$$

$$= SCOD_{in}\text{(mg/l)}x0.30x1.1x1.18$$

$$= SBOD_{in}\text{(mg/l)}x0.30x1.6x1.18$$

The last two equations apply only to influents before oxidative processes such as the activated sludge process.

Estimates of total TOC, COD or BOD removals are needed to estimate the loading on the biological processes which follow chemical treatment. However, this information is not needed for estimating sludge formation in the chemical treatment step.

Chemicals Added to Activated Sludge

The chemical sludge is calculated in the manner described in Section 11.3.1.1.

Addition of chemicals to the activated sludge process may produce a significant increase in BOD and suspended solids removals. This is particularly true of a high rate activated sludge plant or an overloaded plant. However, for a well-designed standard rate plant addition of chemicals will produce little additional sludge other than the chemical sludge itself. It is recommended that for a standard rate plant no increase in mass be included for additional suspended solids. For special cases such as an existing high rate plant, judgment will have to be made based on available information.

Chemicals Added After Biological Treatment

The chemical sludge is calculated as in Section 11.3.1.1. If the tertiary treatment follows a standard rate activated sludge plant, additional sludge from improved suspended solids and dissolved BOD removal can be neglected. For a trickling filter plant, however, additional mass removed could be important. For one plant (20) a substantial improvement in BOD and solids removal in a high rate trickling filter plant resulted from alum addition before the final clarifier. In one series of runs, overall BOD removals increased from 73% to 91% and suspended solids removal increased from 70% to 84%.

At another plant (21), under conditions where the final clarifier was not hydraulically overloaded, alum added after a standard rate trickling filter brought BOD and suspended solids from the 15-20 mg/l levels to below 10 mg/l.

These results can be used to calculate additional sludge brought down with the chemical sludge.

11.3.1.5 Illustrative Calculation

Sludge mass is calculated here for a plant utilizing primary treatment with aluminum introduced before clarification, followed by standard rate activated sludge. Details of the calculation are presented in Table 11-6. The existing plant is a properly designed plant operating within its design limits, which is changed only by introduction of a means to inject the chemical and flocculate it before primary clarification.

TABLE 11-6

CALCULATION OF TOTAL SLUDGE MASS WITH METAL ADDITION

Given: Plant utilizes primary clarification followed by standard rate activated sludge. Alum is to be added in a flocculation tank prior to the primary clarifier. Estimate the additional sludge produced by alum addition.

Primary clarifier of the existing plant is designed for 50% suspended solids removal (without chemical). Activated sludge plant is designed for an F/MLVSS ratio of 0.4.

Overall phosphorus removal is to be 90%. Estimate that this can be achieved at an Al/P atomic ratio of 2.0.

Influent to Primary:			
	P	=	10 mg/1
	Total BOD	=	172 mg/1
	Suspended Solids	=	240 mg/1
	Soluble TOC	=	40 mg/1

Sludge from Primary Treatment - No Chemicals

Estimate 50% SS removal, 30% total BOD removal and 0% STOC removal.

240 x 0.50 = 120 mg/l

Sludge from Primary Treatment - Alum Added

Estimate 90% P removal, 85% SS removal, 65% total BOD removal and 30% STOC removal.

Al Dose:	(10/31)(2)(27) =	17.4 mg/1
	10 x 0.90 =	9.0 mg/1 P removal
	9/31 =	0.29 meg (milliequivalent)/1 Al PO_4
	17.4/27 =	0.64 meg/1 Al
	0.64 - 0.29 =	0.35 meg/1 Al excess
	0.29 x 122 =	35.4 mg/1 Al PO_4
	0.35 x 78 =	27.3 mg/1 Al $(OH)_3$
	35.4 + 27.3 =	62.7 mg/1 sludge produced

Chemical Solids: Increase by 35%, per Section 11.3.1.1
62.7 x 1.35 = 84.5 mg/1

Suspended Solids:	240 x 0.85	=	204 mg/l
Dissolved Solids:	40x0.30x2.5x1.18	=	35.4 mg/l
	Total Sludge:		323.9 mg/l

Sludge from Activated Sludge - No Chemicals

Equation from Section 11.3.1.2

$$W_v/F = (0.79) - (0.0714/F/M)$$

F/M is given as 0.4

W_v/F = 0.79 - 0.18 = 0.61 lb VSS/day/lb BOD removed per day

or 0.61 mg VSS/mg BOD removed

BOD in primary effluent is:

172 x 0.70 = 120 mg/l

BOD in final effluent is estimated to be almost 12 mg/l. See reference (6).

BOD removed = 120 - 12 = 108 mg/l

Therefore, 108 x 0.61 = 66 mg/l VSS produced

Assuming the sludge is 70% volatile,

66/0.70 = 94 mg/l waste sludge

Sludge from Activated Sludge - Al to Primary

BOD in primary effluent is:

172 x 0.35 = 60.2 mg/l

This is a very dilute waste. The existing activated sludge plant will be greatly oversized. Air requirements and the contact time may be reduced.

Keeping F/M the same (0.4),

W_v/F = 0.61 (see above)

Estimate BOD in final effluent to be 12 mg/l (see above).

BOD removed = 60.2 - 12 = 38.2 mg/l

Therefore, 38.2 x 0.61 = 23.3 mg/l VSS produced

Assuming sludge is 70% volatile,

23.3/0.70 = 33.3 mg/l waste sludge

Summary of Sludge Quantities Calculated

	Sludge Quantities (mg/l)	
	No Chemical	Al Added
Primary Clarifier		
Sludge from SS	120	204
Sludge from Dissolved Solids	0	35.4
Chemical Sludge	0	84.5
Total Clarifier Sludge	120	323.9
Secondary Clarifier		
Waste Activated Sludge	94	33.3
Total Sludge	214	357.2
Total Sludge Excepting Chemical Sludge	214	272.7

It is clear that chemical addition produces a substantial increase in the total sludge mass. The increase is greater than just the amount of chemical sludge because, in the no-chemical case, there is substantially more conversion of organic substances to carbon dioxide and water.

For a plant in which the chemical is added to the activated sludge process, the increase in mass will be virtually all due to the chemical sludge, since essentially the same amount of biological degradation is occurring as in the no-chemical case.

In the illustration, total sludge production (mass) was increased 67 percent when adding aluminum. The chemical sludge increased sludge mass 40 percent. The additional 27 percent sludge came from organic material which would have been oxidized had it reached the activated sludge aerator. If the chemical had been added to the aerator, sludge mass would have only increased 40 percent. The greater sludge mass which results when the chemical is added to the primary must be balanced against the reduced dimensions of the activated sludge aerator, the reduction in quantity of air needed, any differences in the dewatering properties of the different sludges, and differences in chemical demand which may be dependent on point of addition.

11.3.2 Lime Sludges

Lime is utilized for phosphorus removal in wastewater treatment by adding it just prior to primary clarification or after secondary clarification as a tertiary treatment process. As in the case of metal addition to wastewater, average values for the amount of sludge produced have been tabulated (6). However, there is need for more precise information applied to specific situations.

In response to this need, a procedure has been developed for predicting the mass of sludge produced when wastewater is treated with lime in a tertiary process. Details of this procedure are presented in this section. The method may be applied with minor modification when lime is added just prior to the primary clarifier.

Unlike metal addition, the amount of sludge produced by lime addition depends not only on the degree of phosphate removal desired, but also on alkalinity and other characteristics of the wastewater. In fact, both the amount of sludge produced and the degree of phosphorus removal achieved by raising the pH to a given level both depend on the nature of the water.

The procedure in predicting sludge quantity requires first that pH level be estimated for the desired phosphorus removal, then lime dose is estimated, and finally sludge mass can be calculated. The individual steps are listed below:

a. From desired P-residual and knowledge of the water type, estimate pH level.

b. From pH level and alkalinity, estimate lime dose.

c. From amount of P removed, calculate the hydroxyapatite sludge produced.

d. From magnesium removal, calculate magnesium hydroxide sludge produced.

e. From calcium balance, calculate calcium carbonate sludge produced.

f. From pH and alkalinity, verify that sufficient carbonate is present to supply carbonate in sludge. If not, correct outlet calcium estimate.

g. Include in total sludge production the inerts introduced with the lime.

If lime is added prior to the primary settler, the procedure for estimating sludge for tertiary lime addition may be used with some modification. However, the correlation between pH and the ratio of lime dose to alkalinity was developed from data obtained by adding lime to secondary effluents. There is no assurance it would fully apply to raw wastewater. It would thus be desirable to establish lime dose needed by actual tests.

When lime is added prior to the primary settler, efficiency of the clarifier is increased substantially. From a previous review (12), it is estimated that for well designed plants, suspended solids removal efficiencies are in the order of 85 percent. This figure may be used to estimate additional sludge solids removed in the primary clarifier by lime addition.

The removal of dissolved organic solids by addition of lime prior to the primary appears to be less than obtained with metals. One study (14) reported very small or negative removals. Others (17) reported substantially lower removals with lime than with metal. Considering the relatively large mass of sludge from lime treatment, the small amount of dissolved organic solids brought down in primary treatment can be neglected.

Some prior data (14) have been utilized to estimate whether material balance calculations similar to those described above correlate with measured sludge production. This case is discussed in Section 11.4.2.2. Measured values were about 20 percent higher than the material balance figures. It is probably advisable to utilize a factor of safety of this magnitude in estimating chemical sludge production.

11.3.2.1 Estimation of Lime Dose and Sludge Production in Tertiary Treatment

A quantitative calculation procedure for estimating the required lime dose and the quantity of sludge produced in tertiary treatment of wastewater with lime is presented. On-site testing for obtaining this information is recommended in specific plant designs.

11.3.2.2 Lime Dose

The extent of phosphate removal is a function of a number of variables including chemical composition of the wastewater and the nature of the phosphate compounds contained. The most important variable controlling phosphate removal and quality of effluent is pH rather than quantity of lime added. Figure 11-1, reference (22), shows a correlation of P-removal as a function of pH, utilizing data from several sources.

Another factor of importance is the hardness level of the wastewater. In some cases (23) it was necessary to go to pH in excess of 11 to get good clarification with the low hardness water. Others (24) obtained satisfactory clarification at pH of about 10 in the hard water.

The chemistry of lime-softening of natural waters is sufficiently well developed that the charge of lime required to achieve a given removal of calcium and magnesium can be predicted and the final pH can be closely estimated. With wastewaters, suspended and dissolved organic materials as well as ortho and polyphosphates apparently affect lime dosage required and can cause super-saturation of precipitating species. Typically, lime dose required to reach a given pH is higher than predicted from softening equations. Likewise, calcium and magnesium content of the product water is higher than would be predicted for lime treated natural waters.

A correlation for estimating lime dose needed to bring a secondary effluent to a given pH is presented in Figure 11-2. It is based on data from several sources (23-27). The pH is reasonably well correlated with the ratio of lime dose to initial alkalinity. Other factors which affect the result are ratio of magnesium to calcium, ionic strength, and the phosphate level. The effects caused by these variables can be estimated, and computer programs have been developed to make the necessary calculation (22) (28). However, the accuracy of these calculations is provisional because of the uncertain effects of contaminants in wastewater.

Figure 11-2 can be utilized to predict lime dose if an estimate of the desired pH level can be made. The curve is entered at the ordinate, and the abscissa, the ratio of CaO dose to alkalinity, is read from the graph. If alkalinity of the water is known, active CaO dose can be calculated. Total quicklime dose can be calculated from active CaO dose and the active CaO content of the quicklime.

11.3.2.3 Sludge Production

The principal reactions for estimating lime sludge production are:

FIGURE 11-1

EFFECT OF pH ON TOTAL RESIDUAL PHOSPHORUS

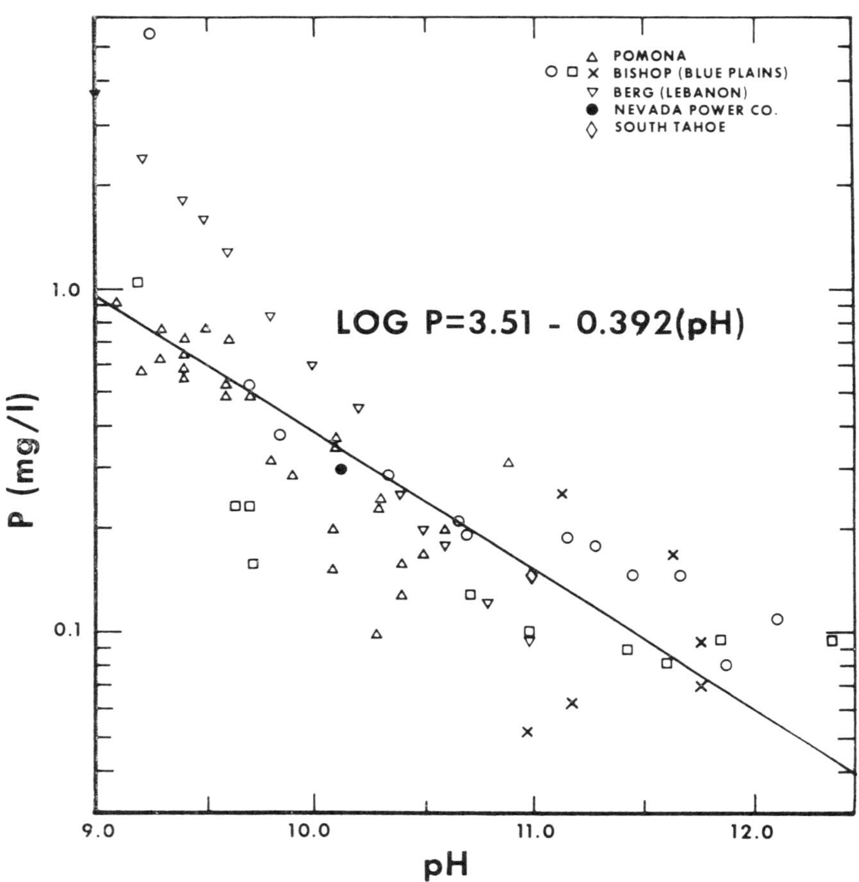

FIGURE 11-2

RATIO OF LIME DOSE TO ALKALINITY (BOTH MG/L) FOR A PRESCRIBED FINAL pH

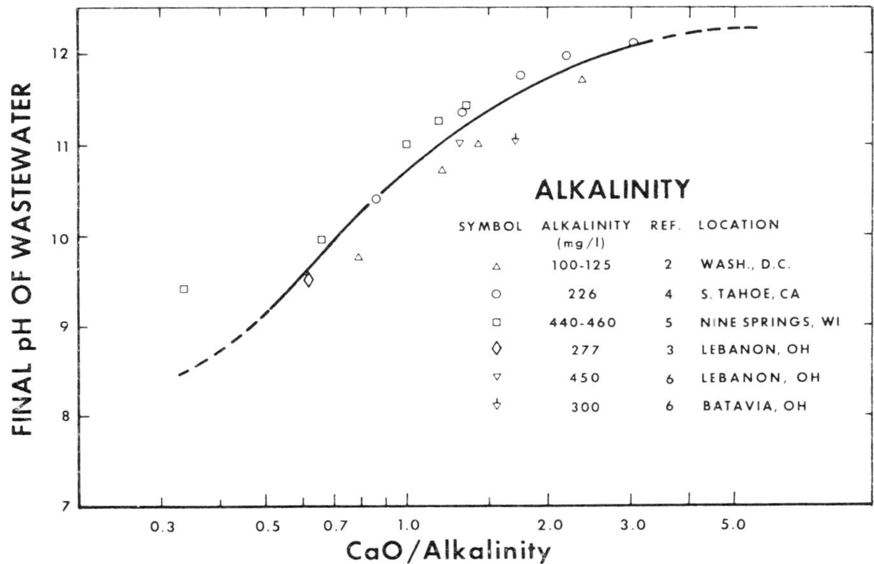

$$5Ca(OH)_2 + 3HPO_4^{--} \longrightarrow Ca_5OH(PO_4)_3 + 6OH^- + 3H_2O \quad (1)$$

$$Mg^{++} + 2OH^- \longrightarrow Mg(OH)_2 \quad (2)$$

$$Ca(OH)_2 + HCO_3^- \longrightarrow CaCO_3 + H_2O + OH^- \quad (3)$$

The information required and the steps to take in estimating the sludge production are listed below. Informational input needs are marked with an asterik.

TABLE 11-7

INFORMATION NEEDED TO CALCULATE LIME SLUDGE MASS

	Information Needed	Determination
A.	P^*_{in}, desired P removal* (sometimes, BOD_{in}, desired BOD removal). Results of jar test or experience*	pH level
B.	Alkalinity of wastewater*, pH level, Figure 11-2	CaO dose
C.	P_{in}, P removal, Equation 1	$Ca_5OH\ (PO_4)_3$(abbrev.:HAP) precipitated
D.	Mg_{in}, from analysis or estimate*, Mg_{out}, from analysis or Figure 11-4, Equation 2	$Mg(OH)_2$ precipitated
E.	Ca_{in}, from analysis or estimate*, Ca_{out}, from analysis or Figure 11-3, Ca from lime dose (see item B), Ca removal in HAP (see item C), Equation 3	$CaCO_3$ precipitated
F.	pH (see item A)	Verify that sufficient CO_3 is present. If not, estimates of Ca_{out} were in error.

G. Active CaO in lime* — Inerts brought in with quicklime.

The procedures for steps A and B have already been presented. Step C requires no explanation. If estimates of Mg_{out} and Ca_{out} are needed for steps D and E, available data (24) can be used as shown in Figure 11-3 and Figure 11-4. They show the calcium and magnesium levels obtained at the clarifier outlet in long-term evaluation of tertiary lime treatment of secondary effluent. Magnesium (Figure 11-4) precipitation commences at pH 10 and is essentially complete at pH 11. The percent in solution could be estimated as declining from 100% at pH 10 to 0 at pH 11.

The calcium concentration (Figure 11-3) is seen to decline precipitously when pH is increased to 9.5. Concentration then increases as pH is increased further. This behavior is rational in light of knowledge of bicarbonate-carbonate equilibria. When pH is increased, bicarbonate ion is converted to carbonate ion. When the solubility product of $CaCO_3$ is exceeded, precipitation occurs and Ca concentration falls. As pH is increased further by more $Ca(OH)_2$ addition, eventually all of CO_3 is removed by $CaCO_3$ precipitation. Calcium concentration then increases as calcium hydroxide dissolves to the limit of its solubility.

Step E outlines the calculation of calcium carbonate precipitation. The calcium carbonate deposited is calculated from a calcium balance. After considering the amount of calcium removed in the hydroxyapatite ($Ca_5OH\ (PO_2)_3$), the $CaCO_3$ is calculated from the net loss of calcium.

A verification step is advisable to determine whether there is sufficient becarbonate to account for the $CaCO_3$ produced. Bicarbonate can be determined from alkalinity. At pH below about 8.3, all of alkalinity measured is due to bicarbonate. At pH below 7, there is more potential carbonate available due to the presence of dissolved carbon dioxide, which is not measured in the alkalinity test. Estimates of dissolved CO_2 can be made by following procedures outlined in "Standard Methods" (29). If the bicarbonate is insufficient to balance with lost calcium calculated from the material balance, calcium in the final effluent was estimated at too low a value. There is little likelihood of this problem at pH levels below about 10.8.

The entire procedure described, including the verification step and the totaling of all of the components of the sludge, is shown in the illustration in Table 11-8. It is evident from the total quantities listed at the end of the table that calcium carbonate makes up about three-quarters of the mass of the sludge. For this reason, substantial errors could occur in estimates of the other sludge quantities without introducing much error into the final result.

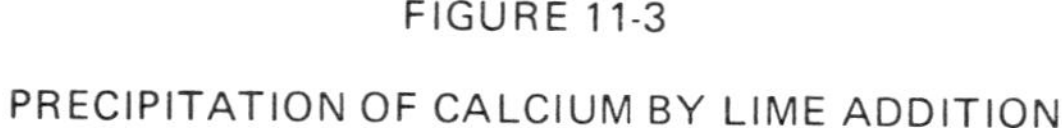

FIGURE 11-3

PRECIPITATION OF CALCIUM BY LIME ADDITION

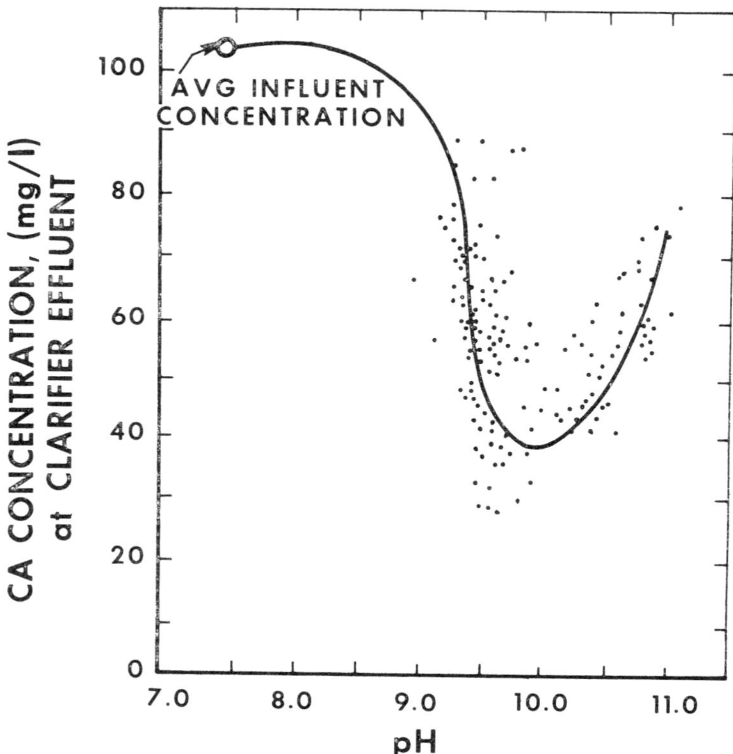

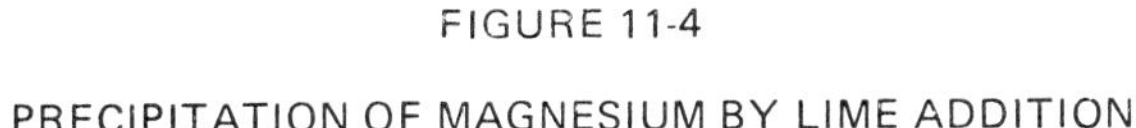

FIGURE 11-4

PRECIPITATION OF MAGNESIUM BY LIME ADDITION

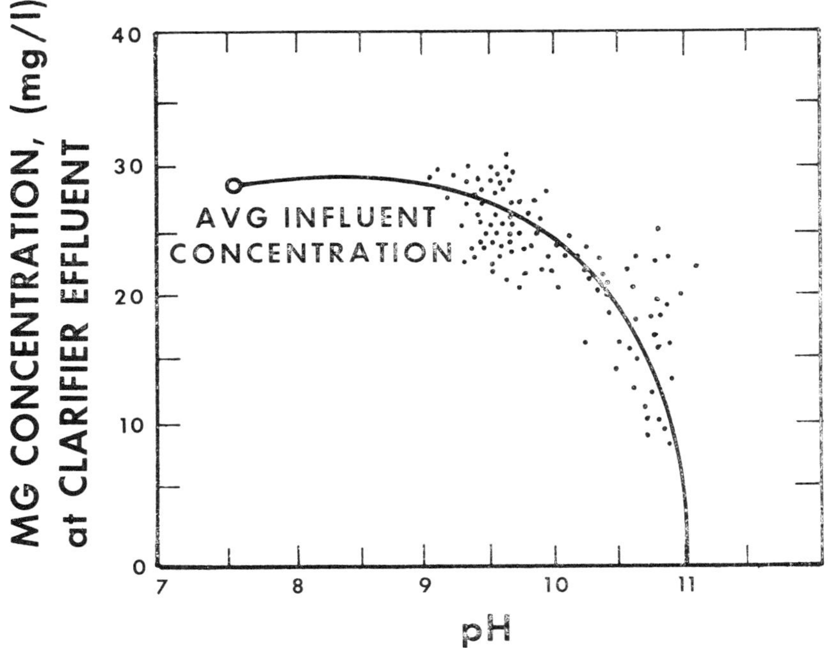

TABLE 11-8

CALCULATION OF LIME SLUDGE MASS

Illustration of Procedure for Estimating Sludge Quantity

Secondary Effluent Characteristics

Phosphate Content	10 mg/l as P
Alkalinity	300 mg/l as $CaCO_3$
Calcium	100 mg/l
Magnesium	20 mg/l

Treatment Conditions

Remove 85% of phosphate by raising pH to 10.5

Final Phosphorus Content	1.5 mg/l as P
Estimate of Ca (Figure 11-3)	50 mg/l
Estimate of Mg (Figure 11-4)	10 mg/l

Lime Dose

From Figure 11-1, 0.92 mg CaO/mg alkalinity at pH 10.5.
Dose = 0.92 x 300 = 276 mg/l CaO

Assume quicklime is 90% active CaO

Total quicklime - 176/0.90 = 307 mg/l

Inerts with quicklime = (0.10)(307) = 30.7 mg/l

Sludge from Phosphate

(10 - 1.5)(502/93) = 45.8 mg/l $Ca_5(OH)(PO_4)_3$

$Mg(OH)_2$ in Sludge

(20 - 10)(58.3/24.3) = 24.0 mg/l $Mg(OH)_2$

Sludge from $CaCO_3$ Precipitation

Input Ca = 100 from sec. effluent + (276)(40/56) from lime dose

Output Ca = 45.8 x (5)(40)/(502) + 50 + Y

(removed with hydroxyapatite) (remaining in solution) (precipitated as $CaCO_3$)

Y = 100 + 197 - 18.3 - 50 = 129 mg/l

129 x (100/40) = 323 mg/l $CaCO_3$ precipitated

Total Sludge

Inerts in quicklime	30.7
$Ca_5OH(PO_4)_3$	45.8
$Mg(OH)_2$	24.0
$CaCO_3$	323.0
Total =	433.5 mg/l

Verification that Sufficient CO_3 is Present

300 mg/l alkalinity (as $CaCO_3$)

300 x (1/50) = 6.0 g-equivalents (or g-mols) HCO_3^-

$$HCO_3^- + Ca^{++} + OH^- \longrightarrow CaCO_3 + H_2O$$

6.0 g-mols HCO_3^- can produce 6.0 g-mols $CaCO_3$

6.0 x 100 = 600 mg/l $CaCO_3$

The CO_3 available could produce as much as 600 mg/l $CaCO_3$. This is substantially greater than the amount calculated above (323 mg/l) from the calcium balance. Thus, there is ample CO_3 present. If there were insufficient CO_3 (which can happen at ph of 11 and greater), it would indicate that the estimate of Ca^{++} after treatment was too low. The amount of $CaCO_3$ indicated by the calculation from the alkalinity would be correct.

Additional Sludge from Recarbonation

Recarbonation to pH 9.5 will reduce Ca concentration to about 30 mg/l (see Figure 11-3). For this example, the additional calcium carbonate sludge produced is calculated as follows:

$$(50 - 30)(100/40) = 50 \text{ mg/l } CaCO_3$$

11.4 Dewatering at Specific Locations

11.4.1 Tertiary Treatment with Lime

11.4.1.1 South Tahoe

At South Tahoe, California, wastewater is treated in conventional primary, followed by a waste activated sludge plant. Phosphate is removed by tertiary lime treatment to pH 11 with recarbonation. A comprehensive report describing the performance of the Tahoe plant and sludge dewatering aspects has been published (25).

The tertiary lime sludge is dewatered and classified at Lake Tahoe by means of solid bowl centrifuges. Sludge is thickened to 8 percent solids. It then flows at about 20 gpm to a 75 HP Bird 24" x 60" solid-bowl concurrent-flow centrifuge. The centrifuge is deliberately operated with high losses to the centrate in order to reject magnesium hydroxide, hydroxyapatite and organic solids, leaving behind a high concentration of calcium carbonate in the cake. The cake is calcined and is recovered. The centrate is sent to a second centrifuge, polymer is added, and solids are removed as a 30% solids cake.

The fact that a centrifuge tends to reject fine low-density solids may lead to fears that fine solids will build up in the system. This has apparently not occured at South Tahoe.

11.4.1.2 Blue Plains Pilot Plant

At the Blue Plains pilot plant, a substantial amount of information was obtained on the sludges produced by treating secondary effluent with lime. This information will be published soon as a report in the EPA series (30).

Secondary effluent from the District of Columbia's Blue Plains Treatment Plant was treated with lime to a pH range of 11 to 12 for phosphate removal. Instead of utilizing second-stage recarbonation to remove excess Ca ions and recover additional calcium carbonate, sodium carbonate was added to the single clarifier for this purpose. The sludge produced was continuously thickened, vacuum filtered, calcined, slaked and recycled to the tertiary clarifier to supply the bulk of the lime demand. Information on sludge thickening and dewatering was obtained with and without recycle of reclaimed lime.

Most of the data during recycle operations were obtained at 56 percent recovery; that is, 56 percent of the active lime dose consisted of recycled lime. Recycle was approximately 3 pounds of recycled solids to one pound of fresh lime. Lime dose consisted of 300 mg/l of active lime plus 240 mg/l of inerts.

Buildup of inerts did not affect phosphorus removal but increased solids loading through the sludge handling system. Solids content of sludge removed from the clarifier was 20 g/l for no-recycle operation and 32 g/l for 56 percent recycle.

The sludge was thickened in a 6-foot diameter vessel equipped with pickets. Sludge thickened to 15-20% solids for recycle as well as no-cycle operations, despite solids loading rates over 70 percent higher for the recycle case.

Information on filtration was obtained from filter leaf tests. Equations for form rate for the recycle and no-recycle cases were developed. These equations, rearranged slightly to give results in terms of yield and cycle time, are:

No Recycle: $Y = (2.18)\,(f)^{0.5}(C)^{0.84}\,(p)^{0.21}/\,(\phi)^{0.5}$

Recycle: $Y = (3.28)\,(f)^{0.5}\,(C)^{0.80}\,(p)^{0.26}\,/\,(\phi)^{0.5}$

Where Y = yield, lb/hr-sq ft

ϕ = cycle time (min), or reciprocal of rpm

C = Solids concentration, wt %

p = operating vacuum, in. Hg

f = fraction submergence

Calculating filter yield at the following conditions,

$f = 0.25$

C = 15% solids

p = 15" Hg vacuum

ϕ = 1 min. (i.e., 1 rpm)

gives the following result:

No Recycle: Y = 18.7 lb/hr-sq ft

Recycle: Y = 29.0 lb/hr-sq ft

This study reports substantial cake cracking which reduced vacuum across the filter. Nevertheless, solids contents from 30-40% were generally obtained. It was also noted that, after a process upset which caused biological solids to carry over into the tertiary treatment system, filter yield deteriorated drastically. However, filtration was possible without chemical conditioning.

11.4.1.3 Butler, Indiana

At Butler, wastewater is treated in a 0.4 mgd conventional primary plus trickling filter plant. Primary settling precedes two-stage biological filters. Conventional sludge is digested and dried on beds. Phosphorus is removed by tertiary lime in a reactor-clarifier at pH 10. Chemical dose is 150 mg/l quicklime plus ferric chloride and polymer to aid settling. Total phosphorus removal is about 80 percent. Sludge is not dewatered and is pumped to a lagoon.

The Butler plant is somewhat unique in that the wastewater is quite hard (300 mg/l) and its magnesium content is high (30 mg/l). Ordinarily, good phosphorus removal is achieved in water of this type at a lower pH than in soft low-magnesium water. Required phosphorus removal was only 80% and a pH of 10 was sufficient to get satisfactory performance.

Sludges leaving the reactor-clarifier were about 8.4 percent solids. Overnight settling increased this to 9.0 or 9.5 percent. Lagoon sludge, even after only a few days settling, was about 17% suspended solids. Fresh sludge from the reactor-clarifier could be filtered at 5.6 lb/hr-sq ft at a cycle time of 2.9 min and 33% immersion. Higher yield could be achieved at lower cycle times but this produced cakes less than one-eighth inch thick, which caused problems with cake release.

Sludge was mostly mineral in nature. Percent volatile solids was under 10%. Magnesium content of the sludge was low. At higher pH, more magnesium would have been precipitated, which could have produced a poorer filtering sludge.

11.4.2 Lime to Primary Settler

11.4.2.1 Contra Costa

The Contra Costa Sanitary District, Walnut Creek, California, has carried out pilot and plant scale evaluations of an integrated process for handling sludges produced by adding lime to primary wastewater treatment. This work has been reported (31) and is discussed in Chapter 5.

The treatment process comprises lime flocculation with polymer or ferric chloride followed by primary sedimentation, activated sludge under conditions such that nitrification is achieved, denitrification with methane addition, and chlorination. Waste activated sludge is recycled to the raw sewage to be removed from the primary clarifier along with the chemical-primary sludge. Denitrification sludge is also returned to the incoming raw sewage, but its amount is negligible.

In addition to investigations of process performance, studies were made to evaluate (a) predictive methods for estimating sludge production, (b) the degree of thickening achieved in primary clarification, (c) the utility of two-step procedure in which sludge is centrifugally classified to recover calcium carbonate and the centrate is centrifuged again to remove solids, and (d) anaerobic digestion of the first-stage centrate.

Sludge mass predicted from analysis of influent and effluent analyses checked satisfactorily with measured sludge mass. Operation at pH 10.2 produced a clarifier sludge of about 4.0 percent solids. At pH 11, higher sludge solids could be attained. Centrifugal classification achieved an 80% recovery of calcium carbonate in relatively dry cakes of 42-57% solids. Solids remaining in the centrate could be removed by polymer-assisted second-stage centrifugation at recoveries as high as 90%. Results were not as good at pH greater than 11. The centrates from the first centrifugation could be anaerobically digested, although volatile solids reduction was lower than for conventional sludge processing. In addition, precipitated magnesium redissolved in digester supernatant. Return of supernatant to process would increase lime demand and magnesium would be precipitated again. It is believed that handling of supernatant by separate treatment or return to the secondary process would avoid that problem.

11.4.2.2 Eimco at Salt Lake City

Eimco Division of Envirotech has conducted a 100 gpm pilot plant investigation of a wastewater treatment system in which chemicals are added in the primary step and soluble organic material is removed by powdered activated carbon. A progress report on this investigation has recently been published (14).

Comparisons of sludge mass calculated from clarifier input and output analyses with measured mass showed reasonable agreement. When iron was added to the primary, measured mass was 20% higher than the calculated value. However, this additional mass appeared to include removal of soluble organics. When adding alum, calculated value (including soluble organic material) was virtually the same as the measured value. For lime, measured values averaged 20% higher than calculated values.

Thickening and filtration data were incomplete for iron and alum. However, data with lime were developed. After 24 hours thickening, sludges with pH values greater than 11.5 thickened to 2.0 to 9.8% solids, with median near 7%. For pH values less than 11.0, sludge thickened to 9.0 to 18.0% solids, with median near 14%.

Filtration yields were obtained for high and low pH operation. No polymer was needed. For low pH sludge at the 14% thickened solids content, maximum filter yield was 6.0 lb/hr-sq ft. For high pH sludge at the 7% thickened solids content, maximum filter yield was 3.0 lb/hr-sq ft.

It should be noted that Salt Lake City wastewater is high in magnesium. Operation at pH greater than 11.5 precipitates much more magnesium than precipitated at pH less than 11.0. Since magnesium hydroxide is a flocculant precipitate, this could account for the low solids content of thickened sludge from high pH operation.

11.4.2.3 Cleveland Westerly

Extensive pilot plant work was conducted at Cleveland's Westerly plant preliminary to preparing a design for a physical-chemical treatment plant. A report describing this work is in progress (32).

Most of the work was done with lime added prior to the primary to reach a pH of 10.5. Sludge was dewatered using a Bird pilot plant centrifuge and an Eimco belt-type pilot plant vacuum filter.

Sludge solids content leaving the clarifier was in the range of 6% to 14% solids. Thickening prior to centrifugation was not needed. Solids recovery in the centrifuge was poor without polymer. However, polymer doses of less than 2 lb/ton gave recoveries of 95%. Solids content of the sludge cake ranged from 20 to 28 percent.

The lime-primary sludge could be filtered without polymers but performance was improved greatly by addition of small quantities of polymer. Filtration yields in excess of 10 lb/hr-sq ft were recorded. Recoveries of solids were in the order of 95%. Cake solids ranged from 20 to 35% with 28% a typical value.

11.4.2.4 Canadian Experience

The Province of Ontario's Ministry of the Environment has carried out a number of plant-scale testing programs at sewage treatment plants. This work has been summarized in several reports (33-35). Some of that information (34) is summarized here.

The use of lime in activated sludge plants increased sludge production from a normal level of about 180 mg/l to 460 mg/l during addition of 200 mg/l slaked lime. Sludge solids content increased from 4.5% to 9%. In primary plants, sludge production increased from 80 mg/l to 240 mg/l. Sludge solids content increased from 7% to 15%.

Some difficulty was encountered with anaerobic digestion because of erratic dosage with high pH sludge. Subsequent studies have shown that holding the sludge longer in the clarifier results in a fall in sludge pH. Digester operation is then normal except for high solids content of the sludge.

Raw lime sludge was dewatered successfully in a continuous solids bowl conveyor-type centrifuge at the Newmarket/East Gwillimbury plant (36). A detailed investigation showed high recoveries could be achieved with polymer addition. At these high recoveries, phosphorus removal from the centrate was also high (37).

With lime added to the primary of a conventional plant (primary plus activated sludge), filter yield increased from 5.2 to 7.2 lb/hr-sq ft, and conditioning costs (lime plus $FeCl_3$) fell from $16/ton down to $11/ton. Cake solids content was 29%.

11.4.3 Alum and Iron Salts

11.4.3.1 Blue Plains

Alum was added to the 300 mgd wastewater treatment plant at Blue Plains, Washington, D. C., for several months. At this plant, primary treatment is followed by high rate activated sludge. Waste activated sludge is combined with sludge withdrawn from the clarifier, and the mixed sludge is thickened and digested. The digested sludge was elutriated in two stages using polymer to retain solids. The sludge was filtered using polymer and ferric chloride for conditioning. The following information on alum addition tests is based on discussions with plant operating personnel.

Alum was added to the aerator in gradually increasing dose until the average dose was 40 mg/l of alum (9.1% Al). This developed a relatively high level of aluminum in the sludge because the Blue Plains wastewater is weak.

Alum addition did not affect thickening. Sludge leaving the thickeners remained at 7.5% solids. Digestion was likewise unaffected. Suspended solids leaving the digesters remained at 3.5%. Elutriation could be carried out but a high polymer dose was required to retain suspended solids. Chemical costs for elutriation increased from $3/ton before alum addition to $7/ton during the test. Filter yields were about the same as with prior operation, but solids content of the cake dropped from 23% to 20%. Polymer condition cost for filtration, which had been $6/ton, rose to $7/ton.

11.4.3.2 Pomona

At Pomona, California, a physical-chemical plant has been in operation to produce a feed stream for a long-term test of carbon columns. The wastewater has been treated in a primary clarifier with 22 mg aluminum per liter. A field test was conducted to determine the dewatering characteristics of this sludge. That unpublished report (38) is summarized here.

Gravity thickening tests showed that, without polymer, sludge settled slowly. Thickening rate to produce a 4% solids sludge was quite low at 6.6 lb/day-sq ft. With polymer addition, that rate increased to 20 lb/day-sq ft with 2 lb/ton of anionic polymer and to 70 lb/day-sq ft with 6 lb/ton. Filtration rates with 12% lime and polymer were less than 2 lb/hr-sq ft at the low cycle time of 2 min. However, the sludge employed was only 2% solids. Rate would be expected to increase proportionately if thicker sludge were available.

One of the reasons this field test was conducted was to verify other data indicating that sludges produced when metals are added to the primary settle poorly and have lower filter yields than conventional sludge. The Pomona field test corroborated these findings.

11.4.3.3 Grand Rapids

The addition of ferric chloride at various points in the process was studied at Grand Rapids, Michigan. The Grand Rapids plant has a conventional primary followed by activated sludge. The ferric chloride was added in three different modes: to the primary alone, split between the primary and the aerator, and to the aerator alone. An investigation of sludge dewatering was not part of this project. However, it was hoped that the filter yield data routinely collected at this plant could be related back to the chemical dosing conditions.

Monthly average filter yields, cake solids, and chemical costs were supplied by Grand Rapids staff and are presented in Figure 11-5. Large fluctuations in all three filtration variables are evident; however, they are not correlated with ferric chloride dosing. This finding was predicted by Grand Rapids staff who observed that the basic nature of the sludge can change from month to month and that necessary maintenance can cause step changes in performance. Some of the interferences noted at Grand Rapids were:

a. Periodic dumping of lime sludge from a water plant into the wastewater collection system.

b. Performance changes in digestion.

c. Acid cleaning of filter cloths.

d. Cleaning of interior piping of vacuum filters.

e. Replacement of vacuum system.

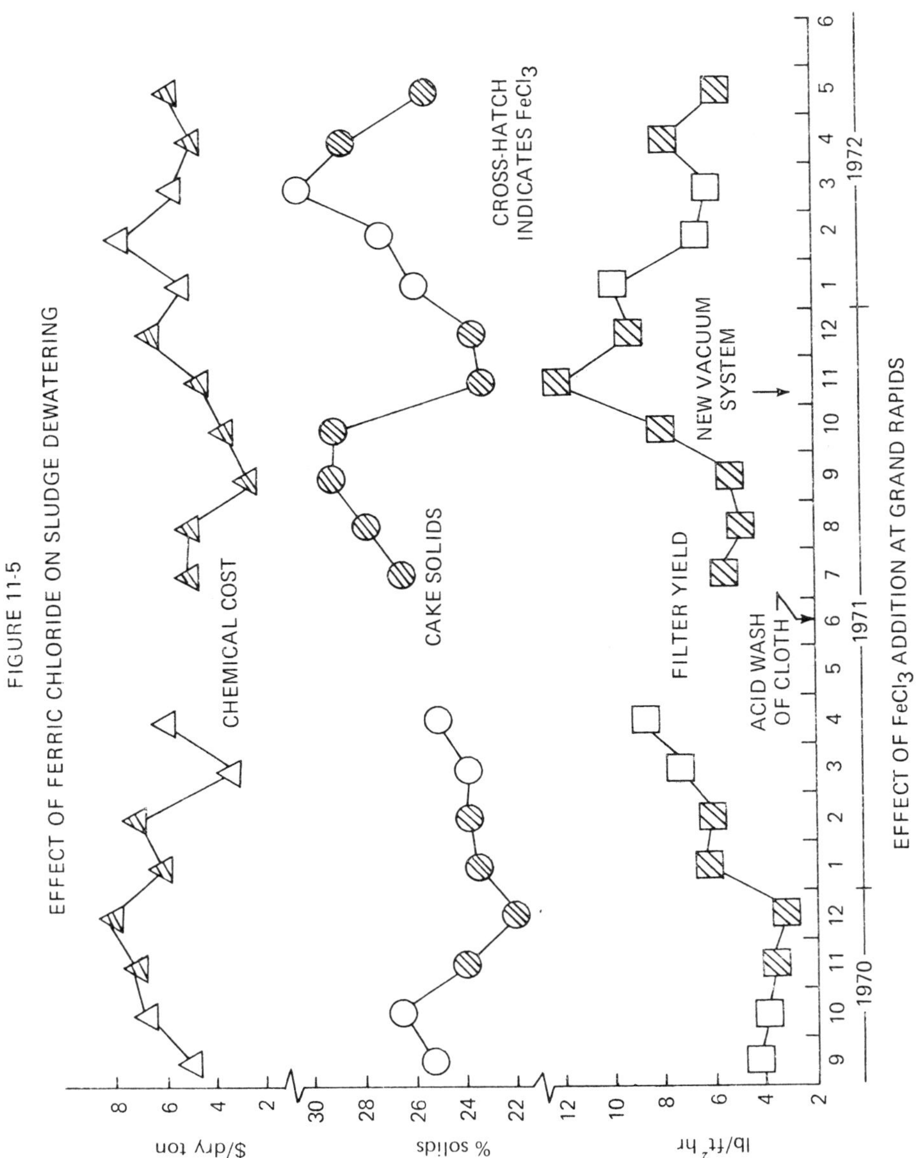

FIGURE 11-5
EFFECT OF FERRIC CHLORIDE ON SLUDGE DEWATERING

f. Periodic switch-over from the incinerators, which had a lower capacity than the vacuum filters and needed a dry cake, to landfill, where maximum output was desired and a dry cake was not so important.

11.4.3.4 Canadian Experience

Reference has been made in Section 11.4.2.4 to the work reported by Ontario's Ministry of the Environment. Additional information from a summary paper (34) is included here.

When metal salts were utilized in activated sludge secondary plants, the dry solids sludge production increased 5% to 25%. Sludge concentration dropped slightly to 3.5 to 5.0% from the normal 4.0 to 5.5%.

In primary plants, sludge production increases substantially because of increased solids capture. An increase of 100% was typical and 20% reduction in sludge solids concentration was reported.

In all cases where metal salts were used to precipitate phosphorus, normal digestion was observed. Problems which developed were the result of increased hydraulic and volatile solids loading.

Filtering rates were approximately halved for a primary plant when alum was added. Lime and ferric chloride conditioning costs increased from \$3.10/ton to \$9.50/ton. Solids content of cake dropped from 31% down to 19%. For a secondary plant with alum added to the primary, filtering rate fell from 5.2 to 4.8 lb/hr-sq ft, and conditioning chemical cost increased slightly. Cake solids was typically 15.9% with alum addition.

11.4.3.5 Chapel Hill

Chapel Hill has a 5 MGD trickling filter plant which is conveniently split into two parallel treatment plants, each with its own primary clarifier, high rate trickling filter, and final clarifier. During an experimental program, the wastewater from one train was dosed with alum for phosphorus removal, just before it entered the final clarifier. A field trip was arranged to this unique facility to determine the impact of alum addition on the wastewater sludge. Unpublished results (39) are summarized here.

Data obtained indicated sludge from the secondary clarifier following the alum-dosed train was essentially unfilterable, even with polymer addition. Since this was a high-rate filter with substantial recirculation before the clarifier, the sludge was very low in organic matter. The control secondary clarifier sludge showed reasonable filter yields. Secondary sludges from both trains were recirculated to their respective primary clarifiers and removed as mixed primary-secondary sludges. Filter leaf tests showed that there was no substantial difference in the filter yields from the two trains.

Data supplied by University of North Carolina staff who operate the plant indicated that 4.75 lb extra sludge was generated per lb of Al added. They also indicated that difficulty was encountered during alum trials in their two-stage digester: the solids content of the supernatant leaving the second digester was unusually high.

11.5 References

1. "Process Design Manual for Sludge Handling and Disposal," Technology Transfer, U. S. Environmental Protection Agency (1974).

2. Smith, R., Eilers, R., "Wastewater Treatment Plant Cost Estimating Program," NTIS PB 219472 (April, 1971).

3. Yeaple, D. S., Barnes, D. A., and DiGiano, F. A., JBF Corp., "A Computer Model for Evaluating Local Phosphorus Removal Strategies," Environmental Protection Agency Contract 68-01-0758, Draft Report (January 3, 1973).

4. Mogelnicki, S., Los Angeles Technology Transfer Session (November, 1972).

5. Teletzke, G. H., "Sludge Dewatering by Vacuum Filtration," Rocky Mountain Sew. & Ind. Wastes Assoc. Annual Meeting, Colorado Springs, CO (October 24-26, 1960).

6. Adrian, D. D., Smith, J. E., Jr., "Dewatering Physical-Chemical Sludges," Vanderbilt Symp., "Applications of New Concepts of Physical-Chemical Wastewater Treatment," in press (September 18-22, 1972).

7. Isgard, E., "Chemical Methods in Present Sewage Purification Techniques," 7th Effluent and Water Treatment Exhib. and Conv., London (June 25, 1971).

8. Mulbarger, M. C., Shifflett, D. G., "Combined Biological and Chemical Treatment for Phosphorus Removal," Chem. Engr. Prog. Symp. Ser. 67, No. 107, 107-116 (1970).

9. Farrell, J. B., "Alum-Phosphate Sludges: Dewaterability and Aluminum Recovery by Lime Treatment," In-House Report, Adv. Waste Treatment Research Lab., NERC, Cincinnati, OH, U. S. Environmental Protection Agency (June, 1973).

10. Smith, R., "Estimating the Rate of Sludge Production in the Activated Sludge Process," Memo (October 10, 1972).

11. City of Austin and University of Texas, "Design Guides for Biological Wastewater Treatment Processes," U. S. Environmental Protection Agency Water Pollution Control Research Series 11010 ESQ (August, 1971).

12. Kreissl, J. F., Westrick, J. J., "Municipal Waste Treatment by Physical-Chemical Methods," Vanderbilt Symp., "Applications of New Concepts of Physical-Chemical Wastewater Treatment," (September 18-22, 1972).

13. "Process Design Manual for Phosphorus Removal," U. S. Environmental Protection Agency (October, 1971).

14. Burns, D. E., Shell, G. L., "Physical-Chemical Treatment of Municipal Wastewater Using Powdered Carbon," Final Report, U. S. Environmental Protection Agency Project No. 17020 EFB (1972).

15. Friedman, L. D., Weber, W. J., Bloom, R., and Hopkins, C. B., "Improving Granular Carbon Treatment," Water Pollution Control Research Series 17020 GDN (1971).

16. Farrell, J. B., unpublished data.

17. Westrick, J. J., Cohen, J. M., "Comparative Effects of Chemical Pretreatment on Carbon Adsorption," 47th Annual Conf., Water Pollution Control Federation, Atlanta, GA (October, 1972).

18. Schaffer, R. B., Van Hall, C. E., McDermott, G. N., Barth, D., Stenger, V. A., Sevesta, S. J., and Griggs, S. H., "Application of a Carbon Analyzer in Waste Treatment," *JWPCF,* 37:11, p. 1545-56.

19. Bunch, R. L., Barth, E. F., and Ettinger, M. B., "Organic Materials in Secondary Effluents," *JWPCF,* 33:2, p. 122-26 (February, 1961).

20. Brown, J. C., "Alum Treatment of High-Rate Trickling Filter Effluent at Chapel Hill, North Carolina," Durham, NC (April, 1973).

21. Derrington, R. E., Stevens, D. H., and Laughlin, J. E., "Enhancing Trickling Filter Plant Performance by Chemical Precipitation," U. S. Environmental Protection Agency, Grant #11010 EGL, (July, 1973).

22. Seiden, L., Patel, K., "Mathematical Model of Tertiary Treatment by Lime Addition," Report No. TWRC-14, FWPCA (September, 1969).

23. O'Farrell, T. P., Bishop, D. F., and Bennett, S. M., "Advanced Waste Treatment at Washington, D. C.," "Water-1969," Chem. Engr. Prog. Symp. Ser. 65, No. 97,

p. 251-58.

24. Berg, E. L., Brunner, C. A., and Williams, R. T., "Single-Stage Lime Clarification of Secondary Effluent," *Water & Wastes Engr.*, (March, 1960).

25. South Tahoe Public Utility District, "Advanced Waste Water Treatment as Practiced at South Tahoe," Water Pollution Control Research Series 17010 ELQ, p. 89 (August, 1971).

26. Malhotra, S. K., Lee, G. F., and Rohlich, G. A., "Nutrient Removal from Secondary Effluent by Alum Flocculation and Lime Precipitation," *Internatl. Jour. Air-Water Poll.*, 8, p. 487-500 (1964).

27. Mulbarger, M. C., Grossman, E. III, Dean, R. B., and Grant, O. L., "Lime Clarification, Recovery, Reuse, and Sludge Dewatering Characteristics," *Jour WPCF,* 41:12, p. 2070-85 (1969).

28. Smith, R., "Calculation of Ionic Equilibria by Means of the Digital Computer," Advanced Waste Treatment Research Lab., Cincinnati, OH (November, 1966).

29. "Standard Methods for the Examination of Water and Wastewater," 13th Edition, APHA, AWWA, WPCF, Washington, D. C. (1971).

30. Bennett, S. M., Bishop, D. F., "Solids Handling and Reuse of Lime Sludge," Environmental Protection Agency Contract #14-12-818, 11010 FYM.

31. Parker, D. S., Zadick, F. J., and Train, K. E., "Sludge Processing for Combined Physical-Biological Sludges," Environmental Protection Agency, R2-73-250 (July, 1973).

32. Zurn Engineers and Battelle/NW, "Westerly Advanced Waste Treatment Facility: Process Development and Engineering Design," Draft Version (September 16, 1971).

33. Van Fleet, G. L., Barr, J. R., and Harris, A. J., "Treatment and Disposal of Chemical Phosphate Sludges in Ontario," Water Pollution Control Federation Meeting, Atlanta, GA (October, 1972).

34. Boyko, B. I., Rupke, J. W. G., "Design Considerations in the Implementation of Ontario's Phosphorus Removal Program," Phosphorus Removal Design Seminar, Toronto, Ontario, Ministry of the Environment (May 28-29, 1973).

35. Knight, C. H., Mondoux, R. G., and Hambley, B., "Thickening and Dewatering Sludges Produced in Phosphorus Removal," Phosphorus Removal Design Seminar, Toronto, Ontario, Ministry of the Environment (May 28-29, 1973).

36. Black, S. A., "Lime Treatment for Phosphorus Removal at the Newmarket/East Gwillimbury WPCF, an Interim Report," Ontario Ministry of the Environment, R.P.W. 2032 (May, 1972).

37. Smith, A. G., "Centrifuge Dewatering of Lime Treated Sewage Sludge," Ontario Ministry of the Environment, R.P.W. 2030 (May, 1972).

38. Hathaway, S. W., Unpublished Report for EPA/NERC (August 27, 1973).

39. Hathaway, S. W., Unpublished Report for EPA/NERC (September, 1972).

APPENDIX A

EVALUATING PHOSPHORUS REMOVAL STRATEGIES

A computer model for evaluating a number of strategies for removing phosphorus in wastewater has been developed for EPA.[1] This model reports to the user the total cost of a selected strategy for removing phosphorus. 22 treatment schemes can be selected and evaluated depending on local conditions. These schemes are:

1. Alum & Polymer Addition to Primary Settler
2. Ferric Chloride & Polymer Addition to Primary Settler
3. Ferrous Chloride & Lime Addition to Primary Settler
4. Lime Addition to Primary Settler
5. Alum & Polymer Addition to Flocculation Basin
6. Ferric Chloride & Polymer Addition to Flocculation Basin
7. Ferrous Chloride & Lime Addition to Flocculation Basin
8. Lime Addition to Flocculation Basin
9. Alum Addition to Aeration Basin
10. Ferric Chloride Addition to Aeration Basin
11. Sodium Aluminate Addition to Aeration Basin
12. Alum Addition to Aeration Basin & Multi-Media Filtration
13. Ferric Chloride Addition to Aeration Basin & Multi-Media Filtration
14. Sodium Aluminate Addition to Aeration Basin & Multi-Media Filtration
15. Alum Addition After Trickling Filter
16. Ferric Chloride Addition After Trickling Filter
17. Alum Addition After Trickling Filter & Multi-Media Filtration
18. Ferric Chloride Addition After Trickling Filter & Multi-Media Filtration
19. Alum Addition to Flocculation Basin After Conventional Secondary Treatment
20. Ferric Chloride Addition to Flocculation After Conventional Secondary Treatment
21. Lime (1-Stage) Addition to Flocculation After Conventional Secondary Treatment
22. Lime (2-Stage) Addition to Flocculation Basin After Conventional Secondary Treatment

Four levels of effluent phosphorus concentration (total unfiltered phosphorus) were assumed to be achievable depending upon the type of treatment provided. These are:

2.0 mg/l (Option 1) Chemical addition to the primary clarifier. (Treatment schemes 1-4)

(Option 2) Chemical addition to a flocculation basin prior to the primary clarifier. (Treatment schemes 5-8)

(Option 3) Chemical addition to the aeration basin. (Treatment schemes 9-11)

[1] "A computer Model for Evaluating Community Phosphorous Removal Strategies," prepared for U.S. EPA by JBF Scientific Corp., Burlington, Mass., Report No. 400/9-73/001 (Oct. 1973).

(Option 4) Chemical addition after the trickling filter. (Treatment schemes 15-16)

0.5 mg/l (Option 5) Chemical addition to the aeration basin plus multi-media filtration. (Treatment schemes 12-14)

or

Chemical addition after the trickling filter plus multi-media filtration. (Treatment schemes 17-18)

(Option 6) Addition of lime to a flocculation basin following conventional secondary treatment. (Treatment scheme 21)

0.3 mg/l (Option 7) Addition of alum or ferric chloride to a flocculation basin following conventional secondary treatment–one stage. (Treatment schemes 19-20)

0.1 mg/l (Option 6) Addition of lime to a flocculation basin following conventional secondary treatment–two stages. (Treatment scheme 22)

The model has been designed specifically for planning purposes and can be used by municipalities, planning agencies and government policy organizations. In order to insure its usefulness, EPA plans to periodically update the model as additional performance data on phosphorus removal processes becomes available.

The phosphorus removal model (REMOVE) is designed to provide an estimate of the cost to a particular community of achieving a designated level of phosphorus removal. The program begins with the entry of data describing the community in terms of population and future trends, the treatment plant in terms of present size and excess capacity, and the influent wastewater characteristics. The data can be entered interactively through a subroutine (SYSTEM).

After input data have been entered, the user has the opportunity to investigate various control strategies for achieving a low phosphorus concentration in the wastewater effluent. Typically, the effects on cost of reducing phosphorus in detergent products, of increasing or decreasing the amount of government financing available, or of using various treatment options can be investigated.

It is important to note that the model has been developed as a simulation only and as such does not achieve a mathematical optimization.

From the description of the community, treatment plant, wastewater characteristics, and desired strategy the model will design a treatment plant incorporating the present plant as described by the user and will add those processes or additional equipment necessary to achieve the desired effluent level. The program converts the sizing information into construction costs, operating man-hours, maintenance man-hours, and total material and supply costs for each process.

The following pages illustrate the type of print-out obtained when using the computer model. The particular cases illustrated here are a 5 mgd trickling filter plant and a 20 mgd activated sludge treatment plant.

5 MGD TRICKLING FILTER PLANT
TREATMENT COSTS IN CENTS/1000 GAL.

P	2	2	2	0.5	0.5	0.5
GOVFF		.8	.8		.8	.8
RESTRICT			yes			yes
SCHNO						
1	13.9	11.9	8.4			
2	10.0	8.4	6.6			
3	17.8	15.0	13.7			
4	13.1	10.9	11.7			
5	14.8	12.3	8.8			
6	11.0	8.8	7.0			
7	18.8	15.3	14.0			
8	14.0	11.2	12.1			
9	12.3	10.6	7.5			
10	10.0	8.4	6.3			
11	15.1	13.5	8.8			
12				18.9	14.3	10.2
13				16.6	12.1	9.1
14				21.7	17.1	11.5
15	7.7	6.7	4.6			
16	8.2	7.3	4.8			
17				13.7	10.0	7.3
18				14.9	11.2	7.7
21				15.3	10.3	11.2

NOTES:
1. P = Phosphorus effluent level in mg/l.
2. GOVFF = Fraction of construction cost financed by federal and state governments.
3. RESTRICT = Restrictions on phosphate detergents.
4. SCHNO = Treatment scheme number.
5. ___ = Lowest treatment cost.

CASE 1

LIQUID TREATMENT SCHEME = 15
AMORTIZATION RATE =0.05 AMORTIZATION LIFETIME =20.0
FRACTION OF CONSTRUCTION COST FINANCED BY GOVERNMENT = .0

PROCESS NAME	CONSTRUCTION COST(1000 $)	SIZING PARAMETER		AMORTIZED COST (1000 $ /TP)	YEAR PROCESS FIRST NEEDED
RAW WSTE PMP	0.0	0.0	MGD	0.0	1994.0
PRE TRETMENT	0.0	0.0	MGD	0.0	1994.0
PRIM SETTLER	0.0	0.0	TSF	0.0	1994.0
PRMYSLGE PMP	0.0	0.0	GPM	0.0	1994.0
TRKLNG FILTR	0.0	0.0	TCF	0.0	1994.0
SECY SETTLER	0.0	0.0	TSF	0.0	1994.0
RECIRC PMPNG	0.0	0.0	MGD	0.0	1994.0
CHLORNE FEED	0.0	0.0	#/D	0.0	1994.0
CHLORNE BASN	0.0	0.0	TCF	0.0	1994.0
GRAV THICKNR	34.824	0.022	TSF	2.794	1974.0
SLGE HLD TNK	15.649	0.987	TCF	1.256	1974.0
VACUUM FILTR	104.751	39.600	SQFT	8.406	1974.0
ALUM FEEDING	49.858	624.750	#/D	4.001	1974.0
FECL FEEDING	45.525	64.565	#/D	3.653	1974.0
LIME FEEDING	15.901	746.367	#/D	1.276	1974.0
TOTALS =	266.507			21.385	

PERIODIC OPERATING COSTS

PROCESS NAME	OPERATING MAN-HOURS (MAN-HOURS/TP)			MAINTAINANCE MAN-HOURS (MAN-HOURS/TP)		
	AVE.	MAX.	MIN.	AVE.	MAX.	MIN.
RAW WSTE PMP	0.	0.	0.	0.	0.	0.
PRE TRETMENT	0.	0.	0.	0.	0.	0.
PRIM SETTLER	0.	0.	0.	0.	0.	0.
PRMYSLGE PMP	0.	0.	0.	0.	0.	0.
TRKLNG FILTR	0.	0.	0.	0.	0.	0.
SECY SETTLER	0.	0.	0.	0.	0.	0.
RECIRC PMPNG	0.	0.	0.	0.	0.	0.
CHLORNE FEED	0.	0.	0.	0.	0.	0.
CHLORNE BASN	0.	0.	0.	0.	0.	0.
GRAV THICKNR	2.	2.	2.	1.	1.	1.
SLGE HLD TNK	18.	18.	18.	11.	11.	11.
VACUUM FILTR	665.	665.	665.	80.	80.	80.
ALUM FEEDING	0.	0.	0.	1473.	1473.	1473.
FECL FEEDING	0.	0.	0.	600.	600.	600.
LIME FEEDING	0.	0.	0.	1578.	1578.	1578.
TOTALS =	685.	685.	685.	3742.	3742.	3742.

PERIODIC OPERATING COSTS

PROCESS NAME	MATERIAL & SUPPLY COST ($/TP)			LABOR COST (AVERAGE) (1000$/TP)	TOTAL 0&M (1000$/TP)
	AVE.	MAX.	MIN.		
RAW WSTE PMP	0.0	0.0	0.0	0.0	0.0
PRE TRETMENT	0.0	0.0	0.0	0.0	0.0
PRIM SETTLER	0.0	0.0	0.0	0.0	0.0
PRMYSLGE PMP	0.0	0.0	0.0	0.0	0.0
TRKLNG FILTR	0.0	0.0	0.0	0.0	0.0
SECY SETTLER	0.0	0.0	0.0	0.0	0.0
RECIRC PMPNG	0.0	0.0	0.0	0.0	0.0
CHLORNE FEED	0.0	0.0	0.0	0.0	0.0
CHLORNE BASIN	0.0	0.0	0.0	0.0	0.0
GRAV THICKNR	5.67	5.67	5.67	0.010	0.015
SLGE HLD TNK	54.57	54.57	54.57	0.133	0.187
VACUUM FILTR	3166.75	3166.75	3166.75	3.382	6.549
ALUM FEEDING	78872.44	78872.69	78872.69	6.693	85.565
FECL FEEDING	2998.12	2998.12	2998.12	2.725	5.724
LIME FEEDING	2855.35	2855.35	2855.35	7.167	10.022
TOTALS =	87952.75	87953.00	87953.00	20.109	108.062

UNIT COST DATA
(CENTS/1000 GALLONS)

PROCESS NAME	AMORTIZATION	LABOR	MATERIAL & SUPPLY	TOTAL
RAW WSTE PMP	0.0	0.0	0.0	0.0
PRE TRETMENT	0.0	0.0	0.0	0.0
PRIM SETTLER	0.0	0.0	0.0	0.0
PRMYSLGE PMP	0.0	0.0	0.0	0.0
TRKLNG FILTR	0.0	0.0	0.0	0.0
SECY SETTLER	0.0	0.0	0.0	0.0
RECIRC PMPNG	0.0	0.0	0.0	0.0
CHLORNE FEED	0.0	0.0	0.0	0.0
CHLORNE BASN	0.0	0.0	0.0	0.0
GRAV THICKNR	0.15	0.00	0.00	0.15
SLGE HLD TNK	0.07	0.01	0.00	0.08
VACUUM FILTR	0.46	0.19	0.17	0.82
ALUM FEEDING	0.22	0.57	4.32	4.91
FECL FEEDING	0.20	0.15	0.16	0.51
LIME FEEDING	0.07	0.38	0.16	0.62
TOTALS =	1.17	1.10	4.82	7.09

SOLID SINKS COSTS

TIME PERIOD	SINK USED	AMOUNT SOLIDS (TONS)	COST ($/TP)
1	1	410.46	10261.50
2	1	410.46	10261.50
3	1	410.46	10261.50
4	1	410.46	10261.50
5	1	410.46	10261.50
6	1	410.46	10261.50
7	1	410.46	10261.50
8	1	410.46	10261.50
9	1	410.46	10261.50
10	1	410.46	10261.50
11	1	410.46	10261.50
12	1	410.46	10261.50
13	1	410.46	10261.50
14	1	410.46	10261.50
15	1	410.46	10261.50
16	1	410.46	10261.50
17	1	410.46	10261.50
18	1	410.46	10261.50
19	1	410.46	10261.50
20	1	410.46	10261.50
TOTALS =		8209.17	205229.12

UNIT COSTS FOR SOLID WASTE DISPOSAL (CENTS/1000 GAL.) = 0.50
TOTAL UNIT COST (CENTS/1000 GAL) = 7.66

20 MGD ACTIVATED SLUDGE PLANT
TREATMENT COSTS IN CENTS/1000 GAL.

P	2	2	2	0.5	0.5	0.5
GOVFF		.8	.8		.8	.8
RESTRICT			yes			yes
SCHNO						
1	11.1	9.9	6.4			
2	7.6	6.5	4.8			
3	16.5	13.9	12.7			
4	9.8	8.6	9.5			
5	11.7	10.1	6.7			
6	8.1	6.7	5.1			
7	17.1	14.2	12.9			
8	10.4	8.9	9.7			
9	8.3	8.0	5.0			
10	6.2	5.9	4.0			
11	11.2	11.0	6.4			
12				13.4	10.9	7.0
13				11.3	8.8	6.0
14				16.3	13.8	8.4
15	7.9	6.4	4.4			
16	8.4	7.0	4.7			
17				12.3	8.9	6.3
18				13.6	10.0	6.8
21				9.4	6.2	7.1

NOTES:
1. P = Phosphorus effluent level in mg/l.
2. GOVFF = Fraction of construction cost financed by federal and state governments.
3. RESTRICT = Restrictions on phosphate detergents.
4. SCHNO = Treatment scheme number.
5. ______ = Lowest treatment cost.

CASE 2

LIQUID TREATMENT SCHEME = 10
AMORTIZATION RATE =0.05 AMORTIZATION LIFETIME =20.0
FRACTION OF CONSTRUCTION COST FINANCED BY GOVERNMENT = .0

PROCESS NAME	CONSTRUCTION COST(1000 $)	SIZING PARAMETER		AMORTIZED COST (1000 $ /TP)	YEAR PROCESS FIRST NEEDED
RAW WSTE PMP	0.0	0.0	MGD	0.0	1994.0
PRE TRETMENT	0.0	0.0	MGD	0.0	1994.0
PRIM SETTLER	0.0	0.0	TSF	0.0	1994.0
PRMYSLGE PMP	0.0	0.0	GPM	0.0	1994.0
AERTN BASIN	0.0	0.0	TCF	0.0	1994.0
MECH AERATN	0.0	0.0	HP	0.0	1994.0
SECY SETTLER	0.0	0.0	TSF	0.0	1994.0
RECIRC PMPNG	0.0	0.0	MGD	0.0	1994.0
CHLORNE FEED	0.0	0.0	#/D	0.0	1994.0
CHLORNE BASN	0.0	0.0	TCF	0.0	1994.0
GRAV THICKNR	38.163	0.546	TSF	3.062	1974.0
SLGE HLD TNK	0.0	0.0	TCF	0.0	1994.0
VACUUM FILTR	98.759	34.950	SQFT	7.925	1974.0
FECL FEEDING	115.133	6446.578	#/D	9.239	1974.0
LIME FEEDING	36.171	2541.481	#/D	2.903	1974.0
TOTALS =	288.226			23.128	

PERIODIC OPERATING COSTS

PROCESS NAME	OPERATING MAN-HOURS (MAN-HOURS/TP)			MAINTENANCE MAN-HOURS (MAN-HOURS/TP)		
	AVE.	MAX.	MIN.	AVE.	MAX.	MIN.
RAW WSTE PMP	0.	0.	0.	0.	0.	0.
PRE TRETMENT	0.	0.	0.	0.	0.	0.
PRIM SETTLER	0.	0.	0.	0.	0.	0.
PRMYSLGE PMP	0.	0.	0.	0.	0.	0.
AERTN BASN	0.	0.	0.	0.	0.	0.
MECH AERATN	0.	0.	0.	0.	0.	0.
SECY SETTLER	0.	0.	0.	0.	0.	0.
RECIRC PMPNG	0.	0.	0.	0.	0.	0.
CHLORNE FEED	0.	0.	0.	0.	0.	0.
CHLORNE BASN	0.	0.	0.	0.	0.	0.
GRAV THICKNR	41.	41.	41.	23.	23.	23.
SLGE HLD TNK	0.	0.	0.	0.	0.	0.
VACUUM FILTR	3013.	3013.	3013.	389.	389.	389.
FECL FEEDING	0.	0.	0.	2884.	2884.	2884.
LIME FEEDING	0.	0.	0.	2009.	2009.	2009.
TOTALS =	3055.	3055.	3055.	5284.	5284.	5284.

PERIODIC OPERATING COSTS

PROCESS NAME	MATERIAL & SUPPLY COST ($ /TP) AVE.	MAX.	MIN.	LABOR COST (AVERAGE) (1000$/TP)	TOTAL O&M (1000$/TP)
RAW WSTE PMP	0.0	0.0	0.0	0.0	0.0
PRE TRETMENT	0.0	0.0	0.0	0.0	0.0
PRIM SETTLER	0.0	0.0	0.0	0.0	0.0
PRMYSLGE PMP	0.0	0.0	0.0	0.0	0.0
AERTN BASIN	0.0	0.0	0.0	0.0	0.0
MECH AERATN	0.0	0.0	0.0	0.0	0.0
SECY SETTLER	0.0	0.0	0.0	0.0	0.0
RECIRC PMPNG	0.0	0.0	0.0	0.0	0.0
CHLORNE FEED	0.0	0.0	0.0	0.0	0.0
CHLORNE BASN	0.0	0.0	0.0	0.0	0.0
GRAV THICKNR	88.40	88.40	88.40	0.290	0.378
SLGE HLD TNK	0.0	0.0	0.0	0.0	0.0
VACUUM FLITR	17365.26	17365.30	17365.30	15.364	32.729
FECL FEEDING	299353.00	299353.06	299353.06	13.101	312.453
LIME FEEDING	9722.82	9722.85	9722.85	9.124	18.847
TOTALS =	326529.44	326529.56	326529.56	37.878	364.407

UNIT COST DATA
(CENTS/1000 GALLONS)

PROCESS NAME	AMORTIZATION	LABOR	MATERIAL & SUPPLY	TOTAL
RAW WSTE PMP	0.0	0.0	0.0	0.0
PRE TRETMENT	0.0	0.0	0.0	0.0
PRIM SETTLER	0.0	0.0	0.0	0.0
PRMYSLGE PMP	0.0	0.0	0.0	0.0
AERTN BASN	0.0	0.0	0.0	0.0
MECH AERATN	0.0	0.0	0.0	0.0
SECY SETTLER	0.0	0.0	0.0	0.0
RECIRC PMPNG	0.0	0.0	0.0	0.0
CHLORNE FEED	0.0	0.0	0.0	0.0
CHLORNE BASN	0.0	0.0	0.0	0.0
GRAV THICKNR	0.04	0.00	0.00	0.05
SLGE HLD TNK	0.0	0.0	0.0	0.0
VACUUM FILTR	0.11	0.21	0.24	0.58
FECL FEEDING	0.13	0.18	4.10	4.41
LIME FEEDING	8.04	0.12	0.13	0.30
TOTALS =	0.32	0.52	4.47	5.31

SOLID SINKS COSTS

TIME PERIOD	SINK USED	AMOUNT SOLIDS (TONS)	COST ($/TP)
1	1	2553.97	63849.36
2	1	2553.97	63849.36
3	1	2553.97	63849.36
4	1	2553.97	63849.36
5	1	2553.97	63849.36
6	1	2553.97	63849.36
7	1	2553.97	63849.36
8	1	2553.97	63849.36
9	1	2553.97	63849.36
10	1	2553.97	63849.36
11	1	2553.97	63849.36
12	1	2553.97	63849.36
13	1	2553.97	63849.36
14	1	2553.97	63849.36
15	1	2553.97	63849.36
16	1	2553.97	63849.36
17	1	2553.97	63849.36
18	1	2553.97	63849.36
19	1	2553.97	63849.36
20	1	2553.97	63849.36
TOTALS =		51079.45	1276985.00

UNIT COSTS FOR SOLID WASTE DISPOSAL (CENTS/1000 GAL.) = 0.87
TOTAL UNIT COST (CENTS/1000 GAL) = 6.18